Advances in Intelligent Systems and Computing

Volume 1173

The series "Advances in Intelligent Systems and Computing" contains publications on theory, applications, and design methods of Intelligent Systems and Intelligent Computing. Virtually all disciplines such as engineering, natural sciences, computer and information science, ICT, economics, business, e-commerce, environment, healthcare, life science are covered. The list of topics spans all the areas of modern intelligent systems and computing such as: computational intelligence, soft computing including neural networks, fuzzy systems, evolutionary computing and the fusion of these paradigms, social intelligence, ambient intelligence, computational neuroscience, artificial life, virtual worlds and society, cognitive science and systems, Perception and Vision, DNA and immune based systems, self-organizing and adaptive systems, e-Learning and teaching, human-centered and human-centric computing, recommender systems, intelligent control, robotics and mechatronics including human-machine teaming, knowledge-based paradigms, learning paradigms, machine ethics, intelligent data analysis, knowledge management, intelligent agents, intelligent decision making and support, intelligent network security, trust management, interactive entertainment, Web intelligence and multimedia.

The publications within "Advances in Intelligent Systems and Computing" are primarily proceedings of important conferences, symposia and congresses. They cover significant recent developments in the field, both of a foundational and applicable character. An important characteristic feature of the series is the short publication time and world-wide distribution. This permits a rapid and broad dissemination of research results.

**** Indexing: The books of this series are submitted to ISI Proceedings, EI-Compendex, DBLP, SCOPUS, Google Scholar and Springerlink ****

More information about this series at http://www.springer.com/series/11156

Wojciech Zamojski · Jacek Mazurkiewicz ·
Jarosław Sugier · Tomasz Walkowiak ·
Janusz Kacprzyk
Editors

Theory and Applications of Dependable Computer Systems

Proceedings of the Fifteenth International Conference on Dependability of Computer Systems DepCoS-RELCOMEX, June 29 – July 3, 2020, Brunów, Poland

Springer

Editors
Wojciech Zamojski
Wrocław University of Science and Technology
Wrocław, Poland

Jacek Mazurkiewicz
Wrocław University of Science and Technology
Wrocław, Poland

Jarosław Sugier
Wrocław University of Science and Technology
Wrocław, Poland

Tomasz Walkowiak
Wrocław University of Science and Technology
Wrocław, Poland

Janusz Kacprzyk
Systems Research Institute
Polish Academy of Sciences
Warsaw, Poland

ISSN 2194-5357 ISSN 2194-5365 (electronic)
Advances in Intelligent Systems and Computing
ISBN 978-3-030-48255-8 ISBN 978-3-030-48256-5 (eBook)
https://doi.org/10.1007/978-3-030-48256-5

This Springer imprint is published by the registered company Springer Nature Switzerland AG
The registered company address is: Gewerbestrasse 11, 6330 Cham, Switzerland

Preface

In this volume, we would like to present proceedings of the Fifteenth International Conference on Dependability of Computer Systems *DepCoS-RELCOMEX* which is scheduled to take place in the Brunów Palace in Poland from 29 June to 3 July 2020.

DepCoS–RELCOMEX is an annual conference series organized since 2006 at the Faculty of Electronics, Wrocław University of Science and Technology, initially by Institute of Computer Engineering, Control and Robotics (CECR) and now by Department of Computer Engineering. The series can be seen as a continuation of other two cycles of events: RELCOMEX (1977–1989) and Microcomputer School (1985–1995) which were organized by the Institute of Engineering Cybernetics (the previous name of CECR) under the leadership of Prof. Wojciech Zamojski, now also the DepCoS Chairman. In this volume, we would like to include results of studies on selected problems of contemporary computer systems and networks, considered as complex systems, related to aspects of their dependability, safety and security. Proceedings of the previous DepCoS events was published (in historical order) by the IEEE Computer Society (2006–2009), by Wrocław University of Technology Publishing House (2010–2012) and since 2011 by Springer in AISC volumes no. 97 (2011), 170 (2012), 224 (2013), 286 (2014), 365 (2015), 479 (2016), 582 (2017), 761 (2018) and 987 (2019). Published by Springer Nature, one of the largest and most prestigious scientific publishers, the AISC series is one of the fastest growing book series in their programme. Its volumes are submitted for indexing in CORE Computing Research & Education database, ISI Conference Proceedings Citation Index (now run by Clarivate), Ei Compendex, DBLP, Scopus, Google Scholar and SpringerLink, and many other indexing services around the world.

DepCoS focus is on issues of dependability and performability—the topics which came naturally as an answer to new challenges in reliability and efficiency evaluation of contemporary computer systems. Being probably the most complex technical systems ever engineered by man, their organization cannot be interpreted only as (however complex and distributed) structures built on the base of technical resources (hardware) but their analysis must take into account a unique blend of interacting people (their needs and behaviours), networks (together with mobile properties, iCloud organization, Internet of Everything) and a large number of users dispersed geographically and producing an unimaginable number of applications. Ever-growing number of research methods being continuously developed for their analysis apply the newest results of artificial intelligence (AI) and computational intelligence (CI). Broad variety of topics in papers selected for this proceedings illustrate diversity of theoretical problems, methodologies and practical tools involved in these fields of human activity. Constant development and research progress have been reflected in evolution of topical range of subsequent DepCoS conferences over the past 14 years.

Preparations of this conference edition and in particular of this proceedings are taking place in a difficult time of SARS-CoV-2 pandemic when the whole world is facing unprecedented threats of both social and economic natures. This challenging period should open up new fields in our research, such as dependable cooperation between people (elements) integrated into one common chain (system structures) created for a global task of saving the population in a pandemic environment. Other new directions are associated with simulation of pandemic processes and the search for factors that can affect its spreading as well as forecasting its time parameters. We believe that we will discuss these problems during next conferences.

We would like to thank everyone who participated in organization of the conference and preparation of this volume—authors, members of the Programme Committee and the Organizing Committee, and all who helped in this difficult time. But especially, this proceedings would not be possible without invaluable contribution of 37 reviewers whose work and detailed comments have helped to select and refine conference submissions. The Programme Committee, organizers and the editors would like to emphasize and gratefully recognize participation in this process of the following experts: Andrzej Białas, Ilona Bluemke, Eugene Brezhniev, Dariusz Caban, Dejiu Chen, Frank Coolen, Mieczysław Drabowski, Zbigniew Gomółka, Alexander Grakovski, Ireneusz Jóźwiak, Igor Kabashkin, Vyacheslav Kharchenko, Artur Kierzkowski, Leszek Kotulski, Alexey Lastovetsky, Henryk Maciejewski, Jan Magott, István Majzik, Jacek Mazurkiewicz, Marek Młyńczak, Yiannis Papadopoulos, Rafał Scherer, Mirosław Siergiejczyk, Czesław Smutnicki, Robert Sobolewski, Janusz Sosnowski, Jarosław Sugier, Kamil Szyc, Przemyslaw Śliwiński, Tadeusz Tomczak, Victor Toporkov, Tomasz Walkowiak, Max Walter, Min Xie, Irina Yatskiv, Wojciech Zamojski and Wlodek Zuberek. Not mentioned anywhere else in this volume, their efforts deserve even more recognition in this introduction.

Finally, we would like to express our thanks to all authors who decided to publish and discuss their research results during the DepCoS Conference. We emphasize our hope that the included papers will contribute to further progress in design, analysis and engineering of dependability aspects of computer systems and networks (theory, engineering and applications), creating a valuable source material for scientists, researchers, practitioners and students who work in these areas.

Wojciech Zamojski
Jacek Mazurkiewicz
Jarosław Sugier
Tomasz Walkowiak
Janusz Kacprzyk

Organization

Fifteenth International Conference on Dependability of Computer Systems DepCoS-RELCOMEX

Brunów Palace, Poland, 29 June – 3 July 2020

Programme Committee

Wojciech Zamojski (Chairman)	Wrocław University of Science and Technology, Poland
Ali Al-Dahoud	AlZaytoonah University of Jordan, Amman, Jordan
Andrzej Białas	Research Network ŁUKASIEWICZ - Institute of Innovative Technologies EMAG, Katowice, Poland
Ilona Bluemke	Warsaw University of Technology, Poland
Wojciech Bożejko	Wrocław University of Science and Technology, Poland
Eugene Brezhniev	National Aerospace University "KhAI", Kharkiv, Ukraine
Dariusz Caban	Wrocław University of Science and Technology, Poland
Dejiu Chen	KTH Royal Institute of Technology, Stockholm, Sweden
Frank Coolen	Durham University, UK
Mieczysław Drabowski	Cracow University of Technology, Poland
Francesco Flammini	University of Naples "Federico II", Italy
Manuel Gill Perez	University of Murcia, Spain
Franciszek Grabski	Gdynia Maritime University, Gdynia, Poland
Aleksander Grakowskis	Transport and Telecommunication Institute, Riga, Latvia

Min Xie	City University of Hong Kong, Hong Kong SAR, China
Irina Yatskiv	Transport and Telecommunication Institute, Riga, Latvia
Włodzimierz Zuberek	Memorial University, St. John's, Canada

Organizing Committee

Chair

Wojciech Zamojski	Wrocław University of Science and Technology, Poland

Members

Jacek Mazurkiewicz	Wrocław University of Science and Technology, Poland
Jarosław Sugier	Wrocław University of Science and Technology, Poland
Tomasz Walkowiak	Wrocław University of Science and Technology, Poland
Mirosława Nurek	Wrocław University of Science and Technology, Poland

Contents

Sequence Mining and Property Verification for Fault-Localization in Simulink Models

Safa Aloui Dkhil[1,3](✉), Mohamed Taha Bennani[2], Manel Tekaya[3], and Houda Ben Attia Sethom[1,4]

[1] Ecole Nationale d'Ingénieurs de Tunis, Laboratoire des Systèmes Electriques, Université de Tunis El Manar, LR11ES15, 1002 Tunis, Tunisia
safa.aloui@enit.utm.tn

[2] Faculté des Sciences de Tunis, Université de Tunis El Manar, 1002 Tunis, Tunisia
taha.bennani@enit.utm.tn

[3] Akeed Solutions, Boulevard de la Terre, Centre Urbain Nord, 1002 Tunis, Tunisia
manel.tekaya@akeed-solutions.com

[4] Ecole Nationale d'Ingénieurs de Carthage, Université de Carthage, 2035 Charguia II, Tunisia
houda.benattia@enicarthage.rnu.tn

Abstract. This paper introduces a novel approach for diagnosing automotive systems and identifying faults at design-time, based on Sequence Mining and Property Verification for Fault Localization (i.e. SMPV4FL). After checking a single property several times, the verifier will generate various failures traces if the model holds a fault. Given their combination, we apply a semantic segmentation method, which extracts the values of the model's variables spanned over the period from the simulation start until the property violation. Then, we infer the root cause of the failure by mining the most frequent sequences of data. We have applied the SMPV4FL approach to a Simulink model of an automatic transmission system from the automotive domain. We have got promising results since we had identified the original fault of the model as well as most of the introduced faults during the evaluation process. Linked to our mutation analysis campaign, we have shown that SMPV4FL kills up to 87% of the mutants (i.e. introduced faults).

Keywords: Fault localization · Temporal property verification · Metric Temporal Logic (MTL) · Sequence mining · Automotive diagnosis systems · Simulink/Stateflow model

1 Introduction

Transportation systems diagnosis is a challenging task that has gained significant consideration in recent years [8, 20]. It consists in identifying the root of a fault which has led to the decay of the system. Therefore, it has to establish a causal link between

This project is carried out under the MOBIDOC scheme, funded by the EU through the EMORI program and managed by the ANPR.

W. Zamojski et al. (Eds.): DepCoS-RELCOMEX 2020, AISC 1173, pp. 1–10, 2020.
https://doi.org/10.1007/978-3-030-48256-5_1

an observed failure and its associated fault. A real-time system's verification process is a crucial task to overcome the fault identification problem. However, it commonly confronts a combinatorial explosion. To back this obstacle, diagnosis at design-time is a promising solution. Typically, the latter process relies on three complementary tasks: fault detection, fault isolation (localization), and fault identification [36]. Fault detection determines whether a fault has occurred, or the system is operating normally. Fault isolation lies in localizing the system's components causing the fault. Finally, fault identification estimates the size, severity, and time-variant behavior of the fault. In the literature, one of the well-used methods for fault localization in diagnosing automotive systems is model checking [4]. It consists of checking the temporal properties of reactive systems. Its major advantage is that it is fully automatic and returns usually a counter-example when the property is not verified. Authors in [23] propose a model checking approach. It uses formal specifications, expressed in Metric Temporal Logic (MTL), and a supervisor to check for correctness of simulation traces against the specification. However, it does not provide helpful information in localizing the failure within the model. As a solution, we present, in this paper, a new approach for fault localization in automotive systems modeled under the Simulink/Stateflow environment and based on temporal property violation. Given a set of traces violating the same property, we synthesize them into complex patterns recording the data-flow between different blocks of the model. Subsequently, we mine recurring sequence patterns that underline the frequent temporal behaviors in these traces. Those sequences will be used to identify the root cause of the fault.

This paper is organized as follows: Sect. 2 reviews and discusses some methods of fault localization proposed in the literature. Section 3 presents the new proposed approach of fault localization in Simulink/Stateflow models. Section 4 introduces an industrial case study that is evaluated in Sect. 5. Finally, Sect. 6 concludes the paper with some challenges and future trends.

2 Related Works

In the last two decades, system fault localization was and still one of the most widespread research areas [1, 4, 15, 19, 24, 26, 31]. It has been applied in many fields such as avionic, robotic, automotive, embedded and electrical systems. Many recent studies deal with fault localization in automotive systems modeled under Simulink/Stateflow environment [4, 19]. For instance, in [4], diagnosis has been performed following three steps: Fault detection step is ensured by checking whether a tested behavior satisfies or violates Signal Temporal Logic (STL) properties [23]. For the second step, it mainly maps the falsified behaviors to the states and transitions of the related Simulink model [30]. The last step [16] identifies the most suitable state/transition to cause the fault using Spectrum based fault-localization (SBFL) procedure. SBFL highlights the most likely element to cause the fault by attributing different scores to each state/transition.

Those scores are given according to their status (activated/deactivated) in failed tests. However, this measure may not reflect the real effect of each element. Indeed, a state/transition can be active during a failed scenario without being part of the fault cause. Authors of [3, 32] underline also the importance of generating test scenarios that violate temporal properties in fault localization when using Simulink/Stateflow models. This approach relies on the robustness metric for evaluating simulation outcomes as a cost function. S TaliRo toolbox [32] uses the same approach to generate random trajectories that violate MTL properties. In the literature, machine learning algorithms are also addressed to localize faults according to three main types of approaches: Classification, fault-injection, and data mining based approaches. Those approaches are applied to complex models in different domains such as photovoltaic [24], automotive [5], bioinformatics [2, 27] and software systems [6, 7, 22]. Classification based approaches [9, 17, 29] deal with fault localization as a simple problem of classification. They have been highly applied for automotive systems [5, 20]. However, their main drawback reside in training databases containing faults with their corresponding failures. It requires even expert intervention or a large set of fault-failure associations. Authors in [11, 28], have discussed fault-injection based approaches. They use targeted fault injection technique to construct the system failures database. Later, when failure is reported, the database is asked to find matched failures generated by fault injection. Information extracted from the matched failures is used as signs to allow the actual root cause identification of the reported failures. This is performed based on the hypothesis that similar faults produce similar failures. Although, the drawback of these approaches is that only the injected faults can be identified, and unexpected troubles cannot be localized. Finally, data mining based approaches deal with the challenges of discovering a huge number of patterns hidden in databases [34]. Authors of [6, 7, 22] prove the efficiency of data mining in localizing faults in software systems by mining the most suitable lines to cause the failed execution. Besides, Sequence pattern mining is a well-used method in diagnosing bioinformatics systems [2, 27]. It proves its efficiency in detecting Type-2 Diabetes [27] and heart diseases [2]. While the first and second approaches have prior knowledge of the faults before monitoring their influence over the system behavior, the third type explores and analyzes large quantities of information to ensure decision-making [10]. Inspired by the advantages of model checking and machine learning approaches, we propose a novel approach for fault localization in Simulink/Stateflow models called SMPV4FL. Indeed, it combines the use of temporal properties in the verification of Simulink/Stateflow models with machine learning-based approaches for fault localization. To the best of our knowledge, there is no research on mining sequence pattern algorithms to identify the root of unknown faults.

3 The Proposed Fault Localization Procedure

Our proposed approach takes as input a Simulink/Stateflow model and a MTL specification that describes the safety behavior of the model. It relies on three main steps: 1) Fault detection, 2) Fault isolation, and 3) Fault identification [36]. In the first step, a list of different execution scenarios (traces) that falsify the MTL property is generated using S TaliRo. We first sample these traces using a predefined step then save them into a multidimensional database D(v, s, t), where v is the number of variables, s is the number of samples and t is the number of traces. This database records the data-flow between the model's blocks during different faulty simulations. In the second step, we browse the lines corresponding to the variables concerned by the property. This process aims to identify the step "moment" when the property was violated. We partition the data-flow recorded into two main segments preceding and succeeding the property violation to process them within the next step. This procedure aims to highlight the behavior of the system that leads to the violation. This latter task is sometimes designated semantic segmentation [37]. Afterward, we label this data. This procedure consists in attaching to each recorded variable a unique id as a prefix. We transform each value into a string to which we add an identifier as a prefix. The output of this process is saved into the database D'(v, s', t) where s' $\leq$ s. This process doesn't affect the identification procedure. However, this will enable the variable's identification. Thereafter, we apply sequential pattern mining [13, 21] to extract the frequent subsequences of data from the labeled database. The input of this process is D'(v, s', t) = t.α which is a set of ordered item-sets denoted as α = h α_1, α_2, .., α_s0 i, α for each trace where α_i is a sample of data, $|\alpha_i|$ = v and $|\alpha|$ = s' defines the number of samples of the segmented database. The output of this process is a set of sequences β^k = h β_1, β_2, .., β_n i where $|\beta_i| \leq$ v and $|\beta^k| \leq$ s'. Those frequent sequences are interpreted as the most likely sub-sequences to model the propagation of the fault from its occurrence until the appearance of the failure. Several sequential pattern mining approaches have been proposed to deal with the efficiency of algorithm improvement [35]. The third step identifies the faulty variable by defining a score function. A score, assigned to each variable, considers the frequency at which the label of each variable appears in the k sub-sequences β^k. The faulty variable is elected with the highest score value. Finally, the data-flow analysis process considers the dependencies of this faulty variable and their scores to highlight the fault propagation in the model. SMPV4FL returns the variable that is at fault with its propagation root.

Trace retrieving, semantic segmentation and labeling functions are developed under the Matlab environment. The sequence mining procedure is a Java function developed in the open-source data mining library Sequential Pattern Mining Framework (SPMF) [12]. The score calculation and data-flow analysis process are currently performed manually.

4 Case Study

This section reveals the applicability of SMPV4FL in fault identification of real systems using a well-referenced benchmark from the literature [14].

4.1 Target System

To apply SMPV4FL, we used the Automatic transmission Controller model proposed in [14]. This model includes five blocks: Engine, Shift-Logic, Threshold Calculation, Transmission, and Vehicle. The Shift logic block which control the transmission ratio contains 19 State/Transitions. The Transmission subsystem is composed by the Torque

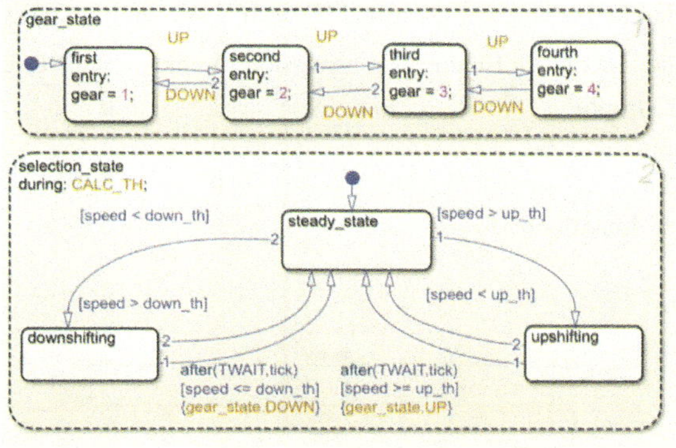

Fig. 1. Automatic transmission model [19]

Converter and the Transmission Ratio blocks. The whole system offers 23 potential mutations as mentioned in Table 1 which are mainly relational operator replacement. The Stateflow diagram depicted in Fig. 1 illustrates the functionality of the Shift-Logic block which implements the gear selection for the transmission [14].

4.2 Fault Detection

This section presents the execution of the first step of SMPV4FL described in Sect. 3. The proper functioning of the target system is resumed by a list of safety specifications. In this paper, we use the requirement that the engine and the vehicle speeds namely Wi and Vi never reach 4500 RPM and 120 km/h, respectively. We formalize this requirement as the MTL specification:

$$\Phi \equiv ((W_i \leq 4500) \wedge (V_i \leq 120))$$

We use S TaliRo to generate 4 traces that falsify the desired specification. Figure 2 illustrates one simulation trace that violates Φ. It comprises 751 samples of the 12 variables Throttle, Break, Speed, RPM, Gear, Ti, Down-Th, Up-Th, N-Out, T-Out, Tiin and Niin for a total simulation time equal to 30 s. For each trace, we extract the data describing the system's behavior and record the dataflow between different blocks in a multidimensional database D(12, 751, 4).

4.3 Fault Isolation

In this step, we first identify the property violation time. Afterward, we label these data. Then, we extract the most frequent sequences of data between different failed traces using BIDE[1] algorithm [33]. This algorithm performs in finding interesting patterns with reduced number of database scans to output patterns without candidate generation.

[1] **BI**-**D**irectional **E**xtension based frequent closed sequence mining algorithm.

Table 1. Mutants identified by SMPV4FL and S TaliRo

Model blocks	Engine	Transmission	Vehicle	Shift-Logic
Catastrophic	0	3	1	0
Abort	1	3	5	2
Silent	1	3	0	4
Total mutations	2	9	6	6

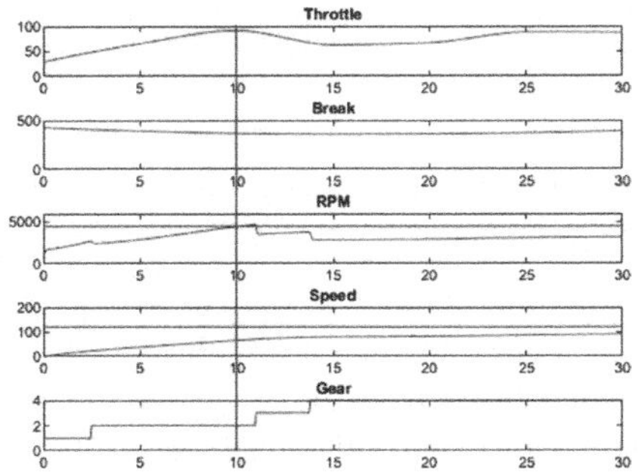

Fig. 2. Simulation trace that falsifies the property Φ

As we monitor 12 variables, $|\alpha_i| = 12$. Therefore, α_{i1} contains the value of throttle at the first sample. The input of this process is the labeled database D'(12, S', 4) where S' = Max(1 $\leq$ i $\leq$ 4) s'i where s'i is the number of samples obtained after performing segmentation in the trace i. For the example illustrated in Fig. 2, as the total simulation time is equal to 30 s sliced into 751 samples and the violation time is equal to 10 s, s'i = 250. The output of this process is 766 sub-sequences β^k.

Table 2. Different variable scores.

Variable	Throttle	Break	Speed	RPM	**Gear**	Ti	Down Th	Up Th	N Out	T Out	Tiin	Niin
Score	4	0	29	23	**439**	66	90	23	23	39	39	23

4.4 Fault Identification

In this step, we compute variable scores Sci $= \sum_{j=1}^{766} (F_{vi})$ where F_{vi} is the frequency of the label of the variable i in β^k. Results of the computation are illustrated in Table 2. We note that, in this case, the Gear, having the higher score, is the variable the most likely to cause a fault. By analyzing the gear selection process described in Fig. 1, we find that the gear depends on four different variables: Speed, Down Th, Up Th, and TWAIT. Results obtained in Table 2 indicate that the Speed, Down Th, and Up Th are not responsible for the fault. Consequently, we conclude that the fault is caused by the waiting time before changing the gear state (TWAIT). To fix this problem, we need to decrease TWAIT until reaching the optimal value that eliminates the violation.

5 Early Evaluation Results

In this section, we empirically evaluate SMPV4FL using experiments applied to the target system. Our evaluation is based on a mutation analysis [25] of the system which consists in replacing one operator in the model by another (Example $< \rightarrow >$). This technique is generally used to evaluate test quality. However, in this paper, we have applied it to introduce faults into the safe model. Our tool, SMPV4FL, will reveal the introduced fault unless the mutant (i.e. introduced faults) is not killed. The assessment process relies on two steps. First, we use a mutation operator to modify the initial model. Only one modification, at a time, is allowed in this paper. Then, we perform SMPV4FL against the mutated model to identify the data related to the modified operator. If SMPV4FL identifies the fault, it means that it has killed the mutant. The Tests were performed by a computer having a processor Intel(R) Xeon(R) CPU ES-2650 v2 @ 2.60 GHz equipped with 32.0 GB RAM.

5.1 Model Instrumentation

When applied to models, mutation analysis makes systematic changes to the model. The fixed target system presented in Sect. 4.1 offers a set of possible mutations as mentioned in Table 1. For each mutation, we generate 4 scenarios of violation, and we apply SMPV4FL to identify the root cause of the failure.

Table 3. Illustration of the results of identification of mutants

Model blocks	Engine	Transmission	Vehicle	Shift-Logic
% of mutants identified by SMPV4FL	100%	50%	67%	100%
% of mutants identified by S TaliRo	0%	17%	17%	0%
Total of identified Mutants	100%	67%	84%	100%

5.2 Observations and Metrics

According to crash observations presented by Koopman in [18], 5 observation levels exist: Catastrophic, Restart, Abort, Silent, and Hindering. By applying the set of mutations presented in Table 1, three observations appeared: The system malfunction is caused by the mutation (Catastrophic), the property violation is caused by the mutation (Abort), the property violation is not influenced by the mutation, so the system operates normally (Silent). The first type blocks the simulation; therefore, we are unable to identify the origin of the fault using SMPV4FL. However, for this type, the error can be returned and its origin is reported by S TaliRo such as a division by zero. The last type is a minor fault that doesn't affect the logic of the system. To test the reliability of our approach, we are interested in the first and second types of observations (Catastrophic and Abort). The metric used to evaluate SMPV4FL is the total percentage of identified mutants using SMPV4FL and S TaliRo respectively. Table 3 illustrates the results of mutants identification. The first line presents the percentage of mutants identified by SMPV4FL. However, the second line presents the first type (Catastrophic) errors returned by S TaliRo. The total identified mutants present the percentage of identified errors in the context of the first and third types of results.

6 Conclusion and Future Works

In this paper, we have introduced a new approach, called SMPV4FL, for localizing faults in Simulink/Stateflow models, which combines property verification with sequence mining. It could help engineers to identify the root of the failures through the design phase of a system. Property violation provides evidence of the model's failure, which would happen when the model contains at least one fault. Through different experiments, we have witnessed that verifying of a single property several times generates different failure traces. We claim that faulty data would keep the same behavior, which could be seen throughout the execution traces. Therefore, we start by extracting the data that span the fault life-cycle of different traces then we mine them to locate which one causes the failure. Our first experiment has identified the fault root of the automatic transmission system from the automotive domain. We have evaluated our approach using mutation analysis, which introduces a single fault into the initial model. Early results are promising since SMPV4FL has identified 87% of the introduced single faults. There are some open questions that we can summarize in three main lines for future work. The first is concerned with the investigation of non-faulty simulations to discriminate routine and faulty behavior before the fault identification process as provided by Bartocci et al. in [4] to upgrade our approach. The second is to extend the evaluation to more models and evaluate for each one to what extent the fault identification we can compute meets the real faults. The last one is concerned with the number of faults our approach may handle at a time.

References

1. Agarwal, P., Agrawal, A.P.: Fault-localization techniques for software systems: a literature review. ACM SIGSOFT Softw. Eng. Notes **39**(5), 1–8 (2014)
2. Bahrami, B., Shirvani, M.H.: Prediction and diagnosis of heart disease by data mining techniques. J. Multi. Eng. Sci. Technol. (JMEST) **2**(2), 164–168 (2015)
3. Balsini, A., Di Natale, M., Celia, M., Tsachouridis, V.: Generation of simulink monitors for control applications from formal requirements. In: 2017 12th IEEE International Symposium on Industrial Embedded Systems (SIES), pp. 1–9. IEEE (2017)
4. Bartocci, E., Ferrère, T., Manjunath, N., Ničković, D.: Localizing faults Insimulink/Stateflow models with STL. In: Proceedings of the 21st International Conference on Hybrid Systems: Computation and Control (part of CPS Week), pp. 197–206. ACM (2018)

5. Bennouna, O., Robin, O., Chafouk, H., Roux, J.: Diagnostic et détection de défauts des systèmes embarqués dans l'automobile. In: Proceeding of the 3SGS 2009 **3** (2009)
6. Cellier, P., Ducassé, M., Ferré, S., Ridoux, O.: DeLLIS: a data mining process for fault localization. In: SEKE, pp. 432–437 (2009)
7. Cellier, P., Ducassé, M., Ferré, S., Ridoux, O.: Multiple fault localization with datamining. In: SEKE, pp. 238–243 (2011)
8. Chen, H., Jiang, B., Chen, W., Yi, H.: Data-driven detection and diagnosis of incipient faults in electrical drives of high-speed trains. IEEE Trans. Industr. Electron. **66**(6), 4716–4725 (2018)
9. Darji, A., Darji, P., Pandya, D.: Fault diagnosis of ball bearing with WPT and supervised machine learning techniques. In: Machine Intelligence and Signal Analysis, pp. 291–301. Springer (2019)
10. Fayyad, U., Piatetsky-Shapiro, G., Smyth, P.: From data mining to knowledge discovery in databases. AI Mag. **17**(3), 37 (1996)
11. Formicola, V., Jha, S., Chen, D., Deng, F., Bonnie, A., Mason, M., Brandt, J.,Gentile, A., Kaplan, L., Repik, J., et al.: Understanding fault scenarios and impacts through fault injection experiments in Cielo. arXiv preprint arXiv:1907.01019 (2019)
12. Fournier-Viger, P., Gomariz, A., Gueniche, T., Soltani, A., Wu, C.W., Tseng, V.S.: SPMF: a java open-source pattern mining library. J. Mach. Learn. Res. **15**(1), 3389–3393 (2014)
13. Fournier-Viger, P., Lin, J.C.W., Kiran, R.U., Koh, Y.S., Thomas, R.: A survey of sequential pattern mining. Data Sci. Pattern Recogn. **1**(1), 54–77 (2017)
14. Hoxha, B., Abbas, H., Fainekos, G.E.: Benchmarks for temporal logic requirements for automotive systems. ARCH@CPSWeek **34**, 25–30 (2014)
15. Isermann, R.: Model-based fault-detection and diagnosis–status and applications. Ann. Rev. Control **29**(1), 71–85 (2005)
16. Jones, J.A., Harrold, M.J.: Empirical evaluation of the tarantula automatic fault localization technique. In: Proceedings of the 20th IEEE/ACM International Conference on Automated Software Engineering, pp. 273–282. ACM (2005)
17. Jung, M., Niculita, O., Skaf, Z.: Comparison of different classification algorithms for fault detection and fault isolation in complex systems. Procedia Manuf. **19**, 111–118 (2018)
18. Koopman, P., Sung, J., Dingman, C., Siewiorek, D., Marz, T.: Comparing operating systems using robustness benchmarks. In: Proceedings of SRDS 1997: 16th IEEE Symposium on Reliable Distributed Systems, pp. 72–79. IEEE (1997)
19. Liu, B., Nejati, S., Briand, L., Bruckmann, T., et al.: Localizing multiple faults in simulink models. In: 2016 IEEE 23rd International Conference on Software Analysis, Evolution, and Reengineering (SANER), vol. 1, pp. 146–156. IEEE (2016)
20. Liu, B., Nejati, S., Briand, L.C., et al.: Effective fault localization of automotive simulink models: achieving the trade-off between test oracle effort and fault localization accuracy. Empirical Softw. Eng. **24**(1), 444–490 (2019)
21. Liu, K., Kim, D., Bissyandé, T.F., Yoo, S., Le Traon, Y.: Mining fix patterns for findbugs violations. IEEE Trans. Softw. Eng. (2018)
22. Maamar, M., Lazaar, N., Loudni, S., Lebbah, Y.: Fault localization using item set mining under constraints. Autom. Softw. Eng. **24**(2), 341–368 (2017)
23. Maler, O., Ničković, D.: Monitoring properties of analog and mixed-signal circuits. Int. J. Softw. Tools Technol. Transfer **15**(3), 247–268 (2013)
24. Mellit, A., Tina, G.M., Kalogirou, S.A.: Fault detection and diagnosis methods for photovoltaic systems: a review. Renew. Sustain. Energy Rev. **91**, 1–17 (2018)
25. Mottu, J.M., Baudry, B., Le Traon, Y.: Mutation analysis testing for model transformations. In: European Conference on Model Driven Architecture-Foundations and Applications, pp. 376–390. Springer (2006)

26. Nandi, S., Toliyat, H.A., Li, X.: Condition monitoring and fault diagnosis of electrical motors—a review. IEEE Trans. Energy Convers. **20**(4), 719–729 (2005)
27. Nithya, P., Maheswari, B.U., Deepa, R.: Efficient sequential pattern mining algorithm to detect type-2 diabetes. Int. J. Adv. Res. Sci. Eng. Technol. **3**(3) (2016)
28. Pham, C., Wang, L., Tak, B.C., Baset, S., Tang, C., Kalbarczyk, Z., Iyer, R.K.: Failure diagnosis for distributed systems using targeted fault injection. IEEE Trans. Parallel Distrib. Syst. **28**(2), 503–516 (2016)
29. Ramos, A.R., García, R.D., Galdeano, J.L.V., Santiago, O.L.: Fault diagnosis in a steam generator applying fuzzy clustering techniques. In: Soft Computing for Sustainability Science, pp. 217–234. Springer (2018)
30. Reicherdt, R., Glesner, S.: Slicing matlab simulink models. In: 2012 34th International Conference on Software Engineering (ICSE), pp. 551–561. IEEE (2012)
31. Triki-Lahiani, A., Abdelghani, A.B.B., Slama-Belkhodja, I.: Fault detection and monitoring systems for photovoltaic installations: a review. Renew. Sustain. Energy Rev. **82**, 2680–2692 (2018)
32. Tuncali, C.E., Pavlic, T.P., Fainekos, G.: Utilizing s-TaLiRo as an automatic test generation framework for autonomous vehicles. In: 2016 IEEE 19th International Conference on Intelligent Transportation Systems (ITSC), pp. 1470–1475. IEEE (2016)
33. Wang, J., Han, J.: BIDE: efficient mining of frequent closed sequences. In: Proceedings, 20th International Conference on Data Engineering, pp. 79–90. IEEE (2004)
34. Yang, H., Hoxha, B., Fainekos, G.: Querying parametric temporal logic propertieson embedded systems. In: IFIP International Conference on Testing Software and Systems, pp. 136–151. Springer (2012)
35. Zaki, M.J.: SPADE: an efficient algorithm for mining frequent sequences. Machinelearning **42**(1–2), 31–60 (2001). https://doi.org/10.1023/A:1007652502315
36. Zaytoon, J., Lafortune, S.: Overview of fault diagnosis methods for discrete event systems. Ann. Rev. Control **37**(2), 308–320 (2013)
37. Zhao, J., Itti, L.: Decomposing time series with application to temporal segmentation. In: 2016 IEEE Winter Conference on Applications of Computer Vision (WACV), pp. 1–9. IEEE (2016)

Handwritten Text Lines Segmentation Using Two Column Projection

Tomasz Babczyński(✉) and Roman Ptak

Department of Computer Engineering, Wrocław University of Science and Technology, Wyb. Wyspiańskiego 27, 50-370 Wrocław, Poland
{tomasz.babczynski,roman.ptak}@pwr.edu.pl

Abstract. Libraries and archives throughout the world keep a great number of historical documents. Many of them are handwritten ones. In many cases they are hardly available. Digitization of such artifacts can make them accessible to the community. But even digitized, they remain unsearchable so the important task is to draw the contents in the computer readable form. One of the first steps on this way is segmentation of the document into the lines. Artificial Intelligence algorithms can be used to solve this problem. In the current paper the projection–based algorithm is presented. Our algorithm finds lines in the left and the right part of the page independently and then associates both sets. Thanks to this, our method can recognize skewed lines better then the algorithms that use global projection. The performance of the algorithm is evaluated on the data-set and with the procedure proposed by the organizers of the ICDAR2009 competition.

Keywords: Document image analysis · Off-line cursive script recognition · Handwritten text line segmentation · Projection profile · ICDAR09 competition

1 Introduction

Document image analysis is one of the important tasks of Artificial Intelligence. Various paper documents and manuscripts including historical primary sources are examined frequently. There are various aims in the document examination. Obviously, the recognition of the manuscript text is the main purpose. In this case, algorithms focus on the obtaining of the contents of handwritten text. Another task related to the manuscript is the writer identification. The writer identification is often carried out for the forensic purposes.

Reliability of a decision-making process based on pattern recognition depends on quality of pre-processing (including segmentation). This applies, for example, to postal codes or addresses on envelopes recognition systems or document (also handwritten) processing (e.g. automatic document sorting or indexing) systems. That is the reason why it is important to perform proper text line localization

W. Zamojski et al. (Eds.): DepCoS-RELCOMEX 2020, AISC 1173, pp. 11–20, 2020.
https://doi.org/10.1007/978-3-030-48256-5_2

and segmentation. The localization is usually the first stage of text line segmentation. In turn, the line segmentation often precedes segmentation into words. The text may be further divided into letters and then the actual recognition can be performed. Therefore, identifications of the verses in the text constitutes the initial step that cannot be omitted. In the present article, we focus on the problem of the text lines segmentation using projection method.

Projection consists of calculating, in a given direction, the number of foreground pixels. Generally horizontal and vertical projections are used for segmentation. In case of line segmentation, horizontal projection is usually employed. Data are examined as the horizontal projection profile of an image (also presented as the histogram). This allows for significant reduction of the amount of information to be analyzed. Values of the histogram represent the density distribution of handwriting. It is possible to perceive areas of letter concentration. In the simplest case a thresholding is used to separate text lines.

The natural way to improve the effectiveness of segmentation algorithm is to calculate histograms piece-wise and examine them independently. Analyzing of many histograms is time consuming so the important question is how many of them is sufficient to give satisfactory results. The aim of the present paper is to check whether splitting of the image into two columns is going to improve the results of analysis and to what extent.

2 Related Work

Problem of text lines segmentation was investigated using many various approaches. A good survey of the methods applied to solve this question is presented in the paper [9]. Comparison of the algorithms can be found in [13]. Various techniques used to line and word segmentation of handwritten document images are compared in the paper [16].

One group of algorithms are these that use projection profiles. They are used for segmentation of printed or handwritten texts and also for other purposes. For example, the article [8] shows color band detection for electrical resistors using vertical projection. Because in many language systems the text is horizontally aligned (the text is running from left to right or right to left), horizontal projection is used to segment the text lines. Fast and rather simple method is global horizontal projection of the pixels followed by the analysis of the obtained histogram. Methods of this kind are quick but not too smart. One of them can be found in the paper [12]. The authors proposed a text line segmentation algorithm based on the projection profile with a variable threshold. The threshold in the method was adaptively tuned and was different for each peak being proportional to its height.

Handwritten text lines can be slanted, undulated and curved in different skew angles. Global projection methods are sensitive to the skew of the lines in the document. In effect segmentation result is not good. To improve results, in piece-wise methods the document is divided into non overlapping vertical parts, and the projection is applied to each one. The partial projections are common methods based on the histogram computing. The parts of document image are also called columns, vertical strips etc.

There are many algorithms applying piece-wise projection profiles, eg. [2–4], and [5].

In work [15], the number of parts, the image may be divided into, is not fixed, but it is defined iteratively checking the segmentation results. In [10] the image is divided in this way that stripe width of a document is computed by statistical analysis of the text height in the document.

The method described in [17] use partial projection combining with partial contour following of every line in the direction of the writing and in the opposite direction. The proposed implementation for Arabic handwriting divide the image of the document into parts of the width approximately equal to the one word width.

In [1], a different approach is used, the image is divided into 20 strips. Smoothed projection profiles are calculated for each of the 5% image strips. The projection profiles of the first 25% of strips are used to determine an initial set of candidate separating lines. The text lines traverse around any obstructing handwritten connected component by associating it to the one of neighbor lines. An association decision is made by modeling the lines as bi-variate Gaussian densities and evaluating the probability of the component under each Gaussian or by the probability obtained from a distance metrics. This method is robust to handwritten documents with lines running into each other but is computationally complicated.

In this work, we propose to divide the image into only two parts in which the projection profiles are calculated.

3 Line Segmentation

The labeling procedure (in segmentation process) consists of several major stages. They are summarized in the algorithm 1 and described below.

Algorithm 1. The algorithm of lines segmentation.

Input: I_{in} {binary image}
{**Parameters:**
t– relative threshold within the interval (0, 1)
w– window size of smoothing histograms }
Output: I_{out} {segmented image}
1: Make projections of left and right part of the image smoothing them using the window of size w
2: Calculate left and right positions of text lines using threshold t
3: Obtain the mapping between both sets of points
4: Compute lines of text
5: Labeling
6: **return** I_{out}

3.1 Half Projections

First, (stage 1) the projections of the left and the right half part of the image are taken. Next, they are smoothed using moving average over the data, i.e. at each point, its value is substituted with the mean value in the $(-w/2 \div w/2)$ window centered at the point. In experiments we also tested smoothing with the Gaussian function $g_{\mu,\sigma}(x)$ with $\mu = 0, \sigma = w/5$. Exemplary result of this stage is shown in the Fig. 1.

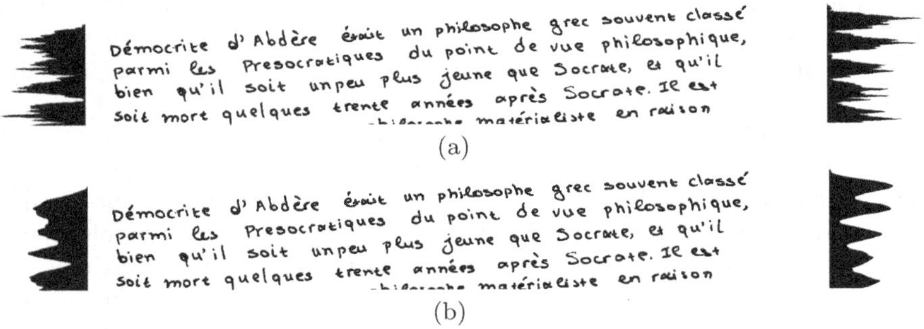

Fig. 1. (a) Projection profiles for the left and the right half of the text, (b) the profiles smoothed using moving average with the window size 31.

Then, each projection profile is analyzed (stage 2) using the variable threshold algorithm from the paper [12]. The points of the profile are processed in a descending sequence starting from the point of the maximum value. Starting at each point, the width of the peak to which it belongs is determined at a certain height t_a. Its value in proportion to the height of the peak h_p is equal to the threshold $t_a = t \cdot h_p$. The $t \in (0, 1)$ is the global parameter of the algorithm. The width of the peak is defined as the size of the range R of arguments having the value greater than the threshold t_a. If a range R does not overlap any of the previously determined ranges, it is accepted as a text line and added to the set lL for the "left profile" or lR for the "right". Otherwise it is rejected to prevent the connection of overlapping ranges which would cause recognition of two or more text lines as the single line. All arguments in the range R are marked as checked. The process terminates when the value of a given point is less than $\alpha = 0.1$ of the maximum value of the profile.

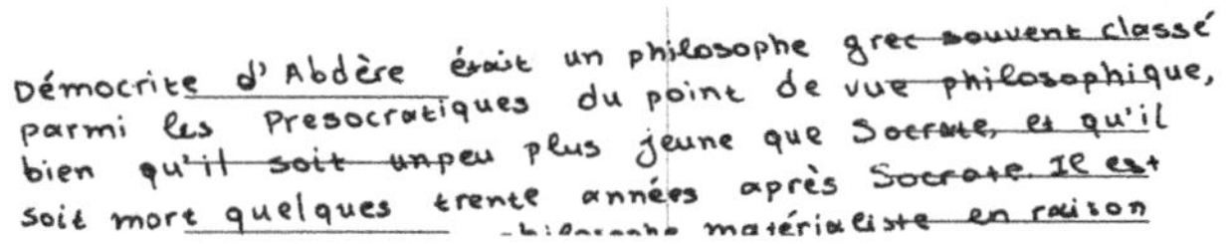

Fig. 2. Lines detected in the left and the right part of the image.

As the result of this stage, two vectors containing the vertical coordinates of the text lines in the left (lL) and the right (lR) part of the image are calculated. This result is shown in the Fig. 2.

3.2 Line Pairing

Now, the two lists lL and lR containing coordinates of lines detected on the left and right side of the document should be paired (stage 3). For each element, the corresponding line at the opposite list should be pointed. Alternatively, it should be stated that a line has no twin.

Our "*jagged combs*" algorithm works as follow. At the beginning, the cross-correlation (Eq. 1) between the two profiles is calculated. The shift (n) here is limited to only $\pm 3/4$ of the mean distance between detected lines for the performance but also for accuracy reasons. The h is the height of the picture or the size of each profile, $|.|$ is the absolute value. Profiles are zero padded at both ends if needed.

$$(f \star g)[n] = \sum_{m=-|n|}^{h+|n|} f[m]g[m+n] \tag{1}$$

The location of the correlation maximum determines the *offset* between the two profiles and, indirectly, the skewness of the text lines. In the Fig. 3 we can see the exemplary chart of the cross-correlation. Next, the actual pairing is performed. For each line from the left set shifted by the offset, the counterpart is looked for. If the distance to the closest line in the right set is less then 1/3 of the mean distance between lines, then this line is counted as the twin of the analyzed one. In the opposite case – the line is counted as unpaired. The same is done with the right set, pairing it with the left one. As the result, two index lists, containing numbers of lines at the opposite side or zeroes in case of unmatched lines, are created.

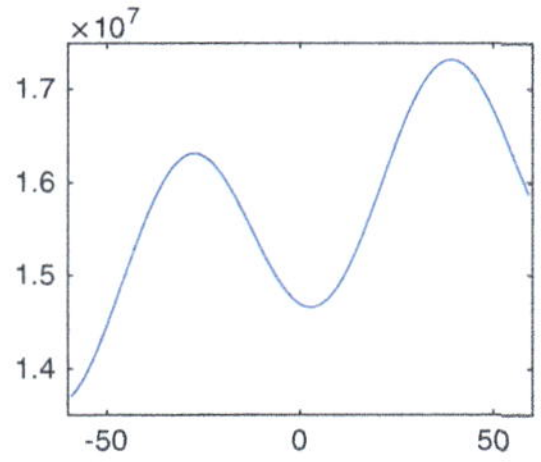

Fig. 3. Limited cross-correlation of the profiles.

The index lists may contain non unique indices when a few lines point the same line at the other side. In this case, only the best matching, i.e. with the smallest distance, is saved. All the others are marked as unpaired.

After this sanity check, two index lists – *indL* and *indR* – together with coordinates *lL* and *lR* can be used to determine the probable course of lines throughout the full width of the text (stage 4). For the paired lines, the left and right coordinates are taken. For the others, the global skewness is assumed. The result of this stage is the numbered list of skewed lines used for text labeling in the next stage. The Fig. 4a presents the lines for the demonstrative example. For some values of parameters the offset is determined incorrectly leading to completely ill results as shown in the Fig. 5a.

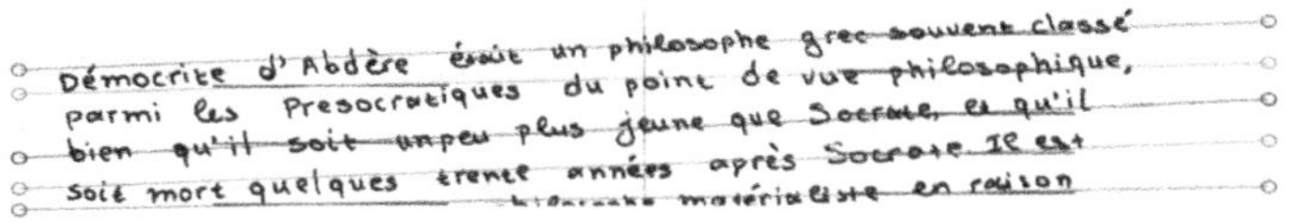

(a)

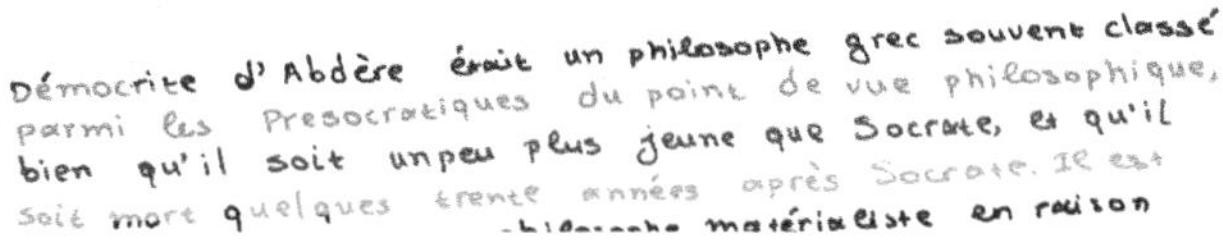

(b)

Fig. 4. (a) The line chains of text, (b) labeled text using the line chains.

3.3 Line Labeling

The labeling (the stage 5) is performed in the following way. The connected regions (in the sense of connected sets of foreground pixels) in the image are identified and analyzed one by one. If the region is touched by one line then all points in this region are labeled with the number of the line. In the opposite case (i.e. no line or more then one line touches the region), each point in the region is labeled with the number of the nearest (using the euclidean distance) line. The results of the labeling are shown in the Figs. 4b and 5b for the good and bad offset value respectively.

The document used here as an example is one of rather poorly recognized documents. Even for properly calculated value of the offset, only one, the third line is correctly labeled. The 2nd line seems good but there is a fragment of the first line recognized as the second one and therefore this line is not counted as one-to-one match by the evaluation procedure described in the next section.

4 Experiments

4.1 Data-Set and Evaluation Methodology

The set of handwritten documents on which we evaluated our algorithm was taken from the materials of the challenge *ICDAR 2009 Handwriting Segmentation Contest* performed during the ICDAR 2009 conference [6]. The data-set

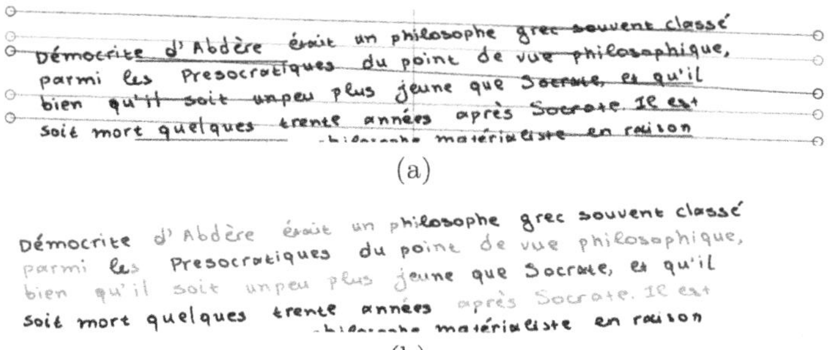

(a)

(b)

Fig. 5. Incorrectly paired lines example: (a) The line chains of text, (b) labeled text using the line chains.

contained 200 1-page handwritten documents in four languages (English, French, German and Greek) written by many writers. The documents were binarized encoding background with zeroes and foreground as ones. The challenge had two parts – line segmentation and word segmentation. We used only data of the first part. Each document was manually annotated by the organizers of the competition to make the *ground truth* which was used to evaluate the participants' results. Each pixel of the image got a label informing to which line it belonged.

The evaluation of the result, during the challenge, was based on the Match-Score table using the one-to-one matching defined in [11]. Detailed description of the performance evaluation is contained in the post-competition report [6]. In the current paper we use the metric applied during the cited competition to compare the results with the attenders of the ICDAR competition.

The MatchScore table (Eq. 2) is constructed as follows. Let I be the set of foreground pixels in the image, R_i the set of pixels recognized as belonged to the ith class, G_j the set of pixels in the jth class of the ground truth. $T(s)$ is a function giving the number of elements in the set s. The MatchScore table takes values in the range $[0, 1]$.

$$\text{MatchScore}(i, j) = \frac{T(R_i \cap G_j \cap I)}{T((R_i \cup G_j) \cap I)} \tag{2}$$

The line i is treated as one-to-one match with the ground truth line j only if the value MatchScore(i, j) is greater then the threshold $T_a = 0.95$. That value was accepted during the ICDAR challenge. Let now the M be the number of recognized lines, N – the number of lines in the ground truth, $o2o$ – the number of one-to-one matches.

The detection rate (DR) and recognition accuracy (RA) metrics are defined in the Eq. 3 as well as the aggregated value FM which was used to range applications during the competition.

$$DR = \frac{o2o}{N},\ RA = \frac{o2o}{M},\ FM = \frac{2 \cdot DR \cdot RA}{DR + RA} \tag{3}$$

The similar competitions were also performed during the conferences ICFHR 2010 [7] and ICDAR 2013 [14] with their own data-sets similar to the above presented one. The goal of next editions of the text segmentation competitions was the baseline detection instead of just segmentation. This problem is harder to resolve and needs more post-processing steps then presented here.

4.2 Experimental Results

The experiments were carried out on the previously described data-set of handwritten documents. We compared the detective performance of the algorithm described in the Sect. 3 for various parameters' values. The threshold t was varied in the range $(0.3 \div 1)$, the smoothing window size w took values from the range $(0 \div 70)$, where 0 meant no smoothing. Additionally, two smoothing patterns were used – rectangular in simple moving average filtering and Gaussian with standard deviation $\sigma = w/5$.

The result for moving average smoothing is shown in the Fig. 6a. It can be seen that for the wide ranges of parameters' values, the FM metric takes values above 95%. The maximum $FM = 95.72\%$ is for the $w = 31, t = 0.7$.

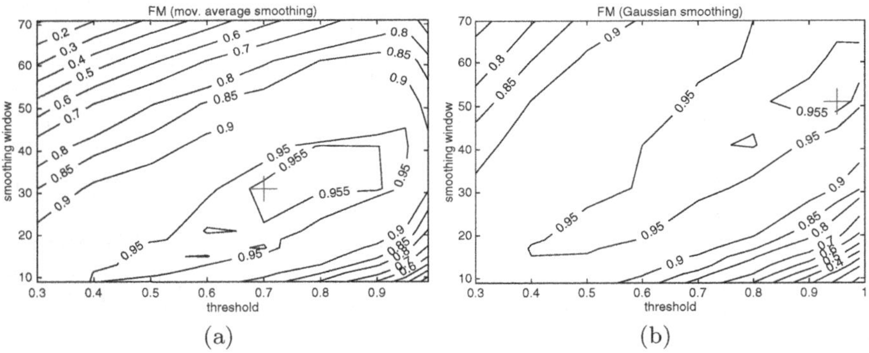

Fig. 6. FM metric for histograms smoothed using (a) moving average, (b) Gaussian shape

In the Fig. 6b, the results for Gaussian smoothing are presented. Here, the area of the peak plateau is even greater than in the previous case. For this case the maximal value of the $FM = 95.79\%$ for the $w = 51, t = 0.95$.

The position of our algorithm (**Proj2M/G**) among the programs of the attenders of the ICDAR competition can be found in the Table 1. Additionally, two algorithm not attending the challenge are presented - one from the paper [12], marked as *ProjFull* and the *Projections* algorithm presented in the post competition report as the reference algorithm. They are marked with italic font and are placed here because they belong to the same group of algorithms. The *ProjFull* uses the same procedure of finding lines in histograms as the presented here but without dividing the text into slices.

Table 1. ICDAR 2009 results.

	DR[%]	RA[%]	FM[%]
CUBS	99.55	99.50	99.53
ILSP-LWSeg-09	99.16	98.94	99.05
PAIS	98.49	98.56	98.52
CMM	98.54	98.29	98.42
Proj2G	95.96	95.63	95.79
Proj2M	95.69	95.76	95.72
CASIA-MSTSeg	95.86	95.51	95.68
PortoUniv	94.47	94.61	94.54
PPSL	94.00	92.85	93.42
LRDE	96.70	88.20	92.25
Jadavpur&Univ	87.78	86.90	87.34
ETS	86.66	86.68	86.67
AegeanUniv	77.59	77.21	77.40
ProjFull	81.45	73.67	77.37
Projections	62.92	57.80	60.25
REGIM	40.38	35.70	37.90

5 Conclusions

Line segmentation in handwritten texts using the projection method is sensitive to irregularities like skew or nonlinearity of the writing. The present paper proves that in some cases a little modification like taking two histograms of the two halves of the text is sufficient. Applying a split into two columns improves the segmentation results (FM) from 77.37% for the global projection algorithm to 95.72% and 95.79%. The proposed algorithm substantially improve the efficiency of the segmentation. This would place the developed algorithms at the fifth and sixth position in ICDAR 2009 results.

The possible directions of future work would be e.g. to evaluate the influence of choosing number of strips in piece-wise algorithms on the line detection capabilities. Another factor that may affect the performance of the algorithm is a method of identifying corresponding lines.

References

1. Arivazhagan, M., Srinivasan, H., Srihari, S.: A statistical approach to line segmentation in handwritten documents. In: Document Recognition and Retrieval XIV, vol. 6500, p. 65000T. International Society for Optics and Photonics (2007)
2. Sakhi, O.B.: Segmentation of heterogeneous document images: an approach based on machine learning, connected components analysis, and texture analysis. Theses, Université Paris-Est, December 2012. https://tel.archives-ouvertes.fr/tel-00912566

3. Gao, Y., Ding, X., Liu, C.: A multi-scale text line segmentation method in freestyle handwritten documents. In: 2011 International Conference on Document Analysis and Recognition, pp. 643–647, October 2011. https://doi.org/10.1109/ICDAR.2011.135
4. Garg, R., Garg, N.K.: An algorithm for text line segmentation in handwritten skewed and overlapped Devanagari script. Int. J. Emerg. Technol. Adv. Eng. **4**(5), 114–118 (2014)
5. Gari, A., Khassidi, G., Mrabti, M.: Novel approach to extract lines of documents for blind and impaired people. Int. J. Comput. Appl. **146**(14), 24–27 (2016). https://doi.org/10.5120/ijca2016910941
6. Gatos, B., Stamatopoulos, N., Louloudis, G.: ICDAR 2009 handwriting segmentation contest. Int. J. Doc. Anal. Recogn. (IJDAR) **14**(1), 25–33 (2011)
7. Gatos, B., Stamatopoulos, N., Louloudis, G.: ICFHR 2010 handwriting segmentation contest. In: 2010 12th International Conference on Frontiers in Handwriting Recognition, pp. 737–742. IEEE (2010)
8. Jung, M.C.: Recognition of resistor color band using a color segmentation in a HSI color model. J. Semicond. Disp. Technol. **18**(2), 67–72 (2019)
9. Likforman-Sulem, L., Zahour, A., Taconet, B.: Text line segmentation of historical documents: a survey. IJDAR **9**(2), 123–138 (2007)
10. Pal, U., Datta, S.: Segmentation of Bangla unconstrained handwritten text. In: Proceedings Seventh International Conference on Document Analysis and Recognition, pp. 1128–1132, January 2003. https://doi.org/10.1109/ICDAR.2003.1227832
11. Phillips, I.T., Chhabra, A.K.: Empirical performance evaluation of graphics recognition systems. IEEE Trans. Pattern Anal. Mach. Intell. **21**(9), 849–870 (1999)
12. Ptak, R., Żygadło, B., Unold, O.: Projection-based text line segmentation with a variable threshold. Int. J. Appl. Math. Comput. Sci. **27**(1), 195–206 (2017). https://doi.org/10.1515/amcs-2017-0014
13. Razak, Z., Zulkiflee, K., Idris, M.Y.I., Tamil, E.M., Noorzaily, M., Noor, M., Salleh, R., Yaakob, M., Yusof, Z.M., Yaacob, M.: Off-line handwriting text line segmentation: a review. Int. J. Comput. Sci. Netw. Secur. **8**(7), 12–20 (2008)
14. Stamatopoulos, N., Gatos, B., Louloudis, G., Pal, U., Alaei, A.: ICDAR 2013 handwriting segmentation contest. In: 2013 12th International Conference on Document Analysis and Recognition, pp. 1402–1406. IEEE (2013)
15. Venturelli, F.: Successful technique for unconstrained hand-written line segmentation. In: Downton, A.C. (ed.) Progress in Handwriting Recognition, pp. 563–568. World Scientific, Singapore (1997)
16. Vishwanath, N.V., Murugan, R., Kumar, S.N.: A comparative analysis of line and word segmentation for handwritten document image. Int. J. Adv. Res. Comput. Sci. **9**(1), 514–519 (2018)
17. Zahour, A., Taconet, B., Mercy, P., Ramdane, S.: Arabic hand-written text-line extraction. In: Proceedings of Sixth International Conference on Document Analysis and Recognition, pp. 281–285, September 2001

Convolutional Neural Networks for Dot Counting in Fluorescence in Situ Hybridization Imaging

Adrian Banachowicz[1], Anna Lis-Nawara[2], Michał Jeleń[2], and Łukasz Jeleń[1(✉)]

[1] Department of Computer Engineering, Wrocław University of Science and Technology, Wybrzeże Wyspiańskiego 27, 50-370, Wrocław, Poland
{235394,lukasz.jelen}@pwr.edu.pl
[2] Department of Immunopathology and Molecular Biology, Wrocław Medical University, ul. Borowska 213, 50-556, Wrocław, Poland
{anna.lis-nawara,michal.jelen}@umed.wroc.pl

Abstract. During breast cancer a small tissue sample is extracted and evaluated to estimate the malignancy of the growth and the possible treatment. When a difficult case is diagnosed, a more accurate diagnosis based on fluorescence in situ hybridization imaging is performed where HER2 and CEN-17 reactions are determined. In this paper we address a problem of immunohistochemistry reaction detection that are often tested when it is difficult to decide on the type of treatment the patient should undergo. Here we describe a segmentation framework adopting convolutional neural networks that are able to classify image pixels into HER2 and CEN-17 reactions respectively. Using the above mentioned framework, the proposed system is able to keep the high segmentation accuracy.

Keywords: FISH · Breast cancer classification · CNN · Image segmentation · Computer aided diagnosis · Breast cancer · HER2 · Dot counting

1 Introduction

Recently, the development of automated systems has influenced most of the areas of human life. The development of computational technology has raised an attention of the medical research and allowed for computer science techniques to be exploited during the diagnosis process [4]. One of the techniques that has influenced the decision making process in medicine is machine learning which allows for efficient and a more accurate diagnosis [4]. According to Yüskel [23] these methods do not always deliver the satisfactory results and deep neural networks are superior. Due to the fact that trained neural networks are able to

W. Zamojski et al. (Eds.): DepCoS-RELCOMEX 2020, AISC 1173, pp. 21–30, 2020.
https://doi.org/10.1007/978-3-030-48256-5_3

detect prevailing patterns in the data, they can be applied to the detection and classification of elements in the image.

According to statistics of the National Cancer Registry, breast cancer is one of the most often diagnosed cancers among middle–age women [1]. Before 2020 there were 17142 diagnosed cases of breast cancer in Poland alone. The number of cases increases year after year. The same records show that out of 17142 cancer cases there were 5651 deaths and that number also increases every year. The key to reduce these rates is to make a diagnosis in the early stage of the disease because cancers in their early stages are vulnerable to treatment [18].

During the diagnosis when a suspicious growth is found in mammography tests a biopsy is taken. At that time a small sample from the questionable breast tissue is extracted and a prognostic factor is assigned during a procedure called a malignancy grading. This grading allows for the description of the type of cancer in detail and estimate its behavior with or without undertaking treatment. When a more accurate diagnosis is required, a set of additional examinations are performed to assess a presence of a HER2 gene and HER2 receptors that stimulate the growth of cancer cells. As described by Hicks and Schiffhaue [11] the treatment is chosen accordingly to human epidermal growth factor receptor 2 (HER2) expression status. To determine the status of HER2 breast cancer biomarker, an Immunohistochemistry (IHC) or Fluorescence In Situ Hybridization (FISH) tests are performed.

Immunohistochemistry helps in the identification of the antigens in cells by staining the HER2 and hormone receptors to be visible on the surface of the cancer cells. This test helps in identification of the antigens in cells. Figure 1a show an example of the immunohistochemistry staining. The final decision is based on the estimation of different markers that may appear within and around the tumor cells [24].

Fluorescence in situ hybridization allows for a visualization of HER2 gene and the determination of its additional copies inside the cell [13]. The rule here is that the more genes one can distinguish, the more HER2 receptors the cells have.

As mentioned by American Society of Clinical Oncologists and College of American Pathologists (ASCO/CAP), the FISH examination requires estimation of chromosome 17 centromere enumeration probe (CEP17 or CEN-17) [20] and the final decision depends on the HER2 to CEN-17 ratio [7]. Here we concentrate on FISH examination.

In literature we can find different attempts of distinguishing the HER2 and CEN-17 reaction within the FISH slides [3,16,19]. This procedure is referred to as dot counting. Proper segmentation of the reactions is the biggest challenge, where the HER2 reactions are visible in the image as red dots while the CEN-17 are detectable as green dots. As a result we need to find the best possible segmentation algorithm that will be able to localize both kinds of dots within the slide. Here we took an opportunity to test the ability of Convolutional Neural Networks to segment the slide regions and classify them as HER2 or CEN-17 reactions.

a) b)

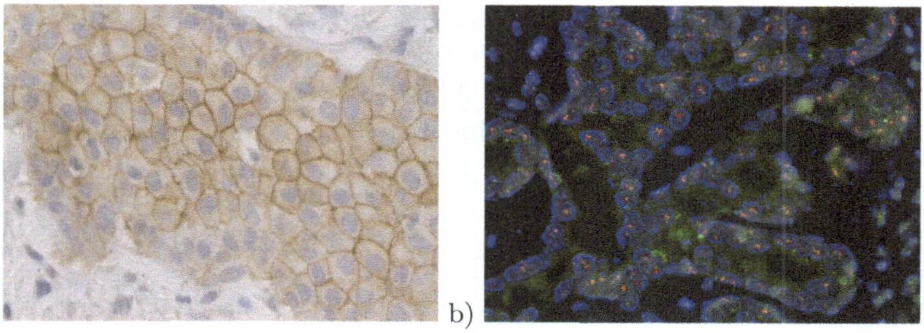

Fig. 1. HER2 slides. a) Immunohistochemistry staining example, b) FISH test example.

Studying the literature on automatic cancer nuclei detection and segmentation from medical images one can find many reports on applying different imaging techniques and segmentation methods [10,17].

Substantial portion of the reports on application of neural networks for segmentation describes methodologies based on self–organized maps (SOMs) as described by Yao *et al.* [22] who segmented sonar images where each pixel of the input image was classified with SOMs. In 2001, Lerner *et al.* [3] described a neural network approach for the detection of fluorescence in situ hybridization images. Method described by authors was able to distinguish between a pair of in– and out–of–focus images with an accuracy of 83–87%. The in–focus images were used for further estimation of the FISH reactions. In 2013, Kiszler *et al.* [13] described a semi–automated approach based on the adaptive thresholding where the final counting was based on the selected areas of the FISH image.

In recent years, deep neural networks are reported to be the very powerful tool in image segmentation [23]. Liu *et al.* [15] described a segmentation method for low-resolution cell images using a convolutional neural network achieving around 96% accuracy. Xia *et al.* [21] on the other hand applied convolutional neural network for retinal vessel segmentation achieving 96.85% accuracy.

It can be easily noticed that Machine Learning is a popular tool for developing support software for supporting a diagnosis process of specialists. This is why as a main contribution of this paper we propose a fully automatic procedure for segmentation of HER2 and CEN-17 reactions based on the convolutional neural networks.

2 Database of FISH Images

The database used in this study consists of 80 fluorescence in situ hybridization images with a size of 1376×1032 pixels. Images were recorded with a resolution of 200 pixels per inch with an Olympus BX61 fluorescence microscope with X-Cite series 120Q EXFO fluorescent system. This system consists of a CCD Olympus XC10 camera with Abbott 30-151332G-Ov2C146747 fluorescent filter.

The magnification of the filter was 60x and 100x. The microscope was working with a Cell-F visualization software.

The images were collected by dr. Anna Lis–Nawara from Wrocław Medical University, Wrocław, Poland according to the procedure described by Cierpkowski *et al.* [6] corresponding to the ISO/IEC 17025:2005 accreditation.

3 Methodology

The framework presented in this paper is a multistage process that includes data preparation, color deconvolution and dot classification as described in subsequent sections.

3.1 Image Preprocessing

The first stage of the proposed segmentation methodology is to estimate the color ranges that represent HER and CEN dots.

To determine the important gene features we performed a color deconvolution of the images into three separate image channels (red, green and blue). Result of the deconvolution is presented in Fig. 2. From the figure one can notice that in the red and green channels the important HER2 genes and CEN-17 centromere are easily visible in individual cells. In blue channel on the other hand (see Fig. 2d), the shape of a cell is evident. Since in this work, we focused only on the dot counting problem, the testing colors of the blue channel are not specified. As we can see, each color space provides different cell information, whee red space shows HER2 gene, green space shows CEN-17 reactions and blue space reveals cell shapes.

3.2 Training Set Preparation

From Sect. 3.1 we can draw a conclusion that working only in one color space is not enough. To obtain important or additional information we need to extract information from each color space separately. This reasoning allowed us to determine the positions of red and green dots in the image.

Depending on the image we noticed that genes dot sizes vary between 7 and 12 pixels in diameter. To automate the process of dot mapping we proposed a simple "Dot mapping" algorithm (see Algorithm 1). The described algorithm makes a mask by zeroing pixels that do not fall into the predefined RGB range that was manually chosen from 20 randomly chosen images with the highest color variations.

With an application of Algorithm 1 we were able to extract red and green dots from a FISH image in a form of 7 to 12 pixels squares. Additionally, to complete the dataset we extracted a background samples at every 500 pixel that didn't represent neither red nor green dot.

Using the above-mentioned procedure we were able to create 36959 images with a size of 9×9 pixels. 5602 of these images represent green CEN-17 reactions, 6954 are the red HER2 reactions and the remaining 24403 images represent background.

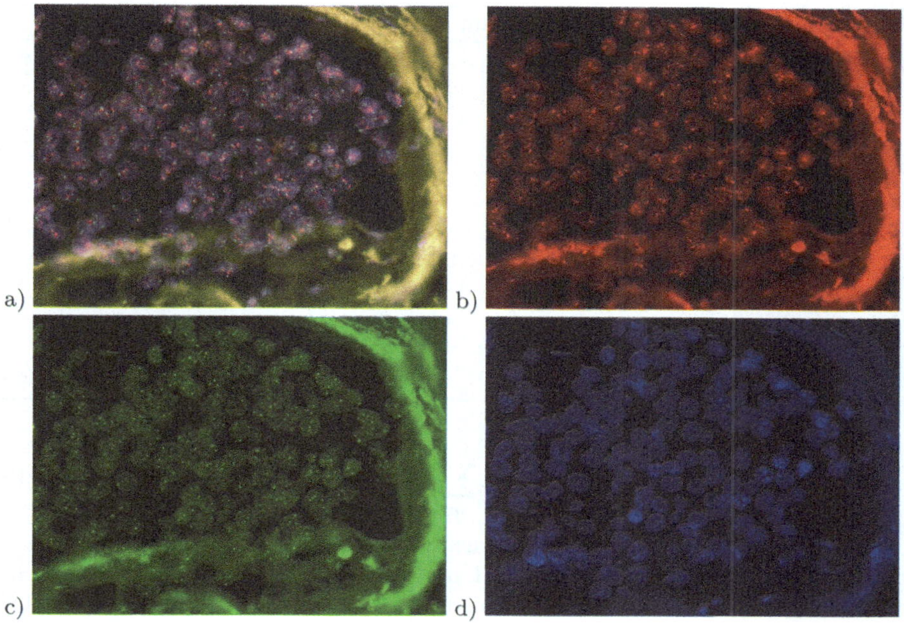

Fig. 2. Color deconvolution of the FISH image.

3.3 Pixel Classification

This section describes a classification methodology that was used for fluorescence in situ hybridization images. As mentioned in Sect. 1 deep learning techniques became an important and powerful tool in image classification and segmentation. In this work we applied convolutional neural networks to classify the pixel regions as background or HER2, CEN–17 reactions.

The model used in this study was originally described by Yann LeCun in 1989 [14] as a neural network that processed data with a known grid-like topology. These networks use at least one convolution operation instead of a traditional matrix multiplication [9].

This is commonly known that the most accustomed form of convolution of two real functions that can be described with Eq. 1.

$$F_m(t) = (x \circledast \omega)(t), \tag{1}$$

In Eq. 1 x argument can be referred to as an *input*, ω is the convolution *kernel* and F_m is a *feature map*. In image processing we always deal with two axes and therefore, the convolution kernel (Ω) used with images (Img) should also be 2–dimensional. In this case the convolution will be described with Eq. 2 or Eq. 3 since convolution is commutative [9].

$$F_M(i,j) = (Img \circledast \Omega)(i,j) = \sum_m \sum_n Img(m,n)\Omega(j-m, i-n), \tag{2}$$

Algorithm 1. Dot mapping algorithm

Input: I – RGB image
Output: HER2, CEN–17 dot map

procedure DotMapping(I)
    $size \leftarrow sizeOf(I)$
    $img[size] \leftarrow 0's$
    $i, j \leftarrow 0$
    for $, j \leftarrow 1, size$ **do** ▷ Loops through each pixel of I
        if img[i,j] < RGBmax and img[i,j] > RGBmin **then**
            $img[i] \leftarrow (255, 255, 255)$
        else
            $img[i] \leftarrow (0, 0, 0)$
    return img ▷ Mapping of HER2, CEN–17 reactions

$$F_M(i,j) = (\Omega \circledast Img)(i,j) = \sum_m \sum_n Img(j-m, i-n)\Omega(m,n). \tag{3}$$

Most of the widely used deep learning libraries uses a similar function to Eq. 3 but without flipping a kernel that maintained the decrease of the kernel when the input indexes increased. This function is described by Eq. 4 and called a cross–correlation function [9].

$$F_M(i,j) = (\Omega \circledast Img)(i,j) = \sum_m \sum_n Img(j+m, i+n)\Omega(m,n). \tag{4}$$

Taking the above into consideration we can build a network consisting of several layers with convolution where each layer will make numerous parallel convolutions to yield a set of linear activation values. Each linear activation value is then pass through Rectified Linear function called ReLu that maintains the non-linearity of the resulting feature map. In the next stage, we need to modify the output of the layer to reduce the output. This is performed with a pooling function Eq. 5 that chooses the maximum output and is called a max–pooling [8].

$$O(x) = max(0, x). \tag{5}$$

Further more, we need to bringing the output vector to the relevant number of dimensions by incorporating a Softmax activation to a Dense layer. The number of dimensions roughly translates to the number of classes. In our case of FISH images there are three dimensions for background $[1, 0, 0]$, red $[0, 0, 1]$ and green $[0, 1, 0]$ dots. The next layer is the regulating Dropout layer that can improve network efficiency by 1–2%. Layer drops in individual pass random neurons that return to work in the next rounds.

Additionally we incorporated a method for stochastic optimization (ADAM) based on categorical cross–entropy that provides a probability that the given input has only one correct solution as described by Kingma *et al.* [12].

4 Training

Training process was performed with Keras [5] and TensorFlow [2] frameworks. To obtain the results presented in this article, we let the network with two convolutional layers learn for 100 epochs to achieve 97.81% effectiveness of the model. The training data was divided to a training set (80%) and validation set (20%).

In the first epochs network reaches the effectiveness above 96%, starting at 81% and increasing at each epoch to achieve nearly 98% at the end.

5 Results

The results obtained in this study include not only the validation metrics but also a comparison with a manual procedure of counting HER2 and CEN–17 reactions. The metric used to measure network's accuracy is relatively simple: how many regions were correctly classified with regard to the number of reactions manually counted. The accuracy of the framework is understood as an average accuracy on test images. The best solution achieves the model accuracy of 97.81% (see Fig. 3a for details).

For comparison purposes we have performed additional test on CNN networks with one and three convolutional layers and without a Dropout layer. The results are presented in Fig. 3. It is easy to notice that for a one layer network during the first 10 epochs, the accuracy is significantly worse than the proposed model (see Fig. 3b). After 100 epochs we were able to achieve 96.1% accuracy which is still not as good as in the proposed framework. When an additional layer was introduced to the model we have noticed that the accuracy dropped significantly which can suggest that the model adapts to the training data (see Fig. 3c). Such over-fitting can be a result of the too complicated architecture. Similar behavior was noticed when a model with two convolution layers but without a dropout layer was used (see Fig. 3d). Looking at Fig. 3 we can draw a conclusion that a network is able to learn to distinguish a background from HER2 and CEN–17 reactions. Furthermore both gene reactions are segmented relatively well. The output of the network applied to the entire FISH image is presented in Fig. 4. The resulting image was achieved by classifying 9 x 9 pixels batches into previously defined regions and the middle pixel of the batch is colored accordingly to the recognition result. To complete the procedure, the dots obtained during segmentation were automatically counted by a computer and the accuracy was verified by manual counting. From Table 1 we can notice that the results of dot counting are very good achieving 98.04% accuracy for both types of reactions.

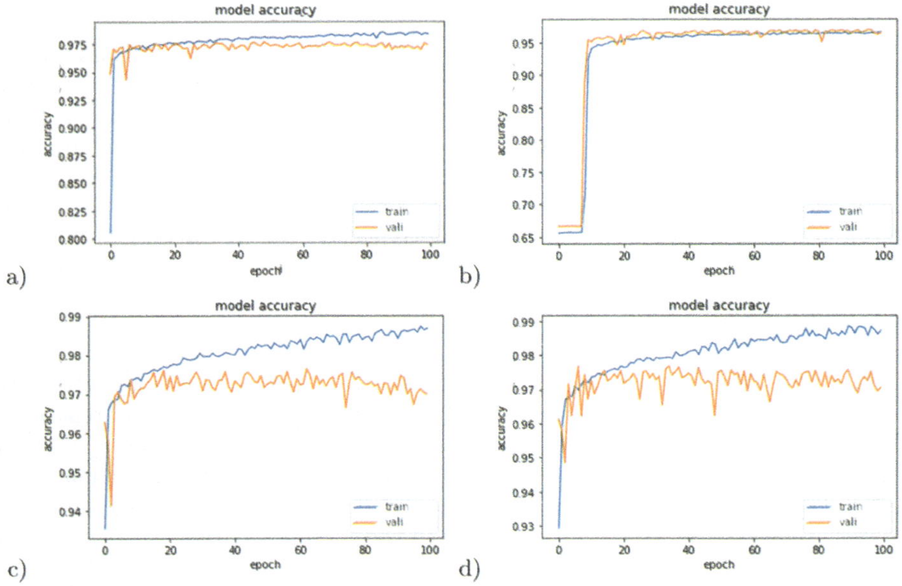

Fig. 3. Model accuracies; a) proposed framework, b) one convolutional layer, c) three convolutional layers, d) no Dropout layer.

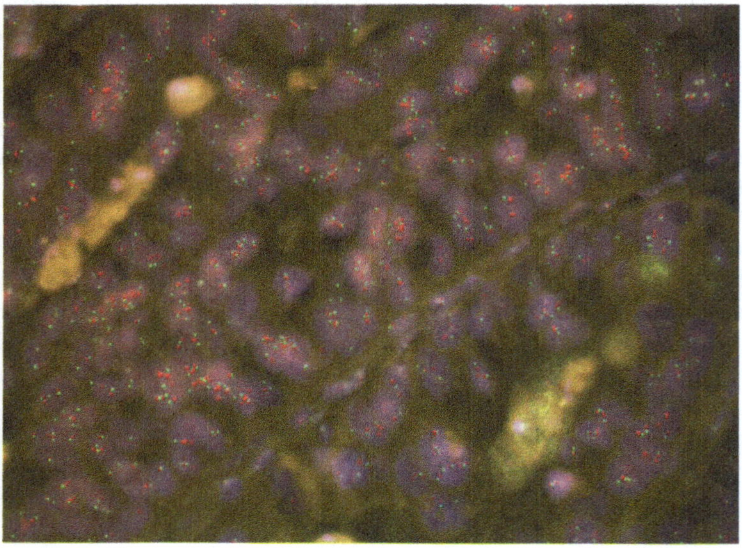

Fig. 4. Example of gene prediction.

Table 1. Dot counting results.

	Proposed framework	Manual counting	Difference	Accuracy
Red dots - HER2	812	819	7	99.15%
Green dots - CEN-17	439	457	18	96.06%
# of dots in the image	1251	1276	25	98.04%

6 Conclusions and Future Work

The main goal of the presented research was to create a convolutional neural network that would be able to segment a fluorescence in situ hybridization images, marking specific regions in the microscopic images. Building and training the convolutional network required to design a simple algorithm to automate the process of training database creation. The trained network model was able to recognize the dots with accuracy as high as 97.81% and applied to the entire FISH image it detected 98.04% of all points in comparison to a manual counting. We can see that CNN can segment HER2 reactions with higher efficiency. Therefore we can draw a conclusion that an automatic dot counting can be used to help diagnosticians in day–to–day work by limiting the number of slides they need to analyze.

In the proposed solution we have mainly focused on the design and implementation of the segmentation algorithm itself. To check further potential of the framework, we plan to increase size of the dataset by annotating more data with the support of a specialist. We also plan to include in the dataset whole slide images from the fluorescent microscope. Further plans would include a creation of the fully operational system that would guide the diagnostician through the entire pipeline of loading the image, counting dots and displaying the results to the user. We believe that an application like that would make a proposed framework more intuitive and useful during the diagnosis.

Acknowledgments. Authors would like to acknowledge the courtesy of prof. Julia Bar from the Department of Immunopathology and Molecular Biology at Wrocław Medical University, Wrocław, Poland for providing the images used during this study.

References

1. National Cancer Registry, December 2013. http://onkologia.org.pl/nowotwory-piersi-kobiet/. Accessed 21 Jan 2020
2. Abadi, M., et al.: Tensorflow: large-scale machine learning on heterogeneous distributed systems (2016). Software available from http://www.tensorflow.org
3. Lerner, B., Clocksin, W., Dhanjal, S., Hultén, M., Bishop, C.: Automatic signal classification in fluorescence in situ hybridization images. Cytometry **43**(2), 87–93 (2001)

4. Chen, A., et al.: Computer-aided diagnosis and decision-making system for medical data analysis: a case study on prostate MR images. J. Manag. Sci. Eng. (2020)
5. Chollet, F., et al.: Keras (2015). Software available from https://keras.io
6. Cierpkowski, P., Lis-Nawara, A., Gajdzis, P., Bar, J.: PDGFRα/HER2 and PDGFRα/p53 co-expression in oral squamous cell carcinoma. Anticancer Res. **38**(2), 795–802 (2018)
7. García-Caballero, T., et al.: Determination of HER2 amplification in primary breast cancer using dual-colour chromogenic in situ hybridization is comparable to fluorescence in situ hybridization: a European multicentre study involving 168 specimens. Histopathology **56**(4), 472–480 (2010)
8. Gomez, R., Gomez, L., Gibert, J., Karatzas, D.: Learning to learn from web data through deep semantic embeddings. In: Computer Vision – ECCV 2018 Workshops, pp. 514–529 (2019)
9. Goodfellow, I., Bengio, Y., Courville, A.: Deep Learning. MIT Press (2016). http://www.deeplearningbook.org
10. Gramacki, A., et al.: Automatic breast cancer diagnostics based on statistical analysis of shape and texture features of individual cell nuclei. In: Stochastic Models, Statistics and Their Applications, pp. 373–383. Springer (2019)
11. Hicks, D., Schiffhaue, L.: Standardized assessment of the HER2 status in breast cancer by immunohistochemistry. Lab Med. **42**(8), 459–467 (2011)
12. Kingma, D.P., Ba, J.: Adam: a method for stochastic optimization (2014)
13. Kiszler, G., et al.: Semi-automatic fish quantification on digital slides. Diagn. Pathol. **8**(1), 1–4 (2013)
14. Lecun, Y.: Generalization and network design strategies. Elsevier (1989)
15. Liu, Y., Yu, N., Fang, Y., Wang, D.: Low resolution cell image edge segmentation based on convolutional neural network. In: 2018 IEEE 3rd International Conference on Image, Vision and Computing (ICIVC), pp. 321–325 (2018)
16. Netten, H., et al.: Fluorescent dot counting in interphase cell nuclei. Bioimaging **4**(2), 93–106 (1996)
17. Piorkowski, A., et al.: Influence of applied corneal endothelium image segmentation techniques on the clinical parameters. Comput. Med. Imaging Graph. **55**, 13–27 (2017)
18. Stachowiak, M., Jeleń, Ł.: Automatic segmentation framework for fluorescence in situ hybridization cancer diagnosis. In: Computer Information Systems and Industrial Management - 15th IFIPTC8 International Conference, CISIM 2016, Vilnius, Lithuania, 14–16 September 2016, Proceedings, pp. 148–159 (2016)
19. Tanke, H.J., et al.: CCD microscopy and image analysis of cells and chromosomes stained by fluorescence in situ hybridization. Histochem. J. **27**(1), 4–14 (1995)
20. Tibau, A., et al.: Chromosome 17 centromere duplication and responsiveness to anthracycline-based neoadjuvant chemotherapy in breast cancer. Neoplasia **16**(10), 861–867 (2014)
21. Xia, H., Zhuge, R., Li, H.: Retinal vessel segmentation via a coarse-to-fine convolutional neural network. In: 2018 IEEE International Conference on Bioinformatics and Biomedicine (BIBM), pp. 1036–1039 (2018)
22. Yao, K., et al.: Unsupervised segmentation using a self-organizing map and a noise model estimation in sonar imagery. Pattern Recogn. **33**, 1575–1584 (2000)
23. Yüksel, M.E.: Accurate disease diagnosis through medical datasets by deep neural networks. J. Biotechnol. **256**, S10 (2017)
24. Zaha, D.C.: Significance of immunohistochemistry in breast cancer. World J. Clin. Oncol. **5**(3), 382–392 (2014)

Classification of Local Administrative Units in Poland: Spatial Approach

Jacek Batóg and Barbara Batóg(✉)

University of Szczecin, Szczecin, Poland
{jacek.batog, barbara.batog}@usz.edu.pl

Abstract. The aim of the paper is the presentation of the modification of the classical discriminant analysis. This proposal is connected with the occurrence of the spatial autocorrelation described by the spatial weights matrix, representing non-measurable spatial relationships between discriminated objects. In the model of spatial discriminant analysis there are non-spatial parameters for diagnostic variables and spatial parameters for the additional variables calculated as the product of spatial weights matrix and diagnostic variables. Such approach is an alternative solution to existing methods based on correction of *a priori* or *a posteriori* probabilities. The empirical verification of the proposed method was conducted for 114 gminas of Zachodniopomorskie voivodship in three years. The results obtained confirmed the improvement of the classification quality in case of spatial discriminant analysis in comparison to classical discriminant analysis.

Keywords: Discriminant analysis · Spatial weight matrix · Administrative units · Level of development

1 Introduction

The results of the existing research indicate the occurrence of spatial inequality of socio-economic development. This development is carried out through sustained processes of its diversification, which lead to the deepening of the division into areas of growth and areas of economic stagnation (Churski 2011). This observation is consistent with the main assumption of the polarization theory, according to which the level of development differentiation depends on the long-term influence of social, economic, cultural and political factors (Dyjach 2013). The consecutive National Development Strategies indicate that the existing inter-regional and intra-regional disparities in the European Union in living conditions and income of the population, which result from the faster growth of the largest urban areas, may become a barrier to maintaining the dynamic development of the whole country (Strategy 2012). The development of the whole country is conditioned by the development dynamics of its individual regions. A similar analogy can be formulated for the relationship between the level of development of the region and the level of development of the local units forming it (Adamowicz and Pyra 2019). In the case of Poland, the local dimension of development refers primarily to the level of gminas, whose level of development depends not only on their specific endogenous resources (Todtling and Trippl 2005), but also, to a

W. Zamojski et al. (Eds.): DepCoS-RELCOMEX 2020, AISC 1173, pp. 31–40, 2020.
https://doi.org/10.1007/978-3-030-48256-5_4

large extent, on their relations with their direct surroundings – neighbouring gminas. This issue is the subject of a conducted study in which a discriminatory analysis was used as a research method. The research hypothesis assumes that gminas with a higher administrative status are not always characterised by the level of socio-economic development warranting their status.

The main objective of the study is to determine the impact of the neighbourhood of administrative local units on the quality of their classification by introducing the spatial autocorrelation factor into the analysis. The effectiveness of the proposed approach will be verified by comparing the results of object classification obtained by means of standard and modified (spatial) discriminant analysis.

The years covered by the research are 2006, 2012 and 2018. The source of statistical data is Local Data Bank provided by Statistics Poland. All calculations were done in the package STATISTICA 13.1.

The rest of the paper is organized as follows. First part presents review of works indicating the importance of taking into account spatial relationships between variables used in the evaluation of the level of socio-economic development. Research method is shortly described in part two. Part three contains data description and empirical results. The paper ends discussion of the main findings and the presentation of recommendations for regional economic policy.

2 Spatial Autocorrelation in the Modelling of Economic Processes

The spatial autocorrelation could occur if there are some specification errors in econometric models (Fischer and Stirbrock 2006, p. 701) or if the administrative boundaries for collecting information do not accurately reflect the nature of the underlying process generating the sample data (LeSage 1999, p. 3). In this case, it is reasonable to assume that "By estimating models with the spatial factor, it is possible to determine the spatial relationship between observations in different locations, and to prove that there is a non-measurable spatial factor differentiating the examined phenomenon between locations" (Kopczewska 2006) Many authors underline the importance of including spatial relationships in regional research e.g. Malina (2004) writes that in most issues, it is only the spatial aspect, i.e. regional disparities and spans, that indicate the characteristic features of socio-economic phenomena that determine the directions of further development and the ability to improve the competitiveness of regions.

The nature of the identified relationships that appear between units in space is determined by a set of factors that can be taken into account in such analyses. As shown, for example, by Nigohosyan and Vutsova (2018), variables used in the analyses of socio-economic development represent diverse concepts with many different typologies. The final set of variables depends not only on the aim of the research but also on availability of statistical data. For example, in the research concerning

Portuguese regions, 5 variables characterising demography, education, income, governance and environment were used (Silva and Ferreira-Lopes 2014). In the evaluation of spatial and temporal changes of the sustainability in Mainland China from 2004 to 2014 twenty one indicators from three areas were used (Qiu et al. 2018). In the analysis of the relationships between the level of development of the region and the local units, Adamowicz and Pyra (2019) used variables characterising seven areas: labour market (6 variables), society (13), economic situation of the households (7), education (10), enterprises (9), research and development and innovation (11) and infrastructure and environment (8). Ikeda et al. (2017) considered 14 variables characterising: produced capital, human capital (education and health), natural capital (agricultural land, timber and non-timber forest, fisheries, minerals), and adjustment factors (resource trade, oil capital gains, CO2 emissions) in examination of regional wealth in 47 prefectures in Japan from 1990 to 2010. Rezende and Sinay (2016) used 88 variables in 7 dimensions: education, culture and sports; housing and transportation; healthcare and social welfare; safety and access to justice; human rights and diversity to compare the level of development of the municipalities in the Baixada Fluminense region in Brasil.

3 Methodology

Discriminant analysis is often applied in classification of regional units according to their economic development (Jaba et al. 2006; Batóg and Batóg 2019). It is also popular in identification of the variables that contribute significantly to the assessment of spatial disparities (El-Hanjouri and Hamad 2015) and bankruptcy prediction (Piasecki and Wójcicka-Wójtowicz 2017). The main aim of discriminant analysis is to examine whether a set of p variables (X1, ..., Xp) is capable to discriminate among *g* groups by means of discriminant functions. These functions are the linear combinations of the discriminant variables (Tacq 2007) and their coefficients (β) satisfy the conditions of maximization of the ratio of between group variance (B) to within group variance (V) (McLachlan 2004):

$$\hat{\beta} = V^{-1}B \tag{1}$$

The following equation represents discriminant functions:

$$Y_{kj} = \beta_{0j} + \beta_{1j}x_{1k} + \ldots + \beta_{pj}x_{pk} \tag{2}$$

where:

Y_{kj} – value of *j*th discriminant function for observation *k*,
x_{ik} – value of *i*th discriminant variable for observation *k*,
$k = 1, \ldots, n$, n – number of observations,
$j = 1, \ldots, r$, r – number of discriminant functions,
$i = 1, \ldots, p$, p – number of discriminant variables,
β_{ij} – parameters of discriminant function

The number of discriminant functions (r) is equal to min(g–1, p).

In order to find estimates of parameters of discriminant function the canonical correlation analysis is applied. The problem is limited to solving the system of equations:

$$\left(V^{-1}B - \lambda I\right)\hat{\beta} = 0 \tag{3}$$

where λ is an eigenvalue, by using the characteristic equation:

$$det\left(V^{-1}B - \lambda I\right) \tag{4}$$

to calculate a maximum value for λ and find the respective vector $\hat{\beta}$.

The introduction of a spatial factor into a classical discriminant analysis is sometimes achieved through an empirical and local choice of the prior class probabilities (Cutillo and Amato 2008) or through modification of the estimated posterior probabilities of group membership produced by automated classifiers (Steele and Redmond 2001).

The current research uses approach involving direct incorporation of spatial relationships to the discriminant function. It results in a correction of the matrix of values of diagnostic variables, taking into account the assumption that, in addition to the original values of variables characterizing the examined objects, the belongingness of a specific object to a given group is also determined by the characteristics of the neighbouring objects. The discriminant function with spatial relationships is given by Eq. 5.

$$D = X\beta + WX\theta + \varepsilon \tag{5}$$

where:

D – vector of values of discriminant functions, $n \times 1$
X – matrix of values of discriminant variables, $n \times r$,
W – spatial weights matrix, $n \times n$,
β – vector of non-spatial parameters, $r \times 1$,
θ – vector of spatial parameters, $r \times 1$,
ε – random error.

In the literature one can find a lot of propositions of spatial weights matrix W (Anselin 1988; Abreu et al. 2004). The current research uses spatial weights matrix based on criterion of neighbourhood (connectivity matrix). If two (different) spatial units have a common border of non-zero length then the entries w_{ij} are equal 1 and 0 otherwise – Eq. 6.

$$w_{ij} = \begin{cases} 1 & \textit{if unit j shares a common boundary with unit i} \\ 0 & \textit{otherwise} \end{cases} \tag{6}$$

Usually the W matrices are standardized to sum unity in each row – Eq. 7 (Abreu et al. 2004).

$$\sum_{j=1}^{n} w_{ij} = 1 \tag{7}$$

where:

w_{ij} – elements of spatial weights matrix,
n – number of units.

4 Empirical Results

The evaluation of the impact of spatial relationships between the objects on the results of their classification was carried out by comparing the results of the classification of 114 gminas in Zachodniopomorskie voivodship obtained with the use of classical discriminant analysis and spatial discriminant analysis for 4 groups of objects: cities with powiat status (CP), urban gminas (U), urban-rural gminas (UR) and rural gminas (R) in 2006, 2012 and 2018. The set of diagnostic variables contains 9 variables characterising selected aspects of functioning these local units (gminas):

- own revenue of gmina budgets per capita (X1),
- national economy entities in the REGON register per 1000 population (X2),
- non-working age population per 100 persons of working age (X3),
- useful floor area of dwellings completed per 1000 population (X4),
- investment property expenditure per capita (X5),
- natural increase per 1000 population (X6),
- employed persons per 1000 population (X7),
- consumption of water from water supply systems in households per capita (X8),
- population density (X9).

All variables were assumed to have the same meaning and no weighting system was introduced.

Table 1 presents canonical discriminant functions obtained in classical discriminant analysis in the research years (only root 1) and basic measures of their quality. Table 2 presents classification quality and classification matrices.

Table 1. Estimation results for classical discriminant analysis (root 1).

Variable	2006		2012		2018	
	Standardized coefficients	Partial Wilk's λ	Standardized coefficients	Partial Wilk's λ	Standardized coefficients	Partial Wilk's λ
X_1	0.050	0.988	−0.588	0.954	−0.150	0.955
X_2	−0.124	0.993	0.215	0.972	0.223	0.972
X_3	0.066	0.979	0.076	0.990	−0.231	0.872
X_4	0.158	0.895	−0.306	0.936	0.181	0.983
X_5	0.028	0.984	0.566	0.949	−0.134	0.970
X_6	0.131	0.966	−0.251	0.869	0.229	0.922
X_7	−0.079	0.919	0.189	0.854	−0.078	0.958
X_8	0.115	0.963	0.091	0.996	−0.001	0.968
X_9	**−0.966**	0.277	**0.929**	0.289	**−0.968**	0.293
Wilk's λ	0.122		0.119		0.117	
χ^2	224.13 (p = 0.000)		226.80 (p = 0.000)		228.59 (p = 0.000)	
F(27, 298)	11.67 (p = 0.000)		11.87 (p = 0.000)		12.00 (p = 0.000)	
Eigenvalue	4.922		4.911		4.730	
Explained variance (%)	92.90		92.29		91.08	

Source: own calculations.

The first discriminant function, which explains more than 90% of variance, discriminates urban gminas and cities with powiat status, and the second discriminant function is responsible for distinguishing urban-rural gminas from rural ones (see Fig. 1).

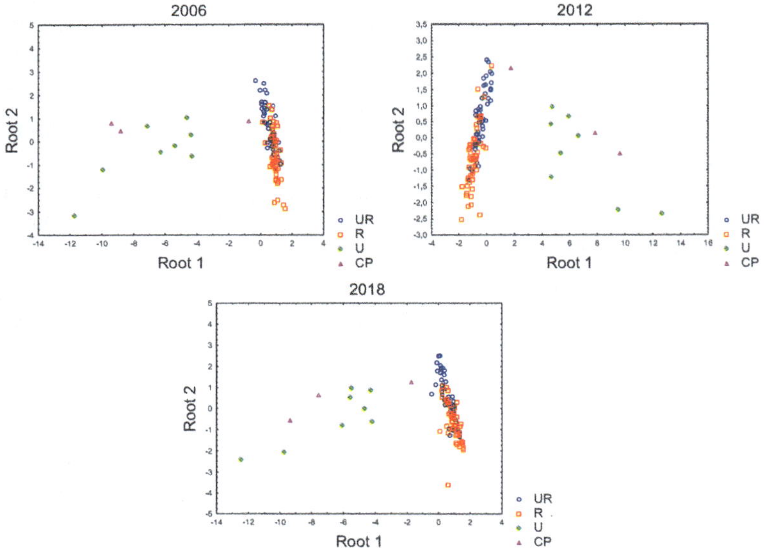

Fig. 1. Gminas in discriminant space in 2006, 2012 and 2018 – classical discriminant analysis.

In all the years analysed, it could be observed the relatively high quality of the classification of urban gminas and the impossibility of recognising the distinct character of cities with powiat status. In the latter case Świnoujście is particularly noteworthy, which is classified in the group of urban-rural gminas. The general level of classification quality may be considered as quite satisfactory.

Table 3 presents quality measures of canonical discriminant functions obtained in spatial discriminant analysis in the research years (only root 1). Table 4 presents classification quality and classification matrices.

Table 2. The classification quality and the classification matrices in classical discriminant analysis in 2006, 2012 and 2018.

Unit	Classification quality (%)	UR	R	U	CP
2006					
UR	68.63	35	16	0	0
R	75.00	13	39	0	0
U	100.00	0	0	8	0
CP	0.00	1	0	2	0
Total	71.93	49	55	10	0
2012					
UR	72.55	37	14	0	0
R	80.77	10	42	0	0
U	87.50	0	0	7	1
CP	0.00	1	0	2	0
Total	75.44	48	56	9	1
2018					
UR	72.55	37	14	0	0
R	76.92	12	40	0	0
U	100.00	0	0	8	0
CP	0.00	1	0	2	0
Total	74.56	50	54	10	0

Source: own calculations.

Table 3. Quality of spatial discriminant function (root 1).

Measure	2006	2012	2018
Wilk's λ	0.051	0.038	0.048
χ^2	304.04 (p = 0.000)	332.77 (p = 0.000)	309.05 (p = 0.000)
F(54, 277)	8.85 (p = 0.000)	10.24 (p = 0.000)	9.08 (p = 0.000)
Eigenvalue	7.229	10.577	7.051
Explained variance (%)	86.71	91.32	85.09

Source: own calculations.

Similarly as in case of classical discriminant analysis, the first discriminant function, which explains more than 90% of variance, discriminates urban gminas and cities with powiat status, and the second discriminant function is responsible for distinguishing urban-rural gminas from rural ones (see Fig. 2).

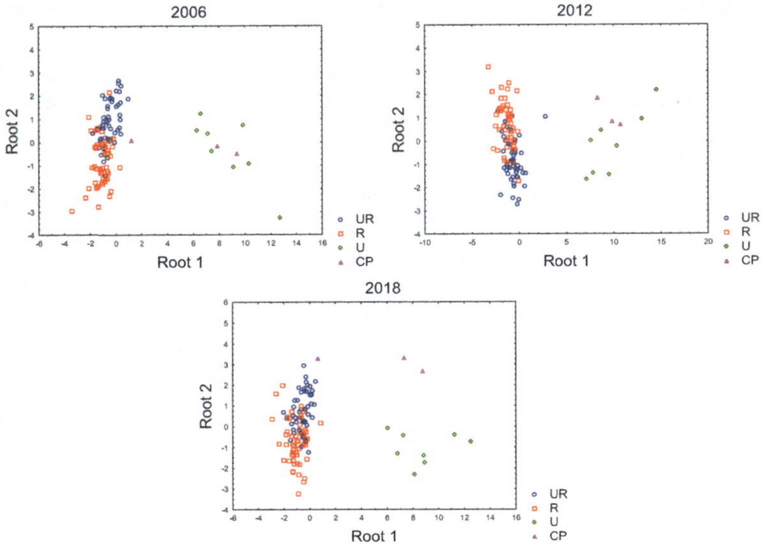

Fig. 2. Gminas in discriminant space in 2006, 2012 and 2018 – spatial discriminant analysis.

Table 4. The classification quality and the classification matrices in spatial discriminant analysis in 2006, 2012 and 2018.

Unit	Classification quality (%)	UR	R	U	CP
2006					
UR	82.35	42	9	0	0
R	84.62	8	44	0	0
U	100.00	0	0	8	0
CP	66.67	1	0	0	2
Total	84.21	51	53	8	2
2012					
UR	80.39	41	10	0	0
R	76.92	12	40	0	0
U	100.00	0	0	8	0
CP	100.00	0	0	0	3
Total	80.70	53	50	8	3
2018					
UR	72.55	37	14	0	0
R	82.69	9	43	0	0
U	100.00	0	0	8	0
CP	66.67	1	0	0	2
Total	78.95	47	57	8	2

Source: own calculations.

In all research years the spatial variant, the classification quality in spatial discriminant analysis is higher than the classification quality in classical discriminant analysis. This applies especially to cities with powiat status. The highest increase in classification accuracy was observed in 2006 (12.28% points). The average increase in the overall quality of the classification for all years was 7.31% points.

5 Conclusions

The results obtained indicate that with spatial relationships between the surveyed objects should be incorporated when discriminant analysis is applied. These spatial relationships allow to reduce the value of classification errors caused not only by the incompatibility of the socio-economic situation of gminas with their administrative status, but also by the influence of neighbouring gminas on their level of development. In the next research, with bigger examined sample, it could be interesting to determine the impact of outliers on the results of classification. The other kinds of spatial weights matrices could be also applied not only to the Polish but also to the European regional and local data.

It seems that the estimated models, in view of their high quality, could have the practical application for determining the kind of local administrative unit in dependence of the characteristics of its level of development. Then it may be used for the creation of new administrative local units or the modification of existing ones.

References

Abreu, M., de Groot, H.L.F., Florax, R.J.G.M.: Space and growth: a survey of empirical evidence and methods. Tinbergen Institute Discussion Paper TI 2004-129/3, Amsterdam (2004)

Adamowicz, M., Pyra, M.: Links between the level of local and regional development – problems of measuring. In: Proceedings of the 2019 International Conference "Economic Science for Rural Development" No 51, Jelgava, LLU ESAF, 9–10 May 2019, pp. 14–22 (2019). https://doi.org/10.22616/esrd.2019.0522019

Anselin, L.: Spatial Econometrics: Methods and Models. Kluwer Academic, Dordrecht (1988)

Batóg, J., Batóg, B.: The application of discriminant analysis to the identification of key factors of the development of Polish cities. Folia Oeconomica. Acta Universitatis Lodziensis **4**(343), 181–194 (2019). https://doi.org/10.18778/0208-6018.343.11

Churski, P.: Obszary wzrostu i obszary stagnacji gospodarczej – kontekst teoretyczny (Areas of Economic Growth and Stagnation – Theoretical Context). In: Churski, P. (ed.) Zróżnicowanie regionalne w Polsce (Regional Differences in Poland). Biuletyn Komitetu Przestrzennego Zagospodarowania Kraju, Polska Akademia Nauk, vol. 248, pp. 9–43 (2011)

Cutillo, L., Amato, U.: Localized empirical discriminant analysis. Comput. Stat. Data Anal. **52** (11), 4966–4978 (2008)

Dyjach, K.: Teorie rozwoju regionalnego wobec zróżnicowań międzyregionalnych (Theories of the Regional Development in View of Interregional Disparities). Annales Universitatis Mariae Curie-Skłodowska. Sectio H Oeconomica **47**(1), 49–59 (2013)

El-Hanjouri, M.M.R., Hamad, B.S.: Using cluster analysis and discriminant analysis methods in classification with application on standard of living family in Palestinian areas. Int. J. Stat. Appl. **5**(5), 213–222 (2015). https://doi.org/10.5923/j.statistics.20150505.05

Fischer, M.M., Stirböck, C.: Pan-European regional income growth and club-convergence. Insights from a spatial econometric perspective. Ann. Reg. Sci. **40**(4), 693–721 (2006)
Ikeda, S., Tamaki, T., Nakamura, H., Managi, S.: Inclusive wealth of regions: the case of Japan. Sustain. Sci. **12**(5), 991–1006 (2017). https://doi.org/10.1007/s11625-017-0450-4
Jaba, E., Jemna, D.V., Viorica, D., Lacatusu, T.: Discriminant analysis in the study of Romanian regional economic development in view of European integration (2006). https://papers.ssrn.com/sol3/papers.cfm?abstract_id=931613
Kopczewska, K.: Ekonometria i statystyka przestrzenna z wykorzystaniem programu R CRAN (Econometrics and Statistics with R CRAN). CeDeWu, Warszawa (2006)
LeSage, J.P.: Spatial Econometrics. University of Toledo, Toledo (1999)
Malina, A.: Wielowymiarowa analiza przestrzennego zróżnicowania struktury gospodarki Polski według województw (A Multi-dimensional Analysis of the Spatial Differentiation of Poland's Economic Structure by Voivodship). Wydawnictwo Akademii Ekonomicznej w Krakowie, Kraków (2004)
McLachlan, G.J.: Discriminant Analysis and Statistical Pattern Recognition. Wiley, Hoboken (2004)
Nigohosyan, D., Vutsova, A.: The 2014–2020 European regional development fund indicators: the incomplete evolution. Soc. Indic. Res. **137**(2), 559–577 (2018). https://doi.org/10.1007/s11205-017-1610-8
Piasecki, K., Wójcicka-Wójtowicz, A.: Capacity of neural networks and discriminant analysis in classifying potential debtors. Folia Oeconomica Stetinensia **17**(2), 129–143 (2017). https://doi.org/10.1515/foli-2017-0023
Qiu, W., Meng, F., Wang, Y., Fu, G., He, J., Savic, D., Zhao, H.: Assessing spatial and tem-poral variations in regional sustainability in Mainland China from 2004 to 2014. Clean Technol. Environ. Policy **20**(6), 1185–1194 (2018). https://doi.org/10.1007/s10098-018-1540-4
Rezende, J.F.D.C., de Sinay, M.C.F.: Methodology for Leading Indicators on Sustainable Regional Development. Revista de Administração Pública **50**(3), 395–423 (2016). https://doi.org/10.1590/0034-7612134163
Silva, R., Ferreira-Lopes, A.: A regional development index for Portugal. Soc. Indic. Res. **118**(3), 1055–1085 (2014). https://doi.org/10.1007/s11205-013-0455-z
Steele, B.M., Redmond, R.L.: A method of exploiting spatial information for improving classification rules: application to the construction of polygon-based land cover maps. Int. J. Remote Sens. **22**(16), 3143–3166 (2001)
Strategia Rozwoju Kraju: Strategy of National Development 2020. Ministerstwo Rozwoju Regionalnego, Warszawa (2012)
Tacq, J.: Multivariate Analysis in Social Science Research. Sage Publications, London (2007)
Todtling, F., Trippl, M.: One size fits all? Towards a differentiated regional innovation policy approach. Res. Policy **34**(8), 1203–1219 (2005)

Development of Methodology for Counteraction to Cyber-Attacks in Wireless Sensor Networks

Olexander Belej[1(✉)], Kamil Staniec[2], Tadeusz Więckowski[2], Mykhaylo Lobur[1], Oleh Matviykiv[1], and Serhiy Shcherbovskykh[1]

[1] Lviv Polytechnic National University, 5 Mytropolyt Andrei Street, Building 4, Room 324, Lviv 79015, Ukraine
Oleksandr.I.Belei@lpnu.ua, loburm@gmail.com, matviykiv@gmail.com, shcherbov@gmail.com

[2] Wroclaw University of Science and Technology, 7/9 Janiszewski Street, 50-372 Wroclaw, Poland
{kamil.staniec, tadeusz.wieckowski}@pwr.edu.pl

Abstract. The study analyzes the physical characteristics of devices that can be targeted. A method of detecting a malicious device with a violation of the physical characteristics of the network node is developed. The proposed method is based on the use of probabilistic functions, the calculation of the confidence interval and the probability of deviation of the current indicators from the confidence interval. Approaches to the control of the output sequences of the encryption algorithm using dynamic chaos and the method of singular spectral analysis are considered. A comparative analysis of the parameters of the input and output sequences of the developed algorithm of encryption based on dynamic chaos and standard algorithms of data encryption is carried out. It is found that the parameters of the output sequences of the encryption algorithm using dynamic chaos and standard encryption algorithms are almost identical. As a result of the study, a method for estimating node load indicators was developed. Evaluation of these indicators using threshold analysis, when the current values of the node fall within the confidence interval, is used to detect deviations in the behavior of the node during a cyber-attack.

Keywords: Attack · Attack detection · Security · Wireless sensor networks · Trust · Encryption · Information · Control · Chaotic mapping · Chaos

1 Introduction

Wireless sensor networks (WSN) have become widely used as a means of monitoring and managing objects from a distance. At the same time, network nodes can be located outside the controlled area and be exposed to an attacker. Also, wireless sensor networks have a large number of vulnerabilities associated with the transmission of data over insecure wireless channels. In this regard, the urgent task is to develop a method that can effectively detect active attacks by an attacker based on an analysis of network

W. Zamojski et al. (Eds.): DepCoS-RELCOMEX 2020, AISC 1173, pp. 41–50, 2020.
https://doi.org/10.1007/978-3-030-48256-5_5

traffic and physical parameters of wireless sensor network nodes. There are two groups of methods for protecting the WSN from active attacks by an attacker: attack and intrusion detection systems and confidence calculation systems [1]. As a way to protect against active attacks by an attacker, this article discusses a method for calculating trust. The methods for calculating trust allow not only detecting abnormal behavior and attacks on the network but also maintaining trusted relationships between nodes, which helps prevent some types of attacks [2]. Today, there are two types of confidence calculation systems: distributed and centralized [3]. Based on the analysis of attacks [4], the most malicious attacks were identified such as Denial of service (DOS) [5], blocking a node, Blocking a node with conditions [6], Tunneling [7], "Attack of Sibylla" [8]. Distributed and centralized systems for calculating trust have their advantages and disadvantages [9], but their common drawback is the inability to counteract the Denial of Service and Sibylla Attacks. Each type of method can use a different mathematical apparatus to calculate confidence. Probabilistic methods combine well with the concept of trust if a trust is defined as the expectation that a network node behaves properly with other nodes and fulfills its obligations when transmitting data, and also does not interfere with the operation of other nodes and the network as a whole [10].

At the current level of development of information technology, issues of information protection in telecommunication systems for various purposes are of great importance. The direction associated with the encryption of information in chaotic systems is developing [11]. The use of dynamic chaos for information protection systems is due to the ability of chaotic mappings to ensure the secrecy of the transmission of encrypted information in block or stream ciphers [12]. The determinism of chaos contributes to the encryption of information, and its randomness makes the system resistant to tampering [13]. Properties such as confusion and sputtering, characteristic of traditional crypto algorithms, are realized in chaotic ones using chaotic mappings and subsequent iterations.

It was shown in [14] that traditional cryptographic systems can be considered within the framework of the synergetic approach, that is, as nonlinear dynamic systems. By cryptosystem then we can understand a dynamic system $\langle F, X, K \rangle$ with a nonlinear function F, state-space X, and parameter space K. The nonlinear function F is specified using the algorithm, X is the set of initial states, the set of keys.

When considering algorithms using dynamic chaos, it is essential to ensure a chaotic mode, which is manifested in obtaining chaotic sequences of the encryption algorithm and is due to the security requirements of the encryption scheme.

In this paper, we propose and consider two approaches for determining the randomness of output sequences of encrypted information. The first of them is based on the approach of nonlinear dynamics methods, which allows determining the parameters of a dynamic system and their changes during encryption. The second option, due to the presence of a deterministic component in the sequences studied, makes it possible to use an approach based on singular spectral analysis with determining the dynamics of the intensity of the main components.

2 Formulation of the Problem

For the research, we used the output sequences obtained for the encryption algorithm we developed using dynamic chaos and the algorithms Advanced Encryption Standard (AES), Data Encryption Standard (DES), as well as input clear-text sequences. The developed encryption algorithm based on dynamic chaos is based on a generalized block symmetric encryption algorithm. The Feistel network is used as the basic transformation, in which the nonlinear function is specified in the form of a chaotic map.

When using the first approach to determining randomness, an analysis of the degree of randomness of the encrypted sequences is carried out by constructing phase portraits and using the delayed coordinate method. The use of the delayed coordinate method, as one of the methods of nonlinear dynamics, allows one to determine quantitative parameters in the form of a correlation dimension d and Kolmogorov entropy K for each of the studied sequences. The correlation dimension determines the region of localization of a dynamic system in phase space or the number of degrees of freedom of a specified system. The Kolmogorov entropy characterizes the stability of the system, as measured by the rate of divergence of its trajectories in phase space. Visual analysis is carried out according to the constructed phase portraits of the system. The construction of phase portraits makes it possible to visually determine the degree of filling of the phase space.

When using the second approach to determining randomness based on the method of singular spectral analysis, a quantitative parameter is estimated - the level of the main components of I. To obtain visual information, phase diagrams are used when various pairs of eigenvectors or principal components are plotted along the x and y axes.

The study aims to develop a method that will effectively detect active attacks by an attacker based on an analysis of network traffic and physical parameters of wireless sensor network (WSN) nodes, as well as quantitative parameters and visualization of encryption algorithm output sequences encrypted using dynamic chaos.

3 Development of an Encryption Algorithm for Messages in Wireless Sensor Network

For research, we used the output sequences obtained for the encryption algorithm and the input plaintext sequences. We developed an algorithm using dynamic chaos, Advanced Encryption Standard, Data Encryption Standard algorithms. The developed encryption algorithm based on dynamic chaos is based on a generalized symmetric block encryption algorithm. The Feistel network is used as a basic transformation in which a nonlinear function is indicated in the form of a chaotic map.

According to the delayed coordinate method, the output sequence of the encryption algorithm is represented as:

$$x_1, x_2, \ldots, x_n, \tag{1}$$

where $x_n = x\ (np)$, p is the sampling step, n is an integer.

This sequence generates m-dimensional vectors lying in the m-dimensional phase space:

$$\overline{x}_i^T = (x_i, \ldots, x_{i+m-1}), \tag{2}$$

where T is the transpose sign.

The state of the system in the reconstructed m-dimensional phase space is determined by m-dimensional points for each implementation $x(p)$:

$$x_i^m = m^{\frac{1}{2}}(x_i, x_{i+1}, \ldots, x_{i+m-1}) \tag{3}$$

The correlation integral $C_m\ (l)$ is a function equal to the probability that the distance between two reconstructed vectors xi is less than l.

The correlation dimension d is determined:

$$d = \lim_{r \to 0} \left[\frac{\lg C_m(r)}{\lg r} \right], \tag{4}$$

where $C_m(r)$ is the correlation integral, r is the size of the partition cell or the similarity coefficient.

The correlation integral is written:

$$C_m(r) = \lim_{N \to \infty} \frac{1}{N^2} \sum_{i,j=1}^{N} \theta(r - |\overline{x_i} - \overline{x}_j|), \tag{5}$$

where $\theta = 0$ at $t \leq 0, \theta = 0.5$, at $t = 0, \theta = 1$, at $t \geq 0$, θ is the Heaviside function, N - the number of points used to estimate the dimension.

It is found that for small values of r the behavior of the function $C_m(r)$ can be described:

$$C_m(r) = r^d, \tag{6}$$

where d is a parameter close to the fractal dimension of the strange attractor, r is the similarity parameter.

To study the realizations of open and encrypted text messages, the method of singular spectral analysis is used, the algorithm of which is reduced to the following.

Let a time series $\{x_i\}_{i-1}^{N}$, formed by a sequence of N equidistant values of some function $f\ (t)$, be given.

1. Scan one-dimensional to multi-dimensional series. As the first row of the matrix *X*, *M* (caterpillar length) values of the sequence are used, starting with the first term. The values of the sequence are used as the second row of the matrix, starting from x_2 to x_{M+1}. The last row of the matrix with the number $k = N - M + 1$ is formed by the last sequence of *M* elements. This matrix whose elements can be considered as an *M*-dimensional time series, which corresponds to an *M*-dimensional trajectory in an *M*-dimensional space of $k - 1$ units.
2. Analysis of the main components: a singular decomposition of a sample covariance matrix. A matrix is calculated that is an off-center covariance matrix $V = (\frac{1}{k})X^T X$. The eigenvalues and eigenvectors of the matrix *V* are determined $V = P \wedge P^T$, its expansion where *L* is the diagonal matrix, on the diagonal of which are descending eigenvalues, and *P* is the orthogonal matrix of eigenvectors of the matrix *V*.
   The matrix P can be considered as a matrix of transition to the principal component $XP = Y = (y_1, y_2, \ldots, y_m)$. If the time series of random numbers is used, then the eigenvalues of the matrix V are sample variances of the corresponding principal components, and their square roots are sample means. To analyze the main components of the series understudy, a graphical representation of the eigenvalues of some functions are used.
3. Taking into account the properties of the matrix *P*, it is possible to represent the matrix of the series *X* in the form $X = YP^T$. We obtain the expansion of the matrix of the series in orthogonal components (main components).
   At the same time, the transformation $y_i = XP_i$ is a linear transformation of the original process using a discrete convolution transformation:

   $$y_i[l] = \sum_{q=1}^{M} x_{lq} p_{jq} = \sum_{q=1}^{M} x_{l+q-1} p_{jq \ldots} \tag{7}$$

   The algorithm generates a set of linear filters tuned to the components of the original process. The eigenvectors of the matrix *V* act as transition functions of the corresponding filters.
   The visual and analytical study of eigenvectors and main components obtained as a result of linear filtering gives information about the structure of the process under study and its properties.
   To obtain visual information, phase diagrams are used when various pairs of eigenvectors or principal components are plotted along the x and y axes. It follows from the orthogonally of the eigenvectors and principal components that the phase shift between such pairs is $\pm\pi/2$.
4. Restoration of a one-dimensional row. The recovery procedure is based on decomposition $X = YP^T$. Recovery is carried out according to the main components if, when applying the formula $X = Y * P$, the matrix $Y*$ is obtained from the matrix *Y* by zeroing out all non-components.
   Thus, we can obtain the approximation of the matrix of the series that we are interested in or the interpreted part of this matrix.

4 Discussion

During a computational experiment using the method of deferred coordinates and the method of singular spectral analysis, the parameters of the output sequences of our encryption algorithm, AES and DES algorithms were obtained. In Fig. 1 shows plots of the correlation dimension d and the Kolmogorov entropy K of input and output sequences. They are obtained by encrypting an arbitrary fragment of plaintext using the developed encryption algorithm and AES.

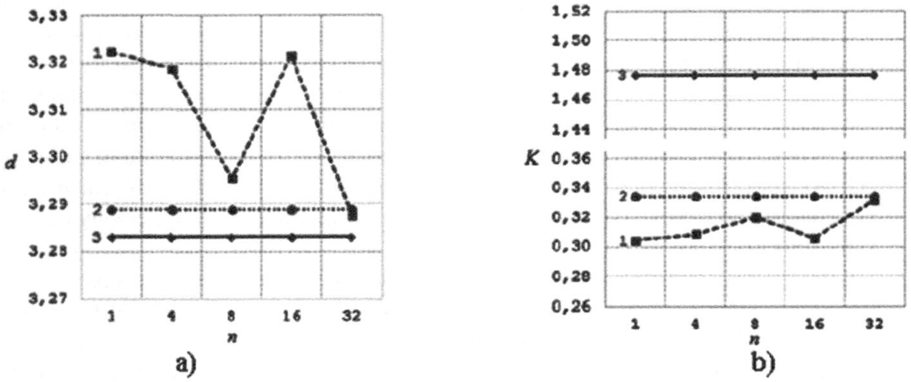

Fig. 1. Graphs of the dependence of the value of the correlation dimension d (a), Kolmogorov entropy K (b) of the output sequences of the encryption algorithm using dynamic chaos (curve 1), the AES algorithm (curve 2), and also the input sequences (curve 3) on the number of rounds of n the base transformation in operating of cipher block grip.

Using information parameters allows you to identify differences in the output sequence of the encryption algorithm relative to the input. In particular, in the WSN operation mode, the values of the correlation dimension for the output sequence exceed the values for the input sequence by 4.0–4.6% and the Kolmogorov entropy values for the output sequence are, respectively, 20.0–21.8% of the values for the entrance.

As can be seen from the graphs shown in Fig. 1, in cipher block grip, for the investigated interval of the number of rounds of the basic transformation (1–32 rounds), the output sequences obtained by the developed encryption algorithm demonstrate a higher degree of randomness than the output sequences obtained by the AES encryption algorithm.

Using these information parameters also allows you to identify areas of determinism in the output sequences that can be formed when the elements of the encryption algorithm are non-chaotic, which cannot be detected by analyzing Lyapunov's indicators. In particular, the behavior of the logistic display in the encryption algorithm for certain values of the control parameter falls into the area with deterministic dynamics, which makes this algorithm vulnerable to attacks based on known plaintext. The values of the correlation dimension and Kolmogorov entropy for the corresponding output sequences will differ significantly from those in the presence of dynamic chaos.

Figure 2 shows phase portraits of plaintext and corresponding encrypted sequences obtained by the developed encryption algorithm and AES algorithm. The number of rounds of the basic transformation of the developed algorithm is 16, and for AES - 14.

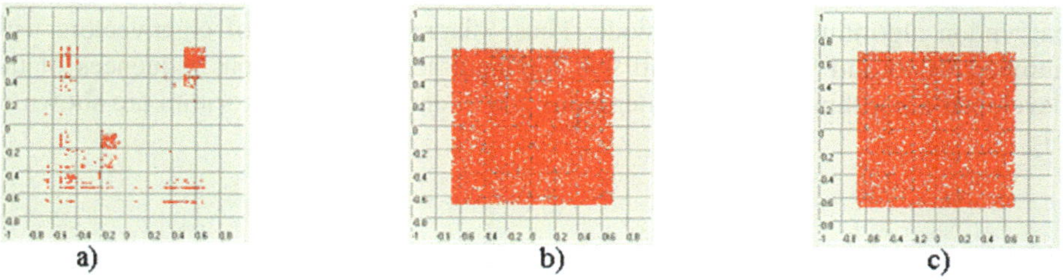

Fig. 2. Phase portraits of input (plain text) (a) and output sequences obtained by the developed encryption algorithm (b) and AES algorithm (c), in the operating of cipher block grip.

The results obtained by the delayed coordinate method and the construction of phase portraits of input (plain text) and output (encrypted) sequences allow us to conclude that: the correlation dimension, Kolmogorov entropy can be used as parameters for determining the degree of randomness of the output sequences, and phase portraits for visual analysis at applying an encryption algorithm using dynamic chaos (Table 1).

Table 1. The level of the main components of I, input sequences (1), output sequences of the developed encryption algorithm using dynamic chaos with the number of iterations z = 8 (2), z = 64 (3), algorithms DES (4) and AES (5) in the operating of cipher block grip.

Main component number	1000	999	998	997	996	995	994	993
1	0,3530	0,3530	0,3036	0,3035	0,2989	0,2979	0,2934	0,2934
2	0,2203	0,2202	0,2122	0,2118	0,2081	0,2080	0,2028	0,2027
3	0,226	0,2258	0,2250	0,2250	0,2109	0,2108	0,2015	0,2014
4	0,1899	0,1893	0,1880	0,1879	0,1841	0,1840	0,1821	0,1820
5	0,2287	0,2287	0,2179	0,2177	0,2135	0,2134	0,2083	0,2081

Using the second approach to determining randomness based on the method of singular spectral analysis, in the process of research, we evaluated a quantitative parameter - the level of the main components of I input sequences, output sequences of the encryption algorithm we developed using dynamic chaos, AES algorithms, DES algorithm. To analyze visual information, phase diagrams are used when different pairs of eigenvectors or principal components are plotted along the x and y axes. The table shows the results of a computational experiment under the following conditions: the length of the analyzed sequences N = 10000, the length of the caterpillar M = 1000, the number of iterations changed z = 8, z = 64.

As can be seen from the table, for the input sequences (1), the level of the main components of I exceeds the values for the indicators of the studied output sequences (2–5) by more than 50%. A comparative analysis shows that for the output sequences of the developed encryption algorithm using dynamic chaos, the level of the main components of I practically coincides with the performance of the AES and DES encryption algorithms in WSN mode. Phase diagrams of pairs of eigenvectors with numbers 1000 and 999, 1000 and 998 for input sequences, output sequences of the developed algorithm (number of iterations z = 8) and the AES algorithm are shown in Fig. 3.

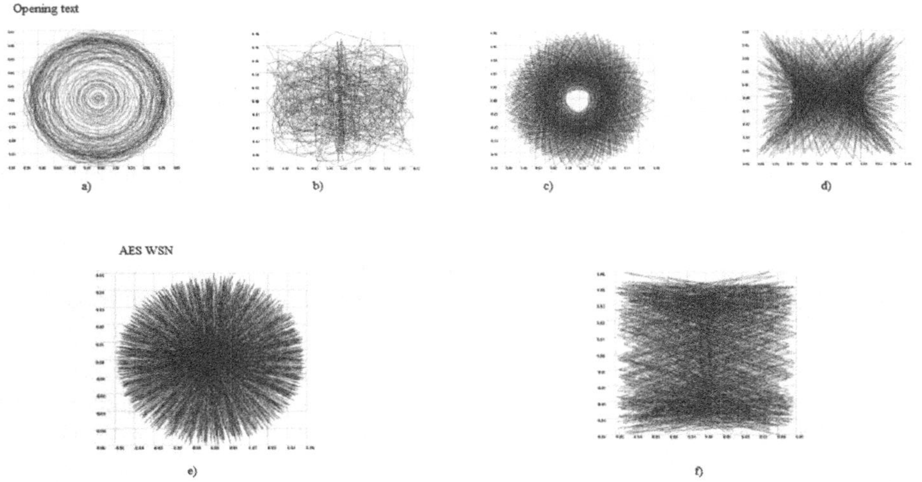

Fig. 3. Phase diagrams of pairs of eigenvectors with numbers 1000 and 999, 1000 and 998: for input sequences (a) and (b); the output sequences of the developed algorithm using dynamic chaos with the number of iterations z = 8 (c) and (d), the output sequences of the AES algorithm (e) and (f) when analyzing sequences by the method of singular spectral analysis.

For phase diagrams of the output sequences of the developed encryption algorithm and the AES cipher (AES WSN) (Fig. 3, e, f), the presence of "noisy" figures is characteristic, in contrast to the plaintext diagrams (Fig. 3, a, b).

Thus, the use of the method of singular spectral analysis as applied to input sequences, as well as to output sequences of an encryption algorithm using dynamic chaos, for example, in WSN mode, allows one to establish qualitative criteria in the form of phase diagrams, as well as a quantitative criterion for the level of principal components to determine randomness of sequences of encryption algorithms.

5 Conclusion

As a result of the studies, it was found that to control the randomness of the output sequences of encryption algorithms using chaotic signals, the delayed coordinate method and the method of singular spectral analysis can be used. It is shown that parameters such as the correlation dimension, the Kolmogorov entropy of the delayed coordinate method can be used as criteria for determining the degree of randomness of output sequences, and phase portraits for visual analysis using an encryption algorithm using chaotic signals. The level parameter of the main components of the method of singular spectral analysis and phase diagrams is recommended to be used as a means for determining the randomness of the output sequences of the encryption algorithm using chaotic signals.

A comparative analysis of the parameters of the methods of delayed coordinate and singular spectral analysis of the input and output sequences of encryption algorithms using chaotic signals, DES and AES showed significant differences in the parameters of the input and output sequences; practical coincidence in terms of the level of the main components in the operating of cipher block grip; improving the parameters of the correlation dimension and Kolmogorov entropy for the encryption algorithm using dynamic chaos, which, in general, can serve as the basis for recommendations on the use of these methods in the development of information security requirements.

The advantage of the developed method for estimating node load indicators is that in WSN there is a high probability of not only attacks from the network, but also attacks aimed at disrupting the physical activity of the node.

Defined error threshold, when the number of malicious nodes is less than 70%, allows us to determine the nodes and block them sufficiently accurately. When the number of malicious nodes is more than 70%, the detection accuracy decreases, and, as a rule, in a real situation when network nodes are located at a sufficiently large distance from each other and their number is measured by thousands of nodes, it is quite difficult for an attacker to exceed the threshold even of 50% of malicious nodes in network. At the same time, distributed and centralized confidence calculation systems are not able to counteract "Sibylla attacks" and "Denial of service" since they analyze only successful/unsuccessful host events, and when implementing these types of attacks, an attacker does not produce unsuccessful.

Therefore, our next study will focus on the development of methods for early diagnosis and countering DDOS attacks.

Acknowledgment. This paper has been written as a result of the realization of the "International Academic Partnerships Program". The project is funded by The Polish National Agency for Academic Exchange (NAWA), the contract for refinancing no. PPI/APM/2018/1/00031/U/001.

References

1. Artyshchuk, I., Belej, O., Nestor, N.: Designing a generator of random electronic message based on chaotic algorithm. In: IEEE 15th International Conference on the Experience of Designing and Application of CAD Systems, Polyana on Proceedings, Ukraine, pp. 1–5 (2019)
2. Belej, O., Artyshchuk, I., Sitek, W.: The controlling of transmission of chaotic signals in communication systems based on dynamic models. In: CEUR Workshop Proceedings, vol. 2353, pp. 1–15 (2019)
3. Govindan, K., Mohapatra, P: Trust computations and trust dynamics in mobile Adhoc networks. In: IEEE Communications Surveys & Tutorials Proceedings, vol. 14, no. 2, pp. 279–298 (2018)
4. Deepali, V., Manas, H., Shringarica, C.: Exponential trust-based mechanism to detect black hole attack in wireless sensor network. Int. J. Soft Comput. Eng. (IJSCE) Proc., 14–16 (2014)
5. Mohammad, M.: Bayesian fusion algorithm for inferring trust in wireless sensor networks. J. Netw. **5**(7), 815–822 (2010)
6. Schoch, E., Feiri, M., Kargl, F., Weber, M.: Simulation of ad hoc networks: ns–2 compared to JiST/SWANS, SIMUTools, Marseille, France (2008)
7. Shelby, Z., Bormann, C.: 6LoWPAN: the wireless embedded internet. Wiley Ser. Commun. Network. Distrib. Syst., 245 (2010)
8. Ho, J.W.: Zone-based trust management in sensor networks. In: IEEE International Conference on Pervasive Computing and Communications Proceedings, pp. 1–2 (2009)
9. Belej, O., Lohutova, T., Banaś, M.: Algorithm for image transfer using dynamic chaos. In: IEEE 15th International Conference on the Experience of Designing and Application of CAD Systems (CADSM) Proceedings, Polyana, Ukraine, pp. 1–5 (2019)
10. Oreku, G.S., Pazynyuk, T.: Security in Wireless Sensor Networks. Springer, Cham (2016)
11. Li, J., Ma, H., Li, K., Cui, L., Sun, L., Zhao, Z., Wang, X.: Wireless sensor networks. In: Revised Selected Papers of 11th China Wireless Sensor Network Conference, CWSN 2017, Tianjin, China (2017)
12. Sun, L., Ma, H., Fang, D., Niu, J., Wang, W.: Advances in wireless sensor networks. In: Revised Selected Papers of the 8th China Conference, CWSN 2014, Xi'an, China. Springer, Heidelberg (2015)
13. Guo, S., Liu, K., Chen, C., Huang, H.: Wireless Sensor Networks. In: Proceedings of the 13th China Conference on Wireless Sensor Networks (CWSN 2019), held in Chongqing, China (2019)
14. Becker, M.: Services in Wireless Sensor Networks. Modeling and Optimisation for the Efficient Discovery of Services. Springer Fachmedien Wiesbaden, Wiesbaden (2014)

The Need to Use a Hash Function to Build a Crypto Algorithm for Blockchain

Olexander Belej[1(✉)], Kamil Staniec[2], and Tadeusz Więckowski[2]

[1] Lviv Polytechnic National University, 5 Mytropolyt Andrei street, Building 4, Room 324, Lviv 79015, Ukraine
Oleksandr.I.Belei@lpnu.ua

[2] Wroclaw University of Science and Technology, 7/9 Janiszewski Street, Wroclaw 50-372, Poland
{kamil.staniec, tadeusz.wieckowski}@pwr.edu.pl

Abstract. The study examines the need and prospects of using blockchain technology for the Internet of Things. The possibility of implementing cryptography for open distributed systems of the Internet of Things without limitation of viewing, but with limitation of unauthorized access is considered. Possibilities of data protection against unauthorized access are shown employing the method of constructing hash function and digital signature in the blockchain itself. This is implemented using a cryptographic algorithm based on a pseudorandom number generator.

Keywords: Blockchain · Hash function · Encryption · Algorithm · Cryptographic · Information · Transformation

1 Introduction

Blockchain technology is one of the most innovative discoveries of our century. We can say this without exaggeration since we are observing the influence that it has had over the past few years and the influence that it will have in the future. To understand the device and the purpose of the blockchain technology itself, we must first understand one of the basic principles of creating a blockchain.

In the simplest sense of the blockchain, there is "an open, distributed registry where transactions between two parties are recorded in an efficient, verifiable and unchangeable way" [1]. Blockchain technology has been around for ten years.

After creating the primary block, the next block is added to each block using a cryptographic hash function [2]. The previous block cannot be changed, replaced or deleted without updating all subsequent blocks in all copies of the chain. Blockchain technology and its most famous implementation of Bitcoin have already become a global phenomenon - the data recorded in Bitcoin blocks correspond to "financial" transactions between the parties, which is an electronic registry. That is why blockchain is sometimes called a distributed registry.

In paper [3] provide a comprehensive, up-to-date discussion of the current state of the art of cryptographic hash functions security and design. The research work [4]

W. Zamojski et al. (Eds.): DepCoS-RELCOMEX 2020, AISC 1173, pp. 51–60, 2020.
https://doi.org/10.1007/978-3-030-48256-5_6

proposes a method for the securing and monitoring of petroleum product distribution records in a decentralized ledger database using blockchain technology.

The research work [5] proposes a method for the securing and monitoring of petroleum product distribution records in a decentralized ledger database using blockchain technology. In research [6] blockchain, as one of the crypto-intensive creatures, has become a very hot topic recently.

To this end, we in this paper conduct a systematic study on the cryptographic primitives in blockchains by comprehensive analysis on top-30 mainstream cryptocurrencies, in terms of the usages, functionalities, and evolutions of these primitives. In other research [7] blockchain is an innovative application model that integrates distributed data storage, peer-to-peer transmission, consensus mechanisms, digital encryption technology, and other computer technologies. It is decentralized, secure, and Information disclosure. The existing security problems of blockchain are analyzed, and the future research direction is expected.

Blockchain and other Distributed Ledger Technologies (DLTs) have evolved significantly in the last years and their use has been suggested for numerous applications due to their ability to provide transparency, redundancy, and accountability [8].

In the future, data from sensors and other smart devices in the Internet of Things (IoT) systems with a complex architecture will be transferred to cloud services for processing. Perhaps the network structure of the blockchain will be an advantage for such data transfer compared to a centralized database in the industrial Internet of things.

2 Problem Formulation

Blockchain security is based on the fact that cryptographic tools and sophisticated algorithms used by users to maintain the integrity of the network replaces the intermediary, which plays the role of a guarantee of trust. Using these tools, you can easily verify the accuracy and correctness of the data. To ensure security and the implementation of the chain, distributed cryptographic techniques such as a cryptographic hash function and electronic digital signature (EDS) are usually used in distributed registry technology.

Traditionally, a hash function [9] is understood as a compressive mapping that translates objects of a set with arbitrarily high power into objects of a set with low power. Such mappings are known to find applications in various application fields, for example, in data retrieval and storage algorithms. In cryptographic applications, hash functions are used to generate integrity codes for transmitted or stored messages, to compress messages when calculating a digital signature, and to authenticate users and the data they transmit.

Currently, existing blockchains use keyless hash functions. It is also interesting to note that in this case, in various implementations of the chain, functions can be used that are based both on the Merkle-Damgard structure that has been well studied and sufficiently rigorously justified at present, and on other principles, the KECCAK-256 function, built on the principle of cryptographic sponges [10].

So, let two users make a transaction in a network using blockchain technology, by which we mean the exchange of data between two parties. The data may include cash, contracts, documents, medical records or any other information that may be digitally presented [11]. Depending on the network parameters, a transaction can be confirmed either instantly or sent to the queue of expected transaction confirmations. In the latter case, the nodes (computers or network servers) check whether the transaction complies with the established network rules. To confirm the transaction, users use an electronic digital signature [12].

Note that the generation and storage of the secret key used in this algorithm requires reliable cryptographic tools formally not related to the implementation of a distributed registry. Network users are divided into two groups: ordinary users who create new transactions, and miners who create blocks from verified records (Fig. 1). Before adding to the chain, blocks must go through the acceptance process by other users or validation. Other network nodes will see changes and simply will not accept the changed block in the network, thus eliminating fraud. Similarly, an incorrectly signed or incorrect transaction will be rejected.

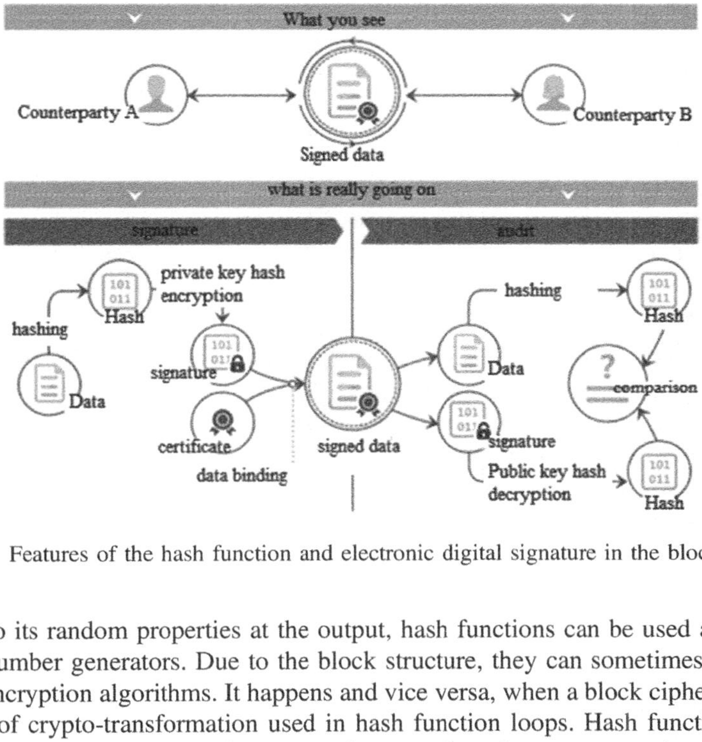

Fig. 1. Features of the hash function and electronic digital signature in the blockchain.

Due to its random properties at the output, hash functions can be used as pseudo-random number generators. Due to the block structure, they can sometimes act as the basis of encryption algorithms. It happens and vice versa, when a block cipher becomes the basis of crypto-transformation used in hash function loops. Hash functions today have become almost the most important element of modern cryptography. They provide security in the ubiquitous SSL secure web connection protocol (Fig. 2).

The Bitcoin protocol contains a set of parameters for the elliptic curve and its final field so that each user uses a strictly defined set of equations. Among the fixed parameters, the curve equation, the value of the field modulus, the base point on the

Fig. 2. The use of a cryptographic algorithm in the formation of an electronic digital signature for the blockchain of Bitcoin.

curve and the order of the base point are distinguished. This parameter is specially selected and is a very large prime number.

Hashing speed means how quickly these hashing operations occur during mining. A high level of hashing means that an increasing number of people and miners are involved in the mining process, and as a result, the system functions normally. If the hash rate is too high, the difficulty level increases proportionally. If the hash speed is too slow, then, accordingly, the difficulty level decreases.

3 The Algorithms of Hash Functions for Blockchain

Inside the operating system, the hash functions are somehow involved in virtually all security functions. In other words, whenever something happens in a computer or network, which implies the protection of information, then at some of the stages a hash function will certainly occur. Today in the science of cryptology there is no strict formal definition that would cover all the properties necessary and sufficient for the hash function to be called "cryptographic". But several properties are initially required for this class of functions:

- Resistance to finding the pre-image: having a digest h, it must be difficult to find such a message-type of m, for which h = hash (m).
- Resistance to finding the second preimage: having the input message m_1, it must be difficult to find the second input m_2 (not equal to m_1), such that hash $(m_1) =$ hash (m_2).
- Collision Resistance: it must be difficult to find two different messages m_1 and m_2 such that hash (m_1) = hash (m_2).

The general set of requirements for cryptographic hash functions is so specific that the ideal hash function should be as boring as possible and have no interesting properties whatsoever. Ideally, it should look like a purely random function - you give anything to the input, and you get a completely random number of fixed lengths at the output. But with the significant difference that the output hash is not random, but a strictly deterministic value, calculated efficiently and quickly.

The hash function *h(x)* is a function that takes as input an information sequence M of arbitrary length and gives an output information sequence (string) of fixed length as a result. The result of hashing the information sequence *M* is called the hash image *h (M)*. The ratio between the lengths of *M* and *h(M)* can be arbitrary, in other words, any relations are possible:

$$|M| > |h(M)|, |M| < |h(M)|, |M| = |h(M)|, \tag{1}$$

Although the first is more common, where $|M|$ is the length of the information sequence *M*. Since the result of a hash function is called a hash image, the data array *M* is sometimes called the prototype (the prototype).

We give a formal definition of the hash function. Let $\{0, 1\}^m$ be the set of all binary strings of length, m, $\{0, 1\}^*$ - the set of all binary strings of finite length. Then the hash function h is called the form transformation:

$$h: \{0, 1\}^* \rightarrow \{0, 1\}^m, \tag{2}$$

where *m* is the digit capacity of the hash image.

In Fig. 3 shows a hashing scheme, where PRNG is a Pseudo-Random Number Generator; Q - PRNG memory elements; h_0 is the initialization vector (IV); $n = |m_i|$ - digit capacity of information sequence blocks, i = 1,..., t:

$$M = m_1, \ldots, m_t, \tag{3}$$

where t is the number of blocks of the sequence M; N is the number of PRNG memory elements. The process of obtaining a hash function can be simply considered as the imposition of a Pseudo-Random Sequence (PRS) on the input transform sequence.

To find a collision in a hash function, it means to find two arbitrary different arrays M1 and M2, such that $h(M_1) = h(M_2)$. In other words, for two different arguments, the hash function values are the same. In Fig. 4 collision occurs when hashing arrays M3 and M4.

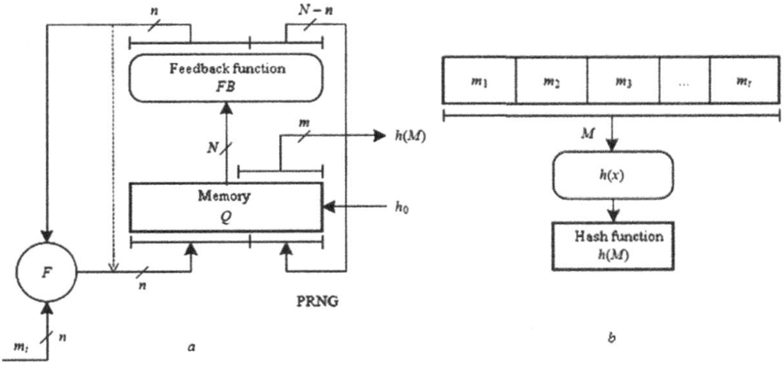

Fig. 3. Hash function: a - the imposition of PRS on the input information sequence; b - a simplified principle of the hash function.

Hash functions together with pseudo-random number generators (PRNG) are the basis of stochastic information protection methods. Stochastic methods are universal and can be used in conjunction with any other protection method, automatically

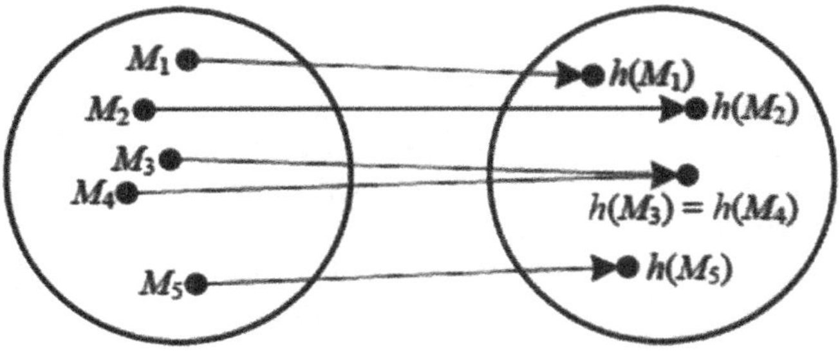

Fig. 4. Sets of prototypes and hash images.

improving its quality. This means that numerous statistical tests designed to assess the quality of PRNG can be used to study the properties of future cryptographic hash functions in the blockchain for IoT.

4 Discussion

Finding the second preimage of the function h(x) means to find, using a given data array M (the first preimage) and its hash image h(M), another array M` $\neq$ M such that h (M) = h (M`).

Any cryptographic hash function h(x) must meet the following requirements:

- The hash image must depend on all bits of the prototype and their mutual distribution;
- When any input information changes, the hash image should change unpredictably, in other words, on average, half the bits of the hash image should change (each bit can change with a probability of 0,5);
- For a given preimage value, the problem of finding a hash image must be computationally solvable;
- For a given value of the hash image, the problem of finding the preimage must be computationally unsolvable, in other words, for a given value h(M) is difficult to calculate the value of M (Fig. 5, a);
- Formally speaking, the hash function h is one-way (One-Way), if for an arbitrary n-bit string $y \in \{0, 1\}^n$ it is computationally difficult to find $x \in \{0, 1\}^*$, such that $h(x) = y$;
- Forgiven values of the hash image h(M) and the first pre-image M, the problem of finding the second pre-image M` $\neq$ M, such that h (M) = h (M`), must be computationally unsolvable (Fig. 5, c) (Second Pre-Image Resistance);
- The task of finding a hash collision, finding two arbitrary messages M1 and M2, such that M` $\neq$ M2, and h (M) = h (M`), must be computationally intractable (Fig. 5, b).

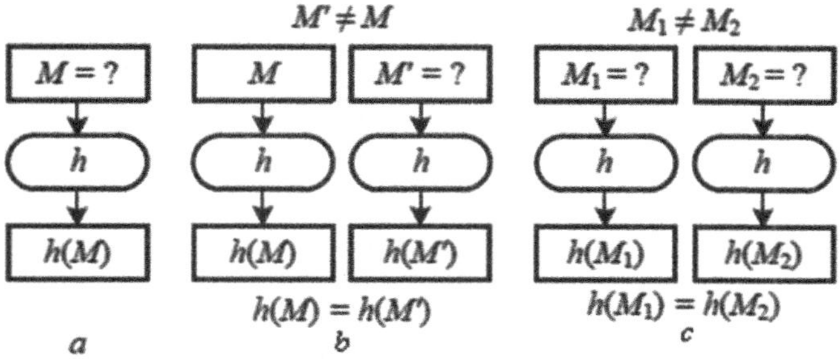

Fig. 5. The tasks of hash function h(x) in cryptographic of blockchain: a - finding the prototype; b - finding the second prototype; c - finding a collision.

In our case, k is the number of attacks on hash functions, N = 365 is the number of used hash functions for the period under consideration. Having solved the problem, find:

$$k \approx 1,18\sqrt{N}\,at\,P = 0,5,\ k \approx 23\,at\,P = 0,5\ \&\ N - 365 \tag{4}$$

Coming out of the fact that there is a random variable, which with equal probability takes any of the *N* possible values, we formulate an algorithm for our problem:

1. Determine the minimum number of realizations of k, at which with probability $P \geq 0,5$ at least one sample turned out to be equal to a predetermined value.
2. Determine the minimum number of realizations of k for which, with a probability $P \geq 0,5$, at least one sample turned out to be equal to the chosen value.
3. Determine the minimum number of realizations of k for which, with a probability $P \geq 0,5$, at least two samples turned out to be equal.
4. Form two sets of random values with k samples in each.
5. Determine the minimum number of realizations of k, at which, with a probability $P \geq 0,5$, at least one sample from the first set turned out to be equal to one sample from the second set.

The solutions to the problems of counteracting an attack on the hash functions of the blockchain are presented in Table 1.

Table 1. The problem of countering an attack on a hash function of blockchain.

Task	Probability	Value *k*	Value *k* at *P = 0,5*	Value *k* at *P = 0,5 & N = 365*
1	$P \approx 1 - e^{\frac{k}{N}}$	$k \approx N\ \ln\frac{1}{1-p}$	$k \approx 0,69N$	253
2	$P \approx 1 - e^{\frac{k-1}{N}}$	$k \approx N\ \ln\frac{1}{1-p} + 1$	$k \approx 0,69N + 1$	254
3	$P \approx 1 - e^{\frac{k(k-1)}{2N}}$	$k \approx \left(2\ln\frac{1}{1-p}\right)^{0,5} N^{0,5}$	$k \approx 1,18N^{0,5}$	23
4	$P \approx 1 - e^{\frac{k2}{2N}}$	$k \approx \left(\ln\frac{1}{1-p}\right)^{0,5} N^{0,5}$	$k \approx 0,83N^{0,5}$	16

In essence, the solution of the first problem is the solution of the problem of finding the first pre-image (Fig. 6, a), the solution of the second problem is finding the second pre-image (Fig. 6, b), the solution of the third and fourth problem is finding the collisions.

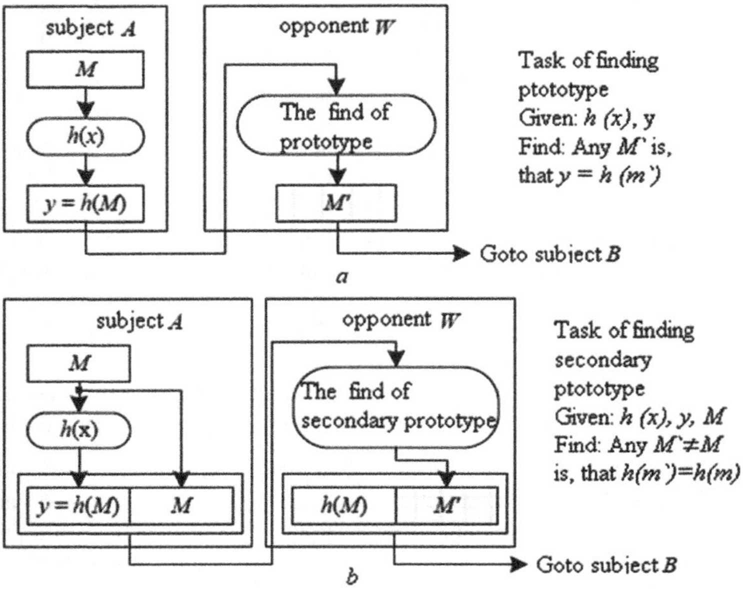

Fig. 6. Attacks on the hash function: a - finding the prototype; b - finding the second type

In Table 2, we present estimates of the complexity of attacks on the hash function, where n is the digit capacity of the hash image.

Table 2. Estimation of the complexity of attacks on the hash function

Attack	Value k at $P = 0,5$	Complexity
Finding the type (1)	$k \approx 0,69 \times 2^n$	2^n
Finding the second type (2)	$k \approx 0,69 \times 2^n + 1$	2^n
Finding a collision (3)	$k \approx 1,18 \times 2^{\frac{n}{2}}$	$2^{\frac{n}{2}}$
Finding a collision (4)	$k \approx 0,83 \times 2^{\frac{n}{2}}$	$2^{\frac{n}{2}}$

When analyzing the security of hash functions, the random oracle model (ROM) is often used. A random oracle is an ideal cryptographic hash function h(x), which gives a random response to each input request, and the same requests will always lead to the same answers of a random oracle no matter when and how many times they were made. There is no formula or algorithm for calculating h(x). There is only one way to find out h(x) - go to the oracle. ROM is an atomic or monolithic entity that cannot be broken apart. However, in practice, the hash function is not monolithic; the process of calculating the hash function is iterative, with the primitive of the next level, called the compression function, used each iteration. So, although the random oracle does not exist in real life, it is to be hoped that a well-designed h(x) will behave like ROM.

As can be seen from the cryptographic algorithm for generating hash functions discussed above, its application is important in the task of protecting blockchain information from third-party access and damage.

5 Conclusion

Cryptography is the basis of the blockchain, which ensures the operation of the system. Blockchain architecture assumes that trust between network participants is based on the principles of mathematics and economics, that is, it is formalized. Cryptography provides security based on the transparency and verifiability of all operations. This limits the visibility of the system to the user. Various cryptographic technologies guarantee the immutability of the blockchain transaction log. They allow solving problems of authentication and access control to the network and data in the blockchain. In our study, we examined cryptographic algorithms for generating hash functions and digital signatures.

The article presents a methodology for constructing cryptographic hash functions for the blockchain, provides algorithms for their use in information protection tasks. It is noted that the task of constructing a high-quality hash function is more complicated than the task of constructing a symmetric block cipher.

Modern peer-to-peer networks built on blockchain technology operate without a centralized trusted third party. This determines their scalability and avoids attracting a trusted third party for a fee. Unfortunately, the transfer of data from sensors to smart contract algorithms in the chain must be through intermediaries. To avoid data monopolization by such oracle intermediaries, an independent algorithm of independent verification of the reliability of the transmitted data by the sensor-video camera is proposed. It is hoped that other sensor manufacturers will also follow this example, which will create a truly decentralized free network that will not require a trusted third party.

The tendency of recent years is noted, namely, the mass appearance of hash functions using multidimensional transformations. In the future, we will consider in more detail the implementation of other cryptographic algorithms for generating hash functions to protect the blockchain for IoT.

Acknowledgment. This paper has been written as a result of the realization of the "International Academic Partnerships Program". The project is funded by The Polish National Agency for Academic Exchange (NAWA), the contract for refinancing no. PPI/APM/2018/1/00031/U/001.

References

1. Iansiti, M., Karim, R.L.: The truth about blockchain. In: Harvard Business Review 1995, no. 1, pp. 118–127 (2017)
2. Drescher, D.: Blockchain basics: a non-technical introduction in 25 steps. In: MITP, p. 17 (2017)
3. Al-Kuwari, S., Davenport, J.H., Bradford R.J.: Cryptographic hash functions: recent design trends and security notions. In: Short Paper Proceedings of 6th China International Conference on Information Security and Cryptology (INS crypt 2010), Science Press of China, pp. 133–150 (2010)
4. Ajao, F.A.D., Agajo, J., Adedokun, A.E., Karngong, L.: Crypto hash algorithm-based blockchain technology for managing decentralized ledger database in oil and gas industry. MDPI **2**, 300–325 (2019)
5. Ajao, L.A., Adedokun, E.A., Nwishieyi, C.P., Adegboye, M.A., Agajo, J., Kolo, J.G.: An anti-theft oil pipeline vandalism detection. Int. J. Eng. Sci. Appl. **2**, 41–46 (2018)
6. Wanga, L., Shena, X., Lib, J., Shaoc, J., Yanga, Y.: Cryptographic primitives in blockchains. J. Netw. Comput. Appl. **127**, 43–58 (2019)
7. Zhai, S., Yang, Y., Li, J., Qiu, C., Zhao, J.: Research on the application of cryptography on the blockchain. In: Journal of Physics, Conference Series, vol. 1168, pp. 1–8 (2019
8. Fernández-Caramès, T.M., Fraga-Lamas, P.: Towards post-quantum blockchain: a review on blockchain cryptography resistant to quantum computing attacks. IEEE Access **8**, 21091–21116 (2020)
9. Liu, Z., Lallie, H.S., Liu, L., Zhan, Y., Wu, K.: A hash-based secure interface on a plain connection. In: 6th International ICST Conference on Communications and Networking in China (CHINACOM), Harbin, pp. 1236–1239 (2011)
10. Dinur, I., Dunkelman, O., Shamir, A.: New Attacks on Keccak-224 and Keccak-256. In: Computer Science, vol. 7549. Springer, Berlin (2012)
11. Xu, M., Chen, X., Kou, G.: A systematic review of blockchain. Finance Innov. **5**, 27 (2019)
12. Setiawan, H., Rey Citra, K.: Design of secure electronic disposition applications by applying Blowfish, SHA-512, and RSA digital signature algorithms to government institution. In: 2018 International Seminar on Research of Information Technology and Intelligent Systems (ISRITI), Yogyakarta, Indonesia, pp. 168–173 (2018)

Common Criteria Vulnerability Assessment Ontology

Andrzej Bialas(✉)

Research Network ŁUKASIEWICZ – Institute of Innovative Technologies EMAG, Leopolda 31, 40-189 Katowice, Poland
andrzej.bialas@ibemag.pl

Abstract. The paper concerns the Common Criteria Evaluation Methodology (CEM) and is focused on the knowledge engineering application for vulnerability assessment. To enable automation of this complex process, better structurization of evaluation activities and data is required. The main finding of the paper is the development of ontology-based data models to be applied in the knowledgebase of a tool supporting the Common Criteria Vulnerability Assessment. The ontology use is exemplified on the vulnerability analysis of a simple firewall. The readers should have basic knowledge about Common Criteria and the ontology development.

Keywords: Common Criteria · Ontology · Security assurance · Security evaluation · Vulnerability assessment

1 Introduction

The paper deals with the security assurance methodology specified in the ISO/IEC 15408 Common Criteria (CC) standard [1, 2]. The assurance is measurable using EALs (Evaluation Assurance Levels) in the range EAL1-EAL7. Thanks to the rigorous development, security independent evaluation and certification, this methodology allows to deliver trustworthy IT products for today's societies and economies. The paper features the preliminary research of the KSO3C project, aiming at establishing a national scheme of IT security evaluation and certification.

The research concerns the Common Criteria Vulnerability Assessment (CCVA) aimed at checking whether exploitable vulnerabilities exist in the IT product and may breach its security. The results are exemplified on a simple firewall (MyFWL).

CC expresses security functional (SFR) and assurance (SAR) elementary requirements (components), grouped by families (ASE_SPD, AVA_VAN, ADV_FSP, etc.), which are, in turn, grouped by classes (ASE, ALC, AVA, ADV, etc.). SFRs express security behaviour and SARs – security assurance. EALs embrace subsets of SARs.

The evaluation evidences are elaborated for an IT product, called TOE (Target of evaluation), according to the EAL claimed for it:

- the Security target (ST), meeting the ASE requirements, presenting the security problem definition (SPD), security objectives as its solution, security requirements and functions;

W. Zamojski et al. (Eds.): DepCoS-RELCOMEX 2020, AISC 1173, pp. 61–70, 2020.
https://doi.org/10.1007/978-3-030-48256-5_7

- the documentation meeting: the ALC (life cycle support), ADV (development), AGD (guidance documents) and the ATE (Tests) classes requirements.

The TOE with evidences are delivered to a security evaluation lab, supervised by a certification authority (scheme). Evaluators check evidences using the Common Criteria evaluation methodology [3], including the vulnerability assessment (AVA_VAN).

CC/CEM are described worldwide, e.g. [2, 4, 5] and in the author's publications, e.g. [6].

The SAR components, in CEM called "sub-activities" (e.g. AVA_VAN.3 "Focused vulnerability analysis"), have the following elements:

- D – evidence should be delivered by the developer, e.g. AVA_VAN.3.1D;
- C – required content and presentation of this evidence, e.g. AVA_VAN.3.1C;
- E – how it will be evaluated, e.g. AVA_VAN.3.1E to AVA_VAN.3.4E; for each E-element a certain number of work units (implied by the D-, C- contents) are specified in CEM to express more precisely the evaluation sub-activity; evaluator's verdicts (Pass/Fail/Inconclusive) are assigned to E-elements.

The vulnerability assessment, a key part of the TOE evaluation, is complex and laborious. Many different factors, specific for the evaluated IT product, and rigour implied by the claimed EAL should be considered. Therefore CCVA is difficult to plan for the given IT product and to automate.

The author proposes to mitigate this problem by deeper formalization and structurization of [7]:

- evaluation activities – by introducing elementary evaluation processes (EEPs);
- input/output data of the EEPs – by the development of ontological models.

The aim of the research presented in the paper is to identify the CCVA-relevant factors, including terms, relations, activities, attacks, vulnerabilities, penetration tests, and to represent them as the CCVA ontology.

Ontologies have recently found application in disciplines where "a common understanding", "a common taxonomy" or "reasoning" are important.

The paper contribution is the knowledge engineering application to the CC vulnerability assessment:

- to help plan the evaluation activity (what kinds of attacks should be considered for the given IT product, proposed investigations, tests scenarios, methods and tools, etc.),
- to identify knowledge, especially reusable, used in the vulnerability assessment,
- to identify main requirements for the software tool supporting the CCVA process.

Section 2 presents the current state of research in the paper domain. The development and validation of CCVAO is discussed in Sect. 3. Section 4 concludes the paper.

2 Current State of Research

The literature review concerns the knowledge engineering application in information security, especially in the CC domain. The author's earlier papers include detailed literature reviews in this domain.

The paper [8] features the results of research dealing with the development of a CC ontology and an ontology-based tool supporting CC knowledge query, mark-up, review, and report functions to better understand CC and enhance the effectiveness of CC certification. This ontology does not embrace the CEM activities, especially concerning CCVA.

The paper [9] presents a tool based on an ontological representation of the Common Criteria components, which is to support the evaluator during the certification process, i.e.: while planning the evaluation process, reviewing relevant documents or making reports. The tool is to decrease the time and costs of certification. This ontology does not express details required for work units, e.g., public sources of information about vulnerabilities, attacks, areas of concerns, attack potential.

The toolset [10], developed in the author's organization, includes the "self-assessment" functionality, which is an ontology-based CEM implementation. Similarly to [9], it is generic and does not embrace details of the vulnerability assessment.

The paper [11] provides a very comprehensive literature survey on "security assessment ontologies" with the following conclusions:

- "Most of works on security ontologies aim to describe the Information Security domain (more generic), or other specific subdomains of security, but not specifically the Security Assessment domain";
- there is "a lack of ontologies that consider the relation of Information Security and Software Assessment fields of research";
- there is "a lack of works that address the research issues: Reusing Knowledge; Automating Processes; Increasing Coverage of Assessment; Identifying Vulnerabilities; Measuring Security; Assessing, Verifying or Testing the Security".

There is neither a method nor tool, especially ontology-based, to support the specific CCVA process, e.g.: searching public information sources for vulnerabilities, evidence-based vulnerabilities search, structurized vulnerability analyses, penetration tests management.

3 Towards the Common Criteria Vulnerability Assessment Ontology

CCVAO is developed as part of the Common Criteria Ontology (CCO) [7]. It is based on the author's earlier elaborated ontologies, e.g. [12], and uses the knowledge engineering principles and the Protégé v.5 tool elaborated at the Stanford Center for Biomedical Informatics Research [13, 14].

The CCVA ontology domain is a vulnerability assessment process which is specified in CC [1], part 3/pp. 311–346, CEM [3]/pp. 182–188 and its Annex B (research input).

The main CCVA process is divided by the author into eleven elementary evaluation processes (EEPs), shown in Fig. 1, and discussed in another submitted paper.

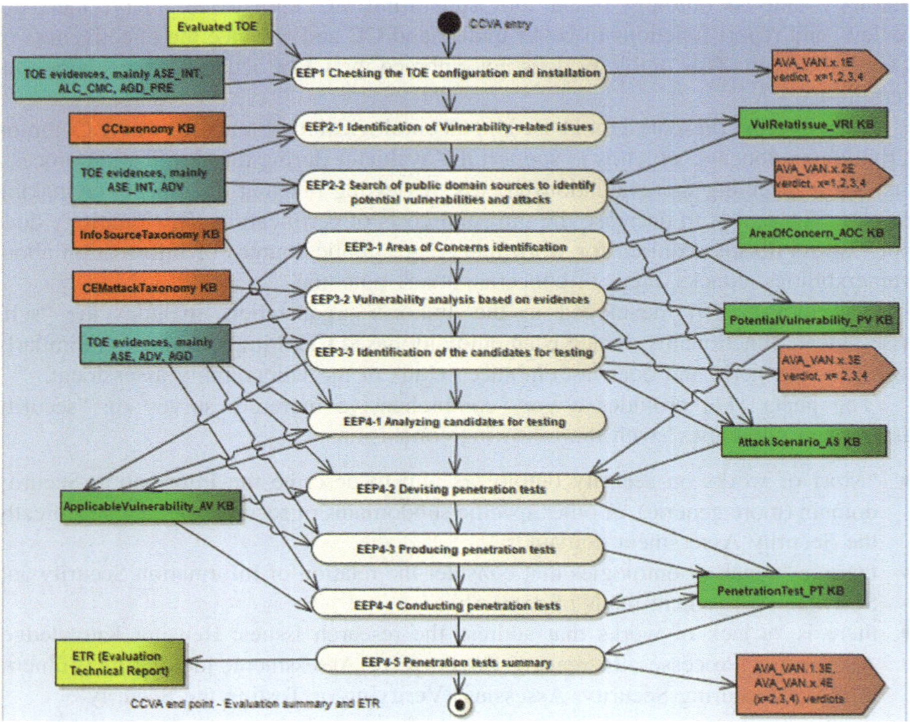

Fig. 1. CCVA elementary processes and their input/output data.

This paper focuses on the input and output data identified for EEPs. Evaluation evidences, turquoise marked objects, represent the documentation delivered to the lab along with the TOE. The elaborated ontology and elementary processes were validated on a simple firewall project (MyFWL) – some examples will be shown below.

3.1 Elementary Evaluation Processes

Particular EEPs [7] correspond to the elements of components, e.g. EEP1 – to AVA_VAN.x.1E, EEP2-1 and EEP2-2 to AVA_VAN.x.2E, etc.

EEP1 checks if the TOE is properly configured and installed, i.e. ready for evaluation experiments and testing.

EEP2-1 identifies the TOE specific characteristics to orientate the potential vulnerability search in public information assets (EEP2-2).

EEP3-n concerns searching of vulnerabilities by the analysis of evaluation evidences (step omitted in the AVA_VAN.1 component) and embraces the identification of areas of concerns, i.e. suspected parts of the TOE design (EEP3-1), orientating the analysis (in EEP3-2). EEP3-3 checks whether the given potential vulnerability is applicable to the TOE in its operational environment – the applicable ones are candidates for penetration testing (EEP4-n).

EEP4-1 preselects the candidates for testing by a deeper analysis, e.g. checks whether the TOE is resistant to the required attack potential (AP). For the preselected pairs: "attack scenario – vulnerability", penetration tests are devised, elaborated, performed, concluded and finally reported in ETR (Evaluation technical report). As a result, exploitable and residual vulnerabilities are identified.

3.2 Data Representation

The paper contribution includes the identification of two kinds of data models used by the elementary processes:

- supporting models, (marked brown in Fig. 1); they represent commonly used information from external services, repositories, data feeds, etc.
- project specific models, (marked green), created and used in the given evaluation project.

Data models are expressed by ontology classes, subclasses, object- and data properties. Classes individuals represent elementary portions of knowledge included in the knowledgebase (KB).

The ontology class *CCtaxonomy*, and its subclasses, complying with the taxonomy of the CC portal, is used here to find products similar to the evaluated one, and finally to detect its potential vulnerabilities. The *CCtaxonomy* subclasses, e.g.: *AccessControlDevicesAndSystems*, *BiometricSystemsAndDevices*, are related to keywords, placed in *VulRelatIssue_VRI*.

CEMattackTaxonomy embraces basic attack categories defined in CEM/B2.1 and refined in the paper [15], like: bypassing, tampering, direct attacks, monitoring, and misuse.

InfoSourceTaxonomy represents identified public information sources used to find TOE-related potential vulnerabilities and attacks. The Information source knowledgebase defines "where to search" for the vulnerabilities.

The ontology classes and individuals related to the supporting data are exemplified in Fig. 2. They concern the MyFWL firewall validation. The ontology classes (yellow circles) are shown on the left. The middle panel presents examples of individuals (violet rhombuses), i.e. "an elementary knowledge". Please note individuals of the *InfoSourceTaxonomy* subclasses expressing different sources to be searched in EEP2-2. For the highlighted *SecRep_OVAL* individual, the data properties (a string type) are shown in the upper right panel. Please note URL links (*infoSourceLink* property) to the external OVAL repository, where potential vulnerabilities are searched. The right lower panel shows the use of the *SecRep_OVAL* individual.

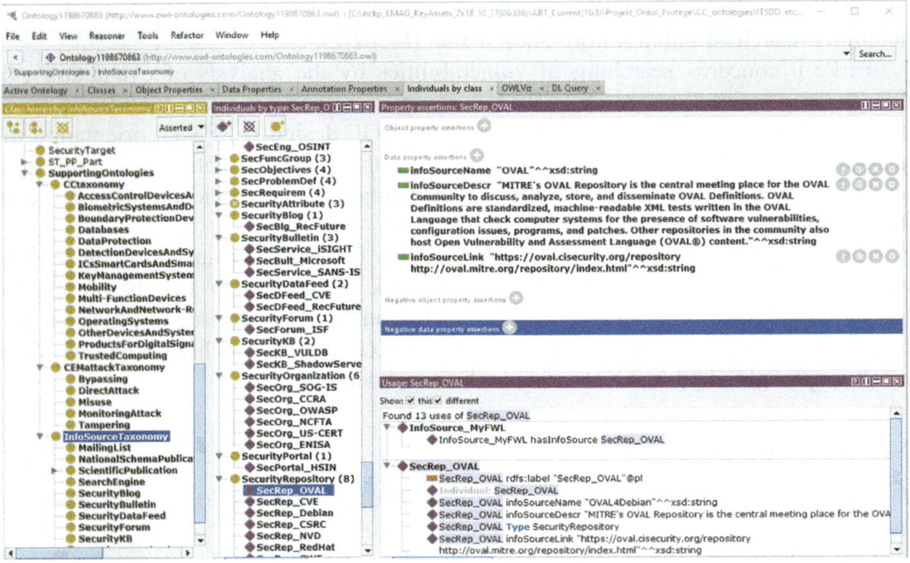

Fig. 2. Supporting data representation in the Protégé ontology editor.

Other identified CCVAO classes concern the project specific data. The first of them, *VulRelatIssue_VRI* (VRI), represents items of different categories characterizing the TOE and its security, orientating the search of the public information sources to identify potential vulnerabilities. The VRI knowledgebase expresses "what to search". The *VulRelatIssue_VRI* class has different keywords (data properties) used while the sources are searched:

- TOE-related keywords derived from the Security target parts:
  - *TOEtypeKeywords* – defined on the "TOE Type" basis,
  - *TOEusageSecFeatureKeywords* – defined on the "Usage and major security features of a TOE" basis,
  - *TOEoperEnvKeywords* – defined on the "Required non-TOE hardware, software, firmware" basis,
  - *targetEnvHardware* – hardware platform, e.g. a microcontroller,
  - *targetEnvSoftware* – software platform, e.g. OS,
- keywords implied by the CC product taxonomy,
  - *CCkeywords*,
- keywords for IT products knowledgebases,
  - *CPEkeywords* – according to (CPE, 2019),
  - *genIT_ProductKeywords* – other keywords defined by the evaluator, concerning the TOE application, operational environment, specific threats, etc.,
  - *similarProducts* – indicating similar IT products.

The *AreaOfConcern_AOC* (AOC) class indicates the "specific portions of the TOE evidence that the evaluator has some reservation about, although the evidence meets the requirements for activity with which the evidence is associated" [3]/B.2.2.2. The AOC class embraces:

- unnecessary complex solutions e.g. functions, interfaces, specifications – potentially vulnerable, expressed by the *complexSolution* data property,
- area in which many known vulnerabilities exist, e.g. input processing, web interfaces, expressed by the *vulnerabArea* data property.

Potential vulnerabilities and attack scenarios are identified during the analysis of:

- public information in the context of VRI (EEP2-2),
- evaluation evidences in the context of AOC (EEP3-2).

The *PotentialVulnerability_PV* concept represents any identified potential vulnerability resulting from the search and analyses of the vulnerability-relevant items. The vulnerability specification is based on the main elements of the Vulnerability Description Ontology (VDO) [16], expressed by data properties:

- *VulnerabProvenance* – like the name of the source which provides information: CWE, CVE [17], NVD [18], CPE [19], CAPEC [20], OWASP [21], etc.;
- *ExtVulnerabID* – unique identifier of a vulnerability in an external source, like a knowledgebase article number, patch number, bug tracking database identifier, or a common identifier such as Common Vulnerabilities and Exposures (CVE) or Common Weakness Enumeration (CWE) [17];
- *VulnerabDescription*, analogously to the VDO "scenario";
- *Product* – software and/or hardware configurations that are recognized as vulnerable;
- *AttackTheater* – area or place from which an attack may occur;
- *EngineeringMethod* – method or mechanism used to manipulate the user into interacting with a malicious resource;
- *VulnerabContext* – entity where the impacts occur from successful exploitation of a security vulnerability;
- *VulnerabilityScore* – according to the Common Vulnerability Scoring System [22];
- *Equipment* – equipment used to exploit a vulnerability;
- *CEMreference* – reference to the CEM attack taxonomy.

Figure 3 presents an example of an identified potential vulnerability *PV_MyFWL_-vul22*. Object properties indicate the related individuals. Data properties present vulnerability details.

The *AttackScenario_AS* (AS) class represents all identified, possible attack scenarios relevant to the given TOE. The attack scenarios are related to a potential vulnerability identified by the object property *attackDealsWith_PV*. AS is characterized by data properties: *AttackDescription* and *AttackProvenance*.

The identified potential vulnerabilities are checked (EEP3-3) whether they are applicable in the operational environment of the TOE (on the ASE, ADV, AGD basis).

The *ApplicableVulnerability_AV* class represents these applicable vulnerabilities preselected as candidates for further analysis and testing. Their data properties show details supplementing information included in *PotentialVulnerability_PV*:

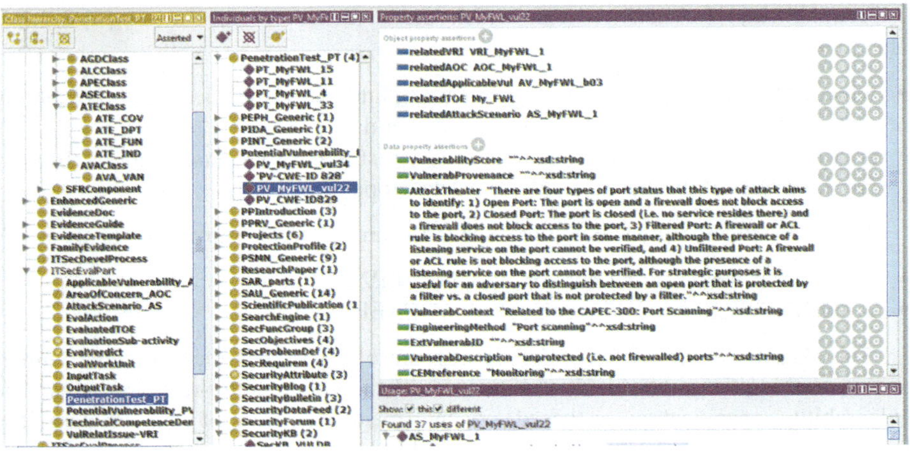

Fig. 3. Potential vulnerability specification in the knowledgebase with the use of Protégé.

- *VulCandidateRationale* – why is it a candidate?
- *AssumedAP* – implied by the EAL (Basic, Enhanced Basic, Moderate, High);
- *CalculatedAP* – calculated Attack Potential (AP) [15];
- *VulSeverity* – characterizing the vulnerability occurrence and severity;
- *VulPriority* – to prepare a prioritized list of vulnerabilities for testing;
- *TestReport* – test report and result (SFRs not met, AP parameters, AP resistance, exploitable or not, etc.);
- *ExploitabilityVerdict* – verdict (not tested, exploitable, non-exploitable, inconclusive).

The *PenetrationTest_PT* class represents any kind of a penetration test planned for the TOE evaluation. The test allows to experimentally confirm if the vulnerability is exploitable or not. Some vulnerabilities have status "residual". They are exploitable, but by attacks whose potential is higher than that considered for the claimed EAL. Figure 4 presents the PT_MyFWL_15 individual of the *PenetrationTest_PT* class.

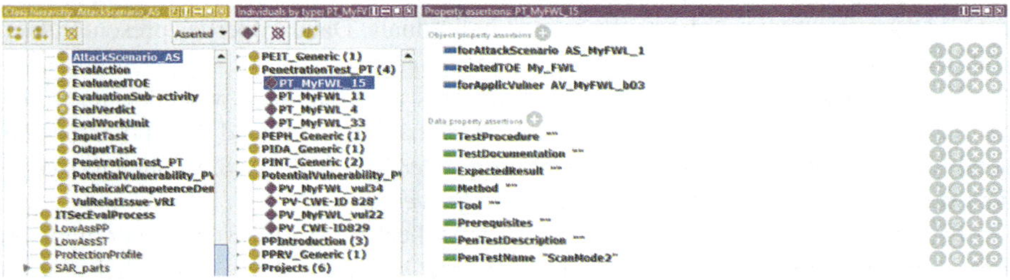

Fig. 4. Penetration test description in Protégé.

Object properties point out the related individuals representing the TOE, applicable to a vulnerability and attack scenario. Data properties express (yet empty) basic information related to the given penetration test in the knowledgebase.

Figure 5 presents the *MyFWL* individual of the *EvaluatedTOE* class.

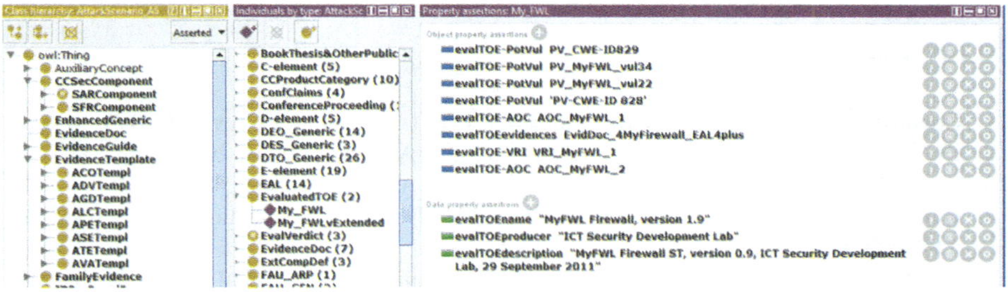

Fig. 5. Evaluated TOE specification in Protégé.

It represents all artefacts of the MyFWL evaluation and is related to the provided evaluation evidences, here the EvidDoc_4MyFirewall_EAL4plus individual.

4 Conclusions

The presented CCVA ontology allows to build a knowledgebase providing all necessary data for the evaluation processes. Data models were presented and exemplified on a simple firewall design. It is the second validation of the CCVA ontology. The first validation concerned sensors [23]. Both validations confirm the possibility to express vulnerability assessment data by ontological models.

Thanks to the applied knowledge engineering methodology, the paper goes one step forward in the structurization and formalization of the vulnerability assessment process comparing to CEM. More structurized operations (EEPs) and data (ontological models of knowledgebase) allow to support the following evaluation activities:

- identification of knowledge orientating the vulnerability search,
- managing knowledge sources about TOE-relevant vulnerabilities,
- supporting vulnerability analyses (e.g. attack potential);
- managing penetration tests designed for the evaluated TOE.

The future automation may bring advantages, like decreasing the cost and time of evaluation, improving the assessment repeatability and quality.

The planned research will focus on:

- more comprehensive validations,
- knowledgebase extension, data query facilities, ontology competency questions,
- establishing a knowledgebase of attacks, potential vulnerabilities, methods and tools for penetration tests,
- requirements for the CCVA tool, knowledgebase preparation and implementation.

Acknowledgements. 1. This work was supported by the Polish National Centre for Research and Development within the programme CyberSecIdent. Grant No. 381282/II/NCBR/2018.

2. This work was conducted using the Protégé resource, which is supported by grant GM10331601 from the National Institute of General Medical Sciences of the United States National Institutes of Health.

References

1. Common Criteria for IT Security Evaluation. part 1-3, version 3.1 rev. 5 (2017)
2. CC Portal. https://www.commoncriteriaportal.org/. Accessed 09 Jan 2020
3. Common Methodology for IT Security Evaluation. version 3.1 rev. 5 (2017)
4. Hermann, D.S.: Using the Common Criteria for IT Security Evaluation. CRC Press, Boca Raton (2003)
5. Higaki, W.H.: Successful Common Criteria Evaluation. A Practical Guide for Vendors, Copyright 2010 by Wesley Hisao Higaki, Lexington, KY (2011)
6. Bialas, A.: Common criteria related security design patterns for intelligent sensors—knowledge engineering-based implementation. Sensors **11**, 8085–8114 (2011)
7. Bialas, A.: Common criteria IT security evaluation methodology – an ontological approach. In: Zamojski, W., et al. (eds.) Advances in Intelligent Systems and Computing, vol. 761, pp. 23–34. Springer, Cham (2019)
8. Chang, S-C., Fan, C-F.: Construction of an ontology-based common criteria review tool. Proc. of the International Computer Symposium (ICS 2010), IEEE Xplore (2010)
9. Ekelhart, A., et al.: Ontological mapping of common criteria's security assurance requirements. In: Venter, H., et al. (eds.) New Approaches for Security, Privacy and Trust in Complex Environments, pp. 85–95. Springer, Boston (2007)
10. CCMODE. http://commoncriteria.pl/index.php/en/. Accessed 09 Jan 2020
11. de Franco Rosa, F., Jino, M.: A survey of security assessment ontologies. In: Rocha, Á., et al. (eds.) Recent Advances in Information Systems and Technologies. WorldCIST 2017. AISC, vol. 569. Springer, Cham (2017)
12. Białas, A.: Ontology based model of the common criteria evaluation evidences. Theoret. Appl. Inform. **25**(2), 69–92 (2013)
13. Musen, M.A.: The Protégé project: A look back and a look forward. AI Matters **1**(4), 4–12 (2015). Association of Computing Machinery Specific Interest Group in Artif. Intelligence
14. Protégé, https://protege.stanford.edu/. Accessed 21 Nov 2016
15. Bialas, A.: Software support of the common criteria vulnerability assessment. In: Zamojski, W., et al. (eds.) Advances in Intelligent Systems and Computing, vol. 582, pp. 26–38. Springer, Cham (2017)
16. Booth, H., Turner, Ch.: Vulnerability Description Ontology (VDO). Draft NISTIR 8138, NIST, Gaithersburg (2016)
17. CWE, CVE. http://cwe.mitre.org/. Accessed 08 Jan 2020
18. NVD. https://nvd.nist.gov/general. Accessed 07 Jan 2020
19. CPE. https://nvd.nist.gov/products/cpe/search. Accessed 05 Jan 2020
20. CAPEC. https://capec.mitre.org/. Accessed 03 Jan 2020
21. OWASP. https://www.owasp.org/index.php/Category:Vulnerability. Accessed 03 Jan 2020
22. CVSS. https://www.first.org/cvss/specification-document. Accessed 05 Jan 2020
23. Bialas, A.: Vulnerability assessment of sensor systems. Sensors **19**, 2518. https://www.mdpi.com/1424-8220/19/11/2518. Accessed 05 Jan 2020

Risk Management Approach for Revitalization of Post-mining Areas

Andrzej Bialas(✉)

Research Network ŁUKASIEWICZ – Institute of Innovative Technologies EMAG, Leopolda 31, 40-189 Katowice, Poland
andrzej.bialas@ibemag.pl

Abstract. The paper concerns the EU RFCS SUMAD (Sustainable Use of Mining Waste Dump) project. It presents the concept of a risk management tool which can be applied to plan the revitalization process of post-mining areas, such as waste dumps. The tool will support decision makers in the selection of the most advantageous revitalization activities for the considered waste dump and the assumed land use. The proposed tool is based on three pillars: Risk Reduction Assessment (RRA), Cost-Benefits Analysis (CBA) and Qualitative Criteria Analysis (QCA) used to work out aggregated information for a decision maker. The RRA user, based on the current risk related to the waste dump, proposes several alternatives of revitalization activities properly reducing this risk. Next, economic parameters of alternatives are analyzed with the use of CBA. Finally, non-financial parameters like: societal, ethical, political, technological, environmental parameters, etc. are considered with the support of QCA. The decision maker gets aggregated information to select the right activities for implementation. The proposed tool concept will be used to orientate the tool design and implementation and to search the project domain for data needed for the tool development.

Keywords: Risk management · Cost-Benefits Analysis · Qualitative Criteria Analysis · Post-mining areas revitalization

1 Introduction

The paper presents research related to the risk management methodology which can be applied to plan the revitalization process of post-mining areas. The results of this preliminary research will be used in the international project SUMAD (Sustainable Use of Mining Waste Dump) to elaborate requirements for the advanced risk management tool and later to implement them in software.

The objective of the SUMAD project is to explore possible future uses of areas which consist of coal-mining spoil with respect to geotechnical, sustainability, environmental, socio-economic, and long-term management challenges. This goal will be achieved by risk management and physical or numerical modeling. These will be applied to different rehabilitation schemes. The focus will be on the technical viability for the development of renewable energy infrastructure. The input will be obtained from tip operators, developers and authorities involved in the project. This will ensure

W. Zamojski et al. (Eds.): DepCoS-RELCOMEX 2020, AISC 1173, pp. 71–81, 2020.
https://doi.org/10.1007/978-3-030-48256-5_8

maximum possible impact of the undertaken operations. The project concepts will be tested on a case-study site.

SUMAD is an interdisciplinary project, embracing geotechnics, geology, ecology and IT. The role of the author's organization is to provide the software component called SUMAD RMT (Risk Management Tool) supporting decision makers in planning revitalization processes of post-mining sites, especially spoil heaps. The revitalization plans should consider different factors representing risk management, financial and non-financial constraints, similarly to these considered in the previously performed projects ValueSec [1] and CIRAS [2]. The author's organization participated in both these projects and acquired knowledge and experience in the field of advanced risk management in different domains of applications. ValueSec concerned: public mass events, mass transportation, air transport/airport security, communal security planning, cyber threats in smart grids, while CIRAS was focused on the critical infrastructure protection.

The author's idea is to adapt and apply the methodology of these projects to quite a new domain of application, i.e. post mining sites revitalization. This idea was included in the SUMAD project, but now it needs to be refined.

The objective of the paper is to work out the SUMAD RMT concept and to orientate the project research on the concept feasibility, data identification and structurization. Section 2 includes an overview of risk management methodologies in domains which partially overlap the SUMAD domain. Section 3 discusses the concept of the tool. Section 4 identifies research issues related to risk management data. Section 5 presents conclusions.

2 Risk Management in the SUMAD Project Domain

SUMAD requires a specific, interdisciplinary approach to risk management, including technical, ecological and geotechnical issues with a view on financial and non-financial limitations of the applied revitalization techniques considered "security measures" here.

Risk management is a continuous process including the identification, analysis, and assessment of potential hazards in a system or hazards related to a certain activity, risk monitoring and communication [3]. Based on the recognized risk picture, risk control measures are proposed to eliminate or reduce potential harm to the people, environment, processes or assets.

Risk management methodologies are very diverse and employed in many domains of application, including the technical domain [4]. Particular domains have developed their own methods and tools which, after adaptation, can be used in other domains. Apart from [4], a comprehensive review of methods, tools and R&D results can be found in the following sources:

- IEC 31010 standard [5] which characterizes about 30 renowned risk assessment methods for different applications;
- [6] Appendix C provides a comparison of the features of about 22 commonly used risk analysis methods;

- ENISA website [7] includes an inventory of risk management/assessment methods, mostly ICT-focused.

Some SUMAD-related risks can be considered a form of ecological risks [8], i.e. risks from natural events (flooding, extreme weather events, etc.), technology, practices, processes, products, agents (chemical, biological, radiological, etc.), and industrial activities that may influence ecosystems, animals and people. Ecological risk assessment embraces a critical review of available data for identification and quantification of the risks associated with a potential threat.

The basic ecological risk assessment methodology is specified as the EPA framework [9], which embraces tree phases: identification of the environmental values to be protected, development of a profile characterizing the ecosystems in which the stressor may occur as well as the biota that may be exposed, and risk characterization which integrates the exposure and effects profiles.

Several ecological risk assessment/management frameworks similar to [9] were compared in the paper [10]. The paper [11] presents experiences with the EPA framework and provides conclusions relevant for the SUMAD RMT development. The goal of the Triad approach [12] is to manage decision uncertainty to increase confidence that project decisions (about contaminant presence, location, fate, exposure, and risk reduction choices and design) are made correctly and cost-effectively [13]. The paper [14] includes guidance for the rehabilitation of brownfield sites, which can be useful for SUMAD.

The report [15] presents an example of a geotechnical risk management methodology. The impact of subsoil condition on a constructed object with the risk/opportunity management view is discussed in the paper [16]. The risk scenarios presented here may be used in planning the revitalization actions in SUMAD.

None of the reviewed approaches considers three kinds of factors: mixed-risk-, financial- and non-financial factors, especially for post-mining waste dumps. Ecological and geotechnical methodologies as well as the research and experiments performed by consortium partners will provide domain data for the tool to predefine threats, vulnerabilities, scenarios, risk measures, and revitalization techniques.

3 Concept of SUMAD Risk Management Methodology and Tool

SUMAD RMT will be based on the well-defined technical risk management approach [3, 5] with elements of environmental management [17]. Knowledge and experience gained from the ValueSec [1] and CIRAS [2] projects will be used for the SUMAD RMT development as well. The similarity of these two projects to SUMAD occurs only on a high level of abstraction. Deep differences between the application domains imply differences in the solutions designed for these domains.

SUMAD RMT is designed to support strategic decisions related to the revitalization objectives of the given site, i.e. the waste dump. Two major revitalization objectives are considered:

- improving the environmental and social properties of the given sites,
- enabling the sustainable exploitation of the sites with a business perspective.

SUMAD RMT is focused on the risk of not achieving the revitalization objectives. The risk can be reduced by applying different revitalization activities (techniques), which have certain costs, bring certain benefits and may have different non-financial constraints. The decision maker selects the most advantageous revitalization techniques for implementation. This decision is based on the results of three types of analyses:

- RRA – Risk Reduction Assessment,
- CBA – Cost-Benefits Analysis,
- QCA – Qualitative Criteria Analysis.

The SUMAD Risk Management Methodology (RMM) embraces the following steps:

1. Preliminary phase (it is necessary, though beyond the SUMAD RMT scope): Select the site and identify a possible revitalization strategy related to the planned land use, constituting a framework for the revitalization process.
2. Use the RRA component and assess the inherent risk (i.e. "risk before" the revitalization techniques application).
3. When the risk value exceeds the risk acceptance level, select a set of revitalization techniques and reassess the risk (i.e. determine the "risk after" the techniques implementation). Compare the new risk value with the risk acceptance level again, and repeat this step until the "risk after" will be acceptable. This way the set of elementary revitalization techniques, properly reducing risk value (i.e. below the risk acceptance level) is identified. It is possible to select several sets of revitalization activities considered further "revitalization alternatives".
4. Use the CBA component to determine the cost-benefits characteristics of alternative sets of revitalization techniques properly reducing the risk. Revitalization alternatives with unsatisfying economic parameters can be disqualified.
5. Use the QCA component to determine the non-financial characteristics of preselected alternative sets.
6. Select one of alternative sets for implementation – make the decision based on the aggregated results of assessments. When no alternative is acceptable, the decision maker looks for new ones repeating all analyses.

The general concept of SUMAD RMT is shown in Fig. 1.

Please note that the site characteristics imply the revitalization strategy and the applicable subsets of revitalization techniques. The pairs threat-vulnerability are analyzed for the given site. On this basis risk scenarios are identified and revitalization actions properly reducing risk are proposed. Next, cost-benefits (financial parameters of applied techniques) are planned and non-financial constraints for techniques are analyzed with the use of the QCA module. The decision maker gets a full picture of the situation to decide about revitalization activities to implement and prepare the revitalization plan.

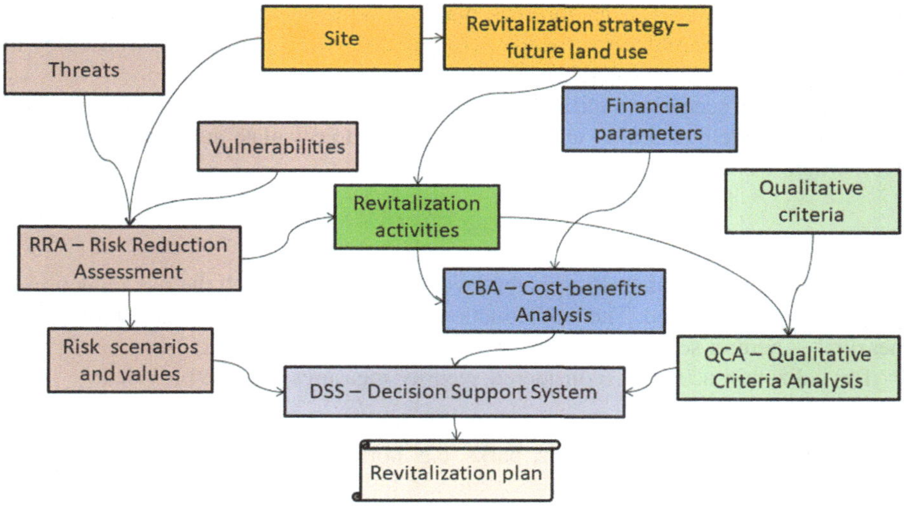

Fig. 1. The general concept of SUMAD RMT.

3.1 Risk Reduction Assessment (RRA) Module Concept

It is assumed that RRA will be based on the consequence-probability method [5] focused on the revitalized site considered "a protected asset", which is understood as "the site and its parts in its desirable, target state" according to the assumed land use. The site, i.e. the given waste dump, is characterized by many parameters, like: name, location, geometrical parameters, soil origin, ingredients and their properties, environmental conditions, thermal activity, and by many physical parameters, such as: cohesion, specific density, compressibility and water content, anisotropy, degradation of inter-particle bonds over time, etc.

Parameters and their values are subject of research performed by the consortium partners. Different sites placed in the countries of the SUMAD participants will be investigated, which will enable to define their profiles, characterized by specific parameters. Apart from these issues, predefined threats and vulnerabilities will be identified.

Threats and hazards embrace everything that might exploit a vulnerability and negatively influence an asset, here: rainfall, tremor, earthquake, uncontrolled human actions, etc. – breaking its stable structure. These are usually external factors.

Vulnerability means a weakness of the protected asset or group of assets that can be exploited by one or more threat agents, e.g.: improper structure/not enough density of the waste material, combustible or toxic waste components, etc. Threat agent is considered a person, organization, thing or entity that acts, or has the power to act in order to cause, carry, transmit, or support a threat.

The risk scenario enforced by the pair: threat-vulnerability specifies how the threat exploiting the vulnerability may breach the protected asset, e.g.: a landslide caused by weather factors with a coincidence of the poor ground condition, low energy production efficiency of wind turbines placed on the site caused by their improper localization, health problems of people living in the site neighbourhood caused by the burst of poisonous chemical substances.

For each risk scenario (a hazardous event) the consequences (C) and likelihood (L), measured in predefined scales, are assessed and used to calculate the overall Risk Value (R_V), where f is a risk function:

$$R_V = f(L, C) \quad (1)$$

Risk Acceptance Value (R_{AV}) is the level of risk above which the risk reduction (or transfer) is recommended. The risk reduction is achieved by implementation of security measures – here: a set of elementary revitalization techniques/activities.

Revitalization activities (considered security measures) can be provided:

- on the waste place, e.g.:
  - establishing recreation areas, parks, e.g. horse racing tracks,
  - flooding to produce lakes,
  - residential purposes (very restricted),
  - photovoltaics/wind farms, ground source heat pumps,
  - coal recovery from dumps (secondary exploitation) – coal is obtained and the risk of spontaneous combustion is reduced,
  - ditches with leachate traps,
  - reclamation of external landfill slopes by covering them with soil and introducing greenery,
- or by using the spoil dump for different purposes, e.g. in:
  - transport infrastructures (railway, roads) on spoil dumps,
  - levelling of degraded areas,
  - highway engineering,
  - cement production,
  - hydraulic stowing (for mines),
  - pavement construction,
  - construction of embankments,
  - levelling the ground.

Generally, revitalization activities can be aimed at the ground improvement (utilization of the appropriate material, use of drainage techniques to eliminate the role of water, both atmospheric rainfall and underground) or the dump stability.

The L and C values will be determined on the basis of expert knowledge. Each of these parameters depends on different factors related to assets, threats and vulnerabilities. Most of them are geotechnical parameters and characteristics. They should be interpreted in terms of the risk management methodology and can be considered:

- site attributes implying risks existing in the site, or
- revitalization technique attributes implying the ability of the technique to reduce the risk.

This issue will be subject of research in the SUMAD project.

3.2 Cost-Benefits Analysis (CBA) Module Concept

The CBA module supports the financial analyses of considered revitalization activities. CBA includes three open, configurable, hierarchical structures of:

- Investment Costs (IC): planning, design, procurement, implementation, setup, integration, etc.,
- Operational Costs (OC): maintenance, end-of-lifetime cost, economic losses, financing, public services disturbances, etc.,
- Future Benefits (FB): reduction of damages, increasing business profits, reduction of insurance fees, etc.,

having categories and nested subcategories defined by the decision maker according to analytical needs, allowing to plan costs/benefits by categories in the considered time horizon, e.g. 10 years.

After preparing cost/benefits distribution in time, deeper analyses are possible. The CBA module will have implemented common economic indicators, like Net Present Value (NPV), Benefit Cost Ratio (BCR), Pay Back Period, Break Even Point, etc.

3.3 Qualitative Criteria Analysis (QCA) Module Concept

The risk picture (from RRA) and the financial picture (from the CBA) of revitalization activities do not constitute a full situation picture for the decision maker. These two pictures should be supplemented by other important constraints for these activities – "soft" factors, sometimes hidden or difficult to express explicitly. The QCA module supports analyses of these non-financial factors related to the revitalization activities. The QCA module should help to identify and to assess factors impairing the implementation of revitalization activities which properly reduce risk and are advantageous from the economic perspective.

QCA is based on an open, configurable list of predefined items representing non-financial issues, called qualitative criteria. The criteria are grouped by categories, such as: societal, economical, ethical, political, technological, environmental, etc.

For the given set of revitalization activities, positive and negative impacts represented by each qualitative criterion are assessed with the use of utility functions configured by the decision maker.

For example, the environmental category embraces the following qualitative criteria:

- natural environment degradation,
- impact on cultural environment (impact on architecture, historic buildings, etc.),
- movement and mobility (whether the activity impacts the mobility/free movement of people),
- aesthetics (sight, smell, sound).

Aggregated results of analyses made with the use of the RRA, CBA and QCA modules constitute a full picture of the situation for the decision maker who selects revitalization activities and elaborates the revitalization plan for the given site according to the assumed land use.

4 Research on Domain Data

Contrary to SUMAD, the ValueSec and CIRAS application domains were well defined and were less-dependent on physical parameters. All risk management data, like assets, threats, vulnerabilities, security measures are very specific in the waste dumps rehabilitation domain, they require identification and right interpretation. All these issues are based on physical properties of the site and its environment. The examples of these data from the previous section are rather superficial and for this reason the research on technical, ecological, geological, geotechnical, financial and other factors is needed. The SUMAD consortium joins experts from these domains, who have at their disposal sites for experimentation, methods and knowledge. Research will be focused on identification and structurization of data processed by SUMAD RMT.

4.1 Asset – a Waste Dump (a Site) Under Rehabilitation

The basic issues are: What is the basic set of characteristics/parameters describing waste dumps in the tool? Is it possible to define profiles for similar dumps? What are the attributes allowing to distinguish such profiles? How to define a data structure specifying the site in the data base?

The author proposes a data structure for the site specification including five segments of attributes (to be extended/refined in the SUMAD research):

General characteristics – basic information to identify the site:

- *ID, name, location, owner,*
- *geometrical parameters, e.g. volume, height, area,*
- *related regulatory actions, e.g. land protection,*

Rehabilitation objectives – considered during risk management:

- *current and planned land use,*
- *objective 1: improvement of the environmental and social properties of the given sites,*
- *objective 2: enabling sustainable exploitation of the sites with a business perspective,*

Internal/inherent factors – implying vulnerabilities:

- *soil origin,*
- *surrounding land,*
- *ingredients and their properties (lithography),*
- *appearance of the site, e.g. odours, thermal activity, aggressive ground chemistry,*
- *geological parameters, e.g. cohesion, anisotropy, specific density, compressibility,*
- *mechanical parameters, e.g. stress path and strain-dependency of stiffness, degradation of inter-particle bonds,*

- *hydrological parameters, e.g. water content,*
- *underground voids,*
- *slopes,*
- *proximity of controlled waters possible to contaminate,*
- *temporary, ad hoc works on revitalization,*

Environmental factors – facilitating identification of threats and hazards

- *condition of the surrounding land,*
- *climate condition,*
- *water condition, proximity of reservoirs, rivers (flood possibility)*

Factors allowing to identify feasible revitalization techniques:

- *current surface features, e.g., vegetation,*
- *current risks and future opportunities,*
- *financial situation.*

It is assumed that the risk will not be referenced to an asset as a whole, but to achieving both above mentioned rehabilitation objectives with respect to the site.

4.2 Threats/Hazards and Vulnerability Structures

These structures are similar. Each includes the following fields: *ID, category, short name, description*. Threats are identified mainly on the basis of environmental factors, but vulnerabilities are identified on the internal factors basis.

4.3 Revitalization Techniques Representation

The examples of revitalization activities mentioned in Sect. 3 are defined on a general level and have different levels of detail. In practice, while choosing a given technique, e.g. arranging a recreation area, one needs to undertake many other elementary revitalization actions. This implies the necessity to distinguish the categories of elementary revitalizations techniques.

A data structure representing an elementary revitalization technique contains: *ID, name, category, description.*

The categories can be grouped in coherent revitalization techniques sets, called alternatives, including: *group ID, list of elementary techniques, CBA assessment record, QCA assessment record.*

4.4 Risk Scenarios Specifying Impacts

Two situations for risk scenarios should be distinguished:

- "risk before", which means the risk picture when no revitalization techniques are applied or only temporary, ad hoc actions were done,
- "risk after", which means the risk picture when preselected sets of elementary revitalization activities were applied (risk reassessment); SUMAD RMT will be

able to consider a certain number (N) of alternatives. For this reason the tool should be able to store each of them.

The proposed structure of a risk scenario includes the following fields:

- *ID, name, general description: "how the threat exploiting the vulnerability may breach the protected asset",*
- *risk before:*
  - *impact description,*
  - *L (Likelihood), C (Consequences),* R_V *(Risk value),*
- *risk after for the alternative i (for i: = 1 to N):*
  - *impact description,*
  - *L (Likelihood), C (Consequences),* R_V *(Risk value),*
  - *Selected for implementation (Boolean).*

5 Conclusions

The paper is focused on the concept of an advanced risk management tool for a very specific domain of application, i.e. waste dumps rehabilitation. Advanced means that revitalization activities are selected not only on the basis of the risk value but also on the financial and non-financial parameters of these activities. The following paper objectives were achieved:

- working out the concept of SUMAD RMT which can be used by the author's team to elaborate the tool requirements and by the consortium to validate the tool before implementation,
- working out basic data structures to orientate research on their refinement and on data identification for the tool.

Please note that the site data structure is of the key meaning for SUMAD RMT as it includes the following segments supporting risk management: general information segment, risk management objectives segment, internal and external properties segments, and a segment implying revitalization techniques.

The paper provides the consortium partners with input to orientate the SUMAD preliminarily planned research so that they could prove the concept before the SUMAD RMT tool has been designed and implemented. Apart from risk factors, the CBA and QCA categories also need research within this specific project domain.

Acknowledgements. The SUMAD project leading to this application has received funding from the EU Research Fund for Coal and Steel under grant agreement No 847227.

References

1. ValueSec. https://cordis.europa.eu/project/rcn/97989/factsheet/en. Accessed Nov 2019
2. Ciras. http://cirasproject.eu/. Accessed Nov 2019
3. ISO 31000:2009 Risk management – Principles and guidelines

4. Rausand, M.: Risk Assessment: Theory, Methods, and Applications. Wiley, Hoboken (2011). Series: Statistics in Practice (Book 86)
5. ISO/IEC 31010:2009 Risk Management – Risk Assessment Techniques
6. Hokstad, P., Utne, I.B., Vatn, J. (eds.): Risk and Interdependencies in Critical Infrastructures: A Guideline for Analysis. Springer, Heidelberg (2012). https://doi.org/10.1007/978-1-4471-4661-2_2. (Springer Series in Reliability Engineering)
7. ENISA. https://www.enisa.europa.eu/topics/threat-risk-management/risk-management/current-risk/risk-management-inventory. Accessed Nov 2019
8. Fargašová, A.: Ecological risk assessment framework. Acta Environ. Univ. Comenianae (Bratislava) **24**(1), 10–16 (2016). https://doi.org/10.1515/aeuc-2016-0002
9. Framework for ecological risk assessment. U.S. Environmental Protection Agency (EPA), Washington (1992)
10. Power, M., McCarty, L.S.: Trends in the development of ecological risk assessment and management frameworks. Hum. Ecol. Risk Assess. **8**(1), 7–18 (2002)
11. Hope, B.K.: An examination of ecological risk assessment and management practices. Environ. Int. **32**(8), 983–995 (2006)
12. TRIAD web page. https://triadcentral.clu-in.org/. Accessed 28 Nov 2019
13. Crumbling, D.M.: Summary of the Triad Approach. Environmental Protection Agency (EPA), Washington, U.S (2004)
14. Mine Site Cleanup for Brownfields Redevelopment - A Three-Part Primer. U.S. Environmental Protection Agency (EPA), Washington (2005)
15. Risk management in geotechnical engineering projects – requirements. Methodology. Swedish Geotechnical Society, SGF Report 1:2014E, 2nd edn (2014)
16. Sondermann, W., Kummerer, C.: Geotechnical opportunity management – subsoil conditions as an opportunity and a risk. In: XVI Danube – European Conference on Geotechnical Engineering, June 2018, vol. 2(2–3), pp. 395–400 (2018)
17. ISO 14001:2015 Environmental management systems—Requirements with guidance for use

CVE Based Classification of Vulnerable IoT Systems

Grzegorz J. Blinowski(✉) and Paweł Piotrowski

Institute of Computer Science, Warsaw University of Technology, Nowowiejska 15/19, 00-665 Warsaw, Poland
g.blinowski@ii.pw.edu.pl

Abstract. Common Vulnerabilities and Exposures database (CVE) is one of the largest publicly available source of software and hardware vulnerability data and reports. In this work we analyze the CVE database in the context of IoT device and system vulnerabilities. We introduce a real-world based classification of IoT systems. Then, we employ a SVM algorithm on selected subset of CVE database to classify "new" vulnerability records in this framework. The subset of interest consists of records that describe vulnerabilities of potential IoT devices of different applications, such as: home, industry, mobile controllers, networking, etc. The purpose of the classification is to develop and test an automatic system for recognition of vulnerable IoT devices and to test completes, sufficiency and reliability of CVE data in this respect.

Keywords: Internet of Things · IoT security · System vulnerability classification · CVE · NVD · SVM

1 Introduction and Background

1.1 IoT Applications an Architecture – An Outline

IoT can be most broadly defined as an interconnection of various uniquely addressable objects through communication protocols. Narrowing down the above said, we can describe it as a communication system paradigm in which the objects of everyday life, equipped with microcontrollers, network transmitters, and suitable protocol stacks that allow them communicate with one another and, via ubiquitous cloud infrastructure and also with users, become an integral part of the Internet environment [1].

IoT is widely, although mostly anecdotally, known as a network of household appliances – from PC equipment and peripherals to fridges, coffee machines, etc. However, the scope of IoT deployments is much wider, and covers the following areas [2–4]:

Smart Cities; Smart environment (monitoring); Smart agriculture and farming; Smart Electric Grid; Manufacturing and Industrial Security and sensing – this range of applications is often referred to as IToT (Industrial IoT) and the systems themselves are referred to as SCADA (Supervisory Control and Data Acquisition); eHealth; Home automation ("Smart homes"). Here, we will consider an IoT model compatible with the

W. Zamojski et al. (Eds.): DepCoS-RELCOMEX 2020, AISC 1173, pp. 82–93, 2020.
https://doi.org/10.1007/978-3-030-48256-5_9

reference architecture model proposed by the EU FP7 IoT-A project [5] and the IoT-A tree structure [6] which consists of three major levels:

- **Perception and execution layer**, which encompasses a wide range of "smart" devices ranging from RFID and NFC enabled tags, environmental sensors and actuators, various home appliances, mobile terminals, smart phones, etc.
- **Network layer**, which provides heterogeneous communication infrastructure based on multiple network standards such as: s Wi-Fi, 3G/LTE, Z-wave, ZigBee, 6LoWPAN, VLC and Ethernet together with the standard internet protocol suite (IPv4/IPv6 and transport layer UDP/TCP stack).
- **Cloud or application layer**, which integrates, manages and analyzes data from IoT devices. The cloud not only gathers data and manages the "things" and "core" layer, but acts as a ubiquitous service provider for end-users, according to the Service Oriented Approach (SOA) paradigm.

1.2 Security Issues with IoT Systems

We can distinguish two general kinds of IoT threats: 1. **threats against IoT** and 2. **threats from IoT**. 1 Threats against IoT occur when a flaw in an IoT device or application, on the perception, network or cloud level is exploited by the hacker, and the device or application is compromised - i.e. a full or limited access to its functions and data is gained by an attacker. In case of threats *from* the IoT, the compromised infrastructure is used to conduct various attacks against IoT or Internet-connected devices. Mirai botnet [7] can serve as an example - when a multitude of compromised webcams and other devices were used to conduct a massive DDoS attack.

In [8] the authors have proposed five "dimensions" relating to IoT security: hardware, operating system/firmware, software, networking and data:

- **Hardware security** is critical when an attacker can physically access the device. Through the hardware backdoors, software level integrity checking can be bypassed by disabling the checking functionality or booting via forged firmware. Almost all IoT devices have hardware vulnerabilities which may be exploited [9]. Microcontrollers (MCUs) which are broadly used in industry applications (SCADA) as well as in automobiles and home automation are also prone to hardware level vulnerabilities.
- **Operating system, firmware and software security and privacy** relates to all three IoT layers: perception, network and cloud. Software security issues are similar to those in the traditional computer systems. Trustworthy operating systems should be used at the perception layer to reduce the risk of remote compromise. However, in practice, this is rarely the case. The controller application is often installed on a PC or a smartphone and software secure measures should be applied in order to prevent the attack against it. The cloud layer security also cannot be blindly trusted, for example: servers installed on Amazon EC2 are secured from the cloud provider's point of view, but not from the point of view of installed application and have to be secured by whoever deploys the servers.
- **Network Security and Privacy** - as a networked system a whole IoT environment has to be secured from end to end. Encryption and authentication should be used

consistently, but often is not. Two security-related functions specific to home IoT devices are pairing and binding. Many attacks relating to their design and implementation have been analyzed and described, for example for: surveillance camera systems [10], wearable devices [11], etc. Week passwords are also a typical security case involved with pairing and binding.

- **Cloud and data** - the cloud collects data from the perception layer, and is responsible for maintaining proper data security. The cloud often handles authentication and associated services and is a peer in end-to-end encryption of transmitted data. Application compromised on the cloud level exposes a significant amount or perhaps the whole of the collected data.

To summarize this section: the majority of security problems emerging in today's IoT systems result directly from buggy, incomplete or outdated software and hardware implementations. A major protocol flaw design error (such as Heartbleed [12] and DROWN [13]) are much rarer. As can easily be verified in public domain vulnerability databases, the number of products reported with serious security flaws is growing year by year.

1.3 Scope of This Work and Related Research

In this work, we propose a classification of device-related (i.e. not "pure software") vulnerability data for IoT and IIoT equipment. We have divided the CVE records from a public database into 7 distinct categories (e.g.: home equipment, SCADA devices, network infrastructure systems, etc.). The database samples were hand-classified by us based on the expert knowledge. We then used support vector machine (SVM) classifier on the device and vulnerability data to predict categories of "new" vulnerabilities – for example data from year 2017 was used to classify 2018's data, etc. The purpose was to predict, and (if possible) prevent and mitigate threats resulting from new vulnerabilities. This is a difficult task given the size of the database and the rate of its growth – each day tens of new records are added to the CVE database alone. Hence, when a new vulnerability or exploit is discovered it is often critical to learn its scope by automatics means, as fast as possible.

There has been some prior research on automatic analysis and classification of vulnerability databases: In [14, 15] models and methodologies of categorizing vulnerabilities from CVE database according to their security types based on Bayesian networks. In [16] Topic Models were used to analyze security trends in CVE database with no prior (expert) knowledge. Huang et al. [17] proposed recently an automatic classification of records from NVD database based on deep Neural Network, the authors compared their model to Bayes and KNN models and found it superior. All of the above cited research was focused on categorizing software aspect of vulnerabilities, with categories such as for example: SQL injection, race condition, cryptographic errors, command injection, etc. According to our knowledge no prior work was done regarding categorizing of the impacted equipment: system or device – our work tries to address this gap.

This paper is organized as follows: in Sect. 2 we describe the contents and structure of the CVE database; we also describe related: CPE (Common Platform Enumeration) and NVD (Network Vulnerability Data) records. In Sect. 3 we introduce our proposed classes of IoT devices; we discuss briefly SVM classifier methods and the measures used by us to test classifiers quality. In Sect. 4 we present the results of the classification. Our work is summarized in Sect. 5.

2 Structure and Contents of CVE Database

2.1 The Common Vulnerability and Exposures (CVE) Database

The Common Vulnerability and Exposures (CVE) database hosted at MITRE is one of the largest publicly available source of vulnerability information [18]. CVE assigns identifiers (CVE-IDs) to publicly known product vulnerabilities. Across organizations, IT-security solutions vendors, and security experts, CVE has become the de facto standard of sharing information on known vulnerabilities and exposures.

In this work we use an annotated version of the CVE database, known as National Vulnerability Database (NVD) which is hosted by National Institute of Standards and Technology (NIST). NVD is created on the basis of information provided by MITRE (and through the public CVE site). NIST adds other information such as structured product names and versions, and also maps the entries to CWE names. NVD feed is provided both in XML and JSON formats structured in year-by-year files, as a single whole-database file and as an incremental feed reflecting the current year's vulnerabilities.

Figure 1 contains a sample (simplified) record from the NVD database. Fields which are relevant for further discussion are as follows:

- **`entry`** contains record id as issued by MITRE, the id is in the form: CVE-yyyy-nnnnn (e.g. CVE-2017-3741) and is commonly used in various other databases, documents, etc. to refer to a given vulnerability;
- **`vuln:vulnerable-configuration`** and **`vuln:vulnerable-software-list`** identifies software and hardware products affected by a vulnerability. This record contains the description of a product and follows the specifications of the Common Platform Enumeration (CPE) standard. Refer to description in Sect. 2.2;
- **`vuln:cvs`** and **`cvss:base_metrics`** describe the scope and impact of the vulnerability. This data allows to identify real-world consequences of the vulnerability;
- **`vuln:summary`** holds a vulnerabilities short informal description.

2.2 Common Platform Enumeration (CPE)

CPE is a formal naming scheme for identifying: applications, hardware devices, and operating systems. CPE is part of the Security Content Automation Protocol (SCAP) standard [19], which was proposed by the National Institute of Standards and Technology (NIST). Here we will refer to the most recent version 2.3 of CPE. The CPE naming scheme is based on a set of attributes called Well-Formed CPE Name (WFN) compatible with the CPE Dictionary format [20]. The following attributes are

```
<?xml version='1.0' encoding='UTF-8'?>
  <nvd xmlns:cvss="http://scap.nist.gov/schema/cvss-v2/0.2" ...>
   <entry id="CVE-2017-3741">
      <vuln:vulnerable-configuration id="http://nvd.nist.gov/">
         . . . </vuln:vulnerable-configuration>
      <vuln:vulnerable-software-list>
        <vuln:product>
         cpe:/a:lenovo:power_management:1.67.12.19</vuln:product>
      </vuln:vulnerable-software-list>
      <vuln:cve-id>CVE-2017-3741</vuln:cve-id>
      <vuln:published-datetime> … </vuln:published-datetime>
      <vuln:cvss>
        <cvss:base_metrics>
          <cvss:score>2.1</cvss:score>
          . . . </cvss:base_metrics>
      </vuln:cvss>
      <vuln:summary>… a local user may alter ... /vuln:summary>
    </entry>
    </nvd>
```

Fig. 1. A single simplified NVD record from NIST CVE feed (some less relevant fields have been abbreviated or omitted).

part of this format: part, vendor, product, version, update, edition, language, software edition, target software, target hardware, and other (not all attributes are always present in the CPE record, very often "update", and the following ones are omitted from the record). Currently, CPE supports two formats: URI and formatted. The CVE database uses URI format and we will only discuss this format further on.

For example: in the following CPE record: **`cpe:/h:moxa:edr-g903:-`** the attributes are as follows: **`part:h`** (indicating hardware device), **`vendor:moxa`**, **`product:edr-903`**, version, and the following attributes are not provided. As a second example let us consider a `vuln:vulnerable-configuration` record – Fig. 2.

```
<vuln:vulnerable-configuration id="http://nvd.nist.gov/">
 <cpe-lang:fact-ref name="cpe:/o:d-link:dgs-1100_firmware:1.01.018"/>
  <cpe-lang:logical-test operator="OR" negate="false">
    <cpe-lang:fact-ref name="cpe:/h:d-link:dgs-1100-05:-"/>
    <cpe-lang:fact-ref name="cpe:/h:d-link:dgs-1100-10mp:-"/>
      ...
  </cpe-lang:logical-test>
</vuln:vulnerable-configuration>
```

Fig. 2. A vulnerable configuration record from CVE – a logical expression build from CPE identifiers.

The record shown on Fig. 2 refers to particular operating systems version (firmware), namely: `cpe:/o:d-link:dgs-1100_firmware:1.01.018` on a list of distinct hardware devices: `cpe:/h:d-link:dgs-1100-05:-,...` etc. CPE does not identify unique instantiations of products on systems, rather, it identifies abstract classes of products. The first component of the CPE descriptor is "part" it can take the following values: a – for application, h – for hardware, o – for operating system.

2.3 Discussion

The NVD database is distributed as XML and JSON feeds, it is also possible to download the whole historical data package (starting from 1999, but records compliant with the current specification are available for data generated since 2002). In addition, there is also an on-line search interface. The database, as at the beginning of 2020 contains over 120 000 records in total, and on the average the number of records increases year by year. Due to historical reasons it is neither completely consistent, nor free of errors. Older records lack some information, there are approximately 900 records without CPE identifier; there exists a large number of records with CPEs inconsistent with the CPE dictionary (approx. 100 000 CPEs). In general, the binding between the vulnerability description and the product concerned may be problematic. Product names containing non-ASCII or non-European characters also pose a problem, as they are recoded to ASCII often inconsistently or erroneously. Lack of record classification on the CVE or CPE level (except for the "application, OS or hardware" attribute in the CPE) is especially cumbersome, because there is no easy or obvious way to differentiate products. Essentially, it is impossible to extract data relating to, for example: web servers, home routers, IoT home appliances, security cameras, cars, SCADA systems, etc. without a priori knowledge of products and vendors.

3 CVE Data Classification and Analysis

3.1 Data Selection

For the classification purposes we have selected only records with the CPE "part" attribute set to "h" (hardware records), namely, the selection criteria was: if any of the records in `vuln:vulnerable-configuration` section contains CPE with part = h then the record was selected for further consideration. Other records were discarded. The reason is the following: all or most of the "hardware" type records refer to devices or systems which can potentially be a component of the perception or network layer of the IoT or IIoT architecture. We have also narrowed down the timeframe to data from years 2010–2019 (data from first quarter of 2019 was taken into account).

Hand analysis of selected vulnerability data led us to a grouping of records into 7 distinct classes as follows:

- **H** – Home and SOHO devices; routers, on-line monitoring.
- **S** – SCADA and industrial systems, automation, sensor systems, non-home IoT appliances, car and vehicles (subsystems), medical devices.

- **E** – Enterprise, Service Provider (SP) hardware (routers, switches, enterprise Wi-Fi and networking) –the network level of IoT infrastructure.
- **M** – mobile phones, tablets, smart watches and portable devices - this constitutes the "controllers" of IoT systems,
- **P** – PCs, laptops, PC-like computing appliances and PC servers (controllers),
- **A** – other, non-home appliances: enterprise printers and printing systems, copy machines, non-customer storage and multimedia appliances.

The reason for the above classification was practical – the key distinction for an IoT component in reference to its security vulnerability is the market and scope of its application (home use, industrial use, network layer, etc.). On the other hand, we are limited by the description of the available data – it would difficult to use a finer-grain classification. Also, it is not practical to introduce to many classes with small number of members, because the automatic classification quality suffers in such case.

3.2 Data Analysis Methodology

We build classifiers by training linear support vector machines (SVM) [21] on the features of "hardware" vulnerability records extracted from the NVD database. The feature vector contains: vendor name, product name and other product data from CPE (if supplied), vulnerability description, and error code (CWE).

The steps of the process of building a classifier are the following: 1. preprocessing of input data (removal of stop-words, lemmatization, etc.); 2. feature extraction, i.e. conversion of text data to vector space; 3. training of the linear SVM.

We use a standard linear SVM, which computes the maximum margin hyperplane that separates the positive and negative examples in feature space. Alternative methods include: k-nearest neighbor, Bayesian classifiers and Neural Nets. We have conducted some experiments with Neural Nets, but finally decided to use SVM, as it proved itself to be: fast, efficient and well suited for text-data classification. With SVM method the decision boundary is not only uniquely specified, but statistical learning theory shows that it yields lower expected error rates when used to classify previously unseen examples [21, 22] – i.e. it gives good results when classifying new data.

We have used Python 3.7.1 with NLTK 3.4.1 [23] to pre-process the text data and SVM and classification quality metrics routines from scikit-learn 0.21.3 [24] libraries.

3.3 Classification Measures

To benchmark classification result we use two standard measures: precision and recall. We define *precision* (Eq. (1)) as the ratio of true positives to the sum of true positives and false positives; we define *recall* (Eq. (2)) as the ratio of true positives the sum of true positives and false negatives (elements belonging to the current category but not classified as such.) Finally, as a concise measure we use the F1 score – Eq. (3). The F1 score can be interpreted as a weighted average of the precision and recall, where an F1 score reaches its best value at 1 and worst at 0.

$$precission = TP/(TP + FP) \tag{1}$$

$$recall = TP/(TP + FN) \tag{2}$$

$$F1 = 2 * \frac{precision * recall}{(prceision + recall)} \tag{3}$$

4 Classification Results

4.1 Data Selection and Classification

We have tested the classifier for historical data in one year intervals. For example, to classify data from 2018 we have used records from the following ranges: 2014–2017, 2015–2017, 2016–2017 and 2017, etc. On we show confusion matrices trained on data ranging from 2014 to 2017 used to classify data for 2018. From a good classifier we would expect a majority of records on the diagonal. Here the classification is not perfect, for example - for training data from year range 2014–2017: 109 H type records were marked as E and 108 as S class; only 62% were correctly classified (recall). When only data from 2017 was used, perhaps surprisingly, the classification is more accurate: 489/85% records of the H type were labelled correctly (recall); however for S and E classes only 55% were correctly identified. For classes with a low number of records (C, M, P) the classification falls below 50% (Fig. 3).

On Fig. 4 both precision and recall for 2018 records are shown based on training data from: 2014–2017, 2015–2017, 2016–2017 and 2017.

Similar trends are visible on Fig. 5 for classified data from the first quarter of 2019 based on SVM trained on records from 2017–2018 range – for E, H and S classes precision is within the range of 70%–90% (with an exception of H class where it is only 44%), and recall falls in the similar range of 70%–90%. The results for years 2015–2016 show similar trends.

Discussion - As we have shown in the previous section, quality of classification results can be summarized as average. We were able to achieve 70–80% of correct labeling for the most populated classes of devices. In some cases the classification falls below 50%. Using more training data, i.e. going back in time does not always improve the classification quality, on the contrary – in most cases it reduces it.

To summarize the classification results for the whole period of 2011–2019 (Q1): Fig. 6 shows the values of *F1* measure weighted by support (the number of true instances for each label). Because of the weighing, this shows the quality of classification for all classes. As shown, the balanced *F1* score varies between 50% and 72%.

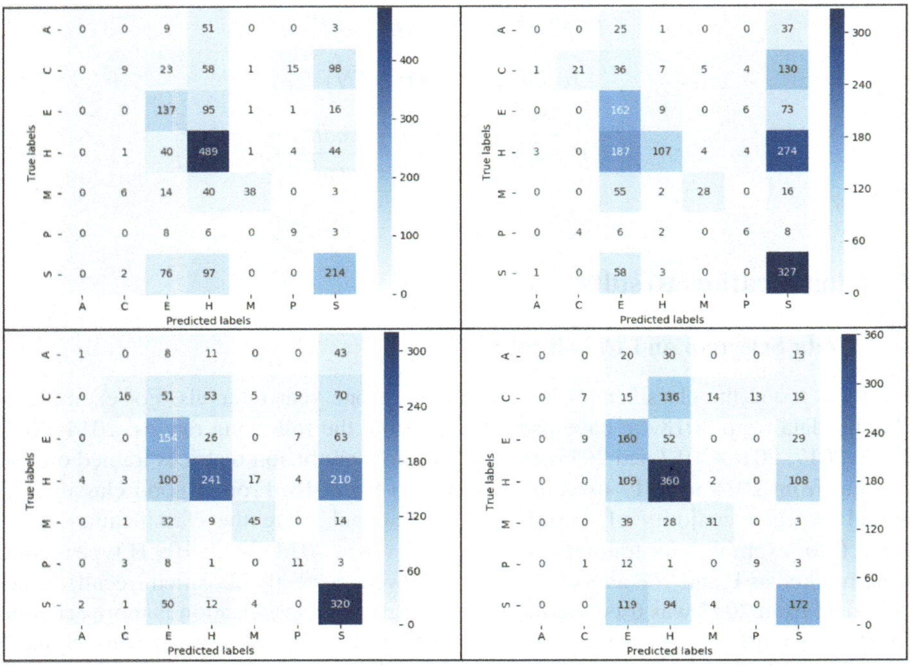

Fig. 3. Classification of records from 2018 based on training data from: 2017, 2016–2017, 2015–2017 and 2014–2017 (left to right, top-bottom). Numbers show the number of records.

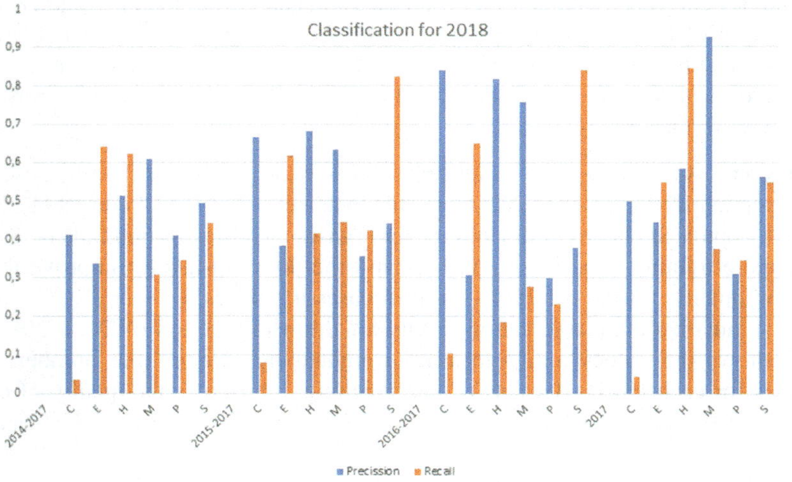

Fig. 4. Precision and recall for 2018 records based on training data from: 2014–2017, 2015–2017, 2016–2017 and 2017 (category "A" was removed).

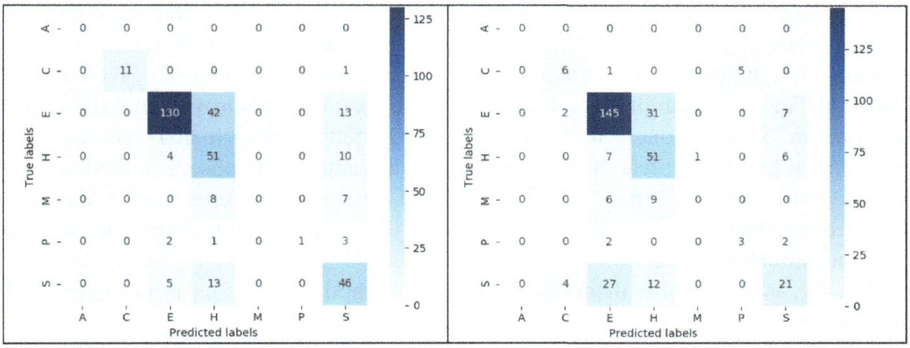

Fig. 5. Classification of records from 1st quarter 2019 based on data from: 2018, 2017–2018 (left to right). Numbers show the number of records.

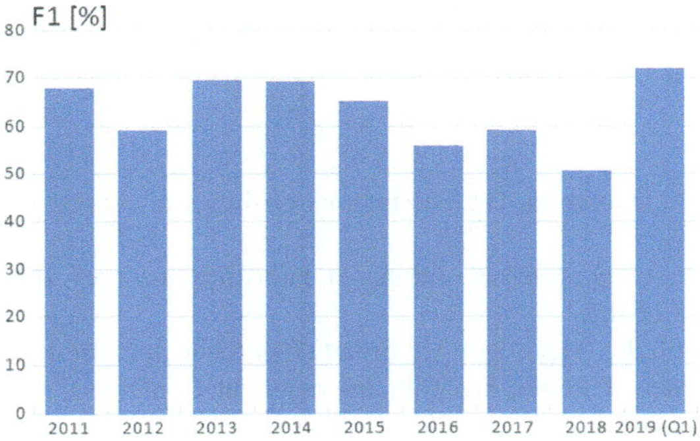

Fig. 6. F1 score for records from 2011 to 2019 (Q1).

5 Summary

We have proposed a classification of IoT device related vulnerability data from the public CVE/NVD database. We have divided vulnerability records into 7 distinct categories: Home and SOHO, SCADA, Enterprise & Networking, Mobile devices, PC devices and other non-home appliances. The hand-classified database samples were used to train a SVM classifier to predict categories of "new" vulnerabilities.

The purpose of the automatic classifier is to predict, and (if possible) in subsequent steps - prevent and mitigate threats resulting from new vulnerabilities. This is not a trivial task to execute by hand given the size of the database and the rate of its growth - when a new vulnerability or exploit is discovered it is often critical to learn its scope by automatics means, as fast as possible.

We have attained classification precision and recall rates of 70–80% for strongly populated categories and of approx. 50% or lower for less numerous categories. This are not ideal results, and in practice they would require further human intervention (verification and possibly reclassification). On the other hand, SVM classifiers have been proved numerous times to be an accurate mechanism for text data classification. The problem in our case lies with the data itself - neither CVE nor CPE contents provide enough specific data for the SVM to discern record categories. We can conclude that the vulnerability ontology should be extended to provide this additional information. Similar conclusions, although not directly related to IoT security, have been drawn by other researchers – e.g. in [25] the authors propose a unified security cybersecurity ontology that incorporates and integrates heterogeneous data and knowledge schemas from different cybersecurity systems, including data about products and product vendors.

Finally, is also worth mentioning that the method used by us is not necessarily limited to CVE database, numerous other on-line vulnerability databases exists, which are managed by companies (e.g. Microsoft Security Advisories, TippingPoint Zero Day Initiative, etc.), national CERTs or by professionals' forums (e.g. BugTraq, Exploit-DB, and others). Information from various sources can be integrated and categorized by the method we proposed in this paper. This should increase the precision of the classification and is a topic of our further research.

References

1. Atzori, L., Iera, A., Morabito, G.: The internet of things: a survey. Comput. Netw. **54**(15), 2787–2805 (2010)
2. Da Xu, L., He, W., Li, S.: Internet of things in industries: a survey. IEEE Trans. Industr. Inf. **10**(4), 2233–2243 (2014)
3. Jalali, R., El-Khatib, K., McGregor, C.: Smart city architecture for community level services through the internet of things. In: 18th International Conference on Intelligence in Next Generation Networks (ICIN), pp. 108–113 (2015)
4. Al-Fuqaha, A., Guizani, M., Mohammadi, M., Aledhari, M., Ayyash, M.: Internet of things: a survey on enabling technologies, protocols, and applications. IEEE Commun. Surv. Tutorials **17**(4), 2347–2376 (2015)
5. The 7th Framework Programme funded European Research and Technological Development from 2007 until 2013; Internet of Things and Future Internet Enterprise Systems. http://cordis.europa.eu/fp7/ict/enet/projects_en.html. Accessed 10 May 2017
6. Architectural Reference Model for the IoT – (ARM). Introduction booklet. http://iotforum.org/wp-content/uploads/2014/09/120613-IoT-A-ARM-Book-Introduction-v7.pdf.. Accessed 10 May 2017
7. Antonakakis, M., April, T., Bailey, M., Bernhard, M., Bursztein, E., Cochran, J., Durumeric, Z., Halderman, J.A., Invernizzi, L., Kallitsis, M., Kumar, D.: Understanding the mirai botnet. In: 26th USENIX Security Symposium, USENIX Security, vol. 17, pp. 1093–1110 (2017)
8. Ling, Z., Liu, K., Xu, Y., Gao, C., Jin, Y., Zou, C., Fu, X., Zhao, W.: IoT security: an end-to-end view and case study. arXiv preprint arXiv:1805.05853 (2018)
9. S. in Silicon Lab: Iot security vulnerability database, August 2017. http://www.hardwaresecurity.org/iot/database
10. Obermaier, J., Hutle, M.: Analyzing the security and privacy of cloud-based video surveillance systems. In: Proceedings of the 2nd ACM International Workshop on IoT Privacy, Trust, and Security, pp. 22–28. ACM (2016)
11. Arias, O., Wurm, J., Hoang, K., Jin, Y.: Privacy and security in internet of things and wearable devices. IEEE Trans. Multi-Scale Comput. Syst. **1**(2), 99–109 (2015)

12. Durumeric, Z., Li, F., Kasten, J., Amann, J., Beekman, J., Payer, M., Weaver, N., Adrian, D., Paxson, V., Bailey, M., Halderman, J.A.: The matter of heartbleed. In: Proceedings of the 2014 Conference on Internet Measurement Conference, pp. 475–488. ACM (2014)
13. Aviram, N., Schinzel, S., Somorovsky, J., Heninger, N., Dankel, M., Steube, J., Valenta, L., Adrian, D., Halderman, J.A., Dukhovni, V., Käsper, E.: DROWN: breaking TLS using SSLv2. In: 25th USENIX Security Symposium (USENIX Security 2016), pp. 689–706 (2016)
14. Wang, J.A., Guo, M.: Vulnerability categorization using Bayesian networks. In: Proceedings of the Sixth Annual Workshop on Cyber Security and Information Intelligence Research, pp. 1–4 (2010)
15. Na, S., Kim, T., Kim, H.: A study on the classification of common vulnerabilities and exposures using Naïve Bayes. In: Proceedings of International Conference on Broadband and Wireless Computing and Applications, pp. 657–662. Springer, Cham (2016)
16. Neuhaus, S., Zimmermann, T.: Security trend analysis with CVE topic models. In: 2010 IEEE 21st International Symposium on Software Reliability Engineering, pp. 111–120. IEEE (2010)
17. Huang, G., Li, Y., Wang, Q., Ren, J., Cheng, Y., Zhao, X.: Automatic classification method for software vulnerability based on deep neural network. IEEE Access **7**, 28291–28298 (2019)
18. MITRE: CVE Common Vulnerabilities and Exposures database (2020). https://cve.mitre.org/. Accessed 02 Jan 2020
19. NIST: Security Content Automation Protocol v 1.3 (2020). https://csrc.nist.gov/projects/security-content-automation-protocol/. Accessed 02 Jan 2020
20. NIST: Official Common Platform Enumeration (CPE) Dictionary (2020). https://csrc.nist.gov/projects/security-content-automation-protocol/. Accessed 02 Jan 2020
21. Vapnik, V.: Statistical Learning Theory. Wiley, New York (1998)
22. Liu, Z., Lv, X., Liu, K., Shi, S.: Study on SVM compared with the other text classification methods. In: Second International Workshop on Education Technology and Computer Science, vol. 1, pp. 219–222. IEEE (2010)
23. NLTK: Natural Language Toolkit. https://www.nltk.org/. Accessed 02 Jan 2020
24. Scikit-learn: Machine learning in Python. https://scikit-learn.org/stable/. Accessed 02 Jan 2020
25. Syed, Z., Padia, A., Finin, T., Mathews, L., Joshi, A.: UCO: a unified cybersecurity ontology. In: Workshops at the Thirtieth AAAI Conference on Artificial Intelligence (2016)

Reliability and Availability Analysis of Critical Infrastructure Composed of Dependent Systems

Agnieszka Blokus(✉) and Przemysław Dziula

Gdynia Maritime University, 81-87 Morska Street, 81-225 Gdynia, Poland
{a.blokus,p.dziula}@wn.umg.edu.pl

Abstract. This paper presents reliability and availability analysis of critical infrastructure consisting of eleven interdependent systems. With use of multi-state approach to analysis, we assumed that the deterioration of one of the systems affects the reliability of other systems and the entire infrastructure. Under this assumption, the critical infrastructure reliability function and basic reliability characteristics were determined. Additionally, the infrastructure availability function was determined, assuming that infrastructure renewal is carried out when its reliability falls below a certain level. Furthermore, we conducted the reliability and availability analysis of critical infrastructure, taking into account additional load on individual infrastructure systems at certain time points. We assumed that due to external factors, the deterioration of reliability and availability of particular systems is caused by crisis situation related to additional load on the system. The summary contains conclusions coming out of the analysis and comparison for various additional load levels.

Keywords: Critical infrastructure · Dependent systems · Interdependencies · Reliability · Infrastructure renewal · Availability · Crisis Management

1 Introduction

Intensive advancement of technologies, that can be observed last years, causes increase of dependences of modern societies, on functioning of the key systems, determined as critical infrastructure (CI). Thus, the most developed countries present activities on appropriate protection of systems identified as belonging to the critical infrastructure [1]. A reference can be made to the European Council Directive 2008/114/EC on the identification and designation of European critical infrastructures and the assessment of the need to improve their protection [2], or US National Infrastructure Protection Plan NIPP 2013 [3]. In addition, even more intensive advancement of new technologies, results in increase of dependencies among particular systems constituting critical infrastructure.

The dependencies can be of different character [4], and appear on various levels. Their nature can be either holistic, arising within the entire system, or more local [5], on particular sectors or systems elements level [6, 7]. For the purpose of critical infrastructure reliability, availability or resilience analysis, they are analysed as system of

W. Zamojski et al. (Eds.): DepCoS-RELCOMEX 2020, AISC 1173, pp. 94–104, 2020.
https://doi.org/10.1007/978-3-030-48256-5_10

systems [7, 8]. Modelling of interdependencies among systems and components forming critical infrastructure, is a key issue in their management. Understanding and identification of the systems relations is essential condition to obtain appropriate functioning and management of infrastructures [9, 10].

Following systems, included in critical infrastructure, have been distinguished for article purposes, basing on Polish Parliament Act on Crisis Management [11]: energy, fuel and energy resources supply; communication; IT networks; financial; food supply; water supply; health protection; transportation; rescue; ensuring the continuity of public administration activities; production, storing and use of chemical and radioactive substances, including pipelines for dangerous substances.

The aim of the article is to present, for exemplary critical infrastructure, that taking into account dependencies among systems constituting the critical infrastructure, and their specific interactions, can be of significant importance for the reliability and availability of the whole infrastructure, and thus for the safety of functioning of the critical infrastructure. Moreover, analysing reliability and availability of the infrastructure and its systems, we very often face additional loads on these systems, resulting in deterioration of their reliability. It can cause appearance of the crisis situation within the systems, and consequently disturbances of proper functioning of whole infrastructure. In the article, by simplified relating to the situation of additional load in the system, we assume deterioration of the reliability and availability of the system, caused by any external factors. The influence of additional deterioration of the conditions of one or more systems, forming the infrastructure, is intensified by interdependencies among the systems, that can cause so called "domino-effect", and result in significant reduction of safety level of functioning of the entire infrastructure.

2 Reliability of Infrastructure Composed of Dependent Systems

We assume that for proper functioning of the entire infrastructure, all included systems are to be in working state. Therefore, analysis of reliability and availability of critical infrastructure, is based on the assumption, that systems S_i, $i = 1, 2, \ldots, 11$, constituting the infrastructure, mentioned in Sect. 1, are connected in series.

Critical infrastructure and its systems are analysed as multistate systems. Following four reliability states are distinguished:

- state 3 of entire reliability – meaning system/infrastructure is fully functional, and all of its subsystems are working without any disruptions,
- state 2 of partial reliability – expressing situation when some disturbances in functioning of the system/infrastructure appear, but it is functioning at appropriate level,
- state 1 of limited reliability – occurring, when the disruptions of the system/infrastructure functioning, cause its exploitation parameters fall below allowed limits,
- state 0 of complete unreliability – indicates system/infrastructure failure, that stops its operation.

We assume that reliability functions of systems S_i, $i = 1, 2,\ldots, 11$, are exponential

$$R_i(t,\cdot) = [1, R_i(t,1), R_i(t,2), R_i(t,3)], t \geq 0, i = 1,2,\ldots,11, \tag{1}$$

and their coordinates $R_i(t,u)$, $u = 1,2,3$, are defined as the probability of system staying in subset $\{u, u + 1,\ldots,3\}$ of reliability states at the moment t, under the assumption that it was at full reliability state (state 3) at the moment $t = 0$, and are given by

$$R_i(t,u) = \exp[-\lambda_i(u)t], u = 1,2,3, \; i = 1,2,\ldots,11, \tag{2}$$

where $\lambda_i(u)$, $u = 1,2,3$, denote the intensities of departure of systems S_i, $i = 1, 2,\ldots, 11$, from the subset $\{u, u + 1,\ldots,3\}$.

Systems S_i, $i = 1,2,\ldots,11$, forming the critical infrastructure are interdependent. Failures and perturbations in one of the systems, affect functioning of other ones. Consequently, failures affect the reliability of entire infrastructure. Hence, we carry out reliability analysis of the infrastructure as a system of series and dependent systems. We assume that the relationships between systems can be unidirectional or bidirectional. Deterioration of the reliability state of one of systems may cause changes in the functioning of other systems, and result in deterioration of their reliability characteristics.

Assuming the local load sharing dependency model for a series system, presented in [12, 13], the magnitude of dependencies between systems is reflected by the influence coefficients $q(\upsilon, S_j, S_i)$, $i,j = 1,2,\ldots,11$, $i \neq j$. Impact between systems does not have to be symmetrical. Therefore, we assume that $q(\upsilon, S_j, S_i)$ determine the effect of changes in reliability state $\{u, u + 1,\ldots,3\}$, $u = 1,2,3$, in S_j, $j = 1,2,\ldots,11$, on lifetimes in the subset $\{\upsilon, \upsilon + 1,\ldots,3\}$, $\upsilon = 1,2$, of S_i, $i = 1,2,\ldots,11$, $i \neq j$. Assuming that systems S_i, $i = 1, 2,\ldots, 11$, have exponential reliability functions, defined by formulas (1)–(2), assuming as described above, the infrastructure reliability function is defined as follows [12, 13]:

$$\boldsymbol{R}_{dep}(t,\cdot) = [1, \boldsymbol{R}_{dep}(t,1), \boldsymbol{R}_{dep}(t,2), \boldsymbol{R}_{dep}(t,3)], t \geq 0, \tag{3}$$

where

$$\boldsymbol{R}_{dep}(t,1) = \exp[-\sum_{i=1}^{11} \lambda_i(2)t] + \sum_{j=1}^{11} \frac{\lambda_j(2)-\lambda_j(1)}{\sum_{i=1}^{11}\lambda_i(2)-\sum_{i=1}^{11}\lambda_i(1)}$$
$$\cdot[\exp[-\sum_{i=1}^{11}\frac{\lambda_i(1)}{1-q(1,S_j,S_i)}t] - \exp[-(\sum_{i=1}^{11}\lambda_i(2) - \sum_{i=1}^{11}\lambda_i(1) + \sum_{i=1}^{11}\frac{\lambda_i(1)}{1-q(1,S_j,S_i)})t]], t \geq 0, \tag{4}$$

$$\boldsymbol{R}_{dep}(t,2) = \exp[-\sum_{i=1}^{11} \lambda_i(3)t] + \sum_{j=1}^{11} \frac{\lambda_j(3)-\lambda_j(2)}{\sum_{i=1}^{11}\lambda_i(3)-\sum_{i=1}^{11}\lambda_i(2)}$$
$$\cdot[\exp[-\sum_{i=1}^{11}\frac{\lambda_i(2)}{1-q(2,S_j,S_i)}t] - \exp[-(\sum_{i=1}^{11}\lambda_i(3) - \sum_{i=1}^{11}\lambda_i(2) + \sum_{i=1}^{11}\frac{\lambda_i(2)}{1-q(2,S_j,S_i)})t]], t \geq 0, \tag{5}$$

$$\boldsymbol{R}_{dep}(t,3) = \exp[-\sum_{i=1}^{11} \lambda_i(3)t], t \geq 0. \tag{6}$$

To illustrate the results of reliability and availability analysis of critical infrastructure, sample expert data for reliability parameters and impact coefficients between systems were adopted. Basic reliability and availability characteristics [14, 15] of critical infrastructure with dependent systems are further determined for the adopted input data.

The mean lifetimes and standard deviations of infrastructure in state subsets {1,2,3}, {2,3}, {3}, counted in years using formulas given in [16–18] and (4)–(6), are:

$$\mu_{dep}(1) = 0.408, \ \mu_{dep}(2) = 0.184, \ \mu_{dep}(3) = 0.133, \tag{7}$$

$$\sigma_{dep}(1) = 0.364, \ \sigma_{dep}(2) = 0.163, \ \sigma_{dep}(3) = 0.133. \tag{8}$$

3 Availability of Renewable Infrastructure

We assume that the critical infrastructure is renewed if probability of its stay in subset of states {2,3} falls below 60%. This probability determines a value of the coordinate $\boldsymbol{R}(t,2)$ of reliability function. Consequently, the coordinate of infrastructure availability function $\boldsymbol{AF}(t,2)$ is defined as the probability of stay of renewable infrastructure in subset of states {2,3}. That coordinate of availability function, assuming regular renewals of infrastructure after exceeding the 60% threshold, is determined as follows:

$$\boldsymbol{AF}_{dep}(t,2) = \begin{cases} \boldsymbol{R}_{dep}(t,2) \text{ if } \boldsymbol{R}_{dep}(t,2) \geq 0.6, t \geq 0 (\text{if } t \leq \tau_{0.6}(2)), \\ \boldsymbol{R}_{dep}(t - \omega \cdot \tau_{0.6}(2), 2) \text{ if } \omega \cdot \tau_{0.6}(2) < t \leq (\omega + 1) \cdot \tau_{0.6}(2), \omega = 1, \ldots, N, \end{cases} \tag{9}$$

where N is the number of infrastructure renewals and $\tau_{0.6}(2)$ is the moment of the first renewal after exceeding the 60% threshold, and $\boldsymbol{R}_{dep}(t,2)$ is given by (5).

Figure 1 shows the coordinate of reliability function in case the systems forming critical infrastructure are dependent $\boldsymbol{R}_{dep}(t,2)$, and to compare - if the systems are functioning independently $\boldsymbol{R}_{indep}(t,2)$. Furthermore, Fig. 1 presents the availability function coordinate of critical infrastructure with dependent systems, that is renewed after exceeding the 60% threshold $\boldsymbol{AF}_{dep}(t,2)$.

3.1 Additional Load on Systems Forming CI

Furthermore, we assume that at certain time points, due to external factors, the reliability of particular systems may deteriorate. The deterioration of reliability, resulting in shortening the system's lifetime in a subset of reliability states, can be caused by various external factors. For simplicity, we will refer to this situation as an additional load on a system afterwards. Obviously, an additional load on one of the systems

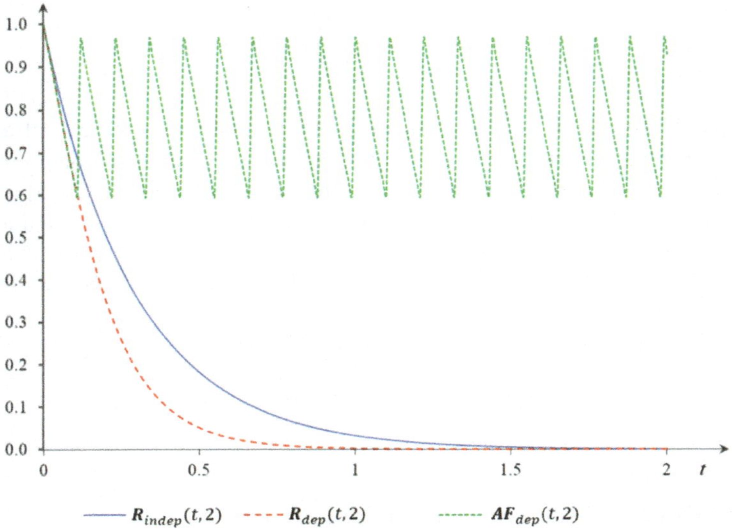

Fig. 1. The coordinates of reliability function of CI with independent systems $\boldsymbol{R}_{indep}(t,2)$, with dependent systems $\boldsymbol{R}_{dep}(t,2)$, and of availability function of CI with dependent systems $\boldsymbol{AF}_{dep}(t,2)$.

forming critical infrastructure, affects the functioning and reliability of the entire infrastructure.

Conducting reliability analysis in case an infrastructure is not renewed, we assume that from the moment an additional load occurs in one of infrastructure systems, the intensity of its departure from a subset of reliability states increases. The coordinate of infrastructure reliability function, impacted by an additional load is then determined as follows:

$$\boldsymbol{R}_{dep}^{S_i\_L}(t, T_L, 2) = \begin{cases} \boldsymbol{R}_{dep}(t,2) \text{ if } 0 \leq t \leq T_L, \\ \boldsymbol{R}_{dep}^{S_i\_L}(t,2) \text{ if } t > T_L, \end{cases} \quad (10)$$

where $\boldsymbol{R}_{dep}^{S_i\_L}(t,2)$ is the reliability function coordinate of infrastructure composed of dependent systems, with increased output intensity from the subset of states {2,3} of the system S_i, i = 1, 2,…, 11, while T_L is the moment when additional load appears in this system. $\boldsymbol{R}_{dep}^{S_i\_L}(t,2)$ is determined similarly to the coordinate $\boldsymbol{R}_{dep}(t,2)$ given by (5), where the intensity $\lambda_i(2)$ of the stay of i-th system S_i, i = 1, 2,…, 11, in the subset of states {2,3}, is replaced by a new value of intensity $\lambda_i^L(2)$, increased due to the additional load L.

Similarly, we determine the availability function of renewable infrastructure, assuming that at the time of additional load in one of systems forming infrastructure, the intensity of its departure from a subset of reliability states increases. However, after infrastructure renewal, its reliability parameters, i.e. the intensity of its systems' departure from subsets of reliability states are the same as at the beginning. We assume that at the initial moment of observation $t = 0$, the infrastructure is in reliability state 3. The infrastructure is renewed after the probability of infrastructure staying in subset of reliability states {2,3}, assuming that systems are dependent but without additional load, falls down below the 60% threshold. For the analyzed critical infrastructure, this means that infrastructure renewals take place every 0.11 years, as shown in Fig. 1.

Under these assumptions, the coordinate of infrastructure availability function having one moment $T_L^{(1)}$ of additional load is determined as follows:

$$\boldsymbol{AF}_{dep}^{S_i\_L}(t, T_L^{(1)}, 2) = \begin{cases} \boldsymbol{R}_{dep}(t,2) \text{ if } 0 \leq t \leq \min\{T_L^{(1)}; \tau_{0.6}(2)\}, \\ \boldsymbol{R}_{dep}^{S_i\_L}(t,2) \text{ if } T_L^{(1)} < t \leq \tau_{0.6}(2), \\ \boldsymbol{R}_{dep}(t - \omega \cdot \tau_{0.6}(2), 2) \text{ if } \omega \cdot \tau_{0.6}(2) < t \leq (\omega + 1) \cdot \tau_{0.6}(2) \\ \text{and } t \leq T_L^{(1)} \text{ or } t > T_L^{(1)} \text{ and } T_L^{(1)} \leq \omega \cdot \tau_{0.6}(2),\ \omega = 1, \ldots, N, \\ \boldsymbol{R}_{dep}^{S_i\_L}(t - \omega \cdot \tau_{0.6}(2), 2) \text{ if } \omega \cdot \tau_{0.6}(2) < t \leq (\omega + 1) \cdot \tau_{0.6}(2) \\ \text{and } t > T_L^{(1)} \text{ and } T_L^{(1)} > \omega \cdot \tau_{0.6}(2),\ \omega = 1, \ldots, N, \end{cases} \tag{11}$$

where N is the number of infrastructure renewals and $\tau_{0.6}(2)$ is the moment of the first renewal of infrastructure, after probability of its staying in subset of states {2,3} falls below 60%.

If an infrastructure, during observation time have M of crisis moments $T_L^{(k)}$, $k = 1, \ldots, M$, in which the load on one of the infrastructure systems increased, the availability function coordinate of renewed infrastructure is determined by:

$$\boldsymbol{AF}_{dep}^{S_i\_L}(t, 2) = \begin{cases} \boldsymbol{AF}_{dep}^{S_i\_L}(t, T_L^{(1)}, 2) \text{ for } t \geq 0 \text{ if } M = 1, \\ \boldsymbol{AF}_{dep}^{S_i\_L}(t, T_L^{(1)}, 2) \text{ if } 0 \leq t \leq (\omega + 1) \cdot \tau_{0.6}(2) \text{ and } T_L^{(1)} > \omega \cdot \tau_{0.6}(2), \\ \text{for } \omega = 0, 1, \ldots, N, M > 1, \\ \boldsymbol{AF}_{dep}^{S_i\_L}(t, T_L^{(k+1)}, 2) \text{ if } \omega \cdot \tau_{0.6}(2) < t \leq (\omega + 1) \cdot \tau_{0.6}(2) \\ \text{and } T_L^{(k)} \leq \omega \cdot \tau_{0.6}(2) < T_L^{(k+1)} \text{ for } \omega = 0, 1, \ldots, N, k = 1, \ldots, M-2, \\ \boldsymbol{AF}_{dep}^{S_i\_L}(t, T_L^{(k+1)}, 2) \text{ if } t > \omega \cdot \tau_{0.6}(2) \text{ and } T_L^{(k)} \leq \omega \cdot \tau_{0.6}(2), \\ \text{for } \omega = 1, \ldots, N, k = M-1, M > 1, \end{cases} \tag{12}$$

where $\boldsymbol{AF}_{dep}^{S_i\_L}(t, T_L^{(1)}, 2)$ is given by (11) and

$$\boldsymbol{AF}_{dep}^{S_i\_L}(t, T_L^{(k)}, 2) = \begin{cases} \boldsymbol{R}_{dep}(t - \omega \cdot \tau_{0.6}(2), 2) \text{ if } \omega \cdot \tau_{0.6}(2) < t \leq (\omega + 1) \cdot \tau_{0.6}(2) \\ \text{and } t \leq T_L^{(k)} \text{ or } t > T_L^{(k)} \text{ and } T_L^{(k)} \leq \omega \cdot \tau_{0.6}(2),\ \omega = 1, \ldots, N, \\ \boldsymbol{R}_{dep}^{S_i\_L}(t - \omega \cdot \tau_{0.6}(2), 2) \text{ if } \omega \cdot \tau_{0.6}(2) < t \leq (\omega + 1) \cdot \tau_{0.6}(2) \\ \text{and } t > T_L^{(k)} \text{ and } T_L^{(k)} > \omega \cdot \tau_{0.6}(2),\ \omega = 1, \ldots, N, \end{cases} \tag{13}$$

for $k = 2, \ldots, M$.

3.2 Exemplary Results and Discussion

To illustrate a crisis situation of additional load on one of the systems, we assume that additional load occurs at the time of 0.25 years in case the infrastructure is not renewed. In case of the availability analysis of renewed infrastructure, additional load occurs at three ($M = 3$) moments of 0.25 year, 0.5 year and after 1 year of observation. The reliability and availability analysis of infrastructure is conducted, assuming that additional load on a system, increases the intensity of its departure from the subset of states {2,3} twice, five times and ten times. Figure 2a presents the results of such an analysis in case of an additional load on the system S_1, while Fig. 2b in case of an additional load on the system S_8. Figures 2a and 2b show both the coordinate of infrastructure availability function, taking into account infrastructure renewals given by (12), and to compare the coordinate of reliability function if the infrastructure is not renewed, given by (10).

Figures 2a and 2b present the graphs of CI reliability function coordinate (10), in case of additional load on the system S_1 and on the system S_8, at critical point T_L = 0.25 year. Additionally, the graphs of CI availability function coordinate (12), in case of additional load on the system S_1 and on the system S_8, are shown in Figs. 2a and 2b. In Figs. 2a and 2b, load increasing the intensity twice is marked in short form by $L2$, load increasing the intensity five times by $L5$, and load increasing the intensity ten times by $L10$. Table 1 contains the values of reliability function coordinate at $t = 0.3$, and the values of availability function coordinate at two exemplary moments, after 0.3 and 0.55 year.

Table 2 contains the values of mean lifetimes in reliability state subset {2,3} of CI with dependent systems in case of additional load on system S_1 and system S_8 compared to mean lifetime of critical infrastructure without additional load.

The obtained results show that disruptions and deteriorations of reliability state of one of systems, can significantly affect the reliability and availability of the entire infrastructure. The impact is of special importance in case the systems constituting the critical infrastructure are interdependent. For example, at the moment $t = 0.3$ years, the infrastructure availability function value can be 9%, 10%, or even 20% worse in case of additional load on one of systems at the moment $T_L = 0.25$ years, related to the availability function value of the system without additional load, decreasing its reliability. If we assume that additional load, influencing on intensity of system departure from set of reliability states, appears at moment $T_L = 0.5$ years, and compare to the availability function values at the moment $t = 0.55$, the difference can even rise to 25% in case of additional load on system S_1, and to 28% in case of additional load on system S_8. Obviously, the value of infrastructure availability function would have decreased more, if the additional load had appeared in a few systems at similar time frames. Similar differences can be observed for infrastructure's mean lifetimes in reliability state subset {2,3}, presented in Table 2. All values specified in Table 2 are related to the reliability of the infrastructure, taking into account dependencies among its systems. If we compare these values to the mean lifetime in subset {2,3} for infrastructure without dependencies among systems, which is 0.294 years, the difference is very significant, even above 40%.

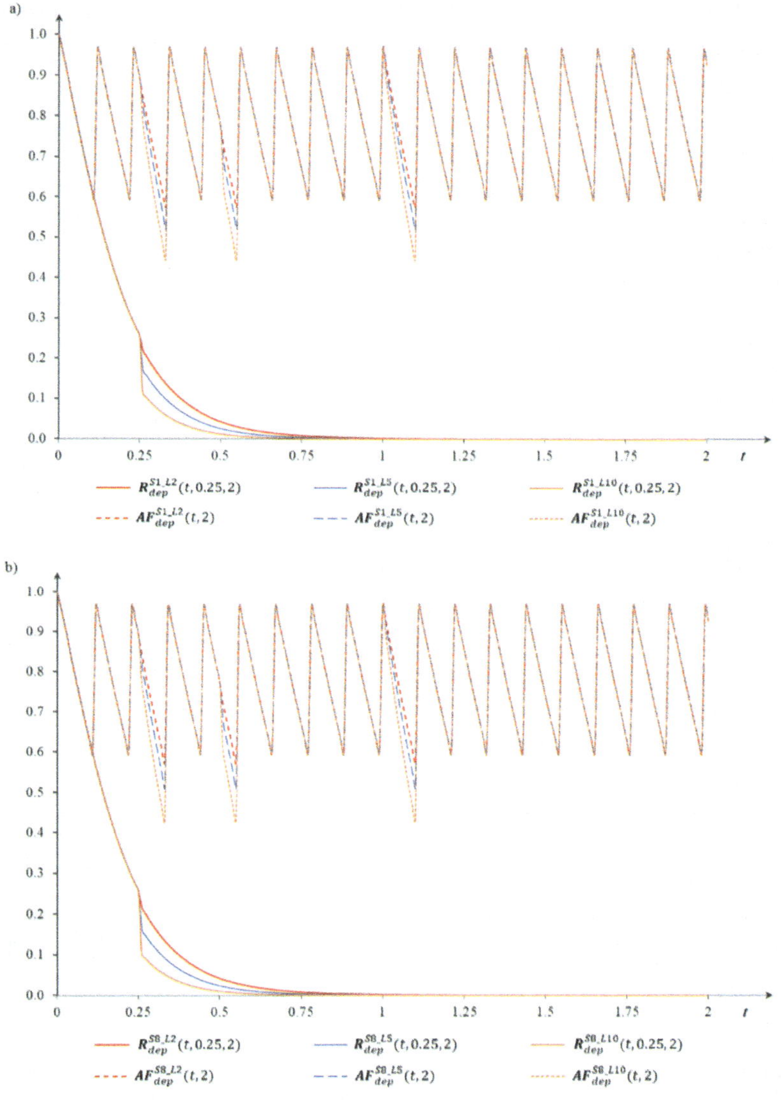

Fig. 2. The coordinates of reliability function and availability function of CI with dependent systems: a) in case of additional load on system S_1, b) in case of additional load on system S_8.

4 Summary

The article introduces analysis of reliability of multistate infrastructure consisting of eleven dependent systems. Then its availability analysis was conducted, basing on the assumption the infrastructure is renewed when probability of its stay in subset of states {2,3} falls below 60%. Moreover, both reliability and availability analyses, take into

Table 1. The values of reliability and availability function of CI with dependent systems, in case of additional load on system S_8, and without additional load, at exemplary time points.

Additional load	$\boldsymbol{R}_{dep}^{S8\_L}(0.3, 0.25, 2)$	$\boldsymbol{AF}_{dep}^{S8\_L}(0.3, 2)$	$\boldsymbol{AF}_{dep}^{S8\_L}(0.55, 2)$
No load	0.1865	0.7012	0.5967
Load increasing the intensity twice	0.1652	0.6832	0.5743
Load increasing the intensity three times	0.1467	0.6659	0.5530
Load increasing the intensity four times	0.1306	0.6492	0.5327
Load increasing the intensity five times	0.1167	0.6330	0.5134
Load increasing the intensity ten times	0.0687	0.5597	0.4294

Table 2. The mean lifetimes in reliability state subset {2,3} of CI with dependent systems, in case of additional load on system S_1 and system S_8, and without additional load [in years].

Additional load	$\mu_{dep}^{S1\_L}(2)$	$\mu_{dep}^{S8\_L}(2)$
No load	0.184	0.184
Load increasing the intensity twice	0.179	0.179
Load increasing the intensity three times	0.175	0.174
Load increasing the intensity four times	0.171	0.170
Load increasing the intensity five times	0.168	0.167
Load increasing the intensity ten times	0.159	0.157

account moments of additional deterioration of reliability parameters of systems constituting the infrastructure, caused by external factors, named "moments of additional loads" for simplicity. Exemplary results, introduced in previous chapter, show significant importance of taking into account dependencies and interactions among infrastructure systems, and additional external factors, for reliability and availability analyses of infrastructure. For the purpose of ensuring of continuity of proper and reliable functioning of infrastructure, and planning of conducting corrective maintenance and renew of the system, it seems to be of key importance to take into account dependencies among infrastructure systems, and their influence on reliability and availability of the entire infrastructure.

The impact of crisis situation in one of systems, on functioning and reliability of the entire critical infrastructure, depends also on the moment of appearance of additional load, related to the moment of infrastructure renewal. Thus, the analysis of reliability and availability, taking into account both interdependencies among systems, and other external factors influencing negatively on systems reliability, is of the key importance for critical infrastructure safety management. The base and necessary effort, leading to ensure proper safety of critical infrastructures, according to Prochazkova [10], is analysis of safety of interdependent infrastructures and their systems in normal, abnormal and critical conditions, and determination of the critical conditions.

Acknowledgements. The paper presents the results developed in the scope of the research project "Safety of critical infrastructure transport networks", WN/2020/PZ/, granted by GMU in 2020.

References

1. Dziula, P.: Selected aspects of acts of law concerning critical infrastructure protection within the Baltic Sea area. Sci. J. Marit. Univ. Szczecin **44**(116), 173–181 (2015)
2. European Union: European Council. Council Directive 2008/114/EC of 8 December 2008 on the identification and designation of European critical infrastructures and the assessment of the need to improve their protection. Brussels (2008)
3. Department of Homeland Security: NIPP 2013: Partnering for Critical Infrastructure Security and Resilience. http://www.dhs.gov. Accessed 06 Jan 2020
4. Saidi, S., Kattan, L., Jayasinghe, P., Hettiaratchi, P., Taron, J.: Integrated infrastructure systems—A review. Sustain. Cities Soc. **36**, 1–11 (2018)
5. Holden, R., Val, D.V., Burkhard, R., Nodwell, S.: A network flow model for interdependent infrastructures at the local scale. Saf. Sci. **53**(3), 51–60 (2013)
6. Rinaldi, S., Peerenboom, J., Kelly, T.: Identifying, understanding and analyzing critical infrastructure interdependencies. IEEE Control Syst. Mag. **21**(6), 11–25 (2001)
7. Rehak, D., Markuci, J., Hromada, M., Barcova, K.: Quantitative evaluation of the synergistic effects of failures in a critical infrastructure system. Int. J. Crit. Infrastruct. Prot. **14**, 3–17 (2016)
8. Eusgeld, I., Nan, C., Dietz, S.: System-of systems approach for interdependent critical infrastructures. Reliab. Eng. Syst. Saf. **96**, 679–686 (2011)
9. Blokus-Roszkowska, A., Dziula, P.: An approach to identification of critical infrastructure systems. In: Simos, T., Tsitouras, Ch. (eds.) ICNAAM 2015, AIP Conference Proceedings vol. 1738(1), pp. 440005-1–440005-4. AIP Publishing, Rhodes (2016). https://doi.org/10.1063/1.4952223
10. Prochazkova, D.: Critical infrastructure safety management. Trans. Transp. Sci. **3**(4), 157–168 (2010)
11. Polish Parliament. Act of 26 April 2007 on Crisis Management. Warsaw (2007)
12. Blokus, A., Kołowrocki, K.: Reliability and maintenance strategy for systems with aging-dependent components. Qual. Reliab. Engng. Int. **35**(8), 2709–2731 (2019). https://doi.org/10.1002/qre.2552
13. Blokus, A., Dziula, P.: Safety analysis of interdependent critical infrastructure networks. TransNav, Int. J. Mar. Navig. Saf. Sea Transp. **13**(4), 781–787 (2019). https://doi.org/10.12716/1001.13.04.10
14. Blokus-Roszkowska, A., Kołowrocki, K.: Reliability analysis of ship-rope transporter with dependent components. In: Nowakowski et al. (eds.) Safety and Reliability: Methodology and Applications – Proceedings of the European Safety and Reliability Conference ESREL 2014, pp. 255–263. Taylor & Francis Group, London (2015)
15. Blokus-Roszkowska, A.: Availability analysis of transport navigation system under imperfect repair. In: Zamojski, W., Mazurkiewicz, J., Sugier, J., Walkowiak, T., Kacprzyk,

J. (eds.) Contemporary Complex Systems and Their Dependability. Advances in Intelligent Systems and Computing, vol. 761, pp. 35–45. Springer, Cham (2019)
16. Kołowrocki, K.: Reliability of Large and Complex Systems, 2nd edn. Elsevier, London (2014)
17. Kołowrocki, K., Soszyńska-Budny, J.: Reliability and Safety of Complex Technical Systems and Processes: Modeling – Identification – Prediction – Optimization, 1st edn. Springer-Verlag, London, Great Britain (2011)
18. Blokus, A., Dziula, P.: Reliability analysis of different configurations of master and back-up systems used in maritime navigation. J. Mar. Sci. Eng. **8**, 34 (2020). https://doi.org/10.3390/jmse8010034

Influence of Component Dependency on System Reliability

Agnieszka Blokus(✉) and Krzysztof Kołowrocki

Department of Mathematics, Gdynia Maritime University, 81-87 Morska St., 81-225 Gdynia, Poland
{a.blokus, k.kolowrocki}@wn.umg.edu.pl

Abstract. This paper presents an analysis of the influence of dependencies among system components on system reliability. The reliability function of aging multistate system with dependent components is determined, in case its components have piecewise Weibull functions. The results obtained are applied for reliability analysis of exemplary four-wheel system. A comparative analysis of this system reliability was carried out for various values of influence coefficients, depending on the distance among components and their reliability states. These coefficients express the influence strength of reliability deterioration of certain components on the other components reliability. The results show that the system lifetime, and more specifically its lifetimes in reliability state subsets, can differ depending on the strength of this influence among components.

Keywords: Multistate aging system · Dependent components · Influence coefficients · Reliability characteristics · Sensitivity analysis

1 Introduction

The reliability analysis of technical systems and estimation of their real lifetime can be important for the safe and reliable functioning of systems. Additionally, by using research on component and system degradation for system maintenance planning can optimize operation costs [1–5]. The authors emphasize that analysis of reliability and system degradation processes can be helpful in creating maintenance strategy. Thus, estimating the system lifetime and predicting wear and fatigue of its components is often crucial, especially in the case of their aging dependence [6–9] considered and developed in the paper.

2 Reliability of System with Dependent Components

We analyze the reliability of a system with series reliability structure, consisting of n components E_i, $i = 1, \ldots, n$. Using a multistate approach to reliability analysis [5, 10, 11], we assume that a system and its components have the states $0, 1, \ldots, z$ ($z \geq 1$), which are ordered from the state of full reliability and operational efficiency i.e. state z,

W. Zamojski et al. (Eds.): DepCoS-RELCOMEX 2020, AISC 1173, pp. 105–114, 2020.
https://doi.org/10.1007/978-3-030-48256-5_11

to the state of complete failure 0. We assume that at initial moment $t = 0$ the system and its components are in the state z.

We assume that components E_i, $i = 1, \ldots, n$, have piecewise Weibull reliability functions:

$$R_i(t,\cdot) = [R_i(t,1),\ldots,R_i(t,z)], t \geq 0, i = 1,2,\ldots,n, \tag{1}$$

with the coordinates

$$R_i(t,u) = \exp[-\lambda_i(u)t^{\beta}], \; t \geq 0, \; u = 1,2,\ldots,z, \; \lambda_i(u) \geq 0, \beta > 0, \tag{2}$$

that define the probability of component E_i, $i = 1,\ldots, n$, staying in reliability state subset $\{u, u + 1, \ldots, z\}$, $u = 1, 2, \ldots, z$, at the moment t. The $\lambda_i(u)$, $i = 1, \ldots, n$, in formula (2), are component intensities of departure from the reliability state subsets $\{u, u + 1, \ldots, z\}$, $u = 1, 2, \ldots, z$.

In that case, the lifetime mean value in the reliability state subset is given by:

$$E[T_i(u)] = (\lambda_i(u))^{\frac{-1}{\beta}} \int_0^{\infty} t^{\frac{1}{\beta}} e^{-t} dt = (\lambda_i(u))^{\frac{-1}{\beta}} \Gamma(1 + \frac{1}{\beta}), \; i = 1,2,\ldots,n, u = 1,2,\ldots,z, \tag{3}$$

where $\Gamma(z) = \int_0^{\infty} t^{z-1} e^{-t} dt, z > 0$.

We conduct a reliability analysis of a multistate system with dependent components [6–8], assuming that the aging of certain components, related to their wear or fatigue, may affect the lifetimes and reliability of other components in the system. Consequently, the interactions among components have an influence on reliability characteristics of the entire system. By applying a multistate approach to reliability analysis [5], the dependence among components affects lifetimes in reliability state subsets of components and the system. More specifically, we assume that after departure one of the components E_j from the subset of reliability states $\{u, u + 1, \ldots, z\}$, $u = 1, 2, \ldots, z$, the system component lifetimes in the subset $\{\upsilon, \upsilon + 1, \ldots, z\}$, $\upsilon = u, u - 1, \ldots, 1$, $u = 1, 2, \ldots, z - 1$, change according to the equation [6, 9]:

$$T_{i/j}(\upsilon) = q(\upsilon, E_j, E_i) \cdot T_i(\upsilon), \; i = 1,2,\ldots,n, j = 1,2,\ldots,n, \tag{4}$$

where $T_{i/j}(\upsilon)$ is the E_i lifetime in the subset $\{\upsilon, \upsilon + 1, \ldots, z\}$ after the E_j departure, and $T_i(\upsilon)$ is the E_i lifetime in this subset before the E_j departure.

We assume a similar equation for mean values of component lifetimes in reliability state subset $\{\upsilon, \upsilon + 1, \ldots, z\}$, $\upsilon = u, u - 1, \ldots, 1$, $u = 1, 2, \ldots, z - 1$, [6, 9]:

$$E[T_{i/j}(\upsilon)] = q(\upsilon, E_j, E_i) \cdot E[T_i(\upsilon)], \; i = 1,2,\ldots,n, j = 1,2,\ldots,n, \tag{5}$$

The coefficients $q(\upsilon, E_j, E_i)$ in (4) and (5) express the strength of impact of component E_j degradation (reliability state change) on lifetimes of other components in the system.

According to the relationship between the lifetime mean value in reliability state subset and the intensity of departure from this subset we get the formula for conditional intensities $\lambda_{i/j}(\upsilon)$, $i = 1, \ldots, n$, $j = 1, \ldots, n$, of the component E_i departure from subset $\{\upsilon, \upsilon + 1, \ldots, z\}$, $\upsilon = u, u - 1, \ldots, 1$, $u = 1, 2, \ldots, z - 1$, after the departure of E_j. Namely, applying (5), in case of components with piecewise Weibull reliability functions (1)–(2), by using (3), we obtain

$$\lambda_{i/j}(\upsilon) = \frac{\lambda_i(\upsilon)}{[q(\upsilon, E_j, E_i)]^{\beta}}. \quad (6)$$

Thus, considering (2) and (4)–(6), the components E_i, $i = 1, \ldots, n$, after the departure of E_j, $j = 1, \ldots, n$, from that subset $\{u, u + 1, \ldots, z\}$, $u = 1,2, \ldots, z$, have the reliability functions with following coordinates:

$$R_{i/j}(t, \upsilon) = \exp[-\frac{\lambda_i(\upsilon)}{[q(\upsilon, E_j, E_i)]^{\beta}} t^{\beta}],\ \upsilon = u, u - 1, \ldots, 1,\ u = 1, 2, \ldots, z - 1, \quad (7)$$

$$R_{i/j}(t, \upsilon) = \exp[-\lambda_i(\upsilon)t^{\beta}],\ \upsilon = u + 1, \ldots, z,\ u = 1, 2, \ldots, z - 1. \quad (8)$$

Applying Proposition 1 from [6] and results given in [9], we determine the reliability function of aging series system with components dependent according to the formulas (4)–(5). Assuming that components have piecewise Weibull reliability functions (1)–(2), the system's reliability function is given by:

$$\boldsymbol{R}(t, \cdot) = [\boldsymbol{R}(t, 1), \ldots, \boldsymbol{R}(t, z)],\ t \geq 0, \quad (9)$$

with the coordinates

$$\begin{aligned}\boldsymbol{R}(t, u) = {} & \exp[-\sum_{i=1}^{n} \lambda_i(u + 1)t^{\beta}] + \sum_{j=1}^{n} \frac{\lambda_j(u+1) - \lambda_j(u)}{\sum_{i=1}^{n} (\lambda_i(u+1) - \lambda_i(u))} \\ & \cdot [\exp[-\sum_{i=1}^{n} \frac{\lambda_i(u)}{[q(u,E_j,E_i)]^{\beta}} t^{\beta}] - \exp[-\sum_{i=1}^{n} (\lambda_i(u + 1) - \lambda_i(u) + \frac{\lambda_i(u)}{[q(u,E_j,E_i)]^{\beta}})t^{\beta}]],\end{aligned} \quad (10)$$

for $u = 1, 2, \ldots, z - 1$, and

$$\boldsymbol{R}(t, z) = \exp[-\sum_{i=1}^{n} \lambda_i(z)t^{\beta}]. \quad (11)$$

3 Exemplary Wheel System Having Dependencies

In Sect. 3 we present the possibility of using the method of determining the system reliability function, proposed in Sect. 2, for research on reliability analysis of a multistate aging technical system consisting of dependently degrading components.

3.1 Assumptions

We consider four-wheel vehicle, with scheme presented in Fig. 1, as an example of a system with dependent components. We assume it is homogeneous multistate aging series system composed of four wheels, i.e. components E_i, i = 1, 2, 3, 4.

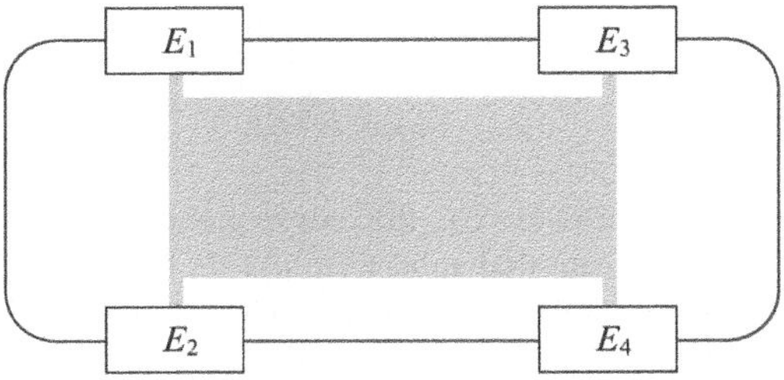

Fig. 1. The scheme of the exemplary four-wheel system structure.

We arbitrarily distinguish following reliability states of the system and its components:

- state 4 – the system/component operation is new and fully effective,
- state 3 – the system/component operation is slightly less effective because of aging, but it operates correctly,
- state 2 – the system/component operation is less effective because of aging, the system/component operates but its functioning may pose a certain danger to the environment and other objects,
- state 1 – the system/component is still operating, but its functioning may pose a serious danger to the environment and other objects; it is not recommended to operate the system in this state,
- state 0 – the system/component is destroyed.

The state 2 (r = 2) was adopted as the critical reliability state of the system and its components. Exceeding this critical state can be dangerous for the environment and other objects.

We assume that components have identical piecewise Weibull reliability functions:

$$R(t,\cdot) = [R(t,1), R(t,2), R(t,3), R(t,4)], t \geq 0, \quad (12)$$

where

$$R(t,u) = \exp[-\lambda(u)t^2], t \geq 0, u = 1,2,3,4. \quad (13)$$

Intensity of departure from the subset of reliability states $\{u, u + 1, \ldots, 4\}$, u = 1, 2, 3, 4, marked by $\lambda(u)$ in (13), is assumed to be of following values for u = 1, 2, 3, 4:

$$\lambda(1) = 0.00125, \lambda(2) = 0.00167, \lambda(3) = 0.00200, \lambda(4) = 0.00250\,year^{-1}. \quad (14)$$

The values of intensities (14) were estimated based on the mean lifetimes of components in reliability state subsets, which were expressed in years.

The influence coefficients express the impact strength of reliability deterioration of one of components on the other components reliability. More specifically, the coefficients $q(\upsilon, E_j, E_i)$ for $\upsilon = u, u-1, \ldots, 1$ and $u = 1, 2, 3$, express the impact of deterioration of component E_j, $j = 1, 2, 3, 4$, on other components E_i, $i = 1, 2, 3, 4$, $i \neq j$. This impact can cause shortening of components E_i, $i = 1, 2, 3, 4$, lifetimes in the reliability state subsets. The influence coefficients take values ranging from 0 to 1, and here are determined by the parameter α, $0 \leq \alpha \leq 0.20$.

Here, we assume that the effect of component aging on other components is stronger for the system in worse reliability states 1 and 2, and for the system in state 3 is less significant. Thus, the coefficients of component E_1 impact take following values:

$$\begin{aligned} q(3, E_1, E_2) = q(3, E_1, E_3) = 1 - 2\alpha,\ q(3, E_1, E_4) = 1 - \alpha, \\ q(\upsilon, E_1, E_2) = q(\upsilon, E_1, E_3) = 1 - 4\alpha,\ q(\upsilon, E_1, E_4) = 1 - 2\alpha,\ \upsilon = 1, 2, \end{aligned} \quad (15)$$

The coefficients of component E_2 impact take following values:

$$\begin{aligned} q(3, E_2, E_1) = q(3, E_2, E_4) = 1 - 2\alpha,\ q(3, E_2, E_3) = 1 - \alpha, \\ q(\upsilon, E_2, E_1) = q(\upsilon, E_2, E_4) = 1 - 4\alpha,\ q(\upsilon, E_2, E_3) = 1 - 2\alpha,\ \upsilon = 1, 2, \end{aligned} \quad (16)$$

The coefficients of component E_3 impact, by assumption, are:

$$\begin{aligned} q(3, E_3, E_1) = q(3, E_3, E_4) = 1 - 2\alpha,\ q(3, E_3, E_2) = 1 - \alpha, \\ q(\upsilon, E_3, E_1) = q(\upsilon, E_3, E_4) = 1 - 4\alpha,\ q(\upsilon, E_3, E_2) = 1 - 2\alpha,\ \upsilon = 1, 2, \end{aligned} \quad (17)$$

The coefficients of component E_4 impact are:

$$\begin{aligned} q(3, E_4, E_2) = q(3, E_4, E_3) = 1 - 2\alpha, q(3, E_4, E_1) = 1 - \alpha, \\ q(\upsilon, E_4, E_2) = q(\upsilon, E_4, E_3) = 1 - 4\alpha,\ q(\upsilon, E_4, E_1) = 1 - 2\alpha,\ \upsilon = 1, 2, \end{aligned} \quad (18)$$

By the assumption $q(\upsilon, E_j, E_j) = 1$, $j = 1, 2, 3, 4$, for $\upsilon = 1, 2, 3$.

3.2 Reliability of Wheel System

For the influence coefficients specified by (15)–(18), and if the components have piecewise Weibull reliability functions (12)–(13), then the reliability function of system with components dependent according to the formulas (4)–(5), using (9)–(11), is as follows:

$$\boldsymbol{R}_\alpha(t, \cdot) = [\boldsymbol{R}_\alpha(t, 1), \boldsymbol{R}_\alpha(t, 2), \boldsymbol{R}_\alpha(t, 3), \boldsymbol{R}_\alpha(t, 4)], t \geq 0, \quad (19)$$

with the coordinates

$$\boldsymbol{R}_{\alpha}(t,1) = \exp[-4\lambda(2)t^2] + \exp[-(\lambda(1) + \frac{\lambda(1)}{(1-2\alpha)^2} + \frac{2\cdot\lambda(1)}{(1-4\alpha)^2})t^2] - \exp[-(4\lambda(2) - 3\lambda(1) + \frac{\lambda(1)}{(1-2\alpha)^2} + \frac{2\cdot\lambda(1)}{(1-4\alpha)^2})t^2], t \geq 0, \tag{20}$$

$$\boldsymbol{R}_{\alpha}(t,2) = \exp[-4\lambda(3)t^2] + \exp[-(\lambda(2) + \frac{\lambda(2)}{(1-2\alpha)^2} + \frac{2\cdot\lambda(2)}{(1-4\alpha)^2})t^2] - \exp[-(4\lambda(3) - 3\lambda(2) + \frac{\lambda(2)}{(1-2\alpha)^2} + \frac{2\cdot\lambda(2)}{(1-4\alpha)^2})t^2], t \geq 0, \tag{21}$$

$$\boldsymbol{R}_{\alpha}(t,3) = \exp[-4\lambda(4)t^2] + \exp[-(\lambda(3) + \frac{\lambda(3)}{(1-\alpha)^2} + \frac{2\cdot\lambda(3)}{(1-2\alpha)^2})t^2] - \exp[-(4\lambda(4) - 3\lambda(3) + \frac{\lambda(3)}{(1-\alpha)^2} + \frac{2\cdot\lambda(3)}{(1-2\alpha)^2})t^2], t \geq 0, \tag{22}$$

$$\boldsymbol{R}_{\alpha}(t,4) = \exp[-4\lambda(4)t^2], t \geq 0. \tag{23}$$

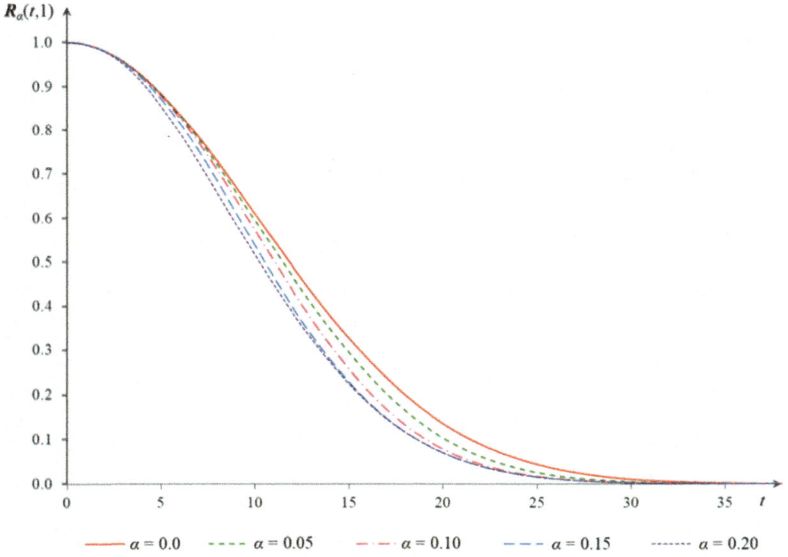

Fig. 2. The graphs of system reliability function coordinate $\boldsymbol{R}_{\alpha}(t,1)$ for different values of parameter α, $0 \leq \alpha \leq 0.20$.

The graphs of system reliability function coordinate $\boldsymbol{R}_{\alpha}(t,1)$ for intensity values of departure from reliability state subsets given by (14), and for different values of parameter α, related to influence coefficients expressing the strength of dependencies among components, are illustrated in Fig. 2. Graphs of this coordinate, determining the probability of system staying in reliability states 1, 2, 3 or 4, are presented for $\alpha = 0$ (meaning independence among components) and for $\alpha = 0.10$, $\alpha = 0.15$, $\alpha = 0.20$ (if the dependence among components is taken into account).

3.3 Discussion of Results

Table 1 shows the mean lifetimes in reliability state subsets {1, 2, 3, 4}, {2, 3, 4}, {3, 4}, {4} of system in case of independent components, when $\alpha = 0$, and for components following the dependency rule described by (4)–(5), with influence coefficients given by (15)–(18), for parameter α ranging from 0.05 to 0.20.

Table 1. The mean lifetimes in reliability state subsets {1, 2, 3, 4}, {2, 3, 4}, {3, 4}, {4} of system in case of independent components and dependent components with different values of influence coefficients reflected by α values (in years).

Dependency strength	$\mu(1)$	$\sigma(1)$	$\mu(2)$	$\sigma(2)$	$\mu(3)$	$\sigma(3)$	$\mu(4)$	$\sigma(4)$
Independent $\alpha = 0$	12.53	6.55	10.84	5.67	9.91	5.18	8.86	4.63
Dependent $\alpha = 0.05$	11.99	5.99	10.53	5.33	9.73	4.98	8.86	4.63
Dependent $\alpha = 0.06$	11.88	5.90	10.47	5.28	9.70	4.95	8.86	4.63
Dependent $\alpha = 0.07$	11.78	5.81	10.41	5.24	9.67	4.91	8.86	4.63
Dependent $\alpha = 0.08$	11.68	5.74	10.36	5.20	9.63	4.88	8.86	4.63
Dependent $\alpha = 0.09$	11.59	5.68	10.31	5.16	9.60	4.85	8.86	4.63
Dependent $\alpha = 0.10$	11.49	5.62	10.25	5.14	9.56	4.82	8.86	4.63
Dependent $\alpha = 0.15$	11.10	5.53	10.04	5.10	9.40	4.68	8.86	4.63
Dependent $\alpha = 0.20$	10.89	5.62	9.93	5.15	9.25	4.59	8.86	4.63

By comparing the results from Table 1, it can be concluded that assuming dependence among components shortens the mean lifetime of system in subset {1, 2, 3, 4} by a value from 4% to 13% depending on the strength of influence expressed by coefficients given by (15)–(18), for parameter α ranging from 0 to 0.20. Mean lifetime in subset {2, 3, 4} is less from around 3% to 8% depending on the value of parameter α, and mean lifetime in subset {3, 4} is shorter by a value between 2% and 7%.

We can notice that although the impact of dependencies among components on mean lifetimes in reliability state subsets of the entire system is moderate, the mean lifetimes of the system in particular reliability states decrease significantly as the strength of dependency among components increases. This is mainly because the system lifetime in state of fully effective operation (state 4) does not depend on the interactions among components, which is illustrated in the results in Table 1. Due to the series reliability structure of the system, the system is in reliability state 4 if all of its components are in this state. Hence, the deterioration of reliability state of one of the components only affects the system lifetime in worse reliability states, i.e. states 3, 2 or 1.

Table 2 contains the mean lifetimes in particular reliability states 1, 2, 3, of system in case of independent components ($\alpha = 0$), and in case of dependent components for different values of parameter α expressing the influence strength between components. The values in Table 2 are given in years and in percentage compared to the results for a system with independent components.

Table 2. The mean lifetimes in reliability states 1, 2, 3, of system in case of independent components and dependent components with different values of influence coefficients reflected by α values (in years).

Dependency strength	$\bar{\mu}(1)$	$\bar{\mu}(1)(\%)$	$\bar{\mu}(2)$	$\bar{\mu}(2)(\%)$	$\bar{\mu}(3)$	$\bar{\mu}(3)(\%)$
Independent $\alpha = 0$	1.69	100%	0.93	100%	1.05	100%
Dependent $\alpha = 0.05$	1.46	86.4%	0.80	86.0%	0.87	82.9%
Dependent $\alpha = 0.06$	1.41	83.4%	0.77	82.8%	0.84	80.0%
Dependent $\alpha = 0.07$	1.37	81.1%	0.74	79.6%	0.81	77.1%
Dependent $\alpha = 0.08$	1.32	78.1%	0.73	78.5%	0.77	73.3%
Dependent $\alpha = 0.09$	1.28	75.7%	0.71	76.3%	0.74	70.5%
Dependent $\alpha = 0.10$	1.24	73.4%	0.69	74.2%	0.70	66.7%
Dependent $\alpha = 0.15$	1.06	62.7%	0.64	68.8%	0.54	51.4%
Dependent $\alpha = 0.20$	0.96	56.8%	0.68	73.1%	0.39	37.1%

By analyzing in more detail the results given in Table 2, it can be noticed that for system with components following the dependency rule described by (4)–(5), in case of parameter value $\alpha = 0.05$ mean lifetime in reliability state 1 is 14% shorter than mean lifetime in this state for a system with independent components ($\alpha = 0$). If the influence of component deterioration on other components' reliability is more intense, this difference reaches up to 43% for $\alpha = 0.20$. By comparing the system mean lifetimes in particular reliability states for different values of parameter α, it can be seen that the difference for mean lifetime in reliability state 3 is the most significant. Namely, for a system with dependent components at the parameter value α of 0.05 this mean lifetime is 17% shorter compared to the mean lifetime of the system with independent components, and for the parameter value α of 0.20 mean lifetime in state 3 it is shorter by up to 63% compared to the mean lifetime for a system with independent components.

Assuming that critical reliability state is $r = 2$, by using definition introduced in [5], the system risk function is given by

$$\boldsymbol{r}_{\alpha}(t) = 1 - \boldsymbol{R}_{\alpha}(t, 2), t \geq 0, \tag{24}$$

where the coordinate of system reliability function $\boldsymbol{R}_{\alpha}(t,2)$ is given by (21). It expresses the probability that a system is in the subset of reliability states worse than the critical reliability state $r = 2$, while it was in the reliability state $z = 4$ at the moment $t = 0$.

Figure 3 illustrates the system risk function for up to 10 years, for different values of parameter α, $0 \leq \alpha \leq 0.20$, and for intensity values of departure from reliability state subsets given by (14).

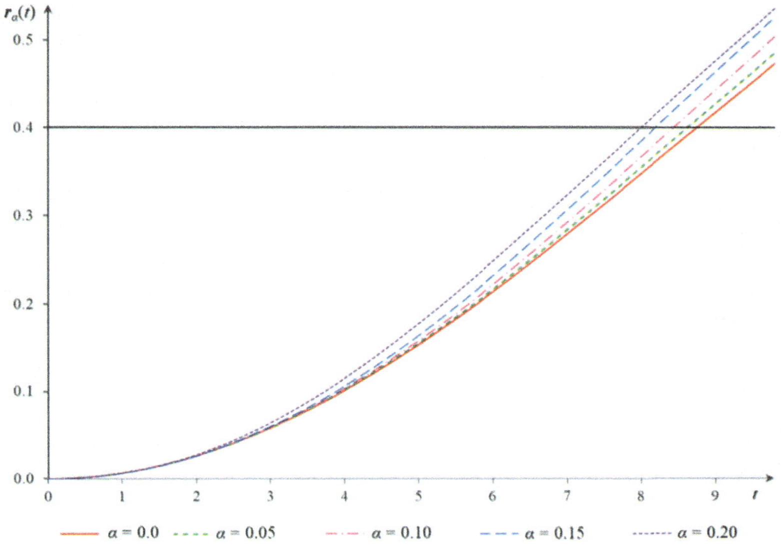

Fig. 3. The graphs of system risk function $r_\alpha(t)$ for different values of parameter α, $0 \leq \alpha \leq 0.20$.

Further, we determine the moment τ_α when the risk function $r_\alpha(t)$ exceeds an acceptable assumed level δ. In Fig. 3, a risk level of 0.4 was marked as an example. Table 3 presents values of moment of exceeding risk level for different values of this level δ and for parameter α ranging from 0 to 0.20.

Table 3. The moments of exceeding an acceptable level δ, ranging from 0.1 to 0.5, of system in case of independent components ($\alpha = 0$) and dependent components with different values of influence coefficients reflected by α values (in years).

Acceptable level	τ_α $\alpha = 0$	τ_α $\alpha = 0.05$	τ_α $\alpha = 0.10$	τ_α $\alpha = 0.15$	τ_α $\alpha = 0.20$
$\delta = 0.1$	3.98	3.96	3.94	3.88	3.73
$\delta = 0.2$	5.78	5.74	5.68	5.55	5.33
$\delta = 0.3$	7.31	7.24	7.11	6.92	6.70
$\delta = 0.4$	8.75	8.62	8.44	8.19	8.00
$\delta = 0.5$	10.19	10.00	9.75	9.46	9.31

Table 3 compares the moments of exceeding an acceptable level δ for the system with dependent components in relation to the values obtained for the system with independent components. Based on results given in Table 3, it can be concluded that depending on the assumed risk level δ, in case of less impact of interactions among components ($\alpha = 0.05$), this difference ranges up to 2%, and for $\alpha = 0.1$ up to 4%. If the coefficient of influence among components is determined by (15)–(18) for $\alpha = 0.15$, this difference ranges from 3% to 7% and for $\alpha = 0.2$ from 6% to 9%.

4 Summary

From the sensitivity analysis carried out in Sect. 3.3 due to parameter α, specifying the strength of relationship between the components, it can be concluded that the dependency among system components and the impact of their aging is significant for the reliability of the entire system and its operational safety. The next stage of our research will be availability analysis of multistate systems, including system maintenance and renewals, and the possibility of using the analysis performed to determine the optimal moment of system repairs or preventive maintenance.

Acknowledgements. The paper presents the results developed in the scope of the research project "Safety of critical infrastructure transport networks", WN/2020/PZ/, granted by GMU in 2020.

References

1. Lin, J., Pulido, J., Asplund, M.: Reliability analysis for preventive maintenance based on classical and Bayesian semi-parametric degradation approaches using locomotive wheel-sets as a case study. Reliab. Eng. Syst. Saf. **134**, 143–156 (2015)
2. Lai, J., Lund, T., Rydén, K., Gabelli, A., Strandell, I.: The fatigue limit of bearing steels – part I: a pragmatic approach to predict very high cycle fatigue strength. Int. J. Fatigue **38**, 155–168 (2012)
3. Blokus, A., Dziula, P.: Reliability analysis of different configurations of master and back-up systems used in maritime navigation. J. Mar. Sci. Eng. **8**(34) (2020). https://doi.org/10.3390/jmse8010034
4. Szymkowiak, M.: Lifetime Analysis by Aging Intensity Functions. Monograph in Series: Studies in Systems, Decision and Control, vol. 196. Springer, Cham (2020)
5. Kołowrocki, K., Soszyńska-Budny, J.: Reliability and Safety of Complex Technical Systems and Processes: Modeling – Identification – Prediction – Optimization, 1st edn. Springer-Verlag, London (2011)
6. Blokus, A., Kołowrocki, K.: Reliability and maintenance strategy for systems with aging-dependent components. Qual. Reliab. Eng. Int. **35**(8), 2709–2731 (2019). https://doi.org/10.1002/qre.2552
7. Blokus-Roszkowska, A., Kolowrocki, K.: Modelling safety of multistate systems with dependent components and subsystems. J. Pol. Saf. Reliab. Assoc. **8**(3), 23–41 (2017). Summer Safety and Reliability Seminars
8. Blokus, A., Dziula, P.: Safety analysis of interdependent critical infrastructure networks. TransNav Int. J. Mar. Navig. Saf. Sea Transp. **13**(4), 781–787 (2019). https://doi.org/10.12716/1001.13.04.10
9. Blokus, A.: Multistate System Reliability with Dependencies, 1st edn. Elsevier Academic Press, United Kingdom (2020)
10. Xue, J.: On multi-state system analysis. IEEE Trans. Reliab. **34**, 329–337 (1985)
11. Xue, J., Yang, K.: Dynamic reliability analysis of coherent multi-state systems. IEEE Trans. Reliab. **4**(44), 683–688 (1995)

Tool for Metamorphic Testing

Ilona Bluemke(✉) and Paweł Kamiński

Institute of Computer Science, Warsaw University of Technology,
Nowowiejska 15/19, 00-665 Warsaw, Poland
I.Bluemke@ii.pw.edu.pl, pawel.kaminski6.stud@pw.edu.pl

Abstract. Metamorphic testing is an approach to test case generation and also to the test result verification. It is a testing technique that can be successfully used in many domains e.g. web services, computer graphics, simulation and even embedded systems. In metamorphic testing checks are performed to verify whether multiple executions of the program under test fulfil certain necessary properties, called metamorphic relations. Since its first publication, many papers on different aspects of metamorphic appeared in the literature but only one tool for this type of testing was described. We decided to design and implement our own tool and with this tool to examine some properties and challenges of metamorphic testing. In this paper we briefly review metamorphic relations and describe our tool. We also present an example of metamorphic testing with our tool.

Keywords: Software testing · Metamorphic testing · Test case

1 Introduction

Software testing, detecting faults, is a very important activity in the software development. It consists of executing a program on some test inputs. To detect faults some procedure is necessary to decide whether the output of the program is correct or not, it is called test oracle. The test oracle compares an expected output value with the observed one. For programs that produce complex output, e.g. numerical simulations, or compilers, predicting the correct output for a given input could be difficult. This problem is called "oracle problem" and it is recognized as one of the challenges of software testing. Metamorphic testing [1] is a technique proposed to mitigate the oracle problem. It is based on the idea that sometimes it is easier to reason about relations between outputs of a program, than it is to fully predict or define its output. The typical example used in many papers e.g. in [7] is program computing the sine function. It is difficult to say what is the value of e.g. sin (30) or if value 0.5 appearing on output is correct. One of mathematical properties of the sine function is that $\sin(x) = \sin(-x)$, and this can be used to check if $\sin(30) = \sin(-30)$ without knowing the exact values. This is an example of a metamorphic relation: a transformation of input that can be used to generate new test cases from existing ones, and an output relation, that compares the outputs produced by a pair of test cases. Metamorphic testing "soothes" the oracle problem and also can be automated. As this technique is not widely used and

W. Zamojski et al. (Eds.): DepCoS-RELCOMEX 2020, AISC 1173, pp. 115–124, 2020.
https://doi.org/10.1007/978-3-030-48256-5_12

recently more and popular in publications e.g. [2–10], we wanted to investigate this domain. However we found one tool for metamorphic testing [11] – JFuzz, we decided to design and implement our own tool because JFuzz implements significantly modified approach and is difficult to use for non programmers.

The paper is organized as follows. More information on metamorphic testing and metamorphic relations is given in Sect. 2. In Sect. 3 the architecture of our tool is briefly described. As the typical application of metamorphic testing are web services our tool is dedicated to this domain. In Sect. 4 an example of metamorphic testing with our tool is presented. Finally, Sect. 5 concludes the paper, highlighting some issues and indicates future research directions.

2 Metamorphic Testing

Metamorphic testing was proposed in [1] and can be used to testing when the expected output of a test case is unknown or hard to compare with the actual output, so when it is difficult to prepare the test oracle. If the program fails a test according to the oracle it implies that the program is not correct on the test case. However, if the program passes a test according to the oracle, it does not imply that the program is correct on the test case. Rather than checking the output of an individual test, metamorphic testing checks whether multiple test executions fulfil certain metamorphic relations. A metamorphic relation is a necessary property of the intended program's functionality that relates two or more input data and their expected outputs. Metamorphic relations are used as the criteria of program correctness.

The basic process of metamorphic testing consists of three steps:

1. *Construction of metamorphic relations*. Properties of the program under test has to be identified and represented as metamorphic relations among multiple test case inputs and their expected outputs. Some method to generate a follow–up test case based on a source test case, must also be given.
2. *Generation of source test cases*. A set of source test cases for the program under test using any traditional testing technique should be generated or selected.
3. *Execution of metamorphic test cases*. Then the metamorphic relations are used to generate follow–up test cases. Source and follow–up test cases are executed and the relations are checked. If the outputs of a source test case and its follow–up test case are not fulfilling the metamorphic relation, the metamorphic test case fails, indicating that the program under test contains an error.

A survey on metamorphic testing was prepared by Segura et al. [7]. They also studied the application domains of this type of testing. The most popular domains are web services and applications (16%), computer graphics (12%), simulation and modelling (12%) and embedded systems (10%). They also found applications of metamorphic testing in financial software, optimization programs and encryption programs. Only 4% of the papers reported results in numerical programs, even though this seems to be the dominant domain used to illustrate metamorphic testing in the literature.

2.1 Metamorphic Relations

A metamorphic relation is a relation over multiple input-output pairs that should be satisfied if the system under test works in accordance with its specification and user's intention. Test cases are executed to see whether metamorphic relations are satisfied. If the system under test works correctly, the relations will be satisfied for all test cases. If the test outcome is not satisfying metamorphic relation, it is possible that there is a defect in the system.

Segura et al. in [6] proposed six Metamorphic Relation Output Patterns (MROPs) that capture the form of common metamorphic relations found in RESTful Web APIs. Describing metamorphic relations we use following notations: let $S = f(x_0)$ be the source output, and $F_i = f(x_i)$ the ith follow-up output.

1. *Equivalence* pattern represents relations where the source and the follow-up outputs are equivalent. Two or more outputs are equivalent if they include the same items although not necessarily in the same order.

$$\forall i \in [1, n] S \equiv F_i \tag{1}$$

2. *Equality* pattern represents relations where the source and follow-up outputs must contain the same items and in the same order.

$$\forall i \in 1, n \ S = F_i \tag{2}$$

3. *Subset* pattern groups relations where the follow-up outputs should be subsets (or strict subsets) of the source output and subsets among them:

$$S \supseteq F_1 \supseteq F_2 \supseteq \ldots \supseteq F_n \tag{3}$$

4. *Disjoint* pattern represents relations where the intersection among the source and follow-up outputs should be empty:

$$\forall i, j \in [1, n], S \cap F_i = \emptyset \wedge (i \neq j) \Rightarrow (F_i \cap F_j) = \emptyset \tag{4}$$

5. *Complete* pattern includes relations where the union of the follow-up outputs should contain the same items as the source output:

$$\mathrm{S} \ = F_1 \cup F_2 \cup \ldots \cup F_n \tag{5}$$

Sometimes it may be necessary to detect duplicated results by checking if the number of items in the source output is equal to the number of items in the follow up outputs:

$$|S| = \sum_{i=1}^{n} |F_i| \tag{6}$$

6. *Difference* pattern includes metamorphic relations where the source output and the follow-up output should differ in a specific set of items D. This pattern is formally defined as:

$$F_1 \backslash S = D \tag{7}$$

2.2 JFuzz

To our best knowledge the only tool supporting metamorphic testing is JFuzz [11]. In JFuzz data mutation and metamorphic testing are combined. Data mutation is a test case generation method proposed in [9]. From the given set of test cases, called seeds, new test cases are generated by modifying the seeds applying a set of data mutation operators. When the modification of the test data is at random, it called fuzz testing. A data mutation operator may be applied to different parts of the input data, if the data are structurally complicated. From a small number of test cases, a large number of test cases can be generated by applying mutation operators shown in [9].

The inputs to JFuzz framework are two Java classes: the class under test (CUT) and a test specification class (TSC). TSC contains attributes for the seed test cases, data mutation operators and methods that are metamorphic relations. The test specification class extends or imports the CUT so that it can access the attributes and methods to be tested. It is compiled before input to the JFuzz tool. Very important in JFuzz is *Metamorphic Relation Class*, with *Assertion* method. An invocation of *Assertion* method implements the mutational metamorphic relation. If the assertion is not satisfied, an error in the CUT is recorded and reported to the tester.

The usage of JFuzz needs some programming skills as some classes have to be written.

3 Meta-Tool

Our goal was to design and implement a metamorphic tool which will be much easier to use than the JFuzz [11] (Sect. 2.2). As the main domain of application of metamorphic testing are web services and web applications we decided to use JSON [12] as the format for data exchange.

The main function of our tool, called *meta-tool*, are following:

- Generates test cases according to Junit 5 standard [13].
- Reads data from XLSX file.
- Writes generated code of test cases as Java file (place and name are given by a user).
- Generates test cases, randomly, based on read data, user determines number of generated test cases.

- Generates test cases, deterministically, based on read data, user sets some properties of generated test cases.
- Determination of the type of metamorphic relation.

These functions are available through user friendly graphical interface.

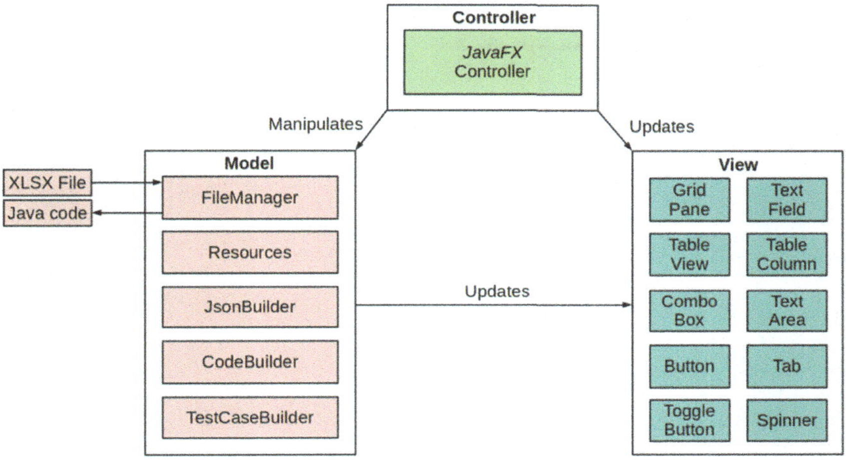

Fig. 1. Architecture of *meta-tool*

The architecture of our tool is shown in Fig. 1. As it can be seen the Model-View-Controller pattern was used. *Controller* is responsible for the flow of information, updating views, capturing user actions and transforming them into calls of appropriate methods.

Module *View* is based on FXML file [14]. All elements of graphical user interface are described by hierarchical structure of markers which is easy to modify. In Fig. 1 in module *View* components of JavaFX, used in building graphical user interface, are shown.

Module *Model* contains five parts. *File Manager* reads input data, which a user can prepare using e.g. Microsoft Excel. It also writes generated Java code in a place determined by user. *Resources* is responsible for storing data, metamorphic relations and data obtained during testing. Module *JsonBuilder* creates JSON messages based on input data, metamorphic relations and types of variables. Module *CodeBuilder* creates Java code based on data generated by *JsonBuilder* and metamorphic relation. Module *TestCaseBuilder* consolidates all above listed modules. On the bases of the information prepared by these modules generates assertions, which are written by *File Manager*.

The whole process of metamorphic testing with our tool is shown in Fig. 2.

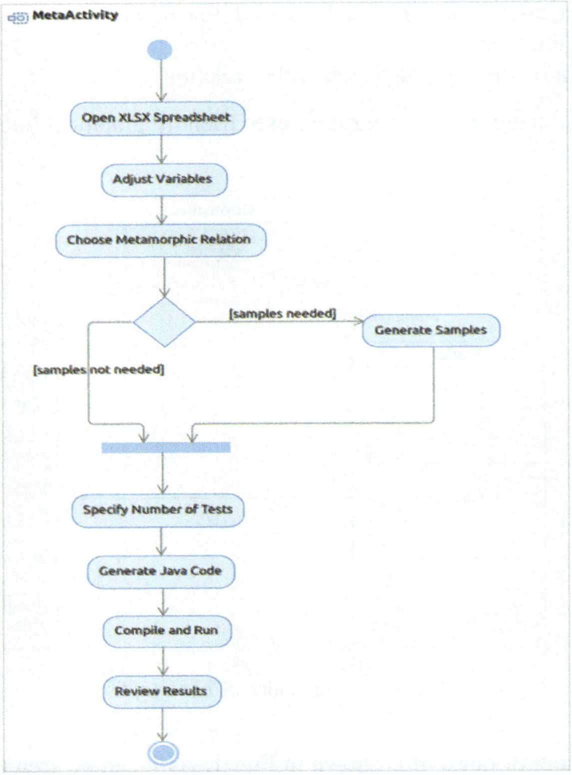

Fig. 2. Metamorphic testing with *meta tool*

4 Example

With our *meta-tool* we performed several experiments. The first one was testing program calculating sin function. This example is used in many papers on metamorphic testing e.g. [7]. As *meta-tool* is dedicated to testing web services and web applications is using JSON [12] as the format for data exchange. A simple application *sin-calc*, using JSON literals for data exchange, was implemented. Standard Java class `Math` was used to calculate sinus and converting angle units.

Input data set is based on two trigonometric identities:

$$\sin(x) = \sin(\pi - x) \tag{8}$$

$$\sin(x) = \sin(2\pi + x) \tag{9}$$

```
{
"argument" : 90
}
```

Fig. 3. Input for *sin-calc* program

In Fig. 3 the JSON format containing argument for calculations of sin function (in degree) is shown while in Fig. 4 the output message with the results of calculation ("value") is presented.

```
{
"value" : 1.0
}
```

Fig. 4. Output from *sin-calc* program

Value x = 30° was used as basic, source, starting test case. It is in the first row of Table 1. In subsequent rows exemplary angles are given, they were obtained using trigonometric identities (8), (9) or composition of both of them. They were used as "follow-up" test cases to the source one. The input data has to be prepared in XLSX file (Fig. 2) and specified to *meta tool* as shown in Fig. 5.

Table 1. Table of input data for *sin-calc*.

Argument
30
150
390
−210
−330

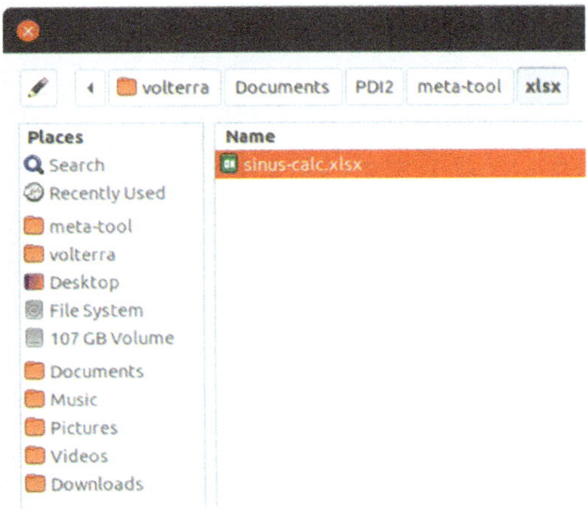

Fig. 5. Specifying input to *meta tool*

In Fig. 6 the configuration of our tool is shown. The *Equality* metamorphic relation is chosen (Sect. 2.1). The variable from the input data set is described as *morph*, which means that field "argument" will be different in each pair of test cases.

Next user has to define directory to store the generated files and run the generation (Fig. 7).

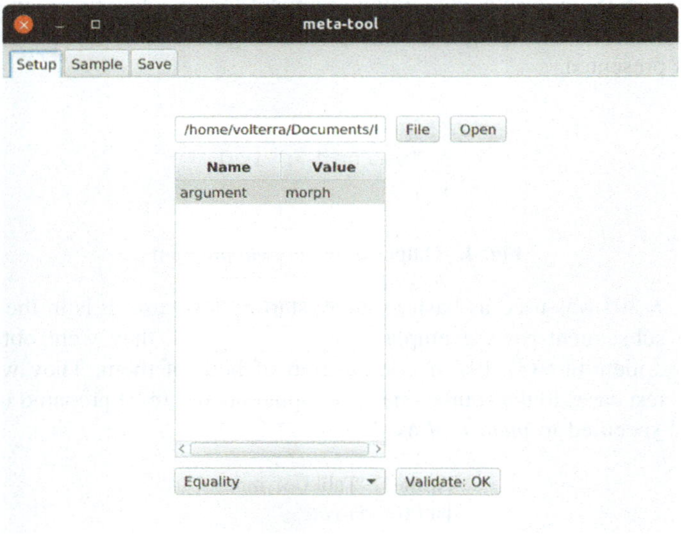

Fig. 6. Configuration of *meta tool* for *sin-calc*

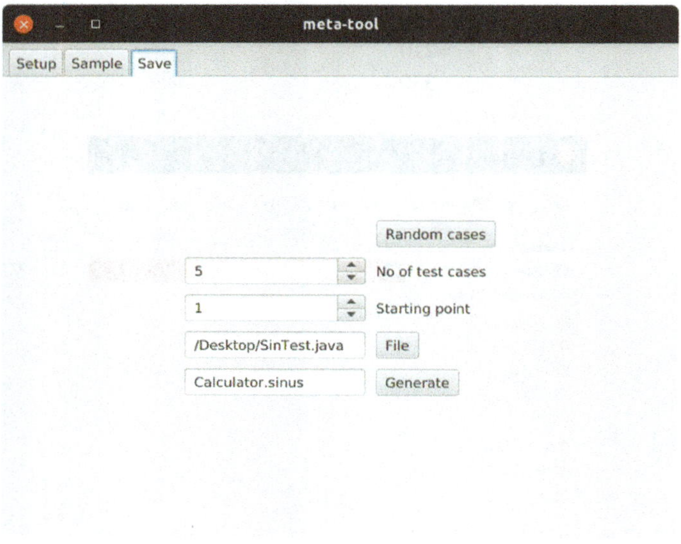

Fig. 7. Specifying number of test cases and directory to store results

As a result of execution of *meta tool* set of test cases was generated. One is presented in Fig. 8.

Method `equalityTestCase1` was generated. Numbers in methods names are automatically generated. Part of generated methods are JSON messages describing source and follow-up test cases. These messages are assigned to `String` objects (lines 3-4). In the final part of method (line 5) the values of sin functions for both arguments are compared. The creation of objects `ArrayList` (before comparison of results) is caused by specifics of *meta tool*. Our tool is dedicated to testing web services which are often returning set of data and *sin-calc* returns only one value. To fulfil equality relation the order of elements in such sets also has to be checked.

```
1 @Test
2 public void equalityTestCase1() throws IOException {
3   String str1 = "{\n \"argument\" : 30\n}";
4   String str2 = "{\n \"argument\" : 150\n}";
5   assertEquals(new ArrayList<>(Calculator.sinus(str1)),
  new ArrayList <>(Calculator.sinus(str2)));
6 }
```

Fig. 8. Exemplary pair of test cases for *sin-calc*

5 Conclusions

We presented a tool for mutation testing with graphical user interface which was designed and implemented at the Institute of Computer Science Warsaw University of technology. To our best knowledge it is the second tool supporting metamorphic testing. It is much easier to use it than JFuzz. More details about our tool can be found in [15]. Building this tool we have shown that it is possible to automate the major steps in metamorphic testing, including test case generation, execution, and verification. The construction of individual test cases appeared to be rather simple. Source test cases can be generated through existing testing methods while the follow-up test cases can be constructed through transformations according to metamorphic relations. Test case execution was also straightforward. It seems to us that the costs of metamorphic testing are lower than in traditional testing techniques. Apart from the metamorphic relations identification process the computational costs for generation and execution of follow-up test cases, and test result verification are low and these steps can be automated. Metamorphic relations identification needs “manual” work but some “manual work” is also necessary in traditional testing.

In future we would like to continue research on metamorphic testing e.g. to compare the effort in metamorphic testing and other traditional methods, to examine the scalability problem when the size of program under test is increasing.

References

1. Chen, T.Y., Cheung, S.C., Yiu, S.: Metamorphic testing: a new approach for generating next test cases. In: Technical Report HKUST-CS98-01, Department of Computer Science, Hong Kong University of Science and Technology (1998)
2. Chan, W.K., Chen, T.Y., Lu, H., Tse, T.H., Yau, S.S.: Integration testing of context-sensitive middleware-based applications: A metamorphic approach. Int. J. Softw. Eng. Know. Eng. **16**(5), 677–704 (2006)
3. Chen, T.Y.: Metamorphic testing: a simple method for alleviating the test oracle problem. In: 2015 IEEE/ACM 10th International Workshop on Automation of Software Test, pp. 53–54 (2015)
4. Hui, Z., Huang, S., Chen, T.Y., Lau, M.F., Ng, S.: Identifying failed test cases through metamorphic testing. In: 2017 IEEE International Symposium on Software Reliability Engineering Workshops (ISSREW), pp. 90–91 (2017)
5. Chen, L., Cai, L., Liu, J., Liu, Z., Wei, S., Liu, P.: An optimized method for generating cases of metamorphic testing. In: 2012 6th International Conference on New Trends in Information Science, Service Science and Data Mining (ISSDM 2012), pp. 439–443 (2012)
6. Segura, S., Parejo, J.A., Troya, J., Ruiz-Cortés, A.: Metamorphic testing of RESTful Web APIs. IEEE Trans. Softw. Eng. **44**(11), 1083–1099 (2018)
7. Segura, S., Fraser, G., Sanchez, A.B., Ruiz-Cortés, A.: A survey on metamorphic testing. IEEE Trans. Softw. Eng. **42**(9), 805–824 (2016)
8. Segura, S., Durán, A., Troya, J., Ruiz-Cortés, A.R.: A template-based approach to describing metamorphic relations. In: 2017 IEEE/ACM 2nd International Workshop on Metamorphic Testing (MET), pp. 3–9 (2017)
9. Lijun, S., Hong, Z.: Generating structurally complex test cases by data mutation: a case study of testing an automated modelling tool. Comput. J. **52**(5), 571–588 (2009)
10. Sun, C., Wang, G., Mu, B., Liu, H., Wang, Z., Chen, T.Y.: Metamorphic testing for web services: framework and a case study. In: 2011 IEEE International Conference on Web Services, pp. 283–290 (2011)
11. Zhu, H.: JFuzz: a tool for automated java unit testing based on data mutation and metamorphic testing methods. In: Second International Conference on Trustworthy Systems and Their Applications, pp. 8–15 (2015)
12. Introducing json. http://www.json.org/. Accessed Jan 2020
13. Junit 5 user guide. https://junit.org/junit5/docs/current/user-guide/. Accessed Jan 2020
14. Fxml documentation. https://docs.oracle.com/javafx/2/api/javafx/fxml/doc-files/introduction_to_fxml.html. Accessed May 2019
15. Kamiński, P.: Metamorphic testing. Bachelor thesis, Institute of Computer Science, Warsaw University of Technology (2019). (in Polish)

Dependability of Web Sites

Dariusz Caban(✉)

Wroclaw University of Science and Technology, 50-372 Wroclaw, Poland
dariusz.caban@pwr.edu.pl

Abstract. The paper presents a view on assessing the dependability of Web sites, abstracting from the technology used in their implementations. It discusses what is dependability in connection to this systems category, what types of faults need to be considered, what are the consequences of error occurrence. It proposes some dependability measures to be used to assess Web sites and discusses the pitfalls connected with their interpretation. It also discusses the different approaches to determining these measures.

Keywords: Dependability · Web sites · Web sites performance

1 Introduction

Web based communication has become the standard for all types of human activity. It is essential that the Web sites provide a high level of assurance that they operate correctly over time. There are numerous reports on how to achieve this by hardening the infrastructure, using various software architectures, providing replication and fallback techniques, employing on-the-fly reconfiguration [4, 7].

The paper abstracts from these techniques, concentrating on the problems of assessing the dependability of the sites. Different analysis methods are already well documented [7], they are usually applied to specific architectures and their assessment. Rather than proposing another approach to the dependability analysis, the paper is focused on the common problems and misconceptions connected with it.

2 Dependability

A. Avizienis, J.C. Laprie and B. Randell proposed that dependability be defined as the capability of systems to deliver service that can justifiably be trusted [2]. It should be noted that:

- The definition relates dependability to the correct performance of a systems, i.e. its ability to provide the functionality in presence of faults.
- The definition relates dependability to justifiable trust, not specifically probability, allowing approaches which are not based on stochastic analysis.

A very important aspect of dependability is a clear distinction between the definition of fault, error and failure:

W. Zamojski et al. (Eds.): DepCoS-RELCOMEX 2020, AISC 1173, pp. 125–132, 2020.
https://doi.org/10.1007/978-3-030-48256-5_13

- Fault corresponds to some system element not working or operating incorrectly. A fault may exist in the system from the beginning of its life cycle (design or production fault, software bugs) or it may occur during its exploitation (natural wear, incidental breakdowns, human errors, security breaches, etc.).
- Error relates to the system operation. A fault may be dormant in the system for any extend of time. When the system makes use of a faulty element during its operation, then the corresponding function is not realized correctly. Then, an error is said to occur.
- Failure relates to the results of system operation. A failure is said to occur if the results of an error occurrence manifest itself in the system not producing output or producing incorrect output.

2.1 System Reliability

As reported by R. E. Barlow [3], the systems reliability was introduced to the engineering community to explain the phenomena occurring in complex military systems. It was often observed that the lifespan of a system was often much shorter than expected by experts on the basis of the quality of the components being used. To understand this and improve the predictions, the structural or system reliability was introduced.

It is defined as the probability of a device performing its purpose adequately for the period of time intended under operating conditions encountered. System reliability depends on the reliability of its components and its reliability structure.

The definition of dependability is very similar to the above, but with significant differences. These differences in the approach address the problems encountered when trying to define the reliability as applying to computer systems, complex fault tolerant digital circuits and, especially, to software. In all these cases the classical definition, identifying faults with errors, cannot be fully upheld. Furthermore, it is very difficult to define reliability structure since the system can perform some functions with a faulty component, while it will fail to perform other functions.

System dependability redefines most of the basic concepts of reliability (while preserving a lot of the developed methodology and techniques). The definition of the operational state is directly connected with the system's ability to deliver service (and not with the state of its components). This reflects the approach developed in 1980's by W. Zamojski, who proposed the reliability-functional analysis of complex systems [6]. Thus, it is possible to have a unified view of system malfunctions, encompassing not only hardware faults, but also software errors, human mistakes and even deliberate user misbehavior.

2.2 Software Reliability

Software is usually regarded as a system component that is not prone to degrade (acquire faults during exploitation). The term "software reliability" was introduced to capture the similarity of software processing to hardware operation. In case of software, the faults do not occur during exploitation, they are dormant in the program from its conception or are introduced when patching or upgrading it. Thus the faults are not random. Instead, when a program is running it activates different parts of the code in a pseudo-random fashion. When a dormant fault is activated, the software fails to operate correctly (thus errors and failures are not distinguished).

Software reliability is defined as the probability of error-free software operation for a specified period of time in a specified environment (see [5]). Software reliability growth models relate this probability to the number of faults dormant in the system and the software lifetime.

Even though this definition is very similar to the definition of reliability, the underlying mechanisms of failure are completely different. In reliability theory a component fault can be either masked or cause system failure when it occurs. Software reliability introduces the concept of systems being operational, regardless faulty components that are not masked. The problem is in the visibility of faults.

2.3 Security

Faults of the hardware are at present relatively infrequent due to the modern design and materials technologies. The main sources of the Web sites failures are faults introduced to the system through the installed software and conscious or unconscious activities of its users. Particularly, the activities of the users can determine the level of the security of the system (viruses, security breaches, attacks on the system).

Security is aimed at preventing incidents, such as vulnerability exploits, malware proliferation, denial-of-service. It also considers the problems of intrusion detection and containment. It hardly ever considers the impact of these incidents: what to do if the preventive measures are breached, how will the system operability be affected. From the point of view of dependability, security vulnerabilities are just another type of faults, whereas security incidents cause operational errors.

2.4 Web Site Faults

As already discussed, dependability is used to assess the capability of systems to deliver service, regardless the potential reason of failure. This is ensured by considering all the potential risks connected with running the system. Thus, the classes of considered faults encompass those defined in reliability, software reliability and security.

In case of Web sites, the faults may occur in the hardware infrastructure (computer hosts and connections), operating systems and Web servers, site specific software, site management activities, and malicious client accesses. The considered faults are classified as:

Host Crash Faults. These hardware faults correspond to failures of the computers hosting the Web site. The software located on the affected hosts must be relocated or the site becomes inaccessible.

Host Performance Faults. The hosts cannot provide the full computational resources, causing some services to fail or increasing their response time above the acceptable limits.

Communication Faults. These faults may cause some hosts and services to become inaccessible preventing the Web site to respond to requests. These faults may also cause increased communication error rate, limiting the system throughput.

Software Crash Faults. The faults are caused by software faults in the operating system or Web server. Some part of the site becomes inaccessible until the software is restarted.

Corruption Faults. The site can produce incorrect or inconsistent responses due to accumulated software errors, effects of transient hardware malfunctions and possible exploitation of vulnerabilities. The Web site becomes unpredictable and misleading to clients.

Exploitation Faults. A special case of the corruption faults, caused by exploitation of security vulnerabilities or proliferation of malware. Unlike corruption faults, this type of fault can seriously damage the business image of the Web site and potentially be harmful to clients (highjacking personal data or compromising the client software).

3 Web Sites Dependability

Dependability expresses the capability of systems to deliver service. Thus, the measures of dependability must reflect the impact of failures on the service performance. To this end, lets consider the performance measures used to characterize Web sites.

3.1 Web Sites Performance Measures

There are multiple tools for testing the performance of Web sites. Most basic is the opensource project Apache JMeter [1]. While there are other more sophisticated tools, both commercial and opensource, the considered measures of performance are similar in all of them.

There are 3 measures that are commonly used, as follows. They all depend on the load of the site, i.e. on the number of concurrently handled clients. The functions, describing this dependency, are considered in whole or some characteristic point measures are derived.

Response Time. This is defined as the time that elapses from the moment that the client sends a request to a Web site until a response is delivered back to it. The response time is not constant, multiple similar requests are handled in differing times. Usually, average or median value is considered. If a request is handled incorrectly (error response) or is not responded to (timeout response), then the response time should not be included in the average.

This measure was very meaningful in classical installations, where most of the processing was done in the backend. It may be misleading in modern architectures, since it does not include the time of response rendering or frontend processing. If it is necessary to compare the performance of sites developed in different architectures, the response time should be redefined to end with complete rendering of the response page, including handling of embedded requests and front-end processing.

A typical average response time function is presented in Fig. 1. The function depends on the number of clients concurrently sending requests to the Web site. It should be noted that there are normally three significant ranges of workload. The first range – normal operation – correspond to the situation where all the incoming requests are immediately handled by concurrent threads or are queued for a short time till a worker thread becomes available. The average response time increases slightly due to increasing number of active worker threads and increasing time of waiting for a free worker.

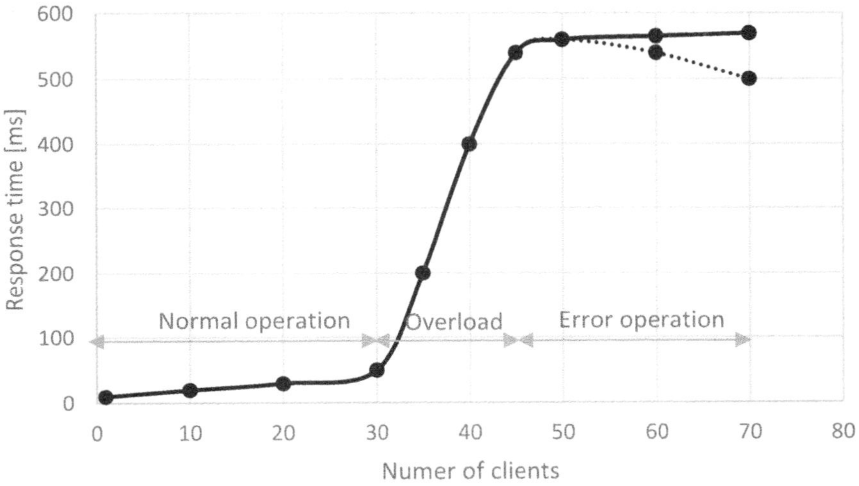

Fig. 1. An example of the average response time function obtained in testbed measurements, dotted line corresponds to a site with custom error responses.

The second range – overload operation – corresponds to the situation when all the active threads used all the time and the waiting queue grows rapidly. The response time increases very quickly with the workload. The site still works correctly, though the response time might not meet the customer quality criteria.

The third range – error operation – is reached when the site is so overloaded that it cannot handle the requests. The response time does not increase since all the requests above the upper limit of overload range are rejected (either by sending an error response or timed out). Thus, the site is actually handling a constant number of requests.

In some experiments, a decrease in the response time is observed in the error operation range (marked by dotted line in Fig. 1). This is an indication that the experiment is corrupt – usually due to incorrect definition of error responses. A lot of Web sites generate custom error responses instead of protocol errors. These may be treated as normal responses by performance measuring tools, not excluded from averaging and biasing the reported value.

Throughput. This is defined as the number of responses that are concurrently correctly handled by the site. Of course, average throughput is proportional to the workload in the normal operation range. Its increase gradually slows down in the overload operation range and becomes almost constant in error operation.

This measure is strongly related to response time, both provide similar information about the site performance. A useful measure related to throughput is maximum throughput. It is a point measure determined from the throughput function – the average throughput at midpoint of the overload range.

Error Rate. This is measured as a percentage of requests that are either not responded to (timed out) or get an error response. The complementary measure (ratio of correctly handled requests) is sometimes referred to as service availability, though this may be misleading in dependability assessment.

As commented when analyzing average response time, there is a problem with the proper defining of an error response. A correct response is defined in HTTP protocol as a response with code from the range 200–299. So, responses with codes above 299 are clearly error responses. HTTP responses with code 500 (server error) usually indicate a software error in the dynamic Web content (programming error). Similarly, errors in TCP protocol (usually timeouts) are treated as error responses. If a Web site is configured to mask errors and respond with some placeholder page (e.g. "site in construction" or "service temporarily not available"), then the available tools treat this as a correct response.

3.2 Dependability Measures

Occurrence of errors when a Web site is running leads to the degradation of its performance (either directly or due to the overhead of error containment procedures). This degradation can be used to quantify the site dependability: the smaller this degradation, the more dependable is the site. The degradation depends on the consequences of the errors and on the likelihood of their occurrence.

Two measures appear most promising when considering the dependability of Web sites:

Throughput Degradation. This measure compares the maximum throughput determined for the site where no faults are observed against normal site lifetime occurrence of faults. During the periods that the site is offline due to crashes, its throughput is 0. When there are errors caused by performance or corruption faults, the throughput is decreased. When there are errors caused by exploitation faults, the site is either put offline or its performance is reduced due to the implementation of error containment procedures.

Availability. The availability function $A(t)$ is defined [3] as the probability that the system is operational at a specific time t. The definition views the system as being either up (operational) or down (offline). This is not consistent with the faults that have to be considered. The definition needs to be extended to accommodate performance and corruption faults.

Before considering these extensions, lets note that the availability function is time invariant in case systems operating in steady state (usually true in case of production Web sites). This constant value is represented as a constant availability, denoted as A. The asymptotic property of the steady-state availability A states that:

$$A = \lim_{t \to \infty} \frac{t_{up}}{t}, \quad (1)$$

where t_{up} denotes the total system uptime. Assuming a uniform rate of requests, the property (1) may be transformed to:

$$A = \lim_{n \to \infty} \frac{n_{corr}}{n}, \quad (2)$$

where n_{corr} denotes the number of correctly handled requests in the site uptime.

If the site is exposed to errors that affect its performance, then the assumption of a uniform rate of requests/responses is not met. Still, the measure proposed in (2) can be used, even though strictly speaking it is no longer consistent with the definition of availability. Since some error responses occur even in a fully operational site, the maximum value of A is less than 1 (hence, it is no longer probability).

Furthermore, the error rate in a fully functional site depends on the workload, as discussed in Sect. 3.1. This is also true, though different in values, in regard to performance of a site with performance degraded by faults occurrence. This means that availability determined using (2) depends on the workload.

4 Dependability Analysis of Web Sites

Successful analysis of systems dependability strongly relies on the technology in use. The following subsections highlight the limitations of some common approaches.

Site Monitoring and Testbeds. The most direct method of determining throughput and availability is to compute them from data collected by monitoring the production Web site. In this case, there is no problem with determining the typical workload to be considered. The approach is very lengthy, as it takes a lot of time to capture a representative set of observations, particularly when one takes in account the low likelihood of errors.

The approach is usually enhanced by building testbeds, i.e. experimental sites that have the same architecture, but can be experimented on without affecting the production site. Usually, these are built using virtual machines with production software. These testbed sites are exposed to artificially generated workloads that allow more speedy collection of the data for analysis. There is still a problem with obtaining data on performance when errors occur. To some extent, this is simulated by shutting hosts and servers, limiting resources allocated to virtual machines, seeding software with artificial faults. The results are very questionable for some categories of faults.

Infrastructure Simulation. This approach is very similar to testbed site monitoring. The site infrastructure is simulated, using a computer/network simulator. The production software is run on the simulated hardware instead of hosts or virtual machines. Thus, hardware and communication faults are very easy to simulate. The approach does not solve the problem of observing software and exploitation faults. Furthermore, it requires extensive rescaling of the obtained results to accommodate differences in performance of real and simulated hardware.

High-Level Simulation and Analysis. This approach requires formulation of models of software behavior (both the servers, dynamic content and service orchestration). The infrastructure and software are simulated by customized simulators. There is no problem with simulating all the categories of faults. Analytical studies, using fault tree or state-transition analysis, are also feasible. Unfortunately, the simplifications introduced by software modelling make the results least credible.

5 Conclusions

It is demonstrated that dependability better characterizes the properties of Web sites, combining properties of reliability, software reliability, and security consequences.

The throughput degradation and modified availability are demonstrated as adequate dependability measures of Web sites, encompassing all the effects of site errors and failures. The pitfalls of these measures and techniques of their assessment are discussed.

References

1. Apache JMeter™ Homepage. http://jmeter.apache.org. Accessed 01 Feb 2020
2. Avizienis, A., Laprie, J., Randell, B.: Fundamental concepts of dependability. In: Proceedings of the 3rd IEEE Information Survivability Workshop, Boston, Massachusetts, pp. 7–12 (2000)
3. Barlow, R.E., Proschan, F.: Mathematical Theory of Reliability. Classics in Applied Mathematics. Society for Industrial and Applied Mathematics, Siam, Philadelphia (1987)
4. Caban, D.: Reconfiguration of complex information systems with multiple modes of failure. In: Monographs of System Dependability, Models and Methodology of System Dependability, pp. 36–47. Politechnika Wroclawska, Wroclaw (2010)
5. Musa, J.D.: Software Reliability Engineering: More Reliable Software, Faster Development and Testing. McGraw-Hill, New York (1999)
6. Zamojski, W.: Functional reliability model of a man-computer system. In: Zamojski, W. (ed.) Computer Engineering. WKŁ, Warsaw (2005)
7. Zamojski, W., Sugier, J. (eds.): Dependability Problems of Complex Information Systems. Advances in Intelligent Systems and Computing, vol. 307. Springer, Heidelberg (2015)

Dependability Analysis of Systems Based on the Microservice Architecture

Dariusz Caban and Tomasz Walkowiak(✉)

Faculty of Electronics, Wrocław University of Science and Technology, Wybrzeże Wyspiańskiego 27, 50-320 Wrocław, Poland
{dariusz.caban, tomasz.walkowiak}@pwr.edu.pl

Abstract. The paper presents an approach to dependability analysis of systems using the microservice architecture. The system model is formulated, categories of faults are identified, adequate dependability measures are proposed (based on availability and response time). The concept of reconfiguration graph for this class of systems is described. Lexical Platform is introduced as a practical example of a system based on microservice architecture. Its reconfiguration graph is shown to prevent any single error or multiple errors with a single root fault to cause the system failure.

Keywords: Systems dependability · Microservices · Dependability analysis

1 Introduction

The paper considers a class of information systems built on the basis of microservice architecture style [12]. It gained a big popularity. Micro-services are defined as a set of "cohesive, independent processes interacting via messages" [5]. It follows SOA [2] ideas though it does not imply the use of protocols associated with SOA systems. The system is seen as a set of interacting distributed components (microservices).

We propose to describe the micro-service information system at 3 levels following approach from [3]. On the highest level, it is represented by the interacting service components. At the physical layer, it is described by the hosts or virtual machines, on which the services are deployed, and by the visibility and communication throughput between them (provided by the networking resources and communication). The third element of the system description is the mapping between the first two layers.

2 Microservice System Model

2.1 Service Model

The system is composed of a number of microservices. Interaction between them is based on the client-server paradigm, i.e. one microservice requests service from some others and uses their responses to produce its own results, either output to the end-user or used to respond to yet another request. The client (user) requests are serviced by some microservices, while others may be used solely in the background.

W. Zamojski et al. (Eds.): DepCoS-RELCOMEX 2020, AISC 1173, pp. 133–142, 2020.
https://doi.org/10.1007/978-3-030-48256-5_14

Microservices are described as independent processes that interact only via message exchange. A microservice is entirely deployed on a single web node (host). Multiple microservices may run on the same host – their implementation must ensure that they are run independently, with no conflicting requirements.

The over-all description of the interaction between the service components is determined by the system choreography, i.e. the scenarios of interactions that produce all the possible usages of the system [9]. The component services generate demand on the networking resources and on the computational power of the hosts running them. This is the only limitation on how the services can be deployed, an example of the microservice based system is discussed in Sect. 4.

2.2 Network Infrastructure

The microservices are deployed on a network of computer hosts. This underlying communication and computing hardware is abstracted as a collection of interconnected hosts. In case of systems that perform mission-critical tasks, this underlying infrastructure is usually hardened by using dynamic routing and load balancing that distributes workload between multiple hosts (and microservices running on them).

The network infrastructure is often abstracted by deploying services to a local or public cloud. This simplifies the technical aspects of deployment of services and management of the infrastructure. This may prevent proper analysis of hardware related dependencies that impact the system dependability.

2.3 System Configuration

System configuration is determined by the deployment of microservices onto the hosts. This is characterized by the subsets of services deployed at each location (see Fig. 1). The deployment clearly affects the system performance, as it changes the communication and computational requirements imposed on the infrastructure.

If $\boldsymbol{H}$ denotes the set of computing hosts and S_h the set of microservices deployed on host $h \in \boldsymbol{H}$, then the system configuration is defined as a vector

$$\psi = [S_h : h \in \boldsymbol{H}]. \tag{1}$$

There are multiple possible configurations of the same system. A configuration is said to be permissible, if it ensures that the system is operational. The set of all permissible configurations is denoted as $\boldsymbol{\Psi}_{up}$. Of course, this does not mean that the choice of the permissible configuration is immaterial - quality of the operation is can be affected by it. The various permissible configurations differ in the efficiency of generating the responses to client requests. This may lead to degraded operation of the system.

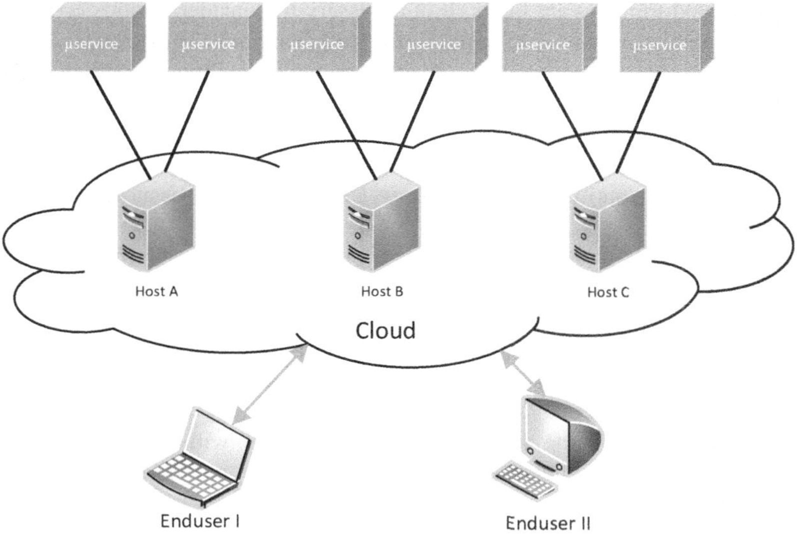

Fig. 1. System configuration – deployment of microservices

3 Reconfiguration

Reconfiguration (change of system configuration) takes place when service deployment is changed, i.e. tasks are redistributed between the hosts. This is fairly easily achieved in case of microservices due to their inherent independence (no conflicting requirements regarding the computing environment, OS and libraries).

Reconfiguration occurs when microservices are redeployed after the administrators observe that some hosts are either over or under-utilized. This is usually a scheduled and carefully planned administrative task connected with optimizing the system performance.

More significant are situations where reconfiguration is used to improve the dependability of systems based on the microservice architecture [5, 9, 12]. Whenever a host becomes inoperational or overloaded due to the occurrence of faults, microservices initially deployed to it are moved to other hosts ensuring continuity of operation. This is a practical approach to the utilization of functional redundancy existing in such systems.

3.1 System Faults

When considering system dependability a number of adverse events must be considered. In fact, dependability is an integrative concept that encompasses: availability (readiness for correct service), reliability (continuity of correct service), safety (absence of catastrophic consequences), confidentiality (absence of unauthorized disclosure of information), integrity (absence of improper system state alterations), maintainability (ability to undergo repairs and modifications). For this reason, these adverse events cannot be limited to hardware faults. It is necessary to consider various errors [3, 10]: transient and persistent hardware faults, software bugs, human mistakes and deliberate attacks on the system (exploitation of software vulnerabilities. draining of limited microservice resources, DOS attacks).

In the considered approach, the hosts are the basic components of the system infrastructure. Thus, all the faults are attributed to either to them (and not to hardware or software components) or to the microservices. This is the basis for the following classification:

Inoperational host faults – the host cannot process services that are located on it, these in turn do not produce any responses to queries from the services located on other nodes.

Impaired performance faults – the host can operate, but it cannot provide the full computational resources, causing some microservices to fail or increasing their response time above acceptable limits.

Connectivity fault – the host cannot communicate with other hosts with the required throughput. In effect, microservices may become unreachable, requiring redeployment.

Service malfunction fault – the microservice can produce incorrect or inconsistent responses due to accumulated software errors, effects of transient hardware malfunctions and possible exploitation of vulnerabilities. The operation of such microservices becomes unpredictable. The fault may propagate to other connected services, producing incorrect responses to user requests. Since microservices are inherently independent from each other, the fault propagation is limited only to those services that use the corrupted responses from the malfunctioning ones. Unlike other Web architectures [12], the propagated malfunction does not persist once the root cause is eliminated.

DOS fault – a special case of a service fault, where the microservice loses its ability to respond to requests. It is usually caused by some exploitation of security vulnerability, often a proliferation of bogus service requests that lock up all its resources. A very important aspect of this class of faults is that the attack may be either host locked or service locked. Reconfiguration is effective only in the first case: moving the affected services to other network addresses can prevent further damage. On the other hand, if a service is moved in case of a service locked attack, then the fault will also be propagated to the new location. In effect, this is a situation when reconfiguration is ineffective.

When considering the dependability of a system it is essential to foresee the potential faults that it may be exposed to. Further, the set of foreseen faults is denoted as $\mathbf{\Theta}$. Of course, multiple faults may be in effect at the same time (if a new fault occurs before the system recovers from a previous one).

3.2 Permissible Configurations

As discussed in Sect. 2.3, the system is operational if it is in any of its permissible configurations $\mathbf{\Psi}_{up}$. If a fault occurs, the set of permissible faults is reduced, i.e. if a host becomes inoperational, then all the configurations that have a microservice deployed to that host are no longer permissible.

If a sequence of faults $\theta_1, \theta_2, \theta_3 \in \boldsymbol{\Theta}$ occurs and the system is not yet restored, then the subset of permissible configurations that remain permissible is denoted as

$$\boldsymbol{\Psi}_{up}/\theta_1, \theta_2, \theta_3 \subset \boldsymbol{\Psi}_{up}. \quad (2)$$

It is a straightforward combinatorial problem to determine all the possible system configurations. A more demanding task is to determine which of the configurations are permissible. This usually is done by testing all the configurations in a testbed. A more feasible approach is to use network simulation tools [4]. Simulation can also be used to automate the process of determining the reduced subsets needed to consider the effects of faults occurrence.

3.3 Reconfiguration Graph

The reconfiguration graph [3] is built to define the possible changes in the configuration, that tolerate the various faults. Set $\boldsymbol{\Psi}_{up}$ is at the root of the graph, since any admissible configuration ensures system being up, if there are no failures. The branches leaving the root correspond to the various faults affecting hosts or services. They point at subsets $\boldsymbol{\Psi}_{up}/\theta_i$ corresponding to effects of single faults occurrence.

Further branches of the graph, corresponding to subsequent faults, are produced by eliminating configurations from $\boldsymbol{\Psi}_{up}/\theta_i$. The procedure is continued until the elimination produce empty sets that correspond to combinations of failures that cannot be tolerated by any reconfiguration. This approach to the reconfiguration graph construction ensures that all the possible configurations are taken into account. An example of such a reconfiguration graph is presented in Fig. 2.

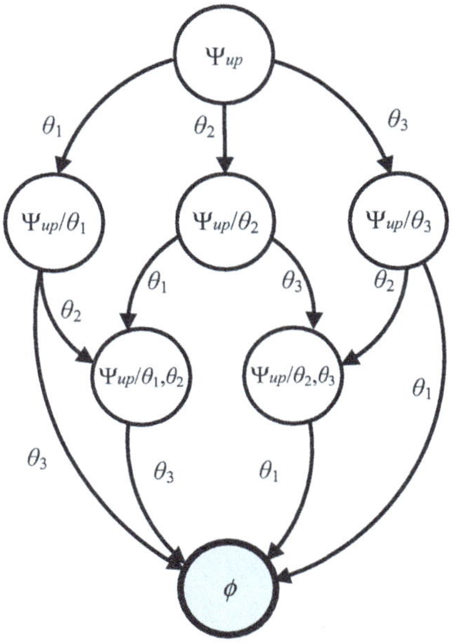

Fig. 2. An example of a simple reconfiguration graph

It should be noted that the reconfiguration graph illustrates all the possible changes in the service deployment that will preserve the system operability.

3.4 Reconfiguration Strategy

System reconfiguration, realized to improve the system dependability, is triggered by the occurrence of a dependability issue, i.e. a fault which causes some services to fail in the current configuration. Reconfiguration is achieved by isolating the faulty hosts and microservices, and then moving the affected services to other hosts.

The reconfiguration strategy should ensure that the target configuration preserves the functionality of all services (continuity of operation), at the same time maintaining their quality at the highest possible level (quality of service).

Continuity of operation is met if all the service components are deployed on unaffected hosts, they do not lead to compatibility issues with other components, and the communication resources ensure their reachability. The reconfiguration strategy is constructed by choosing one configuration from the set corresponding to each node in the reconfiguration graph.

Usually, there are many reconfiguration strategies that can be constructed in this way. Quality of service is ensured by choosing the configuration that ensures the best performance, i.e. with the shortest average response times. This is achieved if there is an efficient tool for predicting the service response time. One of the feasible approaches is to use network simulation.

3.5 Service Availability

Dependability is defined as the capability of systems to deliver service that can justifiably be trusted [1]. This definition implies that there is a clear criterion of service trustworthiness. In the considered case, it is the system ability to respond correctly and timely to user requests.

Any of the considered faults will cause the system to fail if/when they affect its ability to generate correct responses to the client requests. In time, the system is restored after a fault occurs, so its failure is not permanent. Thus the availability function $A(t)$, defined as the probability that the system is operational (provides correct responses) at a specific time t. Most interesting from the practical point of view, the function is usually time invariant, characterized by a constant availability denoted as A.

The asymptotic property of the steady-state availability A states that:

$$A = \lim_{t \to \infty} \frac{t_{up}}{t}, \tag{3}$$

where t_{up} denotes the total system uptime. From the perspective of the request/response systems, the asymptotic property may be transformed, assuming a uniform rate of service requests to:

$$A = \lim_{n \to \infty} \frac{n_{up}}{n}. \quad (4)$$

This yields a common understanding of availability as the number of properly handled requests n_{ok} as a percentage of all the requests n.

Availability does not reflect the quality of service. This has to be analyzed using a different measure. The most natural is to use the average response time, i.e. the time elapsed from the moment of sending a request until the response is completely delivered to the client [3]. The mean value is calculated only on the basis of correctly and timely handled response times. The error and time-out response times are excluded from the assessment (or assessed as a separate average).

4 Lexical Platform

4.1 Overview

Lexical Platform [6] is a web application for a lightweight integration of various lexical resources [6] into one complex (from the perspective of non-technical users). All lexical resources are represented as software web components implementing a minimal set of predefined programming interfaces providing functionality for querying and generating simple common presentation format. A common data format for the resources is not required. Users are able to search, browse and navigate via resources on the basis of anchor elements of a limited set of types. Lexical resources are linked to the platform via components that preserve their identity.

4.2 Anchor System

The Lexical Platform allows users to address its context but a simple anchor system that includes content type, language and content name. Following types are currently defined:

- word forms, i.e.: /orth/pl/domu
- lemma, i.e.: /lemma/pl/dom
- identifiers of synsets [], i.e.: /synset/plwordnet/4782

After providing the component anchor the system checks if any lexical resource includes such entry and provides available information in JSON or HTML format.

Important aspect of Lexicial Platform is the interlinking of resources. Based on the content of the resource the Lexical Platform converts resource internal references into a platform one allowing automatic mapping between references used in the resource and platform references.

4.3 Service Model

Lexical Platform is composed of a set of microservices. Each lexical resource (shaded rectangulars in Fig. 3) is represented by a dedicated mircroservice. It provides a simple

API that allows to: deliver the resource meta-data, checks if an anchored element exists in a given language resources and returns the data in JSON or HTML format.

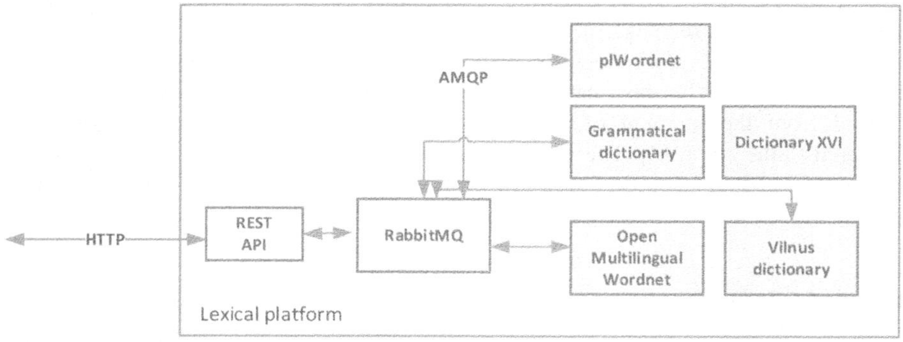

Fig. 3. Lexical Platform architecture

The client (user) requests are serviced by REST microservice and central indexing service that allows interaction between lexical resources. Communication is done asynchronously by the AMQP [8] protocol with a usage of the open source RabbitMQ [7] broker. Each lexical microservice collects tasks from a given queue and sends back messages through REST microservice when results are available.

4.4 Network Infrastructure and System Configuration

The Lexical Platform is hosted by CLARIN-PL Centre of Language Technology (CLT) [11] in a private cloud. Lexical micro-services are deployed on three separate virtual machines running on different hosts. The cloud currently consists of 9 hosts available at CLT. There are no constraints on deployment of virtual machines to hosts.

4.5 Reconfiguration

Due to the usage of virtual environment and backup system, it is very simple to change the deployment of virtual machines to the hosts. This mechanism is the basis of all the reconfigurations required to enact the changes required by the reconfiguration graph.

Inoperational host faults are tolerated by redeploying micro-services to other, still operational hosts. Impaired performance faults are dealt with by moving virtual machines to other host or duplicating the virtual machines and deploying them on multiple hosts. Connectivity faults are tolerated by the private cloud architecture, unless they accumulate to completely break the cloud.

Service malfunction faults are dealt with by restarting the affected micro-services. This approach relies on the assumption that the errors produced by the faulty service do not propagate to other micro-services. This is assured by the discussed architecture of micro-services: the interface between them is designed robust, so illegal requests do not lead to illegal/undefined states of the responding micro-service.

DOS faults are not handled by reconfiguration in the described system. Instead, it was assumed that the cloud networking devices sufficiently eliminate the risk of a successful DOS attack.

Reconfiguration is also used to react to unexpected increase in the demand for service (manifested by a high number of requests). In this case, replication of the virtual machines is also performed.

5 Conclusion

The paper discusses a systematic approach to improving the dependability of micro-service based systems by reconfiguration. This approach is demonstrated by applying it to the Lexical Platform hosted by CLARIN-PL.

The proposed reconfiguration strategy is found to successfully tolerate faults occurring in the system hardware and software. No single fault or group of dependant faults with a common root fault can cause the system to go down for a significant time (longer than the time needed to redeploy a virtual machine). Only the DOS attacks are not dealt with directly – the networking mechanisms were found sufficient to deal with their risk.

References

1. Avizienis, A., Laprie, J., Randell, B.: Fundamental concepts of dependability. In: Proceedings of the 3rd IEEE Information Survivability Workshop, Boston, Massachusetts, pp. 7–12 (2000)
2. Bell, M.: SOA Modeling Patterns for Service-Oriented Discovery and Analysis. Wiley, Hoboken (2010)
3. Caban, D., Walkowiak, T.: Service availability model to support reconfiguration. In: Zamojski, W., Mazurkiewicz, J., Sugier, J., Walkowiak, T., Kacprzyk, J. (eds.) Complex Systems and Dependability. AISC, vol. 170. Springer, Heidelberg (2013)
4. Caban, D., Walkowiak, T.: Risk assessment of web based services. In: Advances in Intelligent Systems and Computing, Theory and Engineering of Complex Systems and Dependability: Proceedings of the Tenth International Conference on Dependability and Complex Systems DepCoS-RELCOMEX, Brunów, Poland, 29 June–3 July 2015, pp. 97–106. Springer, Heidelberg (2015)
5. Dragoni, N., Giallorenzo, S., Lluch-Lafuente, A., Mazzara, M., Montesi, F., Mustafin, R., Safina, L.: Microservices: yesterday, today, and tomorrow. CoRR, vol abs/1606.04036 (2016)
6. Piasecki, M., Walkowiak, T., Rudnicka, E., Bond, F.: Lexical platform – the first step towards user-centred integration of lexical resources. Cogn. Stud. (Études cognitives) **18** (2018)
7. Videla, A., Williams, J.: RabbitMQ in action. Distributed messaging for everyone. Manning (2012)
8. Vinoski, S.: Advanced message queuing protocol. IEEE Internet Comput. **10**(6), 87–89 (2006)
9. Walkowiak, T.: Language processing modelling notation – orchestration of NLP microservices. In: Advances in Dependability Engineering of Complex Systems: Proceedings of the

Twelfth International Conference on Dependability and Complex Systems DepCoS-RELCOMEX, Brunów, Poland, 2–6 July 2017, pp. 464–473. Springer, Heidelberg (2018)
10. Walkowiak, T., Michalska, K.: Functional based reliability analysis of Web based information systems. In: Dependable Computer Systems, pp. 257–269. Springer, Heidelberg (2011)
11. Walkowiak, T., Pol, M.: Dependability aspects of language technology infrastructure. J. Polish Saf. Reliab. Assoc. **9**(3), 101–108 (2018)
12. Wolff, E.: Microservices: Flexible Software Architectures. Addison-Wesley, Boston (2016)

Using Domain Specific Languages and Domain Ontology in Workflow Design in Syndatis BPM4 Environment

Wiktor B. Daszczuk[1(✉)], Henryk Rybiński[1], and Piotr Wilkin[2]

[1] Institute of Computer Science, Warsaw University of Technology, Nowowiejska Str. 15/19, 00-665 Warsaw, Poland
{wbd,hrb}@ii.pw.edu.pl

[2] Syndatis Ltd., Puławska Str. 12a/10b, 02-566 Warsaw, Poland
piotr.wilkin@syndatis.com

Abstract. Defining professional workflows within Workflow Management Systems (WfMS) is not a simple task. Typically, this activity is dedicated to professionals having a high level knowledge and skills in this field, because many aspects of the workflow need to be linked: data model, presentation forms, process flow, synchronization, logical constraints, etc. Therefore in this situation, the work of a specialist is a bottleneck that limits the possibilities of effective workflow creation. The paper is devoted to a new architecture of WfMS, where workflow is defined by means of a set of files describing particular graphically designed various aspects of the workflow, with the use of XML-grounded Domain Specific Languages (DSLs). Each of the aspects has its own XML schema, defining its structure and constraints. An important property is that separate DSLs with their own schemas allow to some extent developing independently individual aspects of workflows (separation of concerns). On the other hand, the process of defining the aspects of the workflow is integrated with an application ontology, which supports automatization of design, and preserves consistency between the schemas, assuring completeness of the workflow.

Keywords: Workflow design · Domain Specific Languages · Domain ontology

1 Introduction

The issue of workflows is closely related to the theory of organization and (discrete) process management in the context of work organization. Practically, a Workflow Management System (WfMS) allows the user to define various workflows for different types of tasks and/or processes, for example, within a manufacturing process, a project document prepared by a designer should be automatically directed to the technical director or to the production engineer. At each stage of the workflow, an employee or a specific group of employees is responsible for a specific task. When the task is completed, WfMS guarantees automatic transfer of information on the completion of the task and its results to other employees involved in the manufacturing process.

W. Zamojski et al. (Eds.): DepCoS-RELCOMEX 2020, AISC 1173, pp. 143–154, 2020.
https://doi.org/10.1007/978-3-030-48256-5_15

Several categories of the WfMS software, aimed at organizing and automating business processes, can be observed on the market. All of them closely refer to the concepts of business process management, such as business rule engines, business process modeling, business monitoring, and finally the flow of human work. Although these individual components of WfMS are often integrated as parts of more general systems, sometimes referred to as IBPMS (Integrated Business Process Management Suites), three basic types of WfMS are distinguished:

- human-centric BPM (concentrating on interaction with people and between people)
- integration-centric BPM (Enterprise Service Bus), especially internet-based and cloud-based, integrating the flows between organizations or inside organizations
- document-centric BPM (Dynamic Case Management), addressed to automate ERP systems (Enterprise Resource Planning) and often being a part of those systems.

The first two types can be categorized as "process-centric", as opposed to document-oriented systems. The best-known document-oriented WfMS is Lotus Notes [1] (and its Domino extension for the WWW environment). Other widely used document-oriented WfMS are: SAP Business Workflow [2], Kissflow [3] and IOnet [4].

Defining professional workflows is not a simple task. Typically, this activity is the task of professionals with a high level of knowledge and skills in this field, because many aspects of the workflow need to be linked: data model, presentation forms, process flow, synchronization, logical constraints, etc. In this situation, the work of a specialist is a bottleneck that limits the possibilities of effective workflow creating.

In the paper, we want to show how to build a workflow as a set of files describing particular aspects, created with the help of graphical specification tools and having a uniform XML form. Each of the aspects has its own schema, defining its structure and constraints. In addition, separate DSLs with their own schemas allow to some extent independent development of individual aspects of workflows (separation of concerns). All aspects are fastened with the clasp of the domain ontology, which determines the objects appearing in the workflow, their attributes and relations between objects (including the hierarchy of abstraction). Semantic links between objects and restrictions are created in a declarative and automated way, which gives more confidence in the consistency and completeness of the workflows.

This approach allows for lowering the threshold of competence and to define workflows by employees with not so high skills, or even by users, and the role of a professional can be limited to advice and consultation.

The paper is composed as follows: Sect. 2 presents related work in the area of Domain Specific Languages (DSL) in the area of workflow design for the needs of WfMS. Then, in Sect. 3 we present our approach to the WfMS, focusing on the features of DSL and the role of ontology in the process modeling activities. Section 4 presents an application example, expressed in our DSL. The paper ends Sect. 5 with conclusions.

2 Related Work

2.1 Domain Specific Languages

Already in 1995, WfMC (Workflow Management Coalition) published a reference architecture of WfMS [5], in which the central position is occupied by the "workflow engine", software that manages workflows in interaction with entities - users and computer systems (and other devices). Important elements of the architecture are the interfaces between the engine and its environment: the application of workflow defining, administration and monitoring tools, client applications, other workflow engines, and other applications. The dependencies between the workflow enactment engine, tools and applications are:

1. The workflow definition interface defines how the workflow modeling tools supplies the workflow engine with process definitions. The most commonly used language is XPDL, BPEL or UML activity diagrams.
2. The client application interface is used to control the implementation and manipulation of the workflow elements in a uniform manner.
3. Interaction standards between the workflow engine and other applications determine the form of adapters for existing applications and provide guidelines for the development of new applications that support workflows.
4. Interoperability standards allow interactions between different organization's engines to perform a complex workflow process.
5. Administration and monitoring standards enable consistent administration of various workflow applications and enable tracking of status changes for workflow elements.

Numerous languages have been designed to implement workflow interfaces. The graphical BPMN (Business Process Model and Notation [6], current version: 2.0.2) language, supported by the OMG organization, is used to define workflows. Its biggest disadvantage, emphasized in many publications, is the lack of formal semantics [7, 8]. Many attempts were made to formalize BPMN, using, for example, Petri nets [9], UML activity diagrams [9], or Event-Driven Process Chains (EPC – [10]).

Separate languages are designed to execute workflows, the most commonly cited in the literature is BPEL [11]. Another approach is to provide specialized libraries for use in programs that deal with defining or executing a workflow. The most familiar is the Windows Workflow Foundation for .NET [12].

The WfMS reference architecture shows the main WfMS interface (1) as uniform, but in fact, there are many aspects of workflow definition that should be passed to the enactment engine. Especially, in document-oriented workflows: data model, its mapping onto a data base, document layout, data constraints and several other aspects should be defined. The workflow itself is typically recorded in one of standard languages (BPMN, YAWL, etc.), and additional information is provided in an internal format of WfMS.

For specific aspects of workflows, Domain Specific Languages (DSL) can be used. Such languages can be defined in terms of the various needs they fulfill. DSL is a programming language or an executable declarative language in which – through appropriate notations and abstractions – the power of expression is focused on, and usually limited to, a specific problem domain. Most often, a declarative DSL is used due to its better application to specific domain models. This is very convenient, especially when the syntax is close to the way of thinking of the specialists in the field. Initial costs associated with the development of DSL tools are often reimbursed many times during WfMS development.

For example, WEE-DSL [13] is used as an intermediate language for workflows, after compiling to the Ruby language. Often, DSLs are designed for graphical interfaces that easily define workflows for specialists in various fields, which are not suitable for BPMN/YAWL languages [14]. In [15] domain languages were used to:

- workflow recording – based on XPDL,
- graphical specification – based on BPMN or Petri nets,
- invoking internet services – the XML schema defines "Web Service endpoint" [16], WSDL URL, site name, information for WS Security, and others [11],
- definition of dialogs – based on W3C Xforms, dynamic behavior in Petri nets,
- data presentation – based on XSL documents.

Our approach is not aimed at creating a DSL for workflows in a given usage domain, like [14]. This is closer to the approach presented in [15], where various DSLs are applied to some aspects of workflows. However, it is different in that it uses the uniform methodology for graphical defining aspects of the scenario in a consistent manner, using a specific schema for the given aspect, and creating XML files on the output. All aspects are immersed in the same domain ontology, which defines objects, their relations, and constraints.

2.2 Ontology

An ontology is a conceptual model (usually represented graphically), constructed by specialists in a given field (domain). It is presented preferably in the form of a graph (or a stratified collection of graphs). Ontology is used in information systems to facilitate work related to the design and implementation, especially where the field of application has a specific and well-defined range of terms used. Recently, ontology is also increasingly used to "generate" (to a significant extent) the functionality of the system. In the last years, ontologies have also become an important element of the development of WfM systems, correlating with the general appearance of ontology applications in information systems [17]. As in our approach, we assign an important role to ontology in the process modeling, in this subsection we review applications of ontologies in WfMS.

In the literature, there are many issues related to ontologies in the context of WfMS. Usually, the use of ontologies aims at improving the application design processes using WfMS systems and improving the quality of these applications. This is evident especially at the design stage because inconsistencies in the model influence the workflow operation. In the design process, many misunderstandings or even errors result from the lack of accurate and systematic knowledge about the business environment, in which the workflows are to be implemented. Semantic ambiguity affects the efficiency of business process modeling and the quality of the obtained specifications. To solve this problem, many approaches based on a formalized ontology have been proposed. In particular, in [18] it is shown how the Resource-Event-Agent ontology (REA) conceptualizes common economic phenomena of a firm independently of application-specific demands. Slightly another approach is shown in [19], where ontologies are used for capturing knowledge about a software system at development time, which may include not only the system architecture, but also its functionality. In this case, ontology is an important part of a methodology known as Ontology-Driven Software Development (ODSD). A step farther is discussed in [20], where ontology is used for semi-automatic generation of various components containing the application description.

One of the main restrictions in using ontologies is the problem of building them. In the area of software engineering, one of the ways is reusing various database schemas or process models. Examples of such an approach are discussed in [18, 19], and [20]. As a matter of fact, in [18] it is shown an important application of DSL for constructing the REA ontology. Similarly, DSL defines an ontology that allows the application structure to be mapped onto elements of the presentation layer [19]. The generation of DSL languages based on OWL ontology saved in Protégé is described in [20]. In addition, in [20] the method of developing textual DSLs, not based on XML, has been presented. In this approach, modularity, associations, and inheritance were applied, which simplifies the development of languages based on existing ones. Also, the concept supports editors generation and template-based code generation.

3 Proposed Solution in Syndatis BPM4

Syndatis BPM4 is a document-oriented workflow management system. Every workflow process is connected with a set of documents. A document consists of a set of fields and some dependencies between them. A stable state of a workflow is defined as a current processing state of the main document defining the workflow (called *process document*). There are also unstable states, which consist in waiting for finalizing some activities, for example, document printing or export operation. The dynamics of workflows is connected with actions on documents, performed by users or initiated by external systems. Process actions change the process state and update the underlying document.

A process model of a workflow is described in the BPMN 2.0 language, modified for Syndatis BPM4 (internally, the BPMN 2.0 process is converted to a jPDL process with additional, Syndatis-specific metadata). All other aspects of a workflow, like data model, document form layout, etc., are described in several Domain Specific Languages. Each DSL determines a format of XML files, with specific schema which defines elements that are specific for a given aspect of the workflow. Every DSL has a graphical editing tool. The main DSLs are presented below.

Data model DSL
It defines document fields, their types and default values. Fields are bound in grouping sections. Restrictions on field values and dependencies between fields are defined as constraints. In addition, the data model specifies the field behavior that depends on the

user's role, document status, and other field values. The constraints in the model define a relation $\lambda \subseteq F \times L \times F$, where F is a set of fields and L is the set of unique constraint labels.

Workflow process DSL

A Workflow process is a graph, which structure reflects changes in a document states. Document states are nodes in the graph, and edges reflect transitions between states. In stable states, documents await users' actions, metastable states are points of synchronization with other processes and waiting points at the end of certain operations in the system (e.g., generating a report).

- Loops in the graph are created only to allow correction of already performed actions. After cutting out corrections from the graph, it becomes acyclic.
- Documents can be grouped by using appropriate "clips". A clip is also a document in the system. Documents can be clipped and unclipped in actions.
- Transitions in a process graph are of type 1:1. Branches occur as decisions in graph nodes. Parallel branches (1:N edge) are rarely used.
- A process can be hierarchical - a state can develop into a subprocess.

Document form DSL

It is used for defining screen layout. In particular, it allows specifying the number of columns, grouping document fields, ordering elements in a group, locating them in a window, rendering and editing rules, etc. The fields are grouped into display sections that correspond to the logical sections of the document specified in the data model.

List DSL - The language is used to define an alternative view of a set of documents in the form of a table. Many actions in processes are defined in lists, for example, a grouping of documents into packages (dynamic sets of documents grouped in order to perform a joint action), accepting, passing to the next stage, etc.

Module DSL - This is the most general element of the system, a form including links to document forms and their lists.

Configuration DSL - A special DSL is dedicated for specifying various elements of a workflow using a *key-value* structure:

- mapping document fields onto columns of SQL tables (or tags of a NoSQL XML file),
- identifier templates, for example, document identifiers,
- resource control; a resource is a certain pool, from which actions in a process can take some parts, without exhausting it completely; the rules determine the control over resource consumption (budget exhaustion, materials usage, etc.).

In addition, there are interface DSLs for describing:

- *Import DSL* – data feed from other systems.
- *Export DSL* – preparation of data for other systems.
- *Reports DSL* – a graphic form of a printout of documents.
- *Acceptance rules DSL* – principles of acceptance of documents by particular roles.

As mentioned above, a special role in Syndatis BPM4 is assigned to ontology. Ontology specifies the concepts (objects) used in a given field of application, usually identified with terms adopted for the determination of concepts. The concepts are accompanied by their attributes and relations between objects. Usually, the ontology is represented in the form of a graph in which the nodes are concepts and their attributes, while the edges are the relationships between the concepts. In another dimension, the hierarchy between objects is constructed. Ontology fully and unambiguously describes the domain, if it is complete, accurate and does not contain redundancy or contradictions. Problems with redundancy and contradictions can be solved in an automated way (see e.g. [21]).

The Syndatis BPM4 system consistently applies the approach of separating individual aspects of defining workflows (*separation of concerns*). However, if the separation is complete, it can lead to inconsistencies in the system. For example, non-compliance of data model elements with process definitions or displayed forms can be achieved. The remedy for this is to incorporate a complete ontology, covering all the domain/application-oriented concepts. With such an ontology one should force using the objects and rules in all aspects of the workflow definition. Consistent usage of ontology, in which the system objects are described coherently, may eliminate contradictions between the specifications, and remove redundancies or ambiguities. The attributes of individual objects are also described in detail, and the relationships between objects are prototypes of relationships in the data (static) or actions in workflows (dynamic). The role of the ontology in the system is to bind the workflow specifications, which speeds up the course of analysts, understanding the workflows by the end-user and shortens the implementation process, as well as facilitates the maintenance.

In Syndatis BPM4, ontology, on the one hand, provides dictionaries with terms that can be used in workflows, and on the other hand, provides the basis for checking static and dynamic consistency of workflows.

The domain languages (DSLs) in the Syndatis BPM4 system, used to design particular aspects of workflows (such as data model, forms, reports, etc.) are essentially XML schemas (Fig. 1). This approach has many advantages, the most important are:

- Many satellite programs for WfMS use XML as input and output: other workflow engines, accounting systems, reporting systems, ERP and CRM systems, etc.
- XML as a text format is legible for designers and allows manual editing (location of errors, analysis of the behavior reported by a user, etc.).
- If the software ignores unknown tags (defining specific sections in the file), XML provides both upward compatibility, as well as downward compatibility, i.e. a new file will be accepted by previous versions of the software.

In order to take full advantage of the approach used, the DSL set used in Syndatis BPM4 was evaluated in accordance with the MDD methodology (Model-Driven Development). The principles of such evaluation were presented in [22]. The multi-criteria evaluation principles relate to linguistic criteria, human factors, software and application engineering. These criteria include, for example: accuracy, flexibility, understandability, level of detail and needed training, simplicity, uniqueness/ orthogonality (lack of redundancy), consistency, space economy.

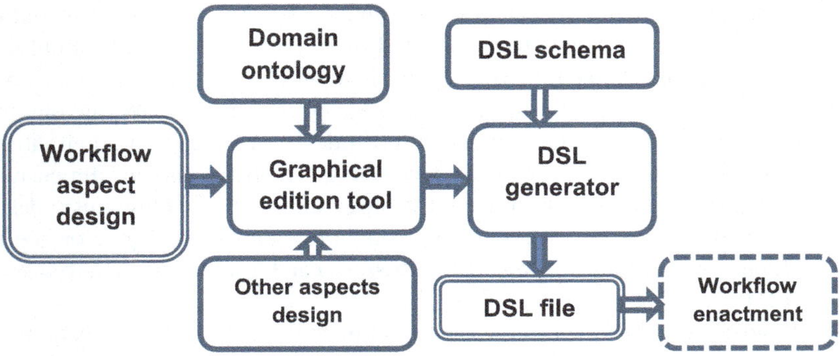

Fig. 1. Generation of DSL files from workflow aspect definition

A well-constructed ontology can be used to automate workflow design. This moves actions related to workflow design towards declarative programming [21]. In Syndatis BPM4, formal ontology supports the process of defining individual aspects of workflows. For example, if there are concepts in the application-oriented ontology whose attribute refers to *person*, then appropriate swimlanes may be created for them in the workflow. The concepts with the "document" attribute can take the form of elements of the data model, and the "document state" attribute can be placed on the swimlanes of those people with whom they are associated in the ontology. If, for example, we have relations such as *consists of*, *accepts* or *sends* between concepts, then these relations translate into actions in the workflow processes. Static relations between concepts can translate into constraints. In this way, the workflow framework can be created, after which more details are filled in by the designer. The construction of ontologies may be oriented to this type of subsequent declarative programming.

System development based on successive approximations of its formal semantic model is called MDD (Model-Driven Development), and the architecture of such a system is known as MDA (Model-Driven Architecture) [10]. Model-oriented systems design methods are called MDE (Model-Driven Engineering). The methodology included in the T□ system (T-Square, [23]) can be used as an example. It allows defining task steps, branching conditions, user interface referring to ontology. The used transformation methods generate the executable software from the abstract process specification [24].

4 Example DSL Specification

The following is a fragment of the schema for data model DSL. It is pretty large and therefore a fragment connected with the example is shown. The tag `<complexType>` for `dataModelType` is a reference for `fields` that have their `fieldType`: they are `fields` with `constraints`, attributed by `name`, optional `parentModel` (to define data model hierarchically) and `version`. Below, the second tag `<complexType>`

for `fieldType` binds it to an object in the ontology (and via ontology to an accompanying java class).

```
<?xml version="1.0" encoding="UTF-8"?>
<schema
  xmlns="http://www.w3.org/2001/XMLSchema"
  targetNamespace="http://www.syndatis.com/documentmodel/DataModel"
  elementFormDefault="qualified"
  xmlns:sdm="http://www.syndatis.com/documentmodel/DataModel">
   <element name="dataModel" type="sdm:dataModelType"></element>
   <complexType name="dataModelType">
    <sequence>
     <element name="fields" type="sdm:fieldsType"
      minOccurs="0" maxOccurs="1">
     </element>
     <element name="constraints" type="sdm:constraintsType"
      minOccurs="0" maxOccurs="1">
     </element>
    </sequence>
    <attribute name="name" type="string"></attribute>
    <attribute name="parentModel" type="string"></attribute>
    <attribute name="version" type="int"></attribute>
   </complexType>
    <!- constraintsType omitted for clarity ->
    <complexType name="fieldType">
     <sequence>
      <element>
       name="metadata" type="sdm:metadataType"
       minOccurs="0" maxOccurs="1">
      </element>
     </sequence>
     <attribute name="name" type="string" use="required"></attribute>
     <attribute name="ontologyType" type="string" use="required">
      </attribute>
     <attribute name="required" type="boolean" default="false">
       </attribute>
    </complexType>
</schema>
```

Using this DSL, a graphical tool can be used to create a data model object based on the presented schema (see Fig. 2 for an example from the prototype of the data model editor). In the picture, two sections of a document are defined (accompanied by "+" buttons): *Header* and *Order*. The *Header* section contains text *Supplier* field while the *Order* contains the Boolean *Sent* field and date-typed *OrderDate* field. For every field, two buttons are shortcuts to edition "E" and deletion "X". The last field *OrderDate* is now opened for edition, in which its name, type and its *obligatory use* status are visible.

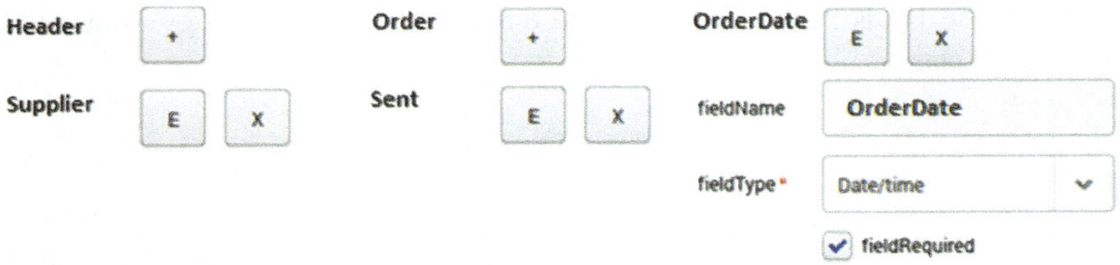

Fig. 2. Editing a sample data model object

Then, a resulting XML file is generated: the definitions shown in Fig. 2 are reflected directly in data model DSL file content.

```
<?xml version="1.0" encoding="UTF-8"?>
<sdm:dataModel
xmlns:sdm=http://www.syndatis.com/documentmodel/DataModel
name="test_data_model"
xmlns:util=http://www.springframework.org/schema/util
xmlns:xsi=http://www.w3.org/2001/XMLSchema-instance
xsi:schemaLocation="http://www.springframework.org/schema/beans
http://www.springframework.org/schema/beans/spring-beans-3.1.xsd">
<sdm:fields>
<sdm:section name="Header">
<sdm:field ontologyType="DescriptiveText" name="Supplier"/>
</sdm:section>
<sdm:section name="Order">
<sdm:field ontologyType="TrueFalseValue" name="Sent"
required="true"/>
<sdm:field ontologyType="DateTimeWithTimezone" name="OrderDate"
required="true"/>
</sdm:section>
</sdm:fields>
</sdm:dataModel>
```

5 Conclusions

Workflow systems have been widely used in information technology solutions which require the automatization of certain repeated, structured business processes. However, the specification of those processes has been usually left to IT specialists and various aspects of the business logic have been coded in imperative programming languages, which limits both the ability to formally verify and validate such processes and the ability of the business user to understand (and verify from a business standpoint) the detailed workings of the workflow.

The new approach in the Syndatis BPM4 system is aimed towards backing the workflow with an underlying formal ontology, which provides a full specification of a large part of the entire workflow ecosystem (as detailed in Sect. 2.1) in a declarative, formally verifiable manner. This approach has three benefits – first of all, it allows for the rapid prototyping and easy deployment of document-centric business processes by relatively non-technical users (certainly without the requirement of proficiency in imperative programming languages). Secondly, it minimizes potential development and implementation turnaround by providing analysts and, to some extent, the end-user with a declarative specification based on the ontology of the target domain, instead of an abstract, technical specification that cannot be understood by non-technical users. Third, it automatizes the design by generating DSL files from ontology, graphical specification of workflow aspects and associated schemas.

Acknowledgment. The research presented in this paper is co-financed by the European Regional Development Fund under the Regional Operational Program of the Lubelskie Voivodeship for 2014-2020 (RPLU.01.02.00-IP.01-06-001/15). Project No. RPLU.01.02.00-06-0048/16

References

1. Nielsen, S.P., Easthope, C., Gosselink, P., Gutsze, K., Roele, J.: Using Domino Workflow, IBM, Poughkeepsie, NY (2000). http://www.redbooks.ibm.com/redbooks/pdfs/sg245963.pdf
2. SAP Business Workflow. https://archive.sap.com/documents/docs/DOC-31056
3. KissFlow. https://kissflow.com/
4. IOnet workflow manager. https://www.ionetsoftware.com/workflow
5. Hollingsworth, D.: The Workflow Reference Model (1995). http://www.wfmc.org/docs/tc003v11.pdf
6. BPMN. http://www.omg.org/spec/BPMN/
7. Poizat, P., Salaün, G., Krishna, A.: Checking business process evolution. In: Kouchnarenko, O., Khosravi, R. (eds.) FACS 2016: Formal Aspects of Component Software, Besançon, France, 19–21 October 2016. LNCS, vol. 10231, pp. 36–53. Springer, Cham (2017). https://doi.org/10.1007/978-3-319-57666-4_4
8. Corradini, F., Muzi, C., Re, B., Rossi, L., Tiezzi, F.: Global vs. local semantics of BPMN 2.0 OR-join. In: Tjoa, A., Bellatreche, L., Biffl, S., van Leeuwen, J., Wiedermann, J. (eds.) 44th International Conference on Current Trends in Theory and Practice of Computer Science, Krems, Austria, 29 January–2 February 2018. LNCS, vol. 10706, pp. 321–336. Edizioni della Normale, Cham (2018). https://doi.org/10.1007/978-3-319-73117-9_23
9. Cai, Y.: Comparative analysis of the workflow modeling. In: 2012 International Conference on Management of e-Commerce and e-Government, Beijing, China, 20–21 October 2012, pp. 226–229. IEEE (2012) https://doi.org/10.1109/icmecg.2012.79
10. Lubke, D., Luecke, T., Schneider, K., Gomez, J.M.: Using event-driven process chains for model-driven development of business applications. Int. J. Bus. Process Integr. Manag. **3**(2), 265–279 (2008). https://doi.org/10.1504/IJBPIM.2008.020974
11. Weerawarana, S., Curbera, F., Leymann, F., Storey, T., Ferguson, D.F.: Web Services Platform Architecture. Prentice Hall, Upper Saddle River (2005). ISBN 978-0-13-148874-8

12. Windows Workflow Foundation. https://docs.microsoft.com/pl-pl/dotnet/framework/windows-workflow-foundation/
13. Sturmer, G., Mangler, J., Schikuta, E.: A domain specific language and workflow execution engine to enable dynamic workflows. In: 2009 IEEE International Symposium on Parallel and Distributed Processing with Applications, Chengdu, China, 10–12 August 2009, pp. 653–658. IEEE (2009). https://doi.org/10.1109/ispa.2009.106
14. Barzdins, J., Cerans, K., Grasmanis, M., Kalnins, A., Kozlovics, S., Lace, L., Liepins, R., Rencis, E., Sprogis, A., Zarins, A.: Domain specific languages for business process management: a case study. In: 9th OOPSLA Workshop on Domain-Specific Modeling (DSM 2009), Orlando, FL, 25–26 October 2009, pp. 34–40 (2009). http://www.dsmforum.org/events/dsm09/papers/barzdins.pdf
15. Freudenstein, P., Buck, J., Nussbaumer, M., Gaedke, M.: Model-driven construction of workflow-based web applications with domain-specific languages. In: 3rd International Workshop on Model-Driven Web Engineering (MDWE 2007), Como, Italy, 17 July 2007, pp. 1–15 (2007). http://ceur-ws.org/Vol-261/paper02.pdf
16. WS endpoint. https://www.techwalla.com/articles/the-definition-of-web-service-endpoint
17. Fanesi, D., Cacciagrano, D.R., Hinkelmann, K.: Semantic business process representation to enhance the degree of BPM mechanization - an ontology. In: 7th International Conference on Enterprise Systems (ES), Kanazawa, Japan, 17–18 November 2015, pp. 21–32. IEEE (2015). https://doi.org/10.1109/es.2015.10
18. Sonnenberg, C., Huemer, C., Hofreiter, B., Mayrhofer, D., Braccini, A.: The REA-DSL: a domain specific modeling language for business models. In: Mouratidis, H., Rolland, C. (eds.) International Conference on Advanced Information Systems Engineering. CAiSE 2011, London, UK, 20–24 June 2011. LNCS, vol. 6741, pp. 252–266. Springer, Heidelberg (2011). https://doi.org/10.1007/978-3-642-21640-4_20
19. Bräuer, M., Lochmann, H.: An ontology for software models and its practical implications for semantic web reasoning. In: Bechhofer, S., Hauswirth, M., Hoffmann, J., Koubarakis, M. (eds.) The Semantic Web: Research and Applications. ESWC 2008, Tenerife, Canary Islands, Spain, 1–5 June 2008. LNCS, vol. 5021, pp. 34–48. Springer, Heidelberg (2008). https://doi.org/10.1007/978-3-540-68234-9_6
20. Ojamaa, A., Haav, H.-M., Penjam, J.: Semi-automated generation of DSL meta models from formal domain ontologies. In: Model and Data Engineering. LNCS, vol. 9344, pp. 3–15. Springer, Cham (2015). https://doi.org/10.1007/978-3-319-23781-7_1
21. Di Ciccio, C., Maggi, F.M., Montali, M., Mendling, J.: Resolving inconsistencies and redundancies in declarative process models. Inf. Syst. **64**, 425–446 (2017). https://doi.org/10.1016/j.is.2016.09.005
22. Mohagheghi, P., Haugen, Ø.: Evaluating domain-specific modelling solutions. In: Trujillo, J., et al. (eds.) Conceptual Modeling – Applications and Challenges. ER 2010. LNCS, vol. 6413, pp. 212–221. Springer, Heidelberg (2010). https://doi.org/10.1007/978-3-642-16385-2_27
23. Rabbi, F., MacCaull, W.: $T_{\square}$: A domain specific language for rapid workflow development. In: France, R.B., et al. (eds.) Model Driven Engineering Languages and Systems. MODELS 2012, Innsbruck, Austria, 30 September–5 October 2012. LNCS, vol. 7590, pp. 36–52. Springer, Heidelberg (2012). https://doi.org/10.1007/978-3-642-33666-9_4
24. Rabbi, F., MacCaull, W.: Model driven workflow development with $T_{\square}$. In: Bajec, M., Eder, J. (eds.) International Conference on Advanced Information Systems Engineering CAiSE 2012, Gdańsk, Poland, 25–26 June 2012. LNBIP, vol. 112, pp. 265–279. Springer, Heidelberg (2012). https://doi.org/10.1007/978-3-642-31069-0_23

GPU Implementation of the Parallel Ising Model Algorithm Using Object-Oriented Programming

Aleksander Dawid(✉)

Department of Transport and Computer Science, WSB University, 1c Cieplaka St., 41-300 Dąbrowa Górnicza, Poland
adawid@wsb.edu.pl

Abstract. The GPU acceleration of the so-called Ising's spins model was investigated in this work. We have implemented the checkerboard algorithm using object-oriented programming techniques. The CUDA enabled graphic card provided by NVIDIA Corporation was used to run MC simulation. The testing code was written in C++ programming language both for GPU and CPU. We have found that the independent multi-threaded calculations provide a huge increase in GPU acceleration. Testing calculations proved that the GPU speed up in comparison to CPU is up to 100 times, on a typical laptop setup. The calculations have been made for six different sizes of thread blocks. The optimal thread block size for Ising model calculations was estimated in this research. We were able to operate only on fixed-point numbers in this solution. The code of this implementation is publicly available for further software development in many fields of science.

Keywords: Ising model · Magnetization · GPU acceleration · OOP method · Checkerboard algorithm · MC simulation

1 Introduction

In the '20s of the 20th century, an unrecognized problem, in theory, was the magnetization of materials. Wilhelm Lenz undertook to solve this problem and develop a mathematical model of magnetic phase transitions. This model is using the fact that magnetization could only take +1 or −1 discrete values. The objects receiving these states were called magnetic spins. This model, also known as Ising model, assumes that an interaction comes only from the nearest neighbors of the selected magnetic spin. The set of spins placed in the nodes of the crystal lattice represents a state of magnetization in the system. The evolution of this state depends on the form of the energy function. The Ising model is also known as a probabilistic graph model. It consists of vertices and edges. Vertices, in this case, can take only two states for simplicity, it is black and white. The edge distribution determines the closest vicinity of a given spin. Generally, it is a graph without cycles. The external parameters of energy function defining the interaction in the Graph. In the case of magnetization, as an external parameters we can consider the temperature and external magnetic field. Applications of the Ising model

W. Zamojski et al. (Eds.): DepCoS-RELCOMEX 2020, AISC 1173, pp. 155–164, 2020.
https://doi.org/10.1007/978-3-030-48256-5_16

depends on the appropriate Graph selection of energy function of external parameters. Ising model has been solved analytically for one-dimensional and two-dimensional cases [1]. For three dimensions, it remains unsolvable exactly, only approximate methods such as Monte Carlo (MC) simulations [2] allow its solution. These simulations for complex systems require quite a lot of computing power. In recent years, the computer games market has been growing considerably intensively. The graphics processing units (GPU) are the most significant part of this market. Their primary purpose is to process and display graphics. They are present in every desktop and most mobile devices. Nowadays, the GPUs are also able to perform general computational tasks. The computing power of a single GPU is many times greater than that of a single central processing unit (CPU). This computational power comes from the fact that GPUs consist of thousands of computing cores, but CPUs usually a dozen. GPUs offer faster floating-point computing speed at a comparable price per unit. The world's fastest supercomputers include GPUs. Due to the strongly multi-core nature of current GPUs, their programming model differs significantly from CPU programming. In 2007 the NVIDIA Corporation introduced the compute unified device architecture (CUDA) for GPU programming [3]. Its basis was C/C++ programming language, but now there are also libraries for programming languages such as Python and Java. This library can be used only on devices equipped with CUDA cores. There are many publications on speeding up calculations in chemistry [4], physics [5, 6], train rescheduling [7], and cognitive computing [8]. GPU acceleration of MC simulations is particularly interesting due to the huge speedup in comparison to CPU calculations. These simulations are applied to many different scientific disciplines. One of them is medicine. The most common use is to estimate the dose in radiotherapy [9, 10]. The GPU acceleration of the 2D and 3D Ising model is also investigated in the past. In the work of Preis et al., we can find GPU based algorithm of solving 2D and 3D Ising model [11]. This solution uses C code and floating-point numbers. The current research in accelerating Ising magnetic spin model calculations concentrates around tensor processing units (TPU) used in artificial intelligent research [12].

In this report, we want to present some practical methods to accelerate the well-known checkerboard algorithm [11, 12] on GPUs. Our main goal is to incorporate the CUDA procedures into an object-oriented programming model based on the C++ language. We want to show that the performance of GPU procedures embedded in a single object is equally effective as that without object definition. Future applications of our library can be used in software development of industry and public transport.

2 The Model

The fundamental element of the Ising model is the energy function. In physics, the energy function is often called the Hamilton function or simply Hamiltonian. In the Ising model, Hamiltonian takes into account interactions between spins located in the nearest nodes in the presence of an external magnetic field, which can be written by the following equation.

$$H = -\sum_{\langle \mathrm{i,j} \rangle}^{N} J_{\mathrm{ij}} S_i S_j - B \sum_{\mathrm{i=1}}^{N} S_i \quad (1)$$

In this equation, S stands for a spin in a single node, and J stands for the magnetic force. The force depends on the dimensions of the system and the network topology. The brackets *<i, j>* mean adding up after the nearest neighbors. In the second term of Eq. 1, B denotes the external magnetic field and, the summation takes place over all nodes. In the case where the number of spins +1 equals the number of spins −1, this member disappears. The same applies if the external field is zero. The algorithm (in steps) for optimizing the system without an external magnetic field can be written as

1. Select the *ith* spin
2. Changing its state $s_i^{(n+1)} = -s_i^{(n)}$
3. Calculate the energy difference $\Delta H = H^{(n+1)} - H^{(n)}$
4. If $\Delta H > 0$ we accept the spin change with appropriate probability by comparing a random variable ξ from the range between 0 and 1 with the Boltzmann factor value $C_B = \exp(-\beta \Delta H)$, where $\beta = k_B/T$. If $\xi > C_B$ than we reject the spin change.

This algorithm is originally known as the Metropolis-Hastings algorithm. The most difficult part of this algorithm can be calculation of the difference $H^{(n+1)} - H^{(n)}$. It is worthy to note that the change of magnetic spin influence only its closest neighbors. Farthest neighbors are no longer affected. Due to the alternation of multiplication $s_i s_j - s_j s_i = 0$, we can treat these contributions as $2 s_i s_j$, so the total energy per node is expressed by the equation

$$H = -2 s_i \sum_{j=1}^{6} s_j \quad (2)$$

To simplify the calculations, we assume that J = 1.

3 Parallel Algorithm

The Metropolis algorithm assumes that the state sequences are dependent only on the state immediately preceding the state. The ideal solution, in the case of GPU calculation, is when each state is assigned to one computing core and also there is no communication between them. Such a situation will be achievable if the set of variables is independent. Such collections represent chaos and are not suitable to describe natural processes. The crucial element is the relation in the data set. On the other side, we may have a system of highly correlated data. In this case, the MC method does not allow the evolution of 2 states simultaneously because each change of a single element changes the state of the entire system. The time evolution of the system must, therefore, carrying out in series. The only thing you can write in parallel is the procedure for determining the value characterizing the state, for example, potential energy. The energy algorithm is usually of the type $\Theta(n^2)$. The more elements in the set, the longer it takes to calculate the energy function. The compromise between these two opposite cases is to limit the interaction to a group of the nearest neighbors of a randomly selected element.

In the case of calculations related to magnetization, there are usually three types of the crystal lattice of magnetic materials (Fig. 1). The algorithm of concurrent calculations for the three-dimensional Ising model is using the simple cubic lattice, shown in Fig. 1a. The distribution of this problem into similar sub-problems consists in the division of spins into two types of colors in a three-dimensional graph. The intersections of two different checkerboards taken from the 3D graph are shown in Fig. 2. This algorithm is a variant of the checkerboard algorithm widely used in this kind of calculation. The black spins interact only with the nearest neighbors marked as white. The colors of the spins change with successive layers alternately. If we now intend to change the value of any of the black spins, then the rest of the black spins will practically not feel this change, because the energy from the nearest neighbors is calculated only for white spins. So you can simultaneously select all the black spins to change if you have so many cores in the graphics processor.

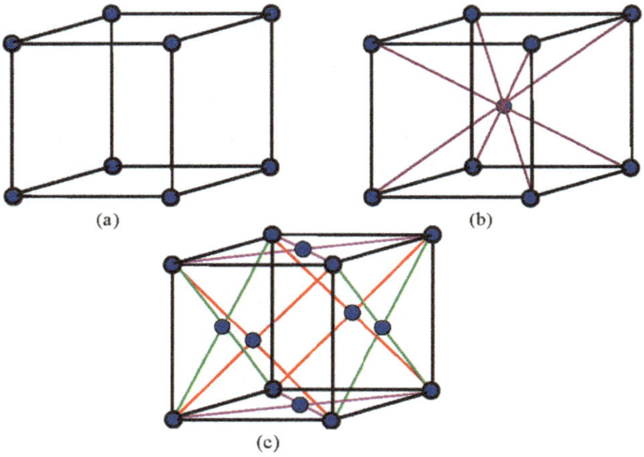

Fig. 1. Crystal lattices: (a) SC - simple cubic, (b) BCC - body centered cubic, (c) FCC - face centered cubic.

After completing the calculations for black spins, we proceed similarly for white spins. In the example presented here, we will only consider the case of minimizing exchange energy without an external magnetic field. For simplicity, we assume that $J_{ij} = 1$ in Eq. 1, so we have to calculate only the exchange energy H (Eq. 2). In this problem, we are dealing with an array of spin values. In our case, it is a three-dimensional array. We can name this array as `Spin[Nx][Ny][Nz]`, where `Nx`, `Ny`, `Nz` are the sizes in each of the dimensions. The memory access in modern computers depends on linear addresses, so we have to translate addresses from 3D array to 1D array. The 1D arrays we can then use to make calculations on GPU. Every good developer knows that we can implement a particular algorithm, in the same

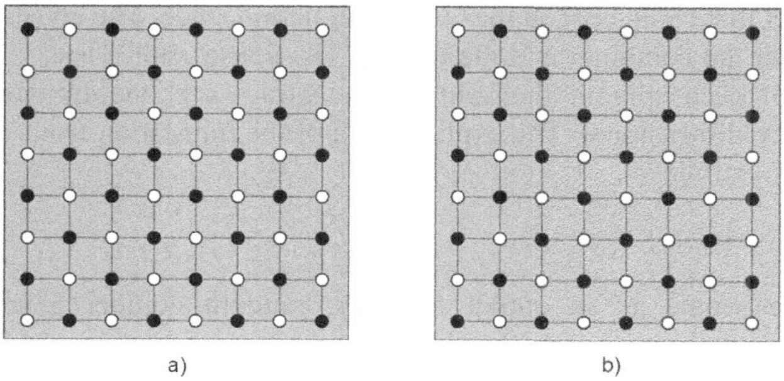

Fig. 2. Layer graph a) 2n i b) 2n + 1 where n = 0, 1, 2, ..., N−1.

programming language, in n-ways. This solution is just one of many possibilities to write such code in C++.

4 Ising Model Class

Currently, computer programs are created based on the object-oriented programming model. The C++ class directive defines an object. In the case of the Ising model of magnetic spins, this class should hold information about the state of spins. This class contains variables and methods that operate only on integers. From the solid-state physics point of view, this solution is not practical, because it is not able to characterize a particular material in terms of its magnetic properties. On the other hand, this computational core will appear in every implementation of the Ising model in material science. In this solution the class name is `IsingModel`. The public variable `N` of this class represents the total number of spins in the system. Private variables `X, Y,` and `Z` define indices in the three-dimensional spin array. The table `w [3]` of type `Vector` contains data describing the unit matrix. It is needed to translate the address from the table in the form of three variables `X, Y, Z,` into a linear address described by one variable `a`. The next variables are related to the preparation of the task for calculation on the graphics processor. The first of them, named `BlockPerGrid`, is responsible for setting up the number of blocks per grid needed to perform the task on the GPU. The second variable named `ThreadPerBlock` determines the number of threads within one block. The product of these two values is the number of threads designated for the GPU. Arrays that collect data about spins are described by succeeding pointers; `*Spin, *SpinNetwork, *SpinTest`. The `SpinTest` array is necessary to store the initial values of the system for their reuse without having to run the procedure assigning spin values multiple times. These tables are allocated based on the number of `N` spins in the `IsingModel` class constructor. The size `N` of the tables is estimated based on the size of the lattice in each directions `N = Nx * Ny * Nz`. The information about the size of the lattice matrix along x, y, and z-axes is important for GPU calculations. We can put it in the global GPU memory of type `__constant__`. Setting the values of these constants is done by the CUDA library function `cudaMemcpyToSymbol()`. These variables for the GPU have different names due to a naming conflict that could arise at the code compilation stage. The constructor allocates memory in both the computer and the device. The constructor's task is also to set the initial value of each spin in the spin tables. Arrays are one-dimensional, but the problem is three-dimensional. The explicit form of the translation from 3 indices into a single index is as follows.

```
nr = z * Nz * Ny + y * Nx + x;
```

Spin values come as an output from the random number generator function `rand() mod 2`. It gives 0 or 1. Recalculating this values by formula $2s - 1$, we can transform the interval [0, 1] into interval [−1, 1]. The spins are then saved to GPU memory using `WriteSpinToGPU()` method. The last constructor instruction creates unit matrix `w[3]`. Dynamically created tables in RAM and GPU memory will exist practically until the end of the program. However, if we want to change the size of the problem during program execution, we must think about the so-called -destructor. Its task is to free memory in the computer and the graphics card. For example, creating an object according to the `IsingModel` class definition for the system consisting of 16 spins in each direction needs the following code.

```
IsingModel MC(16, 16, 16);
```

Now the main task, that optimize such a system, will be using the Metropolis algorithm for the spin system. In the CUDA programming model, the processing of threads occurs in groups inside the blocks. The model assumes the smallest number of iterations in the thread. Imagine that each thread represents a single spin of a given color (black or white). We know that black spins are more than computational cores in the GPU. However, concurrent processing in CUDA architecture prefers a thread stream rather than an instruction stream within a loop. The size of the problem plays an important role in the preparation of concurrent calculations. If we know that every streaming multiprocessor (SM) in our graphics system has 128 cores, we can conclude that 128 threads in the block will be the most appropriate choice. We can set this value that comes from the `IsingModel` class as follows.

```
ThreadPerBlock=128;
```

Now we have to determine the number of blocks. In general, this task is quite complex, but using the simplification in which the number of black and white nodes is the same, we can determine the number of blocks using the following formula.

```
BlockPerGrid = (N / 2) / ThreadPerBlock;
```

Thread synchronization is another important component of concurrent computing. We know that until calculations are not finishing for black (even) nodes, you cannot perform calculations for white (odd) nodes. The CUDA library has an inside mechanism for synchronizing threads within a block. To introduce global synchronization, you need to use not one computing kernel, but two. Returning of control to the CPU synchronizes all the threads in the GPU before performing the next task. The code fragment that performs many calls to the kernel from the graphics system can be written as follows.

```
for(int t=0;t <tN;t++){
EvenSpin <<<BlockPerGrid, ThreadPerBlock>>> (Spin);
//Global synchronization
OddSpin << <BlockPerGrid, ThreadPerBlock >> >(Spin);
}
```

The `for` loop determines how many times the spins in even and odd nodes are to be changed. The particular node in the linear array represents i-index. The value of this index comes from the following dependency.

```
i = thread_in_block + block_size * block_number
```

The maximum number of blocks, in the case of Geforce960M, is equal to over two billion. The code for a specific i-index looks the same as for the serial algorithm. There are also in the GPU definitions of energy and spin changing functions. Declaration of `Vector w[3]` is inside the energy function. This solution turns out to be more efficient due to the faster memory on the level of thread. It is worthy to note here that the index is passed to this function, saving time for recalculation of the index from variables representing thread and block. The calculation kernel for the odd magnetic spins (marked with white nodes) differs practically only by one line responsible for the selection of a specific node.

```
idx = i2 + 1 - a - b;
```

The complete code for this implementation along with the project in the Visual Studio 2017 environment is available on the GitHub platform [13].

5 Results

In this study, we want to show how the performance of the concurrent algorithm scales with the increasing number of spins in the Ising model. In the following experiment, the spins are located in the simple cubic lattice. We have used the reference computer equipped with 8 GB of RAM, Intel i5–6300HQ quad-core processor, and NVIDIA GeForce 960M graphics card. The testing software was developed using Visual Studio 2017 IDE with nvcc compiler from the CUDA toolkit library (version 9.1). In all our MC simulations, the number of steps is constant and equal to 1000. The quantity that was measured here, is the time of application execution for a given problem complexity. In this case, the complexity of the problem is determined by the number of spins in the lattice.

Calculations were made for cubes with edge sizes from 8 to 184 spins, which gives the number of spins between $8^3 = 512$ and $184^3 = 6229504$. We know that the number of threads processed by one computational kernel is half of these limits, i.e., from 256 to 3114752. If we now convert it into blocks of 128 threads, we will receive from 2 to 24335 blocks. Application performance is presented in relative values as the quotient of code execution time on the CPU to code execution time on the GPU.

Table 1. The GPU acceleration against CPU at five different block sizes.

#spins	Speedup by thread block size					
	32	64	128	256	512	1024
512	2.3606	2.2379	1.6473	2.4272	2.3134	1.4504
4096	17.3374	17.5119	17.4182	16.1745	18.2469	14.5452
13824	32.8928	42.9335	38.4795	38.4831	34.0413	34.1045
32768	54.3132	58.4307	59.4651	58.4579	57.0402	51.7657
262144	83.9924	88.1435	89.4465	89.5722	88.5428	84.0677
4096000	79.8162	99.1884	99.2324	98.7915	96.3785	86.5082

Initially, for 512 spins, GPU performance is only, on average, 2.073 times faster than one CPU core (Table 1). So, if we use 4 CPU cores, the CPU calculation time will be lesser than that for the GPU. As the complexity of the problem increases, the graphics card performance becomes more significant. Starting from 4096 spins we can observe the better performance of GPU than CPU. We have found that speedup also depends on the tread block size (Table 1). At small numbers of spins, the worse result is registered for calculations where the block size was set to 1024. For 262144 spins, we can observe the saturation process. This value corresponds to 2048 blocks of 128 threads. Starting from this point, the increasing number of magnetic spins does not improve drastically the speedup of GPU calculations. The second goal of these tests was to check how the efficiency of calculations on the GPU changes when we will changing the block size. Increasing the number of threads in the block twice (up to 256) causes, in general, a decrease of GPU performance in Ising spin model calculations. The drop in performance is small, and it is equal to 0.03%. (Table 1). If we reduce the number of threads in the block by half to 64, the GPU performance for a large number of magnetic spins stays at almost the same level. The only difference is visible for the number of magnetic spins equal to 512 and 13824. Here the performance is about higher than the performance for thread block size equal to 128. The results for 512, 1024, and 32 threads in block shows the drop in the performance of GPU. In the case of 4096000 magnetic spins, the slower GPU computations have been observed for the block size equal to 32 threads (Fig. 3). These testing calculations mostly use fixed-point calculations on GPU and CPU. The floating-point calculations may probably give different accelerations.

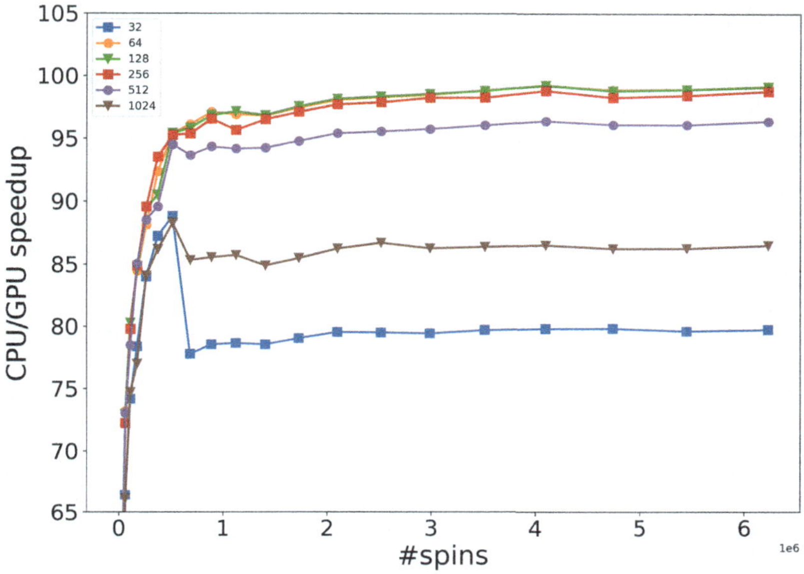

Fig. 3. Dependence of GPU execution speed increase concerning one CPU core on the number of spins in a simple cubic lattice.

The reference systems taken for calculations in this example do not represent the latest CPU and GPU solutions. Calculations on more comprehensive GPU or CPU family could show some development tendency for both the CPU and GPU. We have not observed any drop in performance caused by the OOP method applied in the developing process.

6 Conclusions and Future Work

In summary, this work confirmed that the Ising model of magnetic spins is suitable for calculations on GPUs. The GPU speedup depends on the thread block size. The best choice for block size is 64 and 128. Encapsulating data and functions in one class for processing on the GPU will allow you to quickly run calculations with the possibility of using this class as a library. The future extension of the class functionality by adding any crystal lattice will improve the overall applications.

Acknowledgment. The author wants to acknowledge the professor's Walica funds for supporting this work.

References

1. Onsager, L.: Crystal statistics. I. A two-dimensional model with an order-disorder transition. Phys. Rev. **65**, 117–149 (1944). https://doi.org/10.1103/PhysRev.65.117
2. Metropolis, N., Ulam, S.: The Monte Carlo method. J. Am. Stat. Assoc. **44**, 335–341 (1949). https://doi.org/10.1080/01621459.1949.10483310
3. Wilt, N.: The CUDA Handbook: A Comprehensive Guide to GPU Programming. Addison-Wesley Professional, Upper Saddle River (2013)
4. Klingbeil, G., Erban, R., Giles, M., Maini, P.K.: Fat versus thin threading approach on GPUs: application to stochastic simulation of chemical reactions. IEEE Trans. Parallel Distrib. Syst. **23**, 280–287 (2012). https://doi.org/10.1109/TPDS.2011.157
5. Dawid, A.: GPU-based parallel algorithm of interaction induced light scattering simulations in fluids. Task Q. **23**, 5–17 (2019). https://doi.org/10.17466/tq2019/23.1/a
6. Spiechowicz, J., Kostur, M., Machura, L.: GPU accelerated Monte Carlo simulation of Brownian motors dynamics with CUDA. Comput. Phys. Commun. **191**, 140–149 (2015). https://doi.org/10.1016/j.cpc.2015.01.021
7. Josyula, S.P., Krasemann, J.T., Lundberg, L.: Exploring the potential of GPU computing in train rescheduling. In: 8th International Conference on Railway Operations Modelling and Analysis (ICROMA), RailNorrköping 2019, Norrköping, Sweden, 17th – 20th June, pp. 471–490 (2019)
8. Dawid, A.: PSR-based research of feature extraction from one-second EEG signals: a neural network study. SN Appl. Sci. **1**, 1536 (2019). https://doi.org/10.1007/s42452-019-1579-9
9. Beltran, C., Tseung, H.W.C., Augustine, K.E., Bues, M., Mundy, D.W., Walsh, T.J., Herman, M.G., Laack, N.N.: Clinical implementation of a proton dose verification system utilizing a GPU accelerated Monte Carlo engine. Int. J. Part. Ther. **3**, 312–319 (2016). https://doi.org/10.14338/IJPT-16-00011.1
10. Wang, Y., Mazur, T.R., Green, O., Hu, Y., Li, H., Rodriguez, V., Wooten, H.O., Yang, D., Zhao, T., Mutic, S., Li, H.H.: A GPU-accelerated Monte Carlo dose calculation platform and its application toward validating an MRI-guided radiation therapy beam model. Med. Phys. **43**, 4040–4052 (2016). https://doi.org/10.1118/1.4953198
11. Preis, T., Virnau, P., Paul, W., Schneider, J.J.: GPU accelerated Monte Carlo simulation of the 2D and 3D Ising model. J. Comput. Phys. **228**, 4468–4477 (2009). https://doi.org/10.1016/j.jcp.2009.03.018
12. Yang, K., Chen, Y.-F., Roumpos, G., Colby, C., Anderson, J.: High performance Monte Carlo simulation of Ising model on TPU clusters. arXiv:1903.11714 Phys (2019)
13. Dawid, A.: alex386/IsingModelCpp (2020)

Hydro-Meteorological Change Process Impact on Oil Spill Domain Movement at Sea

Ewa Dąbrowska(✉) and Krzysztof Kołowrocki

Gdynia Maritime University, Morska 81-87, 81-225 Gdynia, Poland
{e.dabrowska,k.kolowrocki}@wn.umg.edu.pl

Abstract. A procedure of oil spill domain movement impacted by changing hydro-meteorological conditions at the sea water area determination, based on a probabilistic approach, is proposed. A stochastic model of the process of changing hydro-meteorological conditions is constructed and identified for Baltic Sea open water area. Prediction procedure of oil spill domain movement at varying in time hydro-meteorological conditions is created and applied to Baltic Sea open water area.

Keywords: Oil spill domain movement · Hydro-meteorological change process · Impact · Modelling · Identification · Prediction

1 Introduction

A very important duty in port activities and shipping is the prevention of oil releases from port installations and ships and the spread of oil spills that often have dangerous consequences for port and sea water areas. Thus, there is a need for methods of oil spill domain movement modelling based on determination of the oil spill central point drift and the oil spill domain probable placement at any moment after the accident. These could be useful tools for increasing the shipping safety and effective port and sea environment protection. Even if, the real oil spill domain movements are slightly different from those determined by the proposed methods, they can be useful in the port and sea environment protection planning and rescue action organizing. This way, the area determined for oil spill can allow us to mark the domain where the actions of mitigating the oil release consequences should be performed. This approach is proposed to make oil releases at the sea prevention and mitigation actions more effective.

The proposed procedure of the oil spill domain determination based on the probabilistic approach may be practically applied in the oil spill modelling, prediction and its consequences mitigation through the search and rescue actions at the sea after the unknown parameters of the used models statistical identification. Research experiments should be organized and performed in order to receive statistical data needed for the statistical methods of the model unknown parameters estimation. Thus, the methods of statistical data collection and evaluation of unknown parameters of the oil spill domain movement should be proposed.

The improvement of the methods of the oil spill domains determination is the main real possibility of the identifying the pollution size and the reduction of time of its

W. Zamojski et al. (Eds.): DepCoS-RELCOMEX 2020, AISC 1173, pp. 165–175, 2020.
https://doi.org/10.1007/978-3-030-48256-5_17

consequences elimination. Therefore, it seems to be necessary to start with the new and effective methods of the oil spill domains at port and sea water areas determination at constant and changing hydro-meteorological conditions. The most important criterion of new methods should be the time of the oil spill consequences minimising. One of the essential factors that could ensure these criteria fulfilment is the accuracy of methods of the oil spill domain determination. Those methods should be the basic parts of the general problem of different kinds of pollution identification, their consequences reduction and elimination at the port and sea water areas to elaborate a complete information system assisting people and objects in the protection against the hazardous contamination of the environment. One of the new efficient methods of more precise determination of the oil spill domains determination and its movement could be a probabilistic approach to this problem presented in this paper based on the models given in [1–3] and preliminarily discussed in [4–7].

2 Process of Changing Hydro-Meteorological Conditions

2.1 Process of Changing Hydro-Meteorological Conditions Modelling

We denote by $A(t)$ the process of changing hydro-meteorological conditions at the sea water area where the oil spill happened and distinguish m its states from the set $A = \{1, 2, \ldots, m\}$ in which it may stay at the moment t, $t \in < 0,T >$, where T, $T > 0$, is the time we are interested in this process analysis. Further, we assume a semi-Markov model [1–3] of the process $A(t)$ and denote by θ_{ij} its conditional sojourn time at the state i while its next transition will be done to the state j, where $i, j \in \{1,2, \ldots, m\}$, $i \neq j$ [4–6]. Under these assumptions, the process of changing hydro-meteorological conditions $A(t)$ is completely described by the following parameters [4–6]:

- the vector of probabilities of its initial states at the moment $t = 0$

$$[p(0)] = [p_1(0), p_2(0), \ldots, p_m(0)], \tag{1}$$

- the matrix of probabilities of its transitions between the particular states

$$[p_{ij}] = \begin{bmatrix} p_{11} & p_{12} & \cdots & p_{1m} \\ p_{21} & p_{22} & \cdots & p_{2m} \\ \vdots & \vdots & \ddots & \vdots \\ p_{m1} & p_{m2} & \cdots & p_{mm} \end{bmatrix}, \tag{2}$$

where $p_{ii} = 0, i = 1, 2, \ldots, m$;

- the matrix of distribution functions of its conditional sojourn times θ_{ij} at the particular states

$$[W_{ij}(t)] = \begin{bmatrix} W_{11}(t) & W_{12}(t) & \cdots & W_{1m}(t) \\ W_{21}(t) & W_{22}(t) & \cdots & W_{2m}(t) \\ \vdots & \vdots & \ddots & \vdots \\ W_{m1}(t) & W_{m2}(t) & \cdots & W_{mm}(t) \end{bmatrix}, t \geq 0, \tag{3}$$

where $W_{ii}(t) = 0, t \geq 0, i = 1, 2, \ldots, m$;

- the matrix of mean values of its conditional sojourn times θ_{ij} at the particular states $[M_{ij}]_{m \times m}$, where

$$M_{ij} = E[\theta_{ij}] = \int_0^\infty t dW_{ij}(t) = \int_0^\infty t w_{ij}(t) dt, i, j \in \{1, 2, \ldots, m\}, \tag{4}$$

and $w_{ij}(t), t \geq 0, i, j \in \{1, 2, \ldots, m\}$, are density function corresponding to distribution functions $W_{ij}(t), t \geq 0, i, j \in \{1, 2, \ldots, m\}$.

2.2 Process of Changing Hydro-Meteorological Conditions at Baltic Sea Open Water Area Identification

Taking into account expert opinions on the process of changing hydro-meteorological conditions $A(t)$ for the Baltic Sea open water area, we distinguished $m = 6$ following states of this process [7, 8], considering two parameters (*wh* – wave height and *ws* – wind speed):

- state 1 – *wh* from 0 up to 2 m and *ws* from 0 m/s up to 17 m/s;
- state 2 – *wh* from 2 m up to 5 m and *ws* from 0 m/s up to 17 m/s;
- state 3 – *wh* from 5 m up to 14 m and *ws* from 0 m/s up to 17 m/s;
- state 4 – *wh* from 0 up to 2 m and *ws* from 17 m/s up to 33 m/s;
- state 5 – *wh* from 2 m up to 5 m and *ws* from 17 m/s up to 33 m/s;
- state 6 – *wh* from 5 m up to 14 m and *ws* from 17 m/s up to 33 m/s.

On the basis of the statistical data collected in Marches (the process depends of the season and is a periodic one) during period of years 1988–1993 [7, 8] and the identification method given in [2, 4], it is possible to evaluate the following unknown basic parameters of the semi-Markov model of the process of changing hydro-meteorological conditions at open waters Baltic Sea area:

- the vector

$$[p_i(0)] = [0.595, 0.349, 0, 0, 0.04, 0.016] \tag{5}$$

of the initial probabilities $p_i(0)$, $i = 1, 2, \ldots, 6$, of the process $A(t)$ of changing hydro-meteorological conditions staying at the particular states i, $i = 1, 2, \ldots, 6$, at the initial moment $t = 0$,

- the matrix of the probabilities p_{ij}, $i, j \in \{1, 2, \ldots, 6\}$, of transitions of the process $A(t)$ of changing hydro-meteorological conditions from the state i into the state j is given as follows

$$[p_{ij}] = \begin{bmatrix} 0 & 0.98 & 0 & 0 & 0.02 & 0 \\ 0.7 & 0 & 0 & 0 & 0.3 & 0 \\ 0 & 0.92 & 0 & 0 & 0.08 & 0 \\ 0 & 0 & 0 & 0 & 1 & 0 \\ 0 & 0.76 & 0.01 & 0 & 0 & 0.23 \\ 0 & 0.41 & 0.15 & 0 & 0.44 & 0 \end{bmatrix}. \quad (6)$$

According to [4], we may verify the hypotheses on the distributions of this process' conditional sojourn times at the particular states. To do this, we need a sufficient number of realizations of these variables, namely, the sets of their realizations should contain at least 30 realizations coming from the experiment.

The sets of the realizations of the conditional sojourn times θ_{12}, θ_{15}, θ_{21}, θ_{25}, θ_{52}, θ_{56}, θ_{62} and θ_{65} of the process $A(t)$ of changing hydro-meteorological conditions were sufficiently large and we verified that they have chimney distributions and gamma distribution (θ_{65}), respectively:

$$W_{12}(t) = \begin{cases} 0, & t<0 \\ 0.016t, & 0 \le t \le 26.52 \\ 0.398 + 0.001t, & 26.52 \le t \le 517.14 \\ 1, & t > 517.14, \end{cases}$$

$$W_{15}(t) = \begin{cases} 0, & t<0 \\ 0.034t, & 0 \le t \le 23.4 \\ 0.608 + 0.008t, & 23.4 \le t \le 46.8 \\ 1, & t > 46.8, \end{cases}$$

$$W_{21}(t) = \begin{cases} 0, & t \le 1.55 \\ 0.036t - 0.056, & 1.55 < t \le 4.45 \\ 0.041t - 0.078, & 4.45 < t \le 13.15 \\ 0.369 + 0.007t, & 13.15 < t \le 94.35 \\ 1, & t > 94.35, \end{cases}$$

$$W_{25}(t) = \begin{cases} 0, & t < 1.35 \\ 0.054t - 0.073, & 1.35 \le t \le 7.95 \\ 0.277 + 0.01t, & 7.95 \le t \le 70.65 \\ 1, & t > 70.65, \end{cases}$$

$$W_{52}(t) = \begin{cases} 0, & t < 1.97 \\ 0.094t - 0.185, & 1.97 \le t \le 8.15 \\ 0.467 + 0.014t, & 8.15 \le t \le 36.99 \\ 1, & t > 36.99, \end{cases}$$

$$W_{56}(t) = \begin{cases} 0, & t \le 1.88 \\ 0.045t - 0.085, & 1.88 < t \le 4.13 \\ 0.098t - 0.304, & 4.13 < t \le 10.88 \\ 0.534 + 0.021t, & 10.88 < t \le 22.13 \\ 1, & t > 22.13, \end{cases}$$

$$W_{62}(t) = \begin{cases} 0, & t < 0 \\ 0.052t, & 0 \le t \le 15 \\ 0.69 + 0.006t, & 15 \le t \le 52.5 \\ 1, & t > 52.5, \end{cases} \quad w_{65}(t) = 0.015 \cdot t^{1.089} \cdot e^{-0.136t}, t \ge 0. \tag{7}$$

The sets of the realizations of the process remaining conditional sojourn times at particular states contained less than 30 realizations. Thus, we assumed that the distribution functions of the process conditional sojourn times θ_{14}, θ_{32}, θ_{35}, θ_{45}, θ_{53} and θ_{63} have the empirical distribution functions, as follows:

$$W_{14}(t) = \begin{cases} 0, & t \le 6 \\ 1, & t > 6, \end{cases} \quad W_{32}(t) = W_{35}(t) = W_{45}(t) = \begin{cases} 0, & t \le 3 \\ 1, & t > 3, \end{cases}$$

$$W_{53}(t) = \begin{cases} 0, & t \le 3 \\ 0.5, & 3 < t \le 9 \\ 1, & t > 9, \end{cases} \quad W_{63}(t) = \begin{cases} 0, & t \le 3 \\ 0.313, & 3 < t \le 6 \\ 0.375, & 6 < t \le 9 \\ 0.5, & 9 < t \le 12 \\ 0.563, & 12 < t \le 15 \\ 0.75, & 15 < t \le 24 \\ 0.875, & 24 < t \le 33 \\ 0.938, & 33 < t \le 51 \\ 1, & t > 51. \end{cases} \tag{8}$$

The remaining distribution functions of the process conditional sojourn times could not be evaluated because of the lack of data. Considering the conditional distributions given by (7), (8), according to (4), the conditional mean values of the sojourn times at the particular states measured in hours are fixed as follows:

$$\begin{aligned} M_{12} \cong 162.97,\ M_{14} = M_{53} = 6,\ M_{15} \cong 16.24,\ M_{21} \cong 32.39, \\ M_{25} \cong 26.96,\ M_{32} = M_{35} = M_{45} = 3,\ M_{52} \cong 12.34, \\ M_{56} \cong 9.2,\ M_{61} \cong 13.33,\ M_{62} \cong 14.25,\ M_{65} \cong 15.38. \end{aligned} \tag{9}$$

3 Prediction of Oil Spill Domain Movement at Varying Hydro-Meteorological Conditions

3.1 Prediction Procedure of Oil Domain Spill Movement at Varying Hydro-Meteorological Conditions

The general prediction procedure of the oil spill domain movement at varying hydro-meteorological conditions, based on the models from [4–6], is given below.

INPUT DATA:

- step of time Δt;
- experiment time T;
- oil spill central point drift trend $K^k : \begin{cases} x^k = x^k(t) \\ y^k = y^k(t), \ t \in <0, T> ; \end{cases}$
- expected values $m_X^{k_i}(t)$, $m_Y^{k_i}(t)$;
- standard deviations $\sigma_X^{k_i}(t)$, $\sigma_Y^{k_i}(t)$;
- correlation coefficient $\rho_{XY}^{k_i}(t)$;
- radius $r^k(t)$;
- $M_{kj\ kj+1} = E[\theta_{kj\ kj+1}]$.

FIX: $s_0 = 0$, $m_X^{k_0}(s_0\Delta t) = 0$, $m_Y^{k_0}(s_0\Delta t) = 0$, $k_i \in \{1, 2, \ldots, m\}$, $m \in$ N;
FOR $i = 1$ TO n,

- *FIX*: $k_i \in \{1, 2, \ldots, m\}$;
- *CHECK*: $(s_i - 1)\Delta t < \sum_{j=1}^{i} M_{kj\,kj+1} \leq s_i\,\Delta t, i = 1, 2, \ldots, n$;
- *SELECT*: $M_{kj\ kj+1}$;
- *CALCULATE*: $s_i\Delta t - s_{i-1}\Delta t$;
  FOR $b_i = 1$ TO $b_i = s_i\Delta t - s_{i-1}\Delta t$,
  FOR $a_i = 1$ TO $a_i = b_i$,
  - $m_X^k(t) := m_X^{k_{i-1}}(s_{i-1}\Delta t) + m_X^{k_i}(a_i\Delta t)$, $m_Y^k(t) := m_Y^{k_{i-1}}(s_{i-1}\Delta t) + m_Y^{k_i}(a_i\Delta t)$;
  - $$\bar{\sigma}_X^k(t) := \bar{\bar{\sigma}}_X^{k_i}(s_{i-1} + a_i\Delta t) = \sigma_X^{k_i}(s_{i-1} + a_i\Delta t) + \sum_{j=1}^{i} r^{k_j}(b_j\Delta t),$$
    $$\bar{\sigma}_Y^k(t) := \bar{\bar{\sigma}}_Y^{k_i}(s_{i-1} + a_i\Delta t) = \sigma_Y^{k_i}(s_{i-1} + a_i\Delta t) + \sum_{j=1}^{i} r^{k_j}(b_j\Delta t);$$
  - PRINT $\bar{\bar{D}}^{k_i}(s_{i-1} + a_i\Delta t) := \bar{D}^k(s_{i-1} + a_i\Delta t)$;

WHILE $s_n\Delta t \leq T$;

OUTPUT: PRINT $\overline{\overline{\boldsymbol{D}}}^{k_1, k_2, \ldots, k_n}(b_i) = \bigcup_{i=1}^{n} \bigcup_{a_i=1}^{b_i} \overline{\overline{\boldsymbol{D}}}^{k_i}(s_{i-1} + a_i\Delta t)$.

3.2 Prediction of Oil Domain Spill Movement at Varying Hydro-Meteorological Conditions at Baltic Sea Open Water Area

On the base of statistical data from Sect. 2.2 and applying the procedure from Sect. 3.1, we can use the probabilistic approach to prediction oil spill domain movement at varying hydro-meteorological conditions at Baltic Sea open water area.

For varying hydro-meteorological conditions, we assume that the process of changing hydro-meteorological conditions *A*(*t*) in succession takes the states k_1, k_2, …, k_n, $k_i \in \{1,2,\ldots,6\}$, i = 1,2, …, n.

Moreover, we arbitrarily assume, that the experiment time T is equal to 48 h and the points $(m_X^{k_i}(t), m_Y^{k_i}(t))$, $t \in \langle 0,48 \rangle$, $k_i \in \{1,2,\ldots,6\}$, i = 1, 2, …, n, for each fixed at varying hydro-meteorological state k_i, create a curve K^{ki}, $k_i \in \{1,2,\ldots,6\}$, i = 1, 2, …, n, called an oil spill central point drift trend which may be described in the parametric form

$$K^{k_i} : \begin{cases} x^{k_i} = t^{k_i} \\ y^{k_i} = t, \quad t \in \ <0,48> , k_i \in \{1,2,\ldots,6\}, i = 1,2,\ldots,n, \end{cases} \tag{10}$$

that varies at different states of the process *A*(*t*), where x and y are measured in meters.

According to the procedure from Sect. 3.1, we arbitrarily assume that the varying in time expected values of the oil spill central point coordinates X and Y respectively are:

$$m_X^{k_i}(t) = t^{k_i}, \ m_Y^{k_i}(t) = t, \ t \in \ < 0,48 > , k_i \in \{1,2,\ldots,6\}, i = \ 1,2, \ldots, n. \tag{11}$$

We arbitrarily fix the standard deviations, the correlation coefficient of the oil spill central point coordinates and the radius of the oil spill domain $\bar{D}^{k_i}(t)$, $t \in \langle 0,48 \rangle$, $k_i \in \{1,2,\ldots,6\}$, $i = \ 1,2, \ldots, n$, as follows:

$$\sigma_X^{k_i}(t) = \sigma_Y^{k_i}(t) = \sigma^{k_i}(t) = 0.1 + 0.2t, \ \rho_{XY}^{k_i}(t) = 0.8, \ r^{k_i}(t) = 0.5 + 0.5t, \tag{12}$$

for $t \in \langle 0,48 \rangle$, $k_i \in \{1,2,\ldots,6\}$, $i = \ 1,2, \ldots, n$.

Further, for a fixed step of time Δt = 1 h, after multiple applying sequentially the procedure from Sect. 3.1, we receive the following sequence of oil spill domains:

- for $t = 1, 2, \cdots, s_1$, at state k_1 we have $\bar{D}^{k_1}(1), \bar{D}^{k_1}(2), \ldots, \bar{D}^{k_1}(s_1)$;
- for $t = s_1 + 1, s_1 + 2, \cdots, s_2$, at state k_2 we have $\bar{D}^{k_2}(s_1 + 1)$, $\bar{D}^{k_2}(s_1 + 2), \ldots, \bar{D}^{k_2}(s_2)$;

 …

- for $t = s_{n-1} + 1, s_{n-1} + 2, \cdots, s_n$, at state k_n we have $\bar{D}^{k_n}(s_{n-1} + 1)$, $\bar{D}^{k_n}(s_{n-1} + 2), \ \ldots, \ \bar{D}^{k_n}(s_n)$,

 where $s_i, i = 1,2,\ldots,n$, are such that

$$s_i - 1 < \sum_{j=1}^{i} M_{k_j k_{j+1}} \le s_i, i = 1, 2, \ldots, n, s_n - 1 \le T = 48 \le s_n, \tag{13}$$

and $M_{k_j k_{j+1}} = E\left[\theta_{k_j k_{j+1}}\right]$ are defined by (4).

We arbitrarily assume that the process of changing hydro-meteorological conditions $A(t)$ in succession takes the states $k_1 = 2$, $k_2 = 1$, $k_3 = 2$, $k_4 = 1$. Thus, $i = 1, 2, 3, 4$:

- $i = 1$;
- for the fixed $k_1 = 2$ and $k_2 = 1$, we select the conditional mean value $M_{21} = 32.39$ of the sojourn time θ_{21};
- we check the condition $(s_1 - 1) = s_0 = 0 < M_{21} = 32.39 \le s_1$;
- hence, $s_1 = 33$ and $s_1 - s_0 = s_1 - 0 = 33$;
- consequently, we draw $b_1 = 1, 2, \ldots, 33$ ellipses;
- we compare s_1 with the experiment time: $s_1 = 33 < 48 = T$, thus, the sequence of the oil spill domains for $a_1 = 1, 2, \ldots, b_1$, $b_1 = 1, 2, \ldots, 33$, is [4–6]

$$\bar{\bar{D}}^{k_1}(b_1) = \bar{\bar{D}}^{k_1}(s_0 + a_1) := \bar{D}^{k}(0 + a_1)$$
$$= \{(x, y) : \frac{1}{1 - (0.8)^2}\left[\frac{(x - m_X^k(a_1))^2}{(\bar{\sigma}^k(a_1))^2} - 2 \cdot 0.8 \frac{(x - m_X^k(a_1))(y - m_Y^k(a_1))}{\bar{\sigma}^k(a_1) \cdot \bar{\sigma}^k(a_1)}\right.$$
$$\left. + \frac{(y - m_Y^k(a_1))^2}{(\bar{\sigma}^k(a_1))^2}\right] \le -2ln(1 - 0.95) \cong 5.99\},$$

where

$$m_X^k(a_1) := m_X^{k_0}(s_0) + m_X^{k_1}(a_1) = (a_1)^2, m_Y^k(a_1) := m_Y^{k_0}(s_0) + m_Y^{k_1}(a_1) = a_1,$$

$$\bar{\sigma}^k(a_1) := \bar{\bar{\sigma}}^{k_1}(s_0 + a_1) = \sigma^{k_1}(s_0 + a_1) + \sum_{j=1}^{1} r^{k_j}(b_j) = \sigma^{k_1}(a_1) + r^{k_1}(b_1)$$
$$= 0.1 + 0.2a_1 + 0.5 + 0.5b_1 = 0.2a_1 + 0.5b_1 + 0.5,$$

- we draw the sequence of the oil spill domains (Fig. 1).

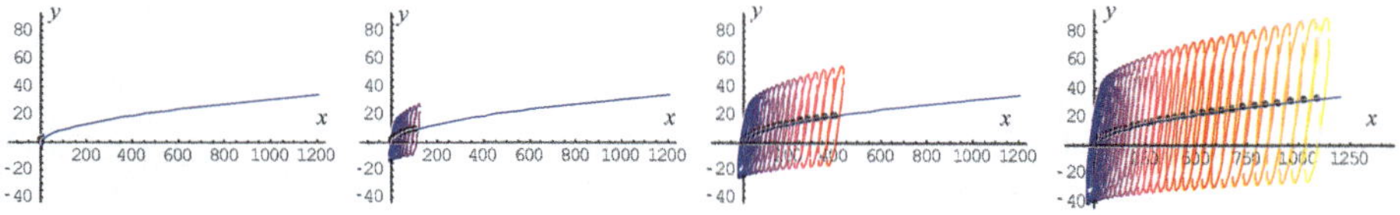

Fig. 1. Oil spill domain at the moments $a_1 = 1, 10, 20, 33$.

- $i = 2$, $s_1 = 33$;
- for the fixed $k_2 = 1$ and $k_1 = 2$, we select the conditional mean value $M_{12} = 162.97$ of the sojourn time θ_{12};
- we have $(s_2 - 1) = s_1 = 33 < M_{21} + M_{12} = 32.39 + 162.97 \leq s_2$;
- hence, $s_2 = 196$ and $196 > 48 = T$, thus $s_2 = 48$ and $s_2 - s_1 = 48 - 33 = 15$;
- consequently, we draw $b_2 = 1, 2, \ldots, 15$ ellipses;
- the sequence of the oil spill domains for $a_2 = 1, 2, \ldots, b_2$, $b_2 = 1, 2, \ldots, 15$,

$$\bar{\bar{D}}^{k_2}(b_2) = \bar{\bar{D}}^{k_2}(s_1 + a_2) := \bar{D}^k(33 + a_2) = \{(x, y) : \frac{1}{1 - (0.8)^2}\left[\frac{(x - m_X^k(33 + a_2))^2}{(\bar{\sigma}^k(33 + a_2))^2}\right.$$

$$\left. -2 \cdot 0.8 \frac{(x - m_X^k(33 + a_2))(y - m_Y^k(33 + a_2))}{\bar{\sigma}^k(33 + a_2) \cdot \bar{\sigma}^k(33 + a_2)} + \frac{(y - m_Y^k(33 + a_2))^2}{(\bar{\sigma}^k(33 + a_2))^2}\right] \leq 5.99\},$$

where

$$\begin{aligned}
& m_X^k(33 + a_2) := m_X^{k_1}(s_1) + m_X^{k_2}(a_2) = 33^2 + a_2, \\
& m_Y^k(33 + a_2) := m_Y^{k_1}(s_1) + m_Y^{k_2}(a_2) = 33 + a_2, \\
& \bar{\sigma}^k(a_2) := \bar{\bar{\sigma}}^{k_2}(s_1 + a_2) = \sigma^{k_2}(33 + a_2) + r^{k_1}(b_1) + r^{k_2}(b_2) \\
& = 0.1 + 0.2(33 + a_2) + 0.5 + 0.5b_1 + 0.5 + 0.5b_2 \\
& = 0.2a_2 + 0.5b_1 + 0.5b_2 + 7.5;
\end{aligned}$$

- the domains are illustrated in Fig. 2.

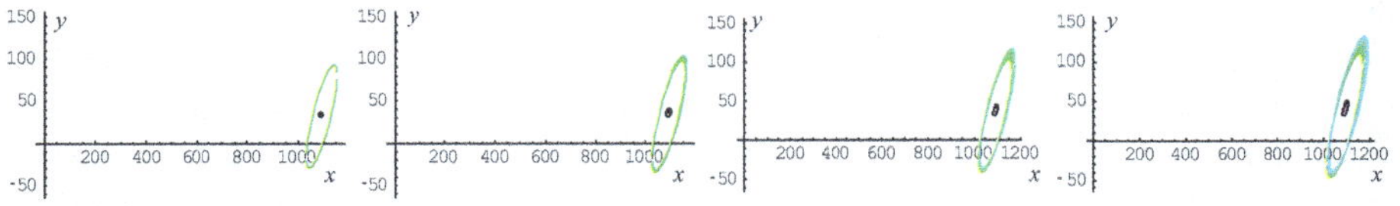

Fig. 2. Oil spill domain at the moments $a_2 = 1, 5, 10, 15$.

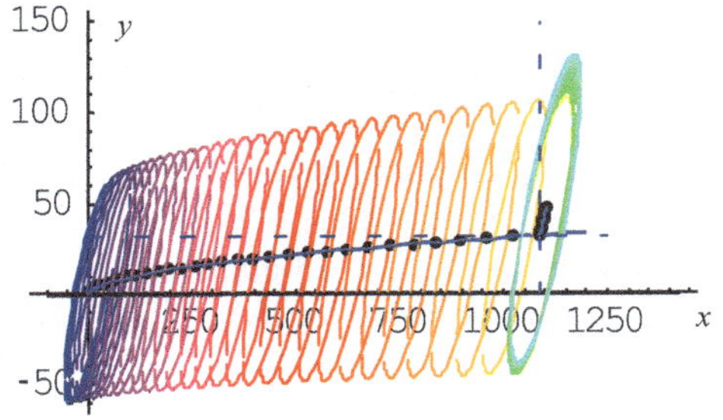

Fig. 3. Oil spill domain movement until the moment $t = 48$.

The oil spill domain movement for the time interval $\langle 0, 48 \rangle$ is illustrated in Fig. 3.

4 Conclusions

The proposed approach to oil spill domains determination and their movement at different hydro-meteorological conditions investigation can be also done for other than oil kind of spills, dangerous for the environment. The purpose of the study is to propose a probabilistic approach to oil spill domains determination and their movement investigation to improve the efficiency of people activities in the environment protection. A weak point of the method is the time and cost of the research experiments necessary to perform at the port and sea water areas in order to identify statistically particular components of the proposed models [4]. Especially, experiments needed to evaluate drift trends and parameters of the central point of oil spill position distributions can consume much time and be costly as they have to be done for different kind of spills [9] and different hydro-meteorological conditions in various areas, e.g. the direction of the wind can be taken into account. South-West wind is dominating direction for moderate and strong winds at the Baltic Sea. A strong and positive point of the method is the fact that the experiments for the fixed port and sea water areas and fixed different hydro-meteorological conditions have to be done only once and the identified models may be used for all environment protection actions at these regions and also transferred for other regions with similar hydro-meteorological conditions. The proposed stochastic approach can be supplemented and developed by considering and applying the approaches discussed in [10] and by the Monte Carlo simulation approach to the spill oil domain movement investigation proposed in [11–15]. These two approaches are original approaches to the oil spill domain determination and its movement propagation which are intended to be significantly developed with the close considering the contents of publications cited in references below.

Acknowledgements. The paper presents the results developed in the scope of the research projects "Safety of critical infrastructure transport networks" and "Impact of hydro-meteorological changes on the movement of oil spills at sea", granted by GMU in 2020.

References

1. Grabski, F.: Semi-Markov Processes: Application in System Reliability and Maintenance. Elsevier, Amsterdam (2014)
2. Kołowrocki, K., Soszyńska-Budny, J.: Reliability and Safety of Complex Technical Systems and Processes: Modeling – Identification – Prediction – Optimization, 1st edn. Springer, London (2011)
3. Kołowrocki, K.: Reliability of Large and Complex Systems, 2nd edn. Elsevier, London (2014)
4. Dąbrowska, E., Kołowrocki, K.: Modelling, identification and prediction of oil spill domains at port and sea water areas. J. Polish Saf. Reliab. Assoc. Summer Saf. Reliab. Semin. **10**(1), 43–58 (2019)
5. Dąbrowska, E., Kołowrocki, K.: Stochastic determination of oil spill domain at gdynia port water area. In: Proceedings of IDT Conference, Zilina (2019)
6. Dąbrowska, E., Kołowrocki, K.: Probabilistic approach to determination of oil spill domains at port and sea water areas. In: Proceedings of TransNav Conference, Gdynia (2020)
7. Kuligowska, E., Torbicki, M.: GMU safety interactive platform organization and possibility of its applications. J. Polish Saf. Reliab. Assoc. Summer Saf. Reliab. Semin. **9**(2), 99–114 (2018)
8. Gdynia Maritime University safety interactive platform (2018). http://gmu.safety.umg.edu.pl/
9. Kurc, B., Chrzanowski, J., Abramoska, E.: Zagrożenia rozlewami szkodliwych chemikaliów oleistych na morzu. Zeszyty Naukowe Akademii Morskiej w Szczecinie **5**(77), 349–359 (2005)
10. Fingas, M.: Oil Spill Science and Technology, 2nd edn. Elsevier, Amsterdam (2016)
11. Dąbrowska, E., Kołowrocki, K.: Monte Carlo simulation applied to oil spill domain at Gdynia Port water area determination. In: Proceedings IEEE of the International Conference on Information and Digital Technologies, 98–102 (2019)
12. Kim, T., Yang, C.-S., Ouchi, K., Oh, Y.: Application of the method of moment and Monte-Carlo simulation to extract oil spill areas from synthetic aperture radar images. In: OCEANS - San Diego, pp. 1–4 (2013)
13. Kuligowska, E.: Monte Carlo simulation of climate-weather change process at maritime ferry operating area. Tech. Sci. Univ. Warmia Mazury Olsztyn **1**(21), 5–17 (2018)
14. Rao, M.S., Naikan, V.N.A.: Review of simulation approaches in reliability and availability modeling. Int. J. Perform. Eng. **12**(4), 369–388 (2016)
15. Zio, E., Marseguerra, M.: Basics of the Monte Carlo method with application to system reliability, LiLoLe (2002)

Subjective Quality Evaluation of Underground BPL-PLC Voice Communication System

Grzegorz Debita[1(✉)], Przemyslaw Falkowski-Gilski[2], Marcin Habrych[3], Bogdan Miedzinski[3], Bartosz Polnik[4], Jan Wandzio[5], and Przemyslaw Jedlikowski[6]

[1] General Tadeusz Kosciuszko Military University of Land Forces, 51-147 Wroclaw, Poland
grzegorz.debita@awl.edu.pl

[2] Faculty of Electronics, Telecommunications and Informatics, Gdansk University of Technology, 80-233 Gdansk, Poland

[3] Faculty of Electrical Engineering, Wroclaw University of Science and Technology, 50-370 Wroclaw, Poland

[4] KOMAG Institute of Mining Technology, 44-101 Gliwice, Poland

[5] KGHM Polska Miedz S.A., 59-301 Lubin, Poland

[6] Faculty of Electronics, Wroclaw University of Science and Technology, 50-370 Wroclaw, Poland

Abstract. Designing a reliable voice transmission system is not a trivial task. Wired media, thanks to their resistance to mechanical damage, seem an ideal solution. The BPL-PLC *(Broadband over Power Line – Power Line Communication)* cable is resilient to electricity stoppage and partial damage of phase conductors. It maintains continuity of transmission in case of an emergency situation, including paramedic rescue operations. These features make it an ideal solution for delivering data, e.g. in an underground mine environment. This paper describes a subjective quality evaluation of such a system. The solution was designed and tested in real-time operating conditions. It involved two types of coupling, namely: induction-inductive and capacitive-inductive, as well as two transmission modes (Mode 1 and Mode 11 operating in the 2–7.5 MHz frequency range). The tested one-way transmission system was designed to deliver clear and easily understandable voice messages. The study involved signal samples in three languages: English (both British and American dialects), German, and Polish, processed in three bitrates: 8, 16, and 24 kbit/s, with the Ogg Vorbis codec. Obtained results confirmed the usefulness of the BPL-PLC technology for voice communication purposes. Results of this study may be of interest to professionals from the mining and oil industry.

Keywords: Communication applications · Signal processing · Quality of service

W. Zamojski et al. (Eds.): DepCoS-RELCOMEX 2020, AISC 1173, pp. 176–186, 2020.
https://doi.org/10.1007/978-3-030-48256-5_18

1 Introduction

Today, communication continuity, along with appropriate QoS *(Quality of Service)* mechanisms, are of major importance, especially in an emergency situation. This applies to e.g. mine disasters, when voice contact with injured or cut-out personnel becomes a key factor. Similar situations may appear in case of black-outs and/or electric energy stoppage. Therefore, in such cases, the use of BPL-PLC *(Broadband over Power Line – Power Line Communication)* technology, in medium voltage (MV) cable networks for data transmission purposes, seems to be a justified solution.

Modern urban cable networks are laid underground, similar as in tunnels and mines. They are robust and resistant to mechanical damage, and can be effectively used as a transmission medium under energy stoppage, even in case of conductor interruption. In order to demonstrate the validity and usefulness of this idea, authors carried out a series of tests for a selected fragment of a 6 kV mining cable network, utilizing a specially developed transmitter and digital receiver dedicated to BPL-PLC modems [1]. The voice quality was evaluated for one-way transmission using two selected modes (Mode 1 from 3 to 7.5 MHz and Mode 11 from 2 to 7 MHz), operating in the 2–7 MHz frequency range. Obtained results confirmed the applicability of BPL-PLC technology in the 6 kV mining cable network for an effective speech transmission system.

2 BPL-PLC Wired Transmission

BPL-PLC is an alternative to other wired technologies for data communication, but the quality of transmission is relatively lower compared to e.g. fiber optic technology. The main disadvantage is the environment's negative impact on range and transmission quality [2, 3]. Therefore, prior to its implementation, respective power cable parameters, including: attenuation, phase constant, characteristic impedance, etc., as well as frequency mode and type of coupling (induction-inductive or mixed capacitive-inductive), must be carefully selected and considered. However, the primary advantage of transmitting carrier-modulated signals over power lines is the possibility of reusing the existing power cable (wired) infrastructure, which does not incur any additional costs, neither operator fees.

In addition, BPL-PLC transmission can be performed successfully under any operation conditions of the electricity grid, even under power outage, what is particularly vital during an emergency situation, e.g. mine disasters or other threats. The transmission channel is then constituted by the armor and/or shield of the electric cable, and of course the battery power supply of the modems. One must note, that in case of dense urban as well as underground environments, cable networks are laid directly underground or run in tunnels, which makes them most resistive to mechanical damage. However, it should be emphasized, that the main task of any power grid is to transmit and deliver electricity at the frequency of 50/60 Hz. Therefore, BPL-PLC technology (of much higher frequency) should not interfere with it, and should be considered as a supplementary technology [4, 5].

In Europe, two frequency bands are assigned to BPL-PLC technologies [6, 7], namely: 3–148 kHz for low bitrates (narrowband frequency range), 2–30 MHz for high bitrates (wideband frequency range), as shown in Fig. 1.

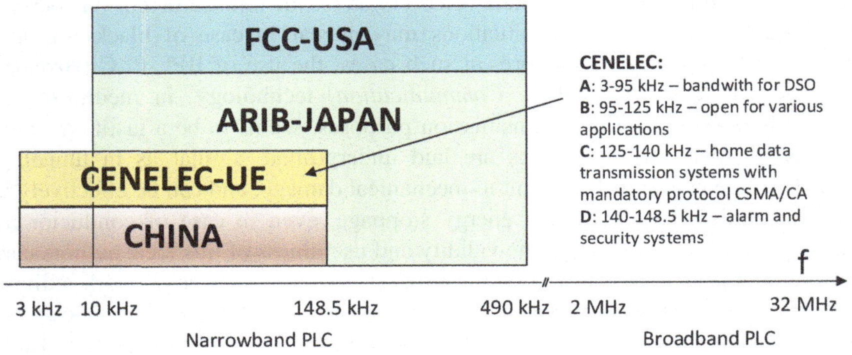

Fig. 1. Frequency bands utilized in BPL-PLC technology.

The advantage of BPL-PLC technology over medium voltage power lines is the possibility to employ additional data services, including voice transmission. This is particularly important under various types of hazards, particularly power supply outage. The power cables that are located underground or in tunnels enable to connect appropriate transmitters and receivers to modems, especially in case of the induction-inductive coupling. This enables to set a secured voice transmission, both pear-to-pear and/or master-slave.

3 Voice Processing and Coding

In order to provide clear and understandable voice information in the BPL-PLC wired system, one needs to know how many bits are sufficient to convey quality content. When it comes to voice transmission systems, the key issue is to provide high-quality audio services over varying bandwidth conditions and heterogeneous networks. In this case packets may be lost or delayed, which is not acceptable for real-time applications. This may cause degradation in quality, observed as either network QoS or perceived user QoE *(Quality of Experience)* [8].

The use of network communication imposes serious restrictions, including bandwidth limitations, associated with available bitrates. In the last two decades, many research efforts have been devoted to the problem of audio compression. Two different compression categories have been of particular interest, namely high performance and low bitrate audio coding [9].

High performance audio coding is aimed to achieve the audio quality as high as possible at a certain bitrate. On the other hand, lossless compression always ensures the highest possible quality, in which the objective redundancy of multimedia content is the only source of compression. Of course, each coding algorithm has a limit for lowest acceptable bitrate. Nevertheless, this limit may be sometimes hard to determine [10].

In most cases, the higher the bitrate, the higher the quality. However, this increase in quality does not resemble a linear scale. In every coding algorithm there is always a break point, when further increase in bitrate does not imply further raise in perceived quality [11]. This breakpoint is highly dependable on the type of transmitted content, as well as type of medium. In our case, we investigated different couplings (induction-inductive and capacitive-inductive), modes (Mode 1 and Mode 11), and speech samples (English, German, Polish), as well as bitrates (8, 16, and 24 kbit/s).

4 About the Study

The main aim of the study was to determine the feasibility of utilizing a predefined BPL-PLC line for voice communication services. In this scenario, the wired medium was located in an underground mine environment, operating in real-time conditions. For the purpose of this test, we have selected a 3-phase cable, about 300 m long, located in the mine shaft headroom. This cable was a part of the tested medium voltage radial network of 6 kV, its total length was equal to approx. 1300 m. This subjective study was a direct continuation of previous work described in [12].

In order to simulate an emergency condition, the tested cable, shown in Fig. 2, was disconnected from the power supply and shorted, as well as earthed at both ends. A specially developed transmitter and digital receiver were connected respectively to the BPL-PLC modems. The best quality and stability of BPL-PLC transmission was obtained for a frequency range of 2–7 MHz.

Fig. 2. Tested wired BPL-PLC medium in an underground mine shaft.

However, it should be emphasized that the BPL-PLC transmission is asymmetrical, showing different values of both throughput (bitrate) and SNR *(Signal-to-Noise Ratio)*, as well as CFR *(Channel Frequency Response)* factors, for reverse direction of transmission. For example, the measured capacity for one direction, which was evaluated in this study, was around 34 Mbit/s. Whereas for the other one, it decreased to approx. 27 Mbit/s.

The tested speech samples were sourced from ITU-T P.501 [13]. In this recommendation, available signal samples consist of two sentences spoken by two female and two male individuals, in different languages. Due to the international character of KGHM Polska Miedz S.A., we have selected samples from 4 sets, namely: American English (AE), British English (EN), German (GE), and Polish (PL). The original signal samples were available in the WAV 16-bit PCM *(Pulse Code Modulation)* format, with sampling frequency set to 32 kHz, typical length equal to 7 s. Next, each sample was coded using the Ogg Vorbis format. We have decided to use this codec due to its openness and full compatibility with the Linux operating system. It should be pointed out, that the custom designed communication system was running on Linux-powered devices. We intended to have as much control over the hardware and software layer as possible. The degraded signal samples were processed in 3 bitrates, namely: 8, 16, and 24 kbit/s, whereas the sampling frequency was set to 44.1 kHz, as in most popular audio systems. After being transmitted in real-time, all samples were recorded on the receiving side for further processing and evaluation purposes. Additional information on low bitrate audio coding may be found in [14–17].

5 Subjective Test Results

In this test, the goal was to investigate how does a BPL-PLC cable perform, when it comes to providing stable and reliable voice services. How does the type of coupling, transmission mode, as well as varying bandwidth conditions, affect the perceived speech quality of samples coded at different bitrates.

The subjective assessment was carried out using Beyerdynamic Custom One headphones in a 5-step MOS *(Mean Opinion Score)* ACR *(Absolute Category Rating)* scale, with no reference signal available, ranging from 1 (bad quality) to 5 (excellent quality). The study involved a group of 16 people, aged between 25–35 years old. Each participant assessed the quality individually, according to [18], and took a training phase before starting the essential study. A single session (2 transmission modes, 4 languages/dialects, 2 male and 2 female speakers, 3 bitrates, 2 types of coupling, each file lasting about 7 s) took approx. 25 min, with a short break in the middle of the study. The results of the subjective quality evaluation, concerning different types of coupling as well as transmission mode, with respect to spoken language and bitrate, are shown in Figs. 3, 4, 5, 6, 7, 8, 9 and 10. Grades for the induction-inductive coupling are shown in Figs. 3, 4, 5 and 6, whereas those for the capacitive-inductive coupling are shown in Figs. 7, 8, 9 and 10.

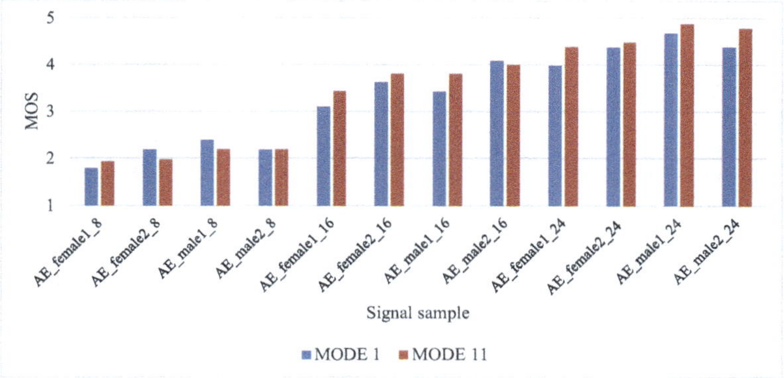

Fig. 3. Induction-inductive coupling – speech samples in American English.

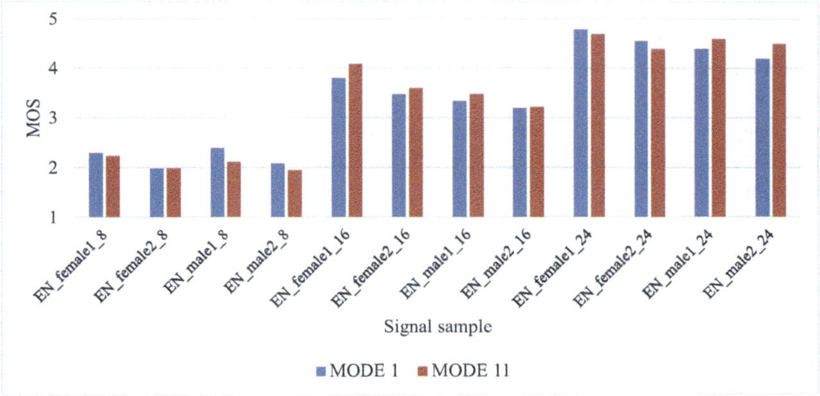

Fig. 4. Induction-inductive coupling – speech samples in British English.

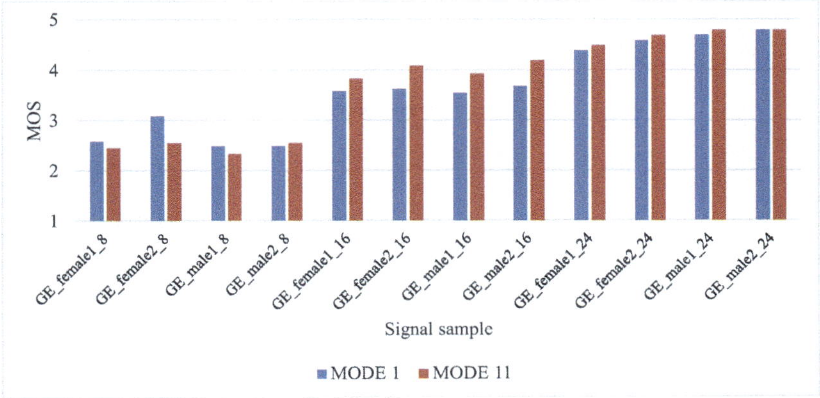

Fig. 5. Induction-inductive coupling – speech samples in German.

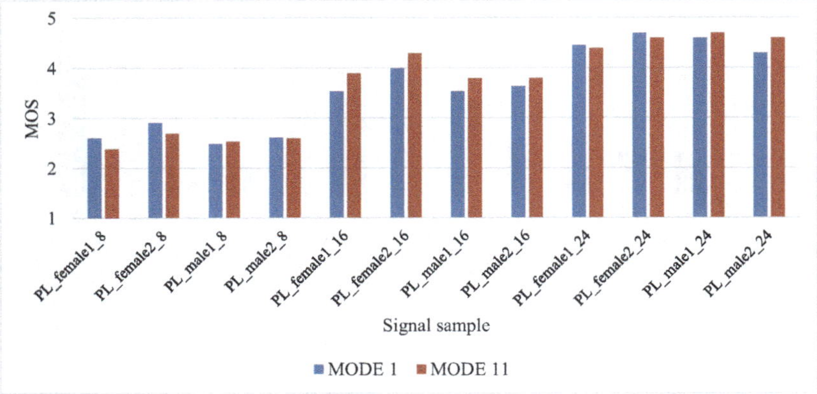

Fig. 6. Induction-inductive coupling – speech samples in Polish.

When statistically processing obtained data, the confidence intervals were set to 5%. In all cases, these ranges were less than 10% of the average values. For the sake of clarity, they were not marked in the diagrams. Additional information on statistical analysis may be found in [19, 20]. Also it should be noted, that neither individual had hearing disorders, and was fluent in both English and German, whereas Polish was their mother tongue.

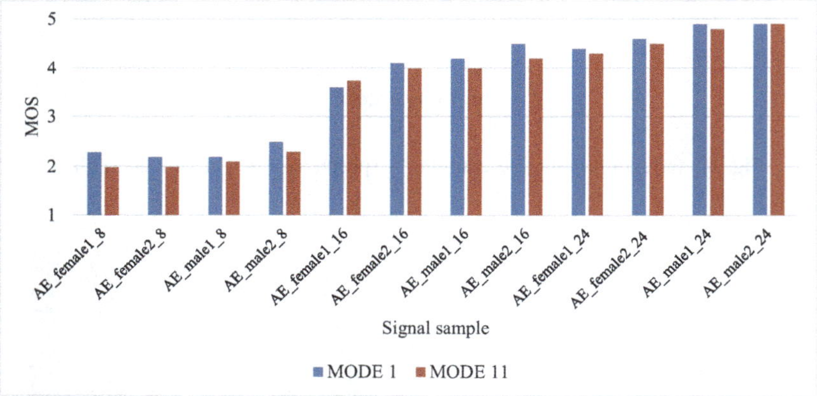

Fig. 7. Capacitive-inductive coupling – speech samples in American English.

While examining these results, one should take into account that any voice transmission system may be considered as of high-quality whenever it receives a MOS score of above 4.0. This breakpoint is sometimes referred to as the broadcast quality criterion. As shown, the lowest bitrate equal to 8 kbit/s proved to be insufficient when it comes to delivering clear and easily understandable voice messages. On the other hand, the medium bitrate of 16 kbit/s was ranked evidently better. Nevertheless, not all samples were perceived as of high quality.

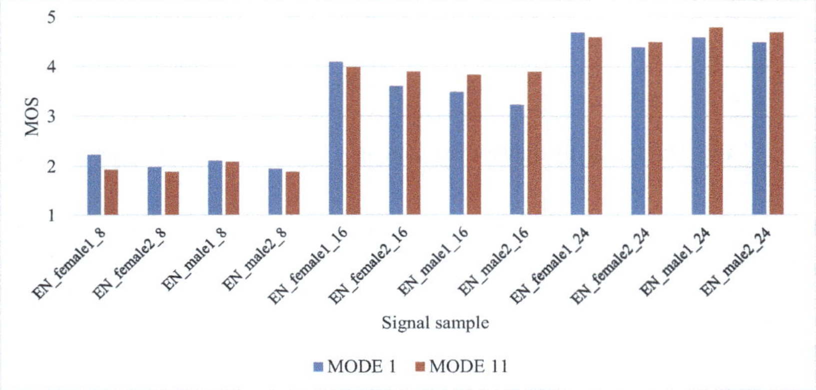

Fig. 8. Capacitive-inductive coupling – speech samples in British English.

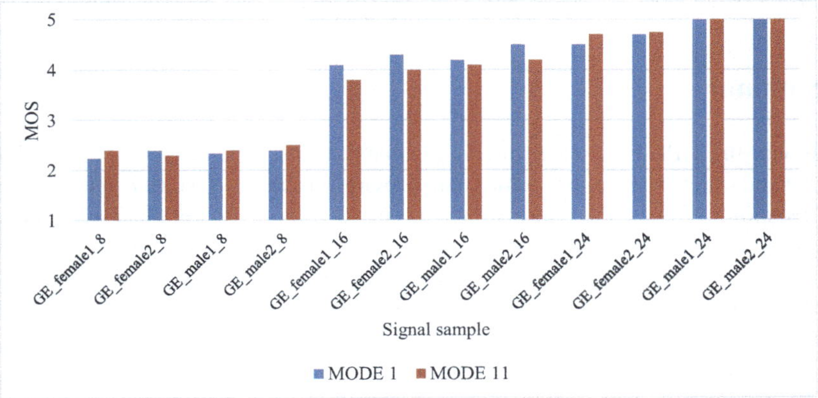

Fig. 9. Capacitive-inductive coupling – speech samples in German.

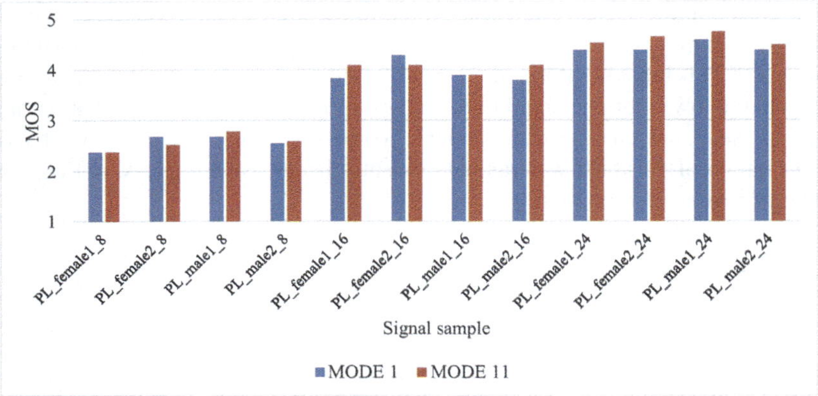

Fig. 10. Capacitive-inductive coupling – speech samples in Polish.

In case of the highest bitrate of 24 kbit/s, all voice messages, whether spoken by a male or female lector, were clear and easily understandable. According to obtained results, the threshold of 24 kbit/s may be viewed as a break point (MOS score of above 4.0 for all signal samples), when further increase in bitrate will not relate to further raise in perceived subjective quality.

When it comes to the background of tested individuals, it is worth mentioning that they were all Polish native speakers, whereas both English and German were the second language of choice (without a clear advantage). Moreover, participants pointed out that sentences spoken by a male lector seemed more appealing. In our case, due to the profile of KGHM Polska Miedz S.A., this feedback becomes an important factor.

6 Summary

As shown, the BPL-PLC wired medium, thanks to its high resistance to mechanical damage and other physical properties, can provide a reliable voice transmission system. This technology, even in a narrowband scenario (bitrates lower than 1 Mbit/s), e.g. caused by bandwidth limitations, severe damage, etc., ensures a stable and reliable connection. Whenever an emergency situation occurs, voice commands, e.g. from a supervisor or paramedic, can help provide maintenance instructions and ease during a rescue operation. Previous random events around the world have shown how important is to maintain contact and communication.

The preliminary technical medium examination shown, that induction-inductive coupling has a clear advantage over mixed capacitive-inductive coupling, when it comes to available throughput. However, when examining results in case of voice transmission, there are situations in which Mode 11 proved to be superior, especially for bitrates of 16 and 24 kbit/s. As shown, the BPL-PLC technology can provide stable and reliable voice transmission services at 24 kbit/s, regardless of the spoken language, or even type of coupling.

Results of this study are characterized by a high degree of usability. They can aid both researchers and scientists during the design and maintenance phase of a wired BPL-PLC voice communication system. Application of such a system may be of interest to professionals from the gas and mining sector, as well as oil industry. The designed solution may help speleologists and paramedics during any rescue operation.

References

1. Pyda, D., Habrych, M., Rutecki, K., Miedzinski, B.: Analysis of narrow band PLC technology performance in low-voltage network. Elektronika ir Elektrotechnika **20**(5), 61–64 (2014)
2. Mlynek, P., Misurec, J., Koutny, M.: Modeling and evaluation of power line for smart grid communication. Przeglad Elektrotechniczny **87**(8), 228–232 (2011)
3. Meng, H., Chen, S., Guan, Y.L., Law, C.L., So, P.L., Gunawan, E., Lie, T.T.: Modeling of transfer characteristics for the broadband power line communication channel. IEEE Trans. Power Delivery **19**(3), 1057–1064 (2004)
4. Lampe, L., Tonell, A.M., Swart, T.G.: Power Line Communications: Principles, Standards and Applications from Multimedia to Smart Grid, 2nd edn. Wiley, Chichester (2016)
5. Carcell, X.: Power Line Communications in Practice. Artec House, London (2006)
6. Habrych, M., Wasowski, M.: Analysis of the transmission capacity of various PLC systems working in the same network. Przeglad Elektrotechniczny **94**(11), 130–134 (2018)
7. CENELEC EN 50065-1: Signalling on Low-Voltage Electrical Installations in the Frequency Range 3 kHz to 148.5 kHz – Part 1: General Requirements, Frequency Bands and Electromagnetic Disturbances (2011)
8. Gilski, P., Stefanski, J.: Subjective and objective comparative study of DAB+ broadcast system. Arch. Acoust. **42**(1), 3–11 (2017)
9. Yang, M.: Low bit rate speech coding. IEEE Potentials **23**(4), 32–36 (2004)
10. Brachmanski, S.: Quality evaluation of speech AAC and HE-AAC coding. In: Proceedings of Joint Conference – Acoustics 2018, pp. 1–4. Polish Acoustical Society – Gdansk Division, Ustka (2018)
11. Falkowski-Gilski, P.: Transmitting alarm information in DAB+ broadcasting system. In: Proceedings of 22nd Signal Processing: Algorithms, Architectures, Arrangements, and Applications Conference (SPA 2018), pp. 217–222. IEEE Poland Section – Circuits and Systems Chapter, Poznan (2018)
12. Debita, G., Habrych, M., Tomczyk, A., Miedzinski, B., Wandzio, J.: Implementing BPL transmission in MV cable network effectively. Elektronika ir Elektrotechnika **25**(1), 59–65 (2019)
13. ITU-T P.501: Test Signals for Telecommunication Systems (2017)
14. Li, T., Rahardja, S., Koh, S.N.: Fixed quality layered audio based on scalable lossless coding. IEEE Trans. Multimedia **11**(3), 422–432 (2009)
15. Griffin, A., Hirvonen, T., Tzagkarakis, C., Mouchtaris, A., Tsakalides, P.: Single-channel and multi-channel sinusoidal audio coding using compressed sensing. IEEE Trans. Audio Speech Lang. Process. **19**(5), 1382–1395 (2011)
16. Helmrich, C.R., Markovic, G., Edler, B.: Improved low-delay MDCT-based coding of both stationary and transient audio signals. In: Proceedings of IEEE International Conference on Acoustic, Speech and Signal Processing (ICASSP 2014), Florence, pp. 6954–6958. IEEE (2014)

17. Lin, J., Min, W., Shengyu, Y., Ying, G., Mangan, X.: Adaptive bandwidth extension of low bitrate compressed audio based on spectral correlation. In: Proceedings of International Conference on Intelligent Computation Technology and Automation (ICICTA 2015), Nanchang, pp. 113–117. IEEE (2015)
18. ITU-R BS.1284: General Methods for the Subjective Assessment of Sound Quality (2003)
19. Mardia, K.V., Jupp, P.E.: Directional Statistics. Wiley, New York (2000)
20. Pearl, J., Glymour, M., Jewell, N.P.: Causal Inference in Statistics: A Primer. Wiley, New York (2016)

Evaluation and Improvement of Web Application Quality – A Case Study

Anna Derezińska(✉) and Krzysztof Kwaśnik

Institute of Computer Science, Warsaw University of Technology University, Nowowiejska 15/19, 00-665 Warsaw, Poland
A.Derezinska@ii.pw.edu.pl

Abstract. Web applications, especially those commonly used by a vast number of clients, should fulfill high quality standards. In this paper, we discuss evaluation and improvement of quality attributes on the example of web applications that support city cards for public transport in urban areas. Their quality has been assessed with a set of autonomous evaluation tools in terms of usability, findability in Internet (SEO), accessibility, design, content, mobile, and performance. Quality of the applications has been compared to quality of a new prototype developed as a Single Page Application (SPA). An improved prototype has been created that referred to suggestions offered by the tools. The case study has also presented effects of the prototype improvement on its quality attributes, and relevance of the tool usage.

Keywords: Web application · Quality improvement · Single Page Application · Web of public transport systems

1 Introduction

User-centered web applications can be devoted to thousands of users with different profiles and capabilities that interact with them daily. Various quality features are of the most importance of such applications. Standards of software application quality, such as SQuaRE series (Systems and Software Quality Requirements and Evaluation) [1, 2] consist of eight general features, subdivided into more precise attributes. However, considering web applications and their interfaces, researchers concentrate on selected characteristics that influence a customary user reception [3–5].

In this paper, we have focused on the city card-based applications, which support modern solutions integrating the region's services related to public transport into one coherent system. We have performed a case study aimed at the following goals:

- Comparison of quality of widely used web applications of similar functionality, on the example of public transport support in big cities of Poland [6–8].
- Review of utilization of tools for evaluation of web application quality [9–13].
- Development of a prototype using Single Page Application (SPA) that fulfills the desired functionality related to public transport in a user-friendly manner.
- Evaluation of the prototype quality and comparison with the real applications.

W. Zamojski et al. (Eds.): DepCoS-RELCOMEX 2020, AISC 1173, pp. 187–196, 2020.
https://doi.org/10.1007/978-3-030-48256-5_19

- Quality improvement in the prototype based on hints suggested by the evaluation tools [9–13] and guidelines on web application improvement [14].
- Evaluation of the impact of the improvements on the web application quality.

Following this introduction, Sect. 2 reviews briefly related work. In Sect 3, quality attributes and their evaluation tools are presented. We discuss the applications used in the case study in Sect. 4. Experiment results are presented and discussed in Sect. 5. Finally, Sect. 6 draws main conclusions.

2 Background and Related Work

Among different models of software quality, one of the most influential is a series of ISO/IEC standards SQuaRE [1, 2] that classify different product characteristics and combine them with several software metrics. Different frameworks try to follow the standard guidelines [15, 16].

While investigating web applications, researchers usually focus on a part of the standard [3], or selected detailed features [4, 5]. Lew et al. [3] proposed a strategy for understanding and improving quality in use and related it to web applications. Kaur et al. [4] reported on evaluation of university websites using automated tools to prepare a website ranking and find their weak spots. Similar features and tools were discussed by Al-Omar [5] after experiments on e-learning websites. The author pointed at unsatisfactory level of the usability and accessibility of the sites in contrast to a better score of their reliability.

Architecture and technology of a web application strongly influences its ergonomic features, performance and security. Two strategies to develop web applications can be selected: multi-page application (MPA) and single-page application (SPA).

In the MPA solution a dynamic HTML is generated by a server. Each time when one moves from one web page to another, a request to a server is generated. A new HTML page is requested. This strategy is usual suitable for bigger applications that have many views. In basic, all page data are updated, regardless the amount of data. However, some technologies can be used (e.g. AJAX requests) do transmit data between a server and a browser only when bigger changes are applied.

An SPA application works within a browser that does not require reloading a page during utility. If a web page is visited for the first time, a source code of the application is loaded. While moving to subsequent views of the web page, a user does not need to wait for a server answer to change a view. Therefore, user presentation views can be altered without reloading the whole page, but only some parts of DOM (Document Object Model).

Some problems of creating an SPA application and methods of tuning its quality attributes were examined by Jadhav et al. in [17]. However, the framework discussed (AngularJS) is different from its successor used in the prototype developed in our case study. SPA applications based on AgularJS were also subjects of investigation performed by Stępniak & Nowak [14]. They analyzed different optimization methods and their impact on the loading time of SPA applications.

3 Evaluation of Website Quality

3.1 Quality Attributes Assessed by Website Quality Evaluation Tools

The following attributes are concerned in the quality evaluation of website applications by automated tools:

- *Usability* relates to an easiness of using a site interface for a given type of users and in a given context.
- *Findability,* often associated with *SEO (Search Engine Optimization),* is specified by a set of rules that could be used by site developers to encourage to visit the site by potentially interested persons. It tends to gain a high position within searching results of web browsers such as Google, Big or Yahoo! for selected content, signs or keywords.
- *Accessibility* is specified by a set of rules that ensure lack of obstacles in using the site by any user. Possible obstacles taken into account may refer to different versions of computer hardware, software, language, localization, technical skills, optical or auditory disability.
- *Mobile* defines how a website is adjusted to mobile devices, such as a smartphone. A monitor resolution should conform to a user needs, regardless using a stationary monitor (e.g. 1900×1200) or a smartphone (e.g. 200×100). An alternative solution is creation of a separate website dedicated to mobile devices. But in the considered case the solution of a site with a single address is preferred.
- *Design* feature defines basic parameters about the site construction, including among others:
  - Type of HTML document.
  - Type of coding, and whether this information is written it the site header.
  - Are Flash animations used by the site?
  - Are Frames and Iframes applied?
  - Are e-mail addresses converted to pictures or JavaScript in order to avoid detection by automata spying the sites?
  - Is the site accessible only via https – an encrypted protocol?
- *Content* is a feature that specifies quality of text presented in a website. The following measures could be taken into account:
  - A ratio of text length to HTML code length.
  - Site headers in HTML source code should be specified from h1 to h6.
  - Word repetitiveness in a site is verified, comparing also additional styles, as bold, underline, title, etc.
  - Availability of Microdata included in the HTML source code.
- *Security* is specified by using an SSL key and encrypting data transferred between a site server and a browser. Such a connection keeps all the data private.
- *Performance* is evaluated by a site content verification and measurement of elements of the site. The following components could be taken into account:
  - Size of the whole starting website.
  - Verification of requests sent during site generation (whether the request number does not influence loading time of a start site).

- Site generation time that is independent of an Internet connection but depends on a server configuration and source code of the site.
- Data compression used to optimize data to be downloaded by a user.
- Verification of JavaScript location. Scripts should be placed at the site end to perform site before script interpretation. It influences loading time in a browser.
- Style verification.
- Verification of a predefined CSS class.

3.2 Web Quality Evaluation Tools

There are different tools that evaluate quality of web sites. They present results of various sets of quality attributes. Moreover, they apply different detailed strategies to calculate the same attributes and different approaches to assess the overall quality. Therefore, we have selected a number of tools to be used in experiments.

A) *Site Analyzer* [9] evaluates SEO, performance, design, content and accessibility assigning up to 100 points to each attribute, and returning the overall quality (also to 100 points).
B) *Quality Validator (Qualidator)* [10] performs about 60–70 automatic tests to assess usability, SEO, and accessibility. Attributes and the overall quality are presented as percent values.
C) *Website Grader* [11] focuses on SEO, mobile, security, and performance assigning up to 30, 30, 10, and 30 points for each attribute accordingly. The overall quality is counted as a sum of these attributes (up to 100 points).
D) *SEO Web Page Analyzer* [12] takes into account SEO, content, design and performance of a website. Analysis result is presented as an overall score (up to 100 points).
E) *Chrome DevTools* [13] can perform a quality audit that comprises SEO, accessibility, design (regarded as application of a set of best practices) and performance. Each quality attribute is normalized to max 100 points. No overall quality measure is provided.

Additionally, the tools provide good and weak points of an analyzed website and/or give suggestions about website changes that might improve its quality outcomes.

4 Case Study Overview

4.1 Web Applications Used in Public Transport

In several provinces of Poland, independent computer systems of public transport have been developed and applied. Three systems with their websites have been examined in the case study of concern. These systems have been commonly used by many clients of different urban areas for several years.

- MKA - Małopolska Agglomeration Card [6]. MKA is dedicated for the citizens and tourists of Krakow and other areas of the Małopolskie Province. The site gives information about MKA card, tickets and services, public transport, parking lots etc.

A client can be registered and use an account with e-mail identifier and password or log in using a Facebook or Twitter account. A history of transactions is supported by a client account. The most important facilities refer to buying various kinds of tickets. The tickets are paid by means of Web schemes. Additionally, tWallet can be used for paying tickets. Apart from services accessible via Client Portal of the MKA website, a mobile application iMKA can be used.
- PEKA [7] It is the Card of the Poznan urban area. Ticket regulations are different from those of MKA, but general functionalities of the web application are similar. There are also client accounts, various ticket selecting and purchasing, tWallet, etc.
- Waw Card [8] is a card associated with a public transport system of the Warsaw urban area. The website also presents necessary information and serves the Client Panel to register clients and sell tickets that are applied to the personal card. Other simple tickets can be bought by other mobile applications delivered by cooperating companies.

The discussed systems were developed using the MPA strategy, do not keep up with current trends, and are scarcely adapted to individual or business requirements. The number of activities to be performed by a client in order to obtain desired transport rights is quite high. In PEKA, purchasing of a long term ticket is divided into several steps, in which some data are loaded in many views. Moreover, in some cases, e.g. no tariff with a number of days selected by a client, a ticket cannot be configured. In the case of the Warsaw system, information about purchasing tickets via the Internet is not easily visible neither in the main site nor in the Client Panel for a logged in client.

4.2 Prototype Requirements

There has been developed a prototype that covers the most important functionalities of the applications supporting a card purchasing in public transport. Therefore, its main functional requirements refer to the following activities:

- *Registration* of new clients using a registration form, in which all text areas should be verified and a password treated accordingly secure.
- *Login* by a client panel that appropriately handles personal data, password and activity time.
- *Language support* for at least two variants that could be easily switched.
- *Edition* of personal data by a logged client.
- *Purchasing of tickets* of different types and configurations.
- *Complaint or comment submission* via a form by a logged client
- *Transaction history* that recalls purchased tickets with their configurations and prices and allows downloading of corresponding VAT invoices.
- *Logout* from a current client session.

The prototype has been aimed at fulfilling a set of non-functional features:

- *Clarity* – a limited number of information should be placed on the main website or client panels.
- *Simplicity* – a website should be loaded with all views, and ought to be legible and without complex navigation.

- *Accessibility* – could be accessed by many clients.
- *Performance* - results should be returned to a client within max 7 s (max acceptable delay according to [18]).
- *Lightness* – application size, including pictures, styles, text files, should be small as far as possible.
- *Intuitive* – all vital elements that lead to a ticket purchasing should be easy performed without any instruction, also by elder or visually impaired people.
- *Smooth transition between views* – transition between view should not require downloading additional data from a server. A next view appears in place of a faded previous one.

4.3 Experiment Set-up

The prototype architecture includes a web application, a data base, and an Internet service in a cloud. The application consists of three components:

- a view created by the Vue.js framework,
- an intermediate layer based on the MVC pattern, which gets data from the data base project, preforms the data and delivers to the view,
- a data base project that communicates with the data base.

The application has been implemented using App Service available in Microsoft Azure Portal delivered in a cloud server.

Experiments on the application quality have been performed according to the following scenario:

1. Quality evaluation of three external websites supporting public transport (MKA, PEKA, and Waw) using five tools A)–E).
2. Quality evaluation of the prototype using the same tools A)–E).
3. Prototype refactoring in order to improve its quality.
4. Quality evaluation of the improved version of the prototype using tools A)–E).

The measurements of the public transport websites correspond to their versions commonly available in May 2019.

5 Experiment Results

5.1 Outcome of Experiments

Evaluation results given by the tools include different attributes and can be expressed in different scopes of values. Therefore, outcomes of applications measured by any tool are presented in separate tables (Tables 1, 2, 3, 4 and 5).

In the upper three rows, results of the public transport websites (Sect. 4.1) are shown. In the case of Qualidator (tool B), evaluation of the Warsaw web application returned an error. Therefore, these results are missing (Table 2). Results of the developed application are shown in the bottom rows: the preliminary version of the prototype (Prototype 1st v.) and the prototype after improvement (Prot. Improved).

Table 1. Comparison of results by Site Analyzer (max 100 points each).

Application	SEO	Accessibility	Content	Design	Performance	Overall
MKA	48.0	61.3	43.9	78.3	61.0	59.3
PEKA	62.2	49.6	27.4	55.7	61.0	58.8
Waw Card	56.5	41.1	63.6	65.9	73.2	55.4
Prototype 1st v.	28.8	41.1	9.5	55.7	73.2	44.1
Prot. improved	43.7	56.0	9.5	62.0	73.2	54.4

Table 2. Comparison of results by Qualidator.

Application	Usability	SEO	Accessibility	Overall
MKA	83.9%	82.0%	86.9%	81.7%
PEKA	77.2%	72.0%	76.6%	76.4%
Waw Card	–	–	–	–
Prototype 1st v.	78.7%	66.5%	83.6%	72.3%
Prot. improved	78.7%	76.6%	83.6%	77.5%

Table 3. Comparison of results by Website Grader.

Application	SEO	Mobile	Security	Performance	Overall
MKA	25/30	30/30	10/10	12/30	77/100
PEKA	15/30	0/30	10/10	12/30	37/100
Waw Card	15/30	0/30	10/10	14/30	39/100
Prototype 1st v.	10/30	0/30	10/10	24/30	44/100
Prot. improved	15/30	30/30	10/10	30/30	85/100

Table 4. Comparison of results by SEO WebPage Analyzer.

Application	Overall
MKA	50/100
PEKA	41/100
Waw Card	46/100
Prototype 1st v.	53/100
Prot. improved	62/100

Table 5. Comparison of results by Chrome DevTool.

Application	SEO	Accessibility	Design (Best Practices)	Performance
Prototype – 1st v.	78/100	49/100	71/100	7/100
Prot. improved	100/100	59/100	93/100	91/100

Results of examination performed by the developer tools included in Google Chrome could only refer to an application under development, hence only two versions of the prototype are compared in Table 5.

5.2 Prototype Improvement

Based on the experiment results and methods to optimize SPA applications [14], the prototype has been updated. The following methods have been used to improve it:

- Concatenation of resources via combining of elements in order to reduce a number of requests to be sent.
- Adjustment of the site content and appearance to resolution of a device.
- Data compression - text files of selected types have been compressed using gzip.
- Removal of unused elements, such as unnecessary rules from a CSS file.
- Minification of files applied to all CSS and JavaScript files using UglifyJS [19].
- Minimization of the number of server queries.
- Selection of scripts to be asynchronously loaded in order not to block the site parsing (application of *async* and *defer* attributes in JavaSript).
- Compression of image files using WebP [20].
- Use of the https protocol instead of htpp.
- Use of HTTP/2.0 protocol instead of HTTP.
- Saving of files in a cache memory. It requires also changes in the application configuration read by the cloud server.
- Tuning the application to mobile devices

5.3 Discussion of Results

Comparing the results of the public transport applications with the prototype, we can observe that the best results were assigned to the MKA application by all tools except SEO WebPage Analyzer. The further judgments are not so unambiguous, as the tools take into account various attributes. Moreover, for the same attribute (e.g. SEO) MKA got the best results from Qualidator, PEKA was better according to Site Analizer, while Website Grader gave the same score to both applications. All in all, we can get an overview of possible weak points of an application and its position among others.

After improvements introduced in the prototype, several quality attributes evaluated with higher values. Based on Site Analyzer, all considered attributes except Content got more points, and the highest rise was for SEO (14.9/100). Similarly, SEO was raised according to Qualidator (10.1/100), although remaining attributes achieved the same outcomes. Website Grader reported also improvement in SEO, but as well in Mobile (from 0 to max) and Performance. Security attribute did not change, but it had already the max score before. SEO WebPage Analyzer finished with an overall rise of about 9/100. Finally, all quality attributes measured by the Chrome development tools resulted in considerable higher values; with the biggest increase in performance.

Summing up all applications, we can observe that the improved prototype achieved the best results according to SEO WebPage Analyzer and Website Grader. Other tools (A and B) counted the Krakow site as the best one. However, Qualidator (B) did not take Performance into account, which could lower the overall score of the prototype. Site Analyzer (A) did not notice any rise in Performance, in the contrary to other tools (C, D, E). It seems to indicate, that this attribute might be measured not precisely enough by this tool.

A great benefit of the tool usage was provision of hints about the application improvement. Especially valuable we have found detailed and accurate suggestions put forward by Website Grader, SEO WebPage Analyzer and the tools from Chrome. Whereas comments received from Qualidator or Site Analyzer were often too general or without practical significance. It is difficult to improve an application quality when no data about its bottlenecks or its critical features are available. It is hardly possible for an application that has many business-dependent views. Thus, accurateness of suggestions offered by the evaluation tools is of the highest importance.

Considering threats to validity, it should be noted that even the same quality attributes measured by tools (A–D) gave different values, as the evaluation criteria are different. However, measurements carried out on the same application by the same tool were deterministic, and always gave the same results independent on other factors.

6 Conclusions

It has been shown that evaluation of quality attributes using available tools can help in verification of a web application improvement, despite some ambiguity in results of the tools. It could be worthwhile to use different tools, although the most beneficial are those offering accurate suggestions apart from giving numerical quality estimates.

Moreover, the experiments have pointed at the necessity for more careful development of the professional web applications used in big cities of Poland. However, it should be stressed that the Warsaw application, which got the overall score of a medium or the worst position, has been reconfigured and its appearance has become more user-friendly since the time of the case study completion. More experiments are also planned on the prototype performance, dealing with loading times, other time constraints and stress testing using other dedicated tools. Comparison of SPA applications created with other frameworks, like React.js or Angular.js, to the prototype based on Vue.js could help in finding solutions of the best quality and performance.

References

1. ISO/IEC 25010: 2011 Systems and software engineering. Systems and software quality requirements and evaluation (SQuaRE) System and software quality models (2010)
2. ISO/IEC 25023: 2016 Software engineering: software product quality requirements and evaluation (SQuaRE) Measurement of system and software quality (2015)
3. Lew, P., Olsina, L., Becker, P., Zhang, L.: An integrated strategy to systematically understand and manage quality in use for web applications. Requirements Eng. **17**, 299–330 (2012). https://doi.org/10.1007/s00766-011-0128-x
4. Kaur, S., Kaur, K., Kaur, P.: Analysis of website usability evaluation methods. In: Proceedings of 3rd International Conference on Computing for Sustainable Global Development, INDIACom, pp. 1043–1046. IEEE, New York (2016)

5. Al-Omar, K.: Evaluating the internal and external usability attributes of e-learning websites in Saudi Arabia. Adv. Comput. Int. J. **8**(3/4) (2017). https://doi.org/10.5121/acij.2017.8401
6. MKA - Małopolska Karta Aglomeracyjna (Małopolska Agglomeration Card). https://mka.malopolska.pl/en. Accessed 08 Jan 2020
7. PEKA - Poznańska Elektroniczna Karta Aglomeracyjna (Card of Poznan urban area). https://www.peka.poznan.pl/SOP/login.jspb. Accessed 08 Jan 2020
8. Warsaw City Card. https://www.wtp.waw.pl/en/. Accessed 08 Jan 2020
9. Site Analyzer. https://www.site-analyzer.com. Accessed 22 Jan 2020
10. Qualidator. https://www.qualidator.com/Wqm/en/default.aspx. Accessed 22 Jan 2020
11. Website Grader. https://website.grader.com. Accessed 22 Jan 2020
12. SEO Web Page Analyzer. http://seowebpageanalyzer.com. Accessed 22 Jan 2020
13. Chrome DevTools. https://developers.google.com/web/tools/chrome-devtools/. Accessed 22 Jan 2020
14. Stępniak, W., Nowak, Z.: Performance Analysis of SPA Web Systems. In: Borzemski, L., et al. (eds) Proceedings of 37th International Conference on Information Systems Architecture and Technology – ISAT 2016. Advances in Intelligent Systems and Computing, vol. 521, pp. 235–247. Springer, Cham (2017). https://doi.org/10.1007/978-3-319-46583-8_19
15. Nakai,H., Tsuda, N., Honda, K., Washizaki, H., Fukazawa, Y.: A SQuaRE-based software quality evaluation framework and its case study. In: 2016 IEEE Region 10 Conference (TENCON), pp. 3704–3707. IEEE, New York (2016). https://doi.org/10.1109/tencon.2016.7848750
16. Martinez-Fernandez, S., et al.: Continuously assessing and improving software quality with software analytics tools: a case study. IEEE Access **7**, 68219–68239 (2019). https://doi.org/10.1109/ACCESS.2019.2917403
17. Jadhav, M.A., Sawant, B.R., Deshmukh, A.: Single Page Application using AngularJS. Int. J. Comput. Sci. Inf. Technol. **6**(3), 2876–2879 (2015)
18. Dennis, A.R., Taylor, N.J.: Information foraging on the web: the effects of "acceptable" internet delays on multi-page information search behavior. Decis. Support Syst. **42**(2), 810–824 (2006). https://doi.org/10.1016/j.dss.2005.05.032
19. UglifyJS 3: Online JavaScript minifier. https://skalman.github.io/UglifyJS-online/. Accessed 31 Jan 2020
20. WebP A new image format for the Web. https://developers.google.com/speed/webp. Accessed 31 Jan 2020

Scheduling Tasks with Uncertain Times of Duration

Dariusz Dorota(✉)

Cracow University of Technology, Warszawska 24, 31-155 Cracow, Poland
ddorota@pk.edu.pl

Abstract. This paper describes methods of scheduling depend tasks. Algorithm based on effective previous research of author work that consider scheduling of task with divisibility attribute with higher level of dependability. As a specification was used as a acyclic task graph which specifies the tasks and their dependencies. In this work duration of task is specified using probabilistic method, especially normal distribution of probability. The research focuses on scheduling tasks under selected uncertainty conditions.

Keywords: Scheduling task · Multiprocessor-Processor task · Divisibility of the task · Uncertain times of duration

1 Introduction

The continuous development of embedded systems forces the development of new, more effective methods of creating systems. As in the area of embedded systems one should focus on the development of not only software but also hardware, it is necessary to apply methods that perform this task from the class of NP-complete problems [27]. One of the most important problems associated with creating a system is the development of an effective algorithm in the maximum way using the equipment for the created software.

The constant development of technology and equipment, especially in the area of IoT and IoV, forces the increase of performance of such systems, that its time of execution and reliability [3], which is associated with the need to develop new and effective task scheduling methods, especially multi-processor tasks. These types of tasks are aimed at introducing and ensuring greater reliability of the system being created, as well as performing complex processes in general, using "real parallelism" [31].

An important aspect is the use of tasks with uncertain response time, which may mean the mutual use of the computing power of devices connected to one IoT network [18]. This concept translates directly into the response time determined from probability. It can be assumed that the system implementation times will be defined as the probability distribution [4]. Normal probability distribution is proposed in the paper.

By following the rules for determining normal distribution and the three sigma principle, you can determine with what probability the system response will be obtained, as outlined in the following chapters. A proposed here concept is to create a target system with an increased level of reliability, obtained through the mutual testing

W. Zamojski et al. (Eds.): DepCoS-RELCOMEX 2020, AISC 1173, pp. 197–209, 2020.
https://doi.org/10.1007/978-3-030-48256-5_20

of tasks [14], which, due to the fact that the time of the whole system is obtained with a certain probability, is also determined in a certain probability range. Applying the principles of mathematics, namely the three sigma principles in the field of probability, we can determine the probability range of system reliability for 40%, 68.7%, 95.4% or 99.7% [5]. This work focuses on using the proposed MCD (Muntz-Coffman Dorota) algorithm [14] for scheduling tasks with specific probabilities.

2 Scheduling Description

This chapter presents current state of knowledge, the scheduling issue, especially in the context of using scheduling methods in the presented work including the concept of using probabilistic methods.

The most common notation is the so-called three-field notation called Graham's notation. In this notation parameters α, β, γ can be distinguished [24, 26]. Considering the Grahams notation, can specify the type of problem (the so-called machine park) as the alpha parameter. Then, the field, which is optional, determines the number of production processors and field describe type and number of transport processors [27]. β field means technological requirements in problem of scheduling tasks (tasks characteristic). When field β is empty it means there are no restrictions. Last parameter γ determines goal function form. Multiprocessor systems allow increasing the time of execution of created embedded systems, while applying other constraints such as the demand for system power or used area [19, 23, 27–29]. The constant development of computer sciences as well as all kinds of computer systems forces the development of increasingly new task scheduling methods, whose task is to use equipment as efficiently as possible for specified tasks. Among the algorithms of scheduling, deterministic and stochastic algorithms can be distinguished [24, 27]. Each scheduling problem can be described as a pair (X, F), where is set of sensible solutions, while $F : X \rightarrow R$ is goal function (optimization criterion). The task scheduling problem boils down to determining a globally optimal solution such that, where x is the solution to the problem. Regardless of what the system's input specification is, each of the tasks and the target system must have specific task parameters and resources to be used for scheduling.

Usually in hard real-time systems where time constraints cannot be violated, design with WCET (Worst Case Execution Time) can be considered. In this case each of its jobs is guaranteed to finish be-fore its deadline. Such a task is initially tested and is released only if it is possible to guarantee its completion before the deadline, otherwise the task is rejected. In soft real-time systems tasks are often accepted in the system without any form of guarantee [1]. Task scheduling methods, especially those based on the use of a graph, allow defining the BCCT (Best-Case Completion Time) and the WCCT (Worst-Case Completion Time) of each task. This formulation of the problem allows you to determine whether the time requirements have been met and the relaxation time [10].

2.1 Concept of Using Probability

Uncertain times are used in scheduling method proposed in paper [2], where author consider periodic and semi-periodic tasks. In that work probabilistic methods are used for providing schedulability guarantees for the semi-periodic tasks [20]. A common approach for scheduling with uncertainty conditions, requires the use of probabilistic models to describe parameters of uncertainty [17]. In probabilistic approach usually is defined probability correlated with events, that means probability of frequency of event. The probability distribution of the variable X can be uniquely described by its probability distribution function. Probability density function F(x) may by any function that determines the probability density in terms of the input variable x. The uncertainty is usually modeling by using discrete probability distributions or where uncertainties is used for describe equipment breakdown, failure of process operations are generally described with discrete parameters [21]. Often when is proposed parameter of uncertainty in system are considered so-called robust system, where priority is reliability of created system [17, 21]. The probabilistic methods are used to guaranteeing a desired QoS. The aim of scheduling with QoS is to ensure that a task will finish within soft deadline with probability. To achieve QoS is necessary to describe all tasks parameters in a probabilistic way [10].

Probability was used to describe the duration of the task. One of the most significant distribution of probability is the normal distribution, also called the Gaussian distribution. Normal distribution can be determined in several ways, including using the density function or the distribution function [30]. If the random variable can be described using this distribution and, the distribution is called the standard normal distribution, whose density function is given by the formula:

$$f(x) = \frac{1}{\varsigma\sqrt{2\pi}} e^{-i\omega t - \frac{(x-\mu)^2}{2\varsigma^2}}, \tag{3}$$

In principle, the author assumes that all variables in the system can be treated as variables described by the normal distribution of probability. Thus, it can be assumed that the arithmetic mean is calculated for all tasks and then the standard deviation. Therefore, when providing task specifications, they should be treated as the number of data that will be used to calculate the probability of scheduling. If it is assumed that the system scheduling time can be described using the normal probability distribution, we can also apply the three sigma principle. Thus, the resulting scheduling time for a specific multiprocessor architecture is time with 40% probability. In order to obtain a higher probability of ordering tasks, it is necessary to appropriately increase the range by adding standard deviations successively to the basic ordering time.

Therefore, in order to obtain a scheduling time with a higher level of probability, it is necessary to assume a the range of time by calculating $\pm\varsigma$ (sigma) from the basic scheduling time, thus obtaining a ranking in the range of 68.3%. Then, to achieve a 95.4% level of probability scheduling time, two standard deviations from the mean should be taken. However, according to the aforementioned principle, assuming three standard deviations, we obtain a scheduling at level of 99.7%.

By using the described approach, apart from the average value, we obtain a time of ranking in the specified range. This range strictly depends on the assumed level of probability, according to how it is adopted in the Gaussian distribution, for a greater probability a larger time discrepancy must occur.

3 Problem Formulation

3.1 Description of Algorithm

The MC (Muntz Coffman) algorithm was proposed in [6] as the optimal scheduling of tasks on two processors. That algorithm is the result of the previous research of the authors. In MC it assumes independent tasks in which the scheduling takes place depending on the priorities set. This approach allows for any scheduling of tasks so as not to exceed the predetermined time for the entire system. The issue of preempted dealt with in earlier author's works is extremely interesting. Preempted in scheduling tasks for two processors was presented in [7]. The algorithm proposed by MC is the optimal algorithm for scheduling tasks divided on two processors [6].

In this chapter was presented modified algorithm for multiprocessor tasks. Originally for scheduling task in two processor architecture was used MC algorithm. In previous work author proposed extension and modification of algorithm. MCD (Muntz Coffman Dorota) algorithm schedule multiprocessor tasks in NoC (Network on Chip) architecture. The new approach is to scheduling with a certain level of probability, both in terms of the time needed to implement the system and the level of system reliability. As set V = {v1, v2, v3, v4} is denoted range of scheduling time for different level of level of probability, using the three sigma rule.

The novel approach proposed by author is to describe time of task using standard normal distribution of probability. Author used MCD algorithm with that assumption, uncertainty was considered as describing the tasks using the probabilistic methods described above. The procedure of the modified MCD algorithm for three-processor tasks is given below:

The procedure of the modified MC algorithm for three-processor tasks is given below:

1. Load the system specification in the form of a task graph
2. If chosen probability is v1 (equals 40%) then go to step 3, else go to step 8
3. Set levels for all non-completed tasks:
   1.1 The task level P_x is the sum of the task execution times ti on the longest path for the selected task $P_x = A_x = max \sum_{i=x}^{n} (t_i)$, where $A_x = \max(A_{i=1}^{K})$, is the time of the longest path for task x, for the initial task x is selected $\max(A_1, \ldots, A_k)$
   1.2 If the selected T_i task at time t_i is a i-processor task:

      1.2.1 $_x = t_x * i$

4. Select the task with the highest level
   3.1 If there are two (or more) tasks at the same level:
   The priority is a dual-processor task, (the exception are dependencies which condition the execution of subsequent tasks, i.e. enabling the implementation of successors, e.g. the need to perform a task/tasks not to exceed the time limits T (Rest) X, where x is the number of the next time limit)
   3.2 If there are two (or more) two-processor tasks at the same level, the first task should be higher in the hierarchy (or with the lower number of the task)
   3.3 Delete the selected task from the set $\{P_a, P_b, \ldots, P_z\}$,
5. Simulate the task execution on the selected processor/processors per unit of time
6. If after the simulation the processor is available in the selected time unit: Select the next task (go to step 1)
7. If there are still unassigned tasks in the graph, go back to step 3
8. Chose next level of probability from set V = {v1, v2, v3, v4}
   1. If minimum level of probability is available chose it and remove it from set V
   2. Else If maximum level of probability is available chose it and remove it from set V
   3. Go to step 3
9. If V is not empty go to step 8
10. For each solution calculate level of dependability
11. Exit the algorithm.

3.2 Description of Graph as an Input with Probabilistic Approach

In scheduling methods we may talk about dependent and independent tasks. If we consider independent tasks then scheduler can map those independently, means without keeping any order. Opposite are tasks which we can schedule as dependent tasks, which are executed in order according to precedence.

This work presents an algorithm for dependent tasks. There are several ways to represent the second type of tasks, including the task graph, SystemC, etc. [16, 18]. In this work, an acyclic task graph was used for the system specification. The input tasks are specified by the graphs proposed by the author based on TGFF graphs [11]. TGFF provides a standard method for generating random allocation and scheduling problem. Randoming graphs which are generated based on this method have an additional task multiprocessor sign. The number of multiprocessor tasks and the multiprocessor designation is generated randomly for the entire graph. In this work designation for multiprocessor tasks are proposed as follows notation for: 1-, 2- and 3-processor tasks.

Acyclic directed graph G = (V, E) can be used to describe the proposed embedded system, where V mean nodes and E mean edges. Each node in the graph represents a task, while the edge describes the relationship between related tasks. Each of the edges in the graph is marked by a label where each of its indices defines the tasks that connect. The graphs of the tasks used for the system specification are shown in Fig. 1.

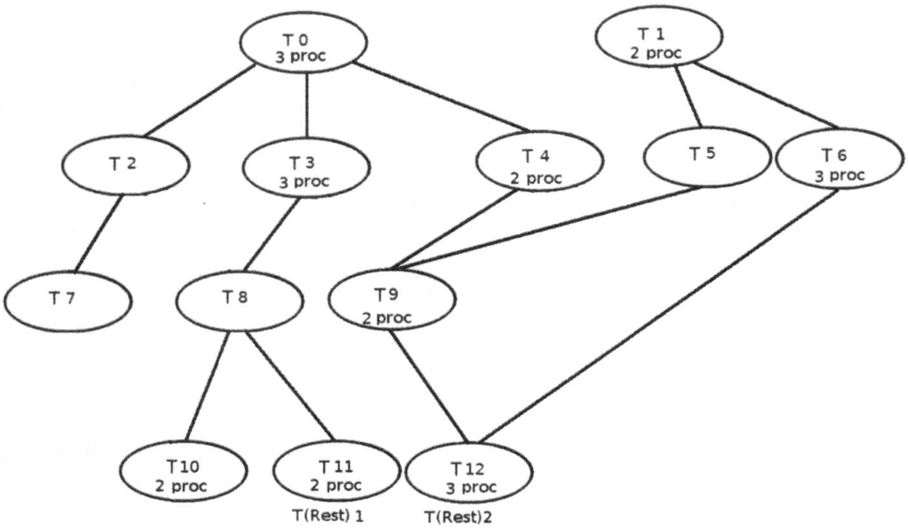

Fig. 1. Exemplary graph 1

4 Model of Dependability

By using the approach used in this paper presented in the article is the use of one-, two- and three-processor tasks, i.e. the need to perform a specific task/tasks on two processors at the same time. The first aspect that will be considered is system reliability. Such reliability can be achieved by using redundancy for critical tasks and thus testing the correctness of the implementation of the selected task on the second processor and if it's necessary adding third processor as arbiter. In the proposed approach, only selected tasks are two- or three-processor tasks in order not to significantly increase the cost of the entire system. Reliability plays a particularly important role in systems requiring a high degree of operational correctness, especially in real-time systems, used in industries, where both temporal determinism and correct operation of the whole system play a significant role. The motivation to enter uncertain task time, which is determined using probabilistic methods, may be the unknown response time of the device to an external or internal event that allows the task to be executed. With this assumption, it should be assumed that not only the initialization task but also the triggered one may have uncertain time of both notification and execution.

Due to the introduction of three-processor tasks, a system with an increased degree of reliability is ultimately created. This concept is based on the previously presented proposal of task redundancy and their mutual testing, if necessary testing with an arbitrator [14].

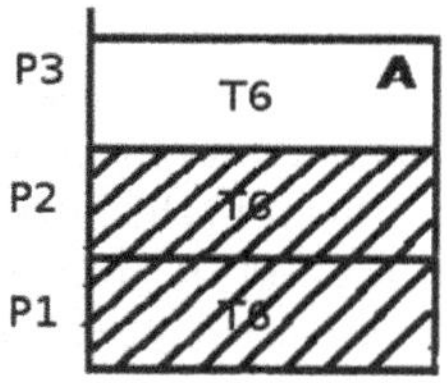

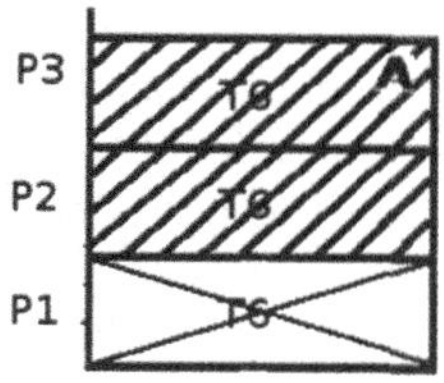

In this article, the condition of uncertainty is also considered, which is understood as the ranking time with a normal probability distribution. Considering the system reliability in this context, one should consider obtaining the reliability parameter also using a certain probability measure.

Based on the effects of the work presented above, as well as the author's previous experience with the issue of scheduling, three-processor tasks with the attribute of divisibility were proposed here [12, 13]. In the algorithm presented in the work, scheduling tasks at the beginning are determined paths in the graph using the A* algorithm. The proposed algorithm for tasks scheduling was based on previous work [14]. To streamline and speed up the operation of the algorithm, omitting the so-called multiprocessor ratio checking. As a result of the conducted research, it was proposed to use the coefficient for both single and multiprocessor tasks. This factor is equal to the number of processors necessary to execute the task. The modification was also proposed in terms of the use of multiprocessor tasks. This algorithm allows you to create a system with higher-level of reliability. The concept of a multiprocessor task comes from tests of scheduling in multiprocessor computer systems, testing of VLSI (ang. Very Large Scale Integration) or other devices [28]. Testing processors by each other requires task action on at least two processors simultaneously [18].

An additional parameter that was consider is probability. that factor has been added to the proposed formula.

To streamline and speed up the operation of the algorithm, omitting the so-called multiprocessor ratio checking. As a result of the conducted research, it was proposed to use the coefficient for both single and multiprocessor tasks. This factor is equal to the number of processors necessary to execute the task. The modification was also proposed in terms of the use of multiprocessor tasks. This algorithm allows you to create a system with higher-level of reliability. The concept of a multiprocessor task comes from tests of scheduling in multiprocessor computer systems, testing of VLSI (Very Large Scale Integration) or other devices [28]. Testing processors by each other requires task action on at least two processors simultaneously [14]. Dependability for the entire system was calculated using the formula:

$$D_x = \frac{\sum_{i=x}^{n}(p_i)}{\sum_{i=x}^{n}(t_i)}, \tag{4}$$

$$P_p(D_x, \varsigma) = P_p(\frac{\sum_{i=x}^{n}(p_i)}{\sum_{i=x}^{n}(t_i)}, \varsigma), \tag{5}$$

Where denote level of probability, is result of dependability function, is standard deviation in probability, denote level of dependability, is task priority and is task time. According to the proposed formula, reliability levels were calculated for each of the examples for which the scheduling was performed. Similarly as time of scheduling we calculate as a probability in case of level of dependability with effort of probability. Therefore, in order to calculate the level of reliability with a higher probability, it must be given to a certain extent as illustrated in Table 2.

5 Representation of Tasks

In this work was used task graph as input data for the system under consideration. The only change in relation to the one-processor tasks is to add annotations in the task graph indicating that the task must be completed as a two-processor or three-processor task. This was accomplished by adding the description "2 Proc" or "3-Proc" in the representation of a given task, which means the necessity of its simultaneous implementation on three machines. Assuming that Ti is a task with a number "i" in a task graph, then the task $T1_i$, $T2_i$, $T3_i$, means that the task has a number and is one-, two-, and three-processor. Of course, here only the annotation about what type of processor is used, which in fact translates into the use of tasks that require one or more processors to operate. Graphical notation is the same like in previous work [23, 27].

The process of scheduling three-processor tasks follows the principles introduced in the MC algorithm with the changes presented in this article [14]. In the approach using the DMC algorithm, it was proposed to use the appropriate multiplier for multiprocessor tasks directly depending on the multiprocessor coefficient of the selected task. The use of this approach allows for the promotion of three-processor tasks from other

Table 1. Exemplary prioritization for tasks shown in Fig. 1.

Task	Time	Level of task/Time to ending										
T0	30	195	195/20	195/5	0	0	0	0	0	0	0	0
T1	10	155	0	0	0	0	0	0	0	0	0	0
T2	5	30	30	30	30	30	0	0	0	0	0	0
T3	10	80	80	80	80	80	80	80	0	0	0	0
T4	15	105	0	0	0	0	0	0	0	0	0	0
T5	20	110	110	110/5	0	0	0	0	0	0	0	0
T6	25	135	135	135	135/10	135/5	0	0	0	0	0	0
T7	25	25	25	25	25	25	25	25	25	25/10	0	0
T8	10	50	50	50	50/5	0	0	0	0	0	0	0
T9	15	90	90	90	90	0	0	0	0	0	0	0
T10	20	40	40	40	40	40	40	40/15	40/5	0	0	0
T11	10	20	20	20	20	20	20	20	20	20	20	0
T12	20	60	60	60	60	60	60	60	60/15	0	0	0

tasks, of course, the order conditions and time constraints must be met here. As an example, you can specify task prioritization according to the modified MC algorithm:

6 Result and Discussion

This chapter presents the results obtained during experiments that reach task scheduling with the attention of one, two and three-processor tasks with probability. Experiments were carried out on a selected number of graphs for systems which included multi-processor tasks in three-, four- and fife-processor architecture.

Exemplary ordering of tasks in the target system is shown in Fig. 2. The target system is implemented on a predefined multiprocessor architecture based on the NoC network. The target architecture is generated in the first step of creating the system, assuming that along with the specification given in the form of a graph of tasks, the number of computational elements on which the system is to be implemented is determined. The Table 1 presents the calculated priorities of tasks in individual steps that directly affect the way of ordering tasks in the target system, and thus have an impact on the duration of the entire system.

The experiments were carried out on graphs generated on the basis of TGFF with the number of tasks from 10 to 50. For each number of tasks in the graph, several

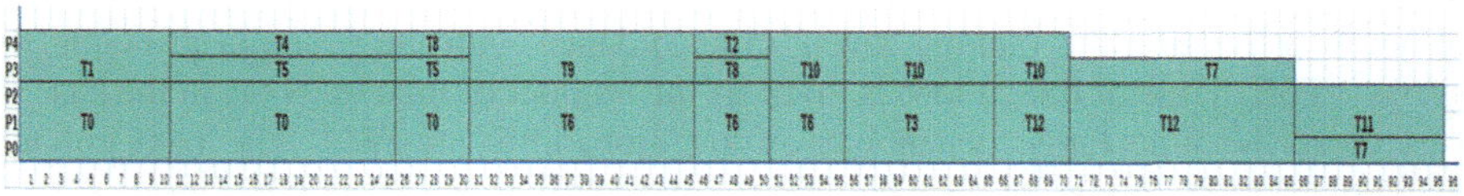

Fig. 2. Scheduling task in 5-processor architecture from specification in Fig. 1

Table 2. Exemplary task scheduling times for selected 3, 4 and 5 processor architectures

Graph	Count of tasks	One-proc tasks	Two-proc tasks	Three-proc tasks	Count of proc.	ToE (3 proc)	ToE (4 proc)	ToE (5 proc)	LoD (5 proc)
G1	9	4	2	3	3	85	70	55	2,29
G2	12	5	4	4	3	170	152	95	2,32
G3	14	5	5	5	3	185	168	105	2,35
G4	18	5	7	7	3	245	224	130	2,40
G5	21	6	6	10	3	305	270	205	2,66
G6	25	8	7	6	3	335	290	240	2,61
G7	31	11	8	13	3	440	355	100	2,68
G8	35	10	8	18	3	640	589	400	2,70
G9	45	12	10	24	3	825	780	780	2,62
G10	50	12	14	25	3	925	800	630	2,70

Table 3. Scheduling with uncertain times

Graph	ToE (5 proc)v1	ToE (5 proc)v2	ToE (5 proc)v3	ToE (5 proc)v4	LoDv1
G1	55	49–61	44-66	38–72	2,29
G2	95	86–105	80–110	73–117	2,32
G3	105	99–111	93–117	87–123	2,35
G4	130	124–136	118–142	112–148	2,40
G5	205	195–215	185–225	175–235	2,66
G6	240	231–249	221–259	212–268	2,61
G7	100	88–112	76–124	64–136	2,68
G8	400	387–413	374–426	360–439	2,70
G9	780	764–796	748–812	732–828	2,60
G10	630	615–645	601–659	585–674	2,62

experiments were carried out to authenticate the scheduling results for graphs of different structure and identical number of tasks. For each scheduling result were calculate probability for four level of probability. The list of scheduling results has been placed in Table 2 and with uncertain tasks time in Table 3.

As shown in Table 3, four results are obtained for one input graph (specifying the system). The obtained scheduling's do not give unequivocal times depending on the assumed level of probability, the exception is v1 (40%). Table 3 presents the elements of set V presenting the ranking levels in accordance with the 3 sigma principle. In Table 3 ToE (Time of Execution) is given as the number of time units (as in the case of v1), or as the time range depending on the probability assumed. The next column marked as LoD (Level of Dependability) reliability is calculated for tasks with a probability of 40%. Result of scheduling for 5-processors architecture and 40% level of dependability is shown on Gant chart in Fig. 2. Of course for tasks scheduling, with another level of probability, is prepared two charts that shows minimum and maximum scheduling.

The conducted research provides promising results for the application of three-processor tasks and achieve optimal scheduling of tasks in the system, so that resources allocated to the implemented system will be used to the greatest possible extent. An additional factor of uncertainty in the form of the normal probability distribution used allows to rank tasks with time at different levels of probability. The use of this uncertainty factor may be relevant especially when scheduling tasks for IoT or IoV. In addition, the target system is constructed with a certain level of reliability also dependent on the probability distribution used. To simplify the model, transmission times between jobs are omitted.

Subsequent research will also focus on taking into account first for all another probability distribution [9, 18] and the transmission between tasks, as they can also affect the execution time of the entire system. The proposed algorithm allows obtaining satisfactory results for both three-, four- and five-processors architectures.

7 Summary

Please Description of result and future work. Conception of sum of probabilistic and consideration of time of transmission in future work, that will allow the development of a more accurate and detailed mathematical model of the system. The presented work was considered the problem of scheduling tasks in multiprocessor embedded systems based on the NoC network with imprecise duration of tasks. The system specification is presented using the task graph, generated by author based on TGFF. Like the previous ones, this author's work also considers the attribute of divisibility of tasks and additionally time defined by using normal distribution of probability. The proposed scheduling algorithm named MCD was based on modification of the approach used in the MC algorithm.

The novel approach proposed by author is to describe time of task using standard normal distribution of probability. A novelty in relation to scheduling works is the consideration of multi-processor tasks and factor of uncertainty here in the form of a probability distribution. This approach with multiprocessors tasks is to ensure the reliability of the system with some level of probability. The obtained results of the experiments confirm that the introduction of three-processor tasks only affects the execution time of the entire system and to achieve higher level of probability of execution time must be using time range ranking. The proposed measure calculates reliability taking into account the multiprocessor factor, allows to create a target system with an increased level of reliability. Because the level of credibility of the system depends here strictly on the multiprocessing of the task, using the three-processor tasks, the system is more reliable than the system using only two-processor tasks. What's more, the reliability of the system also closely depends on the percentage share of three-processor tasks. In addition, it is new to obtain a reliability factor with a certain level of probability that depends on what level of probability we want to get the time to rank the system.

Future work will focus on exploring other probability distributions that better reflect the time uncertainty of tasks specified as system entry. Additionally future work will concern the introduction of n-processor tasks as well as consideration of inter processor transmissions in the NoC network for which the ordering is performed. Additionally future work will concern the introduction of n-processor tasks as well as consideration of inter processor transmissions in the NoC network for which the system is performed. This will helps introducing a more realistic system model.

References

1. Abeni, L., Buttazzo, G.: QoS guarantee using probabilistic deadlines. In: Proceedings of 11th Euromicro Conference on Real-Time Systems. Euromicro RTS 1999, pp. 242–249. IEEE (1999)
2. Tia, T.S.: Probabilistic performance guarantee for real-time tasks with varying computation times. In: Proceedings Real-Time Technology and Applications Symposium, Chicago, IL, USA, pp. 164–173 (1995)

3. Ang, L.M., Seng, K.P., Ijemaru, G.K., Zungeru, A.M.: Deployment of IoV for Smart Cities: Applications, Architecture, and Challenges. Institute of Electrical and Electronics Engineers (IEEE) (2019)
4. Arnheiter, E.D., Maleyeff. J.: The integration of lean management and six sigma. In: Research and Concepts, Connecticut, USA, vol. 17, no. 1 (2005)
5. Bi, S., Zhuang, Z., Xia, T., Mo, H., Min, H., Luo, R.: Multi-objective optimization for a humanoid robot walking on slopes. In: 2011 International Conference on Machine Learning and Cybernetics (ICMLC), vol. 3, pp. 1261–1267, July 2011
6. Błażewicz, J., Cellary, W., Słowiński, R., Węglarz, J.: Badania operacyjne dla informatyków. WNT, Warszawa (1983)
7. Błażewicz, J., Drabowski, M., Węglarz, J.: Scheduling multiprocessor tasks to minimize schedule length. IEEE Trans. Comput. **5**, 389–393 (1986)
8. Błażewicz, J., Drozdowski, M., Guinand, F., Trystam, D.: Scheduling a divisible task in a two-dimensional toroidal mesch. Discrete Appl. Math. **94**, 35–50 (1999)
9. Bożejko, W., Rajba, P., Wodecki, M.: Stable scheduling of single machine with probabilistic parameters. Bull. Pol. Acad. Sci. Tech. Sci
10. Caranevali, L., Pinzuti, A., Vicaro, E.: Compositional verification for hierarchical scheduling of real-time systems. IEEE Trans. Softw. Eng. **39**(5), 638–657 (2012)
11. Dick, R.P., Rhodes, D.L., Wolf, W.: TGFF: task graphs for free. In: Proceedings of the 6th International Workshop on Hardware/Software Codesign (CODES/CASHE 1998), pp. 97–101. IEEE Computer Society, Washington, DC (1998)
12. Dorota, D.: Dual-processor tasks scheduling using modified Muntz-Coffman algorithm. In: International Conference on Dependability and Complex Systems, pp. 151–159. Springer, Cham (2018)
13. Dorota, D.: Scheduling tasks in embedded systems based on NoC architecture using simulated annealing. In: Advances in Dependability Engineering of Complex Systems, pp. 131–140. Springer, Cham (2017)
14. Dorota, D.: Scheduling tasks in a system with a higher level of dependability. In: International Conference on Dependability and Complex Systems, pp. 143–153. Springer, Cham(2019)
15. Drozdowski, D.: Selected Problems of Scheduling Tasks in Multiprocessors Computer Systems, Poznań (1997)
16. Eles, P., Peng, Z., Kuchcinski, K., Doboli, A.: System level hardware/software partitioning based on simulated annealing and tabu search. Des. Autom. Embed. Syst. **2**(1), 5–32 (1997)
17. Garvey, A., Lesser, V.: Design-to-time scheduling with uncertainty. Department of Computer Science (1995)
18. Gubbi, J., Buyya, R., Marusic, S., Palaniswami, M.: Internet of Things (IoT): a vision, architectural elements, and future directions. Future Gener. Comput. Syst. **29**(7), 1645–1660 (2013)
19. Kopetz, H.: Real-time Systems: Design Principles for Distributed Embedded Applications. Springer, Heidelberg (2011)
20. Li, Z., Ierapetritou, M.: Process scheduling under uncertainty review and challenges. Comput. Chem. Eng. **32**(45), 715–727 (2008)
21. Lombardi, M., Milano, M., Benini, L.: Robust scheduling of task graphs under execution time uncertainty. IEEE Trans. Comput. **62**(1), 98–111 (2011)
22. Moselhi, O.; Lorterapong, P.: Fuzzy vs probabilistic scheduling. In: Proceedings of the 12th Conference on Automation and Robotics in Construction (ISARC), Warsaw, Poland (1995)
23. Ost, L., Mandelli, M., Almeida, G.M., Moller, L., Indrusiak, L.S., Sassatelli, G., Moraes, F.: Power-aware dynamic mapping heuristics for NoC-based MPSoCs using a unified model-based approach. ACM Trans. Embed. Comput. Syst. (TECS) **12**(3), 75 (2013)

24. Pinedo, M.L.: Scheduling Theory, Algorithms and Systems. Springer, Heidelberg (2008)
25. Popieralski, W.: Algorytmy stadne w optymalizacji problem przepływowego szeregowania zadań, Ph.D. thesis (2013)
26. Rajesh, K.G.: Co-synthesis of hardware and software for digital embedded systems, Ph.D. thesis, 10 Dec 1993
27. Smutnicki, C.: Algorytmy szeregowania zadań, Oficyna Wydawnicza Politechniki Wrocławskiej, Wrocław (2012)
28. Tarun, K., Aryabartta, S., Manojit, G., Sharma, R.: Scheduling chained multiprocessor tasks onto large multiprocessor system, Wien (2017)
29. Tynski, A.: Zagadnienie szeregowania zadań z uwzględnieniem transportu. Modele, własności i algortmy, Ph.D. thesis, Wrocław (2008)
30. Mardosz, B.: Rozkład normalny, rozkład Gaussa (2008)
31. Zakharov, V.: Parallelism and array processing. IEEE Trans. Comput. **100**(1), 45–78 (1984)

The Concept of Management of Grid Systems in the Context of Parallel Synthesis of Complex Computer Systems

Mieczysław Drabowski(✉)

Department of Theoretical Electrical Engineering and Computing Science, Cracow University of Technology, Warszawska 24, 31-155 Kraków, Poland
drabowski@pk.edu.pl

Abstract. The paper presents the possibilities of using the results of research work on the parallel synthesis of complex systems with an increased degree of dependability in the issues of resource and tasks management of grid systems, including cloud computing and fog computing, too. It is therefore necessary to modify the system model, primarily in terms of performance and optimization criteria, and indicate the main criterion for system operation, which should be the time to obtain the results necessary for the proper operation of the distributed operating system, which manages the sets of resources and tasks in grid. Namely, if the time to obtain results for computer aided design (CAD) procedures is not critical, it is required only that these procedures are polynomial, because the size of these problems completely prevents the use others algorithms, it is obvious that of the execution time of management procedures is critical, because it determines the efficiency and practical usability of the grid.

Thus, the paper will show similarities and differences in the approach to the problems of complex systems synthesis and in the approach to management of distributed grid systems.

Keywords: Grid system · Parallel synthesis · Rapid prototyping · Management of resources and tasks

1 Practical Motivation and Model of Parallel Synthesis of Complex System with Increasing Level of Dependability

1.1 Introduction

The synthesis of complex systems is based on computer-aided procedures that calculate solutions at a high level of abstraction [1, 2]. At this level, the system is modeled, its behavior and structure, modular functionality and connections, and is realized of optimization of its parameters, such as: implementation cost, speed of operation, energy consumption during operation and dependability. In this case a less important parameter is the term of receipt of the calculation results obtained as a result implementation of these procedures supporting the design process, it is important that these results are obtained in practically acceptable time, and it will be such if the procedures will be implemented by algorithms about of polynomial computational complexity;

W. Zamojski et al. (Eds.): DepCoS-RELCOMEX 2020, AISC 1173, pp. 210–220, 2020.
https://doi.org/10.1007/978-3-030-48256-5_21

then the size of the problems solved by these procedures can be fully practically justified.

When modeling a projected complex system, we should be present its specification, i.e. specify a set of requirements implemented by the system functions. We should also present a database of resources with their characteristics that will enable the implementation of the system functions presented in the specification. The database of available resources should dispose the characteristics of these resources, i.e. their cost, speed of operation, connection interfaces, standardization and scalability. This database should include both hardware resources such as electronic devices: processors (universal and specialized), memories, controllers, external devices, as well as program resources such as components containing ready-made drivers, library functions, auxiliary and system programs. The resource base can also contain skeletons of various system structures recommended for realization different applications. The specification of the designed system should also include restrictions imposed on the system being built, e.g. cost, performance, quality or dependability restrictions.

The system design for calculations using CAD procedures should include the definition of the system performance criteria that should be optimized.

CAD procedures should give results (often alternative or varied depending on the preferences of designers) leading to the so-called virtual prototyping. The appropriate resources should stay selected from the resource database, both from the hardware component pool and from software component libraries. Inappropriate resources (e.g. available, but too expensive, or about insufficient reliability characteristics - too small - or energy consumption - too large) and not available should be indicated and given guidelines for their development and implemented.

At this stage, the synthesis carries out modeling and specifications and deals with virtual prototyping and focuses on tasks and resources, on partition, selection, scheduling and allocation.

The next stage of the synthesis is physical prototyping, i.e. determining the structure of the hardware (i.e. in address space) and the behavior of software (i.e. in time space) along with the communication and integration requirements of the entire system.

Virtual and physical prototyping implemented using CAD is the so-called rapid prototyping, which shown Fig. 1.

1.2 Model of Synthesis of Complex Systems

System model, based of deterministic model of task scheduling, for synthesis of complex systems has the form [3] (Eq. 1):

$$\boldsymbol{SYSTEM} = \{\boldsymbol{Resources}, \boldsymbol{Tasks}, \boldsymbol{Criteria}\} \tag{1}$$

where

- *Resources* is a set of hardware resources (processors - general and specialized - as well as additional resources) constituting a pool of components and a set of program resources forming a software library. These sets create database of available resources).

- *Tasks* is a set of tasks (processes) that the system will perform - with attributes of length in time, availability, completion, divisibility, arrival and execution order, deadline and resource requirements, including multiprocessing.
- *Criteria* is a set of optimality criteria for the operation and structure of the system (minimum cost, maximum speed of operation, minimum energy consumption, maximum dependability of implementation) defining the requirements and limitations of the constructed system.

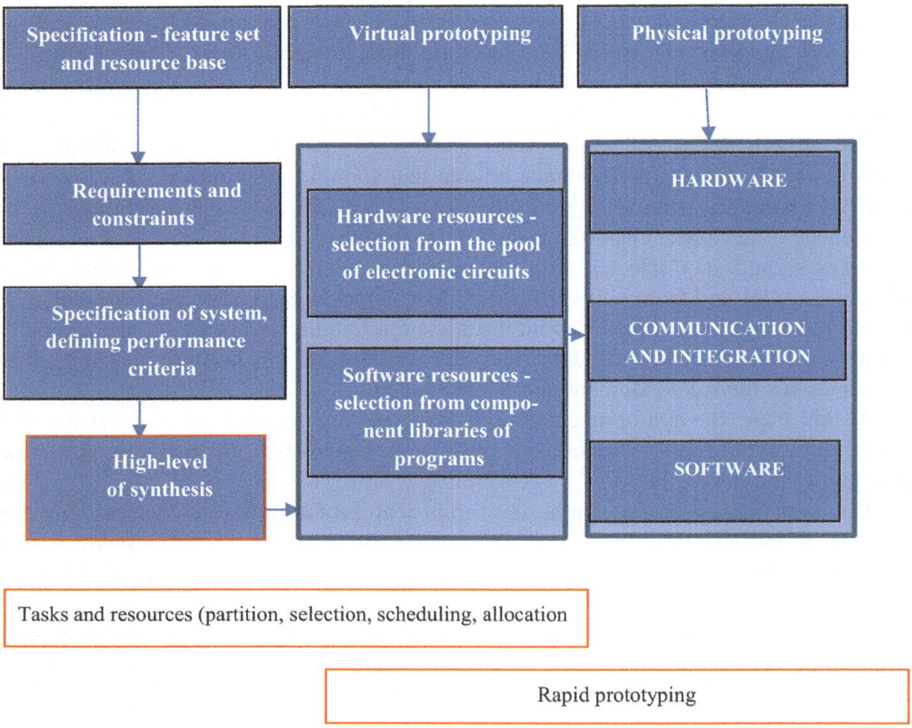

Fig. 1. Practical motivations for high-level synthesis

1.3 Multi-optimization in Synthesis of Complex Systems

Optimizations in synthesis concern the partition of tasks (into hardware and software) and selection of resources for the implementation of optimally arranged tasks allocated to resources and their schedules [4].

As a multi-optimization in system synthesis we will adopt multi-criteria optimization in the sense of Pareto. The solution is optimal in the Pareto sense if it is not possible to find a better solution because of at least one criterion without deterioration due to the others.

Solution *W* can be improved against both criteria C_1 and C_2. There is no such possibility for solutions *P, R* and *Q* - an improvement against one criterion causes deterioration due to the other - they belong to the set of optimal solutions in the *Pareto* sense. Let's assume for example, that we want to optimize a solution of two contradictory requirements: *the Cost* and *Power consumption*, as in Fig. 2.

For bringing multi-criteria to single criteria optimization, one can apply e.g. a method of weighted criteria - surrogate criterion equal to the sum of weighted criteria (Eq. 2):

$$\boldsymbol{MIN}(\boldsymbol{X}) = \sum (\boldsymbol{w_n} \cdot \boldsymbol{C_n}(\boldsymbol{X})), \ \boldsymbol{where}\ 0 \leq \boldsymbol{w_n} \leq 1,\ \boldsymbol{n\ number\ of\ criteria} \qquad (2)$$

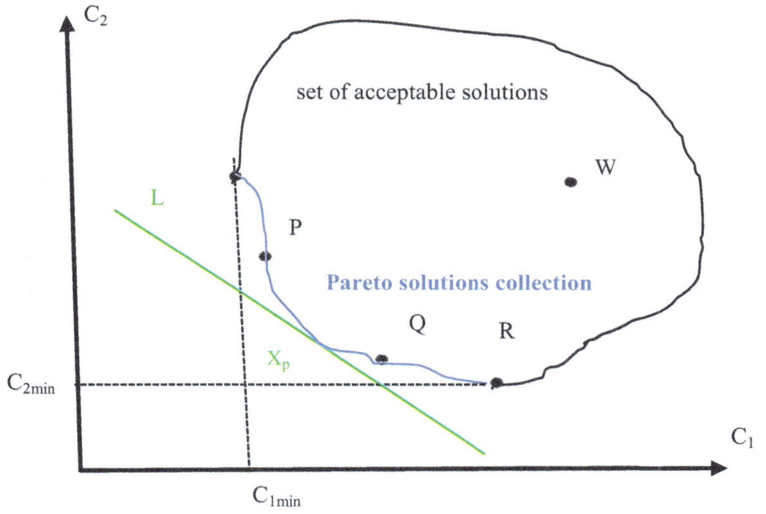

Fig. 2. Practical motivations for high-level synthesis

Graphically, the solution can be presented as the point of intersection of the set of permissible solutions with line $\boldsymbol{L}$ (with the point $\boldsymbol{X_p}$), depending on the values of the criteria weights. Due to the balanced impact of individual criteria, criteria may be standardized [5]. The problem is to choose a priori values of criteria weights, which can lead to different solutions.

The suggested model may be used for defining various synthesis problems for optimum computer systems. Our model of a system in this approach, typical for the theory of task scheduling, consists of a set of requirements and existing relationships between them (related to their order, required resources, time, readiness and completion deadlines, preemptive/non-preemptive [6, 7], priority etc.). The synthesis procedure contains the following phases: identification of hardware and software resources for task implementation, defining the processing time, defining the conflict-free task schedule, defining the degree of concurrency in the performance, allocating the tasks for the recourses and indicating the operations which can be executed concurrent [8].

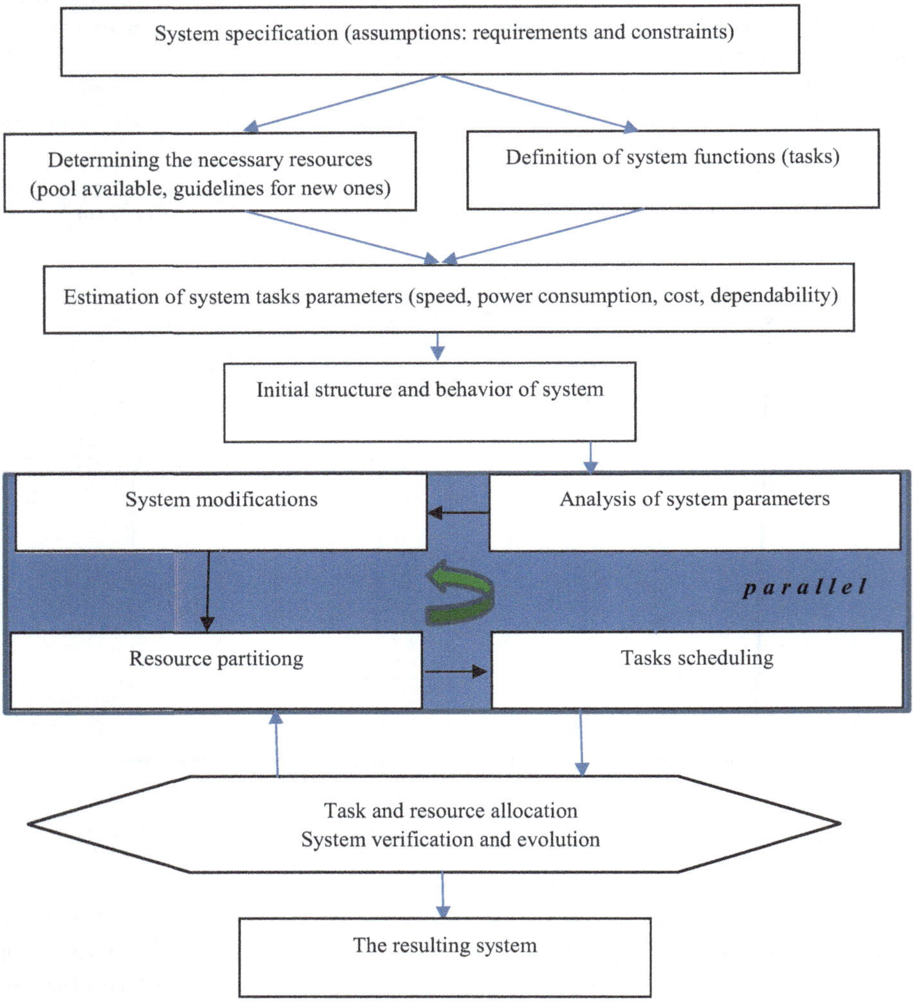

Fig. 3. A diagram of CAD procedures for parallel synthesis of complex systems

1.4 Algorithm of Parallel Synthesis of Complex Systems

A diagram actions operated of CAD procedures perform high-level parallel synthesis of complex systems is shown in Fig. 3 [9, 10].

Modeling the search for the optimum task schedule and resource partition of the designed system into hardware and software parts is fully justified. Parallel consideration of these problems may be useful in implementing optimum solutions, e.g. the cheapest hardware structures along with the shortest schedules. With such approach, the optimum task distribution is possible on the universal and specialized hardware and selected of resources characterized by maximum efficiency. We propose the following schematic diagram of a coherent process of fault tolerant systems synthesis.

The suggested coherent analysis consists of the following steps:

1. Specification of requirements for the system,
2. Specification of tasks,
3. Assuming the initial values of resource set,
4. Defining testing tasks and the structure of system, testing strategy selection,
5. Task scheduling,
6. Evaluating the operating speed and system cost, multi-criteria optimization,
7. The evaluation should be followed by a modification of the resource set, a new system partitioning into hardware and software parts and an update of test tasks and test structure (back to step 4).

In this approach a parallel search for optimal resources partition and optimal tasks scheduling occur. Iterative calculations are executed till satisfactory design results are obtained – i.e. an optimal system structure, a level of dependability and schedules.

2 Goals and Motivations of Management of Grid System

I am convinced that the model and CAD procedures of complex systems may find, after appropriate modifications and adaptations, in solutions for the problems of resource management and tasks of grid systems [11].

As it is known, the grid system is an organized, scattered hardware and software structure which enables cohesive partitioning of and access to computational resources. Thanks to standardized connection protocols of general purpose it provides control over co-sharing of system resources. Yet at the same time individual grid resources are not subject to centralized supervision and operate as autonomous computational units. Central unit only manages cohesive division of the task which is being currently processed by the grid, the distribution of data in the system as well as further data collection and unification.

Individual resources in such systems (these do not have to be computational resources or just such resources) are generally managed according to various strategies, they are dynamically heterogeneous, no uniform access, no cost model, etc. (they constitute a dynamic pool of resources to be used). On the other hand tasks directed to grid have a diversified specification: different quantitative and qualitative requirements on how to utilize individual resources and they can negotiate resource allocation conditions to obtain heterogeneous, dynamically scalable resource environment [12, 13].

The task of grid is to establish such methods of acting in which the user does not have to know its components and can focus only on the merits of the problem she/he is solving. Generally, the problem lies in flexible, secure and co-ordinated resource co-sharing by dynamic collection (set) of tasks, in order to enable various groups ("virtual organisations") the co-sharing of scattered resources for joint solving of complex problems. It is assumed that there is lack of:

- Centralized location of resources,
- Centralized management,
- Knowledge about global system condition,
- Full and mutual trust of users.

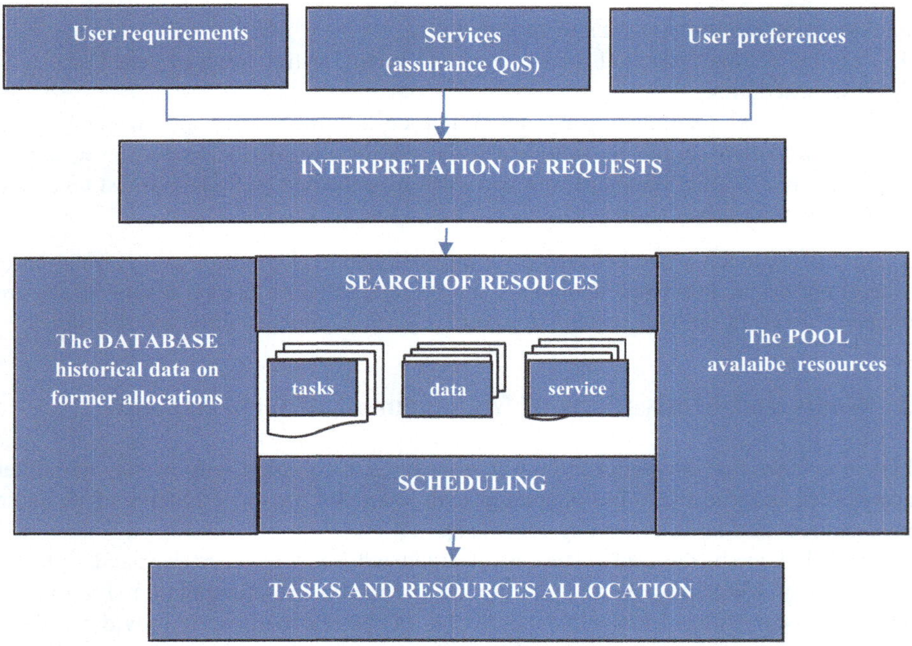

Fig. 4. Practical motivations for grid systems management

From the point of view of operational procedures, the following problems need to be solved:

- Resource partitioning,
- Scheduling of users' tasks,
- Balancing of resource load/burden,
- Multiple criteria assessment of resources, scheduling and allocation,
- Task preference modelling,
- Security,
- Securing appropriate quality level of services,
- Maximum system transparency.

The management system, the grid operating system, is a distributed system [14] whose purpose of calculation is to best meet the demands of users who should be able to determine their preferences regarding the selection of resources. Simultaneously the system should achieve the assumed indicators of service quality, including cost, deadlines, quality and dependability.

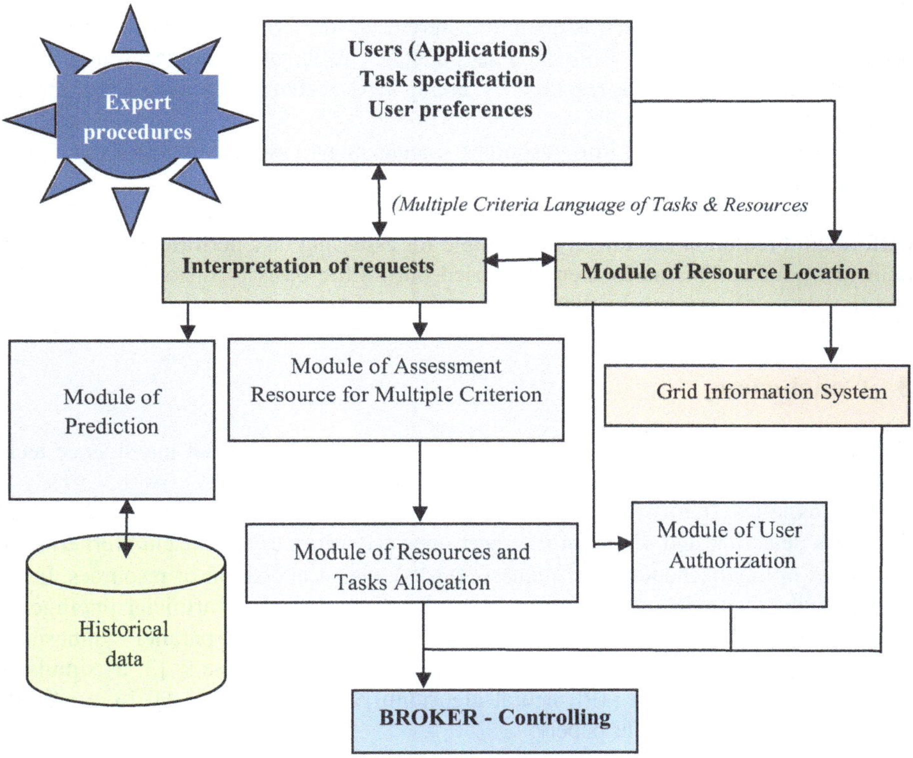

Fig. 5. Grid management

The system, after interpreting the tasks (request, query) of users, searches for a set of resources available in the system. It is, therefore, a database (pool) of resources and it is variable database, resources in this database can be multiplied and these resources can have various performance characteristics, primarily regarding access times.

The system also has a historical database, which contains addresses of resources to which users' tasks were allocated in the past, maybe optimally, suboptimal or not optimally. This database is a knowledge base about past schedules. Such a diagram of the grid system operation is shown in Fig. 4.

Tasks, services, data of task are scheduling by the system and allocated to currently selected and available resources. It is possible to migrate tasks assigned to certain resources and passed on the others, during their performance [15].

The user presents a task described by means of multiple criteria language of tasks and resources specification. The interpreter processes user's requests and calls in Resource Location Module. In next stage Module of Schedule, supported by expert system and on the basis of knowledge about resources and previous task solutions, searches for resources which fulfil basic criteria (resource type and quantity, operational system, etc.). Resources to which the user has no access because of security reasons are rejected (by User Authorization System). In this way a list of potential resources is defined, which is then transferred to the Multiple Criteria Resource Assessment Module [16]. This module generates different schedules (on different resources) and assesses the resource by taking into account both user's and system administrators' preferences.

The general layout of grid resources management system supported by expert system is shown in Fig. 5.

Finally, a set of resources is chosen on which the task will be scheduled and performed. Prediction Module is responsible for assessing task performance times on a defined resource. This assessment is carried out on the basis of historical knowledge and the status of resource load/burden.

3 Conclusions

The expert system of grid management could be based on artificial intelligence technologies e.g. algorithms presented for example [17, 18].

The expert system of grid management must coherently solve the problems of resources detection and selection (the best ones according to optimisation criteria), as well as of optimal scheduling of requests (tasks) and of allocation of resources, tasks.

The expert system of grid management could be based on artificial intelligence technologies e.g. of metaheuristic algorithms used in terms of parallel synthesis of complex systems, so far, such as for example of algorithms genetic [2, 5] simulated annealing [4], tabu search [19], neural algorithm [9] or ant colony [10] etc. already presented in the appropriate papers.

The expert system of grid management must coherently solve the problems of resources detection and selection (the best already according to optimisation criteria), as well as of optimal scheduling of requests (tasks) and of allocation of resources, tasks.

For the management of grid systems it is necessary to computer of solutions in deadlines and in real time (as soon as possible) [20]. If the time to obtain results for computer aided design (CAD) procedures is not critical, it is required only that these procedures are polynomial, because the size of these problems completely prevents the use others algorithms, it is obvious that of the execution times of management procedures are critical, because it determines the efficiency and practical usability of the grid.

So there are some similarities, but there are also significant differences between the problems of systems synthesis and management problems in grid. Algorithms developed for the synthesis of systems after adaptation and necessary modifications could be used in grid system management. It seems that such issues should be currently studied.

References

1. Drabowski, M., Wantuch, E.: Coherent concurrent task scheduling and resource assignment in dependable system design. In: Proceedings of the European Safety and Reliability Conference – ESREL 2005. Advances in Safety and Reliability. Taylor & Francis (2005)
2. Drabowski, M.: Boltzmann tournaments in evolutionary algorithm for CAD of complex systems with higher degree of dependability. In: Zamojski, W., Mazurkiewicz, J., Sugier, J., Walkowiak, T., Kacprzyk, J. (eds.) Theory and Engineering of Complex Systems and Dependability. AISC, vol. 365, pp. 141–152. Springer, Cham (2015)
3. Garey, M., Johnson, D.: Computers and intractability: A Guide to the Theory of NP-Completeness. Freeman, San Francisco (1979)
4. Dick, R.P., Jha, N.K.: MOGAC: a multiobjective genetic algorithm for the cosynthesis of hardware-software embedded systems. In: Proceedings of the International Conference on Computer Aided Design, pp. 522–529 (1997)
5. Dick, R.P., Jha, N.K.: MOGAC: a multiobjective genetic algorithm for hardware-software cosynthesis of hierarchical heterogeneous distributed embedded systems. IEEE Trans. Comput. Aided Des. Integr. Circuits Syst. **17**(10), 920–935 (1998)
6. Błażewicz, J., Drabowski, M., Węglarz, J.: Scheduling independent 2-processor tasks to minimize schedule length. Inf. Process. Lett. **18**, 267–273 (1984)
7. Błażewicz, J., Drabowski, M., Węglarz, J.: Scheduling multiprocessor tasks to minimize schedule length. IEEE Trans. Comput. **C-35**(5), 389–393 (1986)
8. Ziegenbein, D., Richter, K., Ernst, R., Thiele, L., Teich, J.: SPI – a system model for heterogeneously specified embedded systems. IEEE Trans. VLSI Syst. **10**(4), 379–389 (2002)
9. Drabowski, M.: Modification of neural network Tsang-Wang in algorithm for CAD of complex systems with higher degree of dependability. In: Zamojski, W., Mazurkiewicz, J., Sugier, J., Walkowiak, T., Kacprzyk, J. (eds.) Dependability Engineering and Complex Systems. AISC, vol. 470, pp. 121–133. Springer, Cham (2016)
10. Drabowski, M.: Adaptation of ant colony algorithm for CAD of complex systems with higher degree of dependability. In: Zamojski, W., Mazurkiewicz, J., Sugier, J., Walkowiak, T., Kacprzyk, J. (eds.) DepCoS-RELCOMEX 2017. AISC, vol. 582, pp. 141–150. Springer, Cham (2018)
11. Węglarz, J.: Project Scheduling – Recent Models, Algorithms and Applications. Kluwer Academic Publishers, Boston (1999)
12. Yhang, Z., Dick, R., Chakrabarty, A.: Energy-aware deterministic fault tolerance in distributed real-time embedded systems. In: 41st Proceedings Design Automation Conference, Anaheim, California, pp. 550–555 (2004)
13. Dorigo, M., Di Caro, G., Gambardella, L., Gambardella, L.M.: An algorithms for discrete optimization. Artif. Life **5**(2), 137–172 (1999)
14. Błażewicz, J., Ecker, K., Pesch, E., Schmidt, G., Węglarz, J.: Handbook on Scheduling. Springer Verlag, Berlin (2007)
15. Lee, C.Y.: Machine scheduling with availably constraints. In: Leung, J.Y.T. (ed.) Handbook of Scheduling, pp. 22.1–22.13. CRC Press, New York (2004)
16. Schmitz, M.T., Al-Hashimi, B.M., Eles, P.: Energy-efficient mapping and scheduling for DVS enabled distributed embedded systems. In: Proceedings of the Design Automation and Test in Europe Conference, pp. 514–521 (2002)
17. Pricopi, M., Mitra, T.: Task scheduling on adaptive multi-core. IEEE Trans. Comput. **C-59**, 167–173 (2014)

18. Agraval, T.K., Sahu, A., Ghose, M., Sharma, R.: Scheduling chained multiprocessor tasks onto large multiprocessor system. Computing **99**(10), 1007–1028 (2017)
19. Nowicki, E., Smutnicki, C.: An advanced tabu search algorithm for the job shop problem. J. Sched. **8**, 145–159 (2005)
20. Yen, T.Y., Wolf, W.H.: Performance estimation for real-time distributed embedded systems. IEEE Trans. Parallel Distrib. Syst. **9**(11), 1125–1136 (1998)

Performance Assessment of Selected Techniques and Methods Detecting Duplicates in Data Warehouses

Paweł Dymora and Mirosław Mazurek(✉)

Faculty of Electrical and Computer Engineering, Rzeszów University of Technology, al. Powstańców Warszawy 12, 35-959 Rzeszów, Poland
mirekmaz@prz.edu.pl

Abstract. A significant and current research problem, as well as a practical one, is the problem of deduplication in databases. The solution of this problem is applicable, e.g., in the context of the following situations in which are stored apparently different records, which actually refer to the same entity (objects, individuals, etc.) in the real world. In such cases, the purpose is to identify and reconcile such records or to eliminate duplication. The paper describes algorithms for finding duplicates and implements them in the developed data warehouse. Efficiency and effectiveness tests were also carried out for sample data contained in individual tables of the warehouse. The work aims to analyze the existing methodologies for detecting similarities and duplicates in data warehouses, to implement algorithms physically, and to test their effectiveness and efficiency. A large scale of data created by IoT devices leads to the consumption of communication bandwidth and disk space because the data is highly redundant. Therefore, correct deduplication of information is necessary to eliminate redundant data.

Keywords: Deduplication · Duplicate data detection · IoT · Industry 4.0 · Oracle warehouse builder · Match Merge · Jaro Winkler · Levenshtein distance · SoundEx · Double Metaphone

1 Introduction

In recent years, the Internet of Things (IoT) has been widely used and attracted a lot of attention. Since most IoT terminals have limited data storage and processing capabilities, the trend is to outsource data from local processing to cloud computing [1]. Currently, databases are growing to enormous sizes or are integrated with each other. This makes standard database tools insufficient. Then the so-called data warehouses come with help, which allows for the processing of large data sets and their integration when the information comes from many sources. Data warehouses can be defined as massive databases whose data are read-only and not modified. They include not only current but also historical data. Based on these data, analyses are made in order to support business decision making by enterprises.

However, most IoT devices have the disadvantages of limited memory capacity and computing power. With the popularity of IoT technology, content manufacturers need

W. Zamojski et al. (Eds.): DepCoS-RELCOMEX 2020, AISC 1173, pp. 221–232, 2020.
https://doi.org/10.1007/978-3-030-48256-5_22

to collect and process more and more data, which can lead to tight storage space and computational costs for IoT terminals. When integrating data from different sources, i.e., combining individual tables or entire databases, data duplication, redundancy can occur. To further reduce communication bandwidth and storage space, data deduplication must be widely used to eliminate unnecessary data. However, because the data collected in IoT is sensitive and closely related to the personal data of users, protecting the privacy of users' information is also a challenge.

The extensive use of IoT devices has led to a rapid increase in data volumes. However, many studies have shown that a large amount of stored data is duplicated. The duplicated data takes up much space and is a bottleneck in data storage. Deduplication of data [2] has, therefore, become an essential technique for use in storage systems, saving users space and bandwidth. Among the main challenges in this area, we are considering merging or removing duplicate entries.

The article is divided into six parts. Section 2, which follows the introduction, provides an overview of the literature and the latest trends in data deduplication. Section 3 characterizes available data duplication algorithms in databases, including Levenshtein Distance Algorithm, Jaro Winkler Algorithm, and SoundEx/Metaphone and Double Metaphone algorithms. Section 4 presents the logical and physical models of the analyzed data warehouse. Section 5 shows the analysis of the effectiveness of algorithms for detecting duplicates in a data warehouse, tabular and graphical comparison of the results of all methods. The summary, conclusions and scope of future research are presented in Sect. 6.

2 Literature Review

Data quality depends not only on the exact input of data into the system but also on how the data is stored, managed, and accessed. As the amount of data grows exponentially, the possibility of having redundant data increases significantly, which in some cases cannot be acceptable or affordable, which means that a pre-processing phase is required. Unit resolution or record linkage involves detecting records relating to the same object in different databases. This task has many applications in different research areas. For this reason, various terms have been adopted for the same concept, such as duplicate detection, deduplication, reference matching, object identification, combination and cleaning, object consolidation, or reference reconciliation [3].

The problem of unit reconciliation (ER) is one of the fundamentals of data integration. The problem is also known as deduplication, reconciling references, cleaning, and others. ER is a significant and common problem for data cleansing and involves the detection of duplicate data for the same external actors and their aggregation into single representations [4].

Duplicates can lead to erroneous chain decisions that affect profitability and brand. For all these reasons, computer programs are essential to detect and eliminate duplicates. On the other hand, there are different tools for cleaning data based on various criteria, such as single fields or whole documents. The article [3] focuses on the search for duplicate Customers based on the analysis of individual fields of a CRM entry coming from a lexical heterogeneity. Some authors analyze the duplication manually, detecting differences in values of some variables (such as a city or a company) in the customer database. Other authors used different character-based similarity techniques to match fields with string data [3, 5, 6].

A clean database improves performance and then leads to higher customer satisfaction [7]. Good data quality has been shown to increase staff productivity because instead of spending time verifying and correcting data errors, they can focus on their core mission [8, 9]. Since comparing units using a single property is usually not enough to decide whether both entities describe the same object in the real world, the aggregation rule must aggregate the similarity of multiple property comparisons using appropriate aggregation functions.

In [10], a performance analysis of techniques for duplicate data detection in relational databases has been performed. The research focuses on traditional SQL based [11] and new bloom filter techniques to find and eliminate records that already exist in the database while performing bulk insertion operation and data synchronization in multisite databases. The results show that the parallel bloom filter is highly suitable for duplicate detection in the database. Authors in [10] in the SQL based approach for all columns of the record concatenate without key column and then matches with all existing records in the table for a duplicate verification using the Where clause in the Select statement. In our work, unlike [10], we propose to use algorithmic duplicate detection techniques such as Levenshtein Distance Algorithm, Jaro Winkler Algorithm, and SoundEx/Metaphone algorithms, which may prove more promising and innovative approach.

3 Duplicate Detection Algorithms

In order to detect and remove duplicates, dedicated mechanisms are used, which are implemented in data warehouse tools, or predefined functions that can be handled by the user, creating appropriate procedures. There are two approaches to detecting duplicate data. The first one is a more straightforward way. The data is first moved to a single table and then analyzed for duplicate information. The second way is a bit more complicated. It consists of creating a procedure in which two or more tables are loaded into cursors and then checked for duplicate data. The rest of the tables will contain "dirty" information that may contain internal duplicates, as well as duplicate values from the main table. Nonduplicated values should then be moved to the unique data table. The remaining data from the "dirty" tables can then be deleted.

The underlying mechanisms for detecting duplicates are ineffective in such cases. Advanced similarity detection techniques come in support. These algorithms are Levenshtein Distance, Jaro Winkler, Needleman Wunsch, Similarity of character pairs, Trigram comparison. Phonetic algorithms can be added to similarity detection techniques: SoundEx, Refined SoundEx, Metaphone, Double Metaphone.

3.1 Levenshtein Distance Algorithm

This algorithm was created by the Russian scientist Vladimir Levenshtein, who discovered this equation in 1965. It is used to calculate the editing distance, which is why it is often called the Edit Distance algorithm. Levenshtein's distance is a metric sequence to measure the difference between two sequences. Informally, Levenshtein's distance between two words is the minimum number of edits of single characters (i.e., inserts, deletions, or substitutions) required to replace one word with another. Levenshtein's distance can also be referred to as the edit distance, although it can also mean a larger family of distance indicators. It is closely related to the alignment of sequences in pairs [3, 12–14].

The Levenshtein distance between two strings, a and b (of length $|a|$ and $|b|$ respectively), is given by $lev_{a,b}(|a|, |b|)$ where [3, 9–11]:

$$lev_{a,b}(i,j) = \begin{cases} \max(i,j) & if \ \min(i,j) = 0, \\ min \begin{cases} lev_{a,b}(i-1,j) + 1 \\ lev_{a,b}(i,j-1) + 1 \\ lev_{a,b}(i-1,j-1) + 1_{(a_i \neq b_j)} \end{cases} & otherwise. \end{cases}$$

Gdzie: $1_{(a_i \neq b_j)}$ is the indicator function equal to 0 when $a_i \neq b_j$ and equal to 1 otherwise, and $lev\ a, b(i,j)$ is the distance between the first i characters of a and the first j characters of b. The first element in the minimum corresponds to deletion (from a to b), the second to insertion, and the third to match or mismatch, depending on whether the respective symbols are the same [12–14].

3.2 Jaro Winkler Algorithm

The algorithm Jaro Winkler was initially created by Matthew Jaro, and then modified by William Winkler. Its result is a number ranging from 0 to 1, where 0 is a measure of lack of similarity, while 1 is the identity of two words [3, 7, 12–15]. The Jaro-Winkler distance is a measure of the similarity between two strings. This measure works well in matching personal names and entities, and it is widely used in the areas of record linkage, entity linking, information extraction. Taking into account the sequence of queries q, searching for the similarity of Jaro-Winkler's distance finds all the sequences in the data set D, whose similarity of Jaro-Winkler's distance to q is not more significant than the given threshold τ. As the size of the data set increases, an effective search for the similarity of the distance between Jaro-Winkler is getting more difficult [15, 16].

The Jaro-Winkler's distance d_j of two given strings s_1 and s_2 is:

$$d_j = \begin{cases} 0 & if \ m = 0, \\ \frac{1}{3}\left(\frac{m}{|s_1|} + \frac{m}{|s_2|} + \frac{m + \tau}{m}\right) & otherwise. \end{cases}$$

where: m is the number of matching characters, and τ is half the number of transpositions.

3.3 SoundEx, Metaphone and Double Metaphone Algorithms

SoundEx is a phonetic algorithm developed by Robert Russell and Margaret Odell. Unlike the algorithms described above, this algorithm assigns a four-character code to each word, so that different strings of characters can be compared [3, 7, 17].

Metaphone is a phonetic algorithm developed by Lawrence Philips. It is an improvement of the SonudEx algorithm. Metaphone uses word severance information and allows it to standardize many inconsistencies in spelling and pronunciation. Thanks to the applied patches, it enables more accurate coding and distinguishing between many very similar words. Double Metaphone is a new version of the Metaphone algorithm, which can be implemented for many languages, not only English. Most ambiguities in different encoding words have also been improved by introducing additional code in addition to the main code. This makes it possible to distinguish between words that have a common origin from another word [17–19].

4 Logical and Physical Model of the Analyzed Data Warehouse

An Oracle database server, version 11 g Release 2 (11.2.0.1.0), Enterprise Edition (educational license), was installed to research the effectiveness of eliminating duplicates. Installation and implementation of the data warehouse system were performed using a free tool provided by Oracle: Oracle Warehouse Builder (OWB). It is a tool with a graphical interface, thanks to which it is possible to create and manage data warehouses. Figure 1 shows the logical scheme of the created data warehouse.

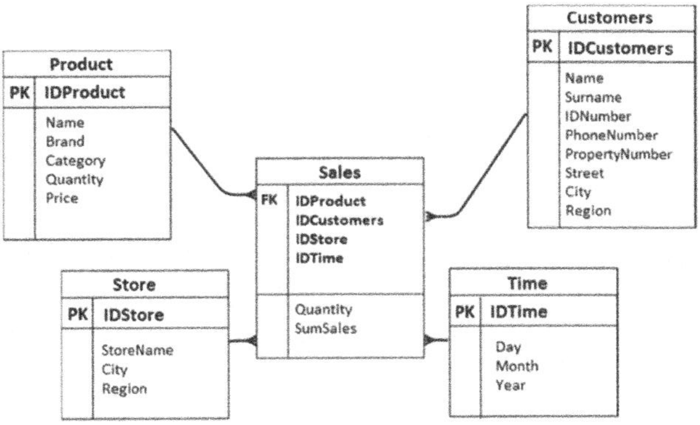

Fig. 1. Logical scheme of the analyzed data warehouse.

The data from the above auxiliary tables are then checked against the data from the main tables (dimension tables) to detect duplicates. The research was carried out on the example of a warehouse prepared, especially for this purpose. All tests consisted of entering 100,000 records into the CUSTOMERS_NEW table and 100 records into the CUSTOMERS table. The dimension tables contain the following information:

- the PRODUCT table contains data on products available for sale. Each product is described by name, brand and category.
- The STORE table contains data about the name of the store, as well as in what city it is located.
- The TIME table has information about the dates on which transactions were carried out in stores.
- The CUSTOMERS table contains information about people who made a purchase in the store. These are personal data such as name, surname, personal identification number, telephone number, property number, street, city and region.

Once the duplicate information has been removed, the unique data is returned to the target structures, which are the dimension tables.

5 Analyzes of the Effectiveness of Algorithms for Detecting Duplicates in a Data Warehouse

The algorithms discussed above were used to search for similar duplicates. They are implemented as pre-defined procedures in Oracle SQL Developer. They are available in the `utl_match` package. The following algorithms are available: Levenshtein Distance and Jaro Winkler. In addition, there are procedures derived from the previously mentioned, which return a percentage of similarity. There is also an implemented SoundEx algorithm, which returns codes that can be compared.

5.1 Analysis of Levenshtein Distance Algorithm (Edit Distance Algorithm) vs. Jaro Winkler Algorithm

The procedure implementing the Levenshtein Distance algorithm works in the following way: data from two tables are extracted to two cursors. The variable RESULT is zeroed. Then, in the cursor loop, individual values of corresponding columns from both tables are compared. If the Edit Distance algorithm returns a value higher than the specified one (ED_PROG variable, e.g., 70), the variable RESULT is incremented. Otherwise, the variable RESULT remains unchanged. After checking all the values of one record, the variable RESULT is compared with a specified threshold value (variable RESULT_PROG). There are eight columns compared to the CUSTOMERS table. The threshold value can be, for example, six. Thus, if the RESULT is more significant than six, i.e., at least six columns contain information in 70% (70 or another value previously determined by the ED_PROG variable) identical in both tables, this record is considered as a duplicate and is removed from the CUSTOMERS_NEW table. At the beginning of each loop, the variable RESULT is zeroed. The records that remain in the CUSTOMERS_NEW table are considered unique and copied to the CUSTOMERS table. At the very end of the CUSTOMERS_NEW table is cleared.

The second test was concerned with performance, i.e., how much time it will take to process 100,000 records. The data set was constructed in such a way that the

CUSTOMERS_NEW table exists only duplicates of similar values contained in the CUSTOMERS dimension.

The efficiency tests for both algorithms were performed for the same variables JW_PROG and RESULT_PROG. The experiments were carried out on the same tables, with the same data sets. The values of parameters for the algorithms are presented in Table 1.

Table 1. Parameters of the Levenshtein Distance/Jaro Winkler algorithms for fifteen runs.

Trial numer	1	2	3	4	5	6	7	8	9	10	11	12	13	14	15
ED_PROG	50	60	70	80	90	50	60	70	80	90	50	60	70	80	90
RESULT_PROG	6	6	6	6	6	7	7	7	7	7	8	8	8	8	8

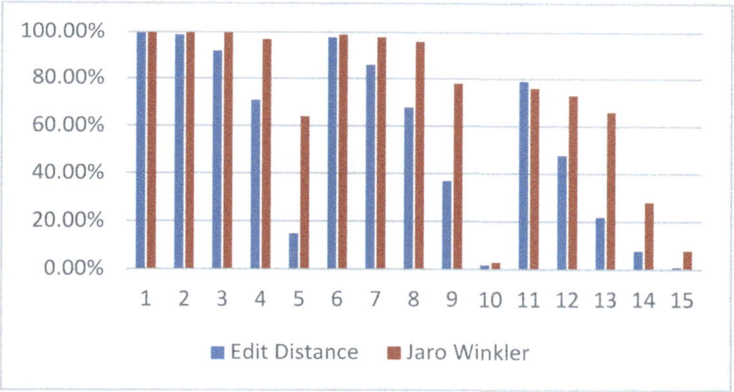

Fig. 2. Effectiveness of eliminating duplicates Edit Distance/Jaro Winkler Algorithms.

Figure 2 shows the results of the survey. The obtained results show how many percent of the duplicate data were detected based on the parameters set according to the above table. The results show that the Edit Distance algorithm works better if one sets less restrictive data similarity requirements. However, there is a risk that unique data that would be very similar to the information already contained in the CUSTOMERS table would be wrongly deleted. Yet, if the requirements are more restrictive, more duplicate data may be added to the main dimension table.

In the case of the Jaro Winkler algorithm, the results show the same behavior of the algorithm concerning the set parameters, as in the case of Levenshtein Distance. It is inversely proportional: the more rigorous the rules are, the fewer duplicates are detected. Comparing the Edit Distance algorithm with Jaro Winkler, we can see that Jaro Winkler managed to discover more duplicates than Edit Distance with identical parameters set.

5.2 The SoundEx Algorithm

The effectiveness analysis was tested for the following values of the RESULT_PROG parameter: 3, 4, 5. The results are shown in Fig. 3. The obtained results show the percentage of duplicate data detected for each value of the set variable. The results of SoundEx algorithm analysis are analogical to the results obtained by previous algorithms. The stricter the requirements, the fewer duplicates the algorithm was able to detect. When the RESULT_PROG variable was changed from 3 to 4, the percentage of duplicates found decreased by 27.5%. With the highest requirements, i.e., all five columns had to contain similar data, the percentage dropped by another 37.5%. The analysis of algorithm performance would also be conducted in the next sections.

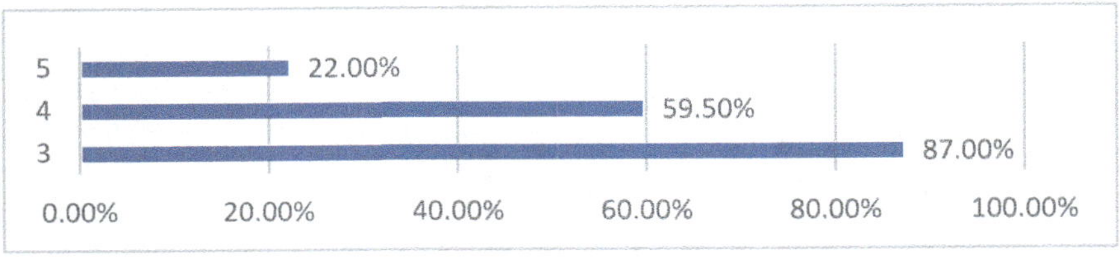

Fig. 3. Rys. 3. The results of the SoundEx algorithm effectiveness in the function of the RESULT_PROG variable.

5.3 Performance Comparison of Levenshtein Distance, Jaro Winkler, and SoundEx Algorithms in Oracle Warehouse Builder Implementation

Oracle Warehouse Builder has implemented the `Match Merge` function, which includes the following algorithms: Jaro Winkler, Levenshtein Distance, SoundEx, and Double Metaphone. These algorithms are selected when configuring the `Match Merge` function in the mapping workspace. All tests were based on inserting 100,000 records into the CUSTOMERS_NEW table and 100 records into the CUSTOMERS table.

By using the `Match Merge` operator, all duplicates have been detected and deleted. The only drawback of this tool is that it has no logic to decide whether the data is correct or not. There are situations when correct data is changed to incorrect data. Most of the available tools are also developed for English language data only, hence their imprecision. It took 5425 s for the program to detect and remove duplicates. It is due to the complexity of actions taken by `Match Merge` operator to identify duplicate data. Nevertheless, it is the best tool for analyzing information for the occurrence of duplicate data.

The Jaro Winkler algorithm proved to be the most effective. In almost each of the tests, which were performed for different parameters, the percentage of detected duplicated data by Jaro Winkler was the highest. The Edit Distance algorithm turned out to be more demanding in the conducted tests; hence the rate of detected duplicate data is lower. The SoundEx algorithm had little configuration possibilities, so it can be used as an auxiliary algorithm when the similarity of the two words is high. The code returned by SoundEx can then be used to qualify the data set as duplicate or unique data. Using Oracle Warehouse Builder and implemented the above three algorithms, as well as the fourth Double Metaphone (with additional options such as comparing data by its initials), 100% of data, duplicates ware detected. With the Edit Distance algorithm, 75% of data duplicates ware detected, and with the use of Jaro Winkler algorithm, 80% of data duplicates ware detected. This is the best result of all carried out efficiency tests.

In order to analyze the performance of the algorithms, the Levenshtein Distance, Jaro Winkler and SoundEx implementation procedures were tested. The performance of `Match Merge` function included in the Oracle Warehouse Builder was also tested. For this purpose, the following parameters were adopted for all algorithms: for Levenshtein Distance, Jaro Winkler: ED_PROG: 75, RESULT_PROG: 7; and for SoundEx RESULT_PROG: 4. The result of the performance test of individual algorithms is presented in Table 2.

Table 2. Results of algorithm performance testing

	Time [s]	Percentage deviation from SoundEx
Edit Distance	1858,62	216%
Jaro Winkler	2126,56	247%
SoundEx	860,47	–
Match Merge	5425	630%

The data was processed most quickly by the procedure implementing the SoundEx algorithm. It was over two times faster than other algorithms implemented in the Oracle Database System. However, this is due to the low complexity of the algorithm. This is due to the lowest complexity of the procedure and less rigorous requirements compared to other algorithms. The procedure using Jaro Winkler needed the most time to detect duplicates, but previous tests showed that it is more effective than other algorithms. From the algorithms implemented in PL/SQL procedures, Jaro Winkler was the slowest.

Nevertheless, it is more efficient (for some parameters even three times) than other algorithms, and this factor gives it an advantage over Edit Distance. The Match Merge operator needed the most time to process 100,000 rows. However, this is due to the multitude of tests performed on the data to detect duplicates. The long processing time, therefore, affects the effectiveness of detecting duplicate information in the data warehouse tables. The tests showed that Jaro Winkler needed the most time to detect duplicates, but was the most successful.

6 Conclusion

The paper presents fundamental issues concerning data warehouses and discusses algorithms responsible for detecting approximate duplicates. In the designed data warehouse, created with the help of Oracle Warehouse Builder, there are implemented methodologies for detecting and removing exact duplicates. Oracle Database System has procedures to support Levenshtein Distance, Jaro Winkler and SoundEx algorithms.

Oracle Warehouse Builder turned out to be the best tool to perform tests for the presence of duplicate data. The advantages are high configuration capabilities and a wide range of activities on data warehouses. Performing data cleaning and standardization is widely developed, but only for data written in English. However, it is possible to add separate tables with correct information, for example, lists of Polish names or cities. This would make it possible to standardize the records in order to analyze them later for the presence of duplicate data. Oracle Warehouse Builder is a graphical tool, so most of the operations are done using wizards. This makes working with a data warehouse much easier and more comfortable. Using Oracle Warehouse Builder and implemented the above three algorithms, as well as the fourth Double Metaphone (with additional options such as comparing data by its initials), 100% of data, duplicates ware detected. With the Edit Distance algorithm, 75% of data duplicates ware detected, and with the use of the Jaro Winkler algorithm, 80% of data duplicates ware detected. The SoundEx algorithm proved to be the fastest.

The data collected in the IoT are sensitive and closely related to users' personal data, so protecting the privacy of users' information becomes a challenge. In times of cyber-terrorism and data theft, security, and privacy issues must be taken seriously as sensor data is transferred from peripheral devices to databases. The amount of data sent affects the encryption time. To protect data privacy, all operations and calculations must be performed in an encrypted form. Therefore, it is necessary to eliminate unnecessary data, which would increase system performance. The data center can eliminate redundancy of collected data, which also allows for a significant reduction of disk space.

In future studies, we would like to compare the obtained results with various implementations of the Bloom filter algorithm (iterative, parallel) in particular concerning the possibility of the appearance of false-positive entries. Also, the use of Delete operations is disputable [10, 11]. It would be advisable to establish a reasonable compromise between the space and false positive, as also additional processing operations in case of Delete operations.

Acknowledgments. We are thankful to the graduate student Andrzej Wilusz of Rzeszów University of Technology, for supporting us in the collection of useful information.

Funding. This work is financed by the Minister of Science and Higher Education of the Republic of Poland within the "Regional Initiative of Excellence" program for years 2019–2022. Project number 027/RID/2018/19, the amount granted 11 999 900 PLN.

References

1. Dymora, P., Mazurek, M.: Anomaly detection in IoT communication network based on spectral analysis and hurst exponent. Appl. Sci. **9**(24), 5319 (2019). https://doi.org/10.3390/app9245319
2. Yan, H., Li, X., Wang, Y., Jia, Ch.: Centralized duplicate removal video storage system with privacy preservation in IoT. Sensors **18**(6), 1814 2018
3. González-Serrano, L., Talón-Ballestero, P., Muñoz-Romero, S., Soguero-Ruiz, C., Rojo-Álvarez, J.L.: Entropic statistical description of big data quality in hotel customer relationship management. Entropy **21**(4), 419 (2019)
4. Bahmani, Z., Bertossi, L., Vasiloglou, N.: ERBlox: combining matching dependencies with machine learning for entity resolution. Int. J. Approx. Reason. **83**, 118–141 (2017)
5. Elmagarmid, A.K., Ipeirotis, P.G., Verykios, V.S.: Duplicate record detection: a survey. IEEE Trans. Knowl. Data Eng. **19**, 1–16 (2007)
6. Pinto, F., Santos, M.F., Cortez, P., Quintela, H.: Data pre-processing for database marketing. In: Data Gadgets, Workshop: Malaga, Spain, pp. 76–84 (2004)
7. Saberi, M., Theobald, M., Hussain, O.K., Chang, E., Hussain, F.K.: Interactive feature selection for efficient customer recognition in contact centers: dealing with common names. Expert Syst. Appl. **113**, 356–376 (2018)
8. Papadakis, G., Svirsky, J., Gal, A., Palpanas, T.: Comparative analysis of approximate blocking techniques for entity resolution. Proc. VLDB Endow. **9**, 684–695 (2016)
9. Lin, M.J., Yang, C.Z., Lee, C.Y., Chen, C.C.: Enhancements for duplication detection in bug reports with manifold correlation features. J. Syst. Softw. **121**, 223–233 (2016)
10. Adil, S.H., Ebrahim, M., Ali, S.S.A., Raza, K.: Performance analysis of duplicate record detection techniques. Eng. Technol. Appl. Sci. Res. **9**, 4755–4758 (2019)
11. Shah, Y.A., Zade, S.S., Raut, S.M., Shirbhate, S.P., Khadse, V.U., Date, A.P.: A survey on data extraction and data duplication detection. Int. J. Recent Innovation Trends Comput. Commun. **6**(5), 77–82 (2018)
12. Guo, L., Wang, W., Chen, F., Tangi, X., Wang, W.: A similar duplicate data detection method based on fuzzy clustering for topology formation. Przegląd Elektrotechniczny (Electr. Rev.) **88**(1), 26–30 (2012). ISSN 0033-2097, R. 88 NR 1b/2012
13. Yujian, L., Bo, L.: A normalized Levenshtein distance metric. IEEE Trans. Pattern Anal. Mach. Intell. **29**, 1091–1095 (2007)
14. Babar, N.: https://dzone.com/articles/the-levenshtein-algorithm-1?source=post_page. Accessed 14 Dec 2019
15. Wang, Y., Qin, J., Wang, W.: Efficient approximate entity matching using Jaro-Winkler distance. In: Bouguettaya, A., et al. (eds.) Web Information Systems Engineering – WISE 2017, WISE 2017. Lecture Notes in Computer Science, vol. 10569. Springer, Cham (2017)
16. Pandya, S.D., Virparia, P.V.: Context free data cleaning and its application in mechanism for suggestive data cleaning. Int. J. Inf. Sci. **1**(1), 32–35 (2011). https://doi.org/10.5923/j.ijis.20110101.05
17. Angeles, M.P., Espino-Gamez, A., Gil-Moncada, J.: Comparison of a Modified Spanish phonetic, Soundex, and Phonex coding functions during data matching process. In: Conference Paper, June 2015. https://doi.org/10.1109/iciev.2015.7334028

18. Mandal, A.K., Hossain, M.D., Nadim, M.: Developing an efficient search suggestion generator, ignoring spelling error for high speed data retrieval using Double Metaphone Algorithm. In: Proceedings of 13th International Conference on Computer and Information Technology (ICCIT 2010) (2010). https://doi.org/10.1109/iccitechn.2010.5723876
19. Uddin, M.P., et. al.: High speed data retrieval from National Data Center (NDC) reducing time and ignoring spelling error in search key based on double Metaphone algorithm. Int. J. Comput. Sci. Eng. Appl. (IJCSEA) **3**(6) (2013). https://doi.org/10.5121/ijcsea.2013.3601

An Overview of DoS and DDoS Attack Detection Techniques

Mateusz Gniewkowski(✉)

Faculty of Electronics, Wrocław University of Science and Technology, Wrocław, Poland
mateusz.gniewkowski@pwr.edu.pl

Abstract. The economic impact of (distributed) denial-of-service attacks is substantial, especially at a time when we rely on web applications more and more often. That is why, it is essential to be able to detect such threats early and therefore react before significant financial losses. In this paper, we focus on techniques, for detecting this type of attacks, that use historical data. We will discuss existing datasets, extracted features and finally the methods themselves. The solutions mentioned in this work are based on supervised learning (k-NN, MLP, DNN), unsupervised learning (mostly modified K-Means) and anomaly detection in time series analysis (ARIMA models family).

Keywords: DoS · DDoS · Anomaly detection · ARIMA · DNN · K-means · Datasets

1 Introduction

The purpose of denial-of-service (*DoS*) attacks is to prevent or disturb users from using internet applications through intentional exhaustion of a given resource (e.g. available sockets, bandwidth or computing power). The distributed version of this type of attack (*DDoS*) is different in that many computers are used to send packets. It is more difficult to perform, but it allows to consume the resource faster and makes it more difficult to react to the threat (it is necessary to filter out multiple connections which often look like normal network traffic).

A typical DDoS attack involves the creation of a so-called botnet - a set of computers on which the attackers took control. Joining new computers to such a network is often based on distributed scanning for hosts with known vulnerabilities and exploiting them, but users usually (unintentionally) install malicious software on their computers themselves (malicious email attachments, suspicious programs from the Internet). Many servers on the network are also constantly subjected to dictionary attacks [25], which also can provide access to the attacker. An interesting phenomenon is a situation in which Internet services cease to function due to naturally increasing interest (e.g. related to sales). Symptoms of such a situation may not always be distinguishable from a real attack.

W. Zamojski et al. (Eds.): DepCoS-RELCOMEX 2020, AISC 1173, pp. 233–241, 2020.
https://doi.org/10.1007/978-3-030-48256-5_23

To better understand the problem that will be addressed in this article, it is good to recall the classification of DoS attacks [17]:

1. Network Device Level – any remote attacks that involve preventing the proper functioning of network devices such as routers or switches,
2. OS level – attacks related to implementation errors of the given protocol in the operating system,
3. Application level – errors in user applications such as a poorly defined interface or buffer overflow errors,
4. Data flood – flooding a single device with a huge amount of data,
5. Protocol feature attack – all the attacks that exploiting protocol features. A good example is the *SYN flooding attack.*

The above classification gives an overview of what type of data can be used in the process of detecting DoS or DDoS attacks. This is primarily intercepted network traffic, but logs and monitoring information can also be helpful, especially when dealing with a new type of attack.

Several works [15,22] proposed a classification of DDoS attacks from both sides: the attacker and the defender perspective. Figure 1 shows the taxonomy of defence mechanisms against DDoS attacks. In this work, we will focus on "Classification by attack detection strategy", in particular on the "NBS-2" group. This group concerns methods that use historical data to detect anomalies: events that are significantly different from the others (in our case an anomaly can be understood as an attack). The main advantage of such a solution is that it allows for a certain generalisation (and therefore, detection of unknown attacks). On the other hand, it has a tendency to misidentify normal user behaviour as an attack.

In the following section, we will shortly describe few datasets that are most commonly used in denial-of-service attack detection techniques. In Sect. 3, we will describe several approaches to the problem. The last section covers conclusions.

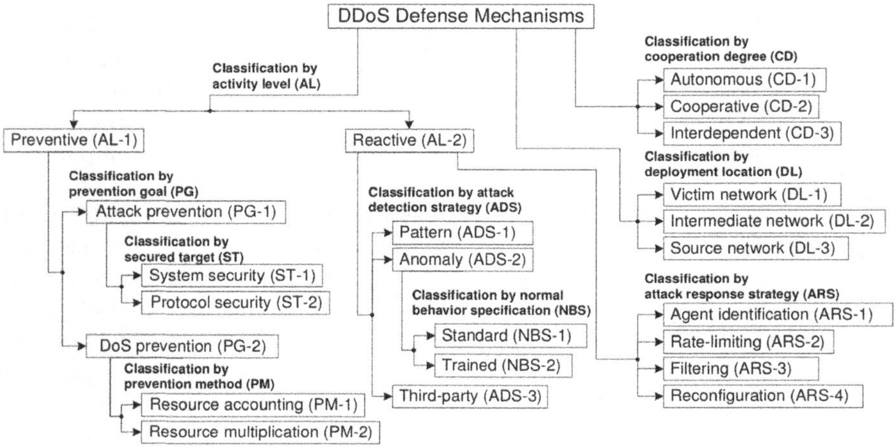

Fig. 1. Taxonomy of DDoS defense mechanisms [22]

2 Datasets

In the process of preparing a decision model, historical data is necessary to allow it to be trained. The problem with DoS-related datasets is that they are most often generated artificially (at least if the dataset is labelled). This requires the creation of statically correct methods of generating background network traffic (not-attacks), which, unfortunately, is a difficult task. In this section, we shortly describe a few most commonly used datasets. We try to focus on their possible criticism. If nothing is said about it, then the dataset is most likely reliable.

2.1 DARPA1998 and DARPA1999

DARPA1998 [1,19] is artificially generated and labelled dataset that contains nine weeks of network sniffing data, audit data (BSM) and full disk dumps from the three UNIX victim machines. It includes four types of attacks: DoS, R2L, U2R and Probing. One year later, a new dataset occurred called DARPA1999 [2,18] - the major differences are the addition of a Windows NT workstation as a victim and expanding the attack list. The complete lists of attacks in DARPA datasets are given in Table 1.

Table 1. DoS attack types used in the DARPA datasets

	Solaris	SunOS	NT	Linux
DARPA1998	apache2	apache2		apache2
	back	back		back
	mailbomb	land		mailbomb
	neptune	mailbomb		neptune
	ping of death	neptune		ping of death
	process table	ping of death		process table
	smurf	process table		smurf
	syslogd	smurf		teardrop
	udp-storm	udp-storm		udp-storm
DARPA1999				apache2
				arppoison
	neptune			back
	pod	arpoison	arppoison	mailbomb
	processtable	land	crashiis	neptune
	selfping	mailbomb	dosnuke	pod
	smurf	neptune	smurf	processtable
	syslogd	pod	tcpreset	smurf
	tcpreset	processtable		tcpreset
	warezclient			eardrop
				udpstorm

Although they are very popular datasets (they have been around for a long time and many researchers, wanting to compare their results with others, use them), they should not be used today. One of the reasons is that they are outdated and therefore less well suited to modern-day attacks and network traffic in general. The main cause is the wide criticism described in [20,21]. In 2000 another dataset appeared [3] in which several DDoS attacks were carried out using specific scenarios, but the background traffic is essentially the same as in 1999.

2.2 KDD1999

KDD [8] is another dataset that appears in many works related to denial-of-service attacks. It is actually a transformed DARPA1998 and should not be used for research due to the mentioned criticism. More about issues with this dataset can be found in [10,29].

2.3 CAIDA2007

The CAIDA2007 [4] is one of many datasets provided by CAIDA organisation. It contains one hour of a sequence of anonymized traffic traces (pcap files) from a real DDoS attack to one victim. Sadly, this dataset is now available only from IMPACT (https://www.impactcybertrust.org/), which means that you can legally download it only from the USA, Australia, Canada, Israel, Japan, The Netherlands, Singapore and UK.

2.4 ISCXIDS2012 and CICIDS2017

The ISCXIDS2012 [7] and CICIDS2017 [5] are two of many labelled datasets provided by University of New Brunswick. The first of them (described in [28]) consist of 7 days of real-like traffic (authors analysed real traces to create agents that generate it). Several attack scenarios were prepared, two of which were related to DoS attacks: "HTTP denial of service" and "distributed denial of service using an IRCBotnet". The second dataset (described in [27]) was generated correspondingly. It consists of 5 days of network traffic and includes following attacks: Brute Force FTP, Brute Force SSH, DoS, Heartbleed, Web Attack, Infiltration, Botnet attack and DDoS.

The datasets provided by the University of New Brunswick seem to be one of the best available to the public. At the end of 2019 another one [6] appeared, which only concerns DDoS attacks. Not many works related to it has yet appeared.

3 Methods Used in DoS and DDoS Attack Detection

In this work, three classes of methods for detection of DoS and DDoS attacks have been distinguished:

1. based on anomaly detection in time series analysis,
2. based on semi-supervised learning or unsupervised learning,
3. based on supervised learning.

This classification is not perfectly separable but allows to better understand the different approaches used in the problem. All of the listed works use at least one of the datasets from the previous section.

3.1 Anomaly Detection in Time Series Analysis

ARIMA [11] is one of the most widely used models for time-series forecasting. Due to the fact that network traffic can be presented as a series of values over time, it is possible to use such a method on it [30]. An anomaly (and thus a potential attack) is a situation in which the predicted value significantly differs from the actual value. Now the question is: How can the input data for the ARIMA algorithm be obtained from the captured network traffic and how exactly can we detect the anomaly?

Probably one of the first works using the ARIMA algorithm is [32]. The authors predict the flow (calculated in MB) of packets in each second. If a certain threshold is exceeded, an alarm is raised. The authors test their solution only for TCP flooding and UDP flooding attacks on their own generated data (they attack their system themselves). It makes it difficult to compare their solution with others, which is a common problem in this type of papers. Relying solely on the flow is also able to detect only a narrow group attacks.

Another example of time series analysis in DDoS detection problem is shown in [13]. Apart from the fact that the AR algorithm was used for forecasting (ARIMA is an AR generalization), the classification method as DDoS traffic has been changed. The solution is based on Lyapunov exponent [31] and [14]. Lyapunov exponent can be defined as follows:

$$\lambda_k \approx \ln(\frac{\Delta x_k}{\Delta x_0}) * \frac{1}{t_k},$$

where Δx_k is the difference between the real and predicted value and t_k is a time range. The researchers state that if $\lambda_k < 0$ the traffic might be a DDoS attack. To evaluate the results, the authors selected three days from the DARPA2000 dataset. The usage of the above equation gave them 71.84% of true positives (positive means anomaly). To improve the results, they trained a back-propagation neural network, which made it possible to achieve the result of 93.75%. The authors, unfortunately, do not give the number of false alarms. Another problem is using a dataset that doesn't have a good reputation.

The latest work [23] based on a similar idea is a work in which the authors managed to achieve results at the level of 98% (sensitivity, but the entire matrix of confusion is also given in this paper). The tests were performed on the fifth Friday of DARPA1998 dataset. The authors argue that this dataset was chosen because the result could be compared with others. The algorithm analysed two time series (the number of packets in one minute and the number of packets in one minute divided by the number of IP source addresses).

The works based on time series analysis are not very precise, because they usually lack accurate evaluation tests (no datasets other than DARPA and selection of only a part of the dataset may be biased). Methods based mainly on deep learning are becoming more and more popular, but it could be worth to ensure whether, for certain specific DoS or DDoS problem, time series analysis algorithms do not perform better.

3.2 Semi-supervised Learning and Unsupervised Learning

Many of DDoS attack detection methods are based on unsupervised learning. Data is usually divided into two clusters where one is designated for regular network traffic and the other for anomalies. Most of the available solutions are based on the classic K-means algorithm. For example, in [26] authors modified the algorithm so that it also iteratively adds and removes additional clusters. This method should allow it to better handle non-spherical distributions. The features were extracted from network traffic using a sliding window algorithm with constant size. Authors use nine of them, among others: number of traffic from the same source IP, number of traffic with "SYN" flag, number of traffic with the same protocol etc. The choice was not justified. For the purpose of evaluation, the DARPA1998 dataset was used. The method obtained 99% of precision and 1.2% of FPR. Apart from the quality of the dataset and the unclear method of testing, the result is quite high and therefore it should be verified. It is worth to notice that this work (and most of the following) does not focus only on DDoS attacks but also classifies all the others available in the dataset.

Another interesting example of work using the K-means algorithm and a hybrid of SVM and ELM algorithms is [9]. Authors separate the training dataset into five categories related to attack types in the dataset (Normal, DoS, Probe, R2L, and U2R). After that, a slightly modified version of K-means algorithm is used in every category to obtain new training datasets. Then the SVM or ELM algorithm is trained on each of the newly received datasets. The prediction process is carried out as shown in Fig. 2. The work, unfortunately, uses the discredited dataset KDD99. Let us remind that this dataset contains already extracted features. The overall performance of the proposed algorithm achieved 95.2% of precision and 1.9% of FPR, also 99.6% of DoS attacks were recognised correctly.

The newer idea based on semi-supervised learning is presented in [16]. The authors attached great importance to the selection of initial features and, based on a review of related works, they selected nine entropy-based features. The proposed algorithm evaluates and selects a subset of those features for a given dataset and performs another version of modified K-Means algorithm. In this work, the initial positions of centroids depend on a labelled sample of data. For

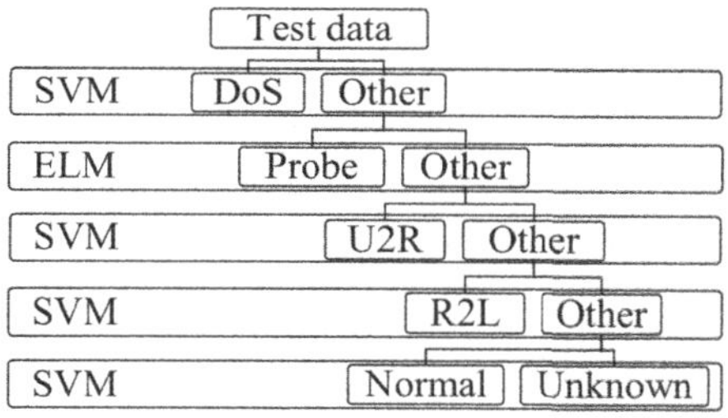

Fig. 2. Multi-level hybrid SVM and ELM [9]

evaluation purposes, they used four different datasets: DARPA2000 (for comparative purposes), CAIDA2007, CICIDS2017 and "Real-world dataset" (their own experiment). They achieved over 99% of precision for each of them.

3.3 Supervised Learning

One of the simplest examples of a supervised learning algorithm is k-NN. It was used among the others in the article [24] in order to classify the network status (normal, pre-attack, attack) rather than the traffic itself. As a distance measure, a weighted cosine formula was applied. The authors used DARPA2000 dataset and mostly entropy-based set of features. In this problem, they managed to achieve 92% of accuracy, but the results are hard to compare with others.

In [10] authors applied decision tree algorithm to classify attacks in KDD99 and they managed to correctly specify 97.1% of DoS attacks. An important contribution that this work brought is that KDD99 is not an appropriate transformation of DARPA1998 dataset, making R2L attacks difficult to classify. The authors introduced a few conditions, that might prevent information losses.

Newer methods often benefit from deep learning and do not bother with elaborated feature extraction. In the example from [12], authors trained two channels CNN network (packet and traffic features) and achieved 98.87% of accuracy for CICIDS2017 dataset and 98.54% for KDD99 dataset. Authors in [33] prepared few variants of LSTM neural network to predict the label for the last packet in a window. They evaluate their work on two days from ISCX2012 dataset and achieved 97.996% and 98.410% of accuracy respectively.

4 Conclusion

In this paper, we shortly discussed several datasets and methods used in DDoS detection problem. Four conclusions are drawn from the overview. First of all, there is a problem with comparing results. Most researchers must refer to the outdated and discredited DARPA dataset. What is more, many of the methods have never been verified on newer datasets. This mainly applies to those based on time series analysis. An interesting and understandable phenomenon is testing solutions on one's servers. However, the results of such an experiment are difficult to evaluate. Maybe researchers should standardise the method of conducting such experiments? This is not an easy task, but it is not impossible. Finally, not many researchers are concerned about the time complexity of their solutions, which may be important, especially for larger networks.

References

1. The 1998 DARPA intrusion detection evaluation dataset. https://www.ll.mit.edu/r-d/datasets/1998-darpa-intrusion-detection-evaluation-dataset. Accessed 05 Dec 2019
2. The 1998 DARPA intrusion detection evaluation dataset. https://www.ll.mit.edu/r-d/datasets/1999-darpa-intrusion-detection-evaluation-dataset. Accessed 05 Dec 2019
3. 2000 DARPA intrusion detection scenario specific datasets. https://www.ll.mit.edu/r-d/datasets/2000-darpa-intrusion-detection-scenario-specific-datasets. Accessed 05 Dec 2019
4. The CAIDA UCSD DDoS attack 2007 dataset. http://www.caida.org/data/passive/ddos-20070804_dataset.xml. Accessed 05 Dec 2019
5. The CICIDS DDoS attack 2017 dataset. https://www.unb.ca/cic/datasets/ids-2017.htm. Accessed 05 Dec 2019
6. DDoS evaluation dataset (CICDDoS 2019). https://www.unb.ca/cic/datasets/ddos-2019.html. Accessed 05 Dec 2019
7. Intrusion detection evaluation dataset (ISCXIDS 2012). https://www.unb.ca/cic/datasets/ids.html. Accessed 05 Dec 2019
8. KDD CUP 1999 data. http://kdd.ics.uci.edu/databases/kddcup99/kddcup99.html. Accessed 05 Dec 2019
9. Al-Yaseen, W.L., Othman, Z.A., Nazri, M.Z.A.: Multi-level hybrid support vector machine and extreme learning machine based on modified k-means for intrusion detection system. Expert Syst. Appl. **67**, 296–303 (2017)
10. Bouzida, Y., Cuppens, F.: Detecting known and novel network intrusions. In: IFIP International Information Security Conference, pp. 258–270. Springer (2006)
11. Brockwell, P.J., Davis, R.A.: Introduction to Time Series and Forecasting. Springer, Cham (2016)
12. Chen, J., Yang, Y.T., Hu, K.K., Zheng, H.B., Wang, Z.: DAD-MCNN: DDoS attack detection via multi-channel CNN. In: Proceedings of the 2019 11th International Conference on Machine Learning and Computing, pp. 484–488. ACM (2019)
13. Chen, Y., Ma, X., Wu, X.: DDoS detection algorithm based on preprocessing network traffic predicted method and chaos theory. IEEE Commun. Lett. **17**(5), 1052–1054 (2013)
14. Chonka, A., Singh, J., Zhou, W.: Chaos theory based detection against network mimicking DDoS attacks. IEEE Commun. Lett. **13**(9), 717–719 (2009)
15. Douligeris, C., Mitrokotsa, A.: DDoS attacks and defense mechanisms: classification and state-of-the-art. Comput. Netw. **44**(5), 643–666 (2004)
16. Gu, Y., Li, K., Guo, Z., Wang, Y.: Semi-supervised k-means DDoS detection method using hybrid feature selection algorithm. IEEE Access **7**, 64351–64365 (2019)
17. Karig, D., Lee, R.: Remote denial of service attacks and countermeasures. Princeton University Department of Electrical Engineering Technical report CE-L2001-002 **17** (2001)

18. Lippmann, R., Haines, J.W., Fried, D.J., Korba, J., Das, K.: Analysis and results of the 1999 DARPA off-line intrusion detection evaluation. In: International Workshop on Recent Advances in Intrusion Detection, pp. 162–182. Springer (2000)
19. Lippmann, R.P., Fried, D.J., Graf, I., Haines, J.W., Kendall, K.R., McClung, D., Weber, D., Webster, S.E., Wyschogrod, D., Cunningham, R.K., et al.: Evaluating intrusion detection systems: the 1998 DARPA off-line intrusion detection evaluation. In: Proceedings DARPA Information Survivability Conference and Exposition. DISCEX 2000. vol. 2, pp. 12–26. IEEE (2000)
20. Mahoney, M.V., Chan, P.K.: An analysis of the 1999 DARPA/Lincoln laboratory evaluation data for network anomaly detection. In: International Workshop on Recent Advances in Intrusion Detection, pp. 220–237. Springer (2003)
21. McHugh, J.: Testing intrusion detection systems: a critique of the 1998 and 1999 DARPA intrusion detection system evaluations as performed by lincoln laboratory. ACM Trans. Inf. Syst. Secur. (TISSEC) **3**(4), 262–294 (2000)
22. Mirkovic, J., Reiher, P.: A taxonomy of DDoS attack and DDoS defense mechanisms. ACM SIGCOMM Comput. Commun. Rev. **34**(2), 39–53 (2004)
23. Nezhad, S.M.T., Nazari, M., Gharavol, E.A.: A novel DoS and DDoS attacks detection algorithm using arima time series model and chaotic system in computer networks. IEEE Commun. Lett. **20**(4), 700–703 (2016)
24. Nguyen, H.V., Choi, Y.: Proactive detection of DDoS attacks utilizing k-NN classifier in an anti-DDoS framework. Int. J. Electr. Comput. Syst. Eng. **4**(4), 247–252 (2010)
25. Pinkas, B., Sander, T.: Securing passwords against dictionary attacks. In: Proceedings of the 9th ACM Conference on Computer and Communications Security, pp. 161–170 (2002)
26. Pramana, M.I.W., Purwanto, Y., Suratman, F.Y.: DDoS detection using modified k-means clustering with chain initialization over landmark window. In: 2015 International Conference on Control, Electronics, Renewable Energy and Communications (ICCEREC), pp. 7–11. IEEE (2015)
27. Sharafaldin, I., Lashkari, A.H., Ghorbani, A.A.: Toward generating a new intrusion detection dataset and intrusion traffic characterization. In: ICISSP, pp. 108–116 (2018)
28. Shiravi, A., Shiravi, H., Tavallaee, M., Ghorbani, A.A.: Toward developing a systematic approach to generate benchmark datasets for intrusion detection. Comput. Secur. **31**(3), 357–374 (2012)
29. Tavallaee, M., Bagheri, E., Lu, W., Ghorbani, A.A.: A detailed analysis of the KDD CUP 99 data set. In: 2009 IEEE Symposium on Computational Intelligence for Security and Defense Applications, pp. 1–6. IEEE (2009)
30. Vafeiadis, T., Papanikolaou, A., Ilioudis, C., Charchalakis, S.: Real-time network data analysis using time series models. Simul. Model. Pract. Theory **29**, 173–180 (2012)
31. Wolf, A., Swift, J.B., Swinney, H.L., Vastano, J.A.: Determining lyapunov exponents from a time series. Physica D **16**(3), 285–317 (1985)
32. Yaacob, A.H., Tan, I.K., Chien, S.F., Tan, H.K.: Arima based network anomaly detection. In: 2010 Second International Conference on Communication Software and Networks, pp. 205–209. IEEE (2010)
33. Yuan, X., Li, C., Li, X.: DeepDefense: identifying DDoS attack via deep learning. In: 2017 IEEE International Conference on Smart Computing (SMARTCOMP), pp. 1–8. IEEE (2017)

Biometric Data Fusion Strategy for Improved Identity Recognition

Zbigniew Gomolka[1(✉)], Boguslaw Twarog[1], Ewa Zeslawska[2], and Artur Nykiel[1]

[1] University of Rzeszow, Pigonia St. 1, 35-959 Rzeszow, Poland
{zgomolka, btwarog, anykiel}@ur.edu.pl
[2] Department of Applied Information, University of Information Technology and Management in Rzeszow, Sucharskiego St. 2, 35-225 Rzeszow, Poland
ezeslawska@wsiz.rzeszow.pl

Abstract. In modern authentication systems, various types of biometric measurements are used to authorize access to protected resources. Constant development of systems using biometric authentication means that they are exposed to hacker methods of stealing users' digital identities. The paper presents a hybrid system for identity identification, which uses various methods of biometric data fusion. Using the MegaMatcher environment and a set of dedicated scanners, an application was designed to allow testing of the authorization process and the impact of biometric data fusion on system security. In the experimental part of the work, two data aggregation strategies were compared, including the False Acceptance Rate (FAR) and False Rejection Rate (FRR) coefficients. The presented methods of biometric data fusion can be applied in authorization systems which use hybrid identity identification.

Keywords: Biometrics · Hybrid fusion · Decision level fusion · Score level fusion

1 Introduction

As technology develops, there is a growing demand for increasingly secure information protection systems. Most common systems can no longer provide a satisfactory level of data protection. Biometric techniques, i.e. one of the most accurate and safest means of authorization, are trying to solve this problem [5, 8, 17, 18]. The main purpose of biometrics is to verify individuals and establish their identity by means of unique biological characteristics. These features are shared by all people, but for everyone they must be unique and should meet such dependencies as: not be subject to significant changes in the time or lifestyle of a given individual; they can be obtained in a simple, quick and non-invasive way; be difficult to replicate and cannot be easily obtained without the knowledge and consent of the person concerned; obtaining them must not conflict with the principles or religion of the society concerned [1, 2, 7, 16].

Two main groups of biometric characteristics suitable for identification of persons are behavioral characteristics related to human behavior and physical characteristics such as: fingerprint, facial geometry, hand geometry, iris of the eye, auricle or blood

W. Zamojski et al. (Eds.): DepCoS-RELCOMEX 2020, AISC 1173, pp. 242–251, 2020.
https://doi.org/10.1007/978-3-030-48256-5_24

vessel system [10–12]. Soft characteristics, such as height or weight, constitute a complementary group, but they change quickly and are not unique, which means that a person's identity cannot be clearly defined. Modern biometric identification is mainly based on human physical characteristics which are unique, one-shot and unchanging over time and therefore considered to be the safest. However, apart from many advantages, there are disadvantages related to the difficulty of providing the same measurement conditions for each biometric test [3, 9, 13, 14]. These differences are caused by the changing condition of measuring devices, lighting or even humidity. Biometric identification systems are therefore exposed to instability due to the large number of incorrectly acquired samples in the system, causing noise and uncertainty in the decision making process. The inefficiency of individual methods has led to the need to combine several biometric characteristics at the same time to identify specific identities. This approach allows the system to maintain correct operation even if the measurement of one or even several characteristics was inaccurate. The accuracy of biometric systems is measured, inter alia, by the following coefficients: FAR (False Acceptance Rate) related to the acceptance of an unauthorized person, more specifically their biometric characteristics into protected resources. It determines the probability of doubtful success as a result of an incorrect comparison of the input template to a mismatched template in the database; FRR (False Rejection Rate), an error in rejecting an authorized person. The probability that the system will not detect a match between the input template and the matched template in the database. Specifies the percentage of valid input data that is incorrectly rejected.

The automatic recognition of a person's template is often difficult because the process can be burdened with technical errors, which are also a determinant of the quality of biometric systems [4, 6, 15]. The accuracy of these estimates depends on a number of test conditions, including the specification of the sensors, the number of entities in the database, the specificity of the population from which the characteristics are taken, selected operating points, etc. The influence of these factors can be reduced or even eliminated by using several biometric techniques simultaneously. Hybrid identification systems are based on the analysis of several sources, thus reducing errors occurring during the identification process, and the probability of falsification of characteristics is low due to the number of analyzed parameters.

2 A Hybrid System Implemented Using the MegaMatcher SDK Environment

Hybrid biometric identification is based on a combination of obtained biometric characteristics to minimize errors. The errors that most often result from inaccurate or incorrect measurements. The combination of characteristics may take place at different stages of the identification process, however, one of the more precise ways of consolidation is decision level fusion (see Fig. 1) and score level fusion (see Fig. 2).

In the decision level fusion approach, the process of identification and assessment of similarity of templates is carried out separately for each of the characteristics. Based on the assessment of similarity and its comparison with the threshold value, a decision of acceptance or rejection is made. The decisions that each feature has obtained are

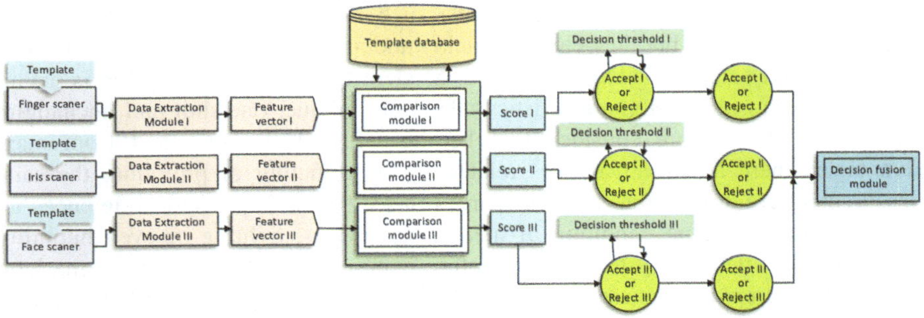

Fig. 1. Decision level fusion

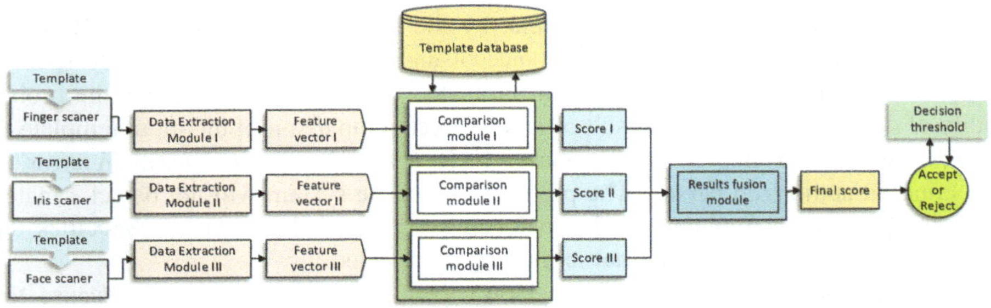

Fig. 2. Score level fusion

taken into account in determining the final decision that will determine whether to accept or reject the entire set of features.

In the score level fusion concept, the similarity assessments of each characteristic in a set are summed up. On the basis of the sum of the similarity assessments obtained by a whole set, a decision is made to accept or reject it.

2.1 Database and Binary Template Processing

The auxiliary database of the established system was divided into four non-relational independent tables, representing respectively analytical approaches that contain individual biometric characteristics. One of them contains collections of biometric characteristic templates. The individual columns of the table contain templates of another biometric characteristic: `fingerToken` (set of fingerprint templates), `irisToken` (set of iris templates), `faceToken` (face geometry templates) (see Fig. 3).

The templates involved in the fusion processes are stored in the templates table in binary form. Templates after being retrieved from the database are eventually processed from a binary form to an `NSubject` type object used by the MegaMatcher SDK environment. A binary array retrieved from the database is passed as a parameter to the

user	fingerToken	irisToken	faceToken
Face + finger	4e54001074070000a014e4654106a0700000a0	*NULL*	4e540010a48c00000a014e4c54109a8c00000a01
Face1	*NULL*	*NULL*	4e540010a48c00000a014e4c54109a8c00000a01
Finger	4e540010750900000a014e4654106b0900000a0	*NULL*	*NULL*
Full scan	4e5400109d0d00000a014e465410930d00000a0	4e5400102c0900000a014e45541022090000a0	4e540010a48c00000a014e4c54109a8c00000a01
Full Scan2	4e540010fd0f00000a014e465410f30f00000a014	4e5400102c0900000a014e45541022090000a0	4e540010a48c00000a014e4c54109a8c00000a01
left iris + finger	4e5400103e0d00000a014e465410340d00000a0	4e5400102c0900000a014e45541022090000a0	*NULL*
lewe	*NULL*	4e5400102c0900000a014e45541022090000a0	*NULL*
Right iris + finger	4e5400100e0c00000a014e465410040c00000a0	4e5400102c0900000a014e45541022090000a0	*NULL*

Fig. 3. Exemplary data in the templates table

constructor of an `NBuffer` type object, which is a transition object as an interface to other end objects (see Listing 1).

Listing 1. Creating a NSubject object from binary data

```
var tempBuffer = new NBuffer((Byte))reader['column'];
list.Add(NSubject.FromMemory(tempBuffer));
```

The step before creating a `NSubject` object is to connect to the database and use the `reader` to retrieve the template from the column of a given characteristic. Each `NSubject` object is assigned to a list which at a later stage is brought down to the form of an `NSubject` object array. Each element of the template array also receives its ID as a name that was assigned to the template in the database (see Listing 2).

Listing 2. The process of creating a list of biometric characteristic templates

```
MySqlCommand command = connection.CreateCommand();
command.CommandText = 'SELECT user FROM templates'
MySqlDataReader usersReader = comm.ExecuteReader();
NSubject[] templates = list.ToArray();
var i = 0;
while (usersReader.Read())
{
if (usersReader['user'] ToString() !=null) {
templates[i].Id = (string)usersReader['user'];
} else {
templates[i].Id = 'Name missing';
}
i++;
}
usersReader.Close();
```

The Listing 3 code fragment shows the process of acquisition of a biometric characteristic and then creating a template object on its basis. This process starts by restoring the initial state for the scanner view, then `NSubject` and `NFinger/NIris/NFace` objects are created depending on the type of characteristic being acquired. The objects of the individual characteristics contain respectively characteristic elements, e.g. `NFinger` contains a set of read minutiae.

Listing 3. The process of loading templates into a hybrid system

```
fingerView.ShownImage = ShownImage.Original;
givenFinger = new NSubject();
fingerEntity = new NFinger();
givenFinger.Fingers.Add(fingerEntity);
fingerView.Finger = fingerEntity;
NBiometricTask task = sdkEngine.CreateTask(
NBiometricOperations.Capture | NBiometricOperations.CreateTemplate,
givenFinger);
sdkEngine.BeginPerformTask(task, AfterCapture, null);
```

2.2 Data Fusion Strategies

Hybrid biometric identification was implemented at two selection levels: through data fusion at the decision level and at the score level. The implementation of the hybrid data connection is possible by combining the average biometric identification values for individual persons. For example, the object responsible for the implementation of the decision level fusion process describes the class presented in Fig. 4.

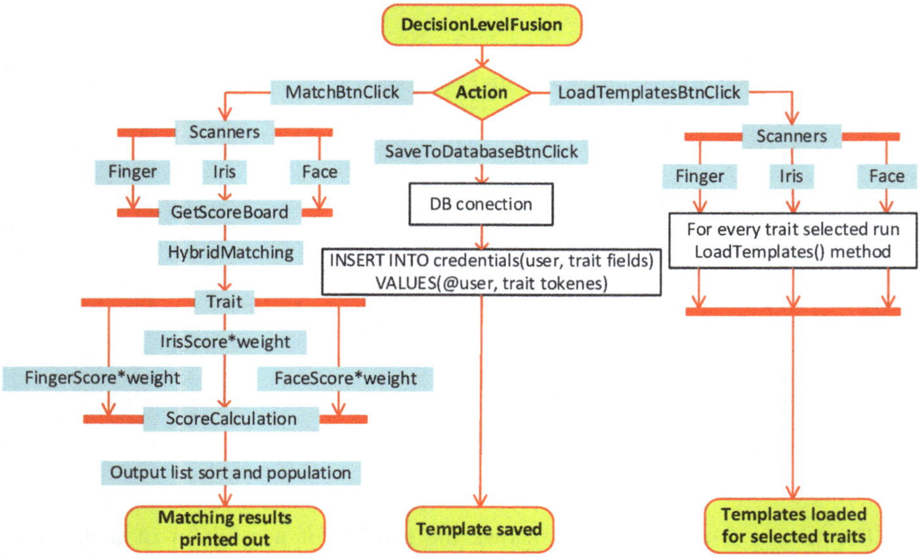

Fig. 4. DecisionLevelFusion class architecture with communication process

By characterizing the process of identification of a set of biometric characteristics in the decision level fusion version, the following design stages can be distinguished:

1. Acquisition of a given biometric characteristic with a scanner, data stored in the `NSubject` object;
2. loading template sets from the database in binary form with conversion to `NBuffer` and finally `NSubject` for each type of characteristics separately;
3. start of the identification process after the transfer of the templates to the MegaMatcher SDK engine:

   - transfer of patterns of individual characteristics to the corresponding identification methods;
   - the scores are multiplied by the weights and then compared with the corresponding threshold values;
   - if all of the characteristics indicated by the user have been accepted, the whole set is accepted, otherwise a set is rejected.

In the case of score level fusion, the process is very similar, but only the total sum of comparisons of all indicated characteristics in the set is compared with the threshold value.

3 Experiments

In order to carry out experiments for the realized hybrid system, a preparatory stage was carried out, which consisted of:

- gathering a group of persons who allowed taking and processing of the scan of their biometric characteristics, after having been informed in advance about any rules for secure processing of sensitive data in relation to the applicable GDPR;
- selecting from the group of participants whose task was to carry out authorization in the system by making both correct and inaccurate measurements of biometric characteristics;
- the tests for the proper functioning of the system consisted in authorizing the group of people 20 times for each characteristic, and the results of the tests allowed to show possible gaps in the system, its deficiencies and possible defects.

By using fusion it is possible to compensate for measurement inaccuracies and significantly reduce the FRR coefficient value, which in turn translates into reliability and safety of the system. Decision level fusion takes place at the final stage, when a decision on acceptance or rejection is set for each characteristic. At the end of the identification process, each attribute receives an appropriate binary value stating whether the similarity assessment it obtained exceeded the threshold value. The system then checks which metrics are required for the whole set to be accepted. The threshold values for fingerprints are 4000, 1000 for the face, 300 for the iris.

The values of the FRR coefficient for single characteristic identification processes and fusion-based processes are shown in Table 1.

Table 1. FRR coefficient for individual scanners

Scanner	Rejections	FRR
Fingerprint	54	0.22
Face geometry	62	0.25
Iris	47	0.19
Decision level fusion	14	0.06
Score level fusion	14	0.06

Identification based on the fusion of characteristics at the decision level allows to reduce the risk of misidentification. By introducing a value for the number of possible characteristics that were rejected, the system could both allow for loose identification and be more stringent. The results of the system based on decision level fusion of characteristics are presented in Table 2.

Table 2. Decision level fusion of characteristics

Person	Fingerprint similarity score	Minimal score 5000	Face geometry score	Minimal score 1000	Iris score	Minimal score 400	Maximum rejection number	Result
1	5219	0	1221	0	621	0	0	ACCEPT
	8310	0	321	1	551	0	1	ACCEPT
	5971	0	451	1	544	0	0	REJECT
2	2305	1	1233	0	213	1	2	ACCEPT
	1499	1	2331	0	364	1	0	REJECT
	4667	1	2671	0	395	1	0	REJECT
3	7110	0	227	1	142	1	1	REJECT
	7767	0	1923	0	312	1	0	REJECT
	10126	0	2213	0	271	1	0	REJECT
4	7331	0	1849	0	546	0	0	ACCEPT
	8772	0	1410	0	487	0	0	ACCEPT
	5612	0	1995	0	517	0	0	ACCEPT
5	11912	0	776	1	215	1	0	REJECT
	6221	0	144	1	189	1	0	REJECT
	5517	0	1722	0	165	1	3	ACCEPT
6	7691	0	2113	0	234	1	0	REJECT
	9201	0	902	1	76	1	2	ACCEPT
	7391	0	512	1	451	1	0	REJECT
7	3178	1	2552	0	612	0	1	ACCEPT
	1023	1	912	1	568	0	0	REJECT
	2215	1	1331	0	504	0	0	REJECT

The fusion of characteristics at the score level requires the indication of weights for each characteristic in order, among other things, to compensate for equipment measurement inaccuracies. The scales are defined as 0.25 for fingerprint, 0.15 for face geometry and 0.10 for iris. By using weights, the impact of inaccuracies in characteristic acquisition could be reduced. This allowed for a significant increase in the effectiveness of the system. The effectiveness of the system was reflected in the FRR coefficient which was only 0.05. The results of the score level are presented in the Table 3.

Table 3. Score level fusion of characteristics

Person	Fingerprint score	Weight 0,25	Face geometry score	Weight 0,15	Iris score	Weight 0,1	Matching sum	Minimal score 2000
1	5219	1304	1221	183	621	62	1549	REJECT
	8310	2077	321	48	551	55	2180	ACCEPT
	5971	1492	451	67	544	54	1613	REJECT
2	2305	576	1233	184	213	21	781	REJECT
	1499	374	2331	349	364	36	759	REJECT
	4667	1166	2671	400	395	39	1605	REJECT
3	7110	1777	227	34	142	14	1825	ACCEPT
	7767	1941	1923	288	312	31	2260	ACCEPT
	10126	2531	2213	331	271	27	2889	ACCEPT
4	7331	1832	1849	277	546	54	2163	ACCEPT
	8772	2193	1410	211	487	48	2452	ACCEPT
	5612	1403	1995	299	517	51	1753	ACCEPT
5	11912	2978	776	116	215	21	3115	ACCEPT
	6221	1555	144	21	189	18	1594	REJECT
	5517	1379	1722	258	165	16	1653	ACCEPT
6	7691	1922	2113	316	234	23	2261	ACCEPT
	9201	2300	902	135	76	7	2442	ACCEPT
	7391	1847	512	76	451	45	1968	ACCEPT
7	3178	794	2552	382	612	61	1237	REJECT
	1023	255	912	136	568	56	447	REJECT
	2215	553	1331	199	504	50	802	REJECT

4 Conclusions

The use of the hybrid approach in the biometric classification process has a very strong impact on improving the performance of person verification and identification systems. The combination of several characteristics showed a highly accurate operation of the identification process, as evidenced by the values of the FAR and FRR coefficients. Presenting the advantages of such an approach, we can see that we have obtained, i.a.: reduction of the FRR and FAR coefficients (see Fig. 5), which directly increases the safety of the system and convenience of use; reduction of the impact of interference from the measuring equipment on the identification process.

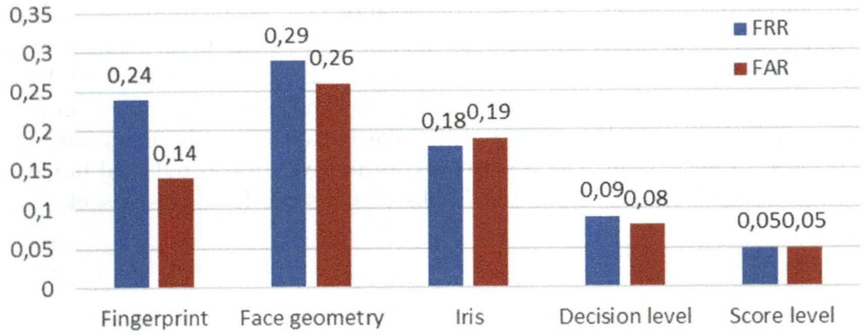

Fig. 5. Diagram of the FAR and FRR coefficients for biometric identification processes for 300 samples

It can be pointed out that the objective property of the proposed approach, which should be considered as its drawback, is the reduced operation speed compared to systems working with one biometric characteristic. The reduced speed is caused by the need to make more template comparisons with the set of scanned characteristics. The use of the hybrid fusion approach in the biometric identification process significantly reduces the risk of errors and thus increases the effectiveness of such systems. The combination of several characteristics allows to eliminate the problem of unauthorized access to a large extent, as evidenced by the obtained FAR coefficient values. It was observed that this coefficient can be as much as about six times lower than for an identification process based on only one characteristic. The values of the FRR coefficient for the biometric characteristics tested: fingerprint, face and iris geometry are 0.24, 0.29 and 0.18, respectively, when for fusion processes it is 0.24, 0.29 and 0.18: 0.09 for decision level fusion and 0.05 for score level fusion. Similar values are achieved by the FAR coefficient. The values of both coefficients directly indicate that the systems using characteristic fusion are much safer and more accurate.

Score level fusion achieved slightly better results compared to decision level fusion. This made it possible to set an overall threshold value for the whole set of characteristics. This value allowed for the final separation of incompatible sets.

The implementation of this research allows to observe the impact of the fusion of biometric characteristics on the security of the biometric identification system. The fusion of characteristics has significantly reduced the chance of unauthorized access to protected resources by reducing the impact of the environment in which the measurements are performed. The chance of obtaining an acceptable degree of similarity of several characteristics simultaneously by an unauthorized person is negligible. An important advantage of the proposed solution is the increase in reliability of the identification process, as hybridization does not cause difficulties of access for authorized persons.

References

1. Advancing Biometric Federal Bureau of Investigation FBI Biometric Specifications. https://www.fbibiospecs.cjis.gov. Accessed 2020
2. Czajka, A.: Iris liveness detection by modeling dynamic pupil features. In: Bowyer, K., Burge, M. (eds.) Handbook of Iris Recognition. Advances in Computer Vision and Pattern Recognition. Springer, London (2016)
3. Gragnaniello, D., et al.: An investigation of local descriptors for biometric spoofing detection. IEEE Trans. Inf. Forensics Secur. **10**(4), 849–863 (2015)
4. Ochocki, M., Kołodziej, M., Sawicki, D.: Identity verification algorithm based on image of the iris, Institute of Theory of Electrical Engineering, Measurement and Information Systems, Warsaw University of Technology (2015). (in Polish)
5. Tanwar, S.: Ethical, legal, and social implications of biometric technologies. In: Biometric-Based Physical and Cybersecurity Systems, pp. 535–569. Springer (2019). ISBN 978-3-319-98734-7
6. Ochocki, M., Kołodziej, M., Sawicki, D.: User verification based on the image of the iris of the eye. Przeglad elektrotechniczny, nr 11, Warsaw (2015). (in Polish)
7. Naidu, M., Govindarajulu, P.: Biometrics hybrid system based verification. Int. J. Comput. Sci. Inf. Technol. **7**(5), 2341–2346 (2016)
8. Li, X., Yin, Y., Ning, Y., et al.: A hybrid biometric identification framework for high security applications. Front. Comput. Sci. **9**, 392–401 (2015). https://doi.org/10.1007/s11704-014-4070-1
9. Dwivedi, R., Dey, S.: A novel hybrid score level and decision level fusion scheme for cancelable multi-biometric verification. Appl. Intell. **49**, 1016–1035 (2019)
10. Meghanathan, N.: Biometric Systems for User Authentication. In: Daimi, K. (ed.) Computer and Network Security Essentials. Springer, Cham (2018)
11. Dasgupta, D., Roy A., Nag A. Biometric authentication. In: Advances in User Authentication. Infosys Science Foundation Series. Springer, Cham (2017)
12. Gomolka, Z., Twarog, B., Zeslawska, E.: The implementation of an intelligent algorithm hybrid biometric identification for the exemplary hardware platforms. In: Contemporary Complex Systems and Their Dependability, DepCoS-RELCOMEX 2018. Advances in Intelligent Systems and Computing, vol. 761, pp. 228–237. Springer, Cham (2019)
13. Mazurkiewicz, J., Walkowiak, T., Sugier, J., Sliwinski, P., Helt, K.: Intelligent agent for weather parameters prediction. In: Proceedings of the Fourteenth International Conference on Dependability of Computer Systems DepCoS-RELCOMEX, Poland, pp. 331–340 (2019)
14. Awad, A., Liu, Y.: Cognitive biometrics for user authentication. In: Biometric-Based Physical and Cybersecurity Systems, pp. 387–399. Springer (2019). ISBN 978-3-319-98734-7
15. Walkowiak, T.: Low-dimensional classification of text documents. In: Proceedings of the Fourteenth International Conference on Dependability of Computer Systems DepCoS-RELCOMEX, Poland, 1–5 July 2019, pp. 534–543 (2019)
16. Bowyer, K., King, M.: Why face recognition accuracy varies due to race. Biometric Technol. Today **2019**(8), 8–11 (2019). ISSN 0969-4765, https://doi.org/10.1016/S0969-4765(19)30114-6
17. Hájek, J. Drahansky, M.: Recognition-based on eye biometrics: iris and retina. In: Biometric-Based Physical and Cybersecurity Systems. Springer (2019). ISBN 978-3-319-98734-7
18. Chi, L., Obaidat, M.: Behavioral biometrics based on human-computer interaction devices. In: Biometric-Based Physical and Cybersecurity Systems. Springer (2019). ISBN 978-3-319-98734-7

Non-homogeneous Four State Semi-Markov Reliability Model of Operation Process

Franciszek Grabski(✉)

Chair of Mathematics and Physics, Polish Naval Academy, Śmidowicza 69, 81-127 Gdynia, Poland
F.Grabski@amw.gdynia.pl

Abstract. Basic concepts, properties and facts concerning homogeneous and non-homogeneous Semi-Markov processes are presented in the paper. The non-homogeneous Semi-Markov reliability model of the operation process of the city transport means is constructed. The model allow to assess some reliability parameters and characteristic of the system.

Keywords: Semi-Markov processes · Non-homogeneous Semi-Markov process

1 Introduction

The semi-Markov processes were introduced independently and almost simultaneously by Levy [17], Smith [22], and Takacs [23] in 1954–1955. The essential development of semi-Markov processes theory were proposed by Pyke [19, 20], Cinlar [3], Koroluk and Turbin [14, 15], Limnios [17]. The non-homogeneous Semi-Markov Process has been applied in the problems relating to life insurance and problems relating to reliability and maintenance [1] Non-homogeneous semi-Markov processes were introduced independently by Iosifescu-Manu [10] and Hoem [6]. The results of Iosifescu-Manu was generalized by Jensen and Dominicis [12]. Theory of discrete time non-homogenous semi-Markov process was developed by Vassiliou and Papandopulu [25].

2 Homogeneous Semi-Markov Processes

We start from brief presentation of concepts and properties of semi-Markov processes theory that are essential in the paper.

A stochastic $\{X(t): t \geq 0\}$ process with a finite or countable state space S, piecewise constant and right continuous trajectory is said to be a homogeneous semi-Markov process if there exist nonnegative random variables $\tau_0 = 0 < \tau_1 < \tau_2 < \ldots$ such that

$$\begin{aligned} & P(\tau_{n+1} - \tau_n \leq \mathrm{t}, X(\tau_{n+1}) = j \,|\, X(\tau_n) = i, \tau_n - \tau_{n-1} \leq \mathrm{t}_n, \ldots, \tau_1 - \tau_0 \leq \mathrm{t}_1) \\ & = P(\tau_{n+1} - \tau_n \leq \mathrm{t}, X(\tau_{n+1}) = j \,|\, X(\tau_n) = i),\ t \geq 0,\ n = 1, 2, \ldots. \end{aligned} \tag{1}$$

W. Zamojski et al. (Eds.): DepCoS-RELCOMEX 2020, AISC 1173, pp. 252–262, 2020.
https://doi.org/10.1007/978-3-030-48256-5_25

Two dimensional sequence $\{(X(\tau_{n+1}), \tau_{n+1} - \tau_n,): n = 0, 1, \ldots\}|$ is said to be the Markov renewal process associated with the semi-Markov process.

The transition probabilities

$$Q_{ij}(t) = P(\tau_{n+1} - \tau_n \leq \mathrm{t}, X(\tau_{n+1}) = j \,|\, X(\tau_n) = i), \tag{2}$$

form a matrix

$$Q(t) = [Q_{ij}(t): i, j \in S] \tag{3}$$

that is called semi-Markov kernel.

To determine semi-Markov process as a model we have to define an initial distribution:

$$\boldsymbol{p}(0) = [p_0(i): i \in S], \quad \text{where} \quad p_0(i) = P(X(0) = i) \tag{4}$$

and all elements of its kernel.

It is easy to notice that the sequence $\{X(\tau_n): n = 0, 1, \ldots\}$ is a homogeneous Markov chain with transition probabilities

$$p_{ij} = P(X(\tau_{n+1}) = j \,|\, X(\tau_n) = i) = \lim_{t \to \infty} Q_{ij}(\mathrm{t}) \tag{5}$$

This random sequence is called an embedded Markov chain in semi-Markov process $\{X(t): t \geq 0\}$.

The function

$$G_i(t) = P(T_i \leq t) = P(\tau_{n+1} - \tau_n \leq t \,|\, X(\tau_n) = i) = \sum\nolimits_{j \in S} Q_{ij}(t) \tag{6}$$

is the CDF distribution of a waiting time T_i denoting the time spent in state i when the successor state is unknown.

The function

$$F_{ij}(t) = P(\tau_{n+1} - \tau_n \leq t \,|\, X(\tau_n) = i, X(\tau_{n+1}) = j) = \frac{Q_{ij}(t)}{p_{ij}} \tag{7}$$

is the CDF of a random variable T_{ij} that is called a holding time of a state i, if the next state will be j. It is easy to see that

$$Q_{ij}(t) = p_{ij} F_{ij}(t) \tag{8}$$

If a set S represents the reliability states of the system, this set may be divided on two subset S_+ and S_- where the first contains the "up" states and the second one contains the failed states ("down" states). Those subset form a partition, i.e.,

$$S = S_+ \cup S_- \text{ and } S = S_+ \cap S_- = \emptyset. \tag{9}$$

From theorem of Korolyuk and Turbin [14], Silvestrov [21], Limnios & Oprisan [17] and also Grabski [5] it follows that for finite state HSMP the following corollary is satisfied:

If $0 < E(T_i) < \infty, i = 1, 2. \ldots, N$ and probability distribution $\pi = [\pi_1, \ldots, \pi_N]$ satisfies a system of linear equations

$$\sum_{i \in S} \pi_i p_{ij} = \pi_j, j \in S, \qquad \sum_{i \in S} \pi_i = 1 \tag{10}$$

then there exist the limiting probabilities

$$P_{ij} = \lim_{x \to \infty} P_{ij}(x) = \lim_{x \to \infty} P_j(x) = \lim_{x \to \infty} P(X(x) = j) = \frac{\pi_j E(T_j)}{\sum_{k \in S} \pi_k E(T_k)} \tag{11}$$

Sufficient conditions of this theorem one can find in [5, 14, 17, 21].

Suppose that $i \in S_+$ is an initial state of the process. Conditional reliability functions of a system are defined by the rule

$$R_i(t) = P(\forall u \in [0, t], X(u) \in S_+ \mid X(0) = i), \ i \in S_+ \tag{12}$$

The conditional reliability functions satisfy system of integral equations [5].

$$R_i(t) = 1 - G_i(t) + \sum_{j \in S_+} \int_0^t R_i(t - x) dQ_{ij}(x), \quad i \in S_+. \tag{13}$$

3 Non-homogeneous Semi-Markov Processes

A two-dimensional Markov chain $\{(\xi_n, \tau_n): n = 0, 1, \ldots\}$ with transition probabilities

$$Q_{ij}(t, x) = P(\xi_{N(t)+1} = j, \ \tau_{N(t)+1} - \tau_{N(t)} \leq x \mid \xi_{N(t)} = i, \ \tau_{N(t)} = \mathrm{t}) \tag{14}$$

A stochastic process $\{X(t): \ t \geq 0\}$ with the piecewise constant and the right continuous sample paths, which is given by

$$X(t) = \xi_{N(t)} \tag{15}$$

is called a non-homogeneous Semi-Markov Process (NHSMP) associated with NHMRP $\{(\xi_n, \vartheta_n): n = 0, 1, \ldots\}$ determined by the initial distribution $\boldsymbol{p} = [p_i(0): \ i \in S]$ and the kernel $Q(t, x) = [Q_{ij}(t, x): \ i, j \in S]$, $t \geq 0$. Recall, that $N(t) = \sup\{n \in : \ \tau_n \leq t\}$ denotes a number of the state changes in a time interval $[0, t]$ and $\{N(t): \ t = 0, 1, 2, \ldots\}$ is a counting process. Recall also that $\tau_{N(t)+1} - \tau_{N(t)} = \vartheta_{N(t)+1}$. The functions

$$p_{ij}(t) = P\left(\xi_{N(t)+1} = j | \xi_{N(t)} = i, \tau_{N(t)} = t\right) = \lim_{x \to \infty} Q_{ij}(t, x),\quad t \in [0, \infty), i, j \in S \tag{16}$$

are called the transition probabilities of non-homogenous embedded Semi-Markov chain in NHSMP. Those functions form a square matrix

$$\boldsymbol{p}(t) = \left[p_{ij}(t)\colon i, j \in S\right]. \tag{17}$$

Similar way like in the case of homogeneous semi-Markov process we can introduce a cumulative distribution function (CDF) of a holding time $T_{ij}(t)\colon t \in [0, \infty)$, $i, j \in S$. The CDF is given by

$$F_{ij}(t, x) = P\left(\vartheta_{N(t)+1} \leq x | \xi_{N(t)+1} = j, \xi_{N(t)} = i, \tau_{N(t)} = t\right),\quad i, j \in S,\ x, t \in [0, \infty). \tag{18}$$

It is easy to show that

$$Q_{ij}(t, x) = p_{ij}(t) F_{ij}(t, x) \tag{19}$$

The cumulative distribution function of a waiting time $\{T_i(t)\colon t \in [0, \infty)\}$ in a state i is given by the formula

$$G_i(t, x) = \sum\nolimits_{j \in S} Q_{ij}(t, x). \tag{20}$$

It means that

$$G_i(t, x) = P(T_i(t) \leq x) = P(\vartheta_{N(t)+1} \leq x \mid \xi_{N(t)} = i,\ \tau_{N(t)} = t). \tag{21}$$

The interval transition probabilities

$$P_{ij}(t, s) = P(X(s) = j \mid X(\tau_{N(t)}) = i, \tau_{N(t)} = t), \quad 0 \leq t < s, \quad i, j \in S. \tag{22}$$

are some of the important characteristics of the non-homogenous semi-Markov process.

Assume that $i \neq j$. The non homogenous semi-Markov process, that starts from a state i at the moment $\tau_{N(t)} = t$ will be in state j at the moment $s > t > 0$ if in an instant $\tau_{N(t)+1}$ the process will pass to a state $k \in S$, and in a time interval $\left(\tau_{N(t)+1}, s\right]$ there takes place at least one change of the state from k to j. Using a memoryless property of a semi-Markov process in the instants $\tau_{N(t)+1}$ and the theorem of the total probability we have:

$$\begin{aligned} P_{ij}(t,s) &= P(X(s)=j|X(\tau_{N(t)})=i, \tau_{N(t)}=t) \\ &= \sum_{k\in S} \int_0^u (P(X(s)=j|X(\tau_{N(t)+1})=k, \vartheta_{N(t)+1}=u)) \\ P(X(\tau_{N(t)+1}) &= k, \vartheta_{N(t)+1}\in du|X(0)=i) \\ &= \sum_{k\in S} \int_0^s P_{kj}(t+u, s-u) Q_{ik}(t,du). \end{aligned} \tag{23}$$

Therefore

$$P_{ij}(t,s) = \sum_{k\in S} \int_0^s P_{kj}(t+u, s-u) Q_{ik}(t,du), \quad i,j\in S, i\neq j, \quad 0\leq t<s.$$

Assume now that $i=j$. The process starting from the state $i\in S$ at the moment $\tau_{N(t)}=t$ will have value $i\in S$ and in the instant $s>t\geq 0$ will have also the same value, if the event $\{\vartheta_{N(t)+1}>s\}$ occurs. Because $P(\vartheta_{N(t)+1}s|X(\tau_{N(t)})=i, \tau_{N(t)}=t)=G_i(t,s)$ then

$$P(\vartheta_{N(t)+1}>s|X(\tau_{N(t)})=i, \tau_{N(t)}=t)=1-G_i(t,s).$$

Hence, for any $i\in S$.

$$P_{ii}(t,s) = 1-G_i(t,s) + \sum_{k\in S} \int_0^s P_{ki}(t+u, s-u) Q_{ik}(t,du), \quad i\in S, \quad 0\leq t<s.$$

Finally we obtain the following system of integral equations

$$\begin{aligned} &P_{ij}(t,s) = \delta_{ij}[1-G_i(t,s)] + \sum_{k\in S} \int_0^s P_{ki}(t+u, s-u) Q_{ik}(t,du), \\ &i,j\in S, 0\leq t<s. \end{aligned} \tag{24}$$

with an initial condition

$$P_{ij}(t,0) = \begin{cases} 1 & if \quad i=j \\ 0 & if \quad i\neq j \end{cases} \tag{25}$$

The non-homogeneous Semi-Markov Process has been applied in the problems relating to life insurance, the medical problems as well as issues of reliability and maintenance.

4 Non-homogeneous Semi-Markov Process as a Model of Transport Means Operation Process

4.1 Description and Assumptions

We assume that the duration of each stopover at the depot space is non negative random variable ξ_t with CDF (cumulative density function)

$$F_{\xi_t}(x) = P(\xi_t \leq x) \tag{26}$$

dependent on t.

The duration of each realization of transport task (carriage of passengers) is non negative random variable variable η_t with PDF

$$f_{\eta_t}(x) = \frac{\kappa_t^2}{2} x e^{-\kappa_t x}, \ \mu > 0, \ x \geq 0 \tag{27}$$

dependent on t.

During functioning period the system can damage. The time to failure during carriage of passenger is a non negative random variable ζ_t with a probability density function $f_{\zeta_t}(x)$, $x \geq 0$ that depends on t. We suppose

$$f_{\zeta_t}(x) = \lambda(t) e^{-\lambda(t)x}, \quad x \geq 0, \quad \lambda(t) > 0 \tag{28}$$

A repair (renewal) time is non negative random variable $\gamma(t)$ that distribution is determined by a probability density function dependent on t.

$$f_{\gamma(t)}(x) = \mu(t)^2 x e^{-x}, \ \mu(t) > 0, \ x \geq 0 \tag{29}$$

Time to a road accident is an exponentially distributed random variable with a probability density function dependent on t:

$$f_{\theta_t}(x) = \alpha(t) e^{-\alpha(t)x}, \ x \geq 0, \ \lambda(t) > 0 \tag{30}$$

The accident causes stopping the operation process for a relatively long time. A repair (renewal) time is non negative random variable variable υ with a probability density function

$$f_{\upsilon}(x) = \nu(t)^2 x e^{-x\nu(t)}, \quad \nu(t) > 0, \ x \geq 0 \tag{31}$$

Moreover we assume that all mentioned above random variable and their copies are identically distributed and mutually independent.

4.2 Model Construction

To consider a stochastic model of the above mentioned operation process we start with introducing the following states of the system, (see Migawa [18]).

s_1 – stopover at the bus depot space
s_2 – realization of transport tasks - carriage of passengers
s_3 – failure during carriage of passanger and repair at the bus depot space or repair by the technical moving support unit
s_4 – accident of the transport mean (bus) and long repair

Possible states changes of the system are shown in Fig. 1.

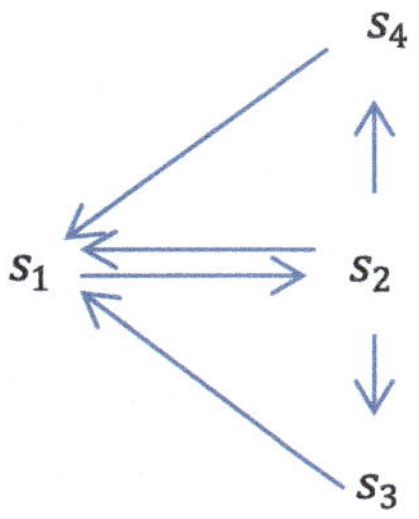

Fig. 1. Possible states changes of the system.

Let $\{X(t): t \geq 0\}$ be a stochastic process describing the transport means operation process. From description and assumptions it follows that we can treat this process as a non-homogenous Semi-Markov process that is determined by the kernel

$$Q(t,x) = \begin{bmatrix} 0 & Q_{12}(t,x) & 0 & 0 \\ Q_{21}(t,x) & 0 & Q_{23}(t,x) & Q_{24}(t,x) \\ Q_{31}(t,x) & 0 & 0 & 0 \\ Q_{41}(t,x) & 0 & 0 & 0 \end{bmatrix}, \tag{32}$$

and an initial distribution

$$\boldsymbol{p}(0) = [1, \quad 0, \quad 0, \quad 0] \tag{33}$$

Now we have to define all elements of the Semi-Markov kernel (32):

$$Q_{12}(t,x) = P(\eta_t \leq x) = \begin{cases} 0 & \text{for} \quad x \in [0,d] \\ 1 & \text{for} \quad x \in (d,\infty) \end{cases} \tag{34}$$

$$Q_{21}(t,x) = P(\xi_t \leq x, \zeta_t > \xi_t, \theta_t > \xi_t) = \int_0^x [1 - F_{\zeta_t}(u)][1 - F_{\theta_t}(u)]\, dF_{\xi_t}(u) \tag{35}$$

For simplification and shortening of equalities we will write and remember that

$$\kappa(t) = \kappa, \quad \lambda(t) = \lambda, \ \alpha(t) = \alpha.$$

From (28) and (30) we obtain

$$\begin{aligned} Q_{21}(t,x) &= P(\xi_t \le x, \xi_t < \zeta_t, \xi_t < \theta_t) \\ &= \int_0^x \kappa(t)^2 u e^{-\kappa(t)x} e^{-\lambda(t)u} e^{-\alpha(t))u} du = \int_0^x \kappa^2 u e^{-(\kappa+\lambda+\alpha)u} du \end{aligned} \tag{36}$$

$$= \frac{\kappa(1 + e^{-(\kappa+\lambda+\alpha)x}(-1 - (\kappa+\lambda+\alpha)x))}{(\kappa+\lambda+\alpha)^2} \tag{37}$$

$$\begin{aligned} Q_{23}(t,x) &= P(\zeta_t \le x, \zeta_t < \xi_t, \theta_t < \zeta_t) \\ &= \int_0^x [1 - F_{\xi_t}(u)]\,[1 - F_{\theta_t}(u)]\lambda e^{-\lambda u} du \\ &= \int_0^x (\lambda(1 + \kappa \mathrm{u})\, e^{-(\kappa+\lambda+\alpha)u} du \\ &= \frac{\lambda((1 - e^{-(\kappa+\lambda+\alpha)x})(\kappa+\lambda+\alpha) + \kappa(1 + e^{-(\kappa+\lambda+\alpha)x}(-1 - (\kappa+\lambda+\alpha)x)))}{((\kappa+\lambda+\alpha)^2} \end{aligned} \tag{38}$$

In a similar way we obtain

$$Q_{24}(t,x) = \frac{\alpha((1 - e^{-(\kappa+\lambda+\alpha)x})(\kappa+\lambda+\alpha) + \kappa(1 + e^{-(\kappa+\lambda+\alpha)x}(-1 - (\kappa+\lambda+\alpha)x)))}{(\kappa+\lambda+\alpha)^2} \tag{39}$$

$$Q_{31}(t,x) = \int_0^x \mu^2 u e^{-\mu u} du = 1 - (1 + \mu x)e^{-\mu x}, \mu > 0 \tag{40}$$

$$Q_{41}(t,x) = \int_0^x v^2 u e^{-vu} du = 1 - (1 + vx)e^{-vx}, v > 0 \tag{41}$$

4.3 Characteristics and Parameters of the Model

From (16) we get the matrix of transition probabilities of the embedded non-homogeneous Markov chain $\{X(\tau_n): n = 0, 1, 2, \ldots\}$:

$$p(t) = \begin{bmatrix} 0 & p_{12}(t) & 0 & 0 \\ p_{21}(t) & 0 & p_{23}(t) & p_{24}(t) \\ p_{31}(t) & 0 & 0 & 0 \\ p_{41}(t) & 0 & 0 & 0 \end{bmatrix}, \tag{42}$$

where

$$p_{12}(t) = 1 \tag{43}$$

$$p_{21}(t) = \frac{\kappa(t)^2}{(\kappa(t) + \lambda(t) + \alpha(t))^2} \tag{44}$$

$$p_{23}(t) = \frac{\lambda(t)}{(\kappa(t) + \lambda(t) + \alpha(t))^2} \tag{45}$$

$$p_{23}(t) = \frac{\lambda(t)(2\kappa(t) + \lambda(t) + \alpha(t))}{(\kappa(t) + \lambda(t) + \alpha(t))^2} \tag{46}$$

$$p_{23}(t) = \frac{\alpha(t)(2\kappa(t) + \lambda(t) + \alpha(t))}{(\kappa(t) + \lambda(t) + \alpha(t))^2} \tag{47}$$

$$p_{31}(t) = 1 \tag{49}$$

$$p_{41}(t) = 1 \tag{50}$$

From (6) it follows that distributions of waiting times $T_i(t)$ are determined by the rules:

$$G_i(t,x) = \sum_{j=1}^{4} Q_{ij}(t,x),\ i = 1,2,3,4 \tag{51}$$

Hence

$$G_1(t,x) = Q_{12}(t,x) = \begin{cases} 0 & \text{for} \quad x \in [0,d] \\ 1 & \text{for} \quad x \in (d,\infty) \end{cases} \tag{52}$$

$$\begin{aligned} G_2(t,x) &= Q_{21}(t,x) + Q_{23}(t,x) + Q_{24}(t,x) \\ &= (-1 - \kappa(t)x + e^{(\kappa(t) + \lambda(t) + \alpha(t))x})(e^{-(\kappa(t) + \lambda(t) + \alpha(t))x}) \end{aligned} \tag{53}$$

$$G_3(t,x) = Q_{31}(t,x) = 1 - (1 + \mu(t)x)e^{-\mu(t)x},\ \mu(t) > 0 \tag{54}$$

$$G_4(t,x) = Q_{41}(t,x) = 1 - (1 + \nu(t)x)e^{-\nu(t)x},\ \nu(t) > 0 \tag{55}$$

Now we calculate expectation of waiting times.

$$E(T_1(t)) = d(t) \tag{56}$$

$$E(T_2(t)) = \frac{2\kappa(t) + \lambda(t) + \alpha(t)}{(\kappa(t) + \lambda(t) + \alpha(t))^2} \tag{57}$$

$$E(T_3(t)) = \frac{2}{\mu(t)} \tag{58}$$

$$E(T_4(t)) = \frac{2}{v(t)} \tag{59}$$

We ought to mention that NHSM process $\{X(t)\colon t \geq 0\}$ is periodic with period of one year. From theorem of Wajda [24] it follows that in this case there exist the stationary probabilities that satisfy system of Eq. (10). As solution we obtain

$$P_1(t) = \frac{E(T_1(t))}{E(T_1(t)) + E(T_2(t)) + E(T_3(t)) + E(T_4(t))} \tag{60}$$

$$P_2(t) = \frac{E(T_2(t))}{E(T_1(t)) + E(T_2(t)) + E(T_3(t)) + E(T_4(t))} \tag{61}$$

$$P_3(t) = \frac{E(T_3(t))}{E(T_1(t)) + E(T_2(t)) + E(T_3(t)) + E(T_4(t))} \tag{62}$$

$$P_4(t) = \frac{E(T_4(t))}{E(T_1(t)) + E(T_2(t)) + E(T_3(t)) + E(T_4(t))} \tag{63}$$

The availability coefficient of the transport means operation process is the function

$$A(t) = P_1(t) + P_2(t), \tag{64}$$

where expectation of waiting times are given by (56)–(59).

To use the model in the real transport system we have to estimate unknown system parameters based on the real data. Next we should find an approximate discrete solution of the equations system (24). Finally we have to find continuous periodic function describing the availability coefficient of the transport means operation process.

References

1. Andrzejczak, K.: Stochastic modelling of the repairable system. J. KONBiN **35**(1), 5–14 (2017)
2. Barlow, R.E., Proshan, F.: Statistical Theory of Reliability and Life Testing. Holt, Rinchart and Winston Inc., New York (1975)
3. Cinlar, E.: Markov renewal theory. Adv. Appl. Probab. **1**(2), 123–187 (1969)
4. Feller, W.: An Introduction to Probability Theory and Its Applications, vol. 2. Wiley, New York (1966)

5. Grabski, F.: Semi-Markov Processes: Applications in Systems Reliability and Maintenance, p. 251. Elsevier, Amsterdam (2015, 2018)
6. Hoem, J.M.: Inhomogeneous semi-Markov processes, select actuarial tables and duration-dependence in demography. In: Greville, T.N.E. (ed.) Population Dynamics, pp. 251–296. Academics Press, Cambridge (1972)
7. Howard, R.A.: Dynamic Programing and Markov Processes. MIT Press, Cambridge (1960)
8. Howard, R.A.: Research of semi-Markovian decision structures. J. Oper. Res. Soc. Jpn. **6**, 163–199 (1964)
9. Howard, R.A.: Dynamic Probabilistic System. Semi-Markov and Decision Processes, vol. II. Wiley, New York (1971)
10. Iosifescu-Manu, A.: Non homogeneous semi-Markov processes. Stud. Lere. Mat. **24**, 529–533 (1972)
11. Iosifescu, M.: Finite Markov Processes and Their Applications. Wiley, Hoboken (1988)
12. Jensen, J., De Dominicisis, R.: Finite non-homogeneous semi-Markov processes. Insur. Math. Econ. **3**, 157–165 (1984)
13. Janssen, J., Manca, R.: Applied Semi-Markov Processes. Springer, New York (2006)
14. Korolyuk, V.S., Turbin, A.F.: Semi-Markov Processes and Their Applications. Naukova Dumka, Kiev (1976). (in Russian)
15. Korolyuk, V.S., Turbin, A.F.: Markov Renewal Processes in Problems of Systems Reliability. Naukova Dumka, Kiev (1982). (in Russian)
16. Lev'y, P.: Proceesus semi-markoviens. In: Proceedings of the International Congress of Mathematicians, Amsterdam, pp. 416–426 (1954)
17. Limnios, N., Oprisan, G.: Semi-Markov Processes and Reliability. Birkhauser, Boston (2001)
18. Migawa, K.: Semi-Markov model of the operation process included in an utilization subsystem of the transport system. Arch. Automot. Eng. **2**, 87–97 (2010)
19. Pyke, R.: Markov renewal processes: definitions and preliminary properties. Ann. Math. Stat. **32**, 1231–1242 (1961)
20. Pyke, R.: Markov renewal processes with finitely many states. Ann. Math. Stat. **32**, 1243–1259 (1961)
21. Silvestrov, D.C.: Semi-Markov Processes with a Discrete State Space. Sovetskoe Radio, Moscaw (1980). (in Russian)
22. Smith, W.L.: Regenerative stochastic processes. Proc. Roy. Soc. London Ser. A **232**, 6–31, 27 (1955)
23. Takács, L.: Some investigations concerning recurrent stochastic processes of a certain type. Magyar Tud. Akad. Mat. Kutato Int. Kzl. **3**, 115–128 (1954)
24. Wajda, W.: Limit theorems for non-homogeneous semi Markov Processes. Aplicationes Mathematicae **21**(1), 1–14 (1991)
25. Vassiliou, P.-C.G., Papadopoulu, A.A.: Non-homogenous semi-Markov systems and maintainability of the state sizes. J. Appl. Probab. **29**, 519–534 (1992)
26. Zajac, M.: Reliability model of the inter-model system. Ph.D. thesis, Wroclaw University of Technology, Wroclaw (2007)

The Efficiency of Energy Storage Systems Use for Energy Cost Mitigation Under Electricity Prices Changes

Alexander Grakovski(✉) and Aleksandr Krivchenkov

Transport and Telecommunication Institute, Lomonosova Street 1, Riga 1019, Latvia
{avg, aak}@tsi.lv

Abstract. The purpose of present research is an analysis of currently promoted energy storage systems based on high-capacity electric batteries from the standpoint of algorithms for intelligent control of their charge and discharge processes. It is discussed the reduction the cost of electricity consumed by the enterprise by the redistributing of energy depending on the variation in tariffs over time. It is based on the use of the Energy Storage System (ESS) and optimal battery charge/discharge schedule. An estimation of savings in consumed energy costs is carried out depending on the power, capacity of ESS, as well as of the period of planned schedule calculating. On base of numerical simulation of battery's charge/discharge control by linear programming optimisation method the efficiency of ESS usage was estimated in the range 10–15% for different periods from 1 up to 5 days of scheduling (planning horizon) respectively.

Keywords: Energy Storage System (ESS) · ESS efficiency · Battery · Electricity tariff · Charge/discharge scheduling

1 Introduction

In different countries or regions over the world, competitive markets have been created for the wholesales of electricity, where the electricity dealers (producers and electric load aggregators) have the opportunity to purchase energy on day-ahead and real-time power markets and sell it to the end users. In the region of Baltic Sea a trading platform is the *Nord Pool Spot*, where the results are available for analysis [1].

It is not possible for individuals and enterprises to buy energy at purely exchange prices, and the tariffs applicable to them contain a significant component of the price of energy transmission and distribution. For Latvia it is a monopoly of ST (*Sadales Tīkli*) [2]. Some part of the energy price, by agreement with the seller, may be proportional to the exchange price. Even if the seller does not take exchange price fluctuations into account, its tariffs are often determined for time zones (day zone or night zone for example).

The problem of enterprise's electricity consumption expenses optimisation became very important due to the growth of electricity consumption, especially for small and

W. Zamojski et al. (Eds.): DepCoS-RELCOMEX 2020, AISC 1173, pp. 263–272, 2020.
https://doi.org/10.1007/978-3-030-48256-5_26

medium enterprises (SME), actively introducing automated and robotic production lines into the production processes.

Under such circumstances SME has the opportunity to redistribute his workload in an organizational or technical way.

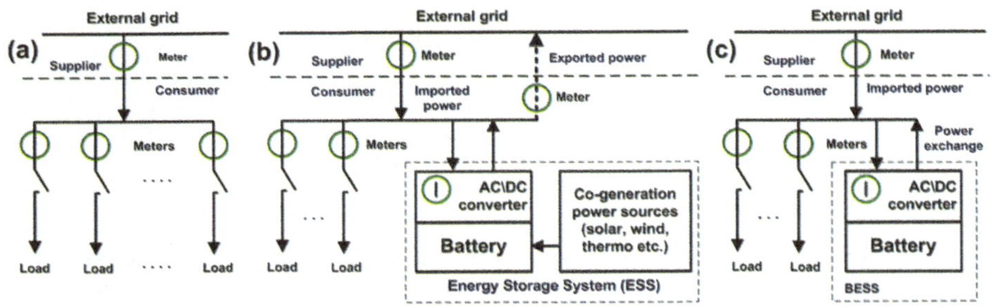

Fig. 1. The architectures for the enterprise's consumption expenses optimization: (a) load manipulation without ESS, (b) full system ESS together with additional energy sources, and (c) battery based energy storage system (BESS) [3].

If an enterprise has the ability to plan the load freely during the day, the easiest way to optimize electricity costs is to disconnect part of the equipment at the hour of the highest price (see Fig. 1(a)), but it usually leads to the transfer of work to night hours and increasing of other costs for the production processes.

The method of saving is also the use of the Energy Storage System (ESS) presented in Fig. 1(b). The 'green' energy sources (solar, wind, hydro and others) significantly increases the possibility of charging the battery and controlling the entire system [5], as well as, with the relevant agreements with the supplier, allows the return (sell) part of the electricity to the grid.

Battery Energy Storage System (BESS), including only the battery and AC\DC converter-inverter (see Fig. 1(c)), seems to be the most suitable for the SME [8].

The problem of optimal control system is reduced to the calculation of a BCDS (Battery Charge/Discharge Schedule) and its implementation in the ESS, which minimizes of the cost of electricity for a certain period of time. The basis for drawing up such a plan are known or forecasted prices for energy and the plan or prediction of the energy consumed by the load.

Main objective of the research is to develop an algorithm for automatic optimal control in the conditions of hourly changes in the market price of electricity, variations in power consumption, and the aging process of sources, formalization of the objective function and limitations of battery charging/discharging control optimization problem, as well as the estimation of the potential economic efficiency of ESS for SME, including not only industry, but also local electricity providers (aggregators). It corresponds to the architecture in Fig. 1(c).

2 ESS and Its Application

Main components of the ESS, which are important in the process of using it to reduce costs, are presented in Fig. 2. Reduced cost, in this case, is achieved by calculation and implementation of the BCDS. In various works offers a variety of algorithms of BCDS [4–7]. The problem of optimal planning and control of ESS was solved for various systems. In [5] a planning and control strategy based on the predictive control model is presented. In the works [6, 7] two BCDS algorithms are given. According to the authors, these algorithms are the simplest and effective. Earlier works have already drawn attention to the fact that the energy price is described to piecewise-continuous function of time, which causes difficulties in optimization, also attempts are made to take more detailed account when planning the properties of batteries [10], which further complicates the planning algorithm. Optimal plan which will ensure maximum savings is formed as follows.

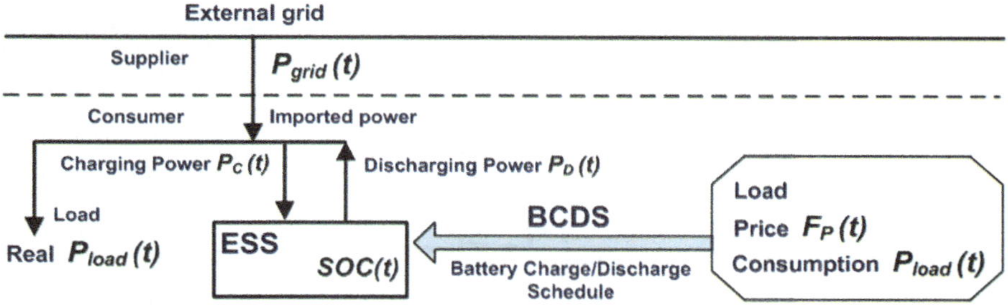

Fig. 2. Main components considered for the costs reduction using ESS [4].

Planning is carried out for a period of time K hours at discrete values of $k = 1, 2, \cdots, K$. Time interval is equal to 1 h and K as an example for a one day is equal to 24. The power consumed from the grid, is equal to:

$$P_{grid}(k) = P_{load}(k) + P_C(k) - \eta_D \cdot P_D(k), \tag{1}$$

where $P_{load}(k)$ is the working plan of the enterprise; the power of charge $P_C(k)$ and discharge $P_D(k)$ to be found in the scheduling process; $\eta_D \leq 1$ is the efficiency coefficient of discharge, and it is determined by the inverter. Condition of power limitation $0 \leq P_{grid}(k) \leq P_{max}(k)$ is applied, i.e. no power is sent to the grid and input power P_{max} is limited by the type of the connection to the grid.

The charge/discharge optimal scheduling is reduced to minimizing the following expression:

$$\sum\nolimits_{k=1}^{K} \left(F_p(k) \cdot P_{grid}(k)\right) \rightarrow min, \tag{2}$$

where $F_p(k)$ is the price for 1 kWh of energy; it can be proportional to the market (on agreement with the seller); in most cases the seller sets this price proportional to some average exchange price; for our consideration it is important only that this is the part of the price that is function of time interval k.

The following conditions are applied to the charge/discharge power and the battery:

$$0 \leq P_C(k) \leq P_{Cmax} \leq P_{max}, \tag{3}$$

$$0 \leq P_D(k) \leq P_{Dmax}, \tag{4}$$

$$SOC(k) = SOC(k-1) + \eta_C \cdot P_C(k) - P_D(k), \tag{5}$$

$$SOC_{min} \leq SOC(k) \leq SOC_{max}, \tag{6}$$

where η_C is the efficiency of charge and it is specified; also there should be a certain range of the battery capacity filling: SOC (State of Charge).

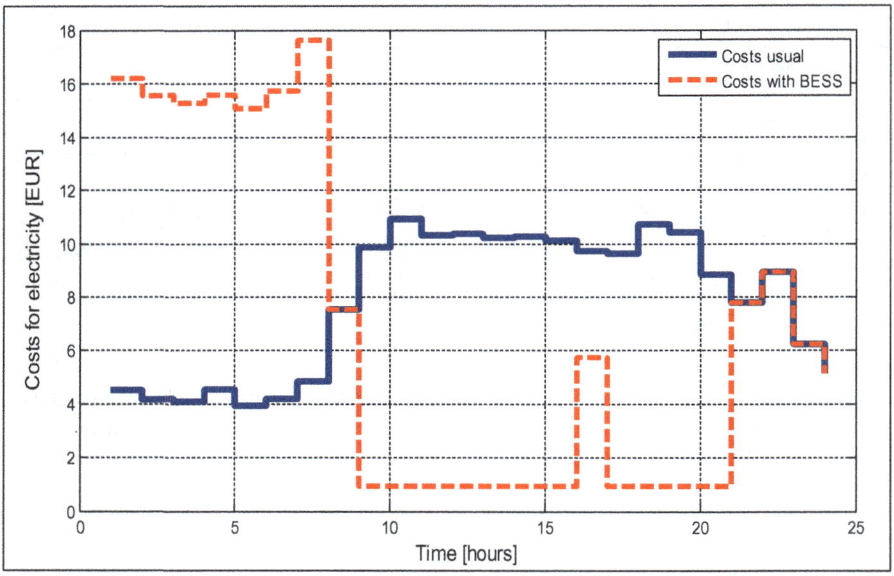

Fig. 3. Cost achieved by Battery ESS charging/discharging according to calculations by linear programming method in Matlab [7]. Hourly average consumption is 85 kW, $\delta C(24) = 13\%$

The scheduling process in this formulation is the solution of the problem of linear programming, the number of variables is $2 \cdot K$. The solution to this problem is quite sensitive to the value of the parameters in (3–6) and planning period K. But it matters how exactly k related parameters in the planning process were known. We estimate the gain in energy costs when implementing the optimal plan. The win in the cost when applying ESS will be:

$$\Delta C(K) = \sum_{k=1}^{K} \left[C(P_{load}(k)) - C\left(P_{grid}(k)\right)\right] = \delta C(K) \cdot \sum_{k=1}^{K} C(P_{load}(k)), \quad (7)$$

where

$$\delta C(K) = \frac{\sum_{k=1}^{K}\left[C(P_{load}(k)) - C\left(P_{grid}(k)\right)\right]}{\sum_{k=1}^{K} C(P_{load}(k))} \quad (8)$$

is a relative gain in the period K. Thus, an absolute win in the cost for the period is proportional to the relative gain and cost of energy for a given load. In [4] it has been shown that for certain prices changes $\delta C(24)$ had the greatest possible value of 18%, and two different planning algorithms gave the different values for $\delta C(24)$ (11% and 7%).

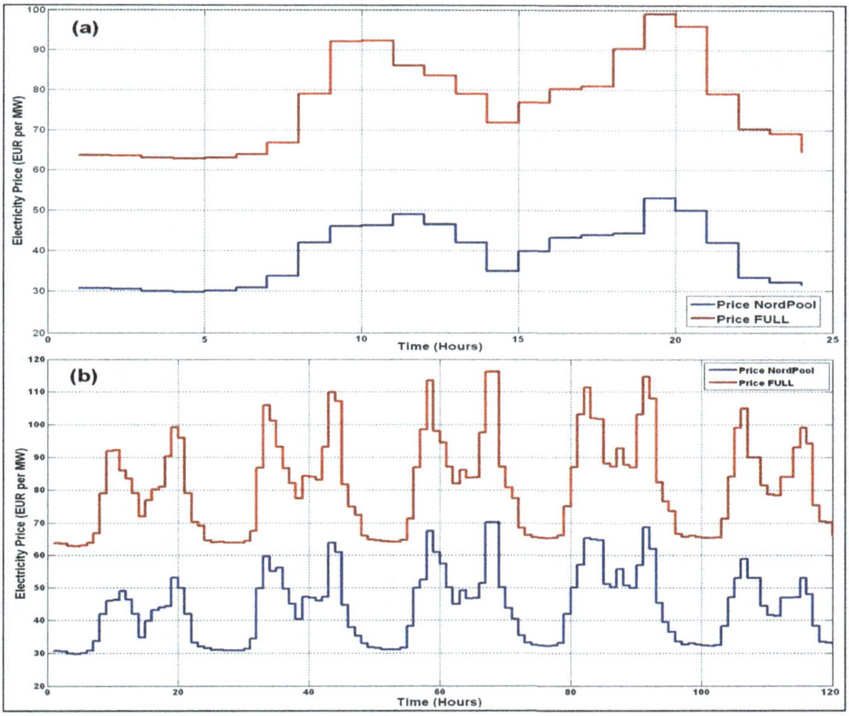

Fig. 4. Changes in electricity prices for: (a) 24 h (1 day: 06-02-2017), and (b) 120 h planning period (5 days), from 06-02-2017 till 10-02-2017 as the example of working week during the winter peak season

This value (8) is the convenient criterion for comparing the effectiveness of scheduling algorithms and we call it later as "efficiency". The estimations of the relative payoff (8) are made in [4] under assumptions that maximum power is fixed; the planning period is equal to 24 (hours); the load is uniformly distributed on time intervals and is equal $P_{load\,avg}$; exchange prices are fixed for 2 zones: for night and day zones, the maximum possible value of (8) has the assessment as 15%. And this value can be reached under conditions [4]:

$$SOC_{max} \geq 16 \cdot P_{load\,avg}; \quad P_{Dmax} \geq P_{load\,avg}; \quad P_{Cmax} \geq \frac{16}{8} \cdot P_{load\,avg}. \tag{9}$$

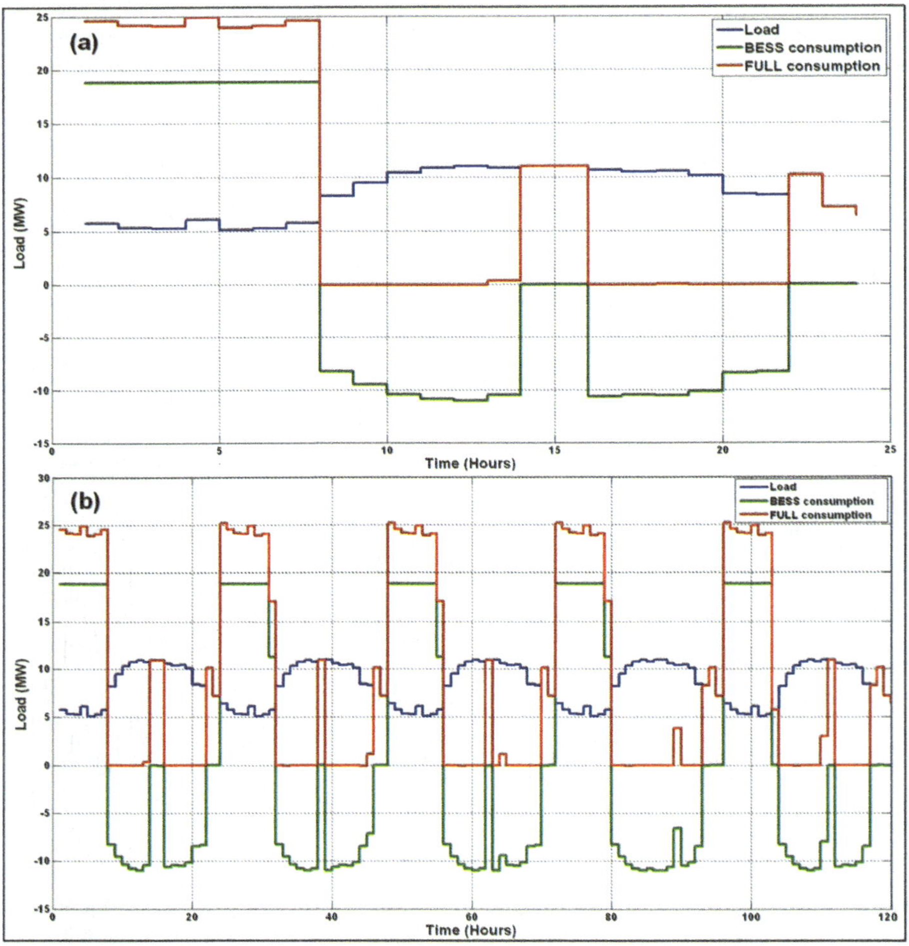

Fig. 5. Typical for the enterprise load and batteries charge/discharge process for: (a) 24 h (1 day), and (b) period of 120 h (5 days)

If the parameters do not meet the requirements for the ESS (9), relative payoff will decrease. Applying specific data on exchange prices and the "typical" load distribution

[5], we have found electricity cost distribution in a time and the value of δC(24) (Fig. 3). The code of MatLab program calculates the optimal BCDS schedule on the base of linear programming [9].

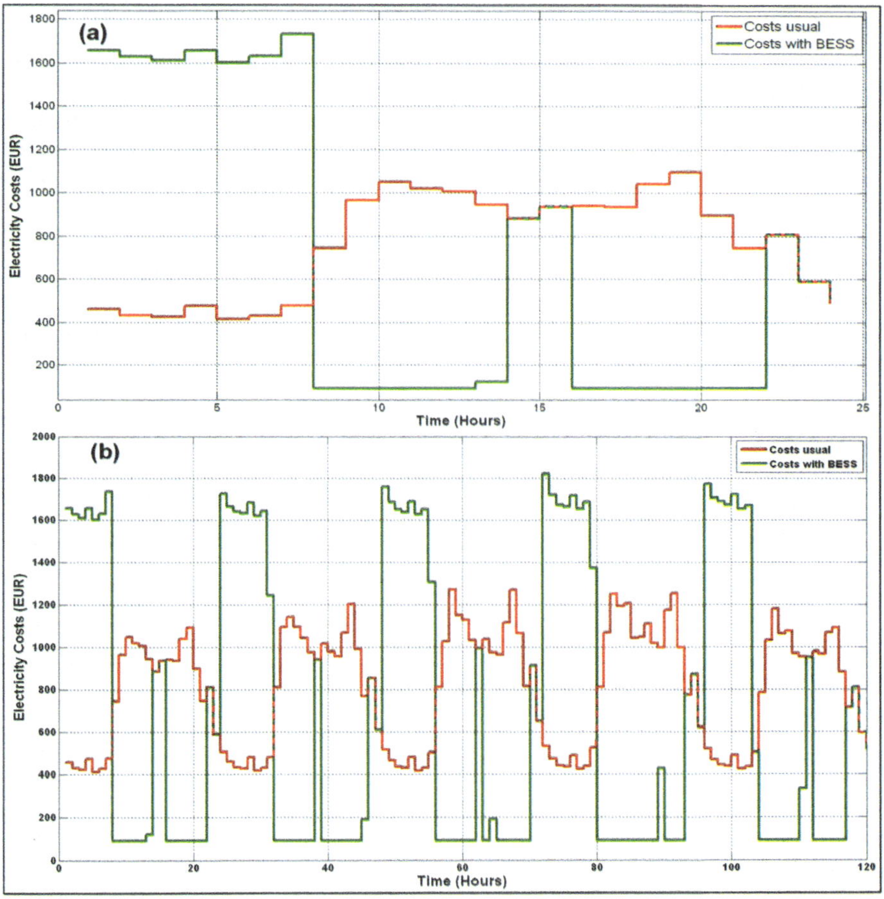

Fig. 6. Costs without and with BESS for: (a) 24 h (1 day) $\delta C(24) = 10\%$, and (b) period of 120 h (5 days) $\delta C(24) = 13\%$

Battery charge/discharge planning can reduce the cost of electricity that is illustrated in Fig. 3. However, the efficiency of BESS remains about 13%, and it can only achieved by applying of high capacity batteries satisfied to the conditions (9).

It was shown that the price of BESS proportional to its power and acceptable power for SME is from 50 kW and more [4]. The capacity of batteries for optimal BSCD algorithm is near to 10 times higher than power and starts from 50 kWh for SME.

3 ESS Efficiency for the Different Planning Periods

So, the changes in electricity tariffs in time and optimal schedule of ESS charge/discharge process in such a way that the most intensive electricity consumption from the grid accounted for the hours of the day with the lowest price, is a way to reduce of energy costs.

ESS parameters play an important role in achieving maximum efficiency (8). It is also assumed that the period for which the optimal plan is calculated will have a significant impact on efficiency. Especially the planning period can have a significant impact in connection with a decrease in the accuracy of forecasting market prices [3] and an increase in the probability of deviation of the planned load from the actual load [5] with an increase in the planning horizon.

The set of optimal schedule BCDS calculations were performed for the time periods of 24, 48, 72, 96 and 120 h. The changes of prices are presented on Fig. 4. The load is taken from [5] and charge/discharge schedule are presented on Fig. 5. The costs of electricity without ESS and with BESS and optimal schedule are presented on Fig. 6. In all cases it was assumed:

$$(SOC_{max} - SOC_{min}) \geq 16 \cdot P_{load\,avg} \tag{10}$$

The efficiency (8) was calculated for every period of scheduling (Fig. 7). Under mentioned before circumstances the increasing in scheduling period leads to some increase of the efficiency but it takes place when possible uncertainties in prices and loads during the period are not taken into account.

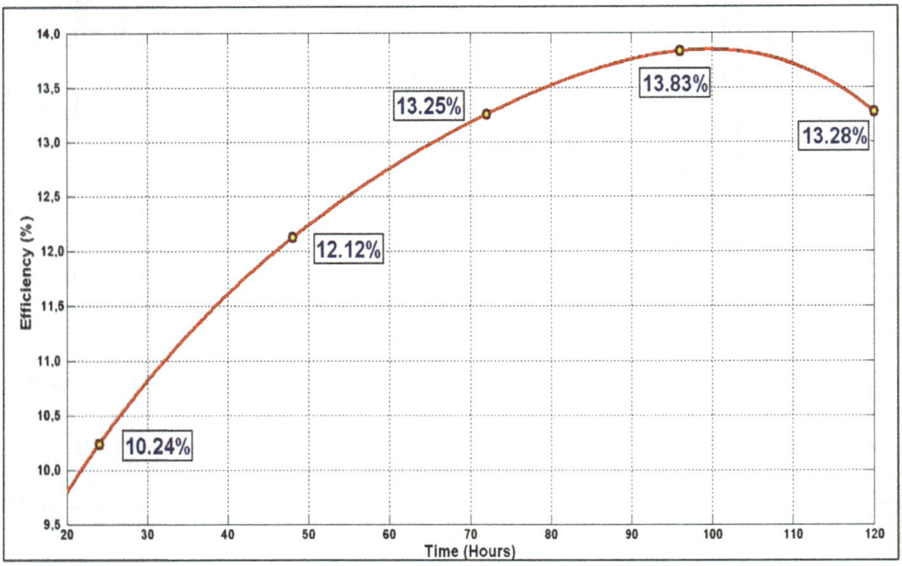

Fig. 7. Efficiency of ESS as a function of the length of scheduling period (24 ÷ 120 h)

The efficiency (8) was calculated for every period of scheduling (Fig. 7). Under mentioned before circumstances the increasing in scheduling period leads to some extend to increase of the efficiency but it takes place when possible uncertainties in prices and loads during the period are not taken into account. The additional analysis shows that deviations in efficiency demonstrated in Fig. 7 are determined by characteristic data changes in the corresponding planning period.

We have applied the linear optimization model because our general assumption is that the BSCD is based on exact knowledge about electricity prices and load demands. The changes in battery capacity along the time will be negligible for the BSCD time periods (for the modern battery types half of the capacity is lost after approximately 7 years) and we include in the algorithm the condition (6).

The number of charge/discharge cycles of battery affects its service life [10] and, consequently, the savings obtained during the operation of the ESS. This effect appears to be relatively small and was not considered in this study.

4 Conclusions

We have considered the use of ESS for energy cost mitigation under electricity prices changes in present research. The first conclusion of the analysis: it's impossible to obtain the reduction in the costs if there are no any changes (volatility) in prices, and the efficiency of BESS is directly dependent on size of price deviations.

The efficiency of ESS usage is estimated according to the relation (8). When the ESS has satisfactory parameters (for an example as in (9)) may be reached the efficiency in the range $10 \div 15\%$ for different periods of scheduling (planning horizon). We have simulated the performance of optimal scheduling algorithm for the periods of 24, 48, 72, 96, and 120 h (5 days) respectively. In all cases the range of efficiency is approximately the same but some uncertainties in loads and possible errors in price forecasting during the period of planning wasn't taken into account. So, the increasing of planning horizon may to decrease the efficiency maximum level of $10 \div 15\%$ only.

In our case, this effect is achieved by calculating and implementing BCDS (Battery Charge/Discharge Schedule) for the conditions of the Latvian legislation on the formation of electricity prices, where more than 50% of the total price is the fixed distribution price of the monopolist company '*Sadales Tīkli*' and state tax (compulsory purchase component of green energy).

In practical applications the accuracy of the optimization model (base of BSCD algorithm) is quite enough for the estimations of ESS efficiency.

Perhaps for other countries with a different legislative principle of electricity price formation, the usage the BESS (Battery Energy Storage System) for costs saving will be more advantageous, but this requires some additional research. In case of Latvia the economy of 15% per day does not allow to reduce the payback period of less than 15-20 years, which is at least twice as much as the warranty battery life [4]. Consequently, the ESS system should be used for its intended purpose: for the accumulation of energy derived from alternative 'green' sources.

Acknowledgements. This research was granted by ERDF funding, project "Optimum planning of an energy-intensive manufacturing process and optimization of its energy consumption depending on changes in the market price (2017–2019)", Contract No 1.1.1.1/16/A/280 (Sub-contract No L-s-2017/12-9).

References

1. Nord Pool Spot market data. Electricity hourly prices. https://www.nordpoolgroup.com/Market-data1/Dayahead/Area-Prices/ALL1/Hourly/. Accessed 27 June 2019
2. Sadales tikls AS, part of Latvenergo AS Group. About tariffs. https://www.sadalestikls.lv/en/to-customers/rates/about-tariffs/. Accessed 27 June 2019
3. Krivchenkov, A., Grakovski, A., Balmages, I.: Required depth of electricity price forecasting in the problem of optimum planning of manufacturing process based on energy storage system (ESS). In: Kabashkin, I., et al. (eds.) RelStat 2018 International Conference. LNNS, vol. 68, pp. 331–342. Springer, Cham (2019)
4. Krivchenkov, A., Grakovski, A., Balmages, I.: Feasibility study on the use of energy storage systems to reduce the enterprise energy consumption costs. In: Kabashkin, I., et al. (eds.) RelStat 2019 International Conference. LNNS. Springer (2020, in Publishing). 10 p.
5. Xu, Y., Xie, L., Singh, C.: Optimal scheduling and operation of load aggregators with electric energy storage facing price and demand uncertainties. In: North American Power Symposium (NAPS), pp. 1–7 (2011)
6. Lebedev, D., Rosin, A.: Modelling of electricity spot price and load. In: Proceedings of 55th International Scientific Conference on Power and Electrical Engineering of Riga Technical University (RTUCON), pp. 222–226. IEEE (2014)
7. Lebedev, D., Rosin, A.: Practical use of the energy management system with day-ahead electricity prices. In: Proceedings of IEEE 5th International Conference on Power Engineering, Energy and Electrical Drives (POWERING), pp. 394–396. IEEE (2015)
8. Varfolomejeva, R., Gavrilovs, A., Iļjina, I.: The regulation possibility of energy-intensive enterprises according to the market price change. In: Proceedings of 2017 IEEE International Conference on Environment and Electrical Engineering and 2017 IEEE Industrial and Commercial Power Systems Europe, Italy, Milan, 6–9 June 2017, pp. 1118–1123. IEEE (2017)
9. Optimisations of Battery Energy Storage System (BESS) daily (24 hours) for 8.49 MW load on base of linear programming (LP) by the interior-point algorithm in MATLAB. https://drive.google.com/file/d/1KVD7YOYL9Ax2jciAdwvDiD9MN8BD7cVA/view?usp=sharing. Accessed 26 July 2019
10. Barnes, A., Balda, J., Geurin, S., Escobar-Mejía, A.: Optimal battery chemistry, capacity selection, charge/discharge schedule, and lifetime of energy storage under time-of-use pricing. In: Proceedings of Innovative Smart Grid Technologies (ISGT Europe), 2nd IEEE PES International Conference and Exhibition, pp. 1–7 (2011)

Capacitated Open Vehicle Routing Problem with Time Couplings

Radosław Idzikowski(✉)

Faculty of Electronics, Wrocław University of Science and Technology, Wybrzeże Wyspiańskiego 27, 50-370 Wrocław, Poland
radoslaw.idzikowski@pwr.edu.pl

Abstract. This paper presents a case study for a real-life Capacitated Open Vehicle Routing Problem with Time Windows problem. The goal function is the sum of travel times of all vehicles. The mathematical model of the problem is presented. In order to take specific road traffic regulations into account, additional time-couplings constraints are formualated. Two heuristic solving methods are proposed: a greedy algorithm and a tabu search metaheuristic. The methods are tested using data from a real-life forwarding company. The results indicate that tabu provides 3.8% improvement compared to the greedy method.

Keywords: Discrete optimization · Tabu search · COVRP · Time windows

1 Introduction

The process of automating the management of a forwarding company can be reduced to the problem of calculating routes for a vehicle fleet in order to deliver orders to customers from a central hub taking into account various types of restrictions. In the literature this problem is referred to as Vehicle Routing Problem (VRP) [2]. VRP can be thought of as a generalization of the classic Traveling Salesman Problem [1], thus it is a NP-hard problems just as TSP is.

VRP is a well-known problem with many existing extensions, like Capacitated VRP (CVRP) [12], which limits the maximal load of vehicles and VRP with Time Windows (VRP-TW) [4], which forces delivery of cargo in a certain time interval. However, such extensions are not always enough to model real-life situations. An example of one such situation are road traffic regulations that forbid vehicle drivers from driving for too long without breaks.

Here we provide a brief overview of a few similar approaches and applications of VRP. In paper by Kok *et al.* a VRP-TW with time constraints, European traffic regulations was considered. The use of Integer Programming in CPLEX Solver resulted in 15% improvement in total travel time. Derigs *et al.* presented an interesting approach of combining road transport with depot time windows with air cargo delivery [3]. The authors used a different goal function, namely

W. Zamojski et al. (Eds.): DepCoS-RELCOMEX 2020, AISC 1173, pp. 273–282, 2020.
https://doi.org/10.1007/978-3-030-48256-5_27

minimization of the number of required vehicles. The authors employed a Local Neighborhood Search method. Finally, Lin and Chou used VRP to model a Public Bicycle Redistribution System, where the goal is to transport bicycles between stations to meet the required bicycle number in each station [9].

This paper presents an approach to real-life Open CVRP-TW problem, where road traffic regulations were modeled as additional time-coupling constraints. Those time constraints force additional calculations to be carried out, making it impossible to use exact algorithms to solve the problem in a reasonable running time. Thus, a greedy algorithm and metaheuristic approach were proposed to find a solution. The problem will be presented on the example of a real map of the United States of America with road traffic regulations take into account.

2 Problem Definition

In the classic VRP we have a set of n clients represented as nodes: $N = \{1, 2, \ldots, n\}$ and a depot called the start/zero node $N_0 = 0$, with a fleet of K identical vehicles. The distances between nodes are described by a cost matrix:

$$D = [d_{ij}]^{(n+1)\times(n+1)} \tag{1}$$

where $i, j \in N \cup \{N_0\}$. Thus, d_{ij} is distance between nodes i and j in miles.

By Open Vehicle Routing Problem (OVRP) we understand a VRP variant in which the vehicles need to start at the depot and deliver the cargo, but without taking into account a return route for the vehicles [8]. Thus, in OVRP we minimize the sum of all route lengths, remembering that all routes start at the depot and each customer should be visited exactly once. The example of OVRP on the map of United States of America is shown in Fig. 1.

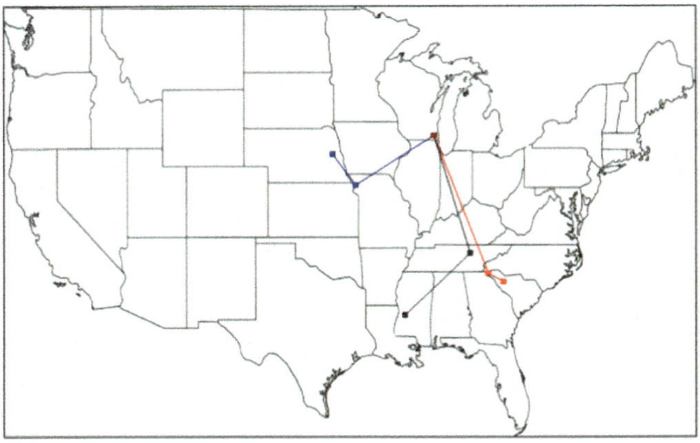

Fig. 1. Example solution for the OVRP

Next, to include Time Windows constraint, first the distances matrix D should be converted to the travel times matrix:

$$T = [t_{ij}]^{(n+1)\times(n+1)}, \tag{2}$$

where $i, j \in N \cup \{N_0\}$. Thus, t_{ij} denotes the time travel from i to j in hours. Due to differences in local and state traffic regulations, an average speed of 45 mph was adopted for distances up to 100 miles and 60 mph for longer distances, assuming that interstate highways and main state roads are used for the latter:

$$t_{i,j} = \begin{cases} d_{i,j}/45 \text{ where } d_{i,j} \leq 100, \\ d_{i,j}/60 \text{ where } d_{i,j} > 100. \end{cases} \tag{3}$$

Each customer has individual order acceptance hours. For all n clients those acceptance hours are described by a set of n time windows:

$$\tau = \{[e_1, l_1], [e_2, l_2], \ldots, [e_n, l_n]\}, \tag{4}$$

where e_i and $l_i > e_i$ mean the earliest and latest possible time to start handling the order during the day, respectively. If the vehicle arrives after l_i, it must wait till the next day when the time window opens.

In the case of the CVRP, the maximum load capacity of the vehicle must not be exceeded. In the considered real-life problem the capacity constraint is twofold, with Q and W being constraints associated with the vehicle length and weight limit respectively. In practice, the sum of order lengths on given vehicle k cannot exceed Q. Similarly, the sum of order weights cannot exceed W.

With this, the set of m orders can be now defined as follows:

$$\Theta = \{[\delta_1, s_1, w_1, p_1], [\delta_2, s_2, w_2, p_2], \ldots, [\delta_m, s_m, w_m, p_m]\}, \tag{5}$$

where:

- δ_i – the customer to deliver order i to,
- s_i – length of order i,
- w_i – weight of order i,
- p_i – handling time of order i.

2.1 Mathematical Model

In this section we present a mathematical model of the problem. We start by creating a full graph $G = (V, E)$ for the classic VRP, where $V = N \cup \{N_0\}$ is the set of n vertices, E is the set of $n(n-1)/2$ non-directed edges, with d_{ij} being the weight of edge (i, j). Structure of graph G for $n = 4$ is shown in Fig. 2.

Using the graph G we can define the feasible solution for the basic VRP can be represented by $n + 1 \times n + 1$-sized matrix of boolean values [11]:

$$X = [x_{ij}]^{(n+1)\times(n+1)}, \tag{6}$$

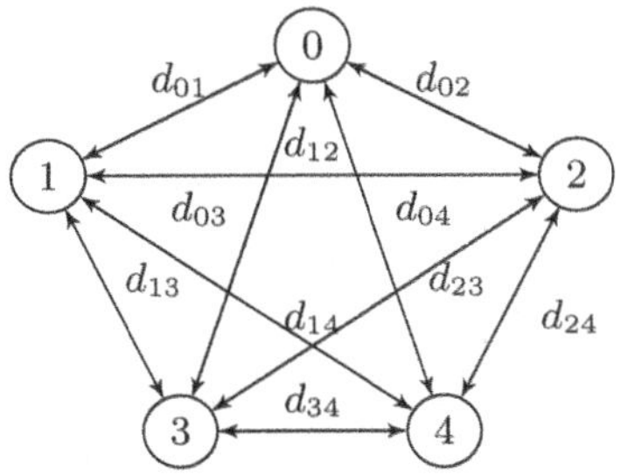

Fig. 2. Graph G for VRP problem for $n = 4$

where:

$$x_{i,j} = \begin{cases} 1 & \text{if the edge is visited,} \\ 0 & \text{otherwise.} \end{cases} \tag{7}$$

Each vertex has to be visited exactly once, therefore:

$$\forall_{i \in N} \sum_{j=1}^{n} x_{ij} = 1 \quad \wedge \quad \forall_{j \in N} \sum_{i=1}^{n} x_{ij} = 1. \tag{8}$$

The number of routes starting at node zero is:

$$\sum_{i=j}^{n} x_{0j} = K. \tag{9}$$

The number of routes ending at node zero is:

$$\sum_{i=1}^{n} x_{i0} = K. \tag{10}$$

Taking into account no return routes in OVRP, the last condition becomes:

$$\sum_{i=1}^{n} x_{i0} = 0. \tag{11}$$

For the classic VRP, the goal function to minimize is the sum of the lengths of all visited paths in the graph which is:

$$F(X) = \sum_{i=0}^{n} \sum_{j=0}^{n} x_{ij} d_{ij}, \tag{12}$$

remembering to keep the route going:

$$\forall_{M \subset N} \sum_{i \in M} \sum_{j \in N \setminus M} x_{ij} \geq 2. \tag{13}$$

However, in our approach we are interested in minimizing the total travel time instead of distance, thus (12) becomes:

$$F(X) = \sum_{i=0}^{n}\sum_{j=0}^{n} x_{ij}t_{ij}. \tag{14}$$

2.2 Transport Constraints

In practice, a simplified description of the load (length and weight) is often used in transport management. In this paper the effective arrangement of the load on the vehicle was not considered. Let us assume that the number of orders equals the number of customers ($n = m$) and each customer has exactly one order assigned to them. Let variable x_{ij}^k represents whether edge (i, j) is on the route of the k-th vehicle. The transport restrictions can now be stated as follows:

- the total length of orders for k-th vehicle can not exceed the maximum length of the vehicle:

$$\sum_{i=0}^{n}\sum_{j=0}^{n} x_{ij}^k s_j \leq Q, \tag{15}$$

- the total weight of orders for k vehicle can not exceed the maximum permissible vehicle weight:

$$\sum_{i=0}^{n}\sum_{j=0}^{n} x_{ij}^k w_j \leq W, \tag{16}$$

where $k \in \{1, 2, \ldots, K\}$.

2.3 Time Constraints

Due to the nature of time restrictions, we can distinguish restrictions regarding (1) customers and (2) vehicles. The order can be shipped to the customer only during customer-specific time interval (time windows). If the vehicle arrives after the time window has closed, then the vehicle has to wait till the next time window which is on the next day. Obviously, such a situation worsens the final time travel considerably. Similar situation occurs if the vehicle arrives before the time window is open, forcing it to wait before the order can be handled.

Traffic regulations forbid drivers from driving for too long, forcing resting breaks. Those restrictions can be summarized in the following 4 points:

a) 8-hour limit: a driver can drive for a maximum of 8 h without taking a break if they took a break of 30 min before driving,
b) 11-hour limit: a driver can drive for a maximum of 11 h without taking a break if they took a break of 10 h before driving,
c) 14-hour limit: a driver can work (including driving, loading and unloading of cargo etc.) for a maximum of 14 h without taking a break if they took a break of 10 h before working,
d) 70-hour limit: a driver can work for 70 h on 8 consecutive days. They may resume after after a 34-hour continuous break.

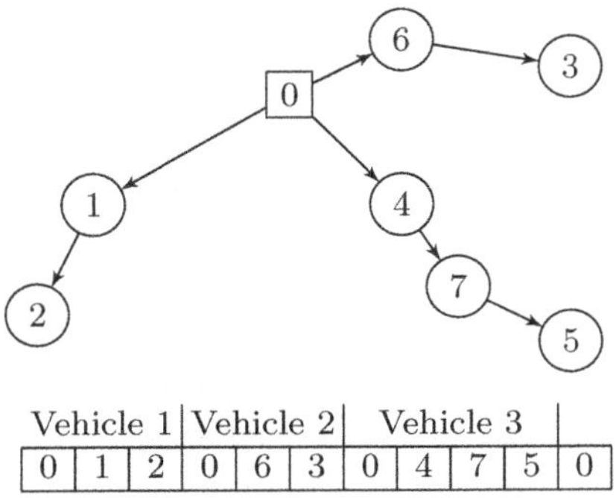

Fig. 3. Giant Tour Representation

3 Algorithms

In this section we decided to implement two inexact solving methods. The first method is a greedy algorithm, which is based on the algorithm currently used in the considered case study transport management company. The algorithm works similar to the K-Nearest Neighbor Search approach [10]. At first, the customer farthest from the depot is chosen. After that we proceed by the nearest-neighbor rule: the customer closest to the previous one is chosen. This is repeated until no more orders can be assigned to this vehicle due to length-weight constraint.

The second proposed algorithm is based on the Tabu Search (TS) metaheuristic [6], which is a local search method. The algorithm requires an initial solution which is usually either random (but must be feasible) or obtained through the greedy algorithm. Then, by performing various so-called moves (e.g. Swap), the neighborhood is searched for the best solution, taking into account the list of banned moves (the Tabu List). Finally, the best solution from the neighborhood is selected. The Tabu List was implemented in the form of a matrix containing the current cadence for all possible moves.

3.1 Giant Tour Representation

In order to check the neighborhood faster, the Giant Tour Representation (GTR) [5] was used. This allows to represent the solution using a single permutation of length $n + K + 1$. In this permutation, all routes are placed sequentially, separated by zeros. A zero value is also placed at the beginning and the end of the permutation. This allows to change the number of customers on a route by changing the positions of the zeros. If two zeros appear side by side, then the route is empty (contains no customers), which is acceptable (i.e. the vehicle does not leave the depot). A sample GTR for $n = 7$ and $K = 3$ is shown in Fig. 3.

3.2 Multi-moves

For the considered problem with capacity limits the well-known swap move introduces many infeasible solutions, e.g.: when trying to swap a longer order with a smaller one when both vehicles are length-full. Due to these, we opted to introduce more complex moves as well, ind order to improve efficiency of the algorithm [7]. Those complex moves are performed after standard swap move, not instead of it.

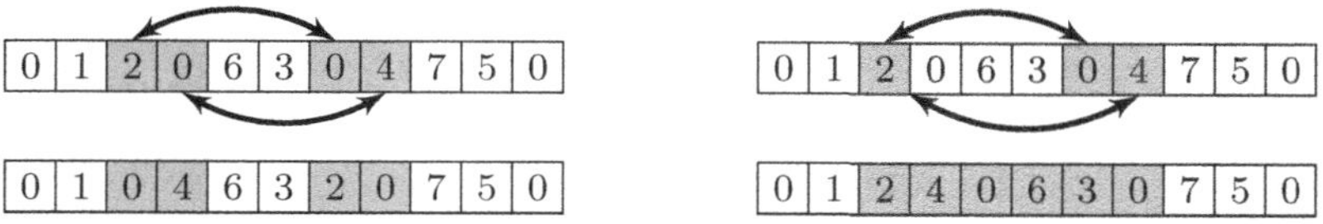

(a) a double swap move

(b) a swap with insert move

Fig. 4. Examples of complex moves

We used double swap and swap with insert complex moves. Double swap is an additional exchange of the pair $(i+1, j+1)$ outside the base pair (i, j) in the neighborhood. The swap with insert movement works analogously to double swap, only instead of the second swap movement, the city at position $j+1$ is cut out and inserted at position $i+1$. The nature of those complex moves is shown in Fig. 4a and 4b respectively.

4 Computational Experiments

The research was carried out on a PC computer equipped with an Intel i7-6700K CPU clocked at 4.00 GHz, 16 GB RAM and an disk SSD working under Microsoft Windows 10 operating system. The algorithms were implemented in C#.

To verify the proposed algorithms, both approaches were implemented. The actual United States of America road network was used. A method in C# was designed to create a cost matrix (neighborhood) for selected cities using the Distance Matrix library from the Google API. Then, for different sizes of the problem, random instances were generated including the demand and load acceptance times for each client. Care was taken that the values were close to those encountered in real-life. The tested instances allowed form more than one order per customer i.e. $m \geq n$. The length and weight limits of vehicles were set to $Q = 53$ *feet* and $W = 45000$ *pounds* respectively.

In order to measure the quality of solutions provided by both algorithms, the Percentage Relative Deviation (PRD) was used, which is defined as:

$$PRD(\pi) = 100\% \frac{F(\pi) - F(\pi^{\text{ref}})}{F(\pi^{\text{ref}})}, \tag{17}$$

where π and π^{ref} are the solutions obtained by the TS and greedy method (as reference) respectively and $F(\pi)$ is the value of the goal function for solution π.

Table 1. PRD and running time results for the COVRP-TW

Dimension	Measures				
m	$t_G[ms]$	$t_{TS}[ms]$	$t_{TSmm}[ms]$	$PRD_{TS}[\%]$	$PRD_{TSmm}[\%]$
5	17.21	6.11	6.25	0	0
10	29.01	21.12	32.53	−0.21	−2.31
15	48.54	62.87	103.49	−1.51	−3.17
20	89.34	133.24	227.39	−1.79	−3.05
25	499.61	241.79	424.30	−1.49	−3.53
50	196.49	1.75×10^3	3.13×10^3	−1.89	−4.42
75	344.51	5.91×10^3	1.08×10^4	−2.32	−3.92
100	553.21	1.34×10^4	2.52×10^4	−2.11	−5.26
Average	–	–	–	−1.41	−3.08

The results, including PRD and running time of algorithms are shown in Table 1 and Fig. 5. Values t_G, t_{TS} and t_{TSmm} are the running times of the greedy, regular TS and TS with Multi-moves methods respectively.

The results indicate that the TS and TS with multi-move methods allow to obtain lower total time travel than the greedy approach by 1.41% and 3.08% on average respectively. We also notice that the PRD value increases with the number of orders m. This implyies that the TS advantage over the greedy methods grows for larger (i.e. more difficult) problem instances.

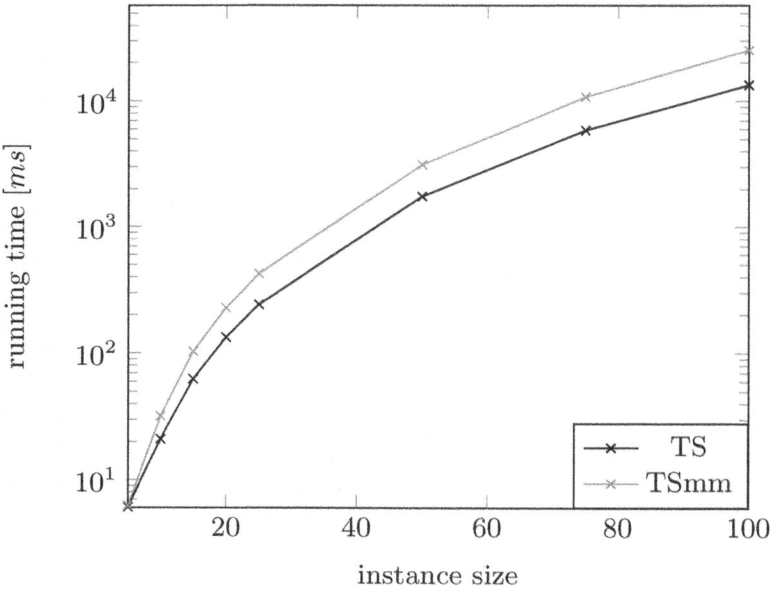

Fig. 5. Comparison of running times for TS and TS with multi-moves

All of the considered methods work in seconds (below 0.5 s for greedy and under half a minute for both TS variants), allowing them to be easily used in real-life situations. We also notice that the TS variant with multi-moves needs twice the computation time of the regular TS variant, but allows to obtain twice the improvement, while still being very fast in practice.

5 Conclusions

In the paper, a case study of a real-life Capacitated Open Vehicle Routing Problem with Time Windows and time couplings was considered. A mathematical model taking into considerations various problem constraints (including vehicle capacity limits, customer time windows and road traffic regulations) was presented. The results of research on effectiveness of the Tabu Search method for the considered problem were presented. The proposed formulation of the problem and the proposed solving algorithm allowed to obtain results better then a greedy heuristic currently employed in the considered real-life company.

In addition, the introduction of complex moves in the Tabu Search method significantly improved the quality of results, with comparable computational complexity to standard Tabu Search. The results could be improved by employing a restricted neighborhood to reduce the number of infeasible neighbors.

Acknowledgements. This work was partially funded by the National Science Centre of Poland, grant OPUS no. 2017/25/B/ST7/02181.

References

1. Dantzig, G.B.: Application of the simplex method to a transportation problem. In: Activity Analysis of Production and Allocation (1951)
2. Dantzig, G.B., Ramser, J.H.: The truck dispatching problem. Manage. Sci. **6**(1), 80–91 (1959)
3. Derigs, U., Kurowsky, R., Vogel, U.: Solving a real-world vehicle routing problem with multiple use of tractors and trailers and eu-regulations for drivers arising in air cargo road feeder services. Eur. J. Oper. Res. **213**(1), 309–319 (2011)
4. Desrochers, M., Desrosiers, J., Solomon, M.: A new optimization algorithm for the vehicle routing problem with time windows. Oper. Res. **40**(2), 342–354 (1992)
5. Funke, B., Grünert, T., Irnich, S.: Local search for vehicle routing and scheduling problems: review and conceptual integration. J. Heuristics **11**, 267–306 (2005)
6. Glover, F., McMillan, C.: The general employee scheduling problem: an integration of MS and AI. Comput. Oper. Res. **13**(5), 563–573 (1986). applications of Integer Programming
7. Grabowski, J., Pempera, J.: Zagadnienie przepływowe z ograniczeniami "bez magazynowania". algorytm tabu search z multiruchami. Automatyka/Akademia Górniczo-Hutnicza im. Stanisława Staszica w Krakowie **T. 9, z. 1-2**, 95–104 (2005)
8. Li, F., Golden, B., Wasil, E.: The open vehicle routing problem: algorithms, large-scale test problems, and computational results. Comput. Oper. Res. **34**(10), 2918–2930 (2007)

9. Lin, J.H., Chou, T.C.: A geo-aware and VRP-based public bicycle redistribution system (2012)
10. Song, Z., Roussopoulos, N.: K-nearest neighbor search for moving query point. In: Jensen, C.S., Schneider, M., Seeger, B., Tsotras, V.J. (eds.) Advances in Spatial and Temporal Databases, pp. 79–96. Springer, Heidelberg (2001)
11. Toth, P., Vigo, D. (eds.): The Vehicle Routing Problem. Society for Industrial and Applied Mathematics, Philadelphia (2001)
12. Toth, P., Vigo, D.: Models, relaxations and exact approaches for the capacitated vehicle routing problem. Discrete Appl. Math. **123**(1), 487–512 (2002)

Mobile Application Testing and Assessment

Marcin J. Jeleński and Janusz Sosnowski(✉)

Institute of Computer Science, Warsaw University of Technology, Warsaw, Poland
j.sosnowski@ii.pw.edu.pl

Abstract. The paper discusses problems related to testing software applications in mobile system environment. We present possibilities and limitations of available test support tools and propose three original programs improving this process. In the experimental studies we analyze test coverage for a set of representative open source projects. This research is enhanced with a detailed exploration of user recommendation reports of these projects to identify improvement and correction needs. Our experience has been discussed with developers of commercial mobile applications which used the developed tools.

Keywords: Software testing · Mobile systems · Application recommendations

1 Introduction

Recently, mobile applications are widely used and many of them should assure high level of dependability. This is especially important in relevance to financial operations, confidentiality, data protection, and other critical issues. In not critical applications we face the problem of user friendly interfaces, lack of annoying errors, low reaction time, high performance. Hence, testing such applications is a crucial issue, the more that mobile environment creates specific requirements not encountered in classical software testing [6–8].

There is a rich literature on software evaluation dealing with general testing schemes or related detailed problems (e.g. [1, 11, 13, 17]) as well as on the software development including issue tracing and source code control [19]. They typically focus on big projects used in stationary systems installed in complex servers or workstations. Moreover, they are targeted at users familiar with the relevant information processing domain. In the case of mobile systems the applications are addressed to a large community of users working with diverse devices and operating system versions. Mobile system environment, software, and devices are quite often replaced by newer ones. This results in a multitude of needed test scenarios.

Another issue is the fact that most mobile applications incorporate extended and specific communication interfaces based on GUI concept which may also differ upon device and system version. Hence, assuring good tests is not a trivial problem. There are many available tools supporting mobile application testing but they have various limitations and drawbacks. Some companies profit the possibility to outsource their testing activities to crowdsourced testers. Such testers may represent diverse skills and use various testing facilities and environments [9]. On the other hand many

W. Zamojski et al. (Eds.): DepCoS-RELCOMEX 2020, AISC 1173, pp. 283–292, 2020.
https://doi.org/10.1007/978-3-030-48256-5_28

applications provide recommendation/review pages where the users can report their comments. This is quite useful source of information neglected in practice [21].

The outline of the paper is as follows. Section 2 describes the context of our research in relevance to the possibilities and limitations of available tools supporting mobile testing. Section 3 focuses on developed new tools for Android environment. Section 4 is devoted to the problem of exploring user review reports. Section 5 concludes our research findings.

2 Testing Problems

In the case of mobile applications we can distinguish 3 classes of tests: *manual tests* – executed by the tester, *script tests* – automatically interacting with the application and checking correctness of responses in relevance to the expected ones, *GUI tests* – focused on user interface. Within the script tests we have: unit tests (covering small program components executed in isolation from other modules); basic and complex integration tests (checking interactions of components). Depending upon the test execution environment we distinguish local and instrumented tests. Local tests are usually executed on a workstation (with JVM) with no access to components of real Android SDK (only a modified version Android.jar can be used here). These tests can be supported with such tools as Mockito and Roboelectric. Instrumented (installed) tests are executed directly in Android system with the installed tested application. Remote handling of the device is assured via *adb* (Android Debug Bridge) and instrumentation classes (e.g. AndroidJunitRunner). Instrumented tests need application installation and its multiple activation which results in long testing time.

A big challenge in mobile testing is GUI [18, 20]. It can be tested partially locally (e.g. using Roboelectric), however the graphical representation has to be tested on a real or virtual device. Such tests can be supported with special tools. Popular Espresso comprises a set of access methods to interface components of the application with activating actions on these components. It assures synchronization with executed operations with the main thread and asynchronous resources (Idling Resources). Similar to Expresso is Robotium, UI Automator is more complex with enhanced functionality, however it lacks synchronization features.

Test automatization and verification of its efficiency are serious problems. In particular, we can use here code coverage metrics (e.g. instruction, method, block, branch, predicate) available is such tools as Emma or JaCoCo (Java Code Coverage). In the case of GUI tests, event coverage is more interesting (intra and inter component), it can be based on event flow graph [15]. Creating tests it is useful to prepare test data, here we can collect and record diverse interactions (signals on changes in components, gest events [7, 14]). This can be provided by Android Accessibility Service. Event replaying during test can be supported with RERUN program [10]. It is also possible to use image recognition techniques for GUI verification (Visual GUI Testing – Eye Automate, Sikulix). Most automatic testing approaches base on exploration techniques [7] which generate randomly such signals as: gests on the screen, system or application state changes (screen rotation, program switching), system events (e.g. notification, SMS messages, connection of earphones, low power signalization). Other exploration schemes base on application models or systematic exploration [16]. There are several tools providing these capabilities, e.g.: Monkey, Dyndroid, MobiGUItar, ACTEve ([2, 3, 12, 14, 18] and references therein).

The available mobile applications are related to many system versions. For Android the used version spectrum is from 2.3.3 (gingerbread) to 8.1 version, with dominating on the market versions 5.1 and 6.1 (about 20% each). Along the system lifecycle we observed changes in available functions, user interface styles, security and power management mechanisms. This increases the risk of errors depending upon the system and device hardware. For not available functions, in older versions, we need to include backward compatibility libraries. Hence, the scope of testing should also cover these issues. Nevertheless, some compromise is needed here. Creating a new application developers specify minSdkVersion and targetSdkVersion, i.e. the oldest and the newest version for which the application can operate, respectively. In Fig. 1 we give box plot of the number of supported versions (y-axis) depending upon the targetSdkVersion. We have derived this with a specially developed script applied to over 1000 applications (compiled using Gradle) available in F-Droid repository. Similar statistics we have generated in function of the number of application installations. Here, the most popular applications supported higher range of system versions.

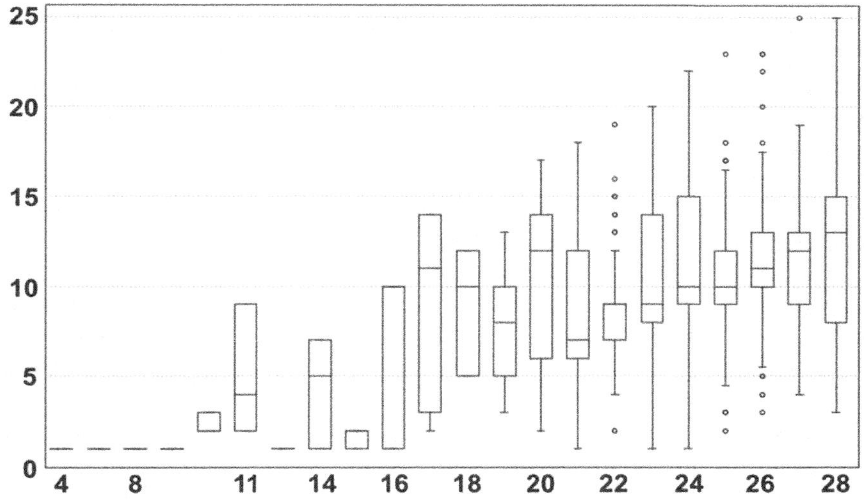

Fig. 1. Box plot showing the number of supported versions (y-axis) depending upon the targetSdkVersion (x-axis) for mobile applications.

Testing a newly introduced application is a challenging problem due to the need of assuring consistency with various system versions and types of devices (including backward conformance). Similarly, application upgrades complicate this problem. Effective tools are required to support test automation and regression testing. Our practical testing experience within several companies resulted in developing special tools targeted at test case creation, automatic exploration and simulation of external resources. All the developed tools can be used both in local and instrumented testing in Android environment. The developed tests do not reveal all application problems. An important issue is to monitor users opinions, e.g. recommendation repositories. These issues were neglected in the literature and are investigated in the sequel.

3 Test Support Tools

Automatic generation of tests attracts practitioners and researchers [8]. In [16] we have a good survey of relevant tools. The key features of these tools are: handling system events; minimizing the number of involved restarts; admitting tester interactions (e.g. to skip test suspension due to login needs); admitting many starting points (achieving some application state may need quite long sequence of events, storing such sequences for future replaying is helpful); test recovery after crash; mimicking (mocking) external environmental behavior; sandboxing (blocking sending harmful or sensitive messages to the external environment, e.g. emails, SMSs); test case portability over various system versions. Test exploration tools reveal new application states, detect diverse errors which lead to application suspension or closing. Unfortunately, they do not verify the correctness of displaying messages and GUI.

Supporting test development it is useful to facilitate description of execution paths for testing (e.g. for regression testing). Having checked this issue in several commercial companies involved in developing mobile application we have found that this has been done manually. The test execution paths should specify such features as: tested object (application, user), required conditions or actions (e.g. state of being logged), expected results. We have developed a special software module (UserCaseCreator) which automatizes the description of the manually executed testing paths. It traces the test execution actions basing on Android Accessibility Services. For the considered application with predefined files subsequent tester interactions are stored in a special file. The contents of this file is derived from texts comprised in application fields for text insertion and content description for the remaining ones (e.g. ikons). The handled events include the following type views: *clicked, long_clicked, object_selected, object_focused text_changed, window_state_changed, hover_enter* (positioning cursor on an object), *hover_exit* (leaving the object), *scrolled* (scrolling the text), *selection_changed*. For an illustration we give an example of a generated description:

1. User clicks 'Sign in', 2. Window context changes [Sign in], 3. User clicks 'Email', 4. User focuses 'Email', 5. User changes text from "to 'login@example.com', 6. User clicks 'Password', 7. User focuses 'Password', 8. User changes text from" to 'secret-password', 9. User clicks 'OK', 10. Window context changes [loading].

The generated test paths constitute a template which can be further extended or modified and used also in exploration tests.

Simulation of external environment is a crucial problem in testing; data changeability and external resource instability create problems. The developed RemoTest (based on *ktor* library) facilitates substituting and simulating REST API interfaces which assure: simple integration with the application code, capability of defining subsequent versions of API, possibility of defining many test cases for responses, test case co-sharing between programmers. RemoTest is a virtual stateless HTTP server, which sends predefined responses for HTTP requests. This can be used to test handling exception situations in applications on the side of services (e.g. exhausted limits of provided resources). RemoteTest can operate on any machine with Java environment. The appropriate configuration can be defined using the program GUI from the browser. RemoTest handles various resources available under different URL addresses relevant to the used services by the application. Resources are defined by a path, name, HTTP method used by the application (GET, POST, DELETE, PUT, etc.), RemoTest case header, expected response code and relevant content (e.g. json). The application sends a request with selected HTTP method (GET) to the appropriate address and receives the requested response. The application or integration test scripts can dynamically specify RemoTest-case header. This assures comprehensive testing of communication layer and observing application reaction for subsequent test cases.

Manually prepared test scripts can be enhanced with developed Monkessso program assuring automatic exploration. It combines and extends capabilities of two other tools: *Monkey* and *Espresso*. As opposed to Monkey [6], for each interaction Monkesso waits for termination of running asynchronous tasks (e.g. http requests) before generating subsequent actions. This assures more predictable and repeatable results of exploration. During the exploration we can detect critical malfunctions (failures) and monitor delays in the main thread. These delays may appear in case of long lasting operations in response to events from user interface, resource allocation, writings to database, file modifications on disc, etc. High delays may be sensitive to many users and limit their interest to the product.

During exploration Monkesso maintains, in the background, a process with a new thread, which cyclically (200 ms) sends a signal to process in the main thread of the application. In the main thread it registers (system diary) the time of response. In case of lacking responses for more than 5 s it registers a warning about potential error (ANR - *application not responding*). The exploration actions comprise random combinations, which include: screen rotation, clicking, double click, long click, introducing random text (1–40 characters), cleaning text field, list scrolling, screen scrolling (left, right, up, down), pressing back button, etc. Monkesso is activated as the installed test so it provides an access to all modules of the tested application (also replacing these modules with others is possible). During in deep searching Monkesso finds all interface elements, which can accept the randomly selected action (listed above). For example for a selected text insertion it looks for all text fields and selects only one of them for inserting the text. The exploration depth depends upon the application complexity and available functions from the level of the interface.

A significant feature of Monkesso (not encountered in other tools) is the possibility to combine execution of predefined test case scripts (e.g. generated by UserCase-Creator) with in depth random exploration. Exploration can be done at specified points in the test with the possibility of replacing components for deeper exploration or extending monitoring functions. Moreover, it can cooperate with Jacoco program to evaluate test coverage. This facilitates finding untested parts of the program, non accessed functions, redundancy in tests, etc. Using Monkesso we have tested 30 applications (appropriately instrumented) from F-Droid repository and verified instruction and path coverage. Path coverage is given in Fig. 2 (minimal, first and third quartile, and maximal values) depending upon the test time. Instruction coverage was higher approaching to 50% in case of the 3rd quartile. This resulted from some limitations of connecting additional devices (e.g. camera), or options requiring providing real correct data (e.g. login data). Nevertheless, the achieved results are comparable with those which we generated for Monkey tool. The performed exploration test detected many exceptions related to standard methods *java.lang*, *kotil* and some for *java.util*, which were skipped by developers. This confirms the usefulness of automatic tests. Moreover, this coverage can be increased by exploring (using Monkesso) test cases obtained with UserTestCreator. The derived tools have been deployed in commercial companies and assessed positively by developers and testers (especially the possibility of combining it with collected test case scripts).

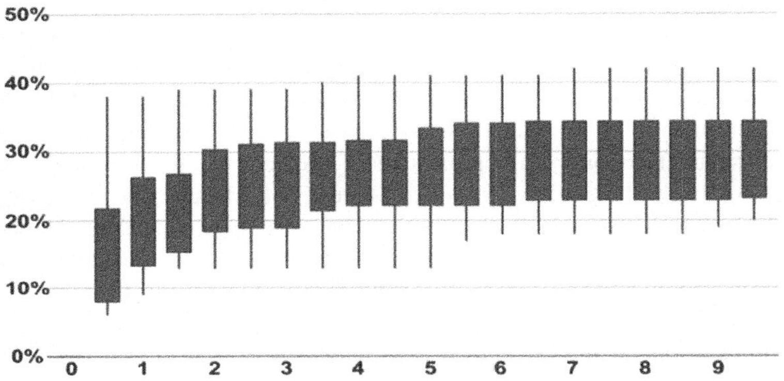

Fig. 2. Program path coverage (y axis) in function of Monkesso testing time (minutes).

4 Analyzing Recommendation Repositories

Quite many applications are assessed within recommendation systems. They provide diverse user opinions, which can be helpful in application improvements, mitigating errors, etc. Usually, these systems specify rules of submitting comments and may provide some additional statistics (e.g. voting related to the stored comments). Unfortunately, the included text is unstructured and freely composed resulting in misspelling, mixing languages, using emoticons, etc. Hence, deriving useful comments for the developer is not a trivial problem. Here, we have to use appropriate text mining tools. In our study we take into account the message text, ranking note and voting score. The primary analysis is focused on general statistics related to description size, ranking categories. More interesting was finding characteristic features by searching for keywords (n-grams) and their correlation with specified problem categories.

Recommendations in *Google play* targeted at smartphone applications (correlated with open source database F-Droid) provide 5 rating notes (1 most negative, 5 most positive) and relevant voting support. Most reports comprised short textual messages with no suggestions. Longer messages seem to be more interesting, especially those with lower rating. Nevertheless, the most positive (note 5) comprised some valuable suggestions for improvement. In case of those with the negative notes some were ambiguous, because they comprised positive terms, e.g. a phrase *"very good"*. We give here some examples of low rank opinions comprising a positive underlined term: *it goes slow please make it a bit faster if you do it ill rate 5 stars: pls make it eaiser make the icon and widget small then i will rate 5*. In practice, we have to take into account larger context (e.g. coexistent phrase *I will, doesent work, fix*). Hence, tuning text mining we have iteratively drilled the text with various features to extract useful data for project improvement. Relying on simple intuitive keywords is not sufficient, and it is reasonable to take into account n-grams and sentiment hints within reports. For this purpose we can use available statistical and text mining tools [4] with appropriate extension and adaptation to analysed problems, e.g. SMILE (*Statistical Machine Intelligence & Learning Engine*).

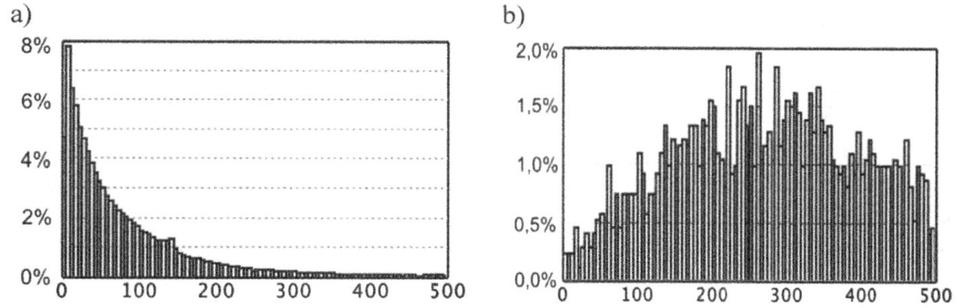

Fig. 3. Distribution of textual comment size (in percent); a) for all comments, b) for comments with voting score 32 or more (x-axis shows he number of characters).

Ranking notes are not sufficient to identify problems or suggestions of improvements. Using text mining techniques based on word frequency and tf-idf (Term Frequency – Inverse Document Frequency [4]) parameter (also extended for n grams) allowed us to extract categories of comments describing errors (failures) or performance drawbacks. We analysed 629 applications from F-Droid catalogue and relevant recommendations in Google play. Having limited maximal number of reports to 24000 per application we have got 433 0000 reports. The distribution of ranking scores was as follows: 6.44, 3.04, 7.06, 15.63 and 67.82% for score 1, 2, 3, 4 and 5 (the best), respectively. Most reports comprised only ranking values. Distribution of reports with additional text comments was as follows 7.04, 6.90, 5.21, 4.74 and 4.6%. We have performed a deep textual analysis using SMILE library and developed special scripts. This analysis was targeted at the following issues: distribution of the number of characters in the textual comment related to note scores, text normalization to the basic forms (stemming), n-gram extraction, finding significance of n-grams and correlating them with ranking values, deriving key features (words, 2–4 grams), comment classification in relevance to problem types. Figure 3a shows the distribution of characters in recommendation reports, Fig. 3b shows this distributions taking into account only recommendations with votes (support) higher than 32. We can find that almost half of recommendations comprise less than 50 characters and 22% no more than 100.

However, for recommendations with high support the contribution of longer comments was higher (Fig. 3b). Moreover, we found that for the most positive notes the comments comprised lower number of characters than for others.

Searching for key features signalling failures or performance problems we have analysed tf-idf parameter values for words and n-grams (up to n = 4). The results were ambiguous. However, relatively good results assured trigrams with high values of tf-idf (given in the brackets), e.g. for negative note 1 we identified such terms as: *don't waste time* (178.49), *app doesn't work* (171.55), but some trigrams included rate *5 star* (119.91). Better outcome (more informative) we obtained taking into account averaged weighed score rate (r_n) for each considered n-gram:

$$r_n = (f_{1,n} + 2f_{2,n} + 3f_{3,n} + 4f_{4,n} + 5f_{5,n}) / (f_{1,n} + f_{2,n} + f_{3,n} + f_{4,n} + f_{5,n})$$

where $f_{k,n}$ is the frequency of considered n-gram appearing in documents with note k. In addition, we calculate standard deviation (σ) of $f_{k,n}$ (over k) for the considered n-gram. For example, word "*crash*" comprised average note r = 2.198 and $\sigma = 0.276$, *doesn't work* r = 1.893 and $\sigma = 0.356$, *best app* r = 4.345 and $\sigma = 0.635$. We found reasonable to take n-grams with σ in the range [0.2–0.8], which resulted in the following keyword features: 22 words (10 for failures, 12 for performance), 1293 bigrams (858 vs 435), 780 trigrams (675 vs 105). As an illustration, we give a sample of n-grams related to performance problems (*slow, drain, battery drain, slow response, huge battery drain, app bit slow*), and failures (*crash, failure, black screen, crash frequently, app force close, latest update crash*). Basing on these characteristic key n-grams, we have analysed 433 000 reports and selected those reporting problems. The classification results are given in tab. 1. Bigram *doesn't work* appeared in 23% of failure reports. It is worth noting that classification based on single words shows also false reports, e.g. reports comprising word "*crash*" may have positive meaning while appearing in the phrase "*doesn't crash*".

Table 1. Statistics of report features used in identifying failures and performance problems.

n- gram category	Reports on failures	Reports on performance drawbacks
Words	15 684	8497
Bigrams	19 811	6114
Trigrams	7187	814

Comments related to failures were mostly shown by n-grams within low rank reports, performance drawbacks dominated in higher rank ones. To give a better view on the performed classification we present some selected comments classified as failures (f1–f3): f1 - *usually crashes immediately the first 4 or 5 times when its opened before it actually works;* f2- *always app crashes during backup android 8 galaxy s7 edge;* f3 - *every time i enter a text in a photo and start the process the app crashes im using redmi 5 plus;* and classified as performance drawbacks: p1 - *after last update drains battery really fast;* p2 - *painfully slow takes forever to load pages.* It is worth noting that the text is clumsy (incorrect spelling, etc.), but interesting for developers.

5 Conclusion

Having analyzed a wide scope of mobile applications we found that they are consistent with many system versions, which imposes additional requirements on efficient testing (including the regression ones). Unfortunately, creating test repositories is still neglected (within open source F-Droid only 20% applications comprised included tests). Having questioned programmers in 6 commercial companies developing mobile applications we found that manual testing dominated, automatic test tools have been used by about 30%–40%, this resulted from difficulties to absorb the functionality of appropriate tools, their usage limitations, problems with simulating environment, low flexibility. The presented new tools mitigate some of these problems and they have been used successfully in three commercial companies.

Moreover, we have demonstrated the possibilities and advantages of analyzing recommendation repositories to trace errors and performance drawbacks. This can be extended by analyzing event logs and other software repositories (e.g. [5, 19]). Further research is targeted at deeper analysis with machine learning techniques.

References

1. Alegroth, E., Feldt, R.: On the long-term use of visual gui testing in industrial practice: a case study. Empir. Softw. Eng. **22**, 2937–2971 (2017)
2. Amalfitano, D., Fasolino, A.R., Tramontana, P., Ta, B.D., Memon, A.M.: MobiGUITAR automated model-based testing of mobile apps. IEEE Softw. **32**(5), 53–59 (2015)
3. Anand, S., Naik, M., Harrold, M.J., Yang H.: Automated concolic testing of smartphone apps. In: ACM SIGSOFT 20th International Symposium on the Foundations of Software Engineering, FSE 2012, pp. 59:1–59:11 (2012)
4. Berry, M.W., Kogan, J.: Text Mining Applications and Theory. Wiley, Chichester (2010)
5. Bhattacharya, P., Ulanova, L., Neamtiu, I., Koduru, S. C.: An empirical analysis of bug reports and bug fixing in open source Android apps. In: 17th European Conference on Software Maintenance and Reengineering, pp. 1–11 (2013)
6. Bie, Y., Bin, S., Sun, G., Zhou, X.: An empirical analysis of Android apps bug and automated testing approach for Android apps. Int. J. Multimedia Ubiquit. Eng. **11**(9), 1–10 (2016)
7. Choudhary, S.R., Gorla, A., Orso, A.: Automated test input generation for Android: are we there yet? arXiv:1503.07217v2 [cs.SE] 31 Mar 2015
8. Cruz, L., Abreu, R., Lo, D.: To the attention of mobile software developers: guess what, test your app! Empir. Softw. Eng. **24**(4), 2438–2468 (2019)
9. Gao, R., Wang, Y., Feng, Y., Chen, Z., Wong, W.E.: Successes, challenges, and rethinking – an industrial investigation on crowdsourced mobile application testing. Empir. Softw. Eng. **24**, 537–561 (2019)
10. Gomez, L., Neamtiu, I., Azim, T., Millstein, T.: RERAN: timing and touch-sensitive record and replay for Android. In: Proceedings of IEEE ICSE Conference, pp. 72–81 (2013)
11. Graham, D., Fewster, M.: Experience of test automation, case studies of test automation. Pearson Education, Inc., London (2012)
12. Hu, Y., Neamtiu, I.: VALERA: an effective and efficient record-and-replay tool for Android. In: IEEE/ACM International Conference on Mobile Software Engineering and Systems, pp. 285–286 (2016)

13. Jorgensen, P.C.: Software Testing: a craftsman's approach, 4th edn. CRC, Boca Raton (2013)
14. Machiry, A., Tahiliani, R., Naik, M.: Dynodroid: an input generation system for Android apps. In: 9th Joint Meeting on Foundations of Software Engineering, ESEC/FSE, pp. 224–234. ACM (2013)
15. Memon, A.M., Soffa, M.L., Pollack, M.E.: Coverage criteria for GUI testing. ACM SIGSOFT Softw. Eng. Notes **26**(5), 256–267 (2001)
16. Moran, K., Linares-Vásquez, M., Bernal-Cárdenas, C., Vendome, C., Poshyvanyk, D.: Automatically discovering, reporting and reproducing android application crashes. In: IEEE International Conference on Software Testing, Verification and Validation, pp. 33–44 (2016)
17. Smidts, C., Mutha, C., Rodriguez, E., Gerber, M.: Software testing with an operational profile: OP definition. ACM Comput. Surv. **46**(3), 39.1–39.39 (2014)
18. Song, W., Qian, X., Huang J.: EHBDroid: beyond GUI testing for Android applications. In: Proceedings of the 32nd IEEE/ACM International Conference on Automated Software Engineering, pp. 27–37 (2017)
19. Sosnowski, J., Dobrzyński, B., Janczarek, P.: Analysing problem handling schemes in software projects. Inf. Softw. Technol. **91**, 56–71 (2017)
20. Zaeem, R.N., Prasad, M.R., Khurshid, S.: Automated generation of oracles for testing user-interaction features of mobile apps. In: IEEE International Conference on Software Testing, Verification, and Validation, pp. 183–192 (2014)
21. Zhang, J., Wang, Y.: Software feature refinement prioritization based on online user review mining. Inf. Softw. Technol. **108**, 30–34 (2019)

Crypto-ECC: A Rapid Secure Protocol for Large-Scale Wireless Sensor Networks Deployed in Internet of Things

Wassim Jerbi[1,2(✉)], Abderrahmen Guermazi[1,2], and Hafedh Trabelsi[2]

[1] CES LAB, National Engineering School of Sfax (ENIS), University of Sfax, Sfax, Tunisia
wassim.jerbi@isetn.rnu.tn
[2] Higher Institute of Technological Studies, 3099 El Bustan Sfax, Tunisia

Abstract. Wireless sensor networks are a new family of computer systems that can observe the world with unprecedented resolution. These systems promise to revolutionize the field of environmental monitoring. However, for the large-scale deployment of sensor network, data security must be ensured. In this context, this paper aims to provide solutions to ensure the authentication of users before having access to services and data collected by the sensor network. In this paper, we seek to accelerate the computation of scalar multiplications by using the paralleling technique which consists in distributing the calculation into several independent tasks that can be processed simultaneously by different nodes. We try to design a secure multicast routing protocol Crypto-ECC that takes into account the constraints of the WSN. Finally, the proposed solution will be evaluated using Telosb sensors.

Keywords: WSNs · Routing protocol · AES · Security · Large scale · Authentication

1 Introduction

In the computer world, the impressive increase in computing power of the processors makes it possible to decipher a message in shorter time. Therefore, in order to increase the security of a system, the most suitable solution is to increase the size of encryption keys and improve the time needed to decipher a message.

In our research, we investigate the possibility of applying ECC in sensor networks in an efficient way, because the operations on elliptic curves are still very complicated for micro-controllers, especially point multiplication, also called scalar multiplication, which is considered as the most expensive operation on the curves.

In this paper, we seek to accelerate the computation of scalar multiplications by using the paralleling technique which consists in distributing the calculation into several independent tasks that can be processed simultaneously by different nodes. The main benefit of a massive node deployment is to have many nodes available that can cooperate closely to achieve a common goal. In addition, nodes are fragile devices that may fail for task processing.

W. Zamojski et al. (Eds.): DepCoS-RELCOMEX 2020, AISC 1173, pp. 293–303, 2020.
https://doi.org/10.1007/978-3-030-48256-5_29

The remainder of this paper is organized as follows: Sect. 2 provides an overview on related works. Section 3 introduces the proposed protocol for asymmetric cryptography. Section 4 is dedicated to simulation results and discussions. Finally, Sect. 5 concludes this paper and outlines the future work.

2 Related Work

Several methods of key management can be cited, for example, a random key pre-distribution method [1, 3], block-based encryption based on symmetric keys [4], using an authentication framework [5]. Each sensor node in S-LEACH [6] has two keys and a key shared with the base station. Base station (BS) can identify malicious cluster head (CH) nodes and credible CH nodes to provide better message authentication. In [7], researchers proposed a security mechanism, allows identifying the malicious nodes by a query processing. This technique provides essential security properties such as confidentiality, integrity, freshness and data authentication.

In [8], a new efficient cryptographic method for better data security in WSN is proposed. The type of encryption used in this method is the symmetric key. The algorithm allows to divide the messages into blocks that are of different sizes and have unique keys for each block. This algorithm presents the feedback. Once the direct step encryption is complete, the entire file is split into two exchanged parts and the encryption method with the comments and a new key will be repeated.

In [9], cryptographic method identifies malicious nodes with high detection rates and low incidence of false detections. However, it focuses on specific attacks related to the package only. Routing protocols are vulnerable to several security attacks including scrambling, impersonation, and replay. Note that for Cluster-based protocols based on their CH, attacks involving CH are the most damaging. If an intruder manages to become a CH, he can make attacks such as radio jamming, flooding and passive listening, thus disrupting the network.

3 Proposed Protocol Crypto-ECC

Elliptic curve cryptography (ECC) includes a set of techniques that allow to secure data by consuming fewer resources. It is attracting more and more attention from researchers all over the world, especially in the field of embedded systems in which electronic devices have very limited computing power. The most important advantage of ECC over other asymmetric cryptography algorithms, for example RSA, is the ability to have a good level of security by using a much shorter key. The performance tests of Gura et al. [10] showed that to have the same level of security as a 1024-bit RSA key, with ECC, you have to use a 160-bit key. Moreover, the use of a shorter key also implies a lower memory, energy consumption and a faster calculation.

The elliptic curve E is an algebraic curve that can be represented by the equation:

$$E : y^2 + a_1xy + a_3y = x^3 + a_2x^2 + a_4x + a_6$$

In order to balance the load between nodes and to accelerate the calculations based on the elliptic curves, we propose, in this paper, to simplify the computation of the scalar multiplications between several nodes of a cluster.

Considering the operation of sensor networks, we decided to break down the calculation of scalar multiplications, the most important operation of elliptic curves, into several completely independent tasks that can be processed simultaneously by cluster members. Once completed, the results will be returned to cluster-head who will combine them together to get the final result. This latter can then be reused in different phases of calculating cryptographic protocols.

Our approach is to divide the tasks and the data between the different members of the CH:

- Data decomposition: A large amount of data is split and distributed to multiple cluster members by executing the same instructions.
- Decomposing tasks: Instead of distributing portions of data, task parallelization simply distributes procedures or threads to the different members of the cluster that execute them in parallel.

In the parallel computation framework, it is possible to share the memory of the CH containing the data circulating between the cluster members. These latter are expected to be able to access data with the same latency and bandwidth. All cluster members participating in parallel calculations can drop and read data, and this is an ideal configuration that allows the system to significantly improve its performance Fig. 1. Because shared memory can provide faster and more reliable data access, synchronization between processors is relatively easier.

The timing management mechanism to set up may be different depending on the type of memory used. For an architecture that uses shared memory, access to memory is easier to monitor and control. In addition, it is more easily to have well-synchronized and consistent data access. When p1 detects that the memory block containing v is momentarily occupied by p2, it is imperative that p1 waits until p2 finishes its current operation as shown in Fig. 2.

The idea of chopping is mainly based on. It offers an efficient data decomposition without requiring shared memory. The goal is to reduce the computation time of scalar multiplications by using the member nodes in the calculation.

We assume that the elliptic curve is loaded on the nodes before deployment, and the generator point G does not change during the lifetime of the network. The scalar multiplication is denoted as: Q = kG, where G is the generating point of a curve defined in a finite prime field, denoted by E (Fp), and k is an integer of length L, which can be represented in binary as the following manner:

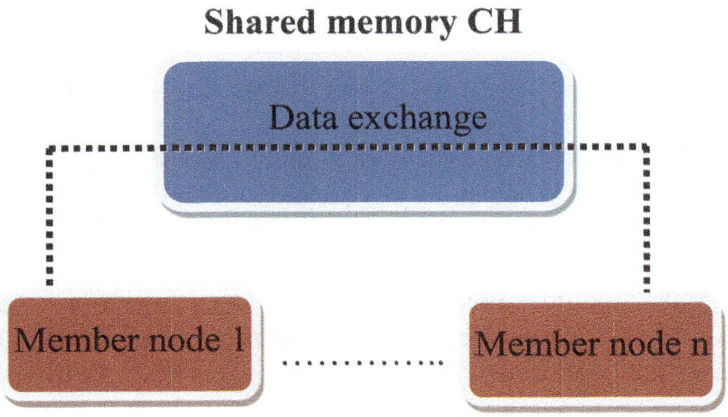

Fig. 1. Data exchange between CH and members of cluster

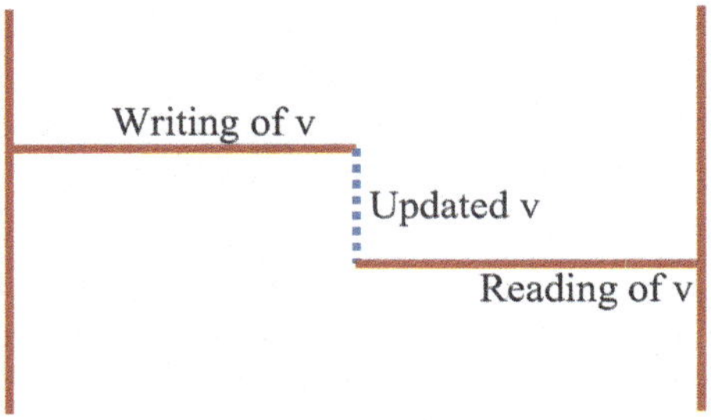

Fig. 2. Synchronization of data access

$$K = \sum\nolimits_{i=0}^{l-1} \left(K_i 2^{\mathrm{i}}\right) \quad (1)$$

The node that decomposes and distributes data is the CH node, and the other nodes used for parallel computing are member nodes. At first, the CH decomposes the integer k into n segments S_i of length b = [l/n], where n is the number of member nodes, and S_i can be represented in binary as the following equation:

$$S_i = \sum\nolimits_{j=ib}^{ib+b-l} \left(K_j 2^{\mathrm{j}}\right) \quad (2)$$

Note that the generator point G is chosen a priori, and it does not change during the lifetime of the network. (The calculation of the points Gi = 2ibG is possible). The calculation of kG can be decomposed as follows:

$$\begin{cases} Q_0 = S_0 G \\ Q_1 = S_1 2^b G \\ \ldots\ldots\ldots\ldots\ldots \\ Q_{n-1} = S_{n-1} 2^{b(n-1)} G \end{cases} \quad (3)$$

Then, the final result $Q = Q0 + Q1 + \ldots Qn - 1$, and each Qi can be calculated independently by a member node.

Suppose, for example, that we have $k = kl - 1kl - 2\ldots k2k1k0$, where $ki \in \{1, 0\}$, and there are (n – 1) members nodes available in the radio range of the CH, the scalar k is therefore decomposed into n segments:

$$\begin{aligned} S_0 &= kb - 1kb - 2\,kb - 3\ldots k2k1k0 \\ S_1 &= k2b - 1k2b - 2\,k2b - 3\ldots kb + 2kb + 1kb \\ S_{n-1} &= knb - 1knb - 2\,knb - 3\ldots k((n-1)b+2)k((n-1)b+1)k(n-1)b \end{aligned} \quad (4)$$

The CH leaves one segment in its local memory, and the other n − 1 segments are distributed to the member nodes. All nodes prepare and store locally a set of points:

$$\begin{aligned} G_0 &= G \\ G_1 &= 2^b G \\ &\ldots\ldots\ldots \\ G_{n-1} &= 2^{(n-1)} G \end{aligned} \quad (5)$$

After receiving the data, each member node calculates a Qi using the formula 3. Once finished, they return Qi to the CH, which combine them with his own local result to get the final result:

$$Q_0 = S_0 G + S_1 2^b G + S_2 2^{2b} G \ldots S_{(n-1)} 2^{(n-1)} G \quad (6)$$

This scalar decomposition method allows splitting the scalar into n parts that can be processed independently, no data exchange is needed between member nodes. In addition, according to the recommended parameters, the size of the body Fp in which the curve is defined is often between 112 and 521 bits, and it is possible to broadcast the set of S_i to the member nodes at same time.

4 Operation of protocol Crypto-ECC

The operation of the protocol begins when the cluster-head detects data revealing a critical event, which must be reported immediately to the base station in a secure manner (Fig. 3):

1. O-LEACH [11], cluster members periodically send collected data to the cluster-head, which is responsible for processing information (aggregation, compression, encryption, etc.);
2. The cluster-head detects an important event, and it wants to use parallelism to accelerate the encryption of sensitive information. To request help from other members, it sends them a request to participate to identify the available nodes in the cluster;
3. After receiving the request, a member returns a response {i, r} where i is its identifier and r is a Boolean value representing its availability;

4. The cluster-head selects the available members which have enough energy and will participate in the parallel computation. The CH first decomposes the computation into n independent parts, and then distributes the calculation tasks to the selected member nodes.
5. The nodes now have the necessary information to perform the task processing. Once completed, the slaves return the results to their CH. This latter combines the returned results to get the final result which can be used to encrypt or sign the message to send.

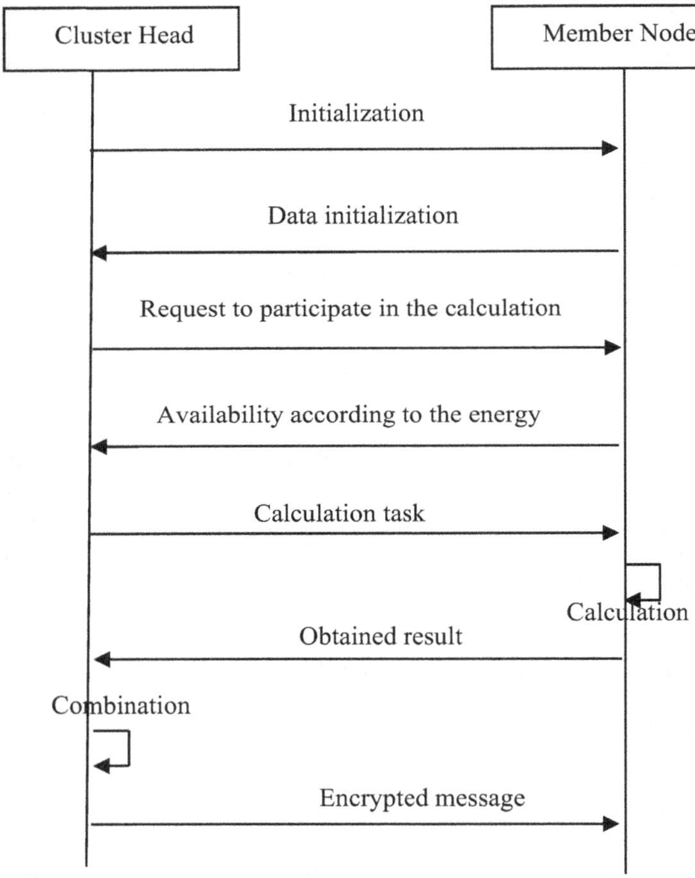

Fig. 3. Deployment of a heterogeneous sensor network

5 Performance and Evaluation of Crypto-ECC

In order to evaluate the performance of our Crypto-ECC fractionation method, we have implemented it on Telosb sensors which is a model designed by Crossbow Technology for research purposes.

We chose nesC [12] as the programming language, which means network embedded system C. It is a language specifically created for event programming. It is also the default development language under Tinyos [13]. One of the advantages of such a program is that its execution is entirely driven by events. If there is no significant event, the execution is automatically paused, and this property allows the node to minimize energy consumption which is considered as one of the crucial factors during the design of an application for sensor networks [14].

The first part of the test consists of evaluating the performance of our parallelism method with recalculated points stored in local memory. The computation times in milliseconds with and without parallel computation are given in Table 1. It is difficult to compare the absolute values of the results with those of the other implementations because of the differences in technique used and test scenario.

We can see in Fig. 4, the percentage of computation time in ms as a function of the numbers of nodes. Show that the computation time decreases progressively when more nodes participate in the parallel computation.

We note that the number of the nodes take part in parallel computation does not exceed the 10 nodes for a cluster. This is due to the load occupied by the CH (an additional cost). We set the percentage of CH at 10% with respect to N nodes for a large scale simulation.

Table 1. Calculation time (ms) of our parallelism protocol

Numbers of nodes	Crypto-ECC	Earnings	AODV	Earnings	KMMR	Earnings
10	16000		10100		12500	
20	12800	20%	8500	15,84%	10300	17,6%
30	9400	41,25%	6750	33,16%	8000	36%
40	7500	53,125%	5800	42,57%	8900	44,8%
50	6600	61,56%	4900	51,48%	5800	53,6%
60	5500	65,63%	3950	60,89%	4700	62,4%
70	4750	70,31%	3550	64,85%	4050	67,6%
80	4500	71,8%	3200	68,31%	3800	70%
100	4150	74,06%	3150	68,88%	3700	70,4%

We have parallelized the calculation of scalar multiplications using up to 100 nodes, and we have obtained a maximum gain of 74.06% compared to the KMMR and AODV protocols.

We assume that the computation time with p nodes is Tp, we can still evaluate the performance of our method with its speedup Sp which is defined in the formula 7, and the speedup results are given in Table 2 and graphically represented in Fig. 5.

We can notice in Fig. 5, a good acceleration when we use more than 100 nodes. The higher the number of member nodes participating in the calculations, the faster the speed increases. The cluster head's task decomposition makes the processing fairly simplified by the member nodes.

Saving time at computing time with a large number of participating nodes reduces energy consumption. Crypto-ECC presents a rapid secure protocol for large-scale Wireless Sensor Networks.

In Fig. 6, we can see that when the calculation is performed on a single node, too much energy is consumed. However, when parallel computing is used, energy consumption is very reduced by a few joules as members nodes participate in parallel calculations, compared to other KMMR and AODV protocols. The member nodes must receive tasks from their CH and return results to them. Comparing the obtained results with a CH that performs the calculations alone, we gained a percentage of up to 74% energy at the CHs level. This gives an extension of the lifetime of the network. Simulation results have shown that parallel computing consumes less energy compared to local computing. Thus, we observe a high consumption on the part of the two protocols AODV and KMMR. The use of Crypto-ECC increases the lifetime of the network, and we therefore propose to use parallel computing in all cases of a normal or urgent communication (the detection of a crucial event).

$$S_p = T_1/T_p \tag{7}$$

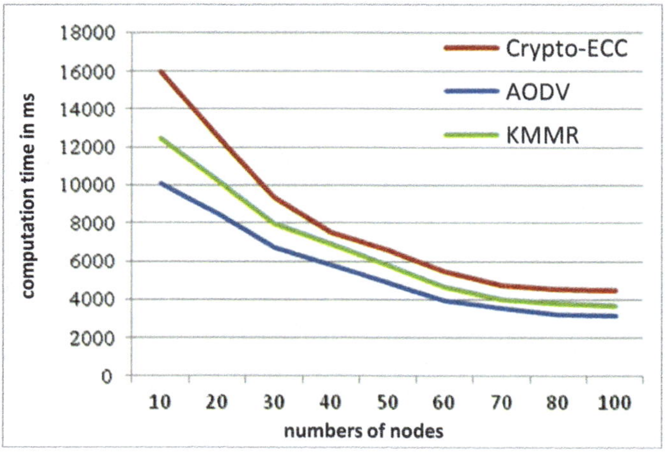

Fig. 4. Computation time of protocol Crypto-ECC

Table 2. Speed of protocol Crypto-ECC

Number of node	20	40	60	80	100
Crypto-ECC	14,3	21,3	34	42,1	51
AODV	11,9	17,4	24,6	29,7	32,5
KMMR	12,1	18,1	26,6	35,7	41

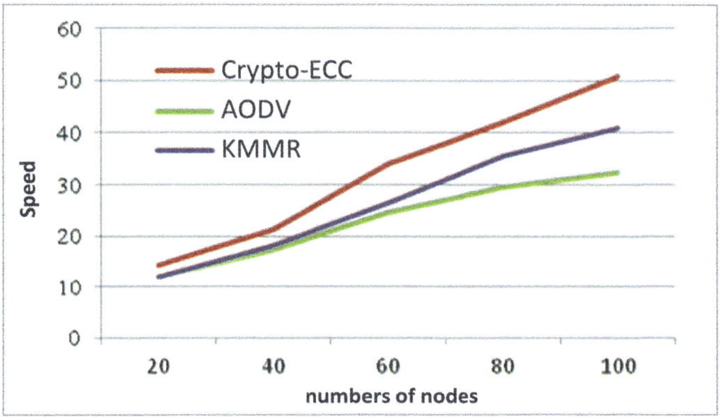

Fig. 5. Speed of protocol Crypto-ECC

We can notice in Fig. 6, a good acceleration when we use more than 60 nodes. The higher the number of member nodes participating in the calculations, the faster the speed increases. The cluster head's task decomposition makes the processing fairly simplified by the member nodes. Saving time at computing time with a large number of participating nodes reduces energy consumption.

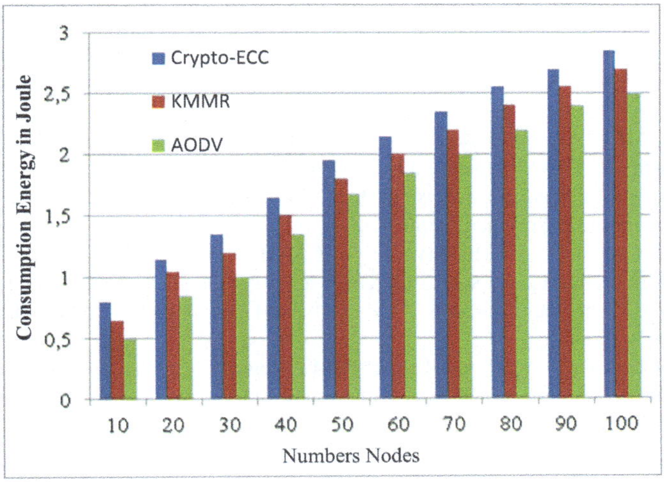

Fig. 6. Consumption energy of protocol Crypto-ECC

6 Conclusion

In this paper, we have applied Crypto-ECC protocol in wireless sensor networks in an efficient and safe manner to ensure the confidentiality of information. The Crypto-ECC protocol can accelerate calculations based on the parallelism. Indeed, a calculation task is divided into several independent parts, which can be processed at the same time by different member nodes. The Crypto-ECC protocol saves a lot of computing time with reduced energy consumption.

References

1. Koblitz, N.: Elliptic curve cryptosystems. Math. Comput. **48**(177), 203–209 (1987)
2. Rivest, R.L., Shamir, A., Adleman, L.: A method for obtaining digital signatures and public-key cryptosystems. Commun. ACM **21**(2), 120–126 (1978)
3. Eschenauer, L., Gligor, V.D.: A key-management scheme for distributed sensor networks. In: Proceedings of the 9th ACM Conference on Computer and Communications Security, pp. 41–47. ACM (2002)
4. Perrig, A., Szewczyk, R., Wen, V., Culler, D., Tygar, J.D.: SPINS: security protocols for sensor networks. Wireless Netw. **8**, 521–534 (2002)
5. Ferreira, A., Vilaça, M., Oliveira, L., Habib, E., Wong, H., Loureiro, A.: On the security of cluster-based communication protocols for wireless sensor networks. In: Networking ICN 2005, pp. 449–458. Springer (2005)
6. Bohge, M., Trappe, W.: An authentication framework for hierarchical ad hoc sensor networks. In: Proceedings of the 2nd ACM workshop on Wireless security, pp. 79–87. ACM (2003)
7. Ghosal, A., Halder, S., Sur, S., Dan, A., DasBit, S.: Ensuring basic security and preventing replay attack in a query processing application domain in WSN. In: LNCS, ICCSA. Springer, Berlin, March 2010
8. Praveena, A., Smys, S.: Efficient cryptographic approach for data security in wireless sensor networks using MES VU. In: 2016 10th International Conference on Intelligent Systems and Control (ISCO). IEEE (2016)
9. Prathap, U., Shenoy, P.D., Venugopal, K.: CMNTS: catching malicious nodes with trust support in wireless sensor networks. In: 2016 IEEE Region 10 Symposium (TENSYMP). IEEE (2016)
10. Gura, N., Patel, A., Wander, A., Eberle, H., Shantz, S.: Comparing elliptic curve cryptography and RSA on 8-bit CPUs. In: Joye, M., Quisquater, J.-J. (eds.) Cryptographic Hardware and Embedded Systems - CHES 2004. Lecture Notes in Computer Science, vol. 3156, pp. 119–132. Springer, Berlin (2004)
11. Jerbi, W., Guermazi, A., Trabelsi, H.: O-LEACH of routing protocol for wireless sensor networks. In: 13th International Conference Computer Graphics, Imaging and Visualization, CGiV 2016, Beni Mlel, Marocco, 29 Mar–01 April, pp. 399–404 (2016). https://doi.org/10.1109/cgiv.2016.84
12. Gay, D., Levis, P., Von Behren, R., Welsh, M., Brewer, E., Culler, D.: The nesC language: a holistic approach to networked embedded systems. In: ACM Sigplan Notices, vol. 38, pp. 1–11. ACM (2003)

13. Levis, P., et al.: TinyOS: an operating system for sensor networks. In: Ambient intelligence, pp. 115–148. Springer (2005)
14. Guermazi, A., Belghith, A., Abid, M., Gannouni, S.: KMMR: an efficient and scalable key management protocol to secure multi-hop communications in large scale wireless sensor networks. In: KSII Transactions on Internet and Information Systems, vol. 11, no. 2, February 2017

Redundancy Management in Homogeneous Architecture of Power Supply Units in Wireless Sensor Networks

Igor Kabashkin(✉)

Transport and Telecommunication Institute, Riga, Latvia
kiv@tsi.lv

Abstract. Wireless sensor networks (WSN) are one of the basic technologies using various Internet of Things applications especially in cyber-physical systems. The cyber-physical system is usually designed for autonomous functioning without direct participation and control by humans. Sensors usually have autonomous power supply from batteries, which is one of the critical factors in the life cycle of a network and requires additional attention of its fault tolerance. In the paper additional method for reliability improving of the sensors in cluster-based WSN with individual and common set of redundant batteries and dynamic management of redundant architecture with two levels of availability is proposed. Mathematical model of the sensor reliability is developed. Comparative analysis of redundancy effectiveness for developed and used structure of backup architecture of batteries in cluster-based WSN is performed.

Keywords: Sensor · Sensor cluster · Wireless sensor networks · Reliability · Battery · Redundancy

1 Introduction

Wireless sensor networks (WSNs) are being actively developed at present. They are one of the basic technologies using various Internet of Things applications especially in cyber-physical systems (CFS). Such applications integrate different technologies and include different network capabilities [1].

In WSN, many interconnected sensor (S) nodes with wireless channels form a spatially distributed system. In large networks, sensors have restrictions on the speed and amount of information processed. To reduce these restrictions on large spatial areas, sensors in networks can be aggregated in spatially distributed groups called clusters. The creation of these clusters can significantly improve the efficiency of network as a whole. The data collected by the sensors of each node are transmitted to the central element of the cluster, which acts as the cluster head element (CH). The base station (BS) collects information from all clusters through their head elements. A cluster-oriented structure of WSN is shown at the Fig. 1.

The cyber-physical system is usually designed for autonomous functioning without direct participation and control by humans. CFS is often designed to collect information and monitor the status of highly responsible and mission critical systems.

W. Zamojski et al. (Eds.): DepCoS-RELCOMEX 2020, AISC 1173, pp. 304–314, 2020.
https://doi.org/10.1007/978-3-030-48256-5_30

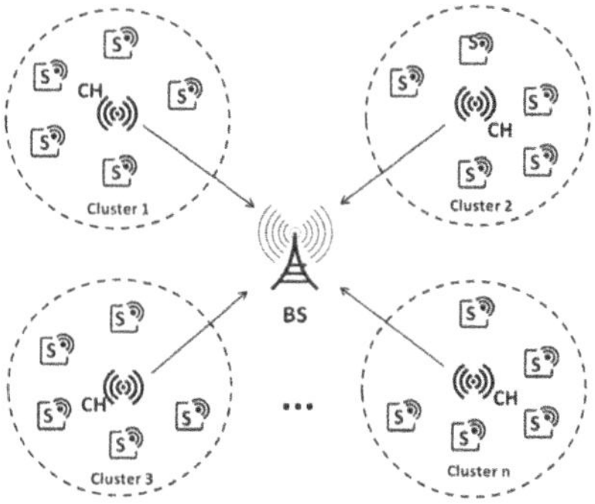

Fig. 1. Architecture of cluster-oriented WSN.

Dependability and fault tolerance is an important attribute of such WSNs. In these networks, requirements for the reliability of the network as a whole and the availability of individual sensor nodes are dominant. In a number of works, the issues of the influence on the reliability of networks of their topology, protocols, and application-level error correction are investigated [2, 3].

Current studies of the reliability of the sensor node mainly investigate the effect on the network efficiency of problems related to equipment reliability and communication problems between network nodes [4]. Sensors usually have autonomous power supply from accumulators or batteries, which is one of the critical factors in the life cycle of a network and requires additional attention from the point of view of its fault tolerance. For sensors that work in wireless networks, the power supply is one of the key elements. Since the sensors are distributed, a typical energy supply structure for them is an individual energy supply using a personal battery (Fig. 2).

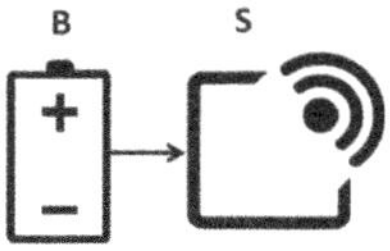

Fig. 2. Traditional sensor with supported battery.

When the battery level drops below a critical level, the sensor cannot perform its functions. In this case, the life cycle of the operation of cyber-physical systems can be extended by introducing the redundancy of the power supply of wireless network sensors due to battery backup [5]. Using this method, each of the m sensors in the cluster works with a pair of batteries, one of which is in the main mode, and the second performs the backup function. The battery communication switch (BCS) performs the functions of monitoring the battery capacity and switching them to standby (Fig. 3).

In this paper we investigate the reliability of selected battery in WSN with common set of redundant batteries and with dynamic management of backup batteries in wireless sensor network.

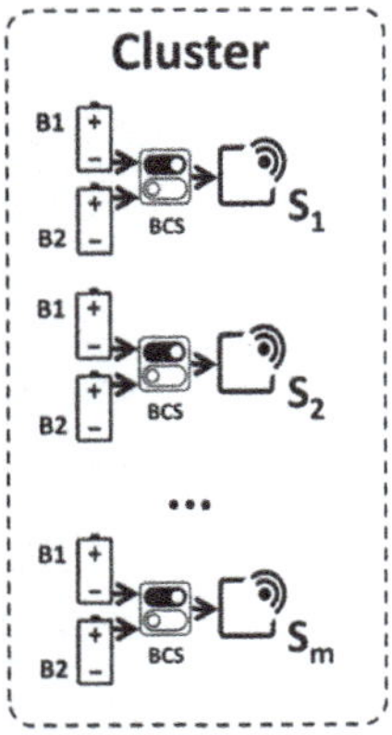

Fig. 3. Architecture of sensor with duplication of batteries

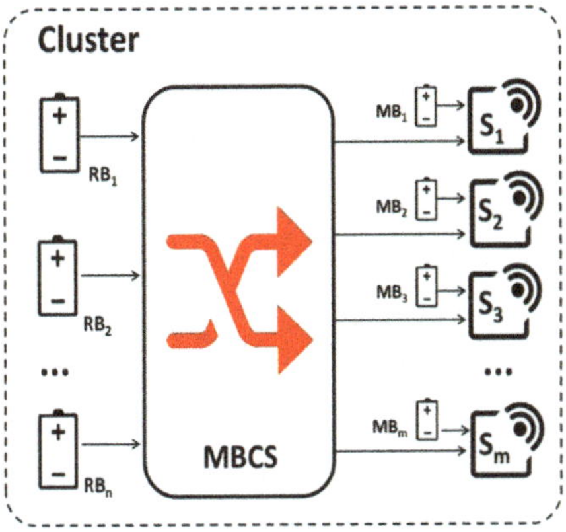

Fig. 4. Architecture of sensor with common set of redundant batteries

The chapter has the following structure. Review of references in the field of dependability of sensor nodes in WSN is carried out in the Sect. 2. The basic definitions and notation used in the chapter are given in Sect. 3. Section 4 describes the proposed approaches to improving fault tolerance of WSN. In the same section, models for the reliability of sensors in cluster-oriented WSN CPS are developed and their comparative analysis is performed. Conclusions are given in the Sect. 5.

2 Related Works

In the paper [7], an overview of existing protocols for the reliability of data transmission in WSN is presented. The paper considers several reliability models. They are focused on the use of various mechanisms for recovering lost data with insufficient reliability of packet transmission, in particular, through the use of retransmission methods and coding redundancy.

Survey of different dependability methods in WSN and specific sensors is presented in paper [8].

Reliability of wireless networks is largely determined by the reliability of sensors. In [9], the sensor was considered at the micro level, its components providing communication are analyzed for influence on the reliability of WSN as a whole. Reliability of network sensors and methods for increasing them is carried out using probabilistic analysis based on Markov models.

In the real world of using wireless sensor networks, various maintenance methods are used to ensure their reliable operation. The influences of the frequency of maintenance on the availability of network nodes, as well as the optimal conditions for its implementation, are studied in [10].

The impact of power outages is especially sensitive in systems critical to safety. New generations of autonomous objects with built-in redundancy of individual subsystems and components, including energy supply sources, are considered in [11].

Duplication of power supply sources of individual nodes of wireless sensor networks as a method of increasing reliability is considered in [5]. For the proposed method, a model of the dependability of the sensor node is proposed and investigation of its availability in comparison with the traditional method of using batteries is made.

The use of duplication of power sources of individual sensor nodes increases their reliability, but at the same time increases the resources to ensure the reliability of their functioning. In [6], an approach was proposed that allows one to find a compromise between these factors based on the use for cluster sensors of a one mutual set of backup energy suppliers. A network reliability model is obtained to determine the availability of the sensor node using the proposed approach. Such method is known as *k-out-of-n* system structure. It is used as method of increase resilience in the systems with uniform elements. The method *k-out-of-n:G* [12–14] describes system with *n* identical elements. The system will fail if more than *n-k* of the elements in the system will fail. The similar model of *k-out-of-n:F* redundancy [15] is a system with *n* identical elements, but the system will fail if more than k of the elements in the system will fail. Applications of *k-out-of-n* structure are very popular for design of fault tolerant telecommunication systems [16].

In real systems, which are critical for security, not only the reliability of the network as a whole is important from the point of view of fault tolerance, but also the availability of individual channels for receiving information. The influence of the used backup methods in such systems was studied in [17–19]. The reliability of switching functions in a redundant switching device is of particular importance in such redundant systems. The architecture of such automatic switching devices allows the possibility of switching failures of two types - "false switching" and "without switching". In [6], models of system reliability for both types of switch failures are considered, and requirements for the probability of each type of failure for highly reliable systems are determined.

In this paper, we discuss additional opportunity to increase efficiency and fault tolerance in WSN cluster nodes with redundancy architecture of batteries (Fig. 4), which provides for each sensor in the cluster a primary energy source MB and a common set of backup energy sources RB. The matrix communications switch MCS replaces any failed battery with a functioning backup battery.

3 Mathematical Background and Main Symbols and Definitions

In the chapter the fault tolerance of cluster-based nodes in IoT sensor networks is investigated. The analysis is performed on the basis of modelling these systems using Markov models of reliability based on the classical mathematical models with construction of Kolmogorov-Chapman differential equations and determining the stationary values of the probabilities of finding the system under study in a stationary state [20].

The following symbols have been used to develop equations for the models:

λ - Failure Rate of battery

μ - Repair Rate of battery

A_0 - Availability of the ideal sensor without test operations

A_1 - Sensor availability with active backup mode of redundant elements

A_2 - Sensor availability with dynamic management of two-stage backup modes

L - Number of repair bodies

V - Factor of reliability improvement

m - Number of sensors in cluster

n - Number of mutual set of redundant batteries in cluster

The reliability of all switching elements in the system is considered ideal; all time intervals have exponential distribution. Redundancy architecture has a homogeneous structure with the same type of energy supply elements.

4 Model Formulation and Solution

The reliability of the operation of the sensor network nodes is determined to the greatest extent by the battery efficiency. The battery efficiency is determined mainly by three factors: the intensity of the cycles of the sensor transmission mode in the network, the temperature of the working environment of the sensor node, and the battery charge level during its recovery.

As an example, Fig. 5 shows the experimental dependences of the battery energy depending on the duty cycle for three different transmission durations (100 ms, 200 ms, 400 ms) [19].

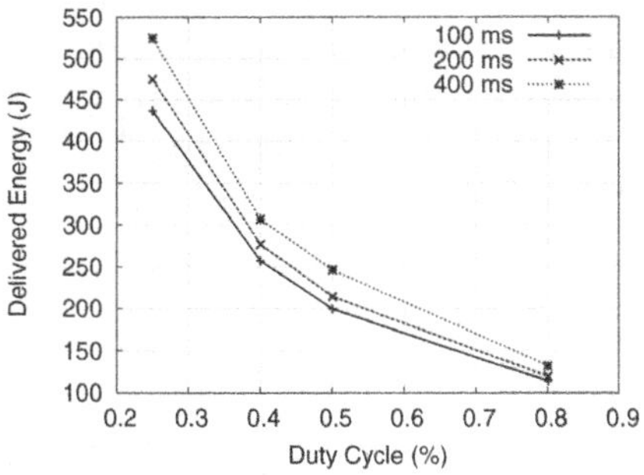

Fig. 5. Impact of duty-cycle on battery delivered energy at different transmission times and sampling intervals [19].

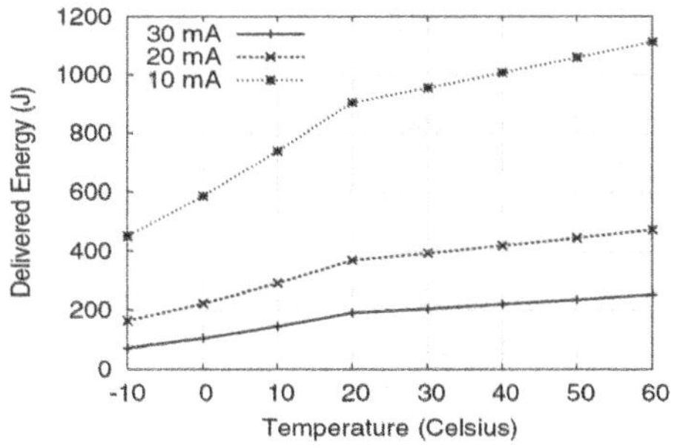

Fig. 6. Impact of ambient temperature on delivered energy for different transmission power levels [19].

Figure 6 shows the experimental dependences of the battery energy depending on the temperature of the working environment of the sensor node and discharge current values [21].

For safety reasons chargers for many batteries cannot exceed some maximum value (for example, 4.20 V/cell for lithium-ions) [22]. While a higher voltage boosts capacity, exceeding the voltage shortens service life and compromises safety. Figure 7 demonstrates cycle count as a function of charge voltage. At 4.35 V, the cycle count of a regular Li-ion is cut in half.

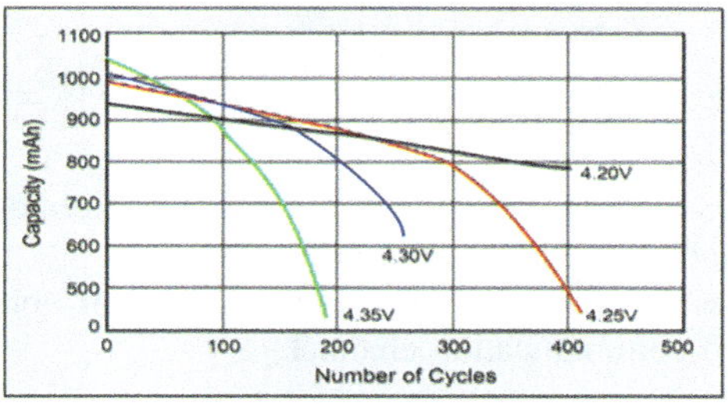

Fig. 7. Effects of elevated charge voltages on cycle life [22].

These examples show that even with the same batteries, their capacity, and therefore the reliability of providing energy to the sensor node, will be different depending on the history of the functioning of these batteries during the life cycle.

As an additional method of improving the reliability of the system we will use a dynamic management of backup modes with two-level modes of redundancy.

The two-level mode backup contains n batteries, of which $r < n$ are at the first mode of availability, and $n - r$ are located on the second level of availability. Redundant batteries with second level are in the cold standby (standby batteries are not under the load and their capacity is practically not consumed, $\lambda_2 = 0$). Redundant batteries at the second level can fail only after their transfer to the first mode of the redundancy. Batteries at the first level are in hot standby (standby batteries in this mode have the same failure rates as the main batteries $\lambda_1 = \lambda$). In case of failure of main battery it without delay is replaced by the backup battery from the first level of availability. Simultaneously the redundant battery from the second level of availability is transferred to the group with first level of availability.

The above two-level model of redundancy required to support readiness on the first level only for z elements, which provide reliable level of operation. In [23] it is shown that if $\lambda_1 > \lambda_2$ at the first level is optimal to have only one no-fault element. In our study this condition is satisfied, so we can accept that $z = 1$.

For the system with $1 \leq L \leq n$ number of repair bodies the behaviour of the examined system is described by the Markov Chain state transition diagram (Fig. 8), where: Hi – state with i failed batteries, but in the selected sensor there is a workable battery; HiI – state with i + 1 failed batteries, in the selected channel there is no a workable battery.

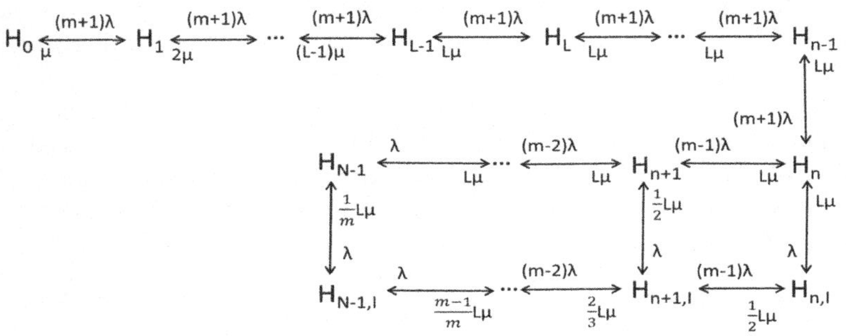

Fig. 8. Markov Chain state transition diagram ($1 \leq L \leq n$).

On the base of this diagram the system of Chapman–Kolmogorov's equations can be writing in accordance with the general rules [20].

By solving the resulting system of equations, we can obtain an expression for availability of selected communication channel:

$$A = 1 - \sum_{\forall i,j} P_{ijl} = \frac{a_1 + a_2}{a_1 + a_2 + a_3}, \tag{1}$$

where

$$a_1 = \sum_{i=0}^{L} \frac{(m+1)^i}{i!} \gamma^i + \frac{L^L}{L!} \sum_{i=L+1}^{n-1} (m+1)^i \omega^i,$$

$$a_2 = \frac{(m+1)^n L^L}{L!} \sum_{i=0}^{m-1} i! \binom{i}{m-1} \omega^{n+i},$$

$$a_3 = \frac{(m+1)^n L^L}{L!} \sum_{i=0}^{m-1} (i+1)! \binom{i}{m-1} \omega^{n+i+1}, \gamma = \frac{\lambda}{\mu}, \omega = \frac{\gamma}{L}.$$

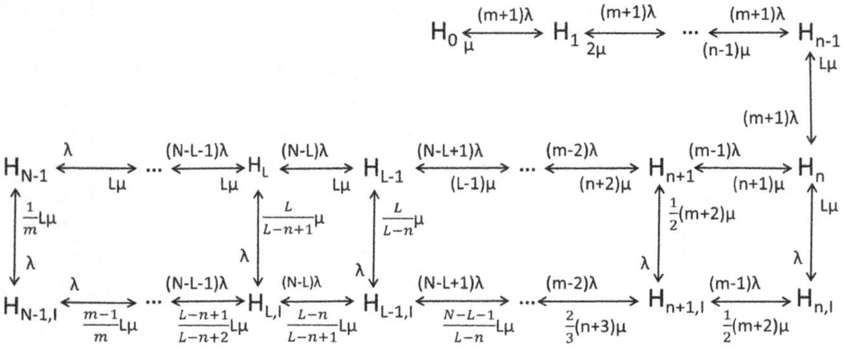

Fig. 9. Markov Chain state transition diagram ($n \leq L \leq N$).

For the system with $n \le L \le N$ number of repair bodies the Markov Chain state transition diagram of the system is shown at the Fig. 9. On the base of this diagram the system of Chapman–Kolmogorov's equations can be writing in accordance with the general rules [20].

By solving the resulting system of equations, we can obtain an expression for availability of selected communication channel in accordance of (1), where

$$a_1 = \sum_{i=0}^{n-1} \frac{(m+1)^i}{i!} \gamma^i,$$

$$a_2 = (m+1)^n \left\{ \frac{\gamma^n}{(n+1)!} \left[n+1+\sum_{i=1}^{L-n} i! \binom{i}{m-1} \gamma^i \right] + \frac{(m-1)! L^L}{L!} \sum_{i=1}^{N-L-1} \frac{\omega^{l+i}}{(N-L-i-1)!} \right\},$$

$$a_3 = (m+1)^n \left[\frac{1}{n!} \sum_{i=0}^{l-n} \binom{i}{m-1} \binom{n}{n+i+1}^{-1} \gamma^{n+i+1} + \frac{(m-1)! L^L}{L!} \sum_{i=1}^{N-L-1} \frac{L-n+i+1}{(N-L-i-1)!} \omega^{L+i+1} \right].$$

Numerical Example

Let us investigate the reliability of selected sensors in cluster-based WSN with common set of standby batteries and with dynamic management of backup modes. It is possible to evaluate the reliability in the proposed model with two-level mode of redundant batteries in comparison to the active backup mode of redundant elements with the help of the factor of reliability improvement V:

$$V = \frac{1 - A_1}{1 - A_2}$$

where the value of A_2 is determined in accordance with the expression (1), equation for the A_1 availability of the system with active backup mode of redundant elements was determined in [17].

At the Fig. 10 the factor of reliability improvement V is shown as function of number m of sensors in cluster with different number n of standby batteries in common redundant set of batteries for mean time between failures of each battery MTBF = $1/\lambda$ = 3000 h, L = 1 and mean time between repairs MTBR = $1/\mu$ = 2 h.

Analysis of the curves at Fig. 10 shows that factor of reliability improvement dramatically increases with increasing number of redundant elements at the second level of redundancy. The intensity of the given functional relation decreases with increasing number of sensors in cluster of WSN.

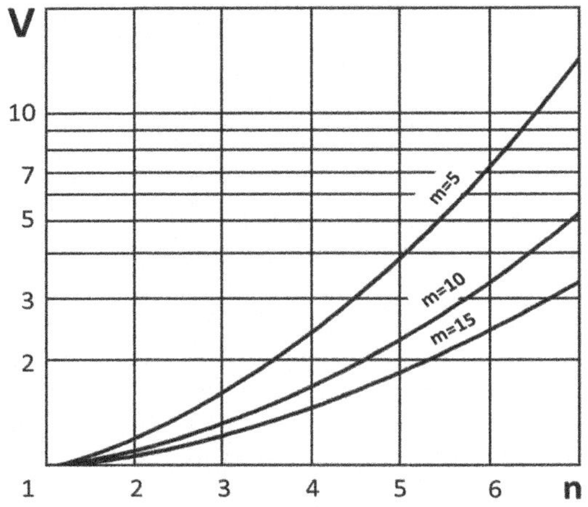

Fig. 10. The factor of reliability improvement.

5 Conclusions

Mathematical model of the sensor reliability in WSN cluster nodes with redundancy architecture of batteries in the real conditions of operation is developed. Cluster-based WSN is one of the basic technologies using various Internet of Things applications especially in cyber-physical systems. As additional method of improving the reliability of the nodes in WSN a dynamic management of backup modes with two-level availability of redundant elements is proposed. Mathematical model of the sensor reliability is developed. Expressions for availability of the dedicated sensor in cluster-based WSN with common set of redundant batteries and dynamic management of backup architecture of batteries with two-level mode of redundant elements are developed. Comparative analysis of redundancy effectiveness for developed and used structure of backup architecture of batteries in cluster-based WSN is performed.

It is shown that optimal redundancy architecture should have only one no-fault element at the first level of two mode backup of power supplies. Sensor reliability of WSN in this case increases with increasing number of redundant batteries at the second level of two mode redundancy architecture. This dependence is more active in the clusters of WSN with a smaller number of sensors.

References

1. Pottie, G.J.: Wireless integrated network sensors (WINS): the web gets physical. In: Frontiers of Engineering: Reports on Leading-Edge Engineering from the 2001 NAE Symposium on Frontiers of Engineering, National Academies Press, p. 78 (2002)
2. Ayadi, A.: Energy-efficient and reliable transport protocols for wireless sensor networks: state-of-art. Wireless Sens. Netw. **3**(3), 106–113 (2011)
3. Sharma, K., Patel, R., Singh, H.: A reliable and energy efficient transport protocol for wireless sensor networks. Int. J. Comput. Netw. Commun. **2**(5), 92–103 (2010)
4. Park, S.-J., Sivakumar, R., Akyildiz, I.F., et al.: GARUDA: achieving effective reliability for downstream communication in wireless sensor networks. IEEE Trans. Mob. Comput. **7**(2), 214–230 (2008)
5. Mahajan, S., Dhiman, P.: Clustering in wireless sensor networks: a review. Int. J. Adv. Res. Comput. Sci. **7**(3), 198–201 (2016)
6. Kabashkin, I.: Reliability of cluster-based nodes in wireless sensor networks of cyber physical systems. Procedia Comput. Sci. **151**, 313–320 (2019). Elsevier
7. Farahani, S.: Battery Life Analysis. In: ZigBee Wireless Networks and Transceivers, pp. 207–224 (2008)
8. Mahmood, M., Seah, W., Welch, I.: Reliability in wireless sensor networks: a survey and challenges ahead. Comput. Netw. **79**, 166–187 (2015)
9. Song, Y., Chen, T., Juanli, M., Feng, Y., Zhang, X.: Design and analysis for reliability of wireless sensor network. J. Netw. **7**(12), 2003–2012 (2012)
10. Kabashkin, I., Kundler, J.: Reliability of sensor nodes in wireless sensor networks of cyber physical systems. Procedia Comput. Sci. **104**, 380–384 (2017)
11. Slovick, M.: Buck-Boost Controller Answers Call for Redundant Battery Systems. Electronic Design, 03 October (2018). https://www.electronicdesign.com/automotive/buck-boost-controller-answers-call-redundant-battery-systems
12. Barlow, R., Heidtmann, K.: On the reliability computation of a k-out-of-n system. Microelectron. Reliab. **33**(2), 267–269 (1993)
13. Misra, K.: Handbook of Performability Engineering. Springer, London (2008)
14. McGrady, P.: The availability of a k-out-of-n: G network. IEEE Trans. Reliab. **R-34**(5), 451–452 (1985)
15. Rushdi, A.: A switching-algebraic analysis of consecutive-k-out-of-n: F systems. Microelectron. Reliab. **27**(1), 171–174 (1987)
16. Ayers, M.: Telecommunications System Reliability Engineering, Theory, and Practice. Wiley-IEEE Press, Piscataway (2012)
17. Kozlov, B., Ushakov, I.: Reliability Handbook (International Series in Decision Processes). Holt Rinehart & Winston of Canada Ltd., New York (1970)
18. Kabashkin, I.: Dynamic redundancy in communication network of air traffic management system. In: Zamojski, W., Mazurkiewicz, J., Sugier, J., Walkowiak, T., Kacprzyk, J. (eds.) Advances in Dependability Engineering of Complex Systems. DepCoS-RELCOMEX 2017. Advances in Intelligent Systems and Computing, vol. 582, pp. 178–185. Springer, Cham (2018)
19. Kabashkin, I.: Dependability of multichannel communication system with maintenance operations for air traffic management. In: Zamojski, W., Mazurkiewicz, J., Sugier, J., Walkowiak, T., Kacprzyk, J. (eds.) Engineering in Dependability of Computer Systems and Networks. DepCoS-RELCOMEX 2019. Advances in Intelligent Systems and Computing, vol. 987, pp. 256–263. Springer, Cham (2020)

20. Rubino, G., Sericola, B.: Markov Chains and Dependability Theory. Cambridge University Press, Cambridge (2014)
21. Park, C. Lahiri, K. Raghunathan, A.: Battery discharge characteristics of wireless sensor nodes: an experimental analysis. In: 2005 Second Annual IEEE Communications Society Conference on Sensor and Ad Hoc Communications and Networks, pp. 430–440. IEEE SECON, Santa Clara (2005)
22. BU-808: How to Prolong Lithium-based Batteries. Battery University (2020). https://batteryuniversity.com/learn/article/how_to_prolong_lithium_based_batteries. Accessed 20 Jan 2020
23. Raikin, I.: Elements of Reliability Theory for Technical Systems. Sov. Radio Publisher, Moscow (1978). (in Russian)

Successive-Interference-Cancellation-Inspired Multi-user MIMO Detector Driven by Genetic Algorithm

Mohammed J. Khafaji and Maciej Krasicki(✉)

Faculty of Computing and Telecommunications,
Institute of Radiocommunications, Poznan University of Technology,
Polanka 3, 61-131 Poznan, Poland
maciej.krasicki@put.poznan.pl

Abstract. Multi-User Multiple-Input Multiple-Output (MU-MIMO) configuration is one of the most promising solutions to the fundamental problem of a telecommunication system: limited bandwidth. According to the MU-MIMO principles, different users transmit their signals concurrently at the same channel. It helps exploit channel capacity to a larger extent, but causes a harmful intra-channel interference at the same time. The receiver's ability to combat the interference and retrieve individual users' signals (Multi-User Detection, MUD) is a measure of the system dependability. In this paper we re-visit the solution to the MUD problem, based on the use of Genetic Algorithm (GA). The novelty of the current contribution is a re-designed method to generate the initial GA population, which improves the performance at no extra computational cost in comparison with the previous proposal.

Keywords: Multi-User Detection · MIMO · Genetic algorithm · Optimization · Zero Forcing

1 Introduction

MU-MIMO configuration is an example of spatial transmission diversity, more and more popular in modern wireless systems, transmitting over uncorrelated Rayleigh-fading channels. The idea behind MU-MIMO is that the signals originating from different transmit stations are carried at the same frequency channel and interfere with each other at the receiver input. At first glance the interference disables detection of any signal, but the truth is that the receiver equipped with a higher number of receive antennas can succeed in resolving individual signals, thereby exploiting high capacity of the MIMO channel. The key to obtain accurate estimates of the transmitted signals is the use of appropriate MUD techniques [1, 2].

According to the optimal MUD Maximum-Likelihood (ML) routine [3], all possible signals must be considered per every transmit station, which makes it too complex for any real-world application in the case of high number of transmit stations and/or medium to high-order modulations (like 16-QAM, 64-QAM, 256-QAM). For example, assuming $N_T = 8$ transmit stations and 16-QAM modulation ($K = 4$ bits per

W. Zamojski et al. (Eds.): DepCoS-RELCOMEX 2020, AISC 1173, pp. 315–324, 2020.
https://doi.org/10.1007/978-3-030-48256-5_31

modulation period), the number of considered transmitted signals' variations would reach $(2^K)^{N_T} \approx 4.3 \cdot 10^9$.

To avoid the ML computational burden, there have been several techniques developed over the years. The most basic linear approach is called Zero Forcing (ZF). It brings the possibility to retrieve individual symbols by reversing the impact of the channel on the transmitted symbols. At the same time, the fading gain is balanced. But, ZF technique is prone to signal deterioration caused by the additive noise. Regarding ZF computational payload, it requires inverting the channel matrix of $N_T \cdot N_R$ size, where N_R represents the number of receiver antennas.

An attractive solution to the MUD problem is interference cancellation [4]. In this paper, a special attention will be paid for Successive Interference Cancellation (SIC). The SIC concept [5, 6] is an iterative procedure, where, at first, the most reliable signal is demodulated and decoded. Basing on the estimated data bits, the impact of that signal is whipped from the mixture of all signals arriving at the receiver antennas. After that, the second most reliable signal undergoes the same procedure, and so on. The criterion of signal reliability can be assessed from the channel gains of individual MIMO subchannels. Decoding and canceling out the signals of adequately strong interferers is a feasible remedy to reduce the level of the effective interference when attempting to detect the intended signal [4]. The main drawback of SIC is the need to demodulate, decode, and encode back all of the interfering signals.

In our previous contribution [7], we proposed to utilize a Genetic Algorithm for MUD purposes. It has appeared that for 16-QAM modulation the use of GA in the scenario of 4×4 MIMO system (4 transmit stations, each with 1 antenna, and one receiver, equipped with 4 antennas) is inferior to the simple ZF detector. However, GA offers a significant SNR gain when used together with ZF detector, not instead of it. In other words, it can improve initial decisions made by ZF detector.

In the current paper we develop another approach to the joined ZF-GA detector. The concept derives its strength from Successive Interference Cancellation principle. It is described in Sect. 3. (Subsect. 3.2 brings the details.) Before that, Sect. 2 is to specify the system model. The performance of the proposed solution is evaluated in Sect. 4. The conclusion is drawn in Sect. 5.

Regarding notation, superscripts $(\cdot)^{-1}$, $(\cdot)^T$, and $(\cdot)^H$ indicate the matrix inverse, transpose, and Hermitian transpose, respectively, while $\|\cdot\|^2$ is the Frobenius norm.

2 System Model

We consider an uplink of the MU-MIMO system, which consists of N_U independent transmit stations transmitting their signals synchronously over the same frequency channel. Each of these stations is equipped with one antenna, so the total number of MIMO channel inputs $N_T = N_U$. The data symbols (signal elements), transmitted in a single modulation period from all the Tx stations, constitute a multi-symbol $\mathbf{x} = [x_i]_{N_T,1}, x_i \in \chi$, where χ is the signal constellation set (e.g., 16-QAM). The multi-symbol is conveyed through an uncorrelated MIMO Rayleigh flat-fading channel, represented by the channel matrix $\mathbf{H} = [h_{ij}]_{N_R,N_T}$, as shown in Fig. 1 for the case of

$N_T = 4$. The elements h_{ij} of the channel matrix are i.i.d. complex Gaussian random variables with zero mean and standard deviation of 1.

At a given modulation period, the signals received through all the receive antennas are represented as $\mathbf{y} = \mathbf{Hx} + \mathbf{n}$, where $\mathbf{n}$ is a zero-mean noise vector with complex Gaussian distribution (standard deviation is the noise power spectral density). The length of $\mathbf{y}$ equals to the number of receive antennas N_R. Throughout the paper it is assumed that $N_R = N_T$.

The role of the multi-user detector is to recover a multi-symbol estimate $\dot{\mathbf{x}}$. If the MUD succeeds, $\dot{\mathbf{x}}$ equals to the actually transmitted multi-symbol $\mathbf{x}$.

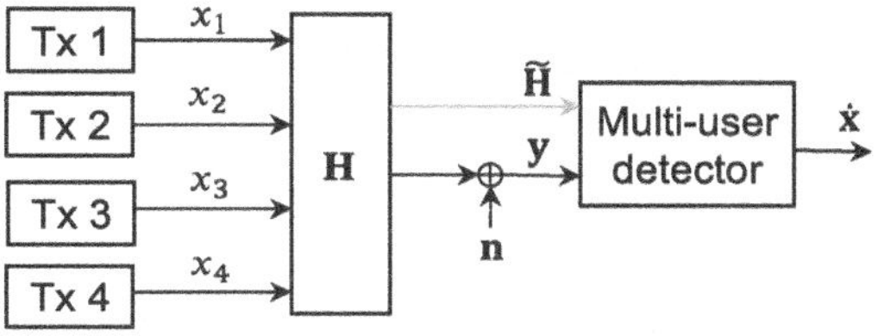

Fig. 1. Multi-user MIMO system model

3 Genetic-Algorithm-Driven Multi-user MIMO Detection

3.1 Essentials

GA is an evolutionary optimization algorithm, which mimics the process of natural selection. The current set of candidate solutions (called individuals) is named the population. Each individual is described by its chromosome and fitness value. The former is (usually binary) representation of given candidate solution, and the latter is its quality measure in terms of the optimization criterion.

According to the considered problem, chromosome $\tilde{\mathbf{b}}$ is defined as a concatenation of N_T labels, denoted by $\tilde{\mathbf{b}}_i$, each having K bits. The labels are assigned to candidate data symbols $\tilde{x}_i$ originating from different Tx stations. Throughout the paper it is assumed that the labels are mapped onto the symbols according to the standard Gray labelling map β. Figure 2 displays an exemplary chromosome in the case of $N_T = 4$ transmit stations (4 sections within the chromosome), and 16-QAM signaling (it results with 4 bits per section, as $K = 4$ for 16-QAM modulation).

The fitness value is the distance of a candidate multi-symbol $\tilde{\mathbf{x}} = [\tilde{x}_1 \ldots \tilde{x}_{N_T}]^T : \tilde{x}_i \in \chi, \forall_i$, and the received symbol $\mathbf{y}$, taking into account the estimated channel state $\tilde{\mathbf{H}}$:

$$f(\tilde{\mathbf{x}}) = \left\| \mathbf{y} - \tilde{\mathbf{H}}\tilde{\mathbf{x}} \right\|^2. \quad (1)$$

Each iterative step in the GA causes the change in the population structure. The new individuals are born from the best fitted parents. Then the weakest population individuals (either parents or children) die to keep the population size constant. To create the offspring, GA applies three independent operators (selection, crossover, and mutation) [8]. The basic GA loop is displayed in Fig. 3. Throughout consecutive iterations, the GA is expected to quickly converge to the global optimum [9]. For more details concerning GAs principles, the reader is referred to [7–10].

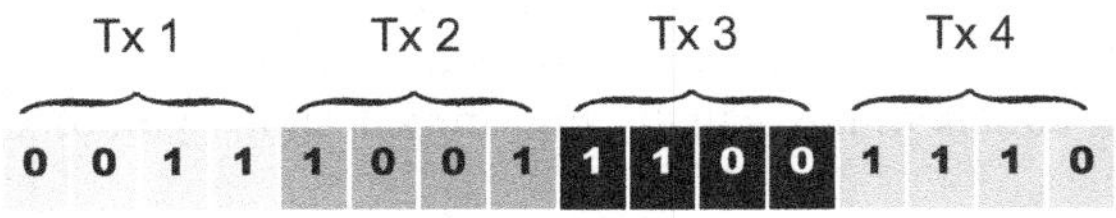

Fig. 2. Chromosome representing a candidate multi-symbol for which $\tilde{\mathbf{b}}_1 = (0011)$, $\tilde{\mathbf{b}}_2 = (1001)$, $\tilde{\mathbf{b}}_3 = (1100)$, $\tilde{\mathbf{b}}_4 = (1110)$

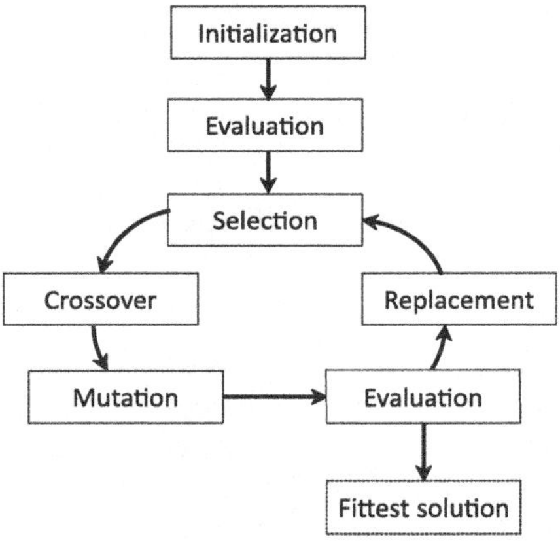

Fig. 3. Basic GA loop

3.2 GA-MUD Initialization

GA initialization has a major impact on its convergence. In vast majority of GA applications, the initial population is constituted randomly. Having many different individuals, the optimization process starts at different localizations in the search space, concurrently, thereby minimizing the risk of stucking at a local optimum [11, 12]. However, from [7] it is clear that assuming random initialization to the GA-MUD problem might result in the lack of any significant convergence of the algorithm, regardless of other GA settings. Thus, in [7] it was proposed to inject one individual representing the ZF solution to the initial population. The fitness of this "superior" individual is massively higher than for any other, so it is very likely to be selected as the parent for many crossover operations, thereby focusing the algorithm onto a promising search region.

3.3 The Novel SIC-Inspired GA-MUD

Taking into account good results of injecting the "superior" ZF individual onto the initial population, reported in [7], in the current contribution we propose another strategy to boost the convergence of the GA-driven MU-MIMO detection algorithm. It is based on the SIC paradigm, summarized in the introductory part.

According to our new proposal, Zero-Forcing detector is still in use to give an initial direction for a good search region. The ZF-based multi-symbol candidate $\dot{\mathbf{x}}_{ZF} = \left[\dot{x}_{ZF,1} \ldots \dot{x}_{ZF,N_T}\right]^T$, used for GA initialization, is obtained as a linear combination of the signals received by different antennas, transformed by the ZF weight matrix $\mathbf{W}_{ZF}$ as follows:

$$\dot{\mathbf{x}}_{ZF} = \mathbf{W}_{ZF}\mathbf{y}, \tag{2}$$

where

$$\mathbf{W}_{ZF} = \left(\widetilde{\mathbf{H}}^H\widetilde{\mathbf{H}}\right)^{-1}\widetilde{\mathbf{H}}^H, \tag{3}$$

and $\mathbf{W}_{ZF}\widetilde{\mathbf{H}}$ gives the identity matrix.

The use of the ZF outcome is quite different than in [7]. Instead of creating a single "superior" individual in the initial population, we judge reliability of the ZF decisions, $\dot{x}_{ZF,1}, \ldots, \dot{x}_{ZF,N_T}$ per every transmit station. The assessment criterion for the ith transmit station is the sum of gains of the subchannels, beginning there:

$$P_i = \sum\nolimits_{j=1}^{N_R} \left|h_{ji}\right|^2; \quad i \in [1, N_T] \tag{4}$$

Without any loss of generality, assume that $P_{i'}$ is the maximum. In such case, the transmit station i' is selected and the ZF decision $\dot{x}_{ZF,i'}$ is demapped onto the binary vector $\dot{\mathbf{b}}_{ZF,i'} = \beta^{-1}\left(\dot{x}_{ZF,i'}\right)$. The last is passed to the GA machine, as shown in Fig. 4, where it accommodates appropriate (the i'th) section of all individuals' chromosomes in the initial population. Remaining bits for all the individuals are generated randomly.

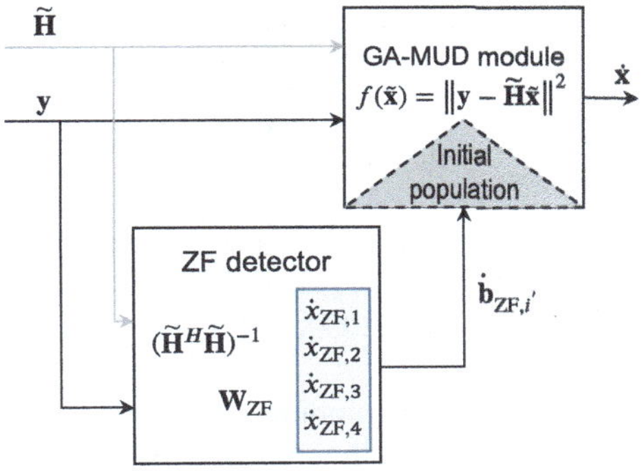

Fig. 4. SIC-inspired GA-MUD block diagram

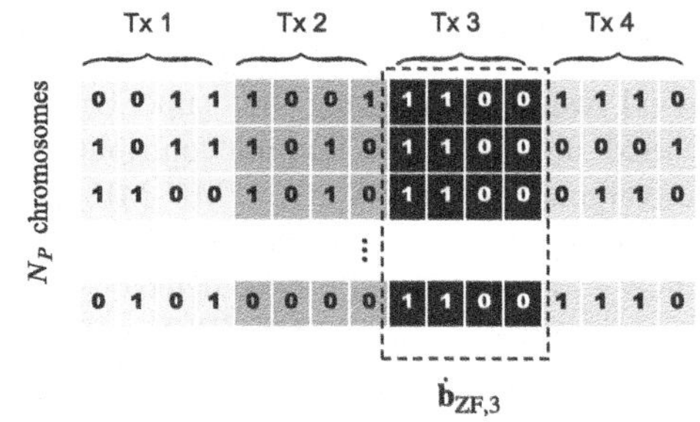

Fig. 5. Initial population in the case of 16-QAM modulation, $i' = 3$, and $\dot{\mathbf{b}}_{ZF,i'} = (1100)$

An exemplary initial population consisting of N_P individuals, given $i' = 3$ is shown in Fig. 5.

4 Simulation Experiment

4.1 Assumptions

To evaluate the performance of the proposed SIC-inspired GA MUD, the simulation experiment is conducted. There are $N_T = 4$ transmit stations, transmitting their 16-QAM signals over the uncorrelated 4×4 MIMO Rayleigh fading channel. It is assumed that the channel state is ideally estimated, so $\widetilde{\mathbf{H}} = \mathbf{H}$. At the receiver, the ZF detector is used at the first stage, and its decision related to the most reliable transmitted symbol is passed to the GA in the way described in Sect. 3.3. GA steps and settings are concisely specified below.

Fitness Evaluation takes into account the fitness measure (1). The lowest value of the objective function corresponds with the fittest chromosome in the generation. Starting from the second generation of the GA algorithm, the worst individuals in the current population are discarded (replacement step) to keep constant population size over the evolutionary processes. The survivors can become the parents for the next generation.

Selection: In the current work, the roulette wheel selection rule is utilized. With this method, all individuals have a chance to be selected with a probability proportional to their fitness [7, 14]. The selection probability of the jth individual is given by:

$$p_j = f_j / \sum_{k=1}^{N_P} f_k, \tag{5}$$

where N_P is the population size. A small number (elite count, ϵ) of the fittest individuals are guaranteed to be alive among the next generation of the individuals. Elitism ensures that the individual quality gained by the GA will not decrease from one generation to the next.

Crossover refers to the process of combining two parents to produce the offspring. The children's chromosomes consist of the fractions of both parents' chromosomes. In this work, the single-point crossover method is applied, *i.e.*, the parents' chromosomes are cut at a random position to create two parts, one of which is exchanged with respective other parent's part. Crossover fraction, p_c, defines the part of the population, other than the elite individuals, that are crossover children.

In order to prevent the GA from converging to a local optimum, some random changes are applied to the genes of the individuals [13]. This operation, so-called **mutation**, consists in replacing 0 by 1 or vice versa on a randomly selected bit position of the chromosome [9]. The mutation rate, p_m, determines the probability that a given individual undergoes mutation. If so, the bit position to be altered is selected according to uniform distribution. The elite individuals avoid mutation of their genes.

Stopping criterion: Every GA needs a mechanism to brake the iterative process judging by some features of the current population. In the current work, the GA stops if there has been no improvement in the best fitness value for a specific number of generations (N_S), called stall generations. If the stopping criterion is met, the individual with the best fitness ever is returned as the final solution.

Settings of the above-mentioned parameters used for the simulation experiment are listed in Table 1 for better clarity. They have been carefully chosen after several preliminary runs.

Table 1. GA parameters

Parameter	Symbol	Value
Population size	N_P	2000
Elite count	ϵ	2
Crossover probability	p_c	0.8
Mutation probability	p_m	0.1
Numb. of stall generations	N_S	20
Selection function		Roulette wheel
Crossover function		Single-point

The choice of the population size $N_P = 2000$ and the number of stall generations $N_S = 20$ is a compromise between the accomplished performance and the computational complexity.

4.2 Simulation Results

The proposed system is evaluated in terms of Bit Error Rate (BER) vs. Signal-to-Noise Ratio (SNR) value performance. It is compared with the following solutions:

- regular ZF detector, which makes final decisions on all users' data,
- simple GA, where all individuals in the initial population are generated randomly,
- ZF-aided GA from [7], wherein one of the initial population's individuals is the outcome of ZF detector.

The results are presented in Fig. 6. From the plot it is clear that the basic genetic algorithm with the initial population generated randomly (the line with diamonds) is inferior to the simple ZF (circle marks) at higher SNRs, and cannot cross an error floor at *ca.* $3 \cdot 10^{-3}$ BER. The solution proposed in [7] (represented by the line with stars) brings a significant improvement: the gain of about 13.5 dB at the level of 10^{-3} can be observed, but the curve is going to merge or cross the one for ZF near SNR of 35 dB. Finally, the new approach, contributed in the current paper offers another 8 dB gain at the level of 10^{-3} BER (the line with squares). What can be also read from the plot, the novel approach is the first to cross the 10^{-4} level.

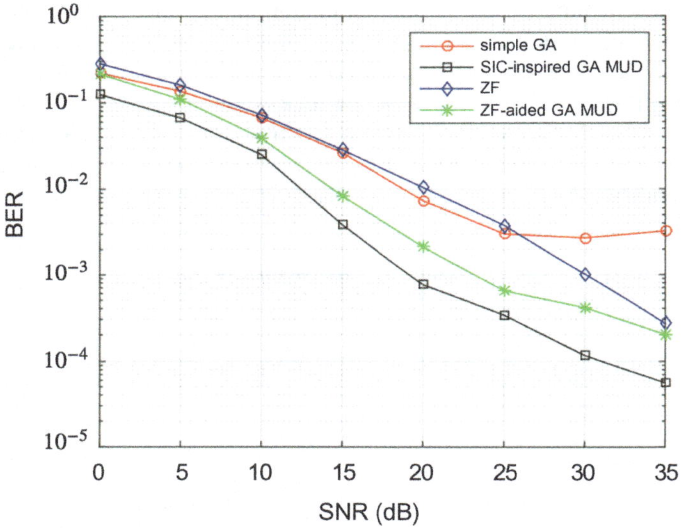

Fig. 6. BER vs SNR for the compared multi-user detectors

4.3 Computational Complexity

The computational complexity of GA increases linearly with N_P, N_G, N_T, and exponentially with K (N_G is the number of populations actually considered in a given algorithm run - it is subject to change, depending on the convergence of the optimization process). To put that in perspective, the computational complexity of ML optimal detector, which increases exponentially with both the number of users (N_T) and the number of bits transmitted per one station per one modulation period (K) [3, 12].

Obviously, the cost of GA procedure is increased by ZF routine from the initialization step. When compared to the solution from [7], the computational payload of the novel SIC-inspired approach is slightly reduced due to the fact that all individuals in the initial population share the same fraction of genes ($\dot{\mathbf{b}}_{ZF,i'}$). As a consequence, the number of products to be computed for the initial population is $(N_T - 1)/N_T$ times the original number.

5 Conclusion

In this work, the performance of GA-driven MIMO multi-user detector has been improved. The new method for population initialization has been proposed. It resembles Successive Interference Cancellation approach, where the most believable signal is detected and decoded first. On the ground of the GA, we use the ZF detector, and the ZF-based decision related to the most reliable signal is reflected in the chromosome of all individuals in the initial population. Respective part of the chromosome, $\dot{\mathbf{b}}_{ZF,\,i'}$, is unlikely susceptible to crossover due to the fact that all of the individuals in the initial population share exactly the same genes on the ZF-decided positions. Obviously, it might appear that the ZF decision related to the most reliable signal is wrong. In such cases, the algorithm can still converge thanks to mutation. Nevertheless, in the light of presented results, it is worth favoring the initial ZF decision to a higher extend than proposed in [7].

The proposed solution needs further investigation. In particular, a higher number of transmit stations, or higher-order modulations should be taken into account.

Acknowledgement. The presented work has been funded by the Polish Ministry of Science and Higher Education under the research grant No. 0312/SBAD/8147.

References

1. Ganti, R., Baccelli, F., Andrews, J.: Series expansion for interference in wireless networks. IEEE Trans. Inform. Theory **58**(4), 2194–2205 (2012)
2. Blomer, J., Jindal, N.: Transmission capacity of wireless ad hoc networks: Successive interference cancellation vs. joint detection. In: IEEE International Conference on Communication, ICC, Dresden. IEEE (2009)
3. Jaldén, J.: Maximum likelihood detection for the linear MIMO channel. Ph.D. dissertation, Royal Institute of Technology, Stockholm (2004)
4. Zhang, X., Haenggi, M.: The performance of successive interference cancellation in random wireless networks. IEEE Trans. Inf. Theory **60**(10), 6368–6388 (2014)
5. He, J., Tang, Z., Ding, Z., Wu, D.: Successive interference cancellation and fractional frequency reuse for LTE uplink communications. IEEE Trans. Veh. Technol. **67**(11), 10528–10542 (2018)
6. Mo, Y., Goursaud, C., Gorce, J.M.: On the benefits of successive interference cancellation for ultra narrow band networks: theory and application to IoT. In: IEEE International Conference on Communication, ICC, Paris. IEEE (2017)
7. Khafaji, M.J., Krasicki, M.: Genetic-algorithm-driven MIMO multi-user detector for wireless communications. In: Zamojski, W. et al. (eds.) Contemporary Complex Systems and Their Dependability: Proceedings of the 13th International Conference on Dependability and Complex Systems DepCoS-RELCOMEX, vol. 761, pp. 258–269. Springer, Cham (2019)
8. Ng, S., Leung, S., Chung, C., Luk, A., Lau, W.: The genetic search approach: a new learning algorithm for adaptive IIR filtering. IEEE Signal Process. Mag. **13**(6), 38–46 (1996)
9. Mitchell, M.: An Introduction to Genetic Algorithms. The MIT Press, Cambridge (1996)

10. Liu, S., St-Hilaire, M.: A genetic algorithm for the global planning problem of UMTS networks. In: IEEE Global Telecommunication Conference, GLOBECOM, Miami. IEEE (2010)
11. Obaidullah, K., Siriteanu, C., Yoshizawa, S., Miyanaga, Y.: Evaluation of genetic algorithm-based detection for correlated MIMO fading channels. In: 11th International Symposium on Communications and Information Technologies, ISCIT 2011, Hongzou, pp. 507–511. IEEE (2011)
12. Yang, C., Han, J., Li, Y., Xu, X.: Self-adaptive Genetic algorithm based MU-MIMO scheduling scheme. In: International Conference on Communication Technology, ICCT 2013, Guilin, pp. 180–185. IEEE (2013)
13. Simon, D.: Evolutionary Optimization Algorithms. Wiley, Hoboken (2013)
14. Lipowski, A., Lipowska, D.: Roulette-wheel selection via stochastic acceptance. Physica A **391**(6), 2193–2196 (2012)

The Availability Models of Two-Zone Physical Security System Considering Cyber Attacks

Vyacheslav Kharchenko[1,2], Yuriy Ponochovnyi[3(✉)], Al-Khafaji Ahmed Waleed[1], Artem Boyarchuk[1], and Ievgen Brezhniev[1,2]

[1] National Aerospace University KhAI, Kharkiv, Ukraine
{V.Kharchenko,a.boyarchuk,e.brezhnev}@csn.khai.edu, eng_ahmed.waleed@yahoo.com
[2] Research and Production Company Radiy, Kropyvnytskyi, Ukraine
[3] Poltava State Agrarian Academy, Poltava, Ukraine
yuriy.ponch@gmail.com

Abstract. Relevance of the paper is confirmed by the need to protect the security systems themselves, not only from physical damage, but also from cyber attacks by intruders. The paper explores the Markov model of the two-zone cyber-physical security system. Evaluation of the functioning of the multi-zone system was carried out taking into account two degrees of degradation (operative condition - the failure state of all zones). The state space of the model (or one fragment) has a dimension of 9 states. In the proposed model, hardware failures caused by vandal attacks on objects of the first zone and software failures due to cyber attacks on the functions of the second zone are considered. The simulation results illustrate different transition intervals of availability indicators of various levels of degradation to a stationary state. For different degrees of degradation, the minimum value of the availability function, the time interval of the transition of the availability function to the stationary mode, and the value of the availability function in the stationary mode are determined. When eliminating software defects and vulnerabilities, the increase in the availability function is 0.23% for a zero level of system degradation.

Keywords: Cyberphysical security system · Availability indicators · Markov model · Degradation levels · Multi-Zone architecture

1 Introduction

Modern physical protection systems have powerful cybernetic components that require full or temporary connection to an open Internet network for full functioning. If before there was a dilemma for ensuring the protection of physical security systems themselves only against physical damage, now such systems are an object for cyber attacks by cybercriminals.

The security system is represented by a set of subsystems, each of which is considered as a separate "zone". Each subsystem contains constituent elements. In aggregate, such a hierarchy is presented in Fig. 1 [1]. Each subsystem is represented by

W. Zamojski et al. (Eds.): DepCoS-RELCOMEX 2020, AISC 1173, pp. 325–333, 2020.
https://doi.org/10.1007/978-3-030-48256-5_32

the failure state spaces of hardware components (HW, hardware) and software/ functions (SW, software), which arise due to the manifestation of physical defects (pf), design defects (df), operator errors (hf), and interaction defects (if).

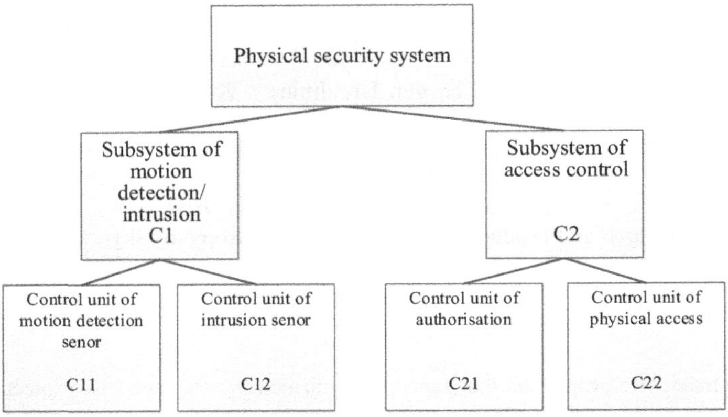

Fig. 1. Two-zone architecture of physical security system

The zonal architecture of physical security systems (PSS, physical security systems), their multifunctionality and functioning in an aggressive external environment require appropriate adequate representation in the construction and analysis of models. The use of the mathematical apparatus of Markov modeling [2, 3], on the one hand, provides a direct assessment of the resulting availability indicator, meets the requirements of standards and normative documents [4, 5]. But, on the other hand, Markov models are limited by assumptions on the simplest flows of events [6], and are also prone to the problem of increasing dimensionality when taking into account a large number of external factors. In [7], a structurally automatic approach to modeling large-dimensional systems is considered. The use of multi-fragment modeling apparatus [8] allows us to study systems with variable parameters, but does not solve the dimensional problem.

In [9], to study systems with variable parameters, it was also proposed to codify Continuous-Time Markov Chains. In [10, 11], Markov and multi-fragment models of hardware and software systems for various purposes for specific architectures are considered. In [12], the Semi-Markov reliability model with variable parameters of event flows was considered.

However, the well-known papers did not consider the influence of zonal architecture on the availability of the system from the standpoint of both reliability and security.

The aim of this study is to develop and analyze the classic Markov model of two-zone PSS availability. The development of the model is based on the determination of many states and mechanisms of interaction taking into account the degree of degradation. Assessment of the availability functions of various degrees of degradation was performed for various sets of input data.

The paper is structured as followed. Next Sect. 2 describes two PSS availability Markov models, their assumptions, states and transitions between them (Subsects. 2.1 and 2.2). The results of the PSS modeling and availability assessment are analyzed in Sect. 3. Section 4 concludes and describes future steps.

2 Development and Research of the Availability Model of Physical Security Systems

2.1 Development of the Initial Model

The availability model of a two-zone cyberphysical security system allows us to study the simultaneous effect of failures of the hardware component of zones and their functions implemented through software. The paper considers a two-zone PSS model (Fig. 1), in which the first zone has an external perimeter and is susceptible to vandal attacks, and the second zone implements access control functions via a remote connection (therefore, it is susceptible to cyber attacks).

The main assumptions of the M_{PSS0} model are:

- the flow of events that transfers the system from one functional state to another has the properties of stationarity, ordinaryness and the absence of aftereffect, the input parameters of the model are assumed to be constant;
- the probability of failure of the cloud service is negligible;
- acts of vandalism (γ_{HW}) are committed on the objects of the first zone, which are located outside the protected perimeter;
- vulnerability attacks (γ_{SW}) are carried out on the functions of the second zone, which are accessible through a public network.

The state space of the model has a dimension of 9 states (Fig. 2), according to combinations of hardware and functional component failures in each of the zones. Also, in Fig. 2, three levels of degradation of the system (0, I and II) are highlighted.

State	Combination of hw/sw failures				Degradation level
S1	1h	2h	1s	2s	0
S2	1h	2h	1s	2s	I
S3	1h	2h	1s	2s	
S4	1h	2h	1s	2s	
S5	1h	2h	1s	2s	
S6	1h	2h	1s	2s	II
S7	1h	2h	1s	2s	
S8	1h	2h	1s	2s	
S9	1h	2h	1s	2s	

Fig. 2. Combinations of zone failures that determine the states of the M_{PSS0} model and degradation levels

In this work, independent failures of hardware and software of two CPSS zones are considered. They are not physically and logically linked and, from the position of Markov modeling, cannot happen simultaneously in an ordinary stream of events.

Figure 3 shows the marked graph of the model, which has end-to-end numbering of states and is developed using the modified grPlot_marker function [13].

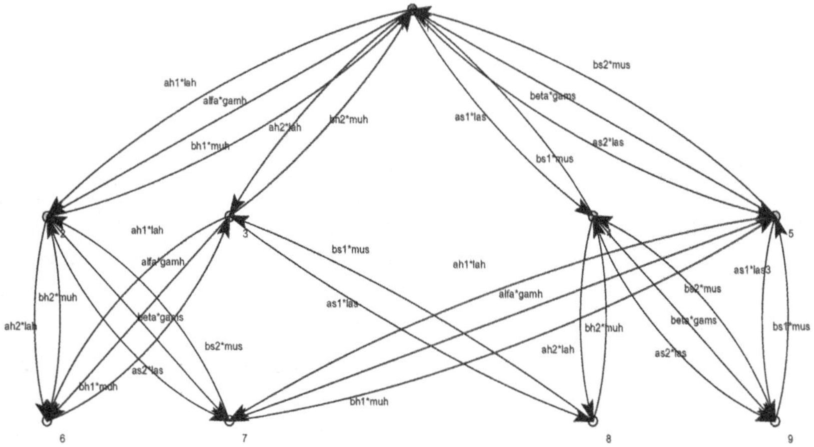

Fig. 3. Marked oriented graph of the two-zone M_{PSSO} model

When constructing the model graph (Fig. 3), a vertical hierarchy of states was used to display the levels of degradation. The upper level is S1 state. It indicates an operative state without failures. The second stage is S2, S3, S4, S5 states. They indicate the state of the first level of system degradation, in which either hardware or functional (software) failure occurred in one of the zones. At the lower level (states S5, S6, S7, S8), the states of complete failure of all zones of the system are indicated. When marking the graph and compiling a system of differential equations, we used the weight coefficients ah1, ah2, as1, as2 to distinguish the failure rates of different zones; and to distinguish the recovery rates, the coefficients bh1, bh2, bs1, bs2 were used.

Hardware failures caused by vandal attacks on objects of the first zone are modeled by transitions - arrows weighted by the index $\alpha^*\gamma_{HW}$. Software failures caused by cyberattacks on the functions of the second zone are modeled by transitions—arrows, weighted by $\beta^*\gamma_{SW}$.

Availability functions for different levels of degradation are defined as:

$$A^{(0)}(t) = P_1(t);\ \ A^{(I)}(t) = \sum_{i=1}^{5} P_i(t) \tag{1}$$

Baseline conditions: t = 0, $P_0(t) = 1$.

2.2 Development of a Multi-fragment Model for Updating Software Functions

In the previous M_{PSSo} model, the assumption was made that the parameters of the functional components of individual zones of the cyberphysical security system are constant. However, modern systems can receive updates and patches of the software component as part of the development and modification cycles. After installing the update or patch, the program code and/or configuration files change, which directly affects the value of the input parameters of the failure flows and cyberattacks. In [8, 11], such changes are modeled using the multi-fragment approach, which is the basis of the M_{PSSm} model.

The assumptions of this M_{PSSm} model are expanded (in comparison with the assumptions of M_{PSSo}):

- the flow of events that transfers the system from one functional state to another one within the same fragment has the properties of stationarity, ordinaryness and the absence of aftereffect, the model parameters within one fragment are assumed to be constant;
- during the upgrade process, the elimination of software defects and vulnerabilities occurs, new defects and vulnerabilities are not introduced.

The state space of the M_{PSSm} model within one fragment, like M_{PSSo}, has a dimension of 9 states (Fig. 2) and three levels of system degradation.

Figure 4 shows a marked graph of three fragments of the M_{PSSm} model. Each fragment of the model is a mapping of the M_{PSSo} graph (Fig. 3), but for compactness, a vertical arrangement of states was performed.

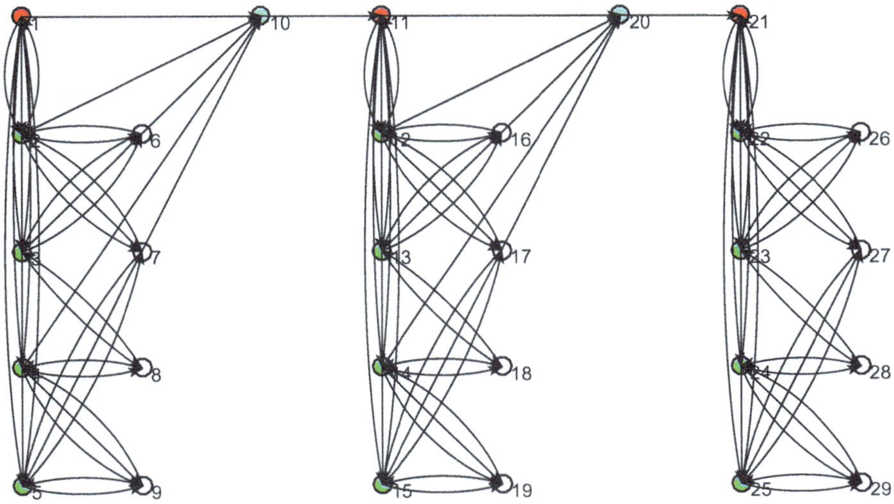

Fig. 4. Marked graph of the three-fragment M_{PSSm} model

When constructing the graph of the model (Fig. 4), the color marking of the states ("Red", "Green", "White") was used to indicate that the states belong to different levels of degradation. Additionally, a "blue" marker was used to highlight SW update states (S10, S20), the system is inoperative in these states.

Availability functions for different levels of degradation are defined as:

$$A^{(0)}(t) = \sum_{i=1}^{Nf} P_{10i-9}(t); \; A^{(1)}(t) = \sum_{i=1}^{Nf} \sum_{j=1}^{5} P_{(10i-10)+j}(t) \quad (2)$$

3 Simulation and Comparative Analysis

The primary input parameters of Markov models were determined on the basis of certification data [1, 5] for the previous CPSS versions samples. Their values are presented in Table 1. To build the matrix of Kolmogorov-Chapman system of differential equations (SDE) in Matlab, matrix A function was used [14].

The Kolmogorov-Chapman SDE can be solved using analytical methods (substitutions, Laplace transforms, etc.). But this approach is applicable for systems of small dimension. In this paper, we consider a single-fragment model of medium dimension (9 states) and a multi-fragment model of large dimension (29 states), therefore, an approach that is universal for both models to the numerical solution of SDEs was chosen.

Table 1. Values of simulation processing input parameters

#	Sym	Parameter	Value
1	λhw	HW failure rate due to unintentional physical and design defects (pf and df)	1e-3 (1/hour)
2	λsw	SW failure rate due to design defects of an unintentional nature (df)	5e-3 (1/hour)
3	γhw	HW failure rate due to intentional actions (if, vandalism)	1e-3 (1/hour)
4	α	The coefficient of "aggression" of physical attackers, depends on external factors	1..100
5	γsw	SW failure rate due to intentional actions (if, viruses, cyberattacks)	5e-3 (1/hour)
6	β	The coefficient of "aggression" of cyber attackers, depends on external factors	1..10
7	μhw	HW recovery rate after failure, averaging is performed in the research and recovery is considered for all causes of failures (pf, df, hf, if)	1 (1/hour)
8	μsw	SW recovery rate after failure, averaging is performed in the research and recovery is considered for all causes of failures (pf, df, hf, if), the recovery does not provide for the elimination of the causes of failure	2 (1/hour)
9	ah1, ah2, as1, as2	The weight coefficients to distinguish the HW failure rates (ah1, ah2) and SW failure rates (as1,as2) of different zones	[1 2 1.5 1.2]
10	bh1, bh2, bs1, bs2	The weight coefficients to distinguish the HW recovery rates (bh1, bh2) and SW recovery rates (bs1,bs2) of different zones	[1 2 1.3 1.25]

In [15], for the Kolmogorov-Chapman SDE solution, it was proposed to use a software implementation of the Runge-Kutta method. In this papers the SDE solution is obtained using the ode15s function [16]. The simulation results are shown in Fig. 5. The graphs of the M_{PSSO} model (Fig. 5, a) illustrate the typical nature of the change in the availability function with a decrease to a stationary coefficient during the first 10 h of operation. Thus, in further analysis of the results, it is necessary to take into account the values of two levels of availability degradation:

- $A^{(0)}M_{PSSO} = 0,9638;$
- $A^{(I)}M_{PSSO} = 0,9997.$

The graphs of the M_{PSSm} model (Fig. 5, b) illustrate the typical nature of the change in the availability function for multi-fragment models [8, 11, 14]. In the initial period of operation, the availability of the system is reduced to a minimum, and then, as the elimination of SW defects and vulnerabilities, strive for a stationary value.

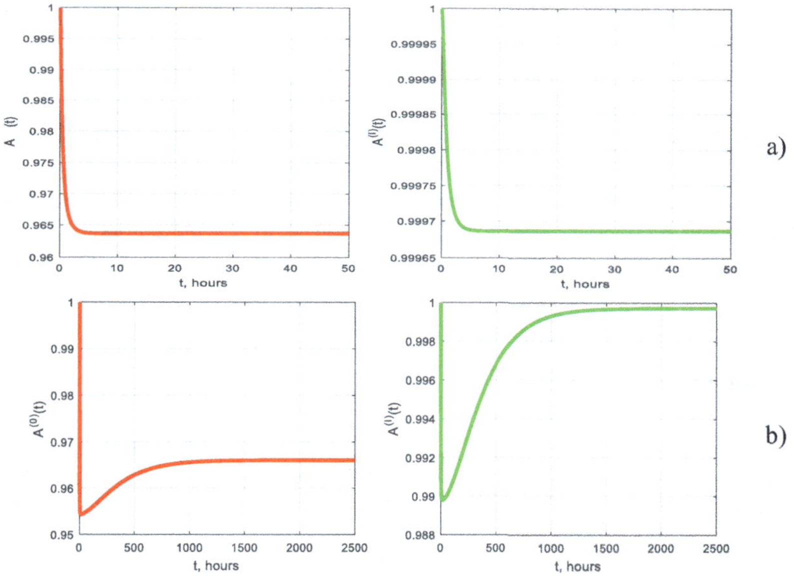

Fig. 5. Results of availability simulations of two-zone CPSS for different levels of degradation: a) model M_{PSSO}, b) model M_{PSSm}

In further analysis of the results, it is necessary to take into account the following groups of values of the resulting indicators for two levels of availability degradation:

a) for zero degradation level $A^{(0)}_{Mpssm}$

- availability function minimum value $A^{(0)}_{Mpssm}\text{min} = 0.9544;$
- availability function value in stationary mode $A^{(0)}_{Mpssm}\text{const} = 0.9661;$

- time interval for the transition of the availability function to the stationary mode $T^{(0)}_{Mpssm} = 3383{,}4$ h.

b) for the first degradation level $A^{(I)}_{Mpssm}$

- availability function minimum value $A^{(I)}_{Mpssm}min = 0.9898$;
- availability function value in stationary mode $A^{(I)}_{Mpssm}const = 0.9997$;
- time interval for the transition of the availability function to the stationary mode $T^{(I)}_{Mpssm}const = 3328{,}3$ h.

4 Conclusion

The article describes two models for assessing the availability of a two-zone physical security system, taking into account vandal and cyber attacks on objects of different zones.

In the M_{PSSO} model, the availability functions of different levels of degradation decrease to stationary values $A^{(0)} = 0{,}9638$ and $A^{(I)} = 0.9997$ during the first 10 h of operation.

In the M_{PSSm} model, the availability function decreases to stationary values $A^{(0)}const = 0.9661$ and $A^{(I)}const = 0.9997$ after 3300 h of operation. Thus, the increase in the availability function while eliminating software defects and vulnerabilities is 0.23% for a zero level of system degradation.

Further research should be directed to the studies of the impact of reducing the failure rate (and recovery) of HW and SW on the resulting indicators, as well as the development and research of both Markov and multi-fragment CPSS availability models, in which the assumption of high reliability of the cloud service is removed.

References

1. Waleed, A., Kharchenko, V., Uzun, D., Solovyov, O.: IoT-based physical security systems: structures and PSMECA analysis. In: 2017 9th IEEE International Conference on Intelligent Data Acquisition and Advanced Computing Systems: Technology and Applications (IDAACS), pp. 870–873. (2017). https://doi.org/10.1109/idaacs.2017.8095211
2. Zheng, Z., Trivedi, K., Wang, N., Qiu, K.: Markov regenerative models of webservers for their user-perceived availability and bottlenecks. IEEE Trans. Dependable Secure Comput. **17**, 92–105 (2020). https://doi.org/10.1109/TDSC.2017.2753803
3. Boano, C., Römer, K., Bloem, R., Witrisal, K., Baunach, M., Horn, M.: Dependability for the Internet of Things—from dependable networking in harsh environments to a holistic view on dependability. e & i Elektrotechnik und Informationstechnik. **133**, 304–309 (2016). https://doi.org/10.1007/s00502-016-0436-4
4. IEC 61508-1: 2010 Functional safety of electrical/electronic/programmable electronic safety-related systems - Part 1: General requirements. https://webstore.iec.ch/publication/5515. Accessed 21 Jan 2020
5. IEC 60050-192: 2015 International Electrotechnical Vocabulary (IEV) - Part 192: Dependability. https://webstore.iec.ch/publication/21886. Accessed 21 Jan 2020

6. IEC 61703: 2016 Mathematical expressions for reliability, availability, maintainability and maintenance support terms. https://webstore.iec.ch/publication/25646. Accessed 21 Jan 2020
7. Volochiy, S., Fedasyuk, D., Chopey, R.: Formalized development of the state transition graphs using the Erlang phase method. In: 2017 9th IEEE International Conference on Intelligent Data Acquisition and Advanced Computing Systems: Technology and Applications (IDAACS). (2017). https://doi.org/10.1109/idaacs.2017.8095255
8. Kharchenko, V., Butenko, V., Odarushchenko, O., Sklyar, V.: Multifragmentation markov modeling of a reactor trip system. J. Nuclear Eng. Radiation Sci. **1**, 031005 (2015). https://doi.org/10.1115/1.4029342
9. Trivedi, K., Bobbio, A., Muppala, J.: Non-homogeneous continuous-time markov chains. Reliabil. Avail. Eng. 489–508 (2017). https://doi.org/10.1017/9781316163047.018
10. Liu, B., Chang, X., Han, Z., Trivedi, K., Rodríguez, R.: Model-based sensitivity analysis of IaaS cloud availability. Fut. Gener. Comput. Syst. **83**, 1–13 (2018). https://doi.org/10.1016/j.future.2017.12.062
11. Kharchenko, V., Ponochovnyi, Y., Abdulmunem, A., Andrashov, A.: Availability models and maintenance strategies for smart building automation systems considering attacks on component vulnerabilities. Adv. Dependab. Eng. Complex Syst. 186–195 (2017). https://doi.org/10.1007/978-3-319-59415-6_18
12. Bobalo, Y., Horbatyi, I., Kiselychnyk, M., Medynsky, I., Melen, M.: Semi-Markov reliability model of functioning of wireless telecommunication system with complex control system. Math. Model. Comput. **6**, 192–210 (2019). https://doi.org/10.23939/mmc2019.02.192
13. Iglin, S. grTheory - Graph Theory Toolbox – File Exchange – MATLAB Central. https://www.mathworks.com/matlabcentral/fileexchange/4266-grtheory-graph-theory-toolbox. Accessed 21 Jan 2020
14. Kharchenko, V., Ponochovnyi, Y., Boyarchuk, A.: Availability assessment of information and control systems with online software update and verification. Inf. Commun. Technol. Educ., Res. Ind. Appl. 300–324 (2014). https://doi.org/10.1007/978-3-319-13206-8_15
15. Yakovyna, V., Seniv, M., Lytvyn, V., Symets, I.: Kolmogorov-chapman differential equation systems software module for reliable design automation. Sci. Bull. UNFU. **29**, 141–146 (2019). https://doi.org/10.15421/40290528
16. Solve stiff differential equations and DAEs – variableorder method – MATLAB ode15s. https://www.mathworks.com/help/matlab/ref/ode15s.html. Accessed 21 Jan 2020

Automatically Created Statistical Models Applied to Network Anomaly Detection

Michał Kierul[1], Tomasz Kierul[1], Tomasz Andrysiak[2(✉)], and Łukasz Saganowski[2]

[1] Research and Development Center, SOFTBLUE S.A., Jana Zamoyskiego 2B, 85-063 Bydgoszcz, Poland
{mkierul, tkierul}@softblue.pl

[2] Institute of Telecommunications and Computer Science, Faculty of Telecommunication, Information Technology and Electrical Engineering, UTP University of Science and Technology, Kaliskiego 7, 85-789 Bydgoszcz, Poland
{andrys, luksag}@utp.edu.pl

Abstract. In this article we present the use of automatically created exponential smoothing models for anomaly detection in networks. We propose the method of parameters estimation and selection by means of model's order obtained by Hyndman-Khandakar algorithm. Optimal values of the model parameters are chosen on the basis of information criteria reflecting a compromise between the consistency model and the size of its estimation error. In the proposed method, we use statistical relationships between the forecasted and real network traffic to determine whether the tested trace is normal or attacked. Efficiency of our method is examined with the use of large set of real network traffic test traces. The experimental results prove resilience and effectiveness of the suggested solutions.

Keywords: Anomaly detection · Exponential smoothing models · Network traffic prediction

1 Introduction

There has been a wide use of security and protection systems of network infrastructure based on previously detected and classified patterns of behavior/threats called signatures. Antivirus software, Intrusion Detection Systems (IDS)/Intrusion Prevention Systems (IPS), or protection against leakage of information are only part of a long and diverse list of this kind of techniques. They have one idea in common, namely, the basis of their operation – they are able to protect the systems and computer infrastructures against known, earlier identified threats described by means of defined patterns [1].

Nevertheless, currently, a more effective solution for protection against new, unknown attacks is a rather radical change in concept/manner of operation. Instead of searching for signatures of abuses in the network traffic, it is necessary to detect and identify abnormal behavior which is a deviation from normal, pattern characteristics of the network. The strength of such an approach is based not on the fact that there is an a

W. Zamojski et al. (Eds.): DepCoS-RELCOMEX 2020, AISC 1173, pp. 334–343, 2020.
https://doi.org/10.1007/978-3-030-48256-5_33

priori knowledge of abuses' signatures, but on what does not comply with given norms, or profiles of the analyzed network traffic [2].

Anomalies are, by definition, irregularities, variations from the accepted norm. Anomalies in the network traffic can mean failure of the device, error of the software, or an attack on the resources and network infrastructure systems. The essence of anomaly detection in computer networks is therefore detection of incorrect behavior or events, especially those which may be a source of potential abuse or attack [3, 4].

Anomaly detection methods have been widely surveyed and analyzed in field articles [3, 5]. Papers discussing the methods proposed techniques including neural networks, clustering techniques, machine learning and expert systems. At present, of all the available methods, the most often developed ones are those which describe the analyzed network traffic by means of statistical models [6].

In this article we suggest automatically created exponential smoothing models for the examined time series which describe the given network traffic. In order to detect anomalies in the network traffic we used differences between the real network traffic and the estimated ExponenTialSmoothing (ETS) model of this traffic. Our experimental results confirm the effectiveness of the suggested method.

This paper is organized as follows. After the introduction, Sect. 2 presents the statistical methods for network anomaly detection. In Sect. 3, there is described in detail the methodology of the statistical models for network traffic prediction. Experimental results and conclusion are presented thereafter.

2 Overview of Statistical Methods for Anomaly/Attacks Detection

Effective protection of information and communications technology (ICT) structures is a serious challenge to all companies providing information technology (IT) services. Next to such threats as viruses, malware, or intrusion into the systems, different kinds of attacks on resources and network services are currently becoming more and more common, e.g. Denial of Service (DoS) or Distributed Denial of Service (DDoS). In simple terms, there are two types of such attacks, i.e. attacks on layers three and four of ISO/OSI model (Open System Interconnection Reference Model) (activities strictly of network type) and attacks on application layer (most often connected with Web services). In practice, intruders skillfully connect different types of attacks correctly assuming that the more of destructive techniques the attack includes, the more effective it will be [7, 8].

DDoS used long-term statistics for attacks detection [9]. The techniques which derive from statistical methods and are used in IDS systems can be divided into two groups. The first group includes methods based on threshold analysis, which examines the frequency of events and surpassing of their limits in a given time unit. When a particular threshold is surpassed, information about an attack is sent. A serious drawback of the above methods is their vulnerability to errors/mistakes caused by violent temporary increase of the legal network traffic and difficulty to define the reference levels and thresholds over which the alarm is activated [1]. The second group includes methods which detect statistical anomalies on the basis of estimated defined parameter profiles of the network traffic. The estimated profiles reflect an average size of internet protocol (IP) packages, an average amount of newly established connections in a given time unit, the quantity proportion of individual network protocols' packages, etc. Other noticeable statistical correlations may result from part of the day (e.g. busier network traffic immediately after the beginning of working hours) or the day of the week. Another apparent element is statistics for individual network protocols (quantity proportion of SYN and FIN packages of transmission control protocol (TCP)). IDS systems which are based on these methods are able to learn a typical network profile. This process lasts from few days to few weeks. The decision whether there is something suspicious occurring in the network or not is based on comparison between the two above mentioned profiles [4, 10].

Nowadays, anomaly detection methods are based on statistical models describing the analyzed network traffic. There are two most often used models, i.e. autoregressive ARIMA and ARFIMA, which allow prospective estimation of the tested network. We can also find hybrid methods which employ elements of prefatory transformation followed by estimation of the transformed signal's statistical parameters. The mentioned prefatory transformation is generally performed in the form of wavelet decomposition [11–13]. In the present article, we propose the use of automatically created exponential smoothing models to perform analysis and forecasting of time series which describe the given network traffic.

3 The Statistical Models for Network Traffic Prediction

3.1 Introduction to Automatic Forecasting of Exponential Smoothing Models

Forecasting is still one of the main tasks of time series analysis. Constructing of the forecasts is usually a multi-stage process involving, among others, matching a pertinent model basing on historical data and evaluation of quality of this matching [14].

Algorithms allowing for automatic construction of forecasts should realize all the stages of time series analysis, i.e. the choice of optimal model for data, parameter estimation and forecast construction. In pursuit for optimal model it is important to apply adequate criteria, protecting against too good matching of the model with the learning data, which may lead to bad quality of forecasts for new periods [15].

One of possible solutions to a such stated problem of automatic forecasting are two models: ExponenTialSmoothing or ErrorTrendSeason (ETS), which belong to a family of adaptive models developed by Hyndman et al. [16]. The family utilizes generalized algorithms of exponential smoothing. An essential advantage of them is their simplicity, relatively quick adaptive matching algorithm and easiness to understand and interpret the results. Common denominator of these methods is (exponential) assigning of weights decreasing with distance in time unit to past observations in the process of defining new forecasts of future observations. This is due to the fact that the classical assumptions of quantitative prediction come down to the postulate of the relative invariability of the development mechanism of the studied phenomena and events.

In methods based on ETS exponential smoothing can be realized with the use of different models, respectively chosen to the processed data.

3.2 The ETS Models – Estimation of Network Traffic Feature

Analyzing the nature and variability of time series, it is easy to notice that they consist of four optional components: trend, seasonal variations, periodical variations and random errors [17]. Periodical variations usually have approximately constant period, while the time of a full cycle of periodical variations is usually variable. In exponential smoothing models, trend is composed of level value v and increment value q. These two components can be connected in four different manners, including attenuation parameter $\varphi \in [0, 1]$. Then, we obtain different types of trends:

$$\mathit{Lack}\ U_p = v, \tag{1.a}$$

$$\mathit{Additive}\ U_p = v + qp, \tag{1.b}$$

$$\mathit{Multiplicative}\ U_p = vq^p, \tag{1.c}$$

$$\mathit{Attenuated}\ U_p = vq^{\left(\varphi + \varphi^2 + \ldots + \varphi^p\right)}, \tag{1.d}$$

where U_p describes the trend's nature, and parameter p is forecast horizon.

Taking three possible variants of combination of seasonal component with trend i.e. lack of seasonality, additive and multiplicative variants, we obtain 12 models of exponential smoothing which can be presented by means of the following formulas:

$$a_t = \alpha F_t + (1 - \alpha)G_t, \tag{2.a}$$

$$b_t = \alpha H_t + (1 - \beta)b_t, \tag{2.b}$$

$$c_t = \gamma I_t + (1 - \gamma)c_{t-m}, \tag{2.c}$$

where a_t is the series level in time t, b_t shows the decrease in time t, c_t denotes the seasonal component of the series in time t, and m is the number of seasons in a given period; the values of F_t, G_t, H_t, and I_t vary according to which of the cells the method belongs to, and α, β, γ, $\varphi \in [0, 1]$ are constants defining the model's parameters [16].

We achieve the method with fixed level (constant over time) by setting the condition $\alpha = 0$, the method with fixed trend (drift) - by setting the condition $\beta = 0$, and the method with fixed seasonal pattern - by setting $\gamma = 0$. It is worth noticing that the additive trend methods are obtained by letting $\varphi = 1$ in the damped trend methods [18, 19].

3.3 The Selection on Model Order and Parameters Estimation

Papers [20, 21] analyze special cases of state space models taking into account a single source of errors, it is easy to see that they can be the basis for some methods of exponential smoothing. Taking under consideration possible character of the errors we can present the state space models for all 12 types of exponential smoothing as:

$$Y_t = s(x_{t-1}) + o(x_{t-1})\vartheta_t, \tag{3.a}$$

$$x_t = f(x_{t-1}) + g(x_{t-1})\vartheta_t, \tag{3.b}$$

where $x_t = [a_t, b_t, c_t, c_{t-1}, \ldots, c_{t-m+1}]^T$ is a state vector, $s(x), o(x), f(x), g(x)$ are continuous functions having continuous derivatives, $\{\vartheta_t\}$ is a Gaussian white noise process with mean zero and variance σ^2, and $\mu_t = s(x_{t-1})$ [21]. Error ϑ_t can be included in the model in the additive or multiplicative way. The model with additive errors has $o(x_{t-1}) = 1$, so that $Y_t = \mu_t + \vartheta_t$. The model with multiplicative errors has $o(x_{t-1}) = \mu_t$ so that $Y_t = \mu_t(1 + \vartheta_t)$. Therefore, the multiplicative model's relative error is $\vartheta_t = (Y_t - \mu_t)/\mu_t$. The models are not unique. Seemingly, any value of $o(x_{t-1})$ will lead to identical point forecasts for Y_t [16, 20].

Out of 12 models of exponential smoothing described with conditions (1–2), after taking into account additivity and multiplicativity of error ϑ_t we obtain 24 adaptive models of states. The choice of proper model of exponential smoothing in a particular prognostic task requires then selection of one of 24 model forms, as well as initialization of vector's x_0 components and parameter estimation $\Psi = [\alpha, \beta, \gamma, \varphi]^T$.

The values of x_0 and the parameter Ψ are obligatory for the forecasting process. It is not difficult to calculate the Likelihood of the Innovations State Space Model (LISSM) by Eq. (4), or to obtain the Maximum Likelihood Estimates (MLE) [16].

$$LISSM(\Psi; x_0) = n log\left(\sum_{t=1}^{n} \frac{\vartheta_t^2}{x_{t-1}}\right) + 2\sum_{t=1}^{n} log|o(x_{t-1})|, \tag{4}$$

where n is the number of observations.

The above can be easily computed by using the recursive equations in [22]. The parameters Ψ and the initial states x_0 can be obtained by minimizing *LISSM*.

The present model is selected on the basis of Akaike Information Criterion (AIC)

$$AIC = LISSM\left(\widehat{\Psi};\, \hat{x}_0\right) + 2k, \tag{5}$$

where k is the number of parameters in Ψ together with the number of free states in x_0, and $\widehat{\Psi}$ and $\hat{x}_0$ define the estimates of Ψ and x_0. From the applicable models, the selected one models minimizes the AIC [23].

The AIC also provides a method for selecting between the additive and multiplicative error models. The two models are characterized by identical point forecasts so that standard forecast accuracy measures, e.g. the mean squared error (MSE) or mean absolute percentage error (MAPE), are unable to choose from the error types [16].

3.4 The Automatic Forecasting Algorithm

On the grounds of the mentioned ideas we obtain an efficient and widely applicable algorithm for automatic forecasting. To conclude, the stages of the performed actions are as follows [15, 22]:

- optimizing the model's parameters by applying all proper models to each of the series (in order to smooth the parameters and the initial stage variables),
- choosing the most efficient model in view of the AIC,
- constructing point forecasts on the basis of the most efficient model (with optimized parameters) for as many stages forward as needed.

All of the above discussed models of exponential smoothing are created in such a way that prediction theory's assumptions are met along with progressive degradation processes (i.e. possible lack of stability of the variable's regularity in time). Great flexibility of these models and their adaptive capability in case of irregular variations of direction or speed of trend, or deformations and shifts of periodical variations make them a comfortable tool of short-term forecasting and prediction. More information concerning the presented algorithm is presented in Hyndman and Khandakar [16, 22].

4 Experimental Results

As a base for experiments we used SNORT IDS [24] with a preprocessor responsible for implementation of the proposed algorithm which utilizes ETS model. We extended standard set of traffic features captured by SNORT to 28 traffic features presented in Table 1.

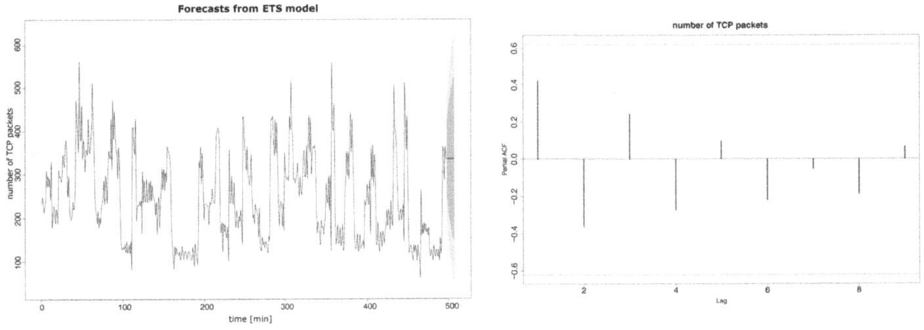

a) b)

Fig. 1. Forecasting interval (10 samples prediction interval) for F_1 traffic feature achieved for ETS model a), Partial Autocorrelation Function (PACF) from ETS model residuals for F_1 feature b).

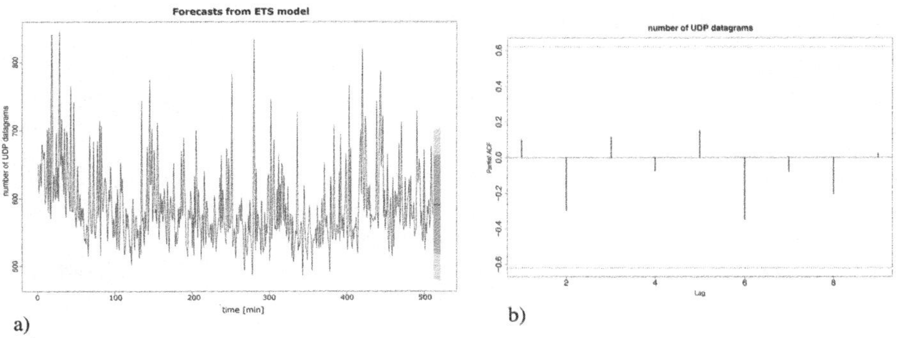

a) b)

Fig. 2. Forecasting interval (10 samples prediction interval) for F_5 traffic feature achieved for ETS model a), PACF from ETS model residuals for F_5 feature b).

Anomalies and attacks were generated based on Kali Linux [25] toolset to simulate abuses that belong, for example, to subsequent class: Application specific DDoS - AppDDos, DDoS, different types of port scanning and spoofing attacks, packet fragmentation, Syn Flooding etc. In order to test usability of the ETS statistical model for a one dimensional univariate time series representing traffic features form Table 1 we calculated Partial Autocorrelation Function from the model residuals. Exemplary PACF graphics representation of calculated for 10 samples prediction intervals are presented in Fig. 1b and 2b (for F_1 and F_5 traffic features respectively). PACF gives us a possibility to measure the model's usage for prediction purposes for a given class of signals. Values of autocorrelations should have low amplitudes constrained by dashed lines presented in Fig. 1b and 2b.

Table 1. Traffic features used for evaluating ETS based anomaly detection algorithm.

Feature	Traffic feature description	Feature	Traffic feature description
F_1	Number of TCP packets	F_{15}	In TCP packets (port 80)
F_2	In TCP packets	F_{16}	Out UDP datagrams (port 53)
F_3	Out TCP packets	F_{17}	In UDP datagrams (port 53)
F_4	Number of TCP packets in LAN	F_{18}	Out IP traffic [kB/s]
F_5	Number of UDP datagrams	F_{19}	In IP traffic [kB/s]
F_6	In UDP datagrams	F_{20}	Out TCP traffic (port 80) [kB/s]
F_7	Out UDP datagrams	F_{21}	In TCP traffic (port 80) [kB/s]
F_8	Number of UDP datagrams in LAN	F_{22}	Out UDP traffic [kB/s]
F_9	Number of ICMP packets	F_{23}	In UDP traffic [kB/s]
F_{10}	Out ICMP packets	F_{24}	Out UDP traffic (port 53) [kB/s]
F_{11}	In ICMP packets	F_{25}	In UDP traffic (port 53) [kB/s]
F_{12}	Number of ICMP packets in LAN	F_{26}	In TCP traffic (port 4444)
F_{13}	Number of TCP packets with SYN and ACK flags	F_{27}	Number of UDP flows per time interval
F_{14}	Out TCP packets (port 80)	F_{28}	Number of TCP flows per time interval

Prediction intervals (10 samples prediction) together with 80% and 95% intervals of prediction variability for F_1 and F_5 traffic features are presented in Fig. 1a and 2a respectively. We used 10 samples prediction interval period because longer prediction interval causes higher errors of prediction represented by, for example, Root Mean Square (RMSE) parameter. Higher RMSE values cause increase of False Positive (FP) indications and decrease of Detection Rate (DR). During online work of the proposed detection method every extracted traffic feature from Table 1 is checked if lie within prediction intervals is achieved for a given traffic feature. If extracted traffic value is outside the statistical model prediction interval variability calculated by ETS, we qualify such an incident as an anomaly.

Table 2. Results achieved for ETS based anomaly/attack detection algorithm.

Feature	DR [%]	FP [%]	Feature	DR [%]	FP [%]
F_1	6.12	4.11	F_{15}	14.14	2.16
F_2	13.15	4.05	F_{16}	0.00	0.00
F_3	13.13	3.94	F_{17}	9.44	0.00
F_4	13.16	4.25	F_{18}	14.53	3.51
F_5	13.14	4.34	F_{19}	14.55	2.35
F_6	0.00	1.37	F_{20}	9.84	3.27
F_7	0.00	3.67	F_{21}	14.77	2.68
F_8	43.34	4.22	F_{22}	0.00	0.87
F_9	97.12	3.86	F_{23}	0.00	2.16
F_{10}	97.46	0.78	F_{24}	0.00	0.00
F_{11}	0.00	1.76	F_{25}	0.00	0.00
F_{12}	92.26	0.20	F_{26}	86.66	0.00
F_{13}	13.11	3.11	F_{27}	95.21	2.45
F_{14}	14.27	4.17	F_{28}	93.33	3.16

Results representing cumulative Detection Rate DR [%] and False Positive [%] for simulated abuses and attacks are presented in Table 2. We achieve highest results of DR [%] for F_{10} traffic feature 97.46% with FP [%] 0.78% and for F_9 traffic feature 97.12% with FP [%] 3.86%. Some set of traffic features gives us zero percent results because simulated abuses did not have an impact on these features.

For anomaly detection class systems, values up to 5% of false positive are considered as a decent level [26–28]. Achieved results are promising and show that the proposed methodology of anomaly and abuses detection can be used for IP, Internet of Things (IoT), or Smart Grid networks traffic, especially where computational resources are constrained and other machine learning methods, like neural networks, cannot be used.

5 Conclusion

Protection of ICT systems' infrastructures against novel, unknown attacks is currently an intensively examined and developed field. One of possible solutions is detection and classification of abnormal behaviors reflected in the analyzed network traffic. An advantage of such an approach is lack of necessity to define and memorize a priori patterns of such behaviors. Thus, in the decision making process, it is only necessary to determine what is and what is not an abnormal behavior in the network traffic in order to detect potential new attack.

In this article we present automatically created statistical ETS models, which are used to estimate behavior of the examined network traffic. Parameters' estimation and identification of the models' order is performed as a compromise between their coherence and size of estimation error. Due to application of the described models, satisfactory statistical estimations were achieved within the examined signals of the network traffic. The process of anomaly detection consisted in comparison of parameters of a regular/normal behavior estimated by means of the mentioned models and parameters of alternation of the real tested network traffic. The results explicitly point that anomalies of signals in network traffic can be efficiently detected by the proposed solution.

The suggested anomaly and abuse detection method based on ETS statistical algorithm gives use promising results (DR [%] 97.46%–97.12% and FP [%] 0.78%–3.86%), especially for purposes where computational resources are constrained, e.g. for IoT devices networks. ETS statistical model gives us acceptable prediction values in forecasting period sufficient for the proposed solution with false positive values up to 5%. The proposed solution was evaluated by means of real world traffic (28 traffic features) taken from SNORT programming interface and the proposed traffic preprocessor where ETS based anomaly detection solution was implemented.

References

1. Esposito, M., Mazzariello, C., Oliviero, F., Romano, S.P., Sansone C.: Evaluating pattern recognition techniques in intrusion detection systems. In: proceedings of the 5th International Workshop on Pattern Recognition in Information Systems, pp. 41–53 (2005)
2. Lakhina, A., Crovella, M., Diot, C.H.: Characterization of network-wide anomalies in traffic flows. In: proceedings of the 4th Conference on Internet Measurement, pp. 201–206 (2004)
3. Chondola, V., Banerjee, A., Kumar, V.: Anomaly detection: a survey. ACM Comput. Surv. **41**(3), 1–72 (2009)
4. Rodriguez, A., Mozos, M.: Improving network security through traffic log anomaly detection using time series analysis. In: Computational Intelligence in Security for Information Systems, pp. 125–133 (2010)
5. Chondola, V., Banerjee, A., Kumar, V.: Anomaly detection for discrete sequences: a survey. IEEE Trans. Knowl. Data Eng. **24**(5), 823–839 (2012)
6. Lim, S.Y., Jones, A.: Network anomaly detection system: the state of art of network behavior analysis. In: proceedings of the 2008 International Conference on Convergence and Hybrid Information Technology, pp. 459–465 (2008)

7. Rajkumar, M., Nene, J.: A survey on latest dos attacks: classification and defense mechanisms. Int. J. Innov. Res. Comput. Commun. Eng. **1**, 1847–1860 (2013)
8. Douligeris, C., Mitrokotsa, A.: DDoS attacks and defense mechanisms: classification and state-of-the-art. Comput. Netw. **44**(5), 643–666 (2004)
9. Scherrer, A., Larrieu, N., Owezarski, P., Borgnat, P., Abry, P.: Non-Gaussian and long memory statistical characterizations for internet traffic with anomalies. IEEE Trans. Dependable Secure Comput. **4**(1), 56–70 (2007)
10. Brockwell, P., Davis, R.: Introduction to Time Series and Forecasting. Springer, Heidelberg (2002)
11. Yaacob, A., Tan, I., Chien, S., Tan, H.: Arima based network anomaly detection. In: proceedings of 2nd International Conference on Communication Software and Networks, pp. 205–209. IEEE (2010)
12. Box, G.E., Jenkins, M.G.: Time Series Analysis Forecasting and Control, 2nd edn. Holden-Day, San Francisco (1976)
13. Andrysiak, T., Saganowski, Ł., Choraś, M., Kozik, R.: Network traffic prediction and anomaly detection based on ARFIMA model. In: proceedings of the 8th International Conference Computational Intelligence in Security for Information Systems, pp. 545–554 (2014)
14. Goodrich, R.L.: The forecast pro methodology. Int. J. Forecast. **16**(4), 533–535 (2000)
15. Ord, K., Lowe, S.: Automatic forecasting. Am. Stat. **50**(1), 88–94 (1996)
16. Hyndman, R.J., Koehler, A.B., Snyder, R.D., Grose, S.: A state space framework for automatic forecasting using exponential smoothing methods. Int. J. Forecast. **18**(3), 439–454 (2002)
17. Gardner Jr., E.S.: Exponential smoothing: the state of the art. J. Forecast. **4**, 1–28 (1985)
18. Gardner, E.S.: Exponential smoothing – the state of the art – part II. Int. J. Forecast. **22**, 637–666 (2006)
19. Archibald, B.C.: Parameter space of the Holt-Winters' model. Int. J. Forecast. **6**, 199–209 (1990)
20. Aoki, M.: State Space Modeling of Time Series. Springer, Berlin (1987)
21. Durbin, J., Koopman, S.J.: Time Series Analysis by State Space Methods. Oxford University Press, Oxford (2001)
22. Hyndman, R.J., Khandakar, Y.: Automatic time series forecasting: the forecast package for R. J. Stat. Softw. **27**(3), 104–133 (2008)
23. Bozdogan, H.: Model selection and Akaike's Information Criterion (AIC): the general theory and its analytical extensions. Psychometrika **52**, 345–370 (1987)
24. SNORT. https://www.snort.org/
25. Kali Linux. https://www.kali.org/
26. Cheng, P., Zhu, M.: Lightweight anomaly detection for wireless sensor networks. Int. J. Distrib. Sens. Netw. **653232**, 2015 (2015)
27. Xie, M., Han, M., Tian, B., Parvin, S.: Anomaly detection in wireless sensor networks: a survey. J. Netw. Comput. Appl. **34**, 1302–1325 (2011)
28. Garcia-Font, V., Garrigues, C., Rifa-Pous, H.: A comparative study of anomaly detection techniques for smart city wireless sensor networks. Sensors **16**, 868 (2016)

Sparse Representation and Dictionary Learning for Network Traffic Anomaly Detection

Tomasz Kierul[1], Michał Kierul[1], Tomasz Andrysiak[2](✉), and Łukasz Saganowski[2]

[1] Research and Development Center, SOFTBLUE S.A., Jana Zamoyskiego 2B, 85-063 Bydgoszcz, Poland
{tkierul,mkierul}@softblue.pl
[2] Institute of Telecommunications and Computer Science, Faculty of Telecommunications, Computer Science and Electrical Engineering, UTP University of Science and Technology, Al. prof. S. Kaliskiego 7, 85-796 Bydgoszcz, Poland
{andrys,luksag}@utp.edu.pl

Abstract. In this article we present the use of sparse representation of signal and dictionary learning method for solving the anomaly detection problem. The signals analysed in the article represented selected features of the network traffic. In the learning process we used modified Method of Optimal Directions in order to find a dictionary resembling correct structures of the network traffic deprived of the influence of possible outlying observations (outliers). A dictionary defined in such a way constituted basis for sparse representation of the analysed signal. Anomaly detection is realised by parameter estimation of the analysed signal and its comparative analysis to network traffic profiles. Efficiency of our method is examined with the use of extended set of test traces from real network traffic. The received experimental results confirm effectiveness of the presented method.

Keywords: Anomaly detection · Signal analysis · Sparse representation · Dictionary learning

1 Introduction

Increasing number of threats and incidents violating safety of systems, computer networks and users of services offered by new information technologies is currently one of the most essential social and civilizational problems. Scope scale and its dynamics concern not only an individual user and small businesses, but also great multinational corporations and governmental institutions and agencies [1].

Antivirus software, intrusion detection and prevention systems, protection against leakage of information are just a few items from a long and diverse list of such techniques. What they all have in common is that they are able to protect systems and computer networks from known threats described by means of previously learned patterns [2].

W. Zamojski et al. (Eds.): DepCoS-RELCOMEX 2020, AISC 1173, pp. 344–354, 2020.
https://doi.org/10.1007/978-3-030-48256-5_34

However, lack of disturbances matching the recognizable signatures does not mean lack of threat. Highly likely, it can be assumed that there can occur changes reflected in the analyzed signals which do not match the given pattern according to the defined classification rule, or there are not yet settled signatures describing given disturbance [3]. In this context, the biggest threat can be posed by the so called *zero-day attacks*, i.e. attacks which so far have not occurred, therefore there is a complete lack of their signatures.

One of the ways to protect against the new, unknown attacks is a rather radical change in concept of action. Instead of searching for signatures of threats it is necessary to seek abnormal behavior which is a deviation from normal, standard characteristics of the analyzed signal. The strength of such an approach lies in solutions which are not based on a priori knowledge of threats' patterns but on what does not comply with particular norms of the analyzed signal [4].

Therefore, an anomaly can be any deviation and irregularities, or deviance from the adopted rule or a selected profile describing "normal" changeability of the analyzed signal. In this context, the mechanisms creating these dependencies should be based exclusively on such signal's features which, in case their tolerance range is breached, will be a symptom of serious disturbance or abuse [5].

Anomaly detection methods have been a topic of numerous scientific research and articles [4, 6, 7]. In the papers describing them, the most often used techniques were those based on machine learning, neural networks or expert systems [8]. Currently, the most intensively developed techniques of anomaly detection are those which use different types of signal's processing and analysis techniques. Most often they are solutions based on adaptive decomposition of the analyzed signal, in particular, using redundant dictionaries [9]. Such approaches allow to estimate profiles of a normal network traffic, and then use them to identify/recognize threats [10].

In the present article we propose utilization of resistant-to-outlier dictionary learning method and signal's sparse representation for given time series describing the analyzed network traffic. Anomaly detection depends on comparison between the parameters of normal behavior (estimated parameters of the sparse representation of a signal) and those of real network traffic.

This paper is organized as follows: after the introduction, in Sect. 2, the motivation and related work is presented. In Sect. 3, the sparse representation of a signal for date traffic prediction is elaborated on. Then, in Sect. 4, the resistant-to-outlier dictionary learning method based on Method of Optimal Directions (MOD) estimation is discussed. Section 5 includes implementation details and experimental results. Conclusions are presented thereafter.

2 Motivation and Related Work

The analyzed signal's linear extensions referring to defined set of basic functions in numerous cases present serious limitations, i.e. they do not particularly exactly represent crucial features of the decomposed signal with the use of a small number of linear extension coefficients [11]. Moreover, if the structure elements of the analyzed signal greatly differ from the scaling factor of the basic function, the coefficients of linear extension do not make optimal representation of the signal [12]. Thus, the signals which change their length require application of basic functions with many different scales. An important restriction of such an approach is, however, connection of frequency and scale parameters [13].

In case of complicated structures of signals, it is impossible to define optimal parameters of the mentioned elements for the given basic functions. In such case, the most effective solution seems to be implementation of more diverse and numerous function sets, known as dictionaries with redundancy [14]. These functions are selected to match best the nature of the analyzed signal. The dictionary can be chosen in two ways: (i) by constructing the dictionary on a mathematical model of data (utilizing already known forms of models such as: wavelets, wavelet packets, contourlets, curvelets or Gabor [14, 15]), or (ii) learning the dictionary on the basis of a training data set (which is the most common part of the processed signal) [16]. The most typical method of dictionary learning is the Method of Optimal Directions. It was introduced by Engan et al. [17] and became one of the first implemented methods, otherwise known as process of sparsification. Another equally popular method, despite being different in character, is K-means Singular Value Decomposition (K-SVD) algorithm, introduced by Ahron et al. [18]. The most important difference between the two methods is the manner of updating the dictionary, i.e. some update all atoms at the same time (MOD), other update atoms one after another (K-SVD).

Implementation of sparse representation as the signal's decomposition with relation to a dictionary requires quick searching and optimal matching of its elements (i.e. atoms) which best resemble given features of the analyzed signal [11]. For such formulated problem, the rate of decrease of the residue standard depends on the correlation between successive signal residues and chosen dictionary's atoms. If the signal is the sum of high-energy components that are dictionary atoms, then the correlation coefficients of the signal and its residues are significant. In such case, their norm rapidly decreases because high-energy components are signal's structural elements well correlated with the given dictionary atoms [15, 16]. Representations of this problem are possible in the form of sparse solutions of equations systems [19] or adaptive greedy approximations [20]. Then, the utilized computational solutions are usually iterative methods, i.e. different variants of Orthogonal Matching Pursuit (OMP) or Basis Pursuit (BP) [21].

Methods and techniques based on idea of sparse representation of signals implemented with the use of dictionaries with redundancy are currently becoming a promising approach towards analysis and anomaly detection [9]. In the process of adaptive decomposition, they allow for separation of essential structural features of the analyzed signal respectively to the character of the exploited dictionary [11, 14].

3 Sparse Representation of a Signal

In many cases, representations of the analyzed signal, performed in form of linear extensions against a definite set of base functions properly located in time and/or frequency, are often described as not precise and optimal enough. Therefore, a better solution is to use dictionaries with redundancy [22] which are sets of numerous, diverse and adjusted to the nature of the signal. In result, more universal and flexible representations are obtained.

Sparse representation of the analyzed signal continuously seeks limited solution of sparse representation coefficients C describing the signal S concerning the over complete dictionary when the residue signal is smaller than the given threshold value δ which we may assume as [20]:

$$\min\|C\|_0 \qquad s.t. \qquad \left\|S-\sum_{m=0}^{M-1} c_m d_m\right\| < \delta, \tag{1}$$

where $\|\cdot\|_0$ describes ℓ_0 norm counting the non-zero entries of a vector, $c_m \in C$ showing a group of decomposition coefficients, $d_m \in D$ are the atoms of the over complete dictionary D, and δ is the constant specifying the exactness of the representation.

The Eq. (1) presents sparse representation of S signal, achieved by means of the minimal number of decomposition coefficients c_m and the d_m atoms of D dictionary corresponding to them (after assuming specific level of δ precision). Optimal representation of the analyzed signal is described as such a subset of dictionary D elements of which linear combination resembles the biggest rate of the signal S energy amongst all the subsets of the same count. Finding such a representation is computationally NP-hard [21]. A suboptimal expansion can be defined by means of greedy algorithms in an iterative procedure, e.g. the orthogonal matching pursuit algorithm.

The Orthogonal Matching Pursuit algorithm is an improved version of the matching pursuit algorithm and is discussed in [23]. The similarity between the two algorithms is that they have a greedy structure but the difference is that OMP algorithm requires all the selected atoms to be orthogonal at every decomposition step.

In each p step of OMP algorithm, subsequent decompositions of S signal are performed by means of projections on elements of D dictionary. In p^{th} step of the decomposition process, we obtain

$$c_p = \left\langle r^{p-1}S, d_{\varphi_p}\right\rangle, \tag{2}$$

where $\langle *, * \rangle$ means scalar product and r^pS constitutes residue being a result of S signal decomposition in the d_{φ_p} direction. For the residue of the obvious dependence $r^0S = S$ occurs. The indices of the chosen p vectors are stored in the index vector $\Phi_p = \{\varphi_1, \varphi_2, \ldots, \varphi_{p-1}, \varphi_p\}$, $\Phi_0 = \emptyset$ and the vectors are stored as the columns of the matrix $D_p = \left\{d_{\varphi_1}, d_{\varphi_2}, \ldots, d_{\varphi_p}\right\}$ and $D_0 = \emptyset$.

The algorithm chooses φ_p in the pth iteration by finding the vector which is best aligned with the residue achieved by projecting r^pS onto the dictionary elements, that is:

$$\varphi_p = arg\ \max_{i \in \Phi_p} |\langle r^pS, d_i\rangle|, \ \varphi_p \notin \Phi_{p-1}. \tag{3}$$

There is no re-selection problem due to the stored dictionary. If $\varphi_p \notin \Phi_{p-1}$ then the index set is updated as $\Phi_p = \Phi_{p-1} \cup \varphi_p$ and $D_p = D_{p-1} \cup d_{\varphi_p}$. Otherwise, $\Phi_p = \Phi_{p-1}$ and $D_p = D_{p-1}$.

The residue $r^p S$ is computed as

$$r^p S = r^{p-1} S - D_p \left(D_p^T D_p \right)^{-1} D_p^T r^{p-1} S, \tag{4}$$

where $D_p^T D_p$ is the *Gram matrix*.

In the next step we calculate the new coefficient c_p described by (2) and new residue $r^p S$ described by (4). Then, we update the set of coefficients $C_p = C_{p-1} \cup c_p$ and set of residues $R^p S = R^{p-1} S \cup r^p S$.

For d_{φ_p} atom, chosen by means of condition (3), residue of S signal is minimized in another step of OMP algorithm. The number of p iterations, within which limitation of residues are performed, depends on the required accuracy of S signal representation and is described by condition (5), which is simultaneously a criterion of stopping OMP algorithm.

The algorithm finishes when residue of signal is below the acceptable limit

$$\| r^p s \| < th, \tag{5}$$

where th is the approximation error.

In our practical implementation of OMP algorithm, a progressive Cholesky upgrading process is used to decrease the effort connected with the matrix inversion [24].

4 Dictionary Learning for Sparse Representations

The question of dictionary learning in the view of sparse representation has been found interesting by numerous researchers [14]. What makes the field of dictionary learning algorithms outstand from other approaches is their adjustment process, i.e. some update all atoms simultaneously (e.g. Method of Optimal Directions [17]), other update atoms one after another (e.g. K-means Singular Value Decomposition [18]).

For these solutions it is essential to search for the best dictionary D (resistant-to-outlier) which will reflect the signal S as sparse representation, as a solution to the Eq. (6)

$$\min_{D,C} \left[\| S - DC - O \|_F^2 + \lambda \| o_i \|_{2,1} \right] \tag{6}$$

where $\| \cdot \|_F^2$ is the Frobenius norm and $O = [o_1, \ldots, o_L]$ is the matrix of outliers, $\| o_i \|_{2,1}$ is the mixed $\ell_{2,1}$ norm of vector $o = \left[\| o_1 \|_2, \ldots, \| o_L \|_2 \right]^T$, where it is defined as the ℓ_1 norm of matrix of outliers, λ is a threshold parameter.

The MOD is used for solving the optimization issue, presented in Eq. (6). This process is done by means of iterative minimization of the objective function over one variable, while the remaining two are fixed. For the first, D and O are initialized. In the following step, minimization over C is performed – the iterative optimization begins. The standard course of initializing D requires the usage of a predefined dictionary, e.g. Gabor's [15], otherwise the dictionary is built of atoms randomly selected from the training signals. The latter solution is not relevant for our process because certain outliers might be mistakenly understood as atoms, this could possibly affect the whole process (in the sequel iterations). O is started by using the zero matrix. To clarify, all the training signals are understood as not "outliers" in the first iteration stage.

The modified MOD algorithm (mMOD), which is resistant to outliers, has three stages [25]:

- **Stage I - Sparse Coding:** in this stage, the decomposition coefficients c_i are collected in the over-complete dictionary D and signal S. The aim of each phase is to find the smallest possible number of coefficients which fulfil the Eq. (7). The given D is known. The OMP algorithm [23] is used to calculate M sparse coefficients c_i for each signal s_i, by estimation of

$$c_i \leftarrow arg\min_{c_i} \|S - Dc_i - o_i\|_2^2 \quad s.t \quad \|c_i\|_0 \leq T, \qquad i = 1, 2, \ldots, M. \tag{7}$$

- **Stage II - Outlier Update:** in this stage, the o outlier vector is updated (every step consists in seeking the minimum quantity of outliers) to complete the Eq. (8).

$$o_i \leftarrow \arg\min_{o_i} \left[\|S - Dc_i - o_i\|_2^2 + \lambda \|o_i\|_2 \right], \qquad i = 1, 2, \ldots, M. \tag{8}$$

To update an outlier vector, it is necessary to solve

$$\min_{o} \left[\|r - o\|_2^2 + \lambda \|o\|_2 \right], \tag{9}$$

where $r = S - Dc$ is the residue vector. When the derivative of the objective function is set to be equal to zero, at the optimal point $\hat{o}$, the outcome is

$$\hat{o} = \left\{ \begin{array}{cc} \left(1 - \frac{\lambda}{2\|r\|_2}\right) r, & if \ \|r\|_2 > \frac{\lambda}{2} \\ 0, & otherwise \end{array} \right\}. \tag{10}$$

Bearing in mind the above, it can be stated that when the sparse representation error (the residue) norm of the training data is beyond the threshold value λ, we obtain an outlier. Otherwise, it is understood as a proper data vector. Another association about the significance of the trade-off factor λ is that the fewer values it has, the larger number of outliers are recognized.

- **Stage III - Dictionary Update:** in this stage, the dictionary D atoms are updated. Alternative values of atoms and decomposition coefficients are calculated to decrease the possibility of an error within the range of the signal S and its sparse representation $D * C$ with outliers.

$$D \leftarrow \arg\min_D \sum_{i=1}^{L} \|S - Dc_i - o_i\|_2^2. \quad (11)$$

In the dictionary update stage, the following problem needs to be solved

$$\min_{D} \|F - DC\|_F^2, \quad (12)$$

where $F = S - O$. We chose the MOD algorithm to calculate the above condition (due to its simplicity) [17], however, any dictionary update algorithm may be utilized in this place.

All dictionary atoms are updated in this manner. Iteration which results from the three above stages enables creation of a dictionary (deprived of outliers) which estimates the signal S in a sparse and concise manner, which then results in the mMOD algorithm being a dictionary D composed of atoms which resemble the examined signal S with reference to its sparse representation $D * C$.

5 Experimental Results

In this section the results obtained for mMOD and KSVD based anomaly detection are compared to SNORT [26] based preprocessor, which was proposed in [27]. The preprocessor utilizes DWT – Discrete Wavelet Transform (Mallat implementation [28]) for detection of anomalies.

Efficiency of anomaly detection algorithm based on mMOD was evaluated by simulation of different real world attacks on test LAN network. Kali Linux [29] distribution was utilized in order to simulate certain attacks, such as: Application specific DDoS, various port scanning, Syn Flooding, DoS, DDoS, pocket fragmentation, spoofing and others.

The same set of attacks as in [27] was used to compare solution based on mMOD algorithms to those based on KSVD and DWT. In order to classify anomalies profiles of normal traffic behavior were created, based on network traffic features with an assumption that the traffic is attack free. 25 traffic features were extracted from network traffic for algorithms' evaluation (see Table 1). The traffic features are shown as one dimensional (1D) univariate time series where every traffic feature (from Table 2) sample arrives in constant period of time. Such a time series is called Tables 2 and 3 present outcomes of DR detection rates and FP false positive, respectively. Noticeably, in the given test, mMOD give better results than anomaly detection methods based on KSVD and DWT.

Detection rate and false positive values strongly depend on the given traffic feature. An attack, on the other hand, has direct influence only on selected traffic features from Table 1. F9 and F10 features give us the best outcomes. DR [%] for F9 and F10 changes in limits of 98.84−97.58 in turn FP [%] changes in the range 5.21−0.45 for mMOD based algorithm. Some features has zero detection rate values (for example F6 and F7) because simulated abuses haven't got influence on these IP traffic parameters.

Table 1. Network traffic features used for experiments.

Feature	Traffic feature description	Feature	Traffic feature description
F_1	Number of TCP packets	F_{14}	Out TCP packets (port 80)
F_2	In TCP packets	F_{15}	In TCP packets (port 80)
F_3	Out TCP packets	F_{16}	Out UDP datagrams (port 53)
F_4	Number of TCP packets in LAN	F_{17}	In UDP datagrams (port 53)
F_5	Number of UDP datagrams	F_{18}	Out IP traffic [kB/s]
F_6	In UDP datagrams	F_{19}	In IP traffic [kB/s]
F_7	Out UDP datagrams	F_{20}	Out TCP traffic (port 80) [kB/s]
F_8	Number of UDP datagrams in LAN	F_{21}	In TCP traffic (port 80) [kB/s]
F_9	Number of ICMP packets	F_{22}	Out UDP traffic [kB/s]
F_{10}	Out ICMP packets	F_{23}	In UDP traffic [kB/s]
F_{11}	In ICMP packets	F_{24}	Out UDP traffic (port 53) [kB/s]
F_{12}	Number of ICMP packets in LAN	F_{25}	In UDP traffic (port 53) [kB/s]
F_{13}	Number of TCP packets with SYN and ACK flags		

Table 2. DR [%] results for three methods of anomaly or attacks generation.

Feature	mMOD	K-SVD	DWT	Feature	mMOD	K-SVD	DWT
F_1	8.87	5.26	5.26	F_{14}	10.72	0.00	5.26
F_2	13.12	5.26	10.52	F_{15}	16.26	0.00	10.52
F_3	13.12	0.00	10.52	F_{16}	0.00	0.00	0.00
F_4	17.45	15.78	10.52	F_{17}	9.41	5.26	5.26
F_5	16.63	10.52	10.52	F_{18}	16.33	10.52	10.52
F_6	0.00	0.00	0.00	F_{19}	10.17	5.26	5.26
F_7	0.00	0.00	0.00	F_{20}	17.42	10.52	5.26
F_8	37.27	25.22	31.58	F_{21}	16.37	12.26	10.52
F_9	98.84	90.73	94.73	F_{22}	0.00	0.00	0.00
F_{10}	97.58	83.68	94.73	F_{23}	0.00	0.00	0.00
F_{11}	12.24	7.24	5.26	F_{24}	0.00	0.00	0.00
F_{12}	88.53	80.42	78.95	F_{25}	10.15	5.26	0.00
F_{13}	16.11	10.52	10.52				

Table 3. FP [%] results for three methods of anomaly or attacks generation.

Feature	mMOD	K-SVD	DWT	Feature	mMOD	K-SVD	DWT
F_1	4.12	5.46	7.43	F_{14}	3.14	4.58	7.48
F_2	4.32	5.17	7.99	F_{15}	3.36	4.86	7.17
F_3	4.45	5.45	7.96	F_{16}	0.02	0.02	0.02
F_4	4.24	5.44	6.06	F_{17}	0.22	0.40	0.39
F_5	4.00	5.64	5.62	F_{18}	3.52	4.80	8.74
F_6	3.02	3.96	4.14	F_{19}	3.78	5.24	8.36
F_7	3.43	5.18	5.33	F_{20}	3.11	4.52	8.50
F_8	3.78	5.24	8.28	F_{21}	3.16	4.23	7.09
F_9	5.21	7.68	9.13	F_{22}	2.47	3.46	3.08
F_{10}	0.45	1.22	0.48	F_{23}	2.77	4.82	3.07
F_{11}	4.12	5.12	12.06	F_{24}	0.00	0.02	0.00
F_{12}	4.16	6.34	4.34	F_{25}	0.03	0.37	0.02
F_{13}	4.38	5.23	7.07				

Additionally, we examined our method with basic traffic base [30] for evaluation of algorithm's performance. Table 4 presents the results of detection rates for two testing days.

Table 4. Detection Rate for Week5, Day5 and Week5, Day1 for DARPA [30] trace.

Traffic feature	DR [%] mMOD W5D5	DR [%] K-SVD W5D5	DR [%] mMOD W5D1	DR [%] K-SVD W5D1
ICMP flows/min.	96.71	64.70	97.43	94.52
ICMP in bytes/min.	94.15	79.14	98.72	93.15
ICMP in frames/min.	94.63	85.29	96.24	93.15
ICMP out bytes/min.	87.35	79.41	98.82	89.04
ICMP out frames/min.	95.67	88.23	96.22	75.34
TCP flows/min.	87.11	48.52	98.76	63.01
TCP in bytes/min.	94.21	55.88	96.12	90.41
TCP in frames/min.	92.14	60.29	98.83	97.26
TCP out bytes/min.	96.56	36.76	97.55	84.93
TCP out frames/min.	87.31	38.23	97.15	89.04
UDP flows/min.	98.73	85.29	96.68	90.41
UDP in bytes/min.	100.00	76.47	98.84	87.67
UDP in frames/min.	98.81	85.29	100.00	68.49
UDP out bytes/min.	98.46	89.70	100.00	98.63
UDP out frames/min.	100.00	91.17	100.00	98.63

6 Conclusion

Monitoring and securing IT systems' infrastructure from new, unknown attacks is currently an issue under intense examination. Development of the network protection systems is forced by growing number of new attacks, globalization of their scope and increasing complexity level. Most often, to ensure safety of these networks, the implemented mechanisms are methods of detection and classification of abnormal behavior spotted in the analyzed network traffic. The advantage of such an approach is that there is no necessity to a priori define and remember patterns of such behaviors (signatures of abuse). Thus, during the decision-making process, the only requirement to define what is and what is not abnormal behavior in the given network traffic in order to detect a possible unknown attack/abuse.

In this article, we present the usage of signal's sparse representation and a dictionary learning method for network traffic analysis. In the learning process we implement the modified Method of Optimal Directions to obtain proper dictionary structure (lacking outliers). Next, classification is performed by means of normal network traffic profiles and sparse representation parameters of the tested signal. An extended set of test traces from real network traffic allow to examine efficiency of our method. The test outcomes clearly prove that abnormal activities included in the network traffic signal can be detected by means of the proposed methods.

Summarizing the attacks/abuse and anomalies created with the use of Kali Linux tools and some of traces from standard DARPA data set, it can be concluded that the proposed mMOD algorithm provides better results than standard K-SVD and DWT spectral analysis of traffic features time series. For the evaluated set of traffic tracks the best outcomes were achieved for F9 and F10 traffic features where DR [%] changes from 98.84−97.58, while FP [%] changes in the range 5.21−0.45 for mMOD based algorithm. The achieved results are promising and may be used for analysis of traffic time series detection of anomalies.

References

1. SANS Institute. Top cyber security risks-zero-day vulnerability trends. http://www.sans.org/top-cyber-security-risks/zero-day.php. Accessed 27 Jan 2020
2. Biggio, B., Fumera, G., Roli, F.: Security evaluation of pattern classifiers under attack. IEEE Trans. Knowl. Data Eng. **26**(4), 984–996 (2014)
3. Duda, R.O., Hart, P.E., Stork, D.G.: Pattern Classification. Wiley, New York (2001)
4. Chondola, V., Banerjee, A., Kumar, V.: Anomaly detection: a survey. ACM Comput. Surv. **41**(3), 1–72 (2009)
5. Chondola, V., Banerjee, A., Kumar, V.: Anomaly detection for discrete sequences: a survey. IEEE Trans. Knowl. Data Eng. **24**(5), 823–839 (2012)
6. Anomalies. IEEE Trans. Dependable Secure Comput. **4**(1), 56–70 (2007)
7. Hodge, V.J., Austin, J.: A survey of outlier detection methodologies. Artif. Intell. Rev. **22**(2), 85–126 (2004)
8. Lim, S.Y., Jones, A.: Network anomaly detection system: the state of art of network behavior analysis. In: Proceedings of the 2008 International Conference on Convergence and Hybrid Information Technology, pp. 459–465 (2008)

9. Adler, A., Elad, M., Hel-Or, Y., Rivlin, E.: Sparse coding with anomaly detection. J. Signal Process. Syst. **79**(2), 179–188 (2015)
10. Garcia-Teodoro, P., Diaz-Verdejo, J., Macia-Fernandez, G., Vazquez, E.: Anomaly-based network intrusion detection: techniques, systems and challenges. Comput. Secur. **2**(8), 18–28 (2009)
11. Białasiewicz, J.T.: Falki i aproksymacje. WNT Warszawa (2004)
12. Gazi, O.: Understanding Digital Signal Processing. Springer, Heidelberg (2018)
13. Mallat, S.G.: A theory for multiresolution signal decomposition: the wavelet representation. IEEE Trans. Pattern Anal. Mach. Intell. **11**(7), 674–693 (1989)
14. Rubinstein, R., Bruckstein, M., Elad, M.: Dictionaries for sparse representation modeling. Proc. IEEE **98**(6), 1045–1057 (2010)
15. Mallat, S., Zhang, Z.: Matching pursuits with time-frequency dictionaries. IEEE Trans. Signal Process. **41**(12), 339–415 (1993)
16. Gribonval, R., Schnass, K.: Dictionary identifiability from few training samples. In: Proceedings of 16th European Signal Processing Conference (2008)
17. Engan, K., Aase, S.O., Husoy, H.J.: Method of optimal directions for frame design. In: proceedings of IEEE International Conference Acoustics, Speech, Signal Process, pp. 2443–2446 (1999)
18. Aharon, M., Elad, M., Bruckstein, A.: K-SVD: an algorithm for designing overcomplete dictionaries for sparse representation. IEEE Trans. Signal Process. **54**(11), 4311–4322 (2006)
19. Bruckstein, A.M., Donoho, D.L., Elad, M.: From sparse solutions of systems of equations to sparse modeling of signals and images. J. SIAM Rev. **51**(1), 34–81 (2009)
20. Davis, G., Mallat, S., Avellaneda, M.: Adaptive greedy approximations. J. Constr. Approx. **13**, 57–98 (1997)
21. Tropp, J.A.: Greed is good: algorithmic results for sparse approximation. IEEE Trans. Inf. Theory **50**(10), 2231–2242 (2004)
22. Elad, M.: Sparse and Redundant Representations: From Theory to Applications in Signal and Image Processing. Springer, Heidelberg (2010)
23. Pati, Y.C., Rezaiifar, R., Krishnaprasad, P.S.: Orthogonal matching pursuit: recursive function approximation with applications to wavelet decomposition. In: proceedings of 27th Asilomar Conference on Signals, Systems and Computers, pp. 40–44 (1993)
24. Seonggeon, K., Uihyun, Y., Jaehyuk, J., Geunsu, S., Jongjin, K., Heung-No, L., Minjae, L.: Reduced computational complexity orthogonal matching pursuit using a novel partitioned inversion technique for compressive sensing. Electronics **7**(206), 2–10 (2018)
25. Amini, S., Sadeghi, M., Joneidi, M., Babaie-Zadeh, M., Jutten, Ch.: Outlier-aware dictionary learning for sparse representation. In: Proceedings of IEEE International Workshop on Machine Learning for Signal Processing, pp. 1–6 (2014)
26. Snort – intrusion detection system. https://www.snort.org/. Accessed 27 Jan 2020
27. Saganowski, L., Goncerzewicz, M., Andrysiak, T.: Anomaly detection preprocessor for snort ids system. In: Proceedings of Image Processing and Communications Challenges, vol. 4, pp. 225–232. Springer (2013)
28. Dainotti, A., Pescap´e, A., Ventre, G.: Wavelet-based detection of dos attacks. In: Proceedings of Global Telecommunications Conference, pp. 1–6 (2006)
29. Kali Linux. https://www.kali.org/. Accessed 27 Jan 2020
30. Defense advanced research projects agency DARPA intrusion detection evaluation data set. http://www.ll.mit.edu/mission/communications/ist/corpora/ideval/data/index.html. Accessed 27 Jan 2020

Changing System Operation States Influence on Its Total Operation Cost

Krzysztof Kołowrocki and Beata Magryta(✉)

Department of Mathematics, Gdynia Maritime University, 81-87 Morska Street, 81-225 Gdynia, Poland
{k.kolowrocki,b.magryta}@wn.umg.edu.pl

Abstract. The operation model of a complex system changing its functional structure and its instantaneous operation costs during the variable at time operation states and linear programming are proposed to optimize the system operation process in order to get the system total operation cost minimal. The optimization method allowing to find the optimal values of the transient probabilities of the complex system operation process at the particular operation state that minimize the system total operation cost mean value under the assumption that the system conditional operation costs mean values at the particular operation states are fixed is presented. The procedure of finding the optimal mean value of system total operation cost during the fixed operation time is applied to the port oil terminal operation cost minimization.

Keywords: Complex system · Operation process · Operation cost · Optimization · Port oil terminal

1 Introduction

To investigate the complex technical system operation process, the semi-Markov model [1] can be used to define its operation states, to introduce its parameters and to determine its characteristics [2, 3]. Having the system operation process characteristics and the system conditional instantaneous operation costs at the particular operation states it is possible to find the mean value of the system total operation cost during the fixed time of the system operation and further to change the system operation process through applying the linear programming [4] in order to minimize the system total operation cost. In the paper, the model for finding and minimizing the mean value of the system total operation cost is created and applied to the port oil terminal.

2 System Operation Cost Model

We assume that the system is operating at v, $v > 1$, operation states z_b, $b = 1, 2, \ldots, v$, that have influence on the system functional structure and on the system operation cost. Assuming semi-Markov model of the system operation process $Z(t)$, $t \geq 0$, it is possible to find this process two basic characteristics:

W. Zamojski et al. (Eds.): DepCoS-RELCOMEX 2020, AISC 1173, pp. 355–365, 2020.
https://doi.org/10.1007/978-3-030-48256-5_35

- the vector of limit values [1, 2]

$$p_b = \lim_{t \to \infty} p_b(t) = b = 1, 2, \ldots, \nu, \tag{1}$$

of transient probabilities

$$p_b(t) = P(Z(t) = z_b),\ t \geq 0,\ b = 1, 2, \ldots, \nu, \tag{2}$$

of the system operation process $Z(t)$ at the particular operation states z_b, $b = 1, 2, \ldots, \nu$,

- the vector $[\hat{M}_b]_{1x\nu}$ of the mean values

$$\hat{M}_b = E[\hat{\theta}_b] \cong p_b\theta,\ b = 1, 2, \ldots, \nu, \tag{3}$$

of the total sojourn times $\hat{\theta}_b$, $b = 1, 2, \ldots, \nu$, of the system operation process $Z(t)$, $t \geq 0$, at the particular operation states z_b, $b = 1, 2, \ldots, \nu$, during the fixed the system operation time θ, $\theta > 0$, where p_b, $b = 1, 2, \ldots, \nu$, are defined by (1)–(2).

Further, we may define the system instantaneous operation cost. Namely, we define the instantaneous system operation cost in the form of the vector

$$\boldsymbol{C}(t) = [[C(t)]^{(1)}, [C(t)]^{(2)}, \ldots, [C(t)]^{(\nu)}], t \geq 0, \tag{4}$$

with the coordinates

$$[C(t)]^{(b)},\ t \geq 0,\ b = 1, 2, \ldots, \nu, \tag{5}$$

that are the coordinates of the system conditional instantaneous operation costs at the system operation states z_b, $b = 1, 2, \ldots, \nu$.

It is natural to assume that the system instantaneous operation cost depends significantly on the system operation state and the system operation cost at the operation states as well. This dependency is also clearly expressed in mean value of the system total operation cost during the system operation time θ, given by

$$\boldsymbol{C}(\theta) = \sum_{b=1}^{\nu} p_b[C(\theta)]^{(b)},\ \theta > 0, \tag{6}$$

where p_b, $b = 1, 2, \ldots, \nu$, are limit transient probabilities at operation states defined by (1)–(2),

$$[C(\theta)]^{(b)} = \int_0^{\hat{M}_b} C[(t)]^{(b)} dt,\ \theta > 0, b = 1, 2, \ldots, \nu, \tag{7}$$

are the mean values of the system total conditional operation costs at the particular system operation states z_b, $b = 1, 2, \ldots, v$, where $\hat{M}_b$, $b = 1, 2, \ldots, v$, are given by (3), and $[C(t)]^{(b)}$, $t \geq 0$, $b = 1, 2, \ldots, v$, are defined by (5).

3 System Operation Cost Minimization

From the linear Eqs. (6), we can see that the mean value of the system total unconditional operation cost $\boldsymbol{C}(\theta)$,$\theta > 0$, is determined by the limit values of transient probabilities p_b, $b = 1, 2, \ldots, v$, of the system operation process at the operation states z_b, $b = 1, 2, \ldots, v$, defined by (1)–(2) and by the mean values $[C(\theta)]^{(b)}$, $\theta > 0$, $b = 1, 2, \ldots, v$, of the system total conditional operation costs at the particular system operation states z_b, $b = 1, 2, \ldots, v$, determined by (7). Therefore, the system operations cost optimization based on the linear programming [3], can be proposed. Namely, we may look for the corresponding optimal values $\dot{p}_b$, $b = 1, 2, \ldots, v$, of the limit transient probabilities p_b, $b = 1, 2, \ldots, v$, of the system operation process at the operation states to minimize the mean value $\boldsymbol{C}(\theta)$ of the system total unconditional operation costs under the assumption that the mean values $[C(\theta)]^{(b)}$, $b = 1, 2, \ldots, v$, of the system total conditional operation costs at the particular system operation states z_b, $b = 1, 2, \ldots, v$, are fixed. Thus, we may formulate the optimization problem as a linear programming model [4] with the objective function of the form given by (6) with the bound constraints

$$\breve{p}_b \leq p_b \leq \hat{p}_b, \; b = 1, 2, \ldots, v, \; \sum_{b=1}^{v} p_b = 1, \tag{8}$$

where

$$[C(\theta)]^{(b)}, \; [C(\theta)]^{(b)} \geq 0, \; b = 1, 2, \ldots, v, \tag{9}$$

are fixed mean values of the system conditional operation costs at the operation states z_b, $b = 1, 2, \ldots, v$, determined according to (7) and

$$\breve{p}_b, \; 0 \leq \breve{p}_b \leq 1 \text{ and } \hat{p}_b, \; 0 \leq \hat{p}_b \leq 1, \; \breve{p}_b \leq \hat{p}_b, \; b = 1, 2, \ldots, v, \tag{10}$$

are lower and upper bounds of the unknown transient probabilities p_b, $b = 1, 2, \ldots, v$, respectively. Now, we can find the optimal solution of the formulated by (6), (8)–(10) the linear programming problem, i.e. we can determine the optimal values $\dot{p}_b$, of the transient probabilities p_b, $b = 1, 2, \ldots, v$, that minimize the objective function given by (6). To do this, we arrange the mean values of the system total conditional operation costs $[C(\theta)]^{(b)}$, $b = 1, 2, \ldots, v$, in non-decreasing order

$$[C(\theta)]^{(b_i)} \leq [C(\theta)]^{(b_2)} \leq \ldots \leq [C(\theta)]^{(b_v)},$$

where

$$b_i \in \{1, 2, \ldots, v\} \text{ for } i = 1, 2, \ldots, v.$$

Next, we substitute

$$x_i = p_{b_i}, \breve{x}_i = \breve{p}_{b_i}, \widehat{x}_i = \widehat{p}_{b_i} \text{ for } i = 1, 2, \ldots, v, \tag{11}$$

and we minimize, with respect to x_i, $i = 1, 2, \ldots, v$, the linear form (6) that after this substitution takes the form

$$\boldsymbol{C}(\theta) = \sum_{i=1}^{v} x_i [C(\theta)]^{(b_i)} \tag{12}$$

with the bound constraints

$$\breve{x}_i \leq x_i \leq \widehat{x}_i, \; i = 1, 2, \ldots, v, \; \sum_{i=1}^{v} x_i = 1, \tag{13}$$

where according to (9)

$$\breve{x}_i, 0 \leq \breve{x}_i \leq 1 \text{ and } \widehat{x}_i, 0 \leq \widehat{x}_i \leq 1, \breve{x}_i \leq \widehat{x}_i, i = 1, 2, \ldots, v, \tag{14}$$

are lower and upper bounds of unknown probabilities x_i, $i = 1, 2, \ldots, v$, respectively.

To find the optimal values of x_i, $i = 1, 2, \ldots, v$, we define

$$\breve{x} = \sum_{i=1}^{v} \breve{x}_i, \hat{y} = 1 - \breve{x} \tag{15}$$

and

$$\breve{x}^0 = 0, \widehat{x}^0 = 0 \text{ and } \breve{x}^I = \sum_{i=1}^{I} \breve{x}_i, \widehat{x}^I = \sum_{i=1}^{I} \widehat{x}_i \text{ for } I = 1, 2, \ldots, v. \tag{16}$$

Next, we find the largest value $I \in \{0, 1, \ldots, v\}$ such that

$$\widehat{x}^I - \breve{x}^I < \hat{y} \tag{17}$$

and we fix the optimal solution that minimize (12) in the following way:

i) if $I = 0$, the optimal solution is

$$\dot{x}_1 = \hat{y} + \breve{x}_1 \text{ and } \dot{x}_i = \breve{x}_i \text{ for } i = 1, 2, \ldots, v; \tag{18}$$

ii) if $0 < \mathrm{I} < v$, the optimal solution is

$$\begin{gathered} \dot{x}_i = \widehat{x}_i \text{ for } i = 1, 2, \ldots, I, \dot{x}_{I+1} = \hat{y} - \widehat{x}^{I} + \breve{x}^{I} + \breve{x}_{I+1} \\ \text{and } \dot{x}_i = \breve{x}_i \text{ for } i = I + 2, I + 3, \ldots, v; \end{gathered} \tag{19}$$

iii) if $I = v$, the optimal solution is

$$\dot{x}_i = \widehat{x}_i \text{ for } i = 1, 2, \ldots, v. \tag{20}$$

Finally, after making the substitution inverse to (11) we get the optimal limit transient probabilities

$$\dot{p}_{b_i} = \dot{x}_i \text{ for } i = 1, 2, \ldots, v, \tag{21}$$

that minimize the mean value of the system total unconditional operation cost, defined by the linear form (6), giving its minimum value in the following form

$$\dot{C}(\theta) = \sum_{i=1}^{i} \dot{p}_b [C(\theta)]^{(b)}. \tag{22}$$

4 Application

4.1 Port Oil Terminal Structure and Operation

We consider the exemplary critical infrastructure [5], which is the port oil terminal, critical infrastructure described in [6, 7]. The main technical assets A_1–A_9 of the port oil terminal critical infrastructure are distinguished in [6, 7]. The asset A_1, the port oil piping transportation system is composed of three pipelines, the subsystems S_1, S_2 and S_3, with the scheme of its functional structure illustrated in Fig. 1.

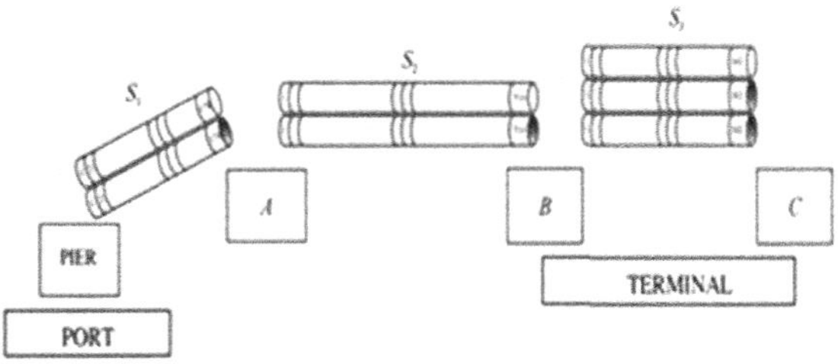

Fig. 1. The scheme of the port oil piping transportation system functional structure.

The asset A_1, the port oil piping transportation system operating at the port oil terminal critical infrastructure consists of three subsystems:

- the subsystem S_1 composed of two pipelines, each composed of 176 pipe segments and 2 valves,
- the subsystem S_2 composed of two pipelines, each composed of 717 pipe segments and 2 valves,
- the subsystem S_3 composed of three pipelines, each composed of 360 pipe segments and 2 valves.

Similarly, as in [6, 7], we assume that the operation of the asset A_1, the port oil piping transportation system, is the main activity of the port oil terminal, involving the remaining assets A_2–A_9 and determining their operation processes. On the basis of the statistical data and expert opinions, it is possible to fix the following basic parameters of the port oil terminal critical infrastructure operation process:

- the number of operation process states, $v = 7$;

and the operation process states:

- the operation state z_1, transport of one kind of medium from the terminal part B to part C using two out of three pipelines of the subsystem S_3 of the asset A_1 and assets A_2, A_4, A_6, A_7, A_9;
- the operation state z_2, transport of one kind of medium from the terminal part C to part B using one out of three pipelines of the subsystem S_3 of the asset A_1 and assets A_2, A_4, A_8, A_9;
- the operation state z_3, transport of one kind of medium from the terminal part B through part A to pier using one out of two pipelines of the subsystem S_1 and one out of two pipelines of the subsystem S_2 of the asset A_1 and assets A_2, A_4, A_5, A_9;
- the operation state z_4, transport of one kind of medium from the pier through parts A and B to part C using one out of two pipelines of the subsystem S_1, one out of two pipelines in subsystem S_2 and two out of three pipelines of the subsystem S_3 of the asset A_1 and assets A_2, A_3, A_4, A_5, A_6, A_7, A_9;
- the operation state z_5, transport of one kind of medium from the pier through part A to B using one out of two pipelines of the subsystem S_1 and one out of two pipelines of the subsystem S_2 of the asset A_1 and assets A_2, A_3, A_4, A_5, A_9;
- the operation state z_6, transport of one kind of medium from the terminal part B to C using two out of three pipelines of the subsystem S_3 and simultaneously transport one kind of medium from the pier through part A to B using one out of two pipelines of the subsystem S_1 and one out of two pipelines of the subsystem S_2 of the asset A_1 and assets A_2, A_3, A_4, A_5, A_6, A_7, A_9;
- the operation state z_7, transport of one kind of medium from the terminal part B to C using one out of three pipelines of the subsystem S_3 and simultaneously transport second kind of medium from the terminal part C to B using one out of three pipelines of the subsystem S_3 of the asset A_1 assets A_2, A_4, A_6, A_7, A_8, A_9.

The port oil terminal critical infrastructure operation process $Z(t)$ main characteristics are the limit values of transient probabilities of the operation process $Z(t)$ at the particular operation states z_b, $b = 1, 2, \ldots, 7$, [2, 3]:

$$p_1 = 0.395, p_2 = 0.060, p_3 = 0.003, p_4 = 0.002, p_5 = 0.20, p_6 = 0.058, p_7 = 0.282. \quad (23)$$

The asset A_1, the port oil terminal system is composed of 2880 components and the number of the system components operating at the various operation states, are different. Namely, there are operating 1086 system components at the operation states z_1, z_2 and z_7, 1794 system components at the operation states z_3 and z_5, 2880 system components at the operation states z_4 and z_6, [2].

4.2 Port Oil Terminal Operation Cost

According to the information coming from experts, the approximate value of the instantaneous operation cost of the single basic component of the asset A_1 used during the operation time interval of $\theta = 1$ year at the operation state $z_b, b = 1, 2, \ldots, 7$, is constant and amounts 9.6 PLN, $t \in <0, 1>$, $b = 1, 2, \ldots, 7$, whereas, the cost of each its singular basic component that is not used is equal to 0 PLN.

Hence, the number of components in a subsystems S_1, S_2, S_3 and their use at particularly operation states imply that the asset A_1 conditional instantaneous operation costs $[C_1(t)]^{(b)}, t \in <0, \theta>$, $b = 1, 2, \ldots, 7$, introduced by (5), are:

$$\begin{gathered} [C_1(t)]^{(1)} = 1086 \cdot 9.6 = 10425.6 \text{ PLN}, [C_1(t)]^{(2)} = 1086 \cdot 9.6 = 10425.6 \text{ PLN}, \\ [C_1(t)]^{(3)} = 1794 \cdot 9.6 = 17222.4 \text{ PLN}, [C_1(t)]^{(4)} = 2880 \cdot 9.6 = 27648 \text{ PLN}, \\ [C_1(t)]^{(5)} = 1794 \cdot 9.6 = 17222.4 \text{ PLN}, [C_1(t)]^{(6)} = 2880 \cdot 9.6 = 27648 \text{ PLN}, \\ [C_1(t)]^{(7)} = 1086 \cdot 9.6 = 10425.6 \text{ PLN}. \end{gathered} \quad (24)$$

Through (3) and (23), the approximate mean values $\hat{M}_b$, of total sojourn times of the port oil terminal at the particular operation states are:

$$\begin{gathered} \hat{M}_1 = 144.175, \hat{M}_2 = 21.9, \hat{M}_3 = 1.095, \hat{M}_4 = 0.73, \hat{M}_5 = 73, \\ \hat{M}_6 = 21.17, \hat{M}_7 = 102.93. \end{gathered} \quad (25)$$

Applying the formula (7) to (24) and (25), we get the approximate mean values $[C_1(\theta)]^{(b)}$, $b = 1, 2, \ldots, 7$, of the asset A_1 total conditional operation costs at the operation state z_b, $b = 1, 2, \ldots, 7$, during the operation time $\theta = 1$ year:

$$\begin{gathered} [C_1(\theta)]^{(1)} = 144.175 \cdot 10425.6 = 1503110.88 \text{ PLN}, \\ [C_1(\theta)]^{(2)} = 21.9 \cdot 10425.6 = 228320.64 \text{ PLN}, \\ [C_1(\theta)]^{(3)} = 1.095 \cdot 17222.4 = 18858.528 \text{ PLN}, \\ [C_1(\theta)]^{(4)} = 0.73 \cdot 27648 = 20183.04 \text{ PLN}, \\ [C_1(\theta)]^{(5)} = 73 \cdot 17222.4 = 1257235.2 \text{ PLN}, \\ [C_1(\theta)]^{(6)} = 21.17 \cdot 27648 = 585308.16 \text{ PLN}, \\ [C_1(\theta)]^{(7)} = 102.93 \cdot 10425.6 = 1073107.008 \text{ PLN}. \end{gathered} \quad (26)$$

The corresponding mean values of the total conditional operation costs for the remaining assets A_2–A_9, during the operation time θ = 1 year, we assume arbitrarily (we do not have data at the moment) equal to 10000 PLN, in all operation states if they are used and equal to 0 PLN if they are not used. Under this assumption, considering the procedure of using assets A_2–A_9 at particular operation states and the total operation costs of asset A_1 given in (26), we fix the total costs of the entire port oil terminal at the particular operation states z_b, $b = 1, 2, \ldots, 7$, given by:

$$\begin{aligned}
[C(\theta)]^{(1)} &= 1503110.88 + 50000 = 1553110.88 \text{ PLN},\\
[C(\theta)]^{(2)} &= 228320.64 + 40000 = 268320.64 \text{ PLN},\\
[C(\theta)]^{(3)} &= 18858.528 + 40000 = 58858.528 \text{ PLN},\\
[C(\theta)]^{(4)} &= 20183.04 + 70000 = 90183.04 \text{ PLN},\\
[C(\theta)]^{(5)} &= 1257235.2 + 50000 = 130735.2 \text{ PLN},\\
[C(\theta)]^{(6)} &= 585308.16 + 70000 = 655308.16 \text{ PLN},\\
[C(\theta)]^{(7)} &= 1073107.008 + 60000 = 1133107.008 \text{ PLN}.
\end{aligned} \tag{27}$$

Considering the values of the total costs $[C(\theta)]^{(b)}$, $b = 1, 2, \ldots, 7$, from (27) and the values of transient probabilities p_b, $b = 1, 2, \ldots, 7$, given by (23), the port oil terminal total operation mean cost during the operation time θ = 1 year, according to (6), is given by

$$\begin{aligned}
C(\theta) &\cong p_1[C(\theta)]^{(1)} + p_2[C(\theta)]^{(2)} + p_3[C(\theta)]^{(3)} + p_4[C(\theta)]^{(4)} + p_5[C(\theta)]^{(5)}\\
&\quad + p_6[C(\theta)]^{(6)} + p_7[C(\theta)]^{(7)}\\
&\cong 0.395 \cdot 1553110.88 + 0.06 \cdot 268320.64 + 0.003 \cdot 58858.528 + 0.002 \cdot 90183.04\\
&\quad + 0.2 \cdot 130735.2 + 0.058 \cdot 655308.16 + 0.282 \cdot 1133107.008 \cong 1013630 \text{ PLN}
\end{aligned} \tag{28}$$

4.3 Port Oil Terminal Operation Cost Minimization

Considering (28) to find the minimum value of the port oil terminal mean cost, we define the objective function given by (6), in the following form

$$\begin{aligned}
C(\theta) &= p_1 \cdot 1553110.88 + p_2 \cdot 268320.64 + p_3 \cdot 58858.528 + p_4 \cdot 90183.04 + p_5 \cdot 130735.2\\
&\quad + p_6 \cdot 655308.16 + p_7 \cdot 1133107.008.
\end{aligned} \tag{29}$$

The lower $\breve{p}_b$, and upper $\widehat{p}_b$ bounds of the unknown optimal values of transient probabilities p_b, $b = 1, 2, \ldots, 7$, respectively are [2]:

$$\begin{aligned}
&\breve{p}_1 = 0.31, \breve{p}_2 = 0.04, \breve{p}_3 = 0.002, \breve{p}_4 = 0.001, \breve{p}_5 = 0.15, \breve{p}_6 = 0.04, \breve{p}_7 = 0.25,\\
&\widehat{p}_1 = 0.46, \widehat{p}_2 = 0.08, \widehat{p}_3 = 0.006, \widehat{p}_4 = 0.004, \widehat{p}_5 = 0.26, \widehat{p}_6 = 0.08, \breve{p}_7 = 0.40.
\end{aligned} \tag{30}$$

Therefore, according to (9)–(10), we assume the following bound constraints

$$0.31 \le p_1 \le 0.46, 0.04 \le p_2 \le 0.08, 0.002 \le p_3 \le 0.006, 0.001 \le p_4 \le 0.004,$$
$$0.15 \le p_5 \le 0.26, 0.04 \le p_6 \le 0.08, 0.25 \le p_7 \le 0.40, \sum_{i=1}^{7} p_b = 1. \quad (31)$$

Now, before we find optimal values $\dot{p}_b$ of the transient probabilities p_b, $b = 1, 2, \ldots, 7$, that minimize the objective function (29), we arrange the mean values of the port oil terminal conditional operation costs $[C(\theta)]^{(b)}$, $b = 1,2,\ldots,7$, determined by (27), in non-decreasing order

$$58858.528 \le 90183.04 \le 130735.2 \le 268320.64 \le 655308.16 \le 1133107.008 \le 1553110.88, \text{ i.e.}$$
$$[C(\theta)]^{(3)} \le [C(\theta)]^{(4)} \le [C(\theta)]^{(5)} \le [C(\theta)]^{(2)} \le [C(\theta)]^{(6)} \le [C(\theta)]^{(7)} \le [C(\theta)]^{(1)}. \quad (32)$$

Further, according to (11), we substitute

$$x_1 = p_3, x_2 = p_4, x_3 = p_5, x_4 = p_2, x_5 = p_6, x_6 = p_7, x_7 = p_1, \quad (33)$$

and

$$\breve{x}_1 = \breve{p}_3 = 0.002,\ \breve{x}_2 = \breve{p}_4 = 0.001,\ \breve{x}_3 = \breve{p}_5 = 0.15, \breve{x}_4 = \breve{p}_2 = 0.04,$$
$$\breve{x}_5 = \breve{p}_6 = 0.04, \breve{x}_6 = \breve{p}_7 = 0.25, \breve{x}_7 = \breve{p}_1 = 0.31,$$

$$\widehat{x}_1 = \widehat{p}_3 = 0.006,\ \widehat{x}_2 = \widehat{p}_4 = 0.004,\ \widehat{x}_3 = \widehat{p}_5 = 0.26,\ \widehat{x}_4 = \widehat{p}_2 = 0.08,$$
$$\widehat{x}_5 = \widehat{p}_6 = 0.08,\ \widehat{x}_6 = \widehat{p}_7 = 0.40,\ \widehat{x}_7 = \widehat{p}_1 = 0.46, \quad (34)$$

and we minimize with respect to x_i, $i = 1,2,\ldots,7$, the linear form (29) that according to (11)–(13) and (33)–(34) takes the form

$$\boldsymbol{C}(\theta) = x_1 \cdot 58858.528 + x_2 \cdot 90183.04 + x_3 \cdot 130735.2 + x_4 \cdot 268320.64 + x_5 \cdot 655308.16 + x_6 \cdot 1133107.008 + x_7 \cdot 1553110.88, \quad (35)$$

with the following bound constraints

$$0.002 \le x_1 \le 0.006, 0.001 \le x_2 \le 0.004, 0.15 \le x_3 \le 0.26, 0.04 \le x_4 \le 0.08,$$
$$0.04 \le x_5 \le 0.08, 0.25 \le x_6 \le 0.40, 0.31 \le x_7 \le 0.46, \sum_{i=1}^{7} x_i = 1. \quad (36)$$

According to (15), we calculate

$$\breve{x} = \sum_{i=1}^{7} \breve{x}_i = 0.793, \;\; \overset{\frown}{y} = 1 - \breve{x} = 1 - 0.793 = 0.207 \tag{37}$$

and according to (16), we find

$$\begin{aligned}
\breve{x}^{0} &= 0, \overset{\frown}{x}^{0} = 0, \overset{\frown}{x}^{0} - \breve{x}^{0} = 0,\\
\breve{x}^{1} &= 0.002, \overset{\frown}{x}^{1} = 0.006, \overset{\frown}{x}^{1} - \breve{x}^{1} = 0.004,\\
\breve{x}^{2} &= 0.003, \overset{\frown}{x}^{2} = 0.01, \overset{\frown}{x}^{2} - \breve{x}^{2} = 0.007,\\
\breve{x}^{3} &= 0.153, \overset{\frown}{x}^{3} = 0.27, \overset{\frown}{x}^{3} - \breve{x}^{3} = 0.117,\\
\breve{x}^{4} &= 0.193, \overset{\frown}{x}^{4} = 0.35, \overset{\frown}{x}^{4} - \breve{x}^{4} = 0.157,\\
\breve{x}^{5} &= 0.233, \overset{\frown}{x}^{5} = 0.43, \overset{\frown}{x}^{5} - \breve{x}^{5} = 0.197,\\
\breve{x}^{6} &= 0.483, \overset{\frown}{x}^{6} = 0.83, \overset{\frown}{x}^{6} - \breve{x}^{6} = 0.347,\\
\breve{x}^{7} &= 0.793, \overset{\frown}{x}^{7} = 1.29, \overset{\frown}{x}^{7} - \breve{x}^{7} = 0.497.
\end{aligned} \tag{38}$$

From the above, since the expression (17) takes the form

$$\overset{\frown}{x}^{I} - \breve{x}^{I} < 0.207, \tag{39}$$

then it follows that the largest value $I \in \{0, 1, \ldots, 7\}$ such that this inequality holds is $I = 5$. Therefore, we fix the optimal solution that minimize linear function (29) according to the rule (19). Namely, we get

$$\begin{aligned}
\dot{x}_1 = \overset{\frown}{x}_1 = 0.006, \dot{x}_2 = \overset{\frown}{x}_2 = 0.004, \dot{x}_3 = \overset{\frown}{x}_3 = 0.26, \dot{x}_4 = \overset{\frown}{x}_4 = 0.08, \dot{x}_5 = \overset{\frown}{x}_5 = 0.08,\\
\dot{x}_6 = \overset{\frown}{y} - \overset{\frown}{x}^{5} + \breve{x}^{5} + \breve{x}_6 = 0.207 - 0.08 + 0.04 + 0.25 = 0.417, \dot{x}_7 = \breve{x}_7 = 0.31.
\end{aligned} \tag{40}$$

Finally, after making the substitution inverse to (33), we get the optimal transient probabilities

$$\begin{aligned}
\dot{p}_2 = \dot{x}_1 = 0.006, \dot{p}_3 = \dot{x}_2 = 0.004, \dot{p}_1 = \dot{x}_3 = 0.26, \dot{p}_5 = \dot{x}_4 = 0.08,\\
\dot{p}_7 = \dot{x}_5 = 0.08, \dot{p}_4 = \dot{x}_6 = 0.417, \dot{p}_6 = \dot{x}_7 = 0.31,
\end{aligned} \tag{41}$$

that minimize the mean value of the port oil terminal total operation cost $\boldsymbol{C}(\theta)$ during the operation time $\theta = 1$ year, expressed by the linear form (28) and according to (22) and (41), its optimal value is

$$\begin{aligned}
\dot{\boldsymbol{C}}(\theta) \cong 0.26 \cdot 1553110.88 + 0.006 \cdot 268320.64 + 0.004 \cdot 58858.528 + 0.417 \cdot 90183.04\\
+ 0.08 \cdot 130735.2 + 0.31 \cdot 655308.16 + 0.08 \cdot 1133107.008 \cong 747513 \text{ PLN}.
\end{aligned} \tag{42}$$

5 Summary

The procedure of using the semi-Markov model of complex technical system operation process [2, 3] and the liner programing [4] is proposed to minimize the system operation cost. Next, this procedure is applied to the optimization of the operation cost of the port oil terminal. The mean value of the port oil terminal total unconditional operation cost for 1-year operation was evaluated and minimize through its operation process modification. Presented in this paper tool can be useful in operation cost optimization of a very wide class of real technical systems operating at the varying conditions that have an influence on changing their functional structures and their operation cost at different operation states.

Acknowledgements. The paper presents the results developed in the scope of the research project "Safety of critical infrastructure transport networks" granted by GMU in 2020.

References

1. Grabski, F.: Semi-Markov Processes: Application in System Reliability and Maintenance. Elsevier, Amsterdam (2014)
2. Kołowrocki, K., Soszyńska-Budny, J.: Reliability and Safety of Complex Technical Systems and Processes: Modeling – Identification – Prediction – Optimization. English/Chinese Edition. Springer, Heidelberg (2011/2015)
3. Magryta, B.: Reliability approach to resilience of critical infrastructure impacted by operation process. J. KONBiN **50**(1), 131–153 (2020)
4. Klabjan, D., Adelman, D.: Existence of optimal policies for semi-Markov decision processes using duality for infinite linear programming. SIAM J. Contr. Optim. **44**(6), 2104–2122 (2006)
5. Lauge, A., Hernantes, J., Sarriegi, J.M.: Critical infrastructure dependencies: a holistic, dynamic and quantitative approach. IJCIP **8**, 6–23 (2015)
6. Kołowrocki, K., Magryta, B.: Port oil terminal reliability optimization. Sci. J. Marit. Univ. Szczecin (2020, to appear)
7. Kołowrocki, K., Soszyńska-Budny, J.: Safety indicators of critical infrastructure application to port oil terminal examination. In: Proceedings of the 29th International Ocean and Polar Engineering Conference – ISOPE 2019, Honolulu, pp. 569–576 (2019). ISBN 978-1 880653 85-2, ISSN 1098-6189

Graph-Based Street Similarity Comparing Method

Konrad Komnata(✉), Artur Basiura, and Leszek Kotulski

Department of Applied Computer Science, AGH University of Science and Technology, Al.Mickiewicza 30, 30-059 Kraków, Poland
{kkomnata,abasiura,kotulski}@agh.edu.pl

Abstract. This paper introduces Street Similarity Graph concept with examples of practical application. A new formal model allows for significant reduction of quantity of lighting situations to calculate during preparation of photometric design.

Keywords: Street Similarity · Graph · Road traffic · Smart city · Photometric calculations

1 Introduction and Motivation

Designers of smart city transformations want cities to be self aware and energy efficient. Our society is consuming more energy yearly. It is predicted, that world energy consumption rises nearly 50% between 2018 and 2050 [1]. Street lighting is one of most energy and finance absorbing aspects of modern smart cities [8]. There are two main ways of approaching this problem:

- Hardware-based approaches, like moving to LED technology, increasing efficiency of luminaires [2],
- Optimization of lighting design and proper, software-based management and dynamic control [16].

As improvement of manufacturing of luminaires is domain of their producers, we can improve software-based approach. Creating optimal photometric design is non trivial task with many cases to be considered by designer. Key for proper street lighting calculations are modern solutions based on graph transformations. Graph transformations are also used for dynamic street lighting control [16]. Applying dynamic street lighting with compliance with ISO 13201 standard [3] is proven to give possibility to save even 40% of energy [15]. Photometric calculations of the outdoor lighting design are very time and resource consuming process, which can take even up to many days to complete. In this paper Street Similarity Graph concept is presented which allows reduction of quantity of lighting situations to calculate, thus improves efficiency of whole process.

W. Zamojski et al. (Eds.): DepCoS-RELCOMEX 2020, AISC 1173, pp. 366–377, 2020.
https://doi.org/10.1007/978-3-030-48256-5_36

2 State of the Art

Photometric calculations are non trivial task, time and resource consuming. For example, having street segments lit with street lights, for each lighting point there are multiple parameters [3] that have to be taken into calculations: 11 pole locations, 6 arm lengths, 10 distances from neighbors, 2 neighboring poles and their influence, 5 fixture inclinations, 5 fixture rotations, 500 fixture models and 76 dimming level, which gives 1 254 000 000 different combinations to consider and verify that everything has been taken into consideration and properly calculated. In a project of lighting modernization in Kraków, Poland, there are 835 street segments. Assuming that every street segment has about 5 lamps lighting it, this gives us 5 235 450 000 000 of situations to calculate. Research shown that using agent environments and graph transformations can significantly improve time of calculations. In [6], authors introduced new multi-agent environment supported by graph transformations, that has been developed further [13] proving that graph transformations are helpful in planning large scale lighting installations [12] and modernization [14]. Utilization of graph-based methods of calculations can lead to significant energy reduction [10] and CO_2 emission reduction [11]. One of main problems with photometric calculations is that every situation has to be calculated separately and independently, so they take much time to complete. Street Similarity Graph brings help to this situation as it reduces quantity of situations to calculate and shortens time of calculating whole project. To properly build Street Similarity Graph, which is described in following sections, Traffic Flow Graph is needed. Traffic Flow Graph (TFG) which has been introduced in [4], models how traffic flows between street segments. Nodes in TFG represent street segments, where edges show traffic flow between them. Edge labels bring information about traffic distribution. Example TFG is presented on Fig. 1 with marked weights of traffic distribution (w_i).

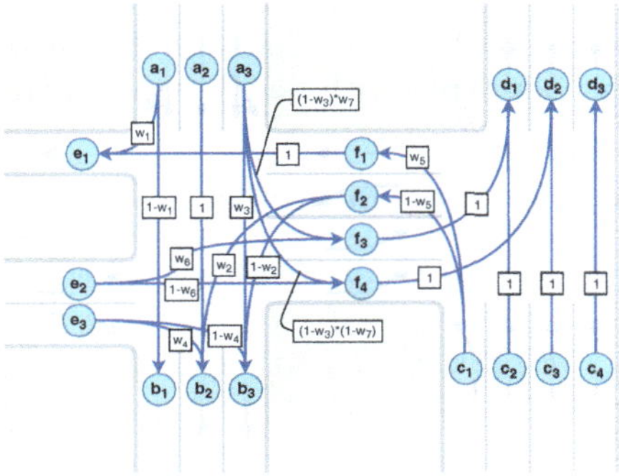

Fig. 1. Traffic Flow Graph example

3 Street Similarity Graph

This chapter presents idea of Street Similarity Graph. At first idea is described, with information about graph structure and its attributes. Next sections describe formal model of Street Similarity Graph and algorithm of its creation.

3.1 General Description

Let us assume street segment to be homogeneous part of the road which has the same lighting parameters on its all length. To distinguish how street segment should be lit, many variables have to be taken into consideration. Street Similarity Graph is designed to represent all variables required by Standard CEN/TR 13201-1:2014 [3].

There are two types of nodes in Street Similarity graph:

- nodes representing street segments,
- nodes representing lighting parameters, these node are having multiple types (Lighting Class, Lanes, Width, etc.).

Also there are two types of edges:

- edges connecting segment nodes with each other, representing possible traffic flow,
- edges connecting lighting parameters nodes with segments.

For every segment, there is only one node. Every possible lighting parameter is represented by single node, with its value. Each lighting parameter node has its type and value type. Parameter node type tells what information is held there, i.e. lighting class, value type informs if node value is picked from finished list of categories or it is numerical. Parameter nodes be connected to multiple segment nodes, i.e. parameter node of type 'Lighting Class' with value 'M2' can be connected to many segments, which tells that these segments have primary lighting class calculated to be M2. Each lighting class has its own node.

Lighting parameter node types with possible values are presented in Table 1.

3.2 Street Similarity Graph Formal Model

Definition 1. *Street Similarity Graph is a directed graph described by a graph grammar:*

$$\Psi = (V, E, \Sigma, \Upsilon, \varrho, \delta, \omega, \lambda, I, \Theta) \quad (1)$$

where:

- *V is a set of nodes, distinguished by indexing function I,*
- *$E \subset V \times V$ is a set of edges,*
- *Σ is a set of node labels,*
- *Υ is a set of edge labels,*
- *ϱ is a set of node attributes,*

Table 1. Parameter nodes description

Node type	Value type	Description	Possible values
Lighting class	Category	Lighting class assigned to this part of road	M2, M3, M4, Me3c, Me4b
Speed limit	Category	Maximum speed allowed for traffic participants	High, Moderate, Low
Traffic volume	Category	Traffic intensity	High, Moderate, Low
Ambient Luminosity	Category	Ambient luminosity	High, Moderate, Low
Carriageway separation	Category	Information if carriageways are separated	True, False
Junction dencity	Category	How often are junctions on this segment	High, Moderate, Low
Navigational task	Category	Information if it is hard to navigate on this street	Easy, Hard
Parked vehicles	Category	Tells if there are parked vehicles by the street	True, False
Traffic composition	Category	Information if traffic consists of only motorized or mixed vehicles	Mixed, motorised-only
Distribution	Category	Distribution of lamps by the street	Unilateral - left, unilateral - right, bilateral
Lanes	Category	Number of lanes	1, 2, 3, 4, 5
Length	Number	Length of road	160 m, 36.5 m
Width	Number	Width of road	5 m, 6.3 m
Area	Number	Area of road	0.3 km^2, 1 km^2
Envelope	Number	Envelope of road	3 km, 2 km

- $\delta : V \longrightarrow \Sigma$ *is the node labeling function,*
- $\omega : V \longrightarrow \varrho$ *is the node attributing function,*
- $\lambda : E \longrightarrow \Upsilon$ *is the edge labeling function,*
- I *is a function from* V *to* δ *indexing nodes,*
- Θ *is an ordering relation in the set of edge labels,*
- $e = (a,b)$, *where* $e \in E$, *understood as a directed edge from* a *to* b,

and following condition is fulfilled: $V, E \neq \emptyset$.

There are two types of node labels - Segment and Property, where Property nodes consist of Category Properties and Numeric Properties. Segment nodes represent part of street which has consistent photometric parameters through all its length and width. Segment node attributes are holding information that identifies street segment (i.e. coordinates). Category nodes represent values that describe photometric parameters of street segments. Category node attributes inform about type of category type (category or numeric), specific node type (i.e. speed limit, length, width) and value (i.e. 50 km/h, 168 m, 15 m). Edges connecting Segment and Category nodes carry information about which segment has which properties. Edges connecting Segments inform, in their attributes, about difference between segments. Only edges connecting segments have attributes, which inform about Measure of Difference between each two segments. This definition is enhanced general graph definition, allowing synchronization with other graph structures.

3.3 Algorithm of Street Similarity Graph Creation

Traffic Flow Graph, which was first introduced in [4], models traffic flow between street segments in a directed graph. Formal model of Street Similarity Graph has been designed to be compatible with Traffic Flow Graph so it can profit from information that Traffic Flow Graph carries. Traffic Flow Graph is needed to fully create Street Similarity Graph as it provides relationships of traffic flow between streets. At first we need to transform information carried in Traffic Flow Graph into table and enhance it with all street attributes, from which we want to build Street Similarity Graph.

Assuming:

- 'Row' as row in Table containing all information about street we want to add to Street Similarity Graph,
- CategoryProperties - as a list of all categorical properties for this way,
- NumericProperty - as a list of all numeric properties for this way, as listed in Table 1,
- ssgWay - as a node representing part of street in Street Similarity Graph,
- tfgWay - as a node representing same part of street in Traffic Flow Graph,
- streetId - as an unique identifier of part of street shared between Street Similarity Graph and Traffic Flow Graph

```
for Row in Table do
    create ssgWay;
    for CategoryProperty in CategoryProperties do
        if CategoryProperty node exists in graph then
            connect to ssgWay with 'hasProperty' relationship;
        else
            create CategoryProperty;
            connect CategoryProperty to ssgWay with 'hasProperty'
            relationship;
        end
    end
    for NumericProperty in NumericProperties do
        if NumericProperty node exists in graph then
            connect to ssgWay with 'hasProperty' relationship;
        else
            create NumericProperty;
            connect NumericProperty to ssgWay with 'hasProperty'
            relationship;
        end
    end
end
for ssgWay in ssgWays do
    tfgNode := tfgNode with the same streetId as ssgWay streetId
    for targetTfgNode in tfgNodes connected to tfgNode do
        targetSsgWay := ssgWay with the same streetId as targetTfgNode;
        connect ssgWay to targetSsgWay with 'trafficFlows' relationship
        and the same direction;
    end
end
```

Algorithm 1: Creation of Street Similarity Graph

4 Similarity Detection

Main value of Street Similarity Graph is that it brings information of similarity between street segments. To properly measure how much street segments differ between each other, we introduce definitions of Measure of Difference, similarity levels and Key Street Segments. As numeric and category properties should be calculated with different manner, there are separate definitions to calculate Measure of Difference of numeric properties (Definition 2) and category properties (Definition 3). Definition 4 defines how to calculate overall Measure of Difference between two street segments.

Definition 2. *Measure of Difference of numeric properties of two street segments in Street Similarity Graph*

$$MDn_{(S1,S2)} = \frac{|S1.n - S2.n|}{\frac{S1.n+S2.n}{2}} * 100\% \tag{2}$$

Where:

- *S1, S2 - compared street segments,*
- *n - numeric property (i.e. length, width).*

Definition 3. *Measure of Difference of category properties of two street segments in Street Similarity Graph*

$$MDc_{(S1,S2)} = 100 - \frac{\sum CCN_{S1,S2}}{CN} \quad (3)$$

Where:

- *c - category properties,*
- *S1, S2 - compared street segments,*
- *CCN - common category nodes, that are shared between S1 and S1 nodes in Street Similarity graph,*
- *CN - quantity of all category types available in Street Similarity Graph, which is constant for instance of Street Similarity Graph.*

Definition 4. $MD_{(S1,S2)}$ *is Measure of Difference of two Street Segments, which is arithmetic average of Measure of Difference of Category Properties and all Measures of Difference of Numeric Properties of two Street Segments.*

Definition 5. *Considering* $MD_{(S1,S2)}$ *as Measure of Difference of street segment nodes in Street Similarity graph, segments S1 and S2 are:*

- *Very Similar when* $MD_{(S1,S2)} < 10\%$*,*
- *Similar when* $20\% < MD_{(S1,S2)} \leq 10\%$*,*
- *Slightly Similar when* $MD_{(S1,S2)} \geq 20\%$*.*

Definition 6. *Key Street Segments in Street Similarity Graph are streets segments that have highest number of exclusive similar street segments.*

4.1 Street Similarity Graph Example

Figure 2 shows example with only part of initial Street Similarity Graph (generated by Algorithm 1), containing 4 segment nodes and 5 Category Property type nodes (CS - carriageway separation, PV - parked vehicles, TC - traffic composition, TV - traffic volume, JD - junction density). Considering Street Similarity Graph build only with those parameters we can calculate Measures of Difference between all segments. Measures of Difference, calculated with formula from Definition 4 are presented in Table 2.

After calculating differences between each segment nodes, Measures of Difference can be added to new edges in Street Similarity Graph which result in creation of Final Street Similarity graph. Result of adding them to graph is presented on Fig. 3.

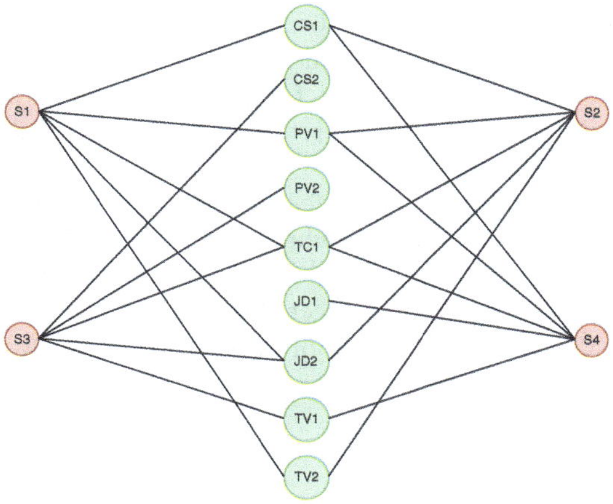

Fig. 2. Initial Street Similarity Graph with similar nodes

Table 2. Measures of Difference between segments in example graph

Segment	S1	S2	S3	S4
S1	x	0%	80%	40%
S2	0%	x	80%	40%
S3	80%	80%	x	80%
S4	40%	40%	80%	x

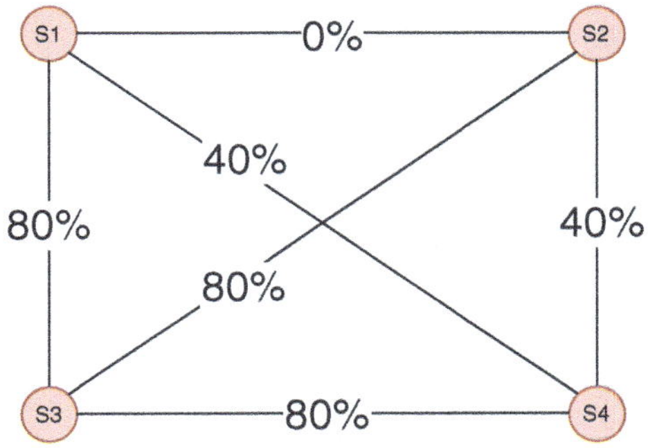

Fig. 3. Final Street Similarity Graph with marked Measures of Difference

5 Photometric Calculations - Real Data Example

Having real and accurate data from part of one of major Polish cities - Kraków, we were able to feed Street Similarity graph model with them. Photometric calculations are very time and resource consuming process. As mentioned in Sect. 2, full calculations can take even years to finish. Considering one calculation to take 1ms, such project would take 166 years to calculate. Publications [9], [5], [7] proved that this time can be shortened to few hours.

We have created Street Similarity Graph to represent situations for this project and verified how much we can reduce situations to calculate with different Measures of Difference thresholds.

We have calculated Measures of Difference for each segment and verified how many segments should be calculated when given thresholds - 1%–20%. To find how many situations are necessary to calculate, we have calculated Key Segments in Street Similarity Graph for this project, for every given Measure of Difference.

Knowing that 835 is total quantity of all segments that we had to calculate, Table 3 shows how many segments had to be calculated with given Measures of Difference.

Table 3. Segments to calculate with various Measures of Difference

Measure of Difference	Segments to calculate
1%	795
2%	707
3%	597
4%	499
5%	393
6%	347
7%	317
8%	289
9%	262
10%	234
11%	208
12%	183
13%	165
14%	147
15%	132
16%	120
17%	112
18%	106
19%	103
20%	89

Looking at Fig. 4, we can notice, that when taking Measure of Difference equal to 5%, which can be considered as very small difference, we had to calculate only 47.07% of all segments. Taking threshold of 10% (see Definition 5), only 28.03% of all segments had to be calculated, so time of photometric verification of all situations shortened to less than 1/3. Figure 4 shows how much we can shorten photometric calculations with properly chosen Measure of Difference threshold.

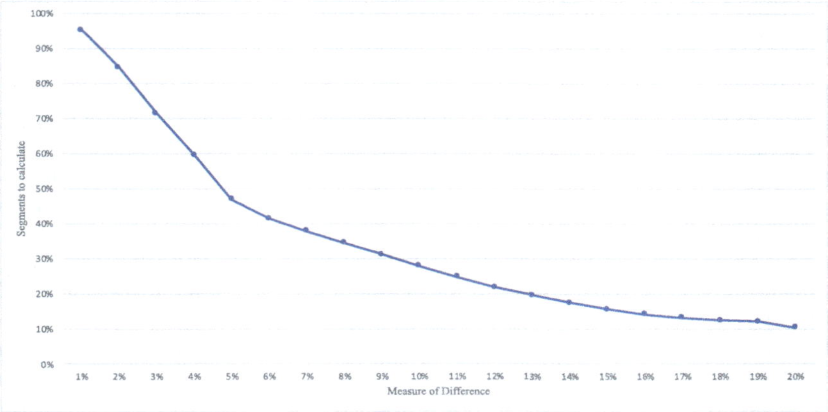

Fig. 4. Percentage of all street segments to calculate with given Measure of Difference

6 Conclusions

Introducing new formal model of Street Similarity Graph allows for significant reduction in the number of photometric calculations in photometric projects. It is especially useful in bigger projects (such as Tbilisi - 100 000 lighting points, Washington - 50 000 lighting points), where calculations lasted several dozen hours for one producer of luminaires. For other we had to repeat. Street Similarity Graph allows for shortening this time to few hours. While the lighting renovations: we prepare design in two stages of the renovation process:

- in audits one: we prepare simplified design to establish parameters for public tender,
- in the exchange one: when the final design of the exchanged lighting luminaries is necessary

In the audits phase the calculations with 20% Measure of Difference is sufficient and the time design preparation will be increased up to 10 times. In the exchange stage more precise calculations (with Measure of Difference $\leq$ 5%), our new model allows shortening calculations by reduction of situations even by 52%.

The acceleration of the computation time is only the one example of the usefulness of Street Similarity Graph concept, the other are:

- the designation of proper traffic parameters in the dynamic control outdoor lighting systems,

– the preparation of the proper designs in the case of partial lack or uncertain quality of inventory data.

Both cases are a subject of the intensive research.

References

1. International energy outlook 2019. Tech. rep. U.S. Energy Information Administration, Office of Energy Analysis, U.S. Department of Energy, Washington, DC 20585 (2019)
2. Burgos-Payan, M., Correa-Moreno, F., Riquelme-Santos, J.: Improving the energy efficiency of street lighting. a case in the south of Spain. In: 2012 9th International Conference on the European Energy Market, pp. 1–8, May 2012
3. CEN: CEN/TR 13201-1:2014, Road lighting – Part 1: Guidelines on selection of lighting classes. Tech. rep. European Committee for Standardization, Brussels (2014)
4. Ernst, S., Komnata, K., Łabuz, M., Środa, K.: Graph-based vehicle traffic modelling for more efficient road lighting. In: Zamojski, W., Mazurkiewicz, J., Sugier, J., Walkowiak, T., Kacprzyk, J. (eds.) Engineering in Dependability of Computer Systems and Networks, vol. 987, pp. 186–194. Springer International Publishing, Cham (2020)
5. Gomez-Lorente, D., Rabaza, O., Estrella, A.E., Pena-Garcia, A.: A new methodology for calculating roadway lighting design based on a multi-objective evolutionary algorithm. Expert Syst. Appl. **40**(6), 2156–2164 (2013)
6. Kotulski, L., Sędziwy, A.: GRADIS - the multiagent environment supported by graph transformations. Simul. Model. Pract. Theory **18**(10), 1515–1525 (2010)
7. Rabaza, O., Pena-Garcia, A., Perez-Ocon, F., Gomez-Lorente, D.: A simple method for designing efficient public lighting, based on new parameter relationships. Expert Syst. Appl. **40**(18), 7305–7315 (2013)
8. Richon, C., Mukish, P.: Led in road & street lighting market analysis, applications, technology trends and industry status. Tech. rep. Yole Développement SA and Lux Fit SAS (2013)
9. Sędziwy, A.: A new approach to street lighting design. LEUKOS **12**(3), 151–162 (2016)
10. Sędziwy, A., Basiura, A.: Energy reduction in roadway lighting achieved with novel design approach and leds. LEUKOS **14**(1), 45–51 (2018)
11. Sędziwy, A., Basiura, A., Wojnicki, I.: Roadway lighting retrofit: Environmental and economic impact of greenhouse gases footprint reduction. Sustainability **10**(11), 3925 (2018)
12. Sędziwy, A., Kotulski, L.: Solving large-scale multipoint lighting design problem using multi-agent environment. Key Eng. Mater. **486**, 179–182 (2011)
13. Sędziwy, A., Kotulski, L., Basiura, A.: Agent aided lighting retrofit planning for large-scale lighting systems. In: Nguyen, N.T., Gaol, F.L., Hong, T.P., Trawiński, B. (eds.) Intelligent Information and Database Systems, vol. 11431. Springer International Publishing, Heidelberg (2019)
14. Sędziwy, A., Kotulski, L., Basiura, A.: Multi-agent Support for Street Lighting Modernization Planning. In: Nguyen, N., Gaol, F., Hong, T.P., Trawiński, B. (eds.) Intelligent Information and Database Systems, vol. 11431, p. 442–452. Springer International Publishing, Cham (2019)

15. Wojnicki, I., Komnata, K., Kotulski, L.: Comparative study of road lighting efficiency in the context of cen/tr 13201 2004 and 2014 lighting standards and dynamic control. Energies **12**(8), 1524 (2019)
16. Wojnicki, I., Kotulski, L.: Improving control efficiency of dynamic street lighting by utilizing the dual graph grammar concept. Energies **11**(2), 402 (2018)

Hybrid Method of the Radio Environment Map Construction to Increase Spectrum Awareness of Cognitive Radios

Krzysztof Kosmowski and Janusz Romanik(✉)

Radiocommunications Department, Military Communication Institute, Warszawska 22A, 05130 Zegrze Poludniowe, Poland
{k.kosmowski, j.romanik}@wil.waw.pl

Abstract. The paper presents the concept of the hybrid method for the Radio Environment Map (REM) construction. REMs are considered as a promising solution for Cognitive Radios (CR) because they can raise the awareness of the electromagnetic environment. This issue is particularly important for the dynamic spectrum management (DSM) since it has an impact on the quality of services provided by the network. The proposed hybrid method combines direct and indirect methods with the aim to achieve higher accuracy of maps. In our previous papers we analyzed the quality of maps created with the use of measurement results and selected direct methods. One of the conclusions was that Kriging interpolation technique is very promising since it offers the highest quality of maps. However, some limitation of the direct methods is the necessity of collecting measurement data from networks with a large number of sensors. The proposed hybrid method uses the results of measurements taken by sensors deployed in a real environment and also the results of the calculations based on the propagation model. The main idea of the method is to use a small number of sensors to adjust the propagation model. In the next step this adjusted propagation model is used to calculate the signal level for the set of points that are treated as virtual sensors and that are applied to increase the quality of maps created with the use of a selected interpolation technique.

Keywords: Cognitive radio · Radio environment map · Spectrum monitoring · Propagation models · Interpolation techniques

1 Introduction

The widespread and rapid development of radio technology and wireless systems results in an increasing demand for spectral resources. Those are very limited and nowadays less and less accessible. Thus, various ideas taking as a goal more effective spectrum utilization are proposed on the civilian and military market. They concern various aspects related to the frequency planning [1], more intelligent radios which are able to choose the best available channel [2, 3] as well as the idea of cognitive radio [4] and dynamic spectrum management [5]. In all of these proposals, at a very early stage some basic assumption on available frequencies has to be made. However, to obtain the effective spectrum utilization in real situations it is required to use some a posteriori

W. Zamojski et al. (Eds.): DepCoS-RELCOMEX 2020, AISC 1173, pp. 378–388, 2020.
https://doi.org/10.1007/978-3-030-48256-5_37

knowledge about utilization of frequency channels in the past. For example, for the development of a cognitive radio which will opportunistically use licensed bands, it is necessary to know how often this band is occupied by licensed users. Such capability can be provided by the Radio Environment Map (REM). In general, the aim of REM is to increase electromagnetic environment situational awareness [8]. However, the question how to use data collected in REM database still remains open and as a result a lot of proposals are presented in the literature. One of them is the problem of locating the transmitter with the use of signals received by sensors [6].

In the paper our attention is focused on the construction of REM with the use of real sensors data and propagation models. The remaining part of the paper is organized as follows: in the second chapter some background information about REM concept and related works is presented and this chapter is also devoted to propagation models that we use as well as the adjustment procedure; Sect. 3 presents the concept of the hybrid method; Sect. 4 shows results in the form of exemplary maps and quality analysis, and finally in Sect. 5 we present our conclusions.

2 State of the Art

2.1 REM Concept

In the literature on the topic of REM different construction techniques are analyzed [7]. The two main categories are *direct methods*, also known as a *spatial statistics based methods,* and *indirect methods*, which are also referred to as *transmitter location based methods*. Hybrid methods combine the two approaches.

In some papers *indirect methods* are preferred due to the fact that they help to construct more accurate maps if the propagation model has been selected properly. On the other hand, *indirect methods* require a set of input data, e.g. transmitter location, the TX power and activity pattern of the transmitter as a minimum.

However, in some systems the TX power is adapted to the channel conditions and what is more, the position of the transceiver may change unpredictably over time, e.g. in MANETs. Moreover, the activity pattern of the transmitter may be unknown and random. In such environment *direct methods* seem to be more appropriate. These methods use input data from the sensor network that monitors the spectrum in a continuous way. Since the sensor network provides the results of measurements that correspond to the points in the space where sensors have been deployed, different interpolation techniques are applied in order to estimate the signal level throughout the whole area. The best known interpolation techniques are Nearest Neighbor (NN), Inverse Distance Weighting (IDW) and Kriging. A more detailed description of the techniques mentioned above is presented in [8]. In some papers the interpolation method based on Kriging is foreseen as promising for REM construction [7, 8] and also the modifications of this technique are proposed with the aim to obtain higher quality of maps [9].

In the literature on the topic the impact of the sensor network density and different arrangements of sensors on the map quality were presented. However, the assumed number of sensors usually ranged from a few dozen up to several hundred. In practice, the number of sensors may be significantly limited.

In [10] the authors presented the results of the experiment conducted in real conditions with the aim to determine the position of a transmitter operating at 800 MHz frequency with the application of the indirect method. The TX antenna was located inside a grid consisting of 49 nodes in a 7×7 arrangement, spaced 5 m apart. In [11] the authors described a method of searching for White Spaces in UHF frequency band (470–900 MHz) and the results of tests that were conducted in a real environment with 100 sensors deployed in the area of 5 km^2. In [12] the authors discussed three methods relevant for REMs creation: path loss based, Kriging based and their own method. To compare the quality of maps constructed with these methods a series of simulations were carried out for the scenario with one transmitting station, 81 sensing nodes and 8 validating nodes (81 sensors and 8 validating nodes did not overlap). All the nodes were deployed on the area of 70 m by 70 m.

In our previous paper [13] we confirmed strong dependency between the sensor network density and the map accuracy. In the experiments conducted in real conditions we used a limited number of sensors ranging from a dozen up to a few dozen. The sensors were deployed on the area of 4 km^2. In general, the increase in the number of sensors from 13 to 26 caused a visible improvement in the quality of maps. In [14] we analyzed the influence of the sensors arrangement on the map accuracy. Our experiments were carried out in real conditions and included two scenarios: the first one with 13 sensors and the other one with 20 sensors. For each scenario we generated 3 tests with different deployments of sensors on the area of 4 km^2. The main conclusion was that an increased number of sensors in the network is beneficial since it causes a drop in the RMSE and thus the quality of maps improves significantly. If the number of sensors in the network is limited, the attention should be paid to the optimum deployment of sensors. We confirmed that even a slight rearrangement of the sensors that were originally randomly deployed may increase the accuracy of maps [14]. In the remaining part of the paper we use the following notation for IDW method: IDW px where x is the power.

2.2 Propagation Models

Radio wave propagation is one of the most important topics in the area of radio communications. A lot of models as well as statistics and reference data are presented in the literature. Variety of aspects related to the propagation issue is particularly well visible in the ITU-R's P-Series of recommendations [16]. In that paper signal strength was obtained on the basis of the Close-in and the Longley-Rice wave attenuation models.

Longley-Rice model is well known and widely discussed in the literature and consequently, it is not necessary to characterize it in detail again. The Close-in model implements a statistical path loss model and can be configured for different scenarios. The default values correspond to an urban macro-cell scenario in a non-line-of-sight (NLOS) environment [15]. Both models have some sets of parameters which have to be defined. For the Longley-Rice model, among others, parameters related to the climate zone and ground characteristics have to be included. The Close-in model takes into account different parameters, e.g. free space reference distance and path loss exponent. In both cases the default sets were used in our experiments. Both models use terrain elevation data. In this case DTED2 (Digital Terrain Elevation Data Level 2) maps were applied. Although we used default parameters in the first step of our experiment, then we adjusted attenuations calculated by theoretical models on the basis of data coming from real measurements. To adjust propagation models the following procedure was adopted: 1) for simulations the transmitter was placed according to coordinates relevant for physical transmitter; 2) signal attenuation was calculated for sensors' coordinates; 3) the median values were calculated for measured data and for theoretically calculated attenuations; 4) the difference between medians was used to modify the calculated attenuation.

This idea is depicted below for exemplary data coming from 39 sensors and for a known transmitter. The differences between signal strengths measured and calculated according to propagation models are depicted in Fig. 1.a.

As one can see both curves are similar and they differ by the values on the y-axis. However, Close-in model produces attenuation approximately 20 dB higher than the other one, which is obvious for an urban scenario.

According to the above procedure the median values were calculated and then corrections were introduced to the original results of the calculations. The median values are as follows: measurements: −95.4260 dBm, Longley-Rice: −56.1336 dBm, and Close-in: −79.4987 dBm. The signal strengths for both models with median correction and the measurement results are presented in Fig. 1.b.

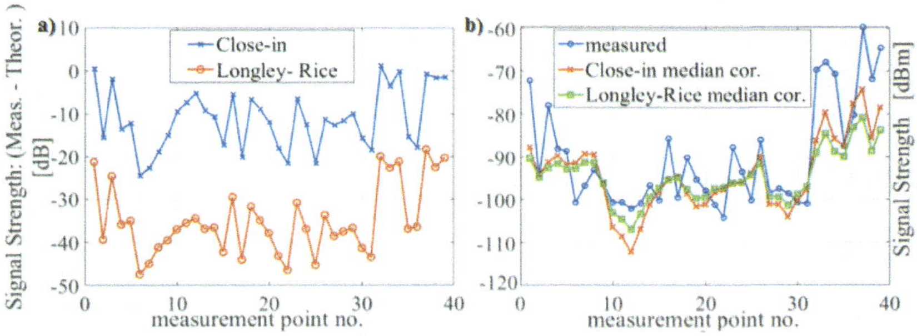

Fig. 1. a) Differences between measured and calculated signal strength, b) measured signal strength for 39 sensors and theoretical values after median correction

3 The Concept of the Hybrid Method

The procedure of adjustment of propagation models presented above requires some information about the transmitter site, i.e. location, power, and antenna height. Measurement data from the sensors are not sufficient for this purpose, however they make it possible to asses probable parameters of the transmitter. To obtain necessary data the optimization algorithm has been created.

In the first step the measurement results from physical sensors are used to create a preliminary map, which is then applied for the selection of the most probable transmitter location. After that, a set of random locations in this area is selected (in our experiment 20 locations were used). For all these probable locations the signal strengths are calculated for selected coordinates (in our case for all the locations of physical sensors). The most probable transmitter location is chosen on the basis of the lowest value of Root Mean Square Error (RMSE) calculated in the following way:

$$RMSE_k = \sqrt{\frac{\sum_{i=1}^{n}(X_i - \bar{X})^2}{n}}, \quad X_i = S_{k,i} - S_i \tag{1}$$

where: k – probable location of the transmitter, $k = 1, 2, \ldots, 20$; n – number of physical sensors; $S_{k,i}$ – calculated signal strength from *k-th* transmitter location at *i-th* sensor location; S_i – measured signal strength at *i-th* sensor location.

In the next step, taking into account only the transmitter location with the lowest RMSE, different sets of the transmitter power and the antenna height are used in simulation models to find the most probable values of these parameters. It is obvious that there are a lot of unknown aspects such as antenna gain, system loss and others. Moreover, the trade-off between the radiated power and the antenna height above the ground has to be considered, i.e. higher antenna placement and lower transmitter power give the same signal strength at some distant point as lower antenna placement with higher power. After this procedure the propagation model is adjusted with the respect to the median values as it was described in Sect 2.2. The last point of the whole algorithm is to generate a set of signal strengths for all the locations of the so-called virtual sensors, which will be used to create REM map. The idea of the application of virtual sensors is presented in the next chapter.

4 Results and Quality Analysis

4.1 Test Scenario

In order to check the efficiency of the hybrid method of REM construction several experiments were conducted. Firstly, measurements were taken in a real environment with the known position of the transmitter and with the sensor network that consisted of 39 sensors. We used 9 sensors for the interpolation process while the remaining 30 sensors served as control sensors. In our measurements we used the frequency of 1997 MHz that was assigned for this purpose with the output power of 40 dBm. As the sensor network we used one mobile sensor that moved within a preliminary area and stopped at certain positions to take measurements (the locations are shown in Fig. 2). The sensors marked with purple dots, bold fonts and named IP_X were used to get the input data for the interpolation process, while the sensors marked with orange stars and named CP_X are control sensors that were used to check the quality of created maps (where X is the sensor ID). It is worth noting that for the interpolation process we intentionally omitted the sensors located very close to the transmitter that measured high signal levels. Instead, only the sensors that measured medium and low levels were taken into account for the interpolation. The position of the TX Antenna is represented by a blue dot.

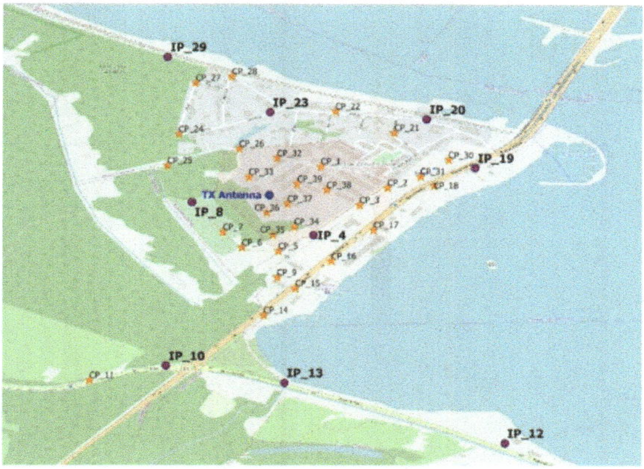

Fig. 2. Area of Zegrze Lake with the position of the TX antenna and deployment of sensors.

In the second phase of our research work we used measurement data from 9 sensors (IP_X) shown in Fig. 2 to: (a) locate the transmitter and estimate the TX power, (b) adjust the propagation model according to our algorithm described in Sect. 3, (c) calculate the signal level in Matlab simulation tool for the whole analyzed area. In the next step of our experiment we created the network of virtual sensors (144 sensors) deployed densely and regularly on the area of 4 km^2, with the signal level taken from the calculations. Then, we constructed maps with the use of selected interpolation techniques on the basis of the virtual sensors only. Finally, we analyzed and compared the quality of maps on the basis of the calculated RMSE for 30 control sensors.

4.2 Exemplary Maps

Exemplary maps created with selected interpolation techniques for the scenario with input data from 9 sensors are shown in Fig. 3.

NN interpolation technique (Fig. 3.a) creates polygons around sensors, which can be identified through different colors representing the signal level measured by the sensor. The size of the polygons and their shape is determined by the number of sensors and the way of their deployment. The signal level changes at the edges of the polygons. Because of the fact that there were no sensors placed close to the TX antenna, the highest signal level represented by yellow color reaches approximately −85 dBm.

For the IDW method the effect of bull's-eye occurs around each sensor. The size of the bull's eye depends on the power p used in the interpolation process. If the power p is set to 1 (Fig. 3.c), the range of the sensor is low and the eye is small. When the power p is set to 3 (Fig. 3.d), the range of the sensor is higher and thus the eye is much bigger. This effect was described in more detail in [8]. The sensors that measured the

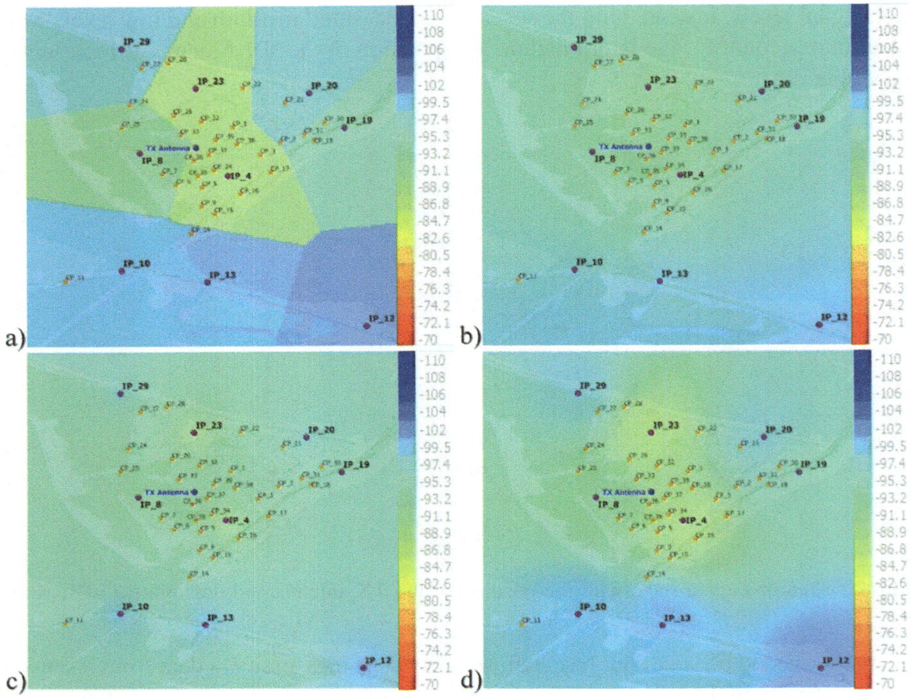

Fig. 3. Exemplary maps for the scenario with 9 sensors used for interpolation (scale in dBm): a) NN method, b) Kriging, c) IDW p1, d) IDW p3.

highest signal level can be identified through the yellow circular areas that surround them (the signal level approx. −85 dBm).

Kriging interpolation technique (Fig. 3.b) creates smooth maps without sudden changes in the signal level or bull's-eye effect. Since there were no sensors located close to the TX antenna and two sensors measured similar signal level, the area with the highest interpolated level (green color – approx. −90 dBm) in that case is extensive.

Maps created with the input data from 144 virtual sensors (Close-in propagation model) that were arranged in a regular way are shown in Fig. 4.

NN interpolation technique (Fig. 4.a) created regular polygons with virtual sensors in the center. The highest signal level, which occurs for only one polygon and reaches −70 dBm, enables us to estimate the position of the TX antenna. The neighboring polygons are marked with orange and yellow color, while distant polygons are blue. There are no sharp changes in the signal level at the borders of the polygons.

For the IDW method with the power p set to 3 (Fig. 4.d), there are four dominating areas with strong signal level. The highest level reaches up to −72 dBm for one bull's-eye, therefore the position of the transmitter can be estimated with quite good accuracy. The areas that are further away from the presumed TX antenna position are represented by yellow color, while the distant ones are blue. When the power p for the IDW method

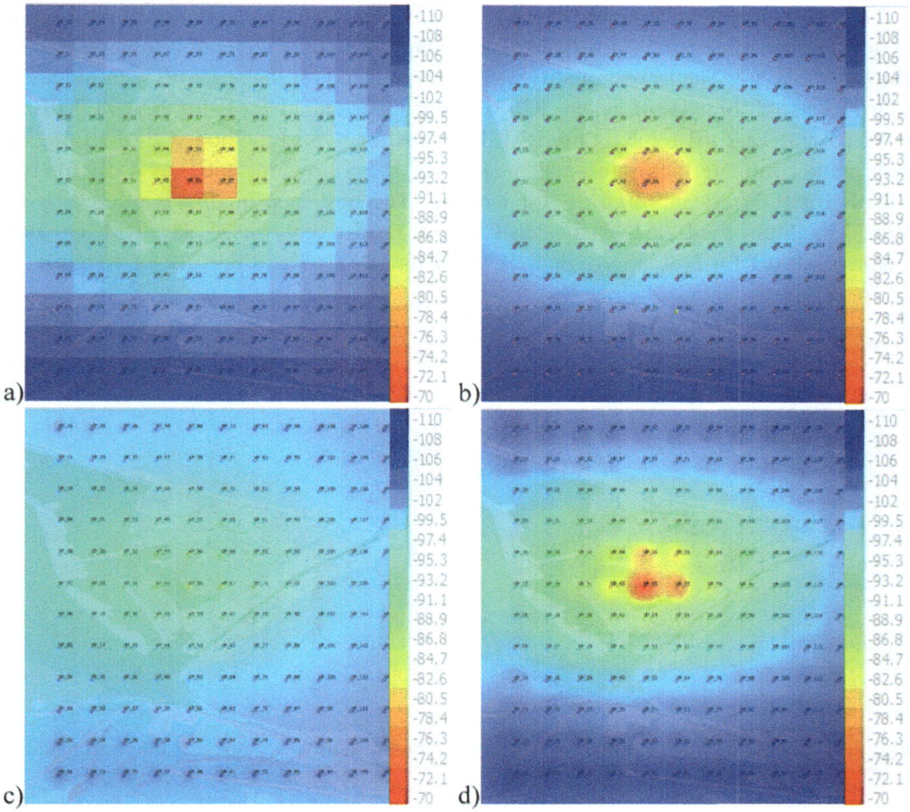

Fig. 4. Maps for hybrid method with 144 virtual sensors and Close-in model (scale in dBm).

takes a small value ($p = 1$), the points with strong signal level are less distinctive (Fig. 3.c). Moreover, between these points the map is blue, which seems to be unnatural.

If the Kriging interpolation technique is applied (Fig. 4.b), the map is smooth all over the area under analysis. The highest level of the signal represented by orange color reaches up to −75 dBm. The orange area in the center of the map enables us to estimate the location of the TX antenna with moderate precision.

4.3 Quality Analysis

For the quality analysis we used the RMSE values for 30 control sensors and for maps constructed with selected interpolation techniques. We compared the quality of the maps for the following three cases: a) input data from 9 sensors; b) hybrid method with the adjusted Longley-Rice model and input data from 144 virtual sensors; c) hybrid method with the adjusted Close-in model and input data from 144 virtual sensors. The results of the comparison are shown in Fig. 5.

The comparison of the calculated RMSE values confirmed the application of the hybrid method is beneficial for REM construction. The only exception to the statement above is IDW p1. The lowest quality maps were created for the scenario with input data from 9 sensors (blue bars), where RMSE values reached 11.5 dB for NN, 11.3 dB for IDW p3 and 12.2 dB for Kriging. RMSE values for hybrid method with the Longley-Rice model (orange bars) are smaller and take the following values: 9.4 dB for NN, 9.9 dB for IDW p3 and 10.6 dB for Kriging. The best map quality is offered by the hybrid method based on the adjusted Close-in model (grey bars). The RMSE values reached 7.8 dB for NN, approximately 9 dB for IDW p3 and 8.9 dB for Kriging.

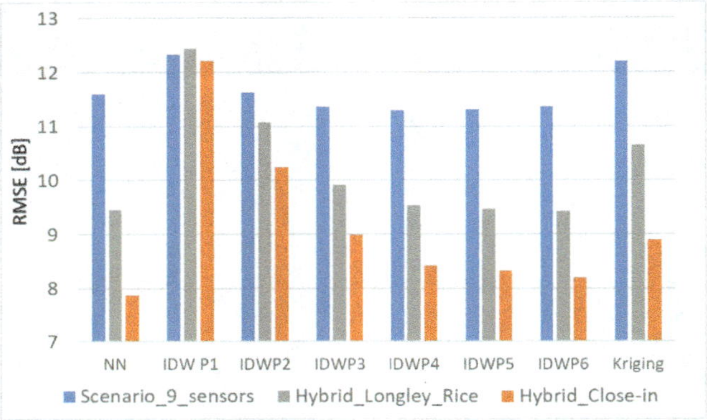

Fig. 5. Calculated RMSE (in dB) for the scenario with 9 sensors for direct and hybrid methods.

5 Conclusions

In this paper we presented the concept of the hybrid method for REM construction, which may be applied in scenarios with a limited number of sensors. In our experiments we used input data from 9 sensors to construct maps with the use of the direct method as well as the hybrid one. In the case of the hybrid method two propagation models were taken into account: Longley-Rice and Close-in. When the hybrid method was used, no matter which propagation model was applied, we noticed a significant improvement in the accuracy of the maps.

The hybrid methods seem to be promising for REM construction, particularly when the number of sensors is limited. However, there are still open issues requiring further analysis, among others, the problem of more accurate adjustment of propagation models, which is possible through a change of some parameters that would enable us to obtain more precise information about the terrain. Other issues include the cases with the transmitter located outside the sensor network and more transmitters operating at the same frequency.

If the quality of maps is high, it is possible to estimate the spectrum usage with better accuracy and consequently, to support the dynamic spectrum management systems in a more efficient way. Finally, it may affect the correctness of the frequency assignment/reassignment and, as a result, the quality of services provided by the cognitive network. The hybrid methods seem to be suitable for this purpose, especially when the sensor network is sparse and irregular.

References

1. Suchanski, M., Matyszkiel, R., Kaniewski, P., Kustra, M., Gajewski, P., Łopatka, J.: Dynamic spectrum management as an anti-interference method. In: Proceedings of SPIE, vol. 10418, Bellingham, WA (2017). https://doi.org/10.1117/12.2269294. ISSN 0277-786X
2. Matyszkiel, R., Polak, R., Kaniewski, P., Laskowski, D.: The results of transmission tests of Polish broadband SDR radios. In: Conference on Communication and Information Technologies (KIT) (2017). https://doi.org/10.23919/KIT.2017.8109462
3. Matyszkiel, R., Kaniewski, P., Polak, R., Laskowski, D.: Selected methods of protecting wireless communications against interferences. In: International Conference on Military Communications and Information Systems (2019). https://doi.org/10.1109/ICMCIS.2019.8842679
4. Bogucka, H.: Cognitive Radio Technology (in Polish: Technologie radia kognitywnego). Wydawnictwo Naukowe PWN (2013)
5. Sliwa, J., Matyszkiel, R., Jach, J.: Efficient methods of radio channel access using dynamic spectrum access that influences SOA services realization - experimental results
6. Kaniewski, P., Golan, E.: Localization of transmitters in VHF band based on the radio environment maps concept. In: 10th International Scientific Conference (KIT), Tatranské Zruby (2019). https://doi.org/10.23919/KIT.2019.8883507
7. Pesko, M., Javornik, T., Košir, A., Štular, M., Mohorčič, M.: Radio environment maps: the survey of construction methods. KSII Trans. Internet Inf. Syst. **8**(11) (2014). https://doi.org/10.3837/tiis.2014.11.008
8. Suchanski, M., Kaniewski, P., Romanik, J., Golan, E.: Radio environment maps for military cognitive networks: construction techniques vs. map quality. In: International Conference on Military Communications and Information Systems (ICMCIS), Warsaw, Poland. IEEE Xplore (2018). https://doi.org/10.1109/ICMCIS.2018.8398723
9. Kliks, A., Kryszkiewicz, P., Kulacz, L.: Measurement-based coverage maps for indoor REMs operating in TV band. In: IEEE International Symposium on Broadband Multimedia Systems and Broadcasting (2017). https://doi.org/10.1109/BMSB.2017.7986162
10. Ezzati, N., Taheri, H., Tugcu, T.: Optimised sensor network for transmitter localisation and radio environment mapping. IET Commun. **10**(16), 2170–2178 (2016). https://doi.org/10.1049/iet-com.2016.0341
11. Patino, M., Vega, F.: Model for measurement of radio environment maps and location of white spaces for cognitive radio deployment. In: IEEE-APS Topical Conference on Antennas and Propagation in Wireless Communications (2018). https://doi.org/10.1109/APWC.2018.8503755
12. Mao, D., Shao, W., Qian, Z., Xue, H., Lu, X., Wu, H.: Constructing accurate radio environment maps with Kriging interpolation in cognitive radio networks. In: Cross Strait Quad-Regional Radio Science and Wireless Technology Conference, CSQRWC 2018 (2018). https://doi.org/10.1109/CSQRWC.2018.8455448
13. Suchanski, M., Kaniewski, P., Romanik, J., Golan, E., Zubel, K.: Radio environment maps for military cognitive networks: density of sensor network vs. map quality. In: Kliks, A., et al. (eds.) Cognitive Radio-Oriented Wireless Networks, CrownCom 2019. LNICST, vol. 291. Springer, Cham (2019). https://doi.org/10.1007/978-3-030-25748-4_15

14. Suchanski, M., Kaniewski, P., Romanik, J., Golan, E., Zubel, K.: Radio environment maps for military cognitive networks: deployment of sensors vs. map quality. In: International Conference on Military Communications and Information Systems (ICMCIS), Budva (2019). https://doi.org/10.1109/ICMCIS.2019.8842720
15. Sun, S., Rapport, T.S., Thomas, T., Ghosh, A., Nguyen, H., Kovacs, I., Rodriguez, I., Koymen, O., Prartyka, A.: Investigation of prediction accuracy, sensitivity, and parameter stability of large-scale propagation path loss models for 5G wireless communications. IEEE Trans. Veh. Technol. **65**(5), 2843–2860 (2016)
16. ITU-R: Publications: Recommendations: P Series. https://www.itu.int/rec/R-REC-P

Group Authorization Using Chinese Remainder Theorem

Tomasz Krokosz(✉) and Jarogniew Rykowski

Department of Information Technology,
Poznań University of Economics and Business, Poznań, Poland
tomasz.krokosz@ue.poznan.pl,
rykowski@kti.ue.poznan.pl

Abstract. The paper presents a new way of group authorization for a population of users of a given IT system. Group authorization is highly desirable in applications where the submission of individual rights is not sufficient: service support (service-owner relationship), anonymous verification of belonging to a particular group (e.g., disabled persons, city inhabitants), mutual authentication of unknown-in-advance persons, discounts for groups (e.g., family travelling), etc. The solution uses the Chinese Remainder Theorem, which for any congruence system (with a collection of pairwise relatively prime moduli for the whole set) allows determining one number, which is its solution. After assigning a congruence (or their collection) to a given person, one can use such a connection to dynamically calculate the congruence system for all group members at a given place and time. The correct result for each member means the successful verification of the rights of the group as a whole. Calculations are repeated after each change in group composition. The proposed solution not only increases the functionality of the authorization system but also increases the degree of anonymity in the authentication process, as the user identification is replaced by the identification of the group of which they are members.

Keywords: Group users authorization · Authorization set of users · Authorization using Chinese Remainder Theorem · Common authorization

1 Introduction

Authorization is a process during which the users confirm their rights to perform specific actions. Authorization is usually based on the identification, which consists of providing a particular identifier and the associated secret - a password. Various methods of implementing this process have been proposed [1], which would protect, among others, password theft, followed by many methods to weaken the requirement of full traceability (e.g., pseudonymization). Successfully completed the authorization process results in granting access to certain resources or functionality. The set of resources/functions granted is declared statically, as an entry in a security database, or dynamically, taking into account the context of the process. Once the authorization process has been passed, the authorized user may present its results to the system to obtain access.

W. Zamojski et al. (Eds.): DepCoS-RELCOMEX 2020, AISC 1173, pp. 389–399, 2020.
https://doi.org/10.1007/978-3-030-48256-5_38

The above situation describes a case where one user at a time is subjected to verification. However, we often have a situation in which many people have to cooperate in order to perform some activities. Thus, the authorization should not be related to a single person, but to the group as a whole. At present, this is done by verifying the rights of each member of the group and then taking into account the assumption that meeting individual requirements by all members of the group implies grating the right for the entire group. However, this approach makes it impossible to grant rights conditionally (e.g. "a child can enter on condition if accompanied by an adult") and to the whole group ("only a group of more than 15 people will receive a discount for visiting the museum"). Anonymization is also important – group-based verification does not have to involve identifying group members, which is necessary for the classic approach.

The purpose of the paper is to present a base for the implementation of the aforementioned situation, i.e., a method of group authorization. We assume that the size of the group or its composition need not be determined in advance, and the whole verification process takes place dynamically as soon as all interested parties have reported themselves at a given place and time. Once the composition of the group is changed, the group permissions are automatically re-verified. To this goal, we applied the so-called Chinese Remainder Theorem based on congruences [17]. Obtaining a solution for a congruence system is possible when the calculations concern a minimum of two congruences (i.e., the authorization covers two users). There is no upper limit while determining the number of users and the congruences in the system; however, it should be noted that for the congruence system and their collection of pairwise relatively prime moduli m_1, m_2, …, m_k, the system solution belongs to the set $Z_{m1*m2*mk} = \{0, 1, \ldots, m_1*m_2*\ldots*m_k\}$, which means that as the group increases, we start to process huge numbers. For this reason, the practical limit of the proposed solution will be applicable to groups of several hundred or, at a maximum, several thousand members.

The remainder of the text is as follows. The second chapter describes some examples of real situations requiring group authorization, supplemented with a need for the anonymization. The third chapter describes the scheme of the proposed group authorization using the Chinese Remainder Theorem. The first part of the chapter presents and explains the computational foundation of the theorem, i.e., congruences. The second part of the chapter is the theoretical basis of the theorem, complemented by an example of an application with a set of required calculations. Finally, the third part of this chapter describes the proper use of this technique to solve the problem of group authorization. Chapter four presents and discusses some examples of similar work related to the Chinese Remainder Theorem and its applications. The last chapter concludes the paper.

2 Examples of Group Authorization

We talk about the need for group authorization in a situation where the sum of individual users' rights does not translate into their combined rights. Imagine an example of how to repair a monitoring system at home. The owner of this system cannot repair it by himself, which is why he must outsource the service to a specialized company. The company representative should authorize himself as a service technician upon arrival. Then the owner, by authorizing himself in the system, confirms the right of the service technician to enter the system. One should note that authorizing access cannot be a simple submission of the rights of both persons. Only simultaneous authorization in a given place and time protects the system against unauthorized access. If the owner had full-service authorization, he could break something due to his ignorance of the system administration. If the service technician had full authority, he could force the system to error or failure without the owner's knowledge. Only the combination of both rights into one group law guarantees that there will be no problem, and both parties can control each other. Another example of group authorization is the granting of group discounts. Let's assume that a public-transport company lowers fares to large families, e.g., 2 + 2. This right cannot be assigned to parents because there is no way to verify the number of children at a given time. Similarly, children cannot be granted such rights because it is not known whether they travel with their parents. Therefore, only creating a group from both parents and children and verifying the rights for such a group ensure that the discount was granted correctly.

If a given user belongs to a broaden group, for example, city residents or physically disabled, instead of individual authorization for such a person, authorization related to the fact of belonging to a group may be used. For example, a disabled person would have the right to park a car at a place reserved for the disabled. Please note that this group authorization method increases the level of anonymity of the system - we know that the parking space is occupied by a person who had the right to do so, but we do not know who this person is. In connection with, e.g., D. Chaum's blind signature [16], we are able to guarantee full individual anonymity. However, typically, as we keep identifiers of the group members in a dedicated database, only the database administrator possesses the full information about the identities of group members. Similarly, one may imagine a situation when a police officer verifies a driving license without knowing the driver identification data; the bank clerk anonymously confirms access to the bank account for an anonymous client, etc. As a part of the semi-anonymous group authorization described above, one should consider a reasonable maximum size of a group, so as not to cause too much demand for computing power when assigning permissions, and to allow efficient administration of identities (and other parameters) of group members. Later in the text, we include a discussion dedicated to this issue.

The last example of group authorization is related to mutual verification. Two or more people who do not know each other often want to know if they are members of the same group, usually during the first meeting. If they show each other some certificates of belonging to the group, they will cease to be anonymous to each other. However, if they confirm their belonging to the same group in the above-described group-verification procedure, they may trust each other without the need for full identification.

Thus, taking into account the above examples and use cases, one may note that there is a strong need to introduce a group authentication mechanism into practice. In the remainder of the text, we describe the theoretical foundations of our solution in this area and then discuss selected aspects of its feasibility and applicability. In the discussion contained in the text, we prove that new methods of group authentication are necessary in some cases when the sum of individual rights is not enough, and additional verification of the group level is required. As already briefly mentioned, we also show that group identification increases the level of anonymity while still observing a high level of security.

3 Group Authorization Using Chinese Remainder Theorem

The first two sections of this chapter stand for a theoretical introduction, which is necessary to present our solution to solve the problem of group authentication. The definition of congruence is the beginning of the theory.

3.1 Congruence

Congruence is a mathematical concept, related to number theory and concerning the theory of finite fields. The concept of congruence, i.e., the equivalence relationship, depends on the target algebraic system. For the purposes of this paper, we assume that congruence is related to integers. According to the definition, two integers a and b are congruent modulo n ($mod\ n$), if difference $a - b$ is divisible by n. This means that, during dividing values a and b by value n, the residues of the division are equal. Form of congruence is as follows:

$$a \equiv b\,(mod\,n) \tag{1}$$

The following equivalence relation stands for an example of congruence in a set of integers:

$$5 \equiv 3\,(mod\,2) \tag{2}$$

which means, in accordance with above-presented definitions, that 2 "completes" difference 5–3 entirely (i.e., the difference between this value and the residue value is equal to zero). This also means that the residue of dividing value a by value n is equal to the residue obtained after dividing b value by the value n. For a set of congruence properties (for integers) belongs among others:

1) $a \equiv a\ (mod\ n)$,
2) $a \equiv b\ (mod\ n) \rightarrow b \equiv a\ (mod\ n)$,
3) if $a \equiv b\ (mod\ n)$ and $b \equiv c\ (mod\ n)$, then $a \equiv c\ (mod\ n)$,
4) if $a \equiv b\ (mod\ n)$ and $c \equiv d\ (mod\ n)$, then $a \pm c \equiv b \pm d\ (mod\ n)$, and $ac \equiv bd$ (mod n) – congruences of the same moduli may be added, subtracted and multiplied by pages,
5) if $a \equiv b\ (mod\ n)$, then $a \equiv b\ (mod\ d)$, for each divisor $d \mid n$,
6) if $a \equiv b\ (mod\ n)$, $a \equiv b\ (mod\ m)$, for relatively prime m and n, $a \equiv b\ (mod\ nm)$.

3.2 Chinese Remainder Theorem

The Chinese Remainder Theorem allows to determine exactly one solution for a set of congruences. Content of theorem may have the following form: if numbers m_1, m_2, m_3, ..., m_k are relatively prime pairs (i.e., $NWD(m_i, m_j) = 1$, for $i \mathrel{!}= j$), then the system of congruences:

$$\begin{aligned} x &\equiv a_1 \ (mod\, m_1) \\ x &\equiv a_2 \ (mod\, m_2) \\ x &\equiv a_n \ (mod\, m_k) \end{aligned} \tag{3}$$

has a common solution, belonging to set $m = \{0, \ldots, m_1\, m_2 \ldots m_k\}$. The example is a way to find a solution for the system of three congruences. For the following system:

$$\begin{aligned} x &\equiv 1 \ (mod\, 2) \\ x &\equiv 2 \ (mod\, 3) \\ x &\equiv 3 \ (mod\, 5) \end{aligned} \tag{4}$$

$a_1 = 1$, $a_2 = 2$, and $a_3 = 3$. Let

$$M_i = m/m_i \tag{5}$$

be the product of all moduli except the i-th. Fulfillment of condition

$$NWD(m_i, M_i) = 1 \tag{6}$$

means that there is such a number N_i that

$$M_i N_i \equiv 1 \ (mod\, m_i) \tag{7}$$

Then x is the common solution:

$$x = \Sigma_i a_i M_i N_i \tag{8}$$

For all i-indexes, the sum components (except i-th) are divisible by m_i, because m_i divides entirely M_j for $j \mathrel{!} = i$, and for each i-index, the received result is

$$x \equiv a_i M_i N_i \equiv a_i \ (mod\, m_i) \tag{9}$$

After applying above definitions for the considered system of congruence, obtained values are: $a_1 = 1$, $a_2 = 2$, $a_3 = 3$, $m_1 = 2$, $m_2 = 3$, $m_3 = 5c$, $m = 30$, $M_1 = 15$, $M_2 = 10$, $M_3 = 6$. In order to find a common solution, one has to calculate (using Euclid's algorithm), values of N_i, that are the inverse of numbers M_i *mod* m_i. After substituting required values for (9), result is equal to $x = 23$, which belongs to the set $Z_{30} = \{0, 1, \ldots, 30\}$. Substituting the calculated x value into a congruence system, one may notice that, for each of the three equivalences, x is the common solution. This value becomes the only solution if some additional restrictions apply (such as x should be the minimum value from the set of all possible values, the value should be smaller than the result of multiplication of all moduli values, etc.).

In theory, a system of congruences may count any number of items. However, while creating groups, one should note that, for large values being prime numbers and acting as congruences moduli, the solution will belong to a set of values ranging from zero to the product of the moduli. This is potentially, for a larger group, a huge number. Thus, our solution is limited only to the group of hundred members, which is the case of small- and middle-size systems. This assumption does not limit the application areas, as one may always limit the set of authorization parameters (and authorization groups) to a reasonable quantity. For example, we plan to address our solution to the set of roles rather than individuals (such as each client is a client, not a person with a unique name; each technician is a member of service team, each disabled person is not identified by name, but rather by reason of disability, etc.).

3.3 Group Authorization

The beginning of the description of the group-authorization solution is the following observation. Let p and q be two prime numbers ($p = 5$, $q = 3$). Table 1 lists the residue factors from dividing numbers in the range 1–15 (product of p and q) by p and q.

Table 1. Residue value from the division of the product by some factors.

N	1	2	3	4	5	6	7	8	9	10	11	12	13	14	15
N mod 3	1	2	0	1	2	0	1	2	0	1	2	0	1	2	0
N mod 5	1	2	3	4	0	1	2	3	4	0	1	2	3	4	0

Each of the N numbers has a different pair of residues of dividing the factor p and q. This relation means that each pair of division residue sets only one N value, which is unique for a given set of residue values. The applied and fulfilled condition of relatively prime numbers guarantees the uniqueness of each pair. If the numbers p and q were not relatively prime, then the value of N could be defined by different pairs. The example (for better readability) concerns two variables (small p, q values); however, the quantity of this set may be increased to include k values.

Taking into account Chinese Remainder Theorem, if the numbers $n_1, n_2, n_3, \ldots, n_k$ are relatively prime numbers, and the numbers $a_1, a_2, a_3, \ldots, a_k$ are any numbers fulfilling the condition $a_i \in \{0, 1, \ldots, n_i\}$, there is one and only one solution for the system of congruences (3), satisfying the condition:

$$0 \leq x \leq n_1 * n_2 * \ldots * n_k \tag{10}$$

The above condition restricts possible set of values obtained according to the computations depicted in the previous section to a single numeric value. The fact that we get a single value for any given set of congruences is fundamental for our solution. We have assumed that each user is associated with a congruence (i.e., a value stored in his/her personal device, repository, or application) or a set of them (if potentially belonging to several groups). During the authorization process, if the requirements described above are met, the number of congruences (i.e., other members of a group) may be of any value. For *k* members, we need at most *k* computations to determine the group status. As a result, if the moduli of each congruence belong to the same set, then we obtain a single value. If a new user joins at any time (or another, currently connected, quits the session), then the above computations should be repeated (if some previous computations are somehow cached – only for the new group members). Thus, the solution (the result of the congruence system) is calculated dynamically and depends on the group of currently detected users.

An example of the computations describes above is shown in Fig. 1.

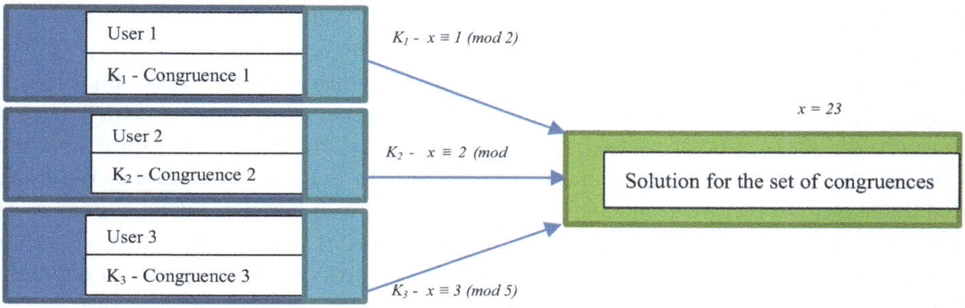

Fig. 1. User communication towards group authentication.

When there are the users in one area who have related collection of pairwise relatively prime moduli, then it becomes possible to obtain a result for a set of congruences. Grouping moduli is also essential for the determination of the set of permissions. If the module meets certain assumptions, then the set of user rights related to this particular congruence may be added to user's profile automatically (and automatically revoked as the user quits the group).

If the computations of congruences for a user are "successful" (i.e., the value stored in personal data repository provides, after some computations, to the well-known value specific for the group), one may imagine the automatic update of this user interface to the system, by adding and revoking specific values, options, menu items, buttons, etc. As a result, system functionality observed by the user may dynamically depend on the neighborhood and other users near-by.

4 Comparison with Similar Work

The Chinese Remainder Theorem has been successfully used in many aspects of cryptography (including e-voting [7, 14]), as well as in, e.g., steganography [3, 9], cloud computing and Internet of Things [13].

The authors of [6] proposed a new method of controlling access to a computer system using the Chinese Remainder Theorem and a time stamp. The text defines the access matrixes, files that associate the user with a key. The key allows insertion, modification, and deletion of the files. The proposal is based on a pair of specialized keys: blocking key, and access key. Each file is related to such a pair, as well as each user, forming a complex matrix of key pairs. Using quite simple computations, one may determine the exact access rights of a user to a file. However, a need to store and maintain a complex matrix of key pairs provokes certain organizational (e.g., while updating access rights of a user, one has to identify and update several key pairs related to the files of this user) and architectural problems, including the need of on-line verification. On the contrary, our solution was not limited by a matrix-like data structure with centralized access rights. Instead, the access rights of a user are to be determined only by his congruence (or a set of congruences possessed by this user). The congruence determined to which group(s) the user belongs to.

Another proposal [8] is related to a scheme for generating and assigning cryptographic keys. The solution is intended for hierarchical structures in which access depends on the rights possessed. The proposal uses the Rabin algorithm and the Chinese Remainder Theorem, and the implementation is quite complicated. We think that the feature of hierarchy management may be narrowed down to the management of many groups, with possible repetitions of the members of these groups. It is a simpler solution for both modeling and implementation. We propose this solution to maintain the hierarchy of users in our solution. If the congruence presented by a user meets certain requirements (i.e., the computations realized with this value lead to expected result value), then the user is granted by additional functions of the application interface. This straightforward approach is repeated to all the congruences for this user, resulting in the possibly more complex and functional interface. It is also possible to determine some multi-parameters describing user roles and privileges in a similar way, such as, e.g., a priority. The solution [4] presents a scheme of secret sharing (and its safe recovery), which consists of dividing information into smaller parts. Collective participation of a fixed number of users allows recovering full information. The scheme, although extremely helpful in building security protocols, is limited in some respects, e.g., it is fragile to an attack of an illegal participant. The solution proposed by the authors of [5] also concerns a combination of a secret-partition scheme and a Chinese Remainder Theorem. Their solution eliminates the problem of intercepting (impersonating the secret holder) a part of the secret thanks to a multi-level, secret division of division thresholds. However, some details are known to the public, such as the quantity of the group. In our solution, due to anonymization, the composition of the group is not possible to determine. The problem of capturing congruence can be partly solved by using biometric mechanisms preceding the group authorization process. Due to the anonymity of the members of the group, the biometric would only be a procedure to confirm their quasi-identity, which would have enabled them to participate (anonymously) in the process of calculating congruence. However, this extension of our proposal requires further study. The authors of [11] have found an application for the Chinese Remainder Theorem in access control. Their solution is based on a periodic sequence, determining access to the communication channel in the access-control layer of the transmission medium (in the ISO/OSI model). Depending on the current sequence, a user sends a data packet or waits for a free transmission channel. Two users broadcasting at the same time cause a collision, and the sequence construction based on the Chinese Remainder Theorem has to protect against it. Our solution is intended for the upper layer and concerns the application layer and group authorization. However, this research confirms the versatility and achievable possibilities of using the Chinese Remainder Theorem.

In [12], the authors used the Chinese Remainder Theorem to aggregate data from the wireless sensor networks (WSN). Their multi-application environment allows for effective aggregation of data from many sensors and easy data extraction. The division of the environment into separate groups allows achieving the intended goal, e.g., the exclusive authorization of sensor operators, each of them responsible for a given sensor group. The above solution is another example of the universality of the Chinese Remainder Theorem we have applied. Another application of the Chinese Remainder Theorem may is related to protect privacy in relation to the location of users [2]. The solution assumes that each user may need different levels of trust and privacy protection depending on the level of trust with other users (close friends, or a public user). This assumption applies strictly to location services and relates to the protection of privacy in geographical, social networks (GSNs). In our solution, other users are not able to identify individual members of the group regardless of location.

Congruences and the Chinese Remainder Theorem are used in solutions that require a high-speed response, as confirmed by the authors of [10]. Their solution is designed for the ad-hoc car networks (VANET - Vehicular ad-hoc network), where there are challenges related to the security and protection of privacy on the one hand, and computational costs and resources, on the other. This feature is also essential for our solution; moreover, what is important is the full automatization of the authorization process.

During the literature review regarding the issue of user group authorization (with dynamically changing cardinality), we did not find any proposals that would coincide with the solution we presented. A direct application of congruences and the Chinese Remainder Theorem directly to the authorization and partial anonymization of group members is not the case for any of the proposals found, even if the Chinese Remainder Theorem is a popular and practical base for a lot of research in the area of security and privacy protection.

5 Conclusions

The paper presents a solution that is dedicated to a group of users for the purpose of common authorization. To this goal, we applied congruences and the Chinese Remainder Theorem. Each user may be associated with different congruences, and in turn, each of them may be used for the authorization within a different group and a different context. The process may be automated in such a way each user observes an interface with specific functionality, with the current set of functions dynamically adjusted to the individual as well as group access rights. In such a way, the targeted functionality depends not only on the user but also on the closed neighborhood, i.e., other users cooperating with the first one. The problem of group authorization is solved by the computations of congruences of all of the members of a group – if the common set of congruences leads to a single solution, then for all of the owners of these congruences certain set of access rights is granted.

Our solution is a proposal for all situations where the authorization of several cooperating users is required. In the paper, we discuss several usage scenarios, including user-service cooperation, group-based discounts, anonymous verification for a group member, etc.

During the literature review, we did not find a similar solution for the problem of simultaneous authorization using congruence systems and the Chinese Remainder Theorem for any number of users. Some proposals referred to the application of the aforementioned theorem for already existing solutions, e.g., the previously mentioned secret partition scheme, or an improvement of the RSA decryption algorithm [15]. The latter is proof of the universality and applicability of the Chinese Remainder Theorem.

References

1. Juyeon, J., Yoohwan, K., Sungchul, L.: Mindmetrics: identifying users without their login IDs. In: Conference Proceedings - IEEE International Conference on Systems, Man and Cybernetics, pp. 2121–2126 (2014). https://doi.org/10.1109/SMC.2014.6974235
2. Karimi, L., Palanisamy, B., Joshi, J.: A dynamic privacy aware access control model for location based services. In: IEEE 2nd International Conference on Collaboration and Internet Computing, IEEE CIC 2016, 6 January 2017
3. Ndoundam, R., Ekodeck, S.: PDF steganography based on Chinese Remainder Theorem. J. Inf. Secur. Appl. **29**, 1–15 (2015). https://doi.org/10.1016/j.jisa.2015.11.008
4. Harn, L.: Secure secret reconstruction and multi-secret sharing schemes with unconditional security. Secur. Commun. Netw. **7**, 567–573 (2014). https://doi.org/10.1002/sec.758
5. Meng, K., Miao, F., Huang, W., Xiong, Y.: Tightly coupled multi-group threshold secret sharing based on Chinese Remainder Theorem. Discret. Appl. Math. **268**, 152–163 (2019). https://doi.org/10.1016/j.dam.2019.05.011
6. Hwang, M.-S., et al.: An access control based on scheme Chinese theorem remainder and time stamp concept (2003)
7. Iftene, S.: General secret sharing based on the Chinese Remainder Theorem with applications in e-voting. Electron. Notes Theor. Comput. Sci. **186**, 67–84 (2007). https://doi.org/10.1016/j.entcs.2007.01.065
8. Chen, T.-S., Chung, Y.-F.: Hierarchical access control based on Chinese Remainder Theorem and symmetric algorithm. Comput. Secur. **21**, 565–570 (2002). https://doi.org/10.1016/S0167-4048(02)01016-7
9. Fridrich, J.: Steganography in digital media: principles, algorithms, and applications. Steganography Digit. Media (2010). https://doi.org/10.1017/CBO9781139192903
10. Alazzawi, M., Chen, K., Yassin, A.A., Lu, H., Abedi, F.: Authentication and revocation scheme for VANETs based on Chinese Remainder Theorem, pp. 1541–1547 (2019). https://doi.org/10.1109/HPCC/SmartCity/DSS.2019.00212
11. Shum, K., Wong, W.S.: Construction and applications of CRT sequences. IEEE Trans. Inf. Theory **56**, 5780–5795 (2010). https://doi.org/10.1109/TIT.2010.2070550
12. Zhou, Q., Qin, X., Liu, G., Cheng, H., Zhao, H.: An efficient privacy and integrity preserving data aggregation scheme for multiple applications in wireless sensor networks, pp. 291–297 (2019). https://doi.org/10.1109/SmartIoT.2019.00051

13. Kavin, B., Sannasi, G.: A secured storage and privacy-preserving model using CRT for providing security on cloud and IoT based applications. Comput. Netw. **151**, 181–190 (2019). https://doi.org/10.1016/j.comnet.2019.01.032
14. Neff, C.: A verifiable secret shuffle and its application to e-voting, pp. 116–125 (2001). https://doi.org/10.1145/501983.502000
15. Wu, C.-H., Hong, J.-H., Wu, C.T.: RSA cryptosystem design based on the Chinese Remainder Theorem. J. Inf. Sci. Eng. **17**, 391–395 (2001). https://doi.org/10.1109/ASPDAC.2001.913338
16. von Solms, S.H., Naccache, D.: On blind signatures and perfect crimes. Comput. Secur. **11**, 581–583 (1992)
17. Fiol, M.A.: Congruences in Z^n, finite Abelian groups and the Chinese Remainder Theorem. Discret. Math. **67**, 101–105 (1987)

Optimal Transmission Technique for DAB+ Operating in the SFN Network

Sławomir Kubal(✉), Michal Kowal, Piotr Piotrowski, and Kamil Staniec

Wroclaw University of Science and Technology, Wyb. Wyspianskiego 27, Wroclaw, Poland
slawomir.kubal@pwr.edu.pl

Abstract. The article presents the issue of signal transmission in the Single Frequency Network SFN network that supports Digital Audio Broadcasting DAB+ digital radio. In the SFN network, the most important issue is the synchronization of transmitters, which is why it is necessary to ensure a stable connection between the multiplexer and the modulator. The paper proposes the use of the optimal, from the SFN point of view, transmission link, and also presents the results of using the Long Term Evolution LTE system as a transmission link that does not meet the requirements of the SFN network. For the proposed solution based on a radio-line, the results of measurements of delay and transmission speed were presented. Also measurements of signal delays in the real SFN network were presented, in which various methods of signal transmission between the multiplexer and transmitters were used.

Keywords: DAB+ · SFN · LTE · Synchronization

1 Introduction – The Single Frequency Network SFN

The Digital Audio Broadcasting DAB+ system has been designed in such way, that it can operate in Single-Frequency Network SFN, in which all transmitters work with the same frequency and send the radio signal in a synchronized mode. Each of the SFN transmitters receives the same stream from the multiplexer, which also contains a synchronization information. The need of synchronization forces the use of highly stable time sources in particular system components. Therefore, the requirements for SFN can be characterized as follows:

- streams received by the transmitters have to be identical, so all transmitters have to receive the stream from the same multiplexer and work in the same distribution network,
- all transmitters have to operate on the same frequency, so it is necessary to ensure a stable frequency source,
- the signal has to be transmitted at the same time from all transmitters, so they must have a stable and synchronous time pattern, and the DAB+ stream must contain a time stamp of the frame.

W. Zamojski et al. (Eds.): DepCoS-RELCOMEX 2020, AISC 1173, pp. 400–409, 2020.
https://doi.org/10.1007/978-3-030-48256-5_39

Despite its invention in 2007 not a lot of attention has been paid to the matter of SFN networks thus far, treating it as mainly an engineering issue. Some items, however, do require scientific effort to answer, these are: provision of limited-range services in wide-area SFN networks or coping with the power and multipath imbalance and finally systematic statistics of the expected SFN gain in terms of range and coverage. Some of the most prominent writings on the former aspects include: e.g. [1, 2] where authors propose a Local Service Insertion (LSI) method allowing to provide local services in wide-area SFN networks (impossible in the original system version) by means of inserting an additional phase reference symbol to the DAB transmission frame. In [3] experimental SFN DVB-T2 network trials were examined for the SFN gain leading to updating empirical minimum C/N (Carrier/Noise) thresholds and providing a thorough discussion on the influence of power imbalance and relative delays occurring in SFN networks on performance in the Rayleigh and Rician channel. Authors of [4] present an original optimization system based on Probabilistic Tabu Search (PTS) and Simulated Annealing (SA) for simultaneously optimizing transmission delays and frequency assignment of single frequency networks.

The SFN network consisting of three DAB+ transmitters is shown in Fig. 1. The signal from the multiplexer is transmitted to modulators in various ways, most often using the Internet for this purpose. The SFN network operates properly under the condition, that all signals reaching the receiver do not have a longer delay than the Guard Interval GI. Otherwise, inter-symbol interferences occur, which may cause, that the correct signal reception is not possible. For the DAB+ standard, the guard interval is 246 μs [5]. The value of the GI determines the maximum distance between particular transmitting stations. The theoretical range results from the propagation time of the signal during the Guard Interval, so the maximum range is obtained for the use of the entire GI value between frames. If the transmitters are out of sync, then the time of the GI is reduced by a synchronization error, and thus the theoretical range of the system is reduced. This phenomenon should be considered during the process of SFN planning.

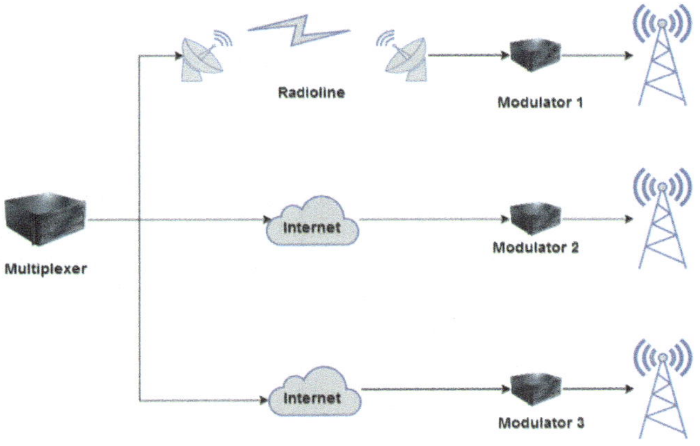

Fig. 1. Example of transmission topology for DAB+ SFN network (own work)

According to [6] four basic types of reference networks were defined for SFN networks:

- reference network 1 for large service-area SFN – contains seven transmitters equipped with a non-directional antenna, one of which is located in a central place. The network coverage area has the shape of a hexagon with a maximum diameter of about 160 km, while the maximum distance between transmitters is about 70 km,
- reference network 2 for small service area and dense SFN – contains three transmitters (equipped with a non-directional antenna), which are located on the vertices of the triangle. The network coverage area has the shape of a hexagon with a maximum diameter of about 50 km, while the maximum distance between transmitters is about 40 km,
- reference network 3 for small service area SFN for urban environment – placement of the transmitters is the same as for the reference network 2, however, due to the propagation environment, greater transmission power is provided for particular transmitters,
- reference network 4 for semi-closed small area SFN – placement of transmitters is the same as for the reference network 2, but transmitters use directional antennas radiating to the center of the triangle,

For the DAB+ standard, it is recommended to use the reference network 3 with a distance of 25 km between transmitters (in order to handle all reception modes mentioned in, i.e. fixed, portable and indoor. Accordingly, for the 'LokalDAB' network developed in Wrocław, a triangular geometry of transmitter lattice has been selected as it provides satisfactory coverage of the city with the smallest number of transmit sites, as advised in Table A.3.6-3 of [6].

2 DAB+ Operating in SFN

DAB+ signal can be sent to transmitters using various techniques and ways. The Internet is the most often used for this purpose, provided that the transmitter is located in a place where an access to this network is provided. For the purpose of effectively operation of the SFN, it is necessary to provide simultaneous or almost simultaneous transmission of individual frames from all transmitters. It can be said, that for a proper operation of the system it is required, that the transmitters maintain 1% accuracy of frequency stability and the time of transmitting frames up to about 10% of the Guard Interval [8]. Maintaining the synchronization in the network can be obtained using Global Positioning System GPS receivers. GPS receivers provide a stable source of second-pulse (1 pps – 1 pulse per second) and 10 MHz frequency. Second pulses are used to synchronize the time of frames transmission while the 10 MHz clock (maintaining a stable carrier frequency) is the reference source for the local oscillator.

As mentioned, the signal can be sent to the transmitters using different routes, which are characterized by different transmission delays. Therefore, time synchronization is obtained by introducing additional delay into the transmission network and/or to particular transmitters. Figure 2 shows the places in the DAB+ SFN system where delays arise or are compensated.

The overall signal delay is the sum of four components [7, 8]:

- network compensation delay - the time, by which the signal is delayed after passing through the distribution network is compensated by the τ_{NC} delay, which matches the total network delay to the largest value of all paths. In this way, the transmitters receive Ensemble Transport Interface ETI frames (standardized output stream for DAB+) with a relatively equal delay,
- transmitter compensation delay – time of frame processing in transmitters is comparable, assuming the same equipment in each transmitter. However, if the delays in individual transmitters arising in the modulator, amplifier and filters are not the same, the delay τ_{TC} should be used to align the processing times. The processing time is usually between 200 and 500 ms,
- transmitter offset τ_{TO} – it is a delay that can be used to compensate delays between transmitters in overlapping coverage areas. In the Open Digital Radio ODR-Dab system (which was used in the developed network) it is a parameter that can be used to compensate all other delays. Its use is also necessary when the coverage areas of particular transmitters are drastically different,
- network padding delay τ_{NP} – it is a delay that can be implemented by the broadcaster to synchronize different networks, e.g. DAB+ and FM of the same broadcaster.

Considering the described delays during designing the SFN network is a condition for maintaining its synchronization.

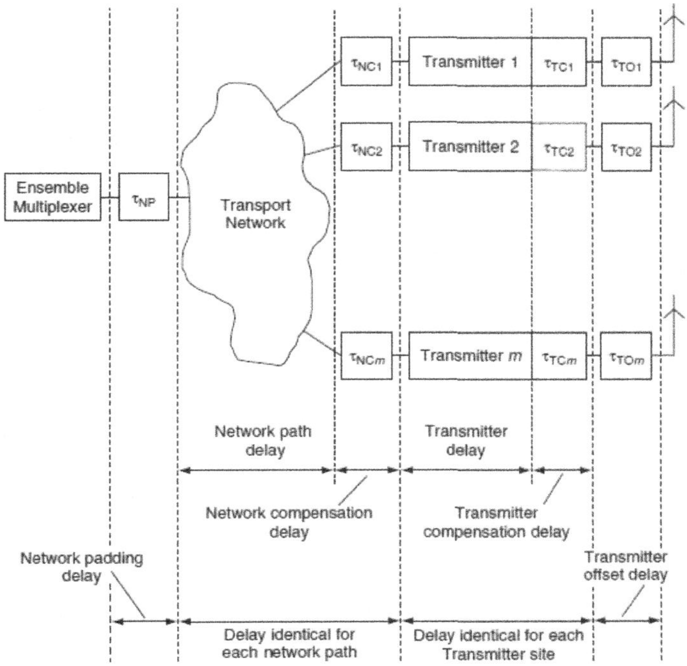

Fig. 2. Transmission delay in SFN network [7]

3 LTE as the Transmission Technique in the SFN

The system built by authors in Wroclaw is based on three transmitters, two of which have been connected to the Internet. Unfortunately, one of the transmitters was designed for installation in a place where there is no fixed connection to the Internet, so it was necessary to develop a dedicated method, other than typical Ethernet networks, of signal transmission from the multiplexer to the modulator. Therefore, it was decided to use the LTE system due to the city-wide coverage of this mobile system. To set up the LTE link, an IDU/ODU 200 modem was used, installed on the transmitter mast. In addition, the network operator configured a dedicated Access Point Name APN for DAB+ transmission. It should be noted that the transmission of the DAB+ stream requires a connection speed of about 1.5 Mb/s, which the LTE network guarantees. In order to check the transmission link based on the LTE network, a series of measurement tests were carried out to measure the basic metrics of the link. The measurement results are shown in Figs. 3 and 4.

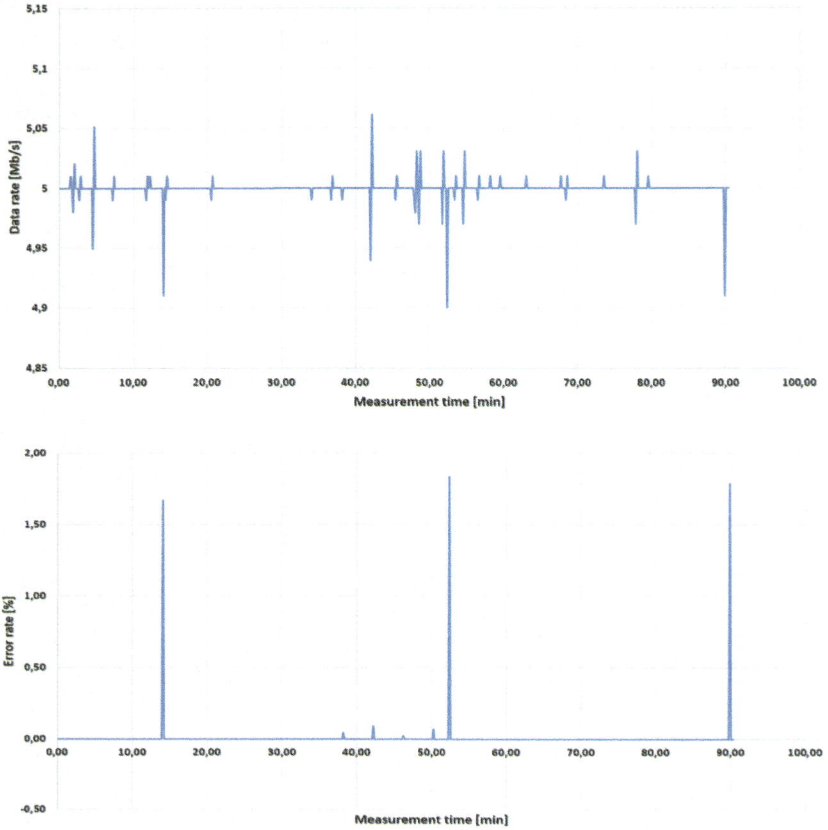

Fig. 3. Data rate (up) and error rate (down) measured for LTE (own work)

To measure link properties in terms of transmission speed and error rate, the Iperf application was used, in which constant one-way traffic at 5 Mb/s was forced. Iperf is an application used to measure link properties based on the client-server architectures. The application allows to generate the traffic with a specific speed and communication protocol (tests have been performed for the UDP protocol). As one can see in Fig. 3, the LTE network provides a constant bandwidth of 5 Mb/s, temporary, slight speed drops do not affect DAB+ transmission, because the lowest measured value exceeds the required minimum, i.e. about 1.5 Mb/s. Unfortunately, when analyzing the link for transmission errors, one can see regular transmission errors of around 2%. Despite the dedicated Access Point Name APN configured for a certain quality of service, the LTE system caused transmission errors at regular intervals. It can be assumed that such a 2% error does not disqualify the LTE link, because such frame dropping would not be detected by the human ear.

However, as was already mentioned, the most important parameter of the transmission link from the point of view of the SFN network is the transmission delay. In order to determine the one-way LTE link delay, a long-term (24 h) measurement of this parameter was performed using a properly configured ping application. The measurement results are shown in Fig. 4.

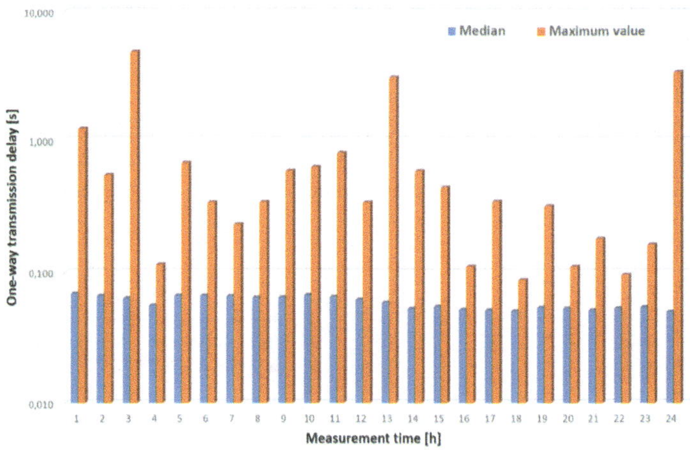

Fig. 4. One-way transmission delay measured for LTE (own work)

One-way LTE link delay measurements have been performed for 24 h, with link statistics in the form of median and maximum value recorded after each subsequent hour. For this test the One-Way Active Measurement Protocol (OWAMP) was used. OWAMP was developed by the Internet Engineering Task Force's (IETF's) IPPM Working Group for latency and delay measurements. OWAMP is a typical client-server application. The owping client contacts the owampd daemon on the peer host to request a specific test. Thanks to strict synchronisation between client and server it is possible to measure one way delay and latency which is valuable since data in DAB system is send mainly from multiplexer to transmitters. As one can see in Fig. 4, the median of the one-way transmission link delay for each hour of measurement does not exceed 100 ms. Unfortunately, the results obtained for the maximum delay of the link disqualify LTE as a DAB+ transmission technique in the SFN network. During the 24-h measurement, values in excess of two seconds were obtained. In the developed SFN network, the maximum allowable time of total delay entered by the transmission link and transmitters that can be corrected in the transmitters is just two seconds. The values obtained for the LTE system would cause desynchronization of the SFN network, and thus would prevent proper reception of the DAB+ signal.

4 Optimal Transmission Link for DAB+ SFN Network

Due to the inability to use a wired Ethernet network, as well as the LTE connection, it was decided to use a dedicated radio-line that sends the signal from the multiplexer to the modulator. For the needs of the DAB+ SFN network, the Ubiquiti AirFiber 5x line operating in the 5 GHz band (at the center frequency of 5480 MHz) was used. One point of the radio-line was placed on the mast with the DAB+ transmitter, while the other on the roof of a 10-floor building on the campus of the Wrocław University of Science and Technology. For radio-lines, it is very important to ensure line-of-sight LOS visibility between antennas. The minimum signal strength ensuring transmission for the used radio-line is −90 dBm, which at 5 GHz band and 30dBm EIRP power, gives an effective range of about 4 km. Figure 5 shows the real value of the transmitted DAB+ stream data rate at the level of about 1.3Mbps, as well as the transmission capacity of the used radio-line. It is worth noting that for the needs of DAB+ transmission the smallest possible bandwidth (10 MHz) was set on the radio-line, which allows to reduce the noise of the receiver, and thus obtaining a greater Signal-Noise-Ratio SNR value at its input.

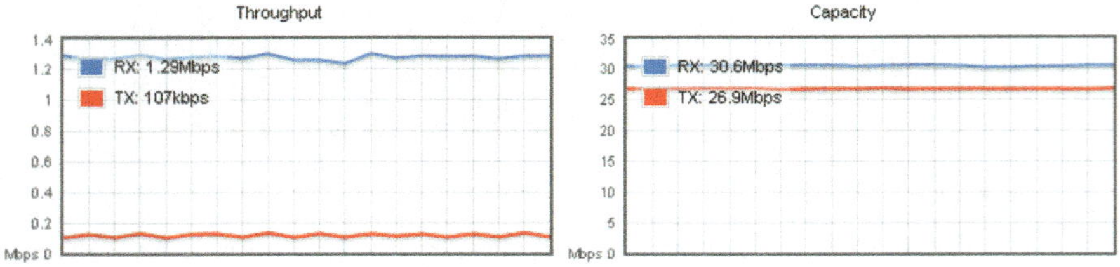

Fig. 5. Real data rate of the DAB+ stream (left) and capacity (right) of the radio-line (own work)

As mentioned earlier, three transmitters were used in the designed DAB+ SFN network. Two of them are connected with the multiplexer through a typical wired Ethernet network, while the connection with the third is based on the radio-line. One-way transmission delay was measured for all three transmitters. For transmitters connected via the Internet, delay values were about 25 and 30 ms, while for a radio-line the delay is about 1 ms. An additional argument for using the radio-line based solution is a permanent supervision over the transmission route. The radio-line is connected to the

LAN network of the Wroclaw University of Science and Technology, in which the multiplexer is also located. For transmitters connected to the Internet, the DAB+ stream is sent over a public network over which the SFN manager has no control. The delay of such network may change e.g. in the case of failure of one of the intermediary points in the network and the need to change the route of packet transmission. Such an uncontrolled delay change can cause desynchronization of the SFN network. The security of the DAB+ transmitted content is also a very important aspect in SFN. Therefore, a Virtual Private Network VPN was established between each of the transmitters and the multiplexer to prevent unauthorized access to transmitted data. In this way, remote access to modulators has been limited and is only possible from the multiplexer. This is especially important for transmitters connected to the public network because they are still exposed to unauthorized access. The LAN network of Wroclaw University of Science and Technology is additionally protected against outside access, which is another argument for using a dedicated connection such as the radio-line.

5 DAB+ SFN in Wroclaw

As was already mentioned, the SFN network built in Wrocław for the needs of DAB+ transmission consists of three transmitters located on the vertices of a triangle. The multiplexer is located in the central part of the triangle at approximately the same distance from each of the transmitters. Detailed deployment of transmitters relative to the multiplexer is shown in Fig. 6. The signal from the multiplexer to Tx2 and Tx3 transmitters is sent using the public Internet network, while to the Tx1 it transmitter is sent via the dedicated radio-line.

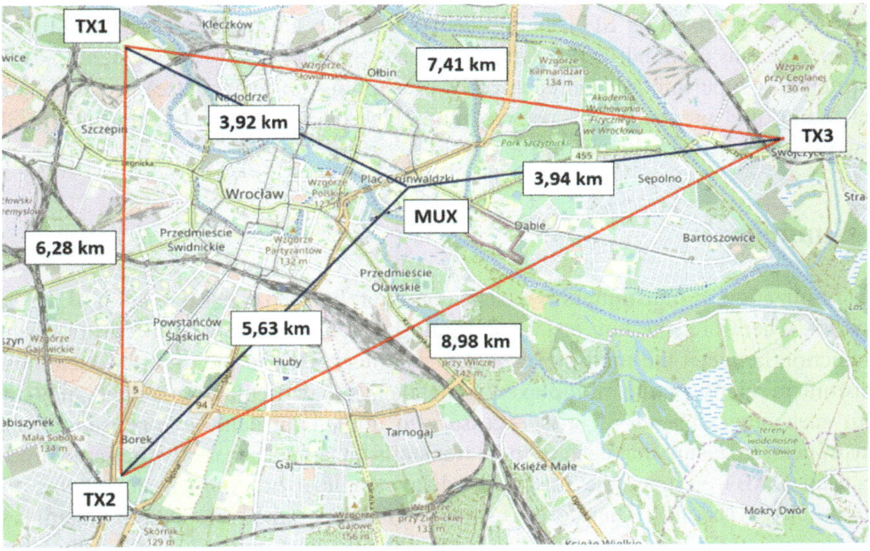

Fig. 6. Architecture of Single Frequency Network in Wroclaw (own work)

In the building where the multiplexer is located, signals from each of the transmitters were measured. The measurement result is presented in Fig. 7.

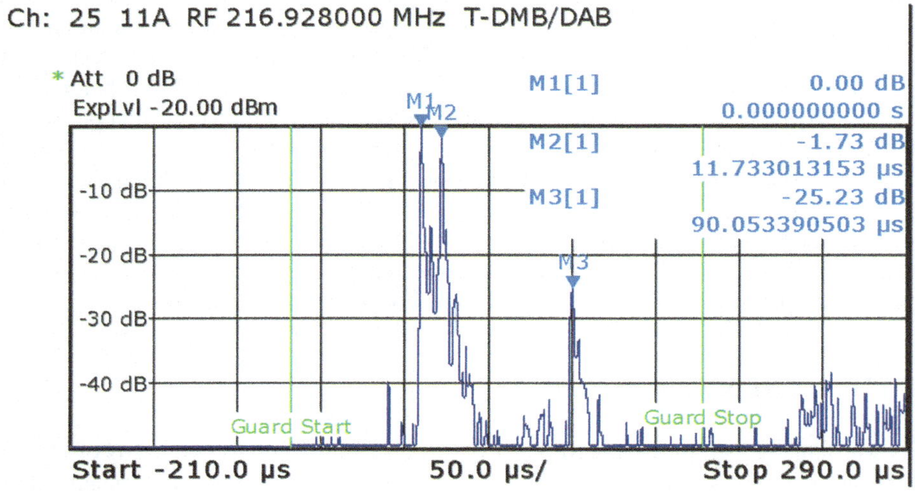

Fig. 7. Delays for DAB+ signals in SFN operating at Wroclaw (own work)

As one can see in Fig. 7, signals from all three transmitters are present and the maximum time difference between DAB+ signals is about 90 µs. This value is smaller than the GI value for the DAB+ standard, which proves the correctness of the SFN network operation. At the measuring point, the correct DAB+ signal reception is possible. When planning SFN, fading analysis is also an important aspect. All transmitters generate a signal at the same frequency, which is why these signals can be destructively superimposed on each other (depending on their phase) causing significant fading and preventing proper reception. The phenomenon of fading in the SFN network is described in detail in [9, 10].

6 Summary

Synchronization of transmitters is a basic condition for the proper operation of the SFN network. Therefore, the key task in designing the SFN network is to ensure a stable connection between the multiplexer and transmitters. Due to the different locations of transmitters relative to the multiplexer, signal delays to particular transmitters may differ from each other. Therefore, it is necessary to implement delay compensation in the transmitters, to ensure that each of them transmits the signal at exactly the same time. In the used DAB+ system, the maximum delay that can be compensated in the SFN network is two seconds. Therefore, as shown in the paper, the LTE cellular system is not suitable for DAB+ stream transmission. Despite adequate network bandwidth, temporary significant delays can cause desynchronization of the SFN network. The article proposes the optimal method of transmission based on the radio-line. This method provides a stable, small delay. In addition, the multiplex administrator has full control over the transmission network, which is impossible to achieve in the case of the public Internet. Of course, the use of the appropriate radio-line depends on the size of the SFN network and the distance between the multiplexer and particular transmitters.

References

1. Schrieber, F.: A backward compatible local service insertion technique for DAB single frequency networks: first field trial results. In: 2018 IEEE International Symposium on Broadband Multimedia Systems and Broadcasting (BMSB), Valencia, pp. 1–5 (2018)
2. Schrieber, F.: A differential detection technique for local services in DAB single frequency networks. In: 2019 IEEE International Symposium on Broadband Multimedia Systems and Broadcasting (BMSB), Jeju, Korea (South), pp. 1–7 (2019)
3. Morgade, J., et al.: SFN-SISO and SFN-MISO gain performance analysis for DVB-T2 network planning. IEEE Trans. Broadcast. **60**(2), 272–286 (2014)
4. Akram, B., Philippe, D., Lhassane, I., René, S., Thierry, S.: Optimizing simultaneously transmission delays and frequency assignment of DVB-T single frequency networks. Int. J. Netw. Commun. (IJNC) **3**, 1–12 (2015)
5. ETSI EN 300 401 V2.1.1 (2017-01) - Radio Broadcasting Systems; Digital Audio Broadcasting (DAB) to mobile, portable and fixed receivers
6. FINAL ACTS of the Regional Radiocommunication Conference for planning of the digital terrestrial broadcasting service in parts of Regions 1 and 3, in the frequency bands 174-230 MHz and 470-862 MHz (RRC-06)
7. Hoeg, W., Lauterbach, T.: Digital Audio Broadcasting - Principles and Applications of Digital Radio, 2nd edn. Wiley, Hoboken (2003)
8. ETSI 300 799 Digital Audio Broadcasting (DAB); Distribution interfaces; Ensemble Transport Interface (ETI)
9. Zielinski, R.J.: Fade analysis in DAB+ SFN network in Wroclaw. In: 2019 International Symposium on Electromagnetic Compatibility - EMC EUROPE, Barcelona, Spain, pp. 106–113 (2019). https://doi.org/10.1109/EMCEurope.2019.8872068
10. Zielinski, R.J.: Analysis and comparison of the fade phenomenon in the SFN DAB+ network with two and three transmitters. Int. J. Electron. Telecommun. **66**(1), 85–92 (2020)

Dynamic Neighbourhood Identification Based on Multi-clustering in Collaborative Filtering Recommender Systems

Urszula Kużelewska(✉)

Faculty of Computer Science, Bialystok University of Technology,
Wiejska 45a, 15-351 Bialystok, Poland
u.kuzelewska@pb.edu.pl

Abstract. This article presents a new approach to collaborative filtering recommender systems that focuses on the problem of an active user's (a user to whom recommendations are generated) neighbourhood modelling. Precise identification of neighbours of an active user is one of the essential problems that recommender systems encounter due to its direct impact on the quality of generated recommendation lists. In classical algorithm, the neighbourhood is modelled by nearest k neighbours, but this approach has poor scalability. Clustering techniques, although improving time efficiency of recommender systems, can negatively affect the quality (precision) of recommendations. It results from inaccurate modelling of object neighbourhood in the case of data located on borders of clusters. A new algorithm based on multi-clustering is proposed in this article. Instead of one clustering scheme, it works on a set of ones, therefore it selects the clustering scheme that models the neighbourhood more precisely. This article presents the experiments confirming these advantages with respect to recommendation quality as well as time efficiency.

Keywords: Recommender systems · Collaborative filtering · Multi-clustering

1 Introduction

Recommender Systems (RSs) are computer applications designed to help users with information overload. The most popular type of RSs are collaborative filtering methods (CF), they are based on the users' behaviour data: search history or visited web sites including time spent on them, and predict a level of interest of users on new, never seen, items [6,8,18]. They search for similar users or items and assume that users with corresponding interests prefer the same items. Collaborative filtering approach has been very successful due to its precise prediction ability [1].

W. Zamojski et al. (Eds.): DepCoS-RELCOMEX 2020, AISC 1173, pp. 410–419, 2020.
https://doi.org/10.1007/978-3-030-48256-5_40

To make recommendations CF are based on either user-based or item-based similarity [17]. The item-based approach usually generates more relevant recommendations since it uses user's ratings - there are identified similar items to a target item, and the user's ratings on those items are used to extrapolate the ratings of the target. This approach is more resistant to changes in the ratings, as well, because usually a number of users is considerably greater than a number of items and new items are less frequently added to the dataset [2].

The article is organised as follows: the first section presents selected collaborative filtering methods with a focus on clustering based solutions in the domain of Recommender Systems, including problems they solve and encounter. Next section describes the proposed multi-clustering algorithm on the background of alternative clustering techniques, whereas the following section contains results of performed experiments with the aim to compare the multi-clustering to the single-clustering approach. The last section concludes the paper.

2 Background and Related Work

The main problem with collaborative filtering methods is their time complexity [14]. Usually, a set of items and users is extremely big, therefore to generate them recommendations in a reasonable time it is appropriate to reduce search space for the candidate objects. k Nearest Neighbors (kNN) is the method that is the most commonly used for this issue [3]. It calculates all the user-user or item-item similarities and identifies k objects (users or items) that are the most similar to the active object as its neighbourhood. Then, the prediction is performed only on objects from the neighbourhood thus reducing the time of calculations. The kNN algorithm is a reference method in order to determine the neighbourhood of an active user for the collaborative filtering recommendation process [6]. Its advantages are simplicity and reasonably accurate results; its disadvantages - low scalability and vulnerability to sparsity in data.

An efficient solution to this problem can be clustering algorithms, that identify clusters for further use as a pre-defined neighbourhood [15]. Recently, clustering algorithms have drawn much attention of researchers and there were proposed new algorithms, particularly developed for recommender systems application [5,11,16]. The efficiency of clustering techniques is related to the fact, that a cluster is a neighbourhood that is shared by all the cluster members, in contrast to kNN approach determining neighbours for every object separately [2]. The disadvantage of this approach is usually loss of prediction accuracy.

The way how clustering algorithms work is the explanation for decreasing recommendation accuracy. A typical approach is based on a single partitioning scheme, which is generated once and then not updated significantly. There are two major problems related to the quality of clustering. The first one is that the clustering results depend on the algorithm's input parameter, and additionally, there is no reliable technique to evaluate clusters before the on-line recommendations process. The other issue is imprecise neighbourhood modelling of the data located on borders of clusters.

The disadvantages described above can be solved by techniques called multi-clustering. They include a wide range of methods which are based on broadly understood multiple runs of clustering algorithms or multiple application of clustering process on different input data.

Some multi-clustering methods, which are often called alternative or multi-view clustering, find partitioning schemes on different data (e.g. ratings and text description) combining results after all [4,12]. The aim of a multi-view partitioning is to generate distinct aspects of the data and to search for the mutual link information among the various views, finally leading to the same cluster structure [7].

Examples of alternative clustering applications in recommender systems are the following. A method described in [13] combines both content-based and collaborative filtering approaches. The system uses multi-clustering, however, it is interpreted as clustering of a single scheme on both techniques. It groups the ratings, to create an item group-rating matrix and a user group-rating matrix. As a clustering algorithm, it uses $k - means$ combined with a fuzzy set theory to represent the level of membership of an object to the cluster. Then a final prediction rating matrix is calculated to represent the whole dataset. In the last step of pre-recommendation process $k - means$ is used again on the new rating matrix to find a group of similar users. The groups represent the neighbourhood of users to limit a search space for a collaborative filtering method.

Another solution is presented in [19]. The authors observed, that users might have different interests over topics, thus might share similar preferences with different groups of users over different sets of items. The method $CCCF$ (Co-Clustering For Collaborative Filtering) first clusters users and items into several subgroups, where the each subgroup includes a set of like-minded users and a set of items in which these users share their interests. The groups are analysed by collaborative filtering methods and the result recommendations are aggregated over all the subgroups.

The multi-clustering method that is described in the next section is different. Although the data as well as attributes on the clustering algorithm's input do not change, the result schemes are different. It means that the data objects can be located more favourably in one cluster than in the other. In the recommendation generation that feature is applied to determine the most appropriate neighbourhood for an active user. It means that the algorithm selects the best cluster from a set of clusters prepared previously (see the following Section).

3 General Description of M-CCF Algorithm

The approach presented in this article, Multi-Clustering Collaborative Filtering ($M-CCF$), defines the multi-clustering process in a different way - as generation of a set of clustering results obtained with the same data on its input. The advantage of this approach is a better quality of the neighbourhood modelling leading to the high quality of predictions, keeping real-time efficiency provided by the clustering methods. This section contains only a general characteristic of the $M - CCF$ technique - a detailed description is in [9,10].

First of all, the set of groups is identified by a clustering algorithm which is run several times with the same or different values of its input parameters. In the experiments described in this article, as a clustering method, $k-means$ was used. Although the set of clusters delivered to the collaborative filtering process is generated with the same parameter k (a number of clusters), the result schemes are different. This step, although time-consuming, has a minor impact on overall system scalability, because it is performed rarely and in an off-line mode.

After neighbourhood identification, the following step, appropriate recommendation generations, is executed. This process requires, despite the great precision, a high time efficiency. The multi-clustering satisfies these two conditions because it can select the most suitable neighbourhood area of an active user and the neighbourhood of all objects is already determined, as well.

One of the most important issues of this approach is to generate a wide set of input clusters that is not very numerous, though it provides a high similarity for every user or item. The other one concerns matching users with the best clusters as their neighbourhood. It can be obtained in the following ways. The first of them compares active user's ratings with the cluster centers' ratings and searches for the most similar one using a certain similarity measure. The other way, instead of cluster centers, can compare an active user with all cluster members and select the one with the highest overall similarity. Both solutions have their advantages and disadvantages, e.g. the first one will work well for clusters of spherical shapes, whereas the second one requires higher time consumption. In the experiments presented in this paper, the clusters for the active users are selected based on their similarity to centers of groups.

Afterwards, the recommendation generation process works as a typical collaborative filtering approach, although the candidates are searched only within the selected cluster of the neighbourhood.

4 Experiments

The dataset examined in the experiments is a subset of benchmark LastFM data [21] containing 10 million ratings. The subset ($100kdata$) consisted of 100 000 entries - 2 032 users and 22 174 artists.

The results obtained with $M-CCF$ were compared with a single-clustering-based recommender system with respect to precision and completeness of recommendation lists generated by the systems as well as a time efficiency. During the recommendation process various similarity indices were taken into consideration (LogLikehood - $LogLike$, cosine coefficient - $Cosine$, Pearson correlation - $Pearson$, Euclidean distance - $Euclid$, CityBlock metrics - $CityBl$ and Tanimoto - $Tanimoto$). The $k-means$ was selected as a clustering algorithm. The influence of a number of clusters for identification in a clustering process was also tested. In the experiments k was equal 20, 50 and 200.

The clustering method, similarity and distance measures were taken from Apache Mahout environment [20]. To achieve a comparable time evaluation, in the implementation of the multi-clustering algorithm, the data models (FileDataModel) and structures (FastIDMap, FastIDSet) derived from Apache Mahout were taken, as well. The following data models were implemented: Clustering DataModel and MultiClusteringDataModel that are in accordance with the interface of DataModel. The appropriate recommender and evaluator classes were implemented, as well.

To provide an objective and appropriate assessment of the compared recommender systems, the quality of recommendations was calculated with $RMSE$ measure (a classical error measure - Root Mean Squared Error) in the following way. Before the clustering, the input data was split into two parts: a training (70%) and testing. Then, in every case of the evaluation, the test set and the items selected for missing ratings evaluation were always the same. To calculate the $RMSE$ measure, the values of ratings from the testing part were removed and estimated by the recommender system. Difference between the original and calculated value was taken for evaluation. For the test set containing N ratings for evaluation, this measure is calculated using (1), where $r_{real}(x_i)$ is a real rating of the user x_i for the particular item i and $r_{est}(x_i)$ is the rating estimated by the recommender system for this user. Although the lower value of $RMSE$ denotes a better prediction ability, there is no maximal value for this measure.

$$RMSE = \frac{\sum_{i=1}^{N} |r_{real}(x_i) - r_{est}(x_i)|}{N} \tag{1}$$

During $RMSE$ calculations there were the cases where estimation of ratings was not possible. It occurs when the item which the calculations are performed to is not present in the clusters to which the items with existing ratings belong. It was assumed that $RMSE$ is taken into consideration if the number of such cases is less than 13 (out of 38 all cases). Table 2 contains a list of the earlier described cases occurred during the evaluation of the item-based collaborative filtering recommender system with the neighbourhood determined by the single $k - means$ clusters. In this table, there are rare examples where the number of such cases equals 13 (e.g. Cosine-based similarity and 4-th and 6-th results of 20 group clustering) and often examples where all cases were not taken into consideration (e.g. Pearson-based similarity for all results of 20 group clustering).

The time efficiency coefficient (t_{av}) has been measured in the following way. A set of M (in the experiments it was taken 10% of users from the input set, so M was equal 203) users was constructed by random users selection from the whole dataset. Then, for each of them, a list of propositions consisted H (in the experiments H was equal to 5) elements was generated. The process was repeated K times (in the experiments K was equal 10) and the final value is the average time of recommendation generation per one user (see 2).

$$t_{av} = \frac{\sum_{i=1}^{K} \frac{\sum_{j=1}^{M} \sum_{k=1}^{H} t_{rec}(x_{jk})}{M}}{K} \tag{2}$$

The results were denoted as significant if it was possible to generate recommendations in more than 80% cases. It means that if the number of users to whom the recommendations were calculated was 203 and for each of them it was expected to generate 5 propositions, therefore at least 800 recommendations should be present in the recommendation lists.

The first experiment was performed on six clustered input data that come from six different runs of the $k - means$ algorithm (in Tables 1, 2 and 3 represented by the number in brackets). The recommender system was built and evaluated on every of six results. Table 1 contains the evaluation of the systems' accuracy run with the following similarity indices: Cosine-based, LogLikelihood, Pearson correlation, Euclidean distance-based, CityBlock distance-based and Tanimoto coefficient. It is noticeable, that the values are different for different input data, even though the number of clusters is the same in the every result. As an example, the recommender system with Cosine-based similarity has $RMSE$ in the range from 0.56 to 0.70. If one takes into consideration a threshold number of cases in which missing ratings were impossible to calculate (see Table 2) the conclusion will be the same: to build a good neighbourhood model for a recommender system, a single run of the clustering algorithm is insufficient. The time of recommendation generation per one user is presented in Table 3. For one case (20(1)) the values are not present due to below the 80% threshold size of the recommendation lists. In the other cases, the time of recommendation generation ranges from 8 to 16 ms per user.

Table 1. RMSE of item based collaborative filtering recommendations with the neighbourhood determined by a single clustering $k-means$ for $k = 20$ (run on sixth different results of the clustering). The best values for every result are in bold.

Number of clusters	Similarity measure					
	Cosine	LogLike	Pearson	Euclidean	CityBlock	Tanimoto
20(1)	**0.65**	**0.65**	0.69	0.66	0.83	0.68
20(2)	0.59	**0.57**	0.57	0.59	0.70	0.60
20(3)	**0.62**	**0.62**	0.52	0.63	0.76	**0.62**
20(4)	0.67	0.65	0.55	0.68	**0.79**	0.69
20(5)	**0.70**	**0.70**	0.62	**0.70**	0.79	0.73
20(6)	0.56	0.54	0.59	0.58	**0.79**	0.58

The overall evaluation of the performance of the recommender built with a neighbourhood model based on a single clustering is presented in Table 4. The tables contain already filtered values with respect to the appropriate thresholds (a number of not calculated missing rating values - for $RMSE$ and the filling of recommendation lists - for the time efficiency). The values of $RMSE$ range from 0.57 to 0.83. It is visible, that the size of the neighbourhood (a number of clusters) does not affect the precision significantly (the lowest values for different numbers of clusters are comparable: 0.57, 0.58, 0.62) and that a greater number of clusters results in the less number of considerate cases (in the case of 200 clusters it was possible to build a valid recommender system only with the similarity based on the CityBlock distance).

Table 2. A number of cases during RMSE calculations that estimation of missing ratings was not possible (an item based collaborative filtering recommender with neighbourhood determined by a single $k-means$ clustering run on sixth different results of the clustering). The significant values are in bold.

Number of clusters	Similarity measure					
	Cosine	LogLike	Pearson	Euclidean	CityBlock	Tanimoto
20(1)	**11**	**11**	22	**11**	**3**	**11**
20(2)	**12**	**12**	23	**12**	**4**	**12**
20(3)	**9**	**9**	20	**9**	**1**	**9**
20(4)	13	13	24	13	**5**	13
20(5)	**10**	**10**	20	**10**	**2**	**10**
20(6)	13	13	22	13	**4**	13

Table 3. Time [ms] of the item based collaborative filtering recommendations with the neighbourhood determined by a single clustering $k-means$ for k=20 (run on sixth different results of the clustering). The best values for every result are in bold.

Number of clusters	Similarity measure					
	Cosine	LogLike	Pearson	Euclidean	CityBlock	Tanimoto
20(1)	–	-	–	–	–	–
20(2)	9	**8**	**8**	**8**	**8**	**8**
20(3)	16	**15**	16	**15**	**15**	**15**
20(4)	15	15	15	15	**14**	**14**
20(5)	16	15	15	16	**14**	**14**
20(6)	13	12	12	12	**11**	12

Table 4. RMSE and time [ms] of item based collaborative filtering recommendations with the neighbourhood determined by a single clustering $k-means$ (range of values for sixth different results of clustering in cases when a number of missing ratings the calculation was greater than 26). The best values result are in bold.

Number of clusters	Similarity measure					
	Cosine	LogLike	Pearson	Euclidean	CityBlock	Tanimoto
20	0.59–0.7	**0.57–0.65**	–	0.59–0.7	0.7–0.83	0.6–0.73
50	0.62	0.6	–	0.63	**0.58–0.78**	0.63
200	–	–	–	–	**0.62**	–
	Time [ms]					
20	9–16	8–15	8–16	8–16	8–15	8–15
50	8–9	8–9	7–9	7–9	7–9	7–9
200	2	2	2	2	2	2

In the following experiment, the $M - CCF$ multi-clustering recommender system was tested. The input dataset, the number of clusters as well as clustering schemes were the same. By an analogy with the previous experiments, different measures were used to determine a level of similarity among vectors of the items in the appropriate recommendation generation process. For every case of the number of clusters were: 20, 50 and 200, all corresponding six clustering schemes from the previous experiment were given for the $M - CCF$ algorithm's input. Table 6 contains the $M - CCF$ algorithm's performance.

Table 5. A number of cases during RMSE calculations that the missing ratings estimation was not possible (the item based collaborative filtering recommender with the neighbourhood determined by the multi-clustering $M-CCF$ method). The significant values are in bold.

Number of clusters	Similarity measure					
	Cosine	LogLike	Pearson	Euclidean	CityBlock	Tanimoto
20	13	**9**	22	13	**4**	**12**
50	14	**12**	24	14	**9**	16
200	25	20	32	26	19	26

The accuracy of the recommender system with the neighbourhood based on the multi-clustering, presented in Table 6 as $RMSE$ values, is greater than in the case of the system with the neighbourhood based on the single-clustering: $RMSE$ ranges from 0.56 to 0.74 (the values are filtered with respect to the threshold from Table 5). The time of recommendation generation is greater: ranges from 34 to 90 ms per user, however, its value can be still considered as a real-time. It is worth to notice, that in all cases the filling of the recommendation lists was not below the 80% threshold.

Similarly to the previous experiments' results - the values of $RMSE$ are not the number of cluster dependent. The value rather depends on the clustering quality. The time of generation of recommendations is lower for the higher number of clusters - the groups contain fewer objects and processing time is shorter, as well.

Taking into consideration all the experiments presented in this article, it can be stated, that the $M - CCF$ multi-clustering recommender system and the technique of dynamic selection the most suitable clusters offers very valuable results with respect to the accuracy of recommendations. However, there is a disadvantage that needs improvement - time efficiency.

Table 6. RMSE and time [ms] of item based collaborative filtering recommendations with the neighbourhood determined by the multi-clustering $M - CCF$ method. The best values are in bold.

Number of clusters	Similarity measure					
	Cosine	LogLike	Pearson	Euclidean	CityBlock	Tanimoto
20	0.63	**0.68**	**0.68**	0.64	0.74	0.69
50	0.49	**0.56**	0.55	0.51	0.62	0.57
200	0.67	0.62	0.82	0.69	0.57	0.63
	Time [ms]					
20	80	80	90	90	90	80
50	70	70	70	70	70	65
200	50	52	50	50	36	34

5 Conclusions

The experiments presented in this article were conducted with the new item-based collaborative filtering recommender system [10]. The neighbourhood of an active user during recommendation generation is modelled by the multi-clustering method $M - CCF$. The multi-clustering delivers a set of clustering schemes obtained from several runs of a clustering algorithm with different values of input parameters. As a result, for an arbitrary object from the input data, its neighbourhood can be identified precisely during the phase of calculation of the most similar users, through selection the most appropriate group from the set of the clustering schemes.

On the contrary to approaches based on k nearest neighbours, application of the algorithms based on a single-scheme clustering characterise high time efficiency and scalability. However, this benefit usually involves decreased accuracy of the prediction feature. It results from inaccurate modelling of objects' neighbourhood in the case of data located on borders of the clusters. The purpose of the $M - CCF$ algorithm was to improve the quality of recommendations, keeping good scalability. Additionally, because $M - CCF$ works on a set of clustering schemes, it is eliminated the issue of selection and evaluation of the most appropriate clustering result.

Acknowledgment. The presented research was performed at Bialystok University of Technology and founded from the resources for research by Ministry of Science and Higher Education.

References

1. Tuzhilin, A., Adomavicius, G.: Toward the next generation of recommender systems: a survey of the State-of-the-Art and possible extensions. IEEE Trans. Knowl. Data Eng. **17**, 734–749 (2005)

2. Aggrawal, C.C.: Recommender Systems. The Textbook. Springer, Cham (2016)
3. Gorgoglione, M., Pannielloa, U., Tuzhilin, A.:Recommendation strategies in personalization applications. Inf. Manage. 23 January 2019. in press
4. Bailey, J.: Alternative Clustering Analysis: A Review, Intelligent Decision Technologies: Data Clustering: Algorithms and Applications, pp. 533–548. Chapman and Hall/CRC (2014)
5. Berbague, C.E., Karabadji, N., Seridi, H.: An evolutionary scheme for improving recommender system using clustering, computational intelligence and its applications, pp. 290–301. Springer, Cham (2018)
6. Bobadilla, J., Ortega, F., Hernando, A., Gutiérrez, A.: Recommender systems survey. Knowl.-Based Syst. **46**, 109–132 (2013)
7. Guang-Yu, Z., Chang-Dong, W., Dong, H., Wei-Shi, Z.: Multi-view collaborative locally adaptive clustering with Minkowski metric. Expert Syst. Appl. **86**, 307–320 (2017)
8. Jannach, D.: Recommender Systems: An Introduction. Cambridge University Press, Cambridge (2010)
9. Kużelewska, U.: Collaborative filtering recommender systems based on k-means multi-clustering. In: Advances in Intelligent Systems and Computing, pp. 316–325. Springer, Cham (2018)
10. Kużelewska, U.: Multi-clustering used as neighbourhood identification strategy in recommender systems. In: Engineering in Dependability of Computer Systems and Networks, pp. 293–302. Springer, Cham (2019)
11. Pireva, K., Kefalas, P.: A recommender system based on hierarchical clustering for cloud e-learning. In: Intelligent Distributed Computing XI, pp. 235–245. Springer, Cham (2018)
12. Mitra, S., Banka, H., Pedrycz, W.: Rough-fuzzy collaborative clustering. IEEE Trans. Syst. Man Cybern. Part B Cybern. **36**(4), 795–805 (2006)
13. Puntheeranurak, S., Tsuji, H.: A multi-clustering hybrid recommender system. In: Proceedings of the 7th IEEE International Conference on Computer and Information Technology, pp. 223–238 (2007)
14. Ricci, F. and Rokach, L. and Shapira, B.: Recommender Systems: Introduction and Challenges, Recommender systems handbook, pp. 1–34. Springer, Heidelberg (2015)
15. Sarwar, B.: Recommender systems for large-scale e-commerce: scalable neighborhood formation using clustering. In: Proceedings of the 5th International Conference on Computer and Information Technology (2002)
16. Selvi, C., Sivasankar, E.: A novel Adaptive Genetic Neural Network (AGNN) model for recommender systems using modified k-means clustering approach. Multimedia Tools Appl. **78**, 1–28 (2018)
17. Schafer, J.B. and Frankowski, D. and Herlocker, J. and Sen, S.: Collaborative Filtering Recommender Systems. In: Brusilovsky, P., Kobsa, A., Nejdl, W. (eds.) The Adaptive Web, pp. 291–324 (2007)
18. Singh, M.: Scalability and sparsity issues in recommender datasets: a survey. Knowl. Inf. Syst. **62**, 1–43 (2018)
19. Wu, Y., Liu, X., Xie, M., Ester, M., Yang, Q.: CCCF: improving collaborative filtering via scalable user-item co-clustering. In: Proceedings of the Ninth ACM International Conference on Web Search and Data Mining, pp. 73–82 (2016)
20. Apache Mahout. http://mahout.apache.org/. Accessed 24 Aug 2019
21. A Million Song Dataset. https://labrosa.ee.columbia.edu/millionsong/lastfm/. Accessed 02 Jun 2019

Increasing the Dependability of Wireless Communication Systems by Using FSO/RF Technology

Robert Matyszkiel[1](✉), Janusz Mikołajczyk[2], Paweł Kaniewski[1], and Dariusz Szabra[2]

[1] Military Communication Institute, Warszawska 22A, 05-130 Zegrze Południowe, Poland
{r.matyszkiel,p.kaniewski}@wil.waw.pl
[2] Military University of Technology, gen. S. Kaliskiego 2, 00-908 Warsaw, Poland
{janusz.mikolajczyk,dariusz.szabra}@wat.edu.pl

Abstract. The article presents some aspects of wireless communication systems dependability used for the army. Free Space Optics (FSO) and optical-radio hybrid (FSO/RF) data transmission technologies were analyzed considering their advantages and disadvantages. In the experimental part, some selected results of the LasBITer research project are described. This project confirmed the significant advantages of the developed hybrid technology. It was shown that FSO links have many advantages, however, they are sensitive to harsh atmospheric conditions, e.g. fog and scintillation. In turn, RF links are characterized by low attenuation in similar weather conditions, but they may be exposed to enemy electronic warfare systems. In such situations, there is need to work in a so-called "radio silence regime" (no radio emission). The combination of both technologies into one FSO/RF hybrid system provides to increase its suitability (up to value of 99.999%), reliability, security of sensitive data transmission, and reduce the probability of detection.

Keywords: FSO/RF · Dependability · Wireless communication

1 Reliability of the FSO/RF Data Link

Free Space Optics is a wireless data transmission technology using optical radiation to transmit data. It can be applied in terrestrial communication systems (between buildings, observation points) as well as for air communication (between airplanes, unmanned aerial vehicles UAVs, high altitude platforms HAP) and space. FSO data link can be integrated with networks implemented through other technologies, e.g. radio, wired and fiber. Increasing use of FSO devices makes it necessary to determine their reliability and availability. To estimate the device reliability, the reliability of each individual element (transmitter, receiving unit and "optical path quality") should be considered. The reliability and availability of the optical path is mainly influenced by local weather conditions, e.g. fog, rainfall, turbulence, as well as the stability of the

W. Zamojski et al. (Eds.): DepCoS-RELCOMEX 2020, AISC 1173, pp. 420–429, 2020.
https://doi.org/10.1007/978-3-030-48256-5_41

optical axes position. If there are used few data links (FSO with RF), the total reliability is determined based on specifications and configurations of each link.

2 Laser Link Availability

2.1 Determinants of Link Availability

The analysis of laser link availability will be limited only to optical path quality analyzes. For transceiver components and network configuration, methodologies of reliability determination are generally known. The "optical path quality" is defined by the transmission properties of the atmosphere. The main limitations of these properties are the phenomena of absorption, dispersion and turbulence. Absorption is determined by the composition and quantity of the gas mixtures in the air. The attenuation impact is minimized by operation in the so-called atmosphere transmission windows. Turbulence can take the form of scintillation, beam wandering, or beam spreading. For the given wavelength, the transmission is usually expressed by an extinction coefficient γ. It contains some components describing the light attenuation caused by absorption and dispersion of particles in the air. The main attenuating factor is scattering by aerosol particles, e.g. fog. This can be described using the Mie model for individual interactions between particles and photons flux. In these calculations, visibility range defines the optical path length at which the transmission drops to 2%. This can be expressed by the Koschmieder equation [1]:

$$Vis = \frac{ln\left(\frac{1}{\tau}\right)}{\gamma_{550\,\mathrm{nm}}} = \frac{ln\left(\frac{1}{0.02}\right)}{\gamma_{550\,\mathrm{nm}}} = \frac{3.912}{\gamma_{550\,\mathrm{nm}}} \tag{1}$$

In the scattering process, there is observed influence of the radiation wavelength. This relationship was empirically measured and described by Kruse [2]:

$$\gamma(\lambda) \cong \frac{3.912}{Vis}\left(\frac{\lambda}{550\,\mathrm{nm}}\right)^{-q} \tag{2}$$

The exponent q depends on the range of visibility. The Eq. (2) should not be used as the determinant but it provides useful information for a preliminary estimation of availability. To make this estimation, it should be assumed that:

- R (T) - system reliability is the probability that it works correctly during the T period in environmental conditions,
- A (t) - system availability is the probability that it is working correctly at the time t.

The conditions for proper FSO device operation are usually determined by the threshold value of bit error rate (BER) for a specific data rate (e.g. BER = 10^{-9} for 100 Mb/s). The BER value is strictly determined by the value of signal to noise ratio (SNR) registered by FSO receiver. It is described by [3]:

$$BER = \frac{1}{2}\left[1 - erf\left(\frac{\sqrt{SNR}}{2\sqrt{2}}\right)\right] \tag{3}$$

where *erf* is the *Gauss error function.*

The availability of the FSO link depends mainly on the power budget and local weather conditions. These conditions also influence on attenuation of registered signal. The basic determinant for the FSO system availability A is expressed:

$$A = \begin{cases} 1 \ for \ SNR \geq SNR_{BER} \\ 0 \ for \ SNR < SNR_{BER} \end{cases} \tag{4}$$

For the scattering, the availability of an FSO link mainly depends on atmospheric phenomena such as fog, drizzle, dust, etc., commonly defined by visibility. Therefore, it is necessary to analyze statistical wheatear data for a given country or area. Figure 1 presents the distribution of visibility changes registered over Warsaw for both the year seasons and the time of day.

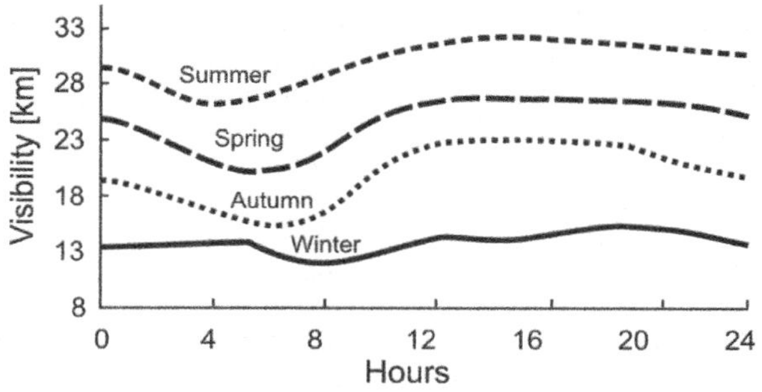

Fig. 1. Visibility changes over Warsaw [4].

In this case, visibility is determined by dust and other particles but not any atmospheric phenomena. However, there is observed that visibility does not generally decrease below 9 km for good wheatear conditions. Mists can be the second factor of visibility deterioration. Figure 2 shows an example of annual distribution of foggy days over the area of Kraków, Bielsko Biała and Katowice.

In this area of Poland, mist has been noticed on average for 53–67 days per year [5]. It was also found that the mist lasts most often from 1 h to 3 h, and the least often over 12 h [6]. Based on the climate analysis, it is possible to calculate the distribution of specific FSO device availability considering their parameters.

2.2 Construction of the Designed Laser Link

The FSO link operating in the LWIR radiation is a unique tool compared to the currently available ones. Due to the use of long infrared radiation, the influence of

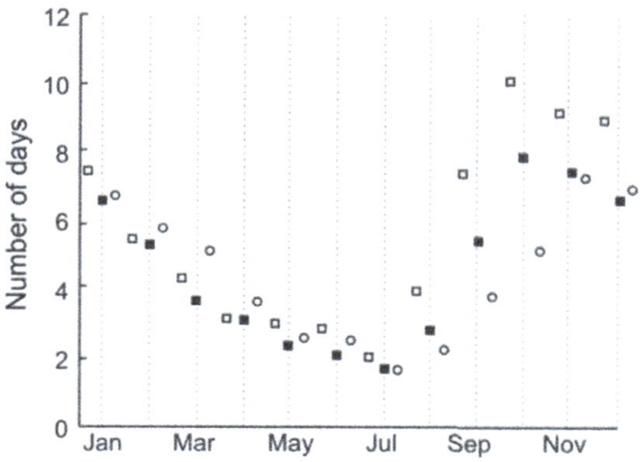

Fig. 2. Annual distribution of foggy days over the area of Krakow, Bielsko Biala and Katowice.

scattering and turbulence (scintillation) on the data range was reduced. Figure 3 presents a view of the constructed FSO transceiver.

Fig. 3. View of the FSO transceiver.

Table 1. Some parameters of the FSO main components.

Parameter	Value
Peak pule power	500 mW
Beam divergence	1 mrad
Transmitter mirror diameter	50 mm
Detectivity	$3\ 10^9$ cm$\sqrt{\text{Hz}}$/W
Active area	0.5×0.5 mm^2
Bandwidth	200 MHz
Receiver mirror diameter	100 mm
Mirror reflectance	90%

A quantum cascade laser emitting light pulses at the wavelength of 9.3 μm. To register these pulses, a detection module equipped with a MCT phoconductive detector was used [7]. Transmitting and receiving optics are off-axis parabolic mirrors. In Table 1, some parameters of the FSO main elements are listed.

The performed test of this FSO link have shown that for good weather conditions (Vis > 5 km and no-noticed scintillation), it can provide a data rate of 10 Mb/s with BER = 10^{-7}. The same tests were performed sending data over the water surface of the Zegrzynskie Lake. When atmospheric conditions were stable, parameters of optical transmission were similar to results obtained during laboratory ones. However, a strong wind creates a water breeze and significantly reduces both the visibility and data transmission range. Some numerical analyzes were performed to determine this phenomenon. Figure 4 shows the SNR values versus link distances for different visibilities. The visibility corresponds to the appropriate atmospheric conditions such as medium fog (Vis = 500 m), light fog (Vis = 1 km), and drizzle (Vis = 2 km).

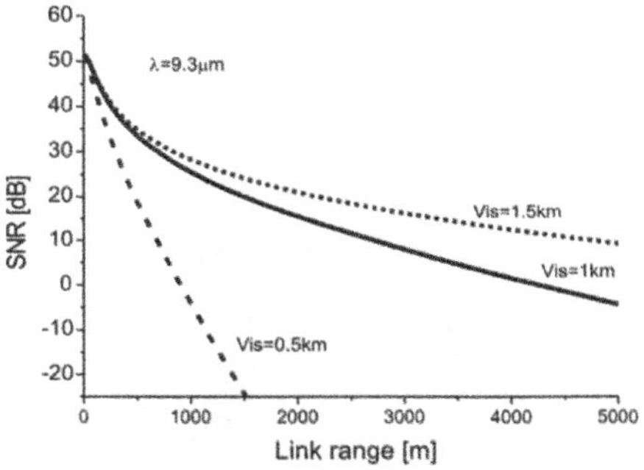

Fig. 4. SNR value vs. data link range of FSO-LWIR system for different visibilities.

However, the developed link is next step of the development of FSO technology considering performances of commercially available systems operating at the wavelengths of 900 nm or 1550 nm. The comparative characteristics of these data links are presented in Fig. 5. It is seen, the developed LWIR link is less sensitive to the effects of fog and drizzle (visibility of 1.5 km).

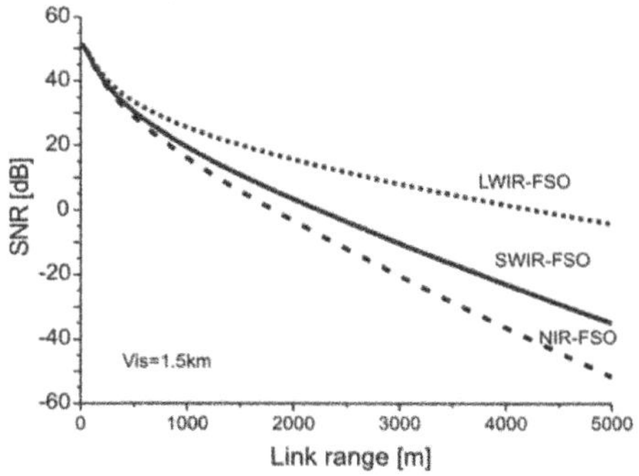

Fig. 5. SNR value vs. data link range for different FSO systems and visibility of 1.5 μm.

The increase in wavelengths of transmitting light provides to obtain better availability of the optical link. NIR-FSO or SWIR-FSO systems operate well at good visibility (data range of a few km), however decrease in visibility can shorten this distance to several dozen meters. In the case of LWIR radiation, this effect is comparatively not so strong. For example, the same level of SNR will be obtained for NIR link with 1.5 km visibility, SWIR link and 1.1 km visibility, and LWIR link and 0.8 km visibility. Analysis of weather conditions showed that the large reduction in visibility (very thick fog) is less common than the appearance of light mists (medium visibility). In such situations, the developed LWIR link is more practical. Estimation of the link availability requires accurate measurements of visibility over a long period of time and at short intervals. The example approximation assumes that the FSO system will work in Poland, where the average number of foggy days is 50 (visibility change to 1.5 km). This fog is rising on average for of 3 h. The availability of the link can be expressed by the formula:

$$A = \left(1 - \frac{N_{FOG}}{N_Y}\right) 100\%, \tag{5}$$

where N_{FOG} - the total number of hours of fog per year, and N_{Y} - the total number of hours in a year.

Using previously defined data, the link availability decreased by fog is 98%. This value is unattainable in practice because it does not consider other phenomena, e.g. rainfall and snowfall, turbulence, strong wind and possible vibrations of the transceiver systems.

2.3 Preliminary Test of the Designed Laser Link

Availability test (*A* parameter) of the laser link was performed at the distance of 800 m. For good weather, the amplitude of the registered pulses was high enough to read data transmitted over this distance. However, worse weather (rain) can cause a large number of errors. Figure 6 illustrates registered changes of both signal amplitude and noise level. These changes are correlated with the observed atmospheric conditions. Determining the SNR value for some time intervals, the BER values (Eq. 3) were determined. Assuming that the threshold level of BER is 10^{-3}, changes in parameter of *A* were also calculated (Eq. 4). In that tests, the designed laser link was not available ($A = 0$) in the case of rain and drizzle.

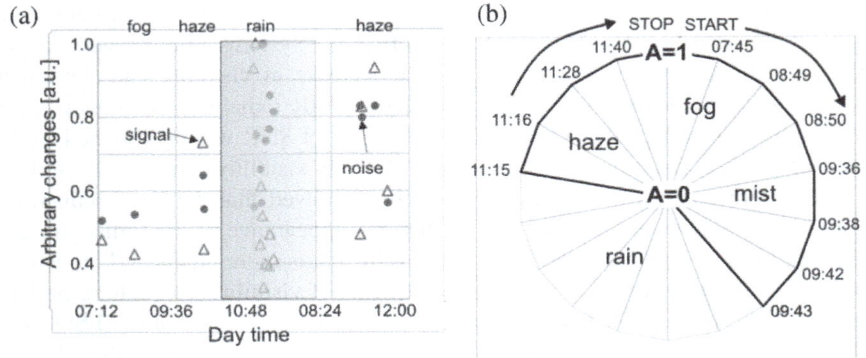

Fig. 6. Arbitrary changes of signal and noise registered by FSO receiver for different weather conditions (a) and their influences on availability A (b).

3 FSO/RF Link Configuration

Laser wireless data transmission is very promising technology. However, the main limitation of FSO systems availability are weather conditions. This limitation can be minimized using backhaul RF data link. The designed FSO link has been functionally connected to the radio link operating at 1.3 GHz. This setup created the so-called FSO/RF hybrid link. It was equipped with a special data transmission management system of two data transmission channels (developed by KenBIT Koenig and Partners). This system defines channel currently responsible for sending information and determines the BER value. It is also possible to setup the automatic operation mode to change transmission channel basing on the BER threshold value. The FSO/RF hybrid link view is shown in Fig. 7.

Fig. 7. View of FSO/RF hybrid data link.

This link was tested at the distance of 700 m. The test results are listed in Table 2.

Table 2. The test results of FSO/RF hybrid data link.

Parameter	Value
Switching time FSO→RF	23 s
Switching time RF→FSO	12 s
Data rate RF	10 Mb/s
Mirror reflectance	90%

The hybrid link provides data transmission with rate of 10 Mb/s using FSO channel or the RF channel. It is also possible to switch channels with a time interval of about 10–25 s. The time difference results directly from the applied data transmission protocols. Such protocols impose strictly defined procedures in advance. However, the switching times are much shorter than the duration of atmospheric phenomena, which can suppress both optical and radio signals. Therefore, using the FSO/RF hybrid configuration can increase the availability of the data link.

4 Assessment of the Reliability of the Wireless Communication Link Using FSO/RF Technology

For the purposes of this article, a technical object (TO) is understood as a means of communication acting in a certain way, i.e. using a certain transmission and reception technique and a given frequency range, which determines the set of physical phenomena associated with the propagation of waves that are the carrier of information. The considered TO functions in a specific environment:

- natural, which is characterized by a number of parameters, e.g. changing weather conditions,
- technical, i.e. as part of a system composed of similar TOs, as well as in the vicinity of other systems of which constituent devices may affect the TO in question.

In the general case, using more than one TO in the communications system to ensure that the same function is performed, increases the available potentiality and the likelihood of the system's fitness (reliability) and thus its dependability. This goal can be achieved by using redundancy in the form of another TO of the same type, as well as with an TO using different principles of operation. TO in a system without redundancy will be designated as TO1, and redundant TO as TO1' if it is of the same type as TO1 or as TO2 in the opposite situation. TO1 together with the redundant TO (TO1' or TO2) will be treated as a complex technical object (CTO).

Let's consider both cases in the situation of TO1 incapacity due to various reasons:

- natural, e.g. climatic and mechanical exposure, atmospheric discharges, electromagnetic pulse;

- dependent on the operator, e.g. exceeding the permissible operating parameters TO, incorrect configuration;
- deliberate or unintentional operation of external systems, e.g. lack of internal compatibility of the system, electromagnetic operation of the opponent.

Some of the above listed reasons are detailed in the literature [8, 9] because they relate to all TO, but many of the reasons for the faults of military equipment, which are counteracted by providing hardware redundancy, are not time-dependent and cannot be described using any probability distribution due to their dependence on variable non-statistical factors, such as the opponent's intentions or technical capabilities of both sides of a conflict.

After analysing the reasons for the failure, it can be concluded that the use of OT1' in case of 3, as well as some of the situations listed in points 1 and 2, will not result in preserving CTO in fitness state, because both TOs of the same type will be equally sensitive to incidental exposures. Depending on the likelihood of different causes of TO1 failure, the change in available potentiality and the probability of CTO's fitness may not occur at all or may be small, disproportionate to the investment potential used.

The specifics of the exploitation of military equipment show that it is particularly important to ensure its fitness during combat operations, i.e. when there is a high probability of occurrence of the events listed in point 3, causing both TO1 and TO1' to fail at the same time. From the point of view of system reliability, it is therefore preferable to use CTO, which beside TO1 includes TO2, not TO1'. In this case it is possible to use the known formula for the reliability of a parallel system:

$$R_{CTO} = 1 - Q_{CTO} = 1 - Q_{TO1} \cdot Q_{TO2}. \tag{6}$$

To achieve a reliability of 0.99999, the product of unreliabilities of the constituent objects ($Q_{TO1} \cdot Q_{TO2}$) should reach the value 0.00001. Assuming that the faults are independent, this requires each TO to provide reliability at the level of 0.003.

5 Conclusions

The recently observed rapid increase in demand for broadband services has caused a rapid increase in demand for the spectral resources, which feature limited availability. There are two ways to solve the problem of spectrum deficit.

The first is Dynamic Spectrum Access – the concept involving sharing of spectral resources and requiring the development of technical capabilities based on cognitive radio, the development of radio environment maps [10, 11] and mechanisms allowing for the reuse of frequencies [12]. This solution requires inter alia legislative changes and is now considered to be achievable in the long term.

The second way to solve the problem of the deficit of spectral resources is to use transmission means that do not use the electromagnetic spectrum in the radio range. Far infrared links can be proposed as one of promising alternatives. There are multiple areas of implementation where they can be successfully used instead of radios.

The use of far-infrared radiation, despite its significant advantages, is, however, sensitive to harsh weather conditions, in particular fog and scintillation.

In order to increase the reliability of the wireless communication system, it was proposed to use a hybrid FSO/RF system combining the advantages of both systems and eliminating disadvantages.

In addition, the performed reliability analysis showed that the FSO/RF link in operational practice can be treated as a parallel system, whose total reliability depends on the product of the unreliabilities of individual components (FSO and RF).

Acknowledgments. We acknowledge support by The National Centre for Research and Development- the grant no. DOB-BIO8/01/01/2016.

References

1. Lee, Z., Shang, S.: Visibility: how applicable is the century-old Koschmieder model? J. Atmos. Sci. **73**(11), 4573–4581 (2016)
2. Corrigan, P., Martini, R., Whittaker, E.A., Bethea, C.: Quantum cascade lasers and the Kruse model in free space optical communication. Opt. Exp. **17**(6), 4355–4359 (2009)
3. Ghatak, A., Thyagarajan, K.: An Introduction to Fiber Optics. Cambridge University Press, Cambridge (1998)
4. Majewski, G., Rogula-Kozłowska, W., Czechowski, P.O., Badyda, A., Brandyk, A.: The impact of selected parameters on visibility: first results from a long-term campaign in Warsaw, Poland". Atmos. (Basel) **6**(8), 1154–1174 (2015)
5. Łupikasza, E., Niedźwiedź, T.: Synoptic climatology of fog in selected locations of southern Poland (1966–2015). Bull. Geogr. Phys. Geogr. Ser. **11**(1), 5–15 (2016)
6. Identyfikacja i ocena ekstremalnych zdarzeń meteorologicznych i hydrologicznych. klimat. imgw.pl
7. Mikołajczyk, J., et al.: Analysis of free-space optics development. Metrol. Meas. Syst. **24**(4), 653–674 (2017)
8. Laskowski, D., et al.: Anthropo-technical systems reliability, safety and reliability: methodology and applications, pp. 399–407. CRC Press-Taylor & Francis Group, ISBN 978-113802681-0 (2015)
9. Lubkowski, P., et al.: Provision of the reliable video surveillance services in heterogeneous networks, safety and reliability: methodology and applications, pp. 883–888. CRC Press-Taylor & Francis Group, ISBN: 978-113802681-0
10. Romanik, J., Golan, E., Zubel, K., Kaniewski, P.: Electromagnetic situational awareness of cognitive radios supported by radio environment maps. In: Signal Processing Symposium, Kraków (2019). https://doi.org/10.1109/sps.2019.8882065
11. Suchański, M., Kaniewski, P., Romanik, J., Golan, E., Zubel, K.: Radio environment maps for military cognitive networks: density of sensor network vs. map quality. In: Lecture Notes of the Institute for Computer Sciences, Social Informatics and Telecommunications Engineering, vol 291. Springer, Cham. https://doi.org/10.1007/978-3-030-25748-4_15
12. Kosmowski, K.: Frequency re-usage in radio planning systems. In: Communication and Information Technologies (KIT) (2019). https://doi.org/10.23919/kit.2019.8883504

Card Game Bluff Decision Aided System

Jacek Mazurkiewicz[1(✉)] and Mateusz Pawelec[2]

[1] Faculty of Electronics, Wrocław University of Science and Technology, ul. Wybrzeże Wyspiańskiego 27, 50-370 Wrocław, Poland
jacek.mazurkiewicz@pwr.edu.pl

[2] Dolby Poland sp. z o.o., ul. Legnicka 48, 54-202 Wrocław, Poland

Abstract. Main goal of the paper is to create and evaluate poker player bluff aided system. To achieve the goal there is a need to create poker game system and learning environment. There are several assumptions and limitations regarding paper subject. Poker game variant is decided to be Texas Hold'em No Limit. This variant was chosen mainly because it is probably most popular poker variant nowadays, it is very entertaining and it has simple rules. Another assumption is that the game is for two players only. This is connected with several factors. Most important is the reduction of computational complexity. Other reasons are: reduction of game system and learning system development complexity, fitness function definition, more complex fitness score assessment. Results evaluation is, due to problem abstraction, limited to analyzing and reasoning of the decisions. AI learning will be performed without external source of knowledge, that is without using historical data containing professional poker players decisions. This system will not use existing poker game solutions as it is an attempt of creating AI with use of live score assessment only. Important thing is, that this system is not meant to model complete poker player but focuses on bluff actions only.

Keywords: Card game analysis · Artificial intelligence · Decision aided system

1 Introduction

Poker is a very well-known card game, played by millions of people every day. Poker-Players Research states that in 2009 40 million people played poker regularly [7]. It is not sure where or when this game came into existence. Poker is believed to have ancient roots that go back nearly 1000 years crossing several continents and cultures. Some historians claim that it origins from domino-card game played by a 10th-century Chinese emperor, others claim that it is a descendant of the Persian card game "As Nas" [11]. There are many variants of poker game. To name couple of the most popular ones: Five-card draw, Seven card stud, Texas hold'em and Omaha hold'em. All poker variants involve betting as an intrinsic part of play, and determine the winner of each hand according to the players cards set. Poker variants vary in the number of cards dealt, number of shared or "community" cards, number of cards that remain hidden and the betting procedures. Poker game itself is not only entertaining to play but it is very interesting when it comes to detailed mechanics of play, modelling player and making

W. Zamojski et al. (Eds.): DepCoS-RELCOMEX 2020, AISC 1173, pp. 430–440, 2020.
https://doi.org/10.1007/978-3-030-48256-5_42

decisions. First of all it is a game of chance since cards are dealt randomly every time. Secondly it is a game of imperfect information, where players decisions, strategy and behavior are more important than the cards. That is in contrast to other games like chess or Go, where both players have the same and complete information. Resolving this type of games is usually based on searching through decision nodes and choosing the best (or with the best potential). Heads-up No-limit Texas Hold'em is said to be solved by DeepStack (described in Sect. 2.3), that is it defeats professional players with statistical significance [5]. This paper is not about creating such solution but about creating artificial intelligence, which will be able to use a game specific action, a "bluff". We wanted to create Artificial Intelligence that player could play with and maybe improve the game skills. This way the system has a potential of becoming really useful tool for poker players who want to improve their heads-up play.

2 Poker Game Description

The game of Texas Hold'em is a poker game where patrons play against each other for "the pot" of money (or chips) on the table. The gambling establishment does not participate in the actual play of the game and has no interest in the outcome of the play [10]. Contrary to Blackjack where players compete against the dealer. Hold'em hand starts with blinds being posted, then all players are dealt two cards facedown, these are known as hole or pocket cards. In No Limit Hold'em players may bet any amount of their chips on the table. First round of betting starts with a player to the left of big blind and follows clockwise. Each player can make one of three decisions: fold, call or raise. Betting round ends when all players that did not fold have put the same amount of chips into the pot, or when only one player is left and others folded. This first stage of game is named preflop. Next stage is the flop, when three cards are dealt on the table (community cards), next round of betting starts with a decision of the player to the left of the dealer button. At this point every player at the table has a unique five-card poker hand consisting of two hole cards and three community cards. Next stages of poker hand are similar, after the flop betting round is completed, another community card is dealt, this is known as the turn. Each of remaining active players has now a six-card poker hand but only five best cards count. Another round of betting is performed. After it is completed a final community card is exposed, the river. Each player has his final hand consisting of the best five cards of seven available. Then the final round of betting begins. The final stage known as showdown happens after final river bets have been placed. The person who initiated the final round of betting is first to show the hand. The action proceeds clockwise and other players muck their hands if they are weaker or show if they are better. Winner is determined by poker hands ranking.

2.1 Playing Strategy

There are several strategies of playing poker, some are more effective, whilst other are less. But in general it always depends on luck and strategies of other players. As stated in previous section, the goal is to win the pot, or more generally to be the last player with chips at the table. Also as stated before, there are two ways of winning the pot.

The obvious one, when we have strong hand. In this situation we focus on winning as much chips as possible. Player has to remember not to make too large bets, so opponents fold, due to the fact that their hand might be not good enough and they don't want to lose more. This way we might lose an opportunity to increase our stack even more when we were lucky having strong hand. Second way to win the pot is to make other people fold. Then it makes no difference which cards were dealt and who had strongest hand since no hole cards are shown. There are two major styles of playing poker: aggressive and passive. Playing aggressive means that we bet or raise even weak hands, this way we take the risk of losing at showdown but we open an opportunity to "steal" the pot even if opponents currently have stronger hands. This type of actions are very effective especially when we make consecutive decisions during the hand. For example if we raise preflop, then on the flop a high card like Ace, King or Queen is drawn, our opponents might suspect that we have just drawn high pair, especially when we make continuation bet. Playing aggressive is less predictable as range of betting hands is wider. This way raise preflop with weak hand (like 9, 7 nonsuit) might result in perfect situation on flop when community cards are low (let's say: 9, 7, 2) and we've drawn top two pairs. This is a perfect opportunity for a trap, when we check on flop (and maybe also turn) just to re-raise our opponent when he decides to make a bet. This kind of situation looks like a bluff for our opponent, because since we raised preflop we would usually have high cards. Playing passive means that we usually tend to call rather than raise. This way we can only win when we have better hand than our opponents. Although this style of playing (with strong hands only), can result in our bluff-raise being more legit, thus opponents might fold more likely. When using this strategy we should remember about its drawbacks. During poker tournaments (e.g. WSOP - World Series Of Poker) passive players tend to reach cash-out stages but rarely win.

2.2 Heads-Up Play

Although the rules remain the same for 2-players Texas Hold'em, the gameplay is very different. Both players at the table are forced to make blind bets so there is no way to fold weak hole cards for free. This implies playing far bigger range of starting cards. Overall probability of drawing strong hand is smaller since players are mainly forced to play weaker cards more often. Heads-up play is more dynamic and gives players more opportunities for bluffing. In heads-up gameplay evaluating your opponent is crucial (especially compared to multi-hands strife). Player's stack is also more important and significantly affects choice of strategy. Less quantity of players determines higher probability of unstable gameplay (bigger losses and wins). One also cannot afford to fold consecutively due to the fact that chips won later would not compensate high amount spent on blinds. Being passive for most of the time reveals when your cards are high and implicates your opponent when to fold. On the other hand it might be a good (but costly) investment for future bluffs.

2.3 DeepStack

DeepStack is one of the latest milestones of the University of Alberta Computer Poker Research Group. This is at the moment one of the best solutions for imperfect information settings. Deepstack combines recursive reasoning, decomposition and a form of intuition learned from self-play. Authors of DeepStack state that this is theoretically sound and is shown to produce more difficult to exploit strategies than prior approaches. DeepStack is a very advanced system, which uses "limited depth lookahead via intuition". It is a general-purpose algorithm for a large class of sequential imperfect information games. It creates possible sequences of public states in the game in form of a public tree with every state having its own subtree [5]. In this system player's strategy define probability distribution over valid actions for each decision point and system can estimate opponent's possible hands when certain public state is reached. The heart of DeepStack is continual resolving, which is a local strategy computation. It uses also sparse lookahead trees to reduce number of actions considered. It takes into account only actions of fold, call, 2 or 3 bet and all-in. This system creates approximately 10^7 decision points and solves them in under five seconds.

3 Poker Game System

3.1 Topology Tuning Idea

Game system we need to have full information about current situation on the table. Game system should also be applicable for described later learning environment. Given these conditions the game system is developed for two players only. The type of a player (bot or human) is still configurable (Fig. 1).

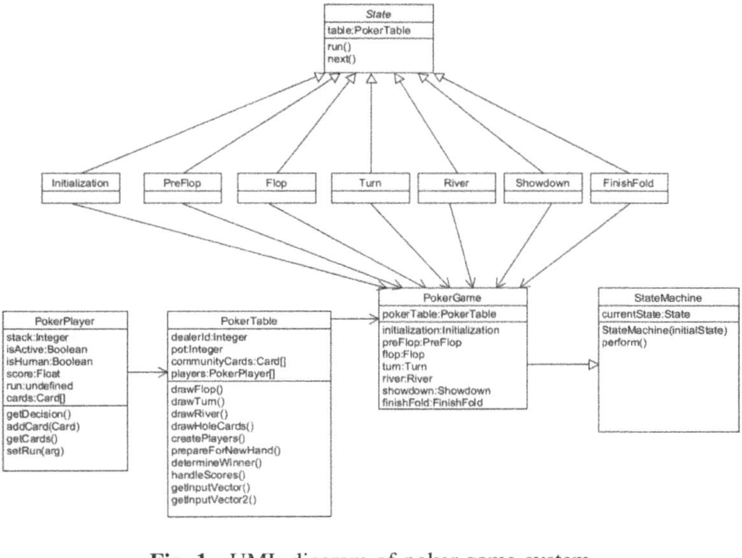

Fig. 1. UML diagram of poker game system

3.2 Historical Data

The database containing more than 10 million poker hands of various types was found at University of Alberta Computer Poker Research Group website [2]. Nonetheless the data was containing hands played by amateur players (and occasionally bots) not for real money. Whole database is divided into variants of play, then each variant into smaller subsets to describe specified state of the table: hole cards, community cards, pot, stacks, previous decisions on given hand. At a glance the database had all we needed. Information derived from the data was supposed to be used for pre-teaching AI for making at least not random decisions. Problem occurred when we realized that in the dataset there was no information on the hands where players folded. This database was created by bots in position of table observer so the only decisions we could teach were raise and call as hole cards were visible only on showdown. This turned out to be major problem for pre-teaching AI as none of the resulting initial models were able to fold anything.

3.3 Learning Procedure

Taking into account difficulties with the dataset usefulness, range of models to choose from was shortened. Multilayer perceptron was selected as very universal, flexible and easily scalable model. The learning algorithm has to work “on-line”. Genetic algorithm was decided to be the best solution for this problem. First of all it is quite natural approach for this subject to create population of players, define score assessment method and let genetic algorithm to produce the best fitted individuals [4]. Second thing is that combining artificial neural network with genetic algorithm is quite natural. When we consider neurons as chromosomes and its weights as genes, it is expected that exchanging it between individuals may give interesting results [1].

Two different inputs for neural network were tested. First one of size 40 consist of hole cards and community cards, each one represented in a form of 17 values [6]. First 13 were count regarding figures, next 4 regarding suit. Then there is a value representing if there is a pair on hand or if the hole cards are suited. Next two Boolean values represent if there is a straight draw and a flush draw. Then there is a float value representing how big is the opponent’s bet in relation to the pot [3]. At the end there are two Boolean values representing opponent’s previous decision (call or raise). Let’s consider a situation when player has Ad, 6d on hand, community cards are Ah, Qd, 7d, 10d (stage of play is the turn). Our opponent bets 30, pot is now 120 and on the flop opponent also raised. Input vector in this case would be:

[0, 0, 0, 0, 1, 0, 0, 0, 0, 0, 0, 0, 1, 2, 0, 0, 0, 0, 0, 0, 0, 0, 1, 0, 0, 1, 0, 1, 0, 1, 3, 1, 0, 0, 1, 0, 1, 0.25, 0, 1]

Second type of input vector has size of 14. It consists of “processed” information derived from the cards. First value represent the hole cards, if there is a pair or cards are suited. Next value is a Boolean representing if hole cards are “connector” (two consecutive figures). Next 9 values are representation of current best poker hand (from high card to straight flush). Next two Boolean variables are as previously a straight and flush draw. The last value represents bet size in similar manner to the first input type.

Considering the same example as for first input type, vector would be: [1, 0, 0, 1, 0, 0, 0, 0, 0, 0, 0, 0, 1, 0.25].

Output of the neural network consists of 4 neurons, three of which are interpreted as a decision (fold, call, raise) and fourth determines bet size when the decision is to raise. Decision is determined by maximum value of first three neurons [9]. Bet size output is interpreted the same as in input vector, it means that it is a fraction of the pot.

3.4 Genetic Algorithm Usage

Score assessment - this is probably the most crucial and most difficult part of creating the genetic algorithm. The process of choosing individuals in selection stage of the algorithm is based on criteria tailored for particular problem. In case of the paper subject - bluffing - there is no universal way of defining if particular decision was good or bad. We have faced many difficulties while defining fitness function. Most important is, as mentioned before, how to assess the score for a decision, since there is no general rule for that. Other problem is that the function has to take into account every decision and cannot be "unbalanced". It means that if the score assessment would favor any particular decision, then the genetic algorithm would produce a population of players who act the same without even taking into account the cards. Let's imagine a player who always folds or always raises. In first case there is no way that this player wins any game. In the second case player might even win couple of hands up until it's opponent has a strong hand and all previous winnings vanish. There were many attempts for defining fitness score. Most of them did not effect in sensible results.

Finally the fitness function was defined as follows: +1 point for a raise when player had better hand and opponent folded; +0.5 point for a fold when opponent had better hand; −1 point for a fold when player had better hand; +2 points for a raise when player had worse hand and opponent folded - reward for bluff; +1 point for a raise when player had better hand and opponent called; −1 point when player called when having worse hand; +1 point for a call when player had better hand; −1 point for a raise when player had worse hand and opponent called - penalty for bad bluff; −0.5 when player checked or called while having better hand; +0.5 point when player checked while having worse hand; +1 point for re-raise (after opponent's bet) when player had better hand; −2 points for re-raise (after opponent's bet) when player had worse hand.

Hands strength for pre-flop stage are determined by probability of winning in 2-players game [8]. For later stages hands strength is determined by best current combination of cards and poker hand ranking. There are two ways of calculating final score of a player: average score for all decisions made; sum of all points. Each of them has a worst-case scenario. For average there is a possibility that player had better hand and raised all the time, when it's opponent also raised the game might finish in just one or two hands. Then the player who had better hand has amazingly good average score but played too few hands to make it reliable. For sum of scores there is a possibility that when two players fold correctly most of the hands, their scores will be disproportionally higher than a good player's when he played with bad opponent and game finished very fast.

The roulette wheel selection of individuals for the new population generation is in use. After parents choosing we go through all of the weights and swap weight between corresponding matrices with predefined probability - denoted as crossover rate [3]. Other way of defining it might be swapping whole columns of weights matrices. The mutation rate was set as very low to prevent population diversity. We have decided to use elitism in form of leaving top 2% of population unchanged between generations [4].

3.5 Results and Discussion

The working point of decisions reasoning system is based on the following parameters: population size: 100, elitism: 2, crossover rate: 20%, mutation rate: 0,01, number of neurons in hidden layer: 20, number of hidden layers: 1, number of epochs: 400. For score representation parameter there was no significant difference in the outcomes. This might be justified by the fact, that each "worst-case scenario" effect could be reduced by the big number of games played. Due to the fact, that each individual in the population played huge number of games, the probability of having weak or strong hands was equal. For input vector method there was actually no statistically significant difference be- tween the final scores. This might be explained with the difference mainly in the form but the information contained was basically the same.

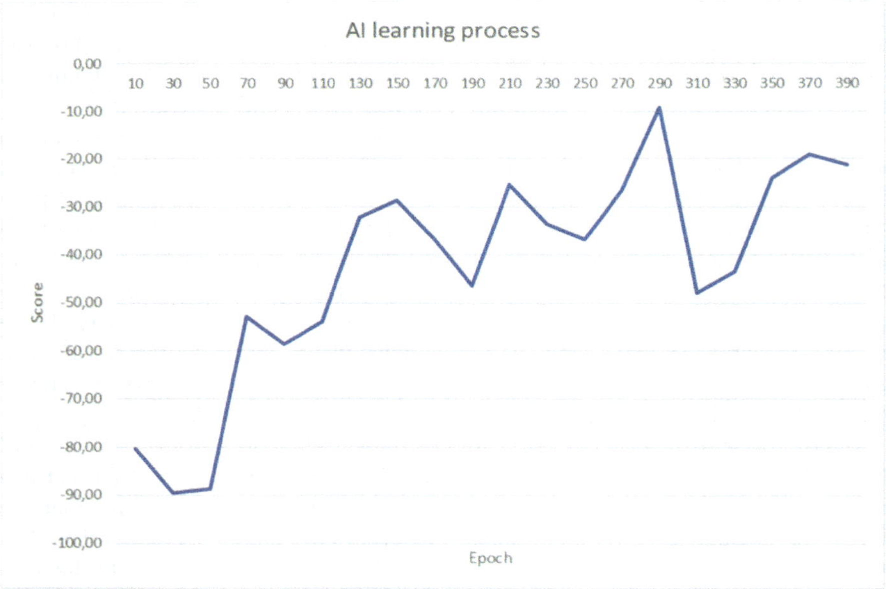

Fig. 2. Learning algorithm quality

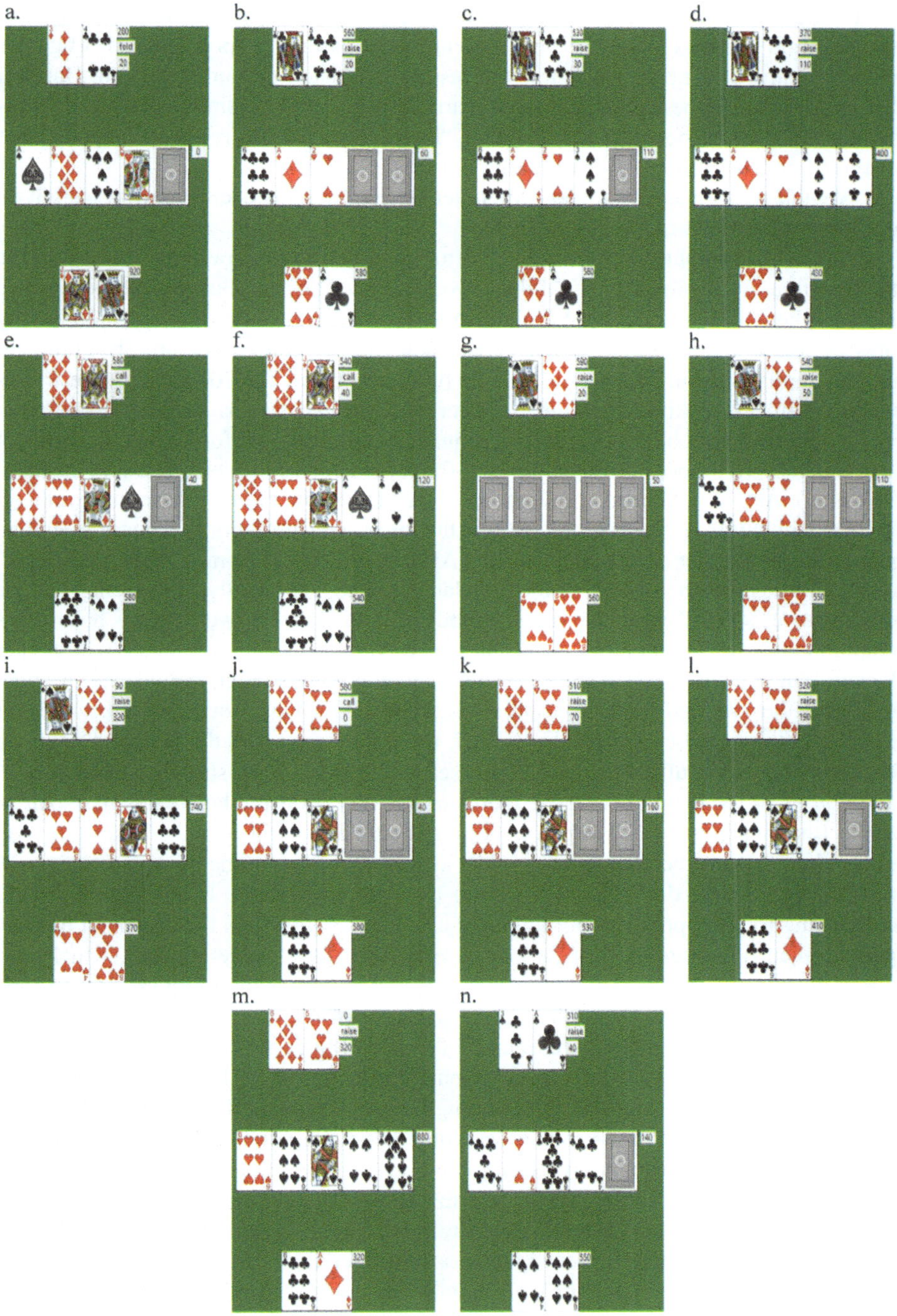

Fig. 3. Computer system in-game decisions with insight on computer player's cards

Figure 2 shows how the score value increased throughout the learning process. The visualization presents 10 of the best individuals in particular epoch played 50 games with the same opponent, which was the result of decision reasoning system. Data for this visualization does not come from learning process but from evaluating the performance of saved best individuals. This "backup" of current results was performed every 10 epochs during the learning process.

We will try to interpret several computer system in-game decisions with insight on computer player's cards.

Figure 3a. computer system folds. It (top) has two cards lower than those on the table, with a 'nut-straight draw' (if there would be 2 of any suit but spade on the river). Human player bet only 20 but AI folded anyway. This was reasonable decision having only 3 "outs" on the river, which gives about 6% of chance. Figure 3b. computer system bluffs without success. It tries to bluff by raising for 20, in this case it was unsuccessful mainly because human player had drawn top pair. Figure 3c. computer system sticks with a bluff. This is a continuation of bluffing performed by AI. The bet size might seem to look like a "value-bet" (bet of small size, it seems like it is meant to be called). Actually AI turns a straight draw (with 4 on the river). Figure 3d. computer system still bluffs at the river. It shows the river, which "pairs the board". Human player decided to bet on the river and the AI decision was to re-raise with "Jack high" only. Holding on to a bluff in the regular gameplay is usually convincing for the opponent, especially with community cards like this. There were many hole cards combinations that beats pair of Aces and which AI player could have: Ace with higher kicker than 7; A, 6; A, 2; A, 3; 3, 4 - which would have straight draw since the flop, pair after the turn and the set on the river; High card (king or queen) with 3 - player flopped nothing and could decide to bluff, then draw a pair on the turn and continue betting with semi-bluff, at the end draw a set on the river; 4, 5 - straight on the turn. In addition to all the mentioned above there are not very probable hole cards like: pair of threes (the nuts in this case) or pair of aces (second nuts) - highly unlikely since human player knows that the opponent would have to draw two remaining cards (3 or A respectively) in the deck. Continuing the bluff throughout the hand might be more convincing for the opponent, but might be also very costly when the opponent has any pair or even high card only. AI's bets sizes looked like "value-betting", which also might be persuasive. To "read" this kind of play correctly, would require knowing the opponent's previous decisions, bets sizes and style of playing. Figure 3e. action goes check-check. Computer system checks on the turn with Jack-high and a straight flush draw. Computer player has 13 outs (9 remaining diamonds for flush and 4 queens for straight). Human player in this situation might have considered betting to steal the pot. Actually we had only 7 4 nonsuit so we would definitely not call. Figure 3f. computer system calls with Jack-high only. Seeing previous check on the turn from the computer we wanted to steal the pot by making pot-size bet of 40. In terms of winning the pot AI made a good decision but in terms of general play and hand strength this kind of action ("hero call") might be very costly and is considered as bad decision. Figure 3g. computer system raises pre-flop with king-seven nonsuit. It raises pre-flop with K,7 nonsuit. In heads-up play raising from small blind position is pretty standard play. Figure 3h. computer system makes a continuation bet on paired board. We called pre-flop with 4, 8 suited to see what will be AI's next moves. Computer player decided to make almost pot-size continuation bet (110 is the pot after the bet) with king-high on paired board. If it was human player we might consider that it flopped a set with high kicker, straight or flush draw. We have a flush draw so we call. Figure 3i. computer system makes a big bet on the river. This is the last betting phase of the hand that started in Fig. 3g. We missed the flush draw on the turn and also on the river. We checked and computer player decided to make another pot-size bet which I could not call with 8, 4. Figure 3j. computer with nothing, human flopped a set with Ace kicker so a very strong hand in heads' up play. Only couple of hands can beat it: pocket queens for full-house queens over sixes or Q, 6 for full-house sixes over queens. Since we know that there is only one six left in the deck and 3 queens it is unlikely that my opponent has any of them. Figure 3k. is a continuation of the hand started in Fig. 3j. We decided to bet 50 with a set and AI tried to raise it to 70 (so increase only 20). This decision does not make any sense in terms of regular play. Figure 3l. shows how computer system re-raises me on the turn. We bet 100 having still very strong hand and computer player re-raised to 190. Although AI has now inside straight draw (with 7 on the river) which gives only 4 potential outs - about 9% probability. Actually there are 3 spades on the table right now which creates a possibility of having a flush. Figure 3m. - there is quite interesting situation. We will consider this situation like we would normally do without seeing opponent's cards. With fourth spade on the river there are possibly many hole cards that computer player could have, which would result in a hand stronger than set of sixes. The most probable with all the decisions made throughout the hand could be something like Ks, 6d, which would result in set of sixes with king kicker since the flop and a king-high flush on the river. Another possibility could be for example pair of fives (with one of them being spade). Any spade in opponent's hole cards would beat set of sixes. But when we take into account actual hole cards that AI has, this is a bad play since the flop. Only the fact that two additional spades were drawn it might just turned out to be perfect bluff since even with a set we would consider folding (if we played with human opposition). In Fig. 3n. we can see that AI bets only about 40% of the pot while having second-nuts with Ace high flush (the nuts would be 6 and 7 of clubs for a straight flush). I had only pair of fours. The bet size could look like a value bet or fainthearted bluff with such a "wet board" (there are several draws for straight or a flush).

4 Conclusions

Creating artificial intelligence playing poker on expert level is a very complicated task. Poker Texas Hold'em No Limit is an extremely complex game, not only in terms of computational magnitude but also in terms of abstract problem definition. Solving Hold'em poker requires combining many different fields of science, most obvious ones like statistics and game theory but also cognitivism and behaviorism for modelling a player. We decided not to use any historical data or poker playing bots but to simplify the problem and define general rules for score assessment. Main goal was to find out how combining ANN with genetic algorithm will perform in creating AI playing poker. Furthermore we decided to solve the problem without any external "teachers", only with fitness score definition. The results obtained are very satisfactory for assumptions made. AI can make somehow reasonable decisions, especially when we take into account that during learning it was rewarded mostly for aggressive style of playing and there was no factor that would consider actual outcome of the game (win or loss). In terms of general poker playing this AI would not stand a chance against human opposition. Main reason for it is that AI after learning is deterministic when it comes to making decisions in particular situations. Above-average human player competing with this AI would notice that it plays aggressive a lot of weak hands and does not know when to let it go. It could be easily exploited for example by seeing couple of flops and betting big when having a strong hand. Good example of this very aggressive game style is a hand showed in Fig. 3k. – Fig. 3n. It might have been successful in this particular case - having four spades on the table - and win over 50% of the opponent's stack with "stone cold bluff". But playing like this in the long run would be very costly.

References

1. Bonarini, A., Masulli, F., Pasi, G.: Soft Computing Applications. Advances in Soft Computing. Springer, Heidelberg (2003)
2. Computer poker research group. http://poker.cs.ualberta.ca/index.html. Accessed 11 Mar 2019
3. Damiani, E.: Soft Computing in Software Engineering. Springer, Berlin (2004)
4. Gómez, F., Quesada, A.: Genetic algorithms for feature selection in data analytics. https://www.neuraldesigner.com/blog/genetic_algorithms_for_feature_selection. Accessed 12 Mar 2019
5. Moravčík, M., Schmid, M., Burch, N., Lisý, V., Morrill, D., Bard, N., Davis, T., Waugh, K., Johanson, M., Bowling, M.: Deepstack: expert-level artificial intelligence in heads-up no-limit poker. Science **356**(6337), 508–513 (2017)
6. Nielsen, M.: Using neural nets to recognize handwritten digits. http://neuralnetworksanddeeplearning.com/chap1.html. Accessed 12 Mar 2019
7. Poker player research ltd. topline findings. http://pokerplayersresearch.com/toplinefindings.aspx. Accessed 10 Mar 2019
8. Probability that your hand will end up being the best hand. http://www.natesholdem.com/pre-flop-odds.php. Accessed 12 Mar 2019
9. Sivanandam, S.N., Deepa, S.N.: Principles of Soft Computing. Wiley, Hoboken (2011)
10. Texas hold'em. https://oag.ca.gov/sites/all/files/agweb/pdfs/gambling/BGC_texas.pdf. Accessed 10 Mar 2019
11. Where did poker originate? https://www.history.com/news/where-did-poker-originate. Accessed 12 Mar 2019

Intelligent Inference Agent for Safety Systems Events

Jacek Mazurkiewicz[1(✉)], Tomasz Walkowiak[1], Jarosław Sugier[1], Przemysław Śliwiński[1], and Krzysztof Helt[2]

[1] Faculty of Electronics, Wrocław University of Science and Technology, ul. Wybrzeże Wyspiańskiego 27, 50-370 Wrocław, Poland
{jacek.mazurkiewicz, tomasz.walkowiak, jaroslaw.sugier, przemyslaw.sliwinski}@pwr.edu.pl

[2] Teleste sp. z o.o, ul. Szybowcowa 31, 54-130 Wrocław, Poland
krzysztof.helt@teleste.com

Abstract. The paper describes the idea of the intelligent agent based on the ontogenic neural networks. The agent is created for inference problem needed in some safety systems. The main goal of the agent is to find if the aggregated data collected from the different sensors should point actual state of the system as alarm. The neural approach to the agent structure allows to combine the diverse nature of the sensors output into single clear answer. The ontogenic type of the neural network is the key to tune the topology to real needs driven by the scale and the safety systems features. The size of the training vectors can be limited as well as the number of the training epochs. Better results of alarm state prediction we can expect when we use the hybrid combination of agents focused on some specific types of events. This way the idea of the Complex Event Processing (CEP) safety systems seems to be sensible where Event Processing Agents (EPA) are also intelligent tools created based on already data sets collected previously.

Keywords: Ontogenic neural network · Safety system · Intelligent agent

1 Introduction

Operation of various contemporary real-time information systems is often based on Detect-Decide-Respond working scheme: in a stream of events which are coming from the environment, functionality which the system must provide can be defined as a detection of some specific temporal and semantic patterns within the event stream, followed by an evaluation of their various characteristics and generation of appropriate reactions as the result of their classification. Such a style of processing can be found e.g. in security or safety monitoring, active diagnostics, predictive processing or business process management. In these cases it can be advantageous to base construction of the whole information system on the event-driven processing paradigm [10]. The events are monitored by the set of sensors which are sensitive for the main features of the events. The outputs of the sensors are specific and it is not trivial to combine them into a single clear answer if the actual state of whole system we find as

W. Zamojski et al. (Eds.): DepCoS-RELCOMEX 2020, AISC 1173, pp. 441–450, 2020.
https://doi.org/10.1007/978-3-030-48256-5_43

alarm or important warning for example. This is the reason why we propose the intelligent agent based on the idea of neural network to produce such global output. The ontogenic type of the neural network is the key to tune the topology to real needs driven by the scale and the safety systems features. The size of the training vectors can be limited as well as the number of the training epochs [8]. Of course we need the set of training vectors with correct outputs given by the teacher to prepare the agent for the normal work [9]. We expect better results of alarm state prediction we can expect when we use the hybrid combination of agents focused on some specific types of events.

Section 2 presents the introduction to ontogenic neural network. Section 3 is focused on the idea of the intelligent agent construction, as well as training and testing of it. Section 4 discusses the results and the test of agent sensitivity for input vector changes. Finally Sect. 5 points the possible paths of next steps in the future.

2 Ontogenic Neural Network

An architecture of the neural network, determines the complexity and capabilities of the entire adaptation model [1]. Models that cannot modify their architecture must have it well determined on prior knowledge before starting the learning process. It is very difficult because usually the complexity of the problem is not known problem. Assuming that the neural network can make proper decisions about the architecture, it may correctly check if the current topology is appropriate to represent problem being solved or is too complex. If the number of connections or neurons of a given neural network will be too small the model is not be able to learn – to create the correct set of weights, however when there are too many connections among the neurons the adaptive cannot reach a satisfactory level of generalization [6]. Therefore, the complexity of the adaptive model should be appropriate the complexity of the problem. The main goal is to find an architecture to achieve the best quality generalization. The network must have some margin freedom, which lies in adaptive parameters and allows to change model states smoothly. Well-chosen margins of freedom along with criteria for controlling complexity model, also allow to fight with the problem of local minima [2]. Model changing its architecture goes to other spaces of adaptive parameters with another error function, in which the learning process continues, and new changes of architecture are possible. In this way, such a learning model can explore different spaces in search of a certain optimum. The methods for checking the complexity of network architectures can be divided into three groups:

- magnifying - these models include algorithms that allow to add new neurons or new connections among neurons;
- reducing - methods that remove unnecessary neurons or connections among neurons or algorithms that can join groups of neurons or connections between neurons;
- cooperative systems - groups of models, each to solve the subtask of the problem and the management system makes final decisions.

3 Intelligent Agent

3.1 Topology Tuning Idea

Our intelligent agent is based on reducing ontogenic neural network. The fully connected three-layer Multilayer Perceptron (MLP) is the starting point [8]. The first possible approach for the reducing procedure is based on significance factor:

$$s_i = E(without\,neuron\,i) - E(with\,neuron\,i) \tag{1}$$

which determines the difference between a network error obtained without and with the participation of a neuron *i*. This method requires considerable calculation costs - to be determined for each coefficient s_i of Eq. (1) an error for the whole training set. Neurons with the less significance factors can be removed.

The similar - also passive – (2) way of the significance coefficients determining has been used in the FSM system (Feature Space Mapping) [1, 3, 4, 10]. Significance coefficients are determined for each hidden layer neuron after interrupting the learning process:

$$Q_i = C_i(\mathbf{X})/|\mathbf{X}| \tag{2}$$

where: $|\mathbf{X}|$ - number of training vectors, $C_i(\mathbf{X})$ – number of correct answers given by neuron *i* from input set $\mathbf{X}$. In FSM type network each neuron from hidden layer is responsible for the class. The neuron with Q_i close to zero is removed.

Methods that reduce the structure of a neural network can often be considered as regularization process. In the weight decay procedure [5] for standard measure of error of model $E_0(f)$

$$E_0(f) = \frac{1}{2}\sum\nolimits_{i=1}^{n}(y_i - f(x_i))^2 \tag{3}$$

the following factor is added [11, 12]:

$$E_w(f) = E_0(f) + \lambda\sum\nolimits_{i=1}^{M}\frac{w_i^2/w_0^2}{1 + w_i^2/w_0^2} \tag{4}$$

where: w_0 – is constant parameter – the experiment shows – should be equal to one, if $|w_i| \gg w_0$ the factor goes to λ, if $|w_i| \ll w_0$ goes to zero. The parameter λ can be tuned during learning process:

- $\lambda = \lambda + \Delta\lambda$ if $E_n < D$ or $E_n < E_{n-1}$
- $\lambda = \lambda - \Delta\lambda$ if $E_n \geq E_{n-1}$ and $E_n \geq D$
- $\lambda = 0.9\lambda$ if $E_n \geq E_{n-1}$ and $E_n < D$

where: E_n – the last epoch error, D – final error for the training process. Finally, the training algorithm called Optimal Brain Damage (OBD) [7] looks as follows:

1. Set the starting topology.
2. Make the training process using classic gradient method until the error is acceptable and the changes are not important.
3. Calculate the significance factors taking into account the regularization parameters.
4. Remove the weights fixed to the extremely low values of significance factors. It means "turn-off" the neurons from the hidden layer.
5. If the weights are reduced go to Step 2.

Of course the reduced number of neurons is acceptable if the network answer is still correct from the functional point of view. If not it is obligatory to come back to the previous version of the topology. The OBD approach does not guarantee correct results of training procedure with the limited number of neurons [11, 12].

3.2 Intelligent Agent as Event Processing Agent

The Event Processing Agents are the essential components which encapsulate actual operations executed on events and the CEP models defines their various types. In general an agent of any type performs a three step processing: filtering (selection of the events from the input streams), matching (finding patterns among the accepted events which will trigger generation of the output) and derivation (forming the output events and evaluating their attributes using the results of the matching step). This generic scheme results in various functionalities (*operators*) of the EPAs which range from simple event filtering based on attribute values (a stateless operation), through transformations (e.g. aggregating a set of input events and composing a derived event from its characteristics), up to different pattern detection schemes which analyze temporal and/or semantical relations between the events. Event is the phenomenon observed pointwise during system operation. Spot observation means that we are able to point a point on the timeline when the phenomenon occurred. Detection of the phenomenon is possible thanks to a set of sensors monitoring the behavior of the system. Sensors are devices with different performance characteristics, usually testing a specific - selective parameter of the monitored system. Certainly, the level of significance for further operation of the system indicated by the sensor of the phenomenon can be very diverse - starting from information having a negligible impact on the need to make decisions related to the further functioning of the system through various levels of warranties requiring preventive or reconfigurable measures to alarm states forcing radical steps to protect system integrity or critical elements of its structure. We decided to create four independent types of intelligent agents to deal with four types of event sensors:

- TS – two-states sensors responsible for simple events – like on-off, open-closed,
- AM – active movement sensors – output: distance to moving object,
- TM – temperature sensors – output: the actual temperature,
- BL – brightness level – output: actual brightness measured using the proper units.

Of course these parameters can be exchanged to actual needs driven by the safety system features. Potential sources of events considered: access monitoring, area monitoring, mass messages, fire risk, internal notifications, video monitoring, communication channels, biometrics, access violation, temperature risk, gas hazard, acoustic threat, biological and medical risks, flood risk, open/close state, assembly/accumulation risk, system operator signal, user defined. A wide spectrum of potential sources of events does not exclude the use of a unified approach to the description of these events. We assume that the source - event generator - will determine the record of the given event in a kind of table. The number of description fields and their types are uniformly defined. Such approach will allow for the initial aggregation of events according to the reasons for their occurrence, as well as subsequent binding events in teaching vectors for intelligent information processing systems that will be used to accurately analyze the situation in the life of the system described by data recorded from many sensors. The package of the following fields is stored in unified structure:

Event_ID, Source_Name, Source_ID, Source_GPS, Object_ID, System_ID, Event_Date, Event_Occurence, Event_Duration, Event_Value, Event_Importance, Event_Probability, Event_Type, Event_Info.

For the set of experiments 1000 records for each of four selected types of sensors were generated covering the wide spectrum of possible input data. The data were used to prepare the training - 70% of population - and testing – 30% of population - vectors of 10 inputs: *Event_ID, Source_ID, Object_ID, System_ID, Event_Occurence, Event_Duration, Event_Value, Event_Importance, Event_Probability, Event_Type*. All introduced data were normalized – reflected from the original scale to [0, 1] range. The training vectors were equipped with the correct answer – gradient methods of neural network training needs "the teacher" – as a source of expected output. We prepared four independent neural networks – four intelligent agents dedicated for each of four types od sensors. The output of each intelligent sensor is the value from 0 to 1 to describe the "importance level" of sensor reaction. The fifth neural network models the intelligent agent – as final voting element. Its input vector is created based on four intelligent agents responsible for sensors' data. This way the final output can be read as aggregated signal of alarm in the system. Of course the final agent training procedure needs the expert answer if the actual input vector looks like the alarm situation. Such hierarchical structure of the single processing element gives a chance to create the flexible solution properly fixed to actual system needs (Fig. 1).

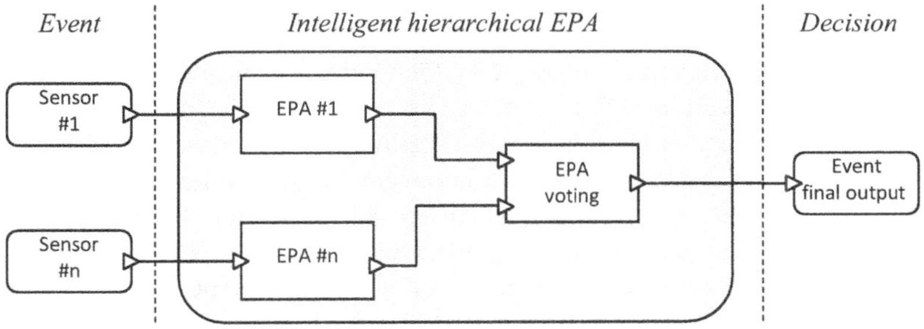

Fig. 1. Intelligent hierarchical event processing agent

3.3 Event Processing Agent Topology and Training

The Event Processing Agents cooperating with the sensors are – initially – fully interconnected Multilayer Perceptrons. The size of the input layer is equal to the size of the input vectors. It means we have 10 neurons there ready to input the float digits representing the components described in the previous subsection. The output layer has only one neuron to generate the answer of the net as the level of importance of the data driven by the input sensor. Of course there is no problem to convert this fraction value to two state using simple threshold mechanism. The size of the single hidden layer is equal to 20 neurons as a starting value, but during the training procedure the number of active neurons is reduced by the Optimal Brain Damage (OBD) mechanism using regularization factors – Sect. 3.1. This way the minimum number of working neurons in hidden layer is only 4. The training procedure was done for each EPA individually using the proper set of input vectors dedicated for each sensor. Each topology created as a result of OBD is trained again. The number of epochs is limited by the "no change" observation taking into account the network error minimizing. The final results for each EPA – and data from each sensor – are presented in Table 1. The initial values of all weights are generated as random from [−1, 1] range. The sigmoid transfer function is applied to all neurons from hidden and output layer. The training is done using Levenberg-Marquardt algorithm [6]. Three different kinds of experimental distance (5) have been used during error of model calculation (4):

$$L_1 = \frac{1}{N}\sum_{i=1}^{N}|y(x_i) - \hat{y}(x_i)| \quad L_2 = \sqrt{\frac{1}{N}\sum_{i=1}^{N}(y(x_i) - \hat{y}(x_i))^2} \qquad (5)$$
$$L_\infty = \max_{i=1,\ldots,N}|y(x_i) - \hat{y}(x_i)|$$

where: $y(x_i)$ – network output, $\hat{y}(xi)$ – desired output.

The voting Event Processing Agents is – initially – also fully interconnected Multilayer Perceptron. The size of the input layer is equal to the size of the input vectors. It means we have 4 neurons there ready to load the products of the EPA collaborating with the data taken from sensors. The output layer has only one neuron to generate the answer: one of two possible states: alarm/no alarm. The size of the single hidden layer is equal to 20 neurons as a starting value, but during the training procedure the number of active neurons is reduced by the Optimal Brain Damage (OBD) mechanism using regularization factors – Sect. 3.1. This way the minimum number of working neurons in hidden layer is only 4. The training procedure was done for using the set of vectors created based on the components aggregated as outputs from EPA cooperating with sensors. The number of epochs is limited by the "no change" observation taking into account the network error minimizing. The final results – the percentage of correct answers - are presented in Table 2. The initial values of all weights are generated as random from [−1, 1] range. The sigmoid transfer function is applied to all neurons from hidden and output layer. The training is done using Levenberg-Marquardt algorithm. Three different kinds of experimental distance (5) have been used during error of model calculation (4):

Table 1. EPA correct answers [%] for different type of sensors and limited number of neurons in hidden layer as OBD mechanism result

Sensor	Distance type	Number of neurons in hidden layer – OBD result								
		4	6	8	10	12	14	16	18	20
TS	L_1	61	65	69	70	70	87	74	71	67
	L_2	56	56	60	62	67	76	71	73	69
	L_∞	47	51	53	57	62	65	61	57	55
AM	L_1	62	68	68	73	73	77	75	72	70
	L_2	55	58	58	63	66	72	72	70	68
	L_∞	45	45	52	52	58	63	63	62	60
TM	L_1	56	57	66	68	70	85	74	70	65
	L_2	56	56	57	62	66	74	69	71	69
	L_∞	43	45	50	55	61	67	61	55	53
BL	L_1	60	65	67	75	71	78	72	72	71
	L_2	52	55	59	65	66	76	71	70	68
	L_∞	43	43	49	51	51	64	63	61	60

Table 2. Voting EPA correct answers [%] for limited number of neurons in hidden layer as OBD mechanism result

Distance type	Number of neurons in hidden layer – OBD result								
	4	6	8	10	12	14	16	18	20
L_1	67	69	72	75	77	82	88	85	81
L_2	58	64	68	70	73	74	75	79	76
L_∞	53	55	58	62	65	71	73	70	70

4 Results and Sensitivity Discussion

The experimental results pointed in Table 1 show that intelligent hierarchical EPA is able to provide correct recognition of the importance level based on data from different types of sensors. Both bi-level and scaled continuous outputs from the sensors can be useful source of data for the proper decision. We find the answer of EPA cooperating with sensor as correct if it equals to required output within [−0.1, +0.1] range. The extremely reduced number of neurons according to OBD mechanism causes insufficient – incorrect from the functional point of view – answer of EPA. It seems natural, but on the other hand – the ontogenic approach to EPA construction – with "on-line" hidden layer tuning – looks very promising. We started with twenty neurons located in this layer, but this value - looking good as a-priori assumption - is not optimal. The best size is 14 neurons for all types of the sensors. Of course this optimal number of neurons can be different if we use other sets of input data or we redefine the input vectors. The ontogenic topology is caused by the data used for training procedure. The correct EPA answers available for all tested types of sensors is a kind of proof that the set of 1000 training vectors sounds sensible to create the required level of "recognition skills" of single EPA. By modeling the training vectors sets we can tune the level of EPA reaction for input as well as we can store in the EPA deeper and more or less detailed "history" of the system life. We know how important is the correct "teacher's" required output during the training process. This output should be based on the expert knowledge to finish the weights setting at the necessary level of details. The EPA outputs for all types of sensors look promising, but better results we find for TS – two-states sensors responsible for simple events and TM – temperature sensors. Maybe these types of data are more convenient for neural modeling or the expert knowledge used during training is better, or other types of sensors need more epochs or more data to establish final values of weights. Table 1 also tells us that for all types of sensors the most classic approach to distance measure L_1 is the best for the task we discuss. It means the easiest implementation in the practical future of the system. The voting EPA results – Table 2 – looks also very promising. The final answer of the hierarchical EPA structure is the best for the topology with 16 neurons in the hidden layer. Again the OBD mechanism allows to reduce the size of this layer to the most suitable size. And the aggregation of the previous layer of EPAs' outputs is absolutely correct using the same L_1 type of distance during training procedure. The hierarchical construction of the intelligent EPA allows to create more sophisticated cascades of EPAs collaborating with sensors with the final decision block. This way we can decide about the components of the voting EPA answer, we can model the influence of the events for the next step of the safety system reaction.

During the last part of the experiment we try to check the EPA sensitivity for the changes of the input vectors. Each unified input vector collects the set of parameters describing the single sensor's actual state. Some of these parameters are constant or almost constant.

The main changes are observable in these components which reflect the environmental feature tested by the sensor. This way we try to find if the EPA answer is really provoked by this leading value from the input vector. Result are presented in Table 3. We can easily notice that greater number of neurons provides the better sensitivity for input data. The net with the greater number of neurons in hidden layer can analyze the input vector in more detailed way. For TM sensors we find better sensitivity than for AM and BL sensors. The previous sentence is kind of analogy to the first part of the experiments and we are not surprised about it. There is no sense to check the sensitivity parameter for TS sensors because inputs are binary. The L_1 distance type is the most suitable.

Table 3. EPA sensitivity [%] for different type of sensors and limited number of neurons in hidden layer as OBD mechanism result

Sensor	Distance Type	Number of neurons in hidden layer – OBD result								
		4	6	8	10	12	14	16	18	20
AM	L_1	10	10	8	7	7	6	6	4	3
	L_2	13	13	11	10	9	7	6	5	4
	L_∞	15	15	12	10	10	9	7	6	5
TM	L_1	8	8	7	7	6	5	5	3	2
	L_2	10	10	11	8	9	6	6	5	4
	L_∞	12	14	12	10	10	8	7	6	5
BL	L_1	10	10	8	7	7	6	6	4	3
	L_2	13	13	11	10	9	7	6	5	4
	L_∞	15	15	12	10	10	9	7	6	5

5 Conclusions

The paper proposes the intelligent agent based on the ontogenic neural networks. The agent is created for inference problem needed in some safety systems. The main goal of the agent is to find if the aggregated data collected from the different sensors should point actual state of the system as alarm. The neural approach to the agent structure allows to combine the diverse nature of the sensors output into single clear answer. The ontogenic type of the neural network is the key to tune the topology to real needs driven by the scale and the safety systems features. The size of the training vectors can be limited as well as the number of the training epochs. Better results of alarm state prediction we can expect when we use the hybrid combination of agents focused on some specific types of events.

This way the idea of the Complex Event Processing (CEP) safety systems seems to be sensible where Event Processing Agents (EPA) are also intelligent tools created based on already data sets collected previously.

Further work is carried out in multiple directions. The first group of activities includes the creation of an expert system, in which a set of rules filled with event descriptions will allow inference leading to the prediction of threats emerging through the aggregation of events observed pointwise. This is possible due to the inclusion in the set of rules of grammars linking the causes and effects of events using intelligent hierarchical EPAs. Perhaps more adequate will be inference ahead, which does not assume the final scenario - predicted “from the beginning” of the resolution of the situation observed in the system. All possible “paths” are tested and the ones that lead to “dead ends” are eliminated.

The second group is intelligent data filtration implemented by self-organizing networks and networks with dynamic feedback - Hopfield network and its related. The effect of this filtration will be the rejection of those features describing events that are in the nature of disturbances and “additions” of missing fragments characterizing the essential features of events.

Acknowledgements. This work was supported by the Polish National Centre for Research and Development (NCBR) within the Innovative Economy Operational Programme grant No. POIR.01.01.01-00-0235/17 as a part of the European Regional Development Fund (ERDF).

References

1. Adamczak, R., Duch, W., Jankowski, N.: New developments in the feature space mapping model. In: Third Conference on Neural Networks and Their Applications, Kule, Poland, pp. 65–70 (1997)
2. Bonarini, A., Masulli, F., Pasi, G.: Soft Computing Applications. Advances in Soft Computing. Springer, Heidelberg (2003)
3. Duch, W., Diercksen, G.H.F.: Feature space mapping as a universal adaptive system. Comput. Phys. Commun. **87**, 341–371 (1994)
4. Duch, W., Jankowski, N., Naud, A., Adamczak, R.: Feature space mapping: a neurofuzzy network for system identification. In: Proceedings of the European Symposium on Artificial Neural Networks, Helsinki, pp. 221–224 (1995)
5. Hinton, G.E.: Learning translation invariant recognition in massively parallel networks. In: Proceedings PARLE Conference on Parallel Architectures and Languages Europe, pp. 1–13. Springer, Berlin (1987)
6. Kung, S.Y.: Digital Neural Networks. Prentice-Hall, Upper Saddle River (1993)
7. Le Cun, Y., Denker, J., Solla, S.: Optimal brain damage. In: Advances in Neural Information Processing Systems 2. Morgan Kauffman. San Mateo CA (1990)
8. Pratihar, D.K.: Soft Computing. Science Press (2009)
9. Sivanandam, S.N., Deepa, S.N.: Principles of Soft Computing. Wiley, Hoboken (2011)
10. Srivastava, A.K.: Soft Computing. Narosa PH (2008)
11. Weigend, A.S., Rumelhart, D.E., Huberman, B.A.: Backpropagation, weight elimination and time series prediction. In: Proceedings of the 1990 Connectionist Models Summer School, pp. 65–80. Morgan Kaufmann (1990)
12. Weigend, A.S., Rumelhart, D.E., Huberman, B.A. Generalization by weight elimination with application to forecasting. In: Advances in Neural Information Processing Systems 3, pp. 875–882. Morgan Kaufmann, San Mateo (1991)

Wi-Fi Communication and IoT Technologies to Improve Emergency Triage Training

Jan Nikodem[1(✉)], Maciej Nikodem[1], Ryszard Klempous[1], and Paweł Gawłowski[2]

[1] Faculty of Electronics, Wrocław University of Science and Technology, Wroclaw, Poland
{jan.nikodem,maciej.nikodem,ryszard.klempous}@pwr.edu.pl
[2] Department of Emergency Medical Service, Wrocław Medical University, Wroclaw, Poland
pawel.gawlowski@umed.wroc.pl

Abstract. The paper presents how Wi-Fi communication and IoT technologies can be used to improve the efficiency of training Emergency Medical Staff in effective emergency triage procedures. To meet training requirements a victim simulator generating vital signs vector, was proposed. Next, the Line of Life, as a time ordered series of vital signs vectors, was created. The Line of Life concept allows to conduct different simulation scenarios corresponding to various severity of injuries modelling real situations of wounded victims. The proposed solution provides an effective communication to organize an appropriate triage and incident command. For wireless on-line communication and immediate feedback at accident site, a Wi-Fi channels and UDP datagrams in IP multicasting mode were proposed. For the purpose of comprehensive debriefing following the training, a client-server architecture was proposed to collect, store and provide triage training data.

Keywords: IoT technologies · Wi-Fi communication · Emergency triage · Victim simulator · UDP datagrams · IP multicasting · TCP/IP client-server

1 Introduction

Emergency triage is a procedure for quick prioritizing for treatment, according to life status of victims injured in mass-scale incidents with large number of casualties. Triage involves assessment and prioritization of casualties, so that activity is addressed first to those who need it the most. Emergency triage is conducted on site by first responders – fire rescues, emergency medical personnel or paramedics who arrive at the scene in tending to medical emergencies in a stressful environment [2,4,5].

W. Zamojski et al. (Eds.): DepCoS-RELCOMEX 2020, AISC 1173, pp. 451–460, 2020.
https://doi.org/10.1007/978-3-030-48256-5_44

Traditional training standards for EMS personnel consist primarily of didactic continuing education lectures that do not adequately address the reality of providing care in an emergent, rural setting. The objective of training exercise is to stage a realistic, in situ simulation. We sought to test the feasibility of deploying Wi-Fi communication [9,10,12] and IoT technologies [3,6,11] in a training exercise [6,8], not only inside university premises, but also in a rural environment and further to provide a live stream via telemedicine technology [7]. From management point of view, objectives were to improve the communication and teamwork skills of the rescue team and the efficiency with which the team performs challenging interventions in a high-stress situation [13]. There are a number of emergency triage procedures [2,4]. In practice, each country develops it's own Proposed National Guideline in which determines when and which procedure should be followed. However, all the procedures have common elements [13] which can be extracted and used as training exercises. This set includes: START, JumpStart, CareFlight and Sieve triage procedures [2,5]. All of them have at least four categories of victims. RED - requiring immediate help, YELLOW - serious but can wait until all reds are transported, GREEN - ambulatory/hold –minor injuries, BLACK - recognized as expectant/deceased.

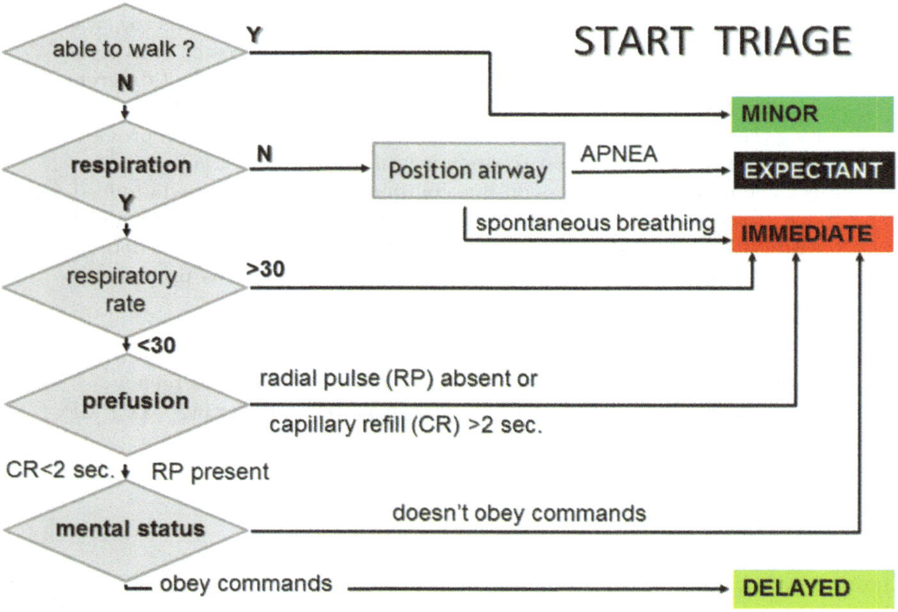

Fig. 1. The Simple Triage And Rapid Treatment (START) block diagram [5].

An emergency triage procedures are based on so-called vital signs which are a strong criteria for prioritization and categorization for treatment of injured people. There are four main substantial criteria:

- **moving**, ability to follow directions and walk,
- **respiratory effort**, clear/open airways, adequate breathing, respiration rate (RR),
- **pulses/perfusion**, control major bleeding, hemorrhaging profusely, radial pulse, capillary refill, saturation of peripheral oxygen, heart rate (HR),
- **mental status**, Glasgow Coma score.

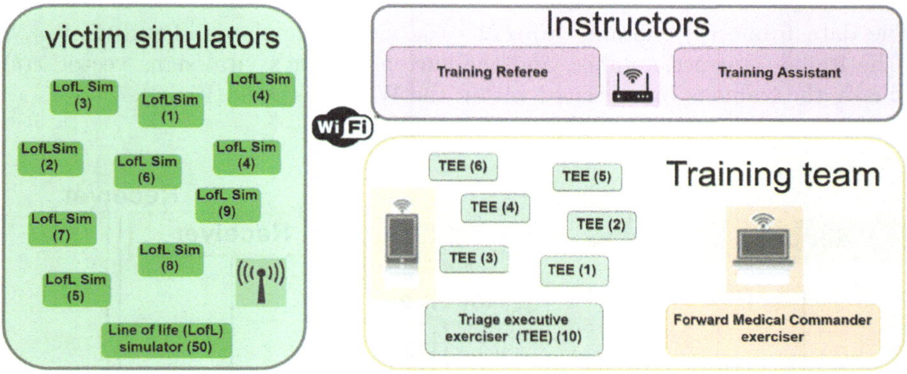

Fig. 2. The Emergency Triage training system - general overview.

2 Emergency Triage Training System

The mass casualty triage requires a cross-cooperation of all first responders – fire rescues, police, emergency medical personnel or paramedics staff, and highlight the need for an effective communication between them. The objective of the proposed training system is to conduct an innovative, simulation training environment for triage exercise. This become possible due to ever-advancing technology enabled miniaturization of computing devices, so they can be used to support mass casualty triage procedures. The growth of information and communication technologies and the prevalence of mobile devices [3,5] make that using IoT for this purpose is a highly topical and relevant issue.

2.1 Victim Simulator

The proposed training system provides dozens victim's line of life simulators, each reproducing different scenario. To meet training requirements, victims simulators provide cyclic generation of vital parameters vector $\overline{x}(t)$, which corresponds to different event scenarios occurring in real cases. A time ordered series of vital parameters vectors creates the Line of Life:

$$LofL(\overline{x},t) = \{\overline{x}(t) \mid t \in [1,2,3,...,T]; \\ \overline{x}(t) = [MA(t), RE(t), RR(t), HR(t), MS(t)]\} \quad (1)$$

were $\overline{x}(t+T) = \overline{x}(t)$ and MA-moving ability, RE-respiratory effort, RR-respiration rate, HR-heart rate, and MS-mental status, are victim's vital signs for START method (Fig. 1).

The $\overline{x}(t)$ vector is generated at every 5 s intervals, and a sequence of $\overline{x}(t)$ vectors forms a Line of Life. Typically LofL is 20 min long. From $\overline{x}(t)$ vector, the data frame is created:

$$SimF(t) = \langle SimID, SeqNo, \overline{x}(t) \rangle \tag{2}$$

The data frame (2) contains $SimID$ - the victim simulator identifier, $SeqNo$ - the frame sequence number, and the current victim's vital signs vector $\overline{x}(t)$. Finally the frame is broadcasted within the Wi-Fi network (Fig. 3).

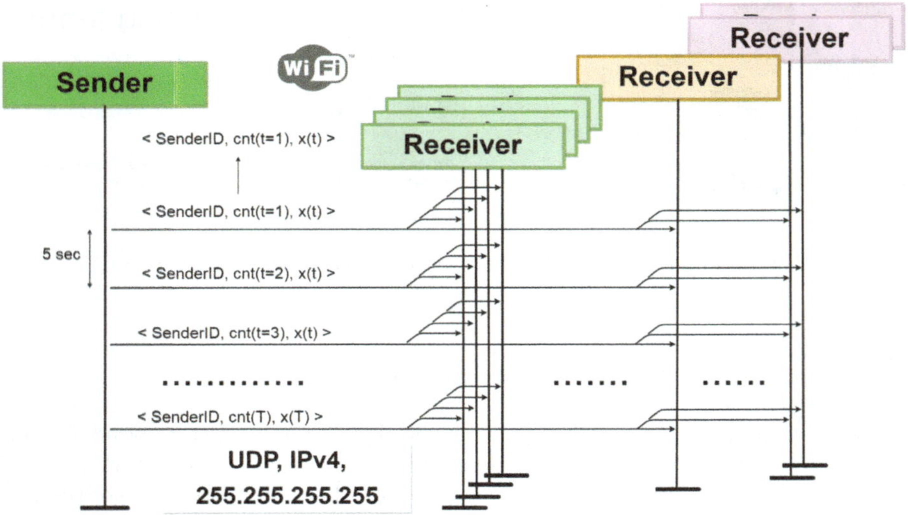

Fig. 3. IP broadcasting of Vital Parameters using UDP datagrams.

For START procedures there are 7 triage scenarios, resulting in four possible categories of triage (7 is the number of path resulting in 4 color categories as shown on Fig. 1). The duration of one scenario varies from 2 to 3 min. These scenario sequences are stored in the victim simulator memory, but the order and moment at which they occur can be modified by training assistant. Therefore, the victim simulator requires communication channel with a training assistant to receive information parameterizing the order and duration of scenarios. In this way, the training assistant can affect the course of the exercise.

2.2 Triage Training Team and Training Instructors

The training team consists of over a dozen person, of which one named Forward Medical Commander performs management functions. The team of Triage

Executive Exercisers (TEEs) triage victims and provide very limited treatment (manually open airways, clear airway with finger sweep, control major bleeding) as necessary, until ambulances arrived at the scene. Triage Executive Exerciser is equipped with a smartphone with Android operating system as a device receiving data frames broadcasted by victim simulators. Each TEE takes 30 s on average, to triage a victim. During this time, he decides which victim he wants to triage and his smartphone selectively receive the frames from the chosen victim simulator. The TEE smartphone displays the current vector of vital parameters and allows exerciser to assign the category. Afterwards the device generates a frame (3) that is send to the Forward Medical Commander (FMC).

$$TriageF = \langle TriaID, TriaPos, TriaTime, TriaCat, SimF \rangle \qquad (3)$$

where $TriaID$ is TEE ID, $TriaPos, TriaTime$ are coordinates of the place and time of the triage, $TriaCat$ is category which was given to the victim and $SimF$ is a vital parameters vector (2), on the basis of which such decision was made.

After FMC confirms reception of the $TraigeF$ frame, the TEE device stops receiving the frames from the currently triaged victim, and continues the exercise looking for the next injured person. A confirmation sent by FMC is received by all the TEE nodes and tells all TEEs not to receive frames from this victim simulator. Next, FMC takes over itself, the monitoring [8] of the frames from this victim simulator.

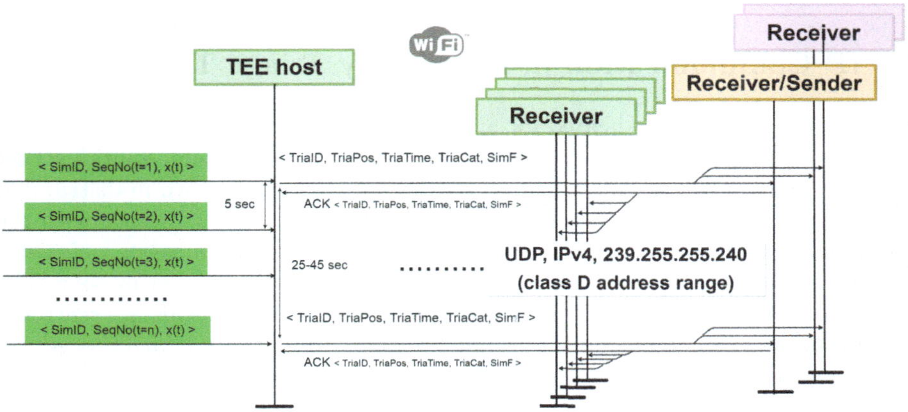

Fig. 4. Sending TEE triage decisions to IP multicasting group.

Forward Medical Commander is a member of training team. He performs management functions and is responsible for the communication and coordination of the emergency triage teamwork activity, and the efficiency with which the team performs challenging interventions in a high-stress situation. In the above scope, his work is evaluated by training instructors conducting the exercise.

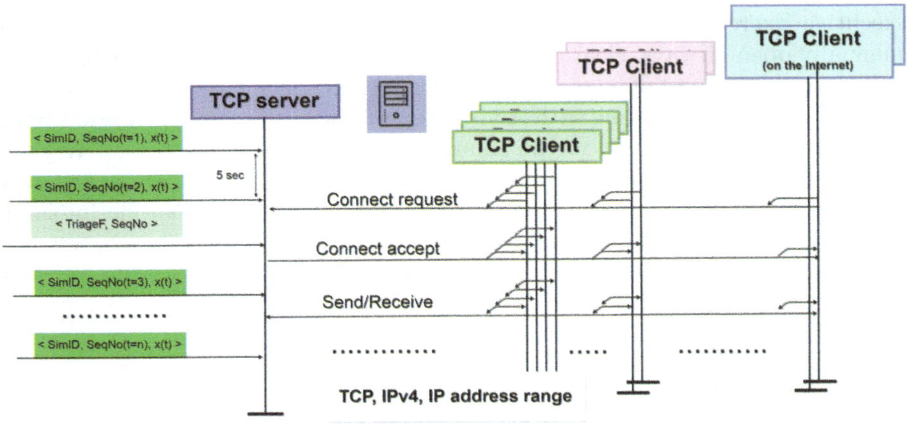

Fig. 5. Serving a triage training data for TCP clients.

Instructors team consists of training assistants and referees who are responsible for the course and student's assessment. Training assistants duties are to arrange the accident scenario, so that they improve triage performance, communication and rescuers teamwork skills the best. Training referees evaluate training team and provide a comprehensive debriefing immediately following the training.

3 IP Communication in Emergency Triage Training

Internet technologies, dedicated for a distributed system, give a number of possible application and can be widely used. They allow to build solutions that acquire data from an area, categorize, classify and aggregate both in distributed or centralized manner [1]. Then, the data is transmitted in accordance with the hierarchy appropriate for the given system, using the communication network.

In the proposed solution, the communication base on Wi-Fi transmission. The choice of wireless transmission is the best solution because victims are scattered within the accident scene and the training team mates perform triage procedures searching this area. Having a choice of several wireless technologies (LoRaWAN, NarrowBand-IoT (NB-IoT), Bluetooth Low Energy and Wi-Fi), we chose Wi-Fi as it simplifies the Internet connection with the use of both mobile LTE communication networks or the possible in the future 5G network.

802.11 standards form the basis of Wi-Fi certificates. This standard covers both Physical (PHY) and Data Link (MAC) Layers in ISO/OIS model. The PHY, MAC protocols are responsible for putting bits on wireless channel and reliable point to point data transfer. Therefore, according to the ISO/OSI reference model, the proposed software focuses on higher 3-Network and 4-Transport layers. From layer 3 the IPv4 protocol, and from layer 4 the UDP and TCP protocols are used [12].

In the Triage Training System (Fig. 2) three functionally different communication mechanisms are used:

- **UDP broadcasting** network (Fig. 3), to propagate vital parameters of wounded victims. All training team members and Emergency Medical Services (EMS) instructors receive broadcasts datagrams [10]. Broadcast datagrams are not sent through routers.
- **UDP multicasting** network (Fig. 4), to communicate and propagate results of triage procedure within training team and EMS instructors. The device must be configured to receive multicast datagrams. Multicast datagrams are sent through routers. The device can send and receive on multiple multicast addresses, moreover device doesn't need to be a member of a group to send multicast datagrams [10] to that group.
- **TCP connection-oriented** network (Fig. 5). All data generated during the exercises are stored on the server and can be streamed to a group of medical students at the University. The entire exercise data can be used for the debriefing, live or after completing of the exercises.

4 Technologies Supporting Triage Training System

The structure of the proposed training system uses a number of modern IoT technologies with communication based on Wi-Fi network [9]. Each of the four system areas (Fig. 2) uses a different type of hardware and software.

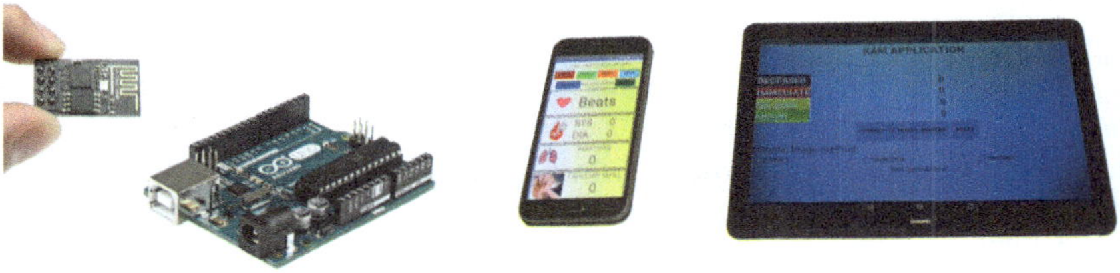

Fig. 6. ESP-01 and Arduino Uno modules (left) and TEE smartphone and FMC tablet (right).

4.1 Victim Simulator

Victim simulators are hardware devces composed of programmable ESP-01 modules. Arduino Uno is used as a programming station (Fig. 6, left). The ESP-01 is a small module (25 mm x 15 mm), produced by Espressif Systems. The module contains ESP8266 [11] chip - low cost Wi-Fi microprocessor with full TCP/IP stack. By default, it can support multiple TCP connections (5 at the same time). The Arduino IDE programming environment is used for software development in C language for direct programming of ESP modules. The module is programming using Hayes-style AT+ commands [11], which allow to connect to Wi-Fi network and setup TCP/IP transmission. Arduino IDE with the sequence of AT+ commands, necessary for setting up ESP-01 module for UDP multicast transmission on Wi-Fi channel, is presented in Fig. 7.

The broadcasting of current vital parameters, using UDP datagrams in IP broadcast mode, is shown in Fig. 3. All broadcasting datagrams are heard by every network user, but exercising TEE (green receiver on Fig. 3) selectively receive only one simulation device at a time. The one being triaged. The FMC (orange receiver on Fig. 3) cumulatively receives transmissions from already categorized victims.

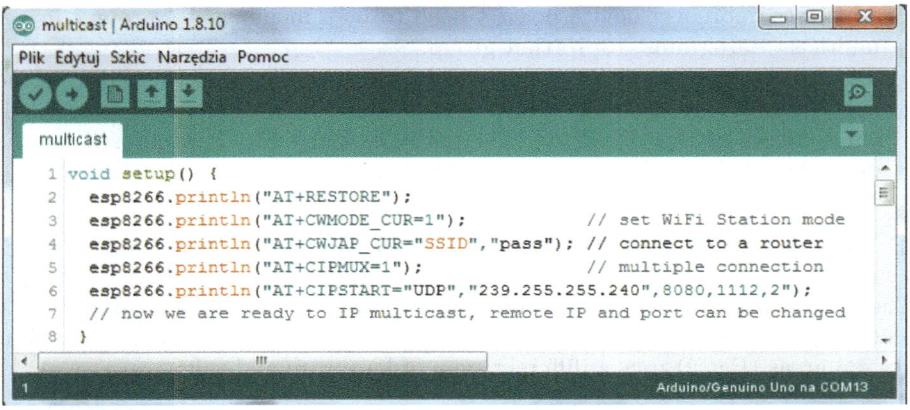

Fig. 7. Sequence of AT+ commands to setup ESP-01 module for UDP multicast transmission on Wi-Fi channel.

4.2 EMS Triage Team and Training Personnel

EMS students training a triage procedures use smartphones (Fig. 6, center), with the Android operating system, on which a triage dedicated application is installed. This equipment is a compromise between the graphical interface sufficient to represent the condition of the wounded victims, and the device that is handy to use in an emergency rescue operation.

The Forward Medical Commander, who manages the action, uses a tablet (Fig. 6, right), which gives better opportunities to present both the progress of the action and coordinate the actions of individual triage team mates. In both cases (smartphone, tablet), the software was developed using integrated programming environment Android Studio [9], in the Windows environment. Java, C++ and XML languages, were used.

The software for FMC tablet and TEE smartphones are functionally different, because of different (as presented in the Sect. 2.2) requirements. However, the communication functions using TCP/IP are almost identical, and the fact that both devices work under the same Android operating system has significantly simplified the software development.

Training Instructors and Assistants use notebooks with Windows operating system, and dedicated programs written in C++ language using Visual Studio programming environment. In the software managing communication sessions, application synchronization, supervising the connection and ensuring the correct direction of data flow, the WinSock2 library was used (Fig. 8).

```
int main()   // receiver of multicast UDP datagrams
{   WSAStartup(MAKEWORD(2, 2), &wsaData);
    mySocket = socket(AF_INET, SOCK_DGRAM, 0);
    setsockopt(mySocket, SOL_SOCKET, SO_BROADCAST, &broadcastPermission, sizeof(broadcastPermission));
    setsockopt(mySocket,SOL_SOCKET, SO_RCVTIMEO,(const char *)&iTimeout,sizeof(iTimeout));
    myAddress.sin_family = AF_INET;  //IPv4
    myAddress.sin_addr.s_addr = inet_addr("239.255.255.240");
    myAddress.sin_port = htons(port);
    bind(mySocket, (SOCKADDR*)&myAddress, sizeof(myAddress));
    while (!and_of_work)
    {   n = recvfrom(mySocket, recMessage, STRLEN, 0, (SOCKADDR*)&myAddress, &server_length);
        if (n <> 0xFFFFFFFF)
        {   WriteToFile(recMessage);
            sendto(mySocket, sendACK, strlen(sendACK), 0, (SOCKADDR*)&myAddress, server_length);
        }
    }
    return 0;
}
```

Fig. 8. Multicast UDP datagrams receiver based on WinSock2 C++ code.

5 Conclusion

We demonstrated the feasibility of using Wi-Fi communication and IoT technologies to improve emergency triage training. The proposed solution of victim vital parameters simulator, uses Wi-Fi channel, generates and broadcast Line of Life that is used in the training exercise. If new functionalities need to be added, using Arduino IDE with ESP-01 module guarantees simple reprogramming process. Moreover, this module is a stand alone microprocessor board with 1MB Flash memory. After programming using Arduino IDE, it can work independently.

The use of smartphones and tablets has several advantages. They offer great development opportunities in the field of visualization and data processing. First responders and emergency medical personnel are familiar with such devices. After all they use them both at work and in every facet of everyday lives. Therefore, during the exercises, they do not waste time for learning how to use the device. Moreover, exactly the same equipment will be used in real rescue operations, as soon as vital parameters sensors for wounded victims [6] will be available on the market.

References

1. Baran, P.: On distributed communications networks. RAND Corporation Papers (1962). https://doi.org/10.7249/P2626

2. Bazyar, J., Farrokhi, M., Khankeh, H.: Triage systems in mass casualty incidents and disasters: a review study with a worldwide approach. Open Access Maced J. Med. Sci. **7**(3), 482–494 (2019). https://doi.org/10.3889/oamjms.2019.119
3. Cook, D.J., Das, S.K.: Smart Environments. Technologies, Protocols and Applications. Wiley, Hoboken (2005). ISBN 0-471-54448-5
4. Lerner, E., Schwartz, R., Coule, P., Weinstein, E., Cone, D., Hunt, R., Sasser, S., Liu, J., Nudell, N., Wedmore, I., Hammond, J., Bulger, E., Salomone, J., Sanddal, T., Lord, G., Markenson, D., O'Connor, R.: Mass casualty triage: an evaluation of the data and development of a proposed national guideline. Disaster Med. Public Health Preparedness **2**(S1), S25–S34 (2008). https://doi.org/10.1097/DMP.0b013e318182194e
5. Nikodem, J., Nikodem, M., Gawłowsk, I.P., Klempous, R.: Training system for first response medical emergency groups to guide triage procedures. In: Bruzzone, A.G., et al. (eds) The 8th International Workshop on Innovative Simulation for Health Care, IWISH, pp. 27-33. Rende : DIME Universitá di Genova ; DIMEG University of Calabria (2019). ISBN 978-88-85741-35-5
6. Niswar, M., Wijaya, A.S., Ridwan, M., Adnan Ilham, A.A., Sadjad, R.S., Vogel, A.: The design of wearable medical device for triaging disaster casualties in developing countries. In: 2015 Fifth International Conference on Digital Information Processing and Communications (ICDIPC), pp. 207–212 (2015). https://doi.org/10.1109/ICDIPC.2015.7323030
7. Sakanushi, K., Hieda, T., Shiraishi, T., Ode, Y., Takeuchi, Y., Imai, M., Higashino, T., Tanaka, H.: Electronic triage system for continuously monitoring casualties at disaster scenes. J. Ambient Intell. Human. Comput. **4**, 547–558 (2011). https://doi.org/10.1007/s12652-012-0130-2
8. Stewart, C., Stewart, M.: Patient-tracking systems in disasters. In: Ciottone, G., et al. (eds.) Ciottone's Disaster Medicine, pp. 344-350. Elsevier (2016). ISBN 9780323286657, https://doi.org/10.1016/B978-0-323-28665-7.00055-8
9. Android Studio Developer Guides: Wi-Fi. https://developer.android.com/guide/topics/connectivity/wifi-scan
10. CISCO technology white papers. IP Multicast Technology Overview (2001). https://www.cisco.com/c/en/us/td/docs/ios/solutions_docs/ip_multicast/White_papers/mcst_ovr.html#wp1008683
11. ESP8266 AT Instruction Set. Version 3.0.2. Espressif Systems (2019). https://www.espressif.com/sites/default/files/documentation/4a-esp8266_at_instruction_set_en.pdf
12. How IPv4 Multicasting Works. Microsoft Docs (2009). https://docs.microsoft.com/en-us/previous-versions/windows/it-pro/windows-server-2003/cc759719(v=ws.10)?redirectedfrom=MSDN
13. World Health Organization: Mass casualty management systems : strategies and guidelines for building health sector capacity. World Health Organization, Geneva (2007). ISBN 9789241596053

Robust Radio Communication Protocol for Traffic Analysis Application

Maciej Nikodem[1(✉)], Tomasz Surmacz[1], Mariusz Slabicki[2], Dominik Hofman[1], Piotr Klimkowski[1], and Cezary Dołęga[3]

[1] Department of Computer Engineering, Wrocław University of Science and Technology, Wybrzeże Wyspiańskiego 27, 50-370 Wrocław, Poland
{maciej.nikodem,tomasz.surmacz}@pwr.edu.pl

[2] Institute of Theoretical and Applied Informatics, Polish Academy of Sciences, ul. Bałtycka 5, 44-100 Gliwice, Poland

[3] Neurosoft Sp. z o.o., ul. Życzliwa 8, 53-030 Wrocław, Poland

Abstract. This article presents a communication protocol designed for low-power radio networks in vehicle traffic analysis application. The protocol is designed for an efficient and robust communication of several vision sensors that detect and track vehicle routes across the intersection. Article presents the protocol designed to ensure reliable transmission over unreliable communication links that satisfies requirements of the traffic analysis application. The paper also presents results of a real live evaluation of the proposed system in a pilot deployment.

Keywords: Traffic analysis · Low-power radio · Robust communication · Dependability

1 Introduction

Detecting, categorizing and counting vehicles is an important part of building efficient transportation systems. Vision based systems can accurately implement these tasks, can efficiently operate in various environmental conditions, and are often used in practice On the other hand, vision based vehicle tracking is less common and limited to single camera systems, usually used in tunnels or highway sections, in order to detect hazardous situations [3].

Vision based traffic analysis, that focuses on individual vehicle traveling across a larger area (e.g. the whole tunnel or intersection), requires cooperation of a number of cameras and is more challenging. This is due to the technical difficulties (e.g. close to real-time image processing, communication between the cameras) and legal restrictions (tracking a vehicle may invade privacy and violate the General Data Protection Regulation - GDPR rules). Recently a new approach to traffic analysis was proposed by Lira et al. [9] who used aerial videos of the intersection. In this approach a drone-mounted camera was used for identifying, tracking and analysing vehicle routes. The authors report issues with

W. Zamojski et al. (Eds.): DepCoS-RELCOMEX 2020, AISC 1173, pp. 461–471, 2020.
https://doi.org/10.1007/978-3-030-48256-5_45

correct detection and tracking of individual vehicles from high altitude and in dense traffic. Fedorov et al. [2] took different approach and used a single camera in a specific setup. In their approach the camera was mounted so that it could observe the whole intersection. Unfortunately, theses approaches are not universally applicable: the first one can operate for a limited period of time (until drone's battery runs out) and depends heavily on weather conditions, the second was tailored to a specific intersection.

Vision based traffic analysis systems are becoming more and more attractive. This is due to low costs of installation (compared to inductive loops in roads and other types of sensors) and reliable operation in various environmental conditions [5]. However, this type of systems require the use of several cameras that can individually identify and jointly track vehicles. This requires cameras to process the video stream in real-time and exchange information about identified vehicles. To comply with GDPR rules the system should not use vehicle specific information but be able to track vehicles based on unique identifiers generated from the vehicle's image (e.g. shape, colour), location and time. Tracking requires the identifiers to be transmitted between cameras as vehicles move across the intersection area. Consequently, traffic analysis requires efficient, real-time and robust wireless communication system.

The goal of this article is to present a proprietary radio communication protocol designed for traffic analysis application. The focus is on the protocol's mechanisms to ensure reliability and correctness of communication in the presence of interference.

2 Related Work

In wireless communication various factors affect packet loss, including interference, background noise, and inadequate choice of parameters for a particular environment. Packet loss affects reliability of data transmission and is crucial for every communication system. Therefore, every communication system implements mechanisms to compensate for packet loss and to ensure efficient communication, even in adverse conditions.

Cattani et al. [1] analysed LoRaWAN communication, and showed that different physical (PHY) layer parameters heavily affect the communication robustness. They argue that using less reliable, faster transmission with retransmission usually gives better results compared to more reliable but slower transmission. Their experiments also show a strong correlation of signal strength (affecting the packet error rate) with ambient temperature.

A retransmission mechanism is often implemented in the communication protocol, as packet loss is inherent characteristics of wireless media. In [6] various policies for packet retransmissions have been evaluated. Although the research was focused on devices with multiple radio interfaces, the underlying universal techniques were also described. These included proper packets scheduling by managing parameters such as round trip time or congestion window.

Packet loss can be alleviated by minimising the risk of collisions using various media access methods, such as carrier sense multiple access (CSMA) or time division multiple access (TDMA). The challenge of clock synchronization, required for the TDMA method, was addressed in [8]. Using hardware support for timers and radio-triggered interrupts, the authors were able to achieve microsecond accuracy of time synchronization in lab environments. However, they suggest to use GPS if devices communicate in 868 MHz band over large distances (i.e. around 1 km).

3 Traffic Analysis System

The system consists of many sensors, which monitor different sections of an intersection. As the exact route vehicles take through the intersection cannot be predicted therefore, information about every detected vehicle is transmitted to all the neighbouring sensors. When a vehicle enters the sensor's field of view (FoV) it is identified and starts being tracked. When it leaves the FoV, a message with vehicle information is broadcasted. Subsequent sensors receive the broadcast message, detect the vehicle, track it and broadcast new message when the vehicle leaves their FoV. This operation is repeated until the vehicle leaves the intersection. The last sensor to track the vehicle gets the information about all the previous sensors that have seen the vehicle and consequently knows the route the vehicle followed through the observed area.

3.1 System Requirements

To be able to determine application requirements, we have recorded traffic in a number of intersections. The recordings were then analysed and used to define

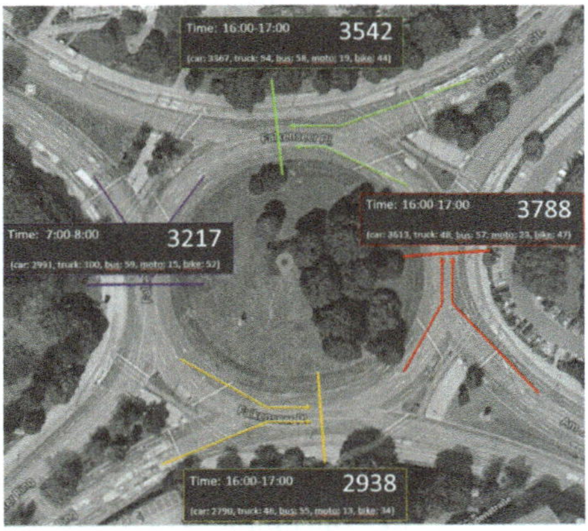

Fig. 1. Sample intersection with 4 measurement areas and the maximum number of vehicles recorded

requirements for the wireless technology and the protocol. Figure 1 presents a sample intersection with peak number of vehicles per hour, during the 24 h period of the recording. Table 1 provides the most important requirements for the communication system.

Table 1. Communication system requirements

Cameras	Number of lanes per camera	Vehicles per lane	Latency	Bandwidth	Range
Up to 16	Up to 3	1 per 2 s	25 ms	50 Kbps	>250 m

During the initial feasibility study various radio communication technologies were considered, including Bluetooth Low Energy in long-range mode, IEEE 802.15.4 radios operating in 2.4 GHz and radios operating in 868 MHz. The experiments aimed at verifying the communication range and packet reception rate in real life. They were conducted at a busy intersection in Wrocław and a parking lot near Komorniki (DK94). Based on their results we decided to use 868 MHz radio and design a proprietary communication protocol.

3.2 The Purpose of Traffic Analysis

The main purpose of the traffic analysis for our system is to assist in traffic surveys and provide additional information on vehicle moving across the area of interest. In a traffic survey an intersection or a particular section of a road is observed continuously for a specified time, usually 24 or 48 h. Nowadays, most of the surveys are conducted manually and are only limited to vehicle counting and classification – people observe the road and note amounts of vehicles and their categories in 3–4 h shifts. The accuracy of such surveys is low, as people get tired by the repetitive task very soon and start averaging instead of counting the vehicles carefully. Currently traffic flows (i.e. routes the vehicles traveled) are very rarely registered. In 2020 nationwide traffic survey is organized by the Polish General Directorate for National Roads and Highways (GDDKiA). According to GDKiA about 50% of 2285 measurement sites will involve video recording and analysis, however only 5% of that will be done automatically [4]. In the remaining sites vehicles will be counted manually and video recording will be only used for the purpose of verification.

Automated surveys need appropriate sensors that can detect vehicles, and ideally classify them at the same time. Various types of measurement devices are available [5,7], such as piezo-electric axle sensors, inductive loops, infrared or laser counters, microwave radar detectors, ultrasonic detectors, and video detection systems. For long-term continuous traffic monitoring, piezo-electric and inductive sensors are the well-established standard, but their deployment is expensive and troublesome as they need to be integrated into the road surface.

Consequently, they are not suitable for short-term surveys. Video image processing is the most flexible solution in this context. Cameras can be mounted on any existing road infrastructure, such as traffic light or street lighting poles. For short-term measurements such devices can even be battery powered. This greatly simplifies installation, as getting an external power supply can be problematic.

4 Reliable Communication for the Traffic Analysis System

The proposed communication system uses radio transceivers operating in 868 MHz radio spectrum (CC 1352P from Texas Instruments). There can be up to 16 sensors deployed to monitor and analyse the traffic. Each sensor is equipped with a single board computer based on Raspberry Pi that contains a GPS receiver, LTE module, a camera to identify and track the vehicles, and is connected to a 868 MHz radio using a serial interface. The LTE module is optional and used mostly in the development phase. The device-to-device communication uses the 868 MHz radio and the proprietary communication protocol designed.

4.1 Communication Protocol

The system uses both broadcast and addressed (directed) communication. Broadcasts are used for transmission of vehicle-related information. Addressed communication is used for setup and management of the system, and for the transmission of some of commands and responses. The sensors are organized in sequences, i.e. each of them has a FoV with predefined entry and exit areas. Each sensor's exit area may be linked to the entry area of another one. As a whole system, the sensors can form a sequence (e.g. a highway or a tunnel), a loop (e.g. a roundabout) or a mesh (typical 4-entry crossing or a parking lot) to allow different real-life scenarios.

The designed protocol uses the following mechanisms to ensure reliable and correct transmission of the information:

Error Detection. The radio modules used, have a built-in Cyclic Redundancy Check (CRC) mechanism and validate CRC on packet reception. As only packets with valid CRC are received, there was no point in using additional error detection or correction mechanism for radio transmission. During the evaluation, it turned out that serial communication between the Raspberry Pi and the radio module is also prone to errors. To eliminate errors CRC for serial communication was implemented. All messages that have incorrect CRC value are dropped.

Time to Live. Vehicle-related information transmitted between the senors has time-limited validity as image processing cannot be delayed by more than tens of seconds. Therefore, if information has expired the message does not need to be transmitted. This may happen due to extremely large traffic or interferences in the communication channel. Each vehicle-related information is assigned a time to live (TTL) that determines how long data is valid. When TTL is exceeded the data is dropped. This adversely affects the data reception rate but improves performance of the communication and shortens recovery time from overloads.

Sequence Numbers. To uniquely identify each vehicle-related information, the transmitting sensors add a sequence number to each transmission. Receiving sensors keep track of the sequence numbers received (individually for each transmitter), can detect when some vehicle-related information is missing and take appropriate actions.

Retransmission Requests. When a receiving sensor detects a missing sequence number it may requests a retransmission. Retransmission request is a command sent directly to the source sensor. To reduce the number of responses the source sensor may delay the retransmission and send the requested information by broadcast. The sensor responds to the maximum number of most recent requests (based on the sequence numbers), and limits the total number of retransmission being simultaneously processed. These methods allow several requesting sensors to be served by a single retransmission and reduces the communication overhead. Moreover, this preserves communication bandwidth for new vehicle-related information that could be otherwise consumed by the retransmission of the expired information.

Fair Spectrum Access. To ensure coexistence between sensors and other radios operating in the same frequency band, the CSMA and TDMA access methods are implemented.

The CSMA is used to ensure that sensors do not start transmission if anyone else is already transmitting. Every sensor samples the radio channel before the transmission and measures the strength of the radio signal in the channel (*received signal strength indicator* – RSSI). If the RSSI is above a predefined threshold the channel is considered busy and the sensor backoffs for a random time, and attempts again after the time elapses. This procedure is repeated until the channel is available or the predefined number of retries is exceeded. The use of CSMA ensures coexistence with all other radios using the same radio spectrum. The disadvantage of the procedure is the delay in the transmission.

The TDMA is used to ensure that sensors do not interfere with each other, use their own time slots for transmission, minimize collisions and retransmissions. Traditional TDMA requires synchronization between nodes and a quite complex procedures for network management resulting in communication overhead that reduces bandwidth available for data transmission. TDMA also increases latency of the transmission as senors need to wait until their slot. In our system the synchronization is based on the radio beacon transmitted every predefined period of time but in future a nanosecond-accurate pulse generated every second by the GPS module in each sensor can be used. The system also uses a proprietary protocol for slot allocation to sensors. It is a distributed two-phase commit procedure, where each sensor selects presumably unallocated slot and requests its acquisition, which can be denied or accepted by other sensors.

The slot selection procedure also uses broadcast communication and was designed so that the number of radio messages transmitted is reduced.

5 Experimental Validation

The working system with 6 devices was deployed at the university campus along one of the internal roads (Fig. 2). Devices have been installed at different heights, on top of university buildings. The maximum distance between devices exceeds 250 m and there are no major obstacles between them. For the purpose of system validation every sensor received all vehicle-related information transmitted by other sensors. The sensors simulated real traffic and generated the information repeatedly with the predefined frequency of 1–6 Hz. The size of a single information was randomly selected from between 10 to 200 bytes, to simulate different number of vehicles detected by the sensor.

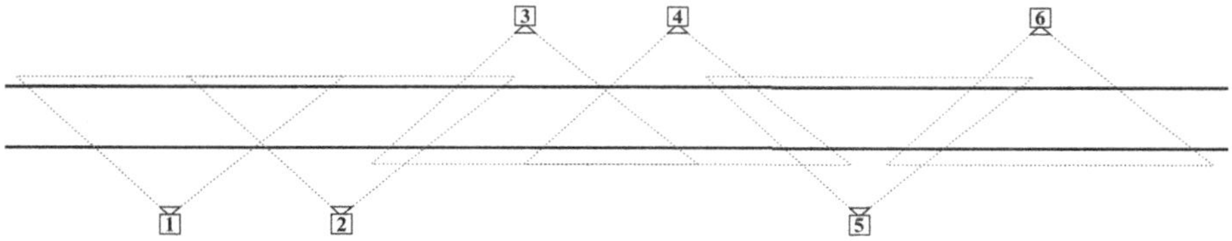

Fig. 2. Outline of the test deployment at the university campus

5.1 Test Scenarios

The main goal of the tests was to verify the operation of the system and robustness of its operation. During the tests a number of parameters were registered and analysed afterwards. The most important ones include:

- data reception rate ($DRR_{\mathtt{dst,src}}$)—is a ratio of the number of vehicle information generated by `src` sensor and successfully received at `dst` sensor vs. the number of all vehicle information generated by `src`,
- packet reception rate ($PRR_{\mathtt{dst,src}}$)—is a ratio of the number of radio packets transmitted by `src` and received at `dst` versus the total number of packets sent by the `src` sensor.
- channel access efficiency—is a distribution of CSMA backoffs and a ratio of CSMA procedure failures versus all the CSMA procedures run,
- duty cycling—the amount of time spent transmitting (time on air) in last 60 min (expressed in percentages).

Table 2 presents differences between test scenarios. The tests were selected to show how operation of the system changes when different reliability mechanisms are used. In tests 1–3 each sensor generates 3 events per second. This corresponds to the heavy traffic situation and thus is expected to be the worst case scenario

Table 2. Settings of the scenarios tested

Scenario	CSMA	TDMA	Retransmissions	No. of vehicles per second
1	ON	OFF	OFF	3
2	ON	OFF	ON	3
3	ON	ON	ON	3
4	ON	ON	ON	6

in real-life. In test 4 the system is stressed with high volume of data to be transmitted which corresponds to extremely heavy traffic. Six vehicle-related information per second almost saturate the available bandwidth of the radio. The test was run to verify that the system is stable in such extreme situation and correctly recovers when the traffic lowers.

5.2 Results

Figure 3 presents a ratio of the number of backoffs in the CSMA procedure for sensor 3, in scenarios 2 and 3. In both scenarios no significant external interference were detected so the number of backoffs resulted from the operation of the sensors. In scenario 2 the number of CSMA backoffs is large because interference are caused by simultaneous transmission of several sensors. The use of TDMA together with CSMA (scenario 3) allows to minimise the number of backoffs because sensors used individual time slots and no simultaneous transmissions were possible. Consequently, the number of backoffs was significantly reduced and over 99% of transmissions did not require any backoff.

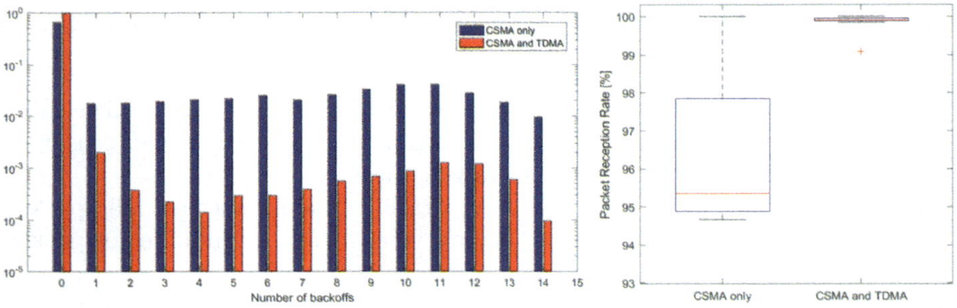

Fig. 3. Distribution of the number of backoffs in CSMA procedure (left), and statistics of PRR for channel access with and without TDMA (right)

The packet reception rate for all the sensors (Fig. 3), when using TDMA and CSMA (scenario 3), is close to 100%. When only CSMA is used (scenario 2) the PRR varies between 95% and 98%. Despite the PRR values, the retransmission mechanism allowed us to keep the DRR above 99% in scenario 2, and at 100% (for all the sensors) in scenario 3.

All sensors had the same CSMA threshold (−95 dBm) but the noise in the radio channel at sensor locations was different. For example, for sensor 6 the noise was only 1–5 dBm lower than the CSMA threshold. Consequently the number of CSMA backoffs for sensor 6 is larger when compared to other sensors (Fig. 4). The use of TDMA in this case does not affect the CSMA procedure as it was the case for sensor 3 (Fig. 3).

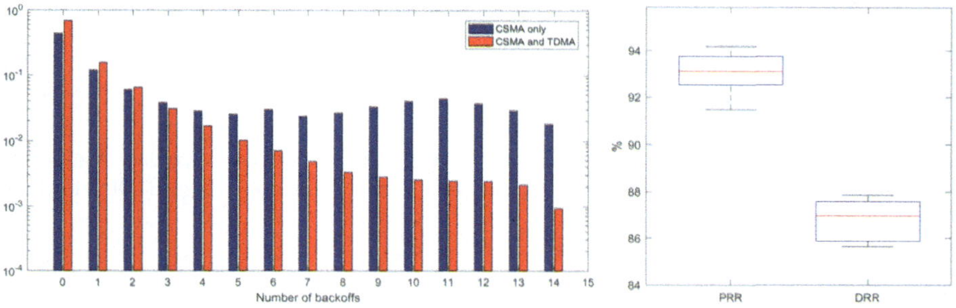

Fig. 4. Distribution of the number of backoffs in CSMA procedure with and without TDMA when CSMA threshold is close to the channel noise (left) and PRR and DRR for the scenario 4 (right)

For extremely heavy traffic (scenario 4) both the PRR and DRR dropped (Fig. 4). The drop in DRR is a result of time-limited validity of vehicle-related information. Limited validity is reflected in: i) dropping information that has expired (TTL), even if it was not sent, and ii) the implementation of the retransmission procedure that restricts the number of simultaneous retransmissions a sensor can proceed. In scenario 4 the limits were exceeded, so some data was dropped (due to TTL) or not retransmitted (despite beeing requested). This policy preserved communication bandwidth for new vehicle-related information but lowered the DRR. This is a trade-off specific to the traffic analysis application.

6 Conclusions

The system was run for over 6 months and tested with various settings, parameters and traffic of different volume. The proposed communication protocol achieves high efficiency and robust transmission of information in unreliable communication channel and meets the time constrains of the application. Different reliability mechanisms can be individually enabled and configured, so that the protocol operation can be adjusted to a particular installation. For example, in small traffic scenarios (the number of vehicles ≤ 3 per second) the TDMA can be disabled without affecting the DRR and improving the latency of data transmission. The communication protocol can be used in other applications where limited number of time-relevant information needs to be transmitted between devices deployed in one area (e.g. parking lot monitoring, tracking of AGV in factories, etc.)

The proposed solution can be further developed. In particular, sensors should not only monitor the environment but also adapt to it. For example, the CSMA threshold should be adjusted to the channel noise, so that the number of backoffs in the CSMA procedure is minimised. Also the threshold of the maximum number of retransmissions and the TTL parameter can be adjusted to the traffic volume and the quality of the communication channel. This should ensure a tradeoff between DRR and time constrains imposed by the application. The TDMA beacon can be generated by each sensor using the GPS receiver. This will eliminate the need for radio transmitted beacons and improve the reliability of the sensors synchronization.

Acknowledgements. This research was jointly funded by the Wroclaw University of Science and Technology and The National Centre for Research and Development (01.01.01-00-1143/17).

References

1. Cattani, M., Boano, C.A., Römer, K.: An experimental evaluation of the reliability of LoRa long-range low-power wireless communication. J. Sens. Actuator Netw. **6**(2), 7 (2017). https://doi.org/10.3390/jsan6020007
2. Fedorov, A., Nikolskaia, K., Ivanov, S., Vladimir, S., Minbaleev, A.: Traffic flow estimation with data from a video surveillance camera. J. Big Data **6** (2019). https://doi.org/10.1186/s40537-019-0234-z
3. Foresti, G.L., Snidaro, L.: Vehicle detection and tracking for traffic monitoring. In: Roli, F., Vitulano, S. (eds.) Image Analysis and Processing - ICIAP 2005, pp. 1198–1205. Springer, Heidelberg (2005)
4. GDDKiA: GPR 2020 – General Traffic Survey, Janaury 2020. https://www.gddkia.gov.pl/pl/3959/GPR-2020
5. Gordon, R.L., Tighe, W.: Traffic Control Systems Handbook, second edition. U.S. Department of Transportation, Federal Highway Administration (2005). https://ops.fhwa.dot.gov/publications/fhwahop06006/fhwa_hop_06_006.pdf. Dunn Engineering Associates and United States. Federal Highway Administration
6. Halepoto, I.A., Khan, U.A., Arain, A.A.: Retransmission policies for efficient communication in IoT applications. In: 2018 IEEE 6th International Conference on Future Internet of Things and Cloud (FiCloud), pp. 197–202, August 2018. https://doi.org/10.1109/FiCloud.2018.00036
7. Klein, L.A., Mills, M.K., Gibson, D.R.: Traffic Detector Handbook. U.S. Department of Transportation, Federal Highway Administration (2006). https://www.fhwa.dot.gov/publications/research/operations/its/06108/06108.pdf

8. Kovácsházy, T., Várallyay, S.: Time-driven sub-GHz wireless communication protocol for real-time cyber-physical systems. In: 2019 20th International Carpathian Control Conference (ICCC), pp. 1–5, May 2019. https://doi.org/10.1109/CarpathianCC.2019.8765929
9. Lira, G., Kokkinogenis, Z., Rossetti, R.J.F., Moura, D.C., Rúbio, T.: A computer-vision approach to traffic analysis over intersections. In: 2016 IEEE 19th International Conference on Intelligent Transportation Systems (ITSC), pp. 47–53, November 2016. https://doi.org/10.1109/ITSC.2016.7795530

Automatic Recognition of Gender and Genre in a Corpus of Microtexts

Adam Pawłowski[1] and Tomasz Walkowiak[2(✉)]

[1] Institute of Information and Library Science, University of Wrocław, Wrocław, Poland
adam.pawlowski@uwr.edu.pl

[2] Faculty of Electronics, Wrocław University of Science and Technology, Wrocław, Poland
tomasz.walkowiak@pwr.edu.pl

Abstract. In this paper, we focus on author's gender and writing genre recognition solely on books titles. We analyse data extracted from the bibliography resources of the National Library of Poland. Within a paper, we compare different methods of text (title) representation and classification. It includes word embedding models such as word2vec, ELMo and classification algorithms such as linear models, multilayer perceptron and bidirectional LSTM. It is shown, that the writing genre (for defined 28 classes) could be automatically recognized based only on the book title with accuracy equal to 0.74. The best results were achieved by fastText methods with word n-grams.

Keywords: Stylometry · Bibliography · Text mining · Feature extraction · Word embedding · fastText · ELMo

1 Introduction

The large-scale bibliographies are maintained by national libraries in many countries. They are used to search for publications, their archiving and to evaluate the scientific results of institutions, disciplines and individuals. They include such information as the author's name, the title, the date and place of publication, and the genre of the text. Their size counted in millions of records and digital form make them valuable objects for text mining researches. The titles have specific features. They are very short compared to other texts being the subject of text mining researches and contain very synthesize information. Titles belong to the class of microtexts [11] that incorporates such texts as SMS, tweets or internet comments.

The research problem investigated by authors is the recognition of the author's gender and the literary genre of the text solely from the titles. Titles are extracted from the National Library of Poland bibliography[1]. The author's

[1] https://www.bn.org.pl/en/catalogues-and-bibliographies.

W. Zamojski et al. (Eds.): DepCoS-RELCOMEX 2020, AISC 1173, pp. 472–481, 2020.
https://doi.org/10.1007/978-3-030-48256-5_46

gender recognition on microtexts available in social media is discussed widely in the literature (i.e. [8,11,12]). The problem of the automatic literary genre recognition is analyzed in the aspect of stylometry [1,2,17]. However, the analysis is done their on entire texts, not just titles. Therefore, the discussed here problem is more close to the subject classification of text documents [15,16] than to stylometry.

The paper is structured as follows, firstly, we describe data sets used in the analysis. In the Sect. 3, we overview used data mining methods (feature generation methods and classifiers). After that, the performed experiments and results are presented. Finally, the conclusions are given.

2 Data Set

2.1 Bibliographic Records

The analyzed corpus consist of bibliographic records of books from the National Library of Poland[2]. The records are stored in the MARC 21 format [13] and consists of ca. 1,850,000 records. It includes bibliographic descriptions of books published in the 20th and 21st centuries, but they could be written earlier. However, contemporary texts are predominant. The records contain, among others, the title, the language of the text (Polish is predominant), authors' names, and writing genre. In further analysis, only books written in Polish were analyzed.

2.2 Gender Recognition Corpus

As mentioned in the introduction, the first aim of the paper is to determine the gender of the author from the title. Therefore, we needed to extract titles and author gender. The task was carried in several stages. Since there is no information about the authors gender in the database, we have to develop a method to determine it based on first names. It was done semi-automatically. The list of all first names was extracted from the database and ordered by occurrence. Only names that occur more there 9 times were further analyzed. Since Polish is an inflected language, there is a common occurrence of the female first name suffix – a. Male names have different endings. Using these simple rules the database of names (ca. 2,700) was automatically tagged by gender. Next, the database was checked manually. Some names, especially foreign ones, are ambiguous, so they were marked us unknown. Finally, we got a simple method that allows identifying gender based on the author's first name, just check the name gender in the tagged list. The analysis was done as follows. First of all, list of authors' names for each book was extracted. Texts where authors were marked as editors were eliminated. In the case of names not on the list or with marked ambiguous gender texts (titles) were rejected. In case of multi-author books, we required that each author has the same gender. Finally, we obtained ca. 855,000 annotated by gender titles where 76.2% authors were male and 23.8% female.

[2] http://data.bn.org.pl/db/bibs-ksiazka.marc.

2.3 Genre Recognition Corpus

The second aim is to recognize the literary genre of the text based on its title. The analyzed bibliographic records were classified into a huge number of classes: i.e. 7,000. However, many of the classes consist of only a few examples. There are ca. 4,000 classes with less than 5 books. The other problem was the similarity between classes. It is mainly due to the polysemy of the language. For example "collections" and "anthologies" are treated in the analyzed records as separate categories, although in reality, they describe the same genre. Therefore, the number of classes had to be limited based on the criterion of the number of elements and similarities between classes. We have selected only records for genre classes that had no less than 5,000 exemplar titles. Finally, we got a corpus that includes ca. 571,000 titles. Resulting 52 writing genres were manually grouped into 28 classes, namely: *textbooks, novels, anthologies, popular publications, guides, biographies, albums, diaries, stories, children's literature, travel guides, textbooks for primary schools, polish journalism, support materials, comics, children's poetry, youth novel, religious considerations and meditations, publications for children, textbooks for high schools, encyclopedias, bibliography, commemorative books, statistical data, youth novel, literature, analysis and interpretation, and informant.*

3 Methods

3.1 Text Classification

Commonly used methods of text classification [14–16] rely on representing documents with feature vectors and use of statistical classifiers to assign documents to defined groups. Classifiers are learned [4] (their parameters are set-up) on trained data set (pairs of texts and class labels). The classical feature vectors are based on the bag-of-words technique [3]. Components of these vectors represent frequencies (weighted) of occurrences of words/terms in individual documents. The state-of-the-art technique is word2vec [7], where individual words are represented by high-dimensional feature vectors (word embedding) trained on large text corpora. In performed experiments, we used pre-trained vectors for Polish language [6][3]. Many statistical classifiers require constant length vector representation of documents. Since texts differ in document length, the feature vectors for document classification could be gained by averaging vector representations of individual words. This approach is known as doc2vec [5]. Within reported experiments, we have used the multilayer perceptron (MLP) [4] and the regularized linear classifier (Elastic Net [19]) with stochastic gradient descent (SGD) [4]. Moreover, we used the fastText algorithm [5], which performs word embedding and linear soft-max classifier learning simultaneously. Words without corresponding entry within built mapping are ignored. All mentioned methods ignores word order since in all cases an average of word embedding is used as a representation of a document (doc2vec). So, they are not-aware of word contexts.

[3] http://hdl.handle.net/11321/606.

3.2 Context-Aware Methods

To incorporate word contexts in the classification, we have replaced the word2vec representation by deep-learning ELMo [9] language model. In ELMo, the word embeddings are defined by the internal states of a deep bidirectional LTSM language network (biLSTM) [10], which is trained on a large text corpus. What is important, ELMo looks at the whole sentence before assigning an embedding to each word in it. Therefore, the embeddings are context-aware.

Moreover, we used biLSTM network to learn long-term dependencies between word2vec word embeddings. It operates forward and backward on the text, allowing detection cues and scopes. The final classification is done by a softmax. We also extended the multiplayer perceptron approach by a usage of a set of multilayer perceptrons (we call this approach hierarchical MLP). Five networks are dedicated to titles of lengths one to five words. In these cases, the title representation (input to the network) consists of concatenated word embedding. The sixth network input consists of the first five-word embeddings and doc2vec representation of the full text (an average of word embedding).

And finally, we use again the fastText algorithm, since, it allows to build the embeddings not only for single words but also for word n-grams making the solution context-aware.

3.3 Standardization

Many machine learning algorithms require features to have a normal distribution, therefore, standardization of data before classification is commonly used. In performed experiments, we tested three methods:

1. no standardization - use of raw data,
2. Z-score normalization - removes the mean and scales to unit variance:

$$\hat{x}_i = \frac{x_i - \overline{X}}{\sigma},$$

   where σ is the standard deviation of the feature and $\overline{X}$ is the mean value of the feature,
3. power transformation [18] - given by equation:

$$\hat{x}_i = \begin{cases} \frac{(x_i+1)^{\lambda}-1}{\lambda}, & \text{if } \lambda \neq 0, x \geq 0 \\ \log(x_i+1), & \text{if } \lambda = 0, x \geq 0 \\ -\frac{(-x_i+1)^(2-\lambda)-1}{2-\lambda}, & \text{if } \lambda \neq 2, x < 0 \\ -\log(-x_i+1), & \text{if } \lambda = 2, x < 0 \end{cases}$$

   where λ is the power parameter that is estimated through maximum likelihood.

3.4 Performance Analysis

For both analyzed tasks, we have randomly divided available data sets into three groups: training (80%), validation (20%) and testing (20%). The training was used to learn used classifiers. The validation data-set was used for regularization by early stopping and setting up hyper-parameters of used methods. We used three metrics calculated on the test data-set to report method performance: accuracy, average f1 score, and weighted f1 score. Accuracy is defined as all correctly classified objects over all objects. The accuracy is not well suited for no-equally sized classes. That is why, we have used f1 measure. However, it is defined (as a harmonic mean of precision and recall) for binary classification. For multi-class problems we can average f1 scores for each class or take into account the support (the number of examples in each class) and make a weighted average.

4 Experiments and Results

4.1 Gender Recognition - Non-context Analysis

In the first set of experiments, we have tested non-context methods in the task of automatic recognition of the author's gender. Results are presented in the Table 1. In the case of all methods, except for the last one, the input to the classifier is calculated using pre-trained vectors for Polish language [6]. Results show that standardization has almost no influence on the analyzed metric in case of the MLP. For the SGD a use of raw feature vectors gives better results. The best results from all experiments were obtained for the MLP with the relu activation function regardless of the version of the f1 score. However, results for the fastText classification method are very close. Estimating these results is not easy. Therefore, we calculated (see Table 2) performance metrics for a random selection of gender (taking into account the proportion of authors gender distribution in the data set) and for selecting always 'man'. It can be seen that the MLP classifier provides better accuracy than a random selection of only 0.03. However, in the case of the f1 scores, it is much better: 0.17 for the average f1 and 0.14 for weighted one.

4.2 Gender Recognition - Context Analysis

Next, context-aware methods (see Sect. 3), i.e. methods that take into account the context of words were taken into account. The results are presented in Table 3. The best results are achieved for the fastText method with word 5-grams. The achieved results outperform the random approach (Table 3) by 0.0648 in accuracy, 0.2309 in the average f1 score, and 0.1776 in the weighted f1 score. The detailed analysis for fastText performance in a function of word n-grams is presented in Fig. 1(top). It allows noticing, that the most effective assignments can be carried out on four-word titles and almost stable values for longer titles.

Table 1. Gender recognition performance for non-context methods

Method	Standardization	Accuracy	f1 score	
			Average	Weighted avg.
MLP (relu)	Power	0.7959	0.6587	0.7720
MLP (relu)	Z-score	0.7965	0.6492	0.7683
MLP (relu)	No	**0.7968**	**0.6725**	**0.7782**
MLP (tanh)	No	0.7933	0.6707	0.7760
SGD	Power	0.7465	0.5536	0.7075
SGD	Z-score	0.7465	0.5536	0.7075
SGD	No	0.7619	0.4324	0.6592
fastText		0.7940	0.6578	0.7710

Table 2. Random recognition of author gender

Method	Accuracy	f1 score	
		Average	Weighted avg.
Random	0.7620	0.5000	0.6375
Select man	0.7620	0.4324	0.659

4.3 Genre Recognition - Non-context Analysis

Next, we have tested non-context methods in the task of writing genre recognition. In this case, we have not analyzed different standardization methods and different activation functions for the MLP. The results are presented in Table 4. The best results are achieved for the fastText algorithm. Table 5 shows the performance metrics for a random selection of genre (taking into account the proportion of genre classes distribution in the data set) and for selecting class with the largest support (i.e. textbooks). It could be noticed that the fastText algorithm gives significantly better results than random selection.

Table 3. Gender recognition performance for context-aware methods

Word2vec	Classifier	Accuracy	f1 score	
			Average	Weighted avg.
Elmo	MLP (relu)	0.7958	0.6432	0.7656
Elmo	SGD	0.7617	0.5894	0.7289
word2vec	BiLSTM	0.7646	0.4706	0.6781
word2vec	Hierachical MLP	0.8067	0.6927	0.7907
Built-in	fastText with 4-grams	**0.8270**	**0.7309**	**0.8150**

Table 4. Writing genre recognition performance for non-context methods

Method	Accuracy	f1 score	
		Average	Weighted avg.
MLP	0.6897	0.5589	0.6767
SGD	0.5633	0.2735	0.5087
fastText	**0.7194**	**0.5875**	**0.6968**

Table 5. Random recognition of writing genres

Method	Accuracy	f1 score	
		Average	Weighted avg.
Random	0.1483	0.0357	0.1483
Select ‘textbooks’	0.3215	0.0174	0.1564

4.4 Genre Recognition - Context Analysis

Finally, context methods for genre recognition were evaluated. The results are presented in Table 6. The best results are achieved for the fastText method with word 2-grams. The fastText performance in a function of word n-grams is presented in Fig. 1(bottom).

Table 6. Genre recognition performance for context aware methods

Word2vec	Classifier	Accuracy	f1 score	
			Average	Weighted avg.
ELMo	MLP (tanh)	0.7083	0.5922	0.6987
ELMo	SGD	0.6276	0.5056	0.6246
ELMo	SGD	0.6423	0.4887	0.6268
fastText	BiLSTM	0.4763	0.2781	0.3945
fastText	Hierachical MLP	0.6933	0.5718	0.6858
Properiaty	fastText with 2-grams	**0.74**	**0.655**	**0.7349**

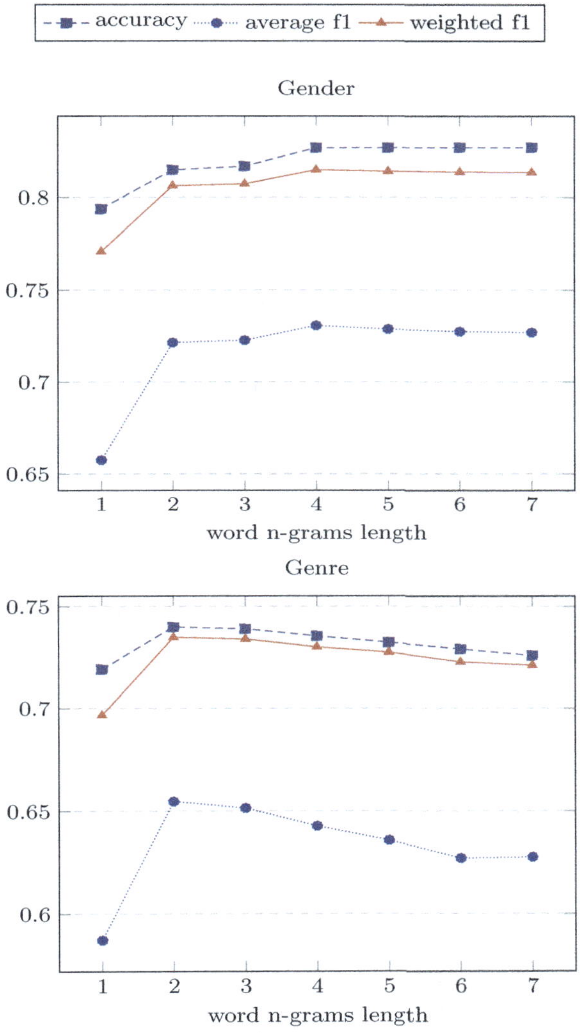

Fig. 1. Performance of fastText in a function of word n-grams length in gender recognition

5 Conclusion

In this paper, we discussed several options for the automatic classification of author gender and book genre based only on the book title. They were tested on the large data set, a bibliography from the National Library of Poland. The experiments show that automatic taxonomy of microtext as short as titles is possible and gives positive results. The recognition of the writing genre for 28 classes gives an accuracy of 0.74 for the fastText method with word 4-grams.

The results for author gender recognition are no so spectacular, but still a bit better (especially for f1-scores) than random selection.

References

1. Baj, M., Walkowiak, T.: Computer based stylometric analysis of texts in polish language. In: Rutkowski, L., Korytkowski, M., Scherer, R., Tadeusiewicz, R., Zadeh, L.A., Zurada, J.M. (eds.) Artificial Intelligence and Soft Computing, pp. 3–12. Springer, Cham (2017)
2. Eder, M., Piasecki, M., Walkowiak, T.: An open stylometric system based on multilevel text analysis. Cogn. Stud. | Etudes cognitives **17**, 267–287 (2017)
3. Harris, Z.S.: Distributional structure. Word **10**(2–3), 146–162 (1954)
4. Hastie, T.J., Tibshirani, R.J., Friedman, J.H.: The Elements of Statistical Learning: Data Mining, Inference, and Prediction. Springer Series in Statistics. Springer, New York (2009). autres impressions: 2011 (corr.), 2013 (7e corr.)
5. Joulin, A., Grave, E., Bojanowski, P., Mikolov, T.: Bag of tricks for efficient text classification. In: Proceedings of the 15th Conference of the European Chapter of the Association for Computational Linguistics: Volume 2, Short Papers, pp. 427–431. Association for Computational Linguistics (2017). http://aclweb.org/anthology/E17-2068
6. Kocon, J., Gawor, M.: Evaluating KGR10 polish word embeddings in the recognition of temporal expressions using BiLSTM-CRF. CoRR abs/1904.04055 (2019). http://arxiv.org/abs/1904.04055
7. Le, Q., Mikolov, T.: Distributed representations of sentences and documents. In: International Conference on Machine Learning, pp. 1188–1196 (2014)
8. Mikros, G., Perifanos, K.: Authorship Attribution in Greek Tweets Using Author's Multilevel N-gram Profiles, pp. 17–23. AAAI Press, Palo Alto (2013)
9. Peters, M.E., Neumann, M., Iyyer, M., Gardner, M., Clark, C., Lee, K., Zettlemoyer, L.: Deep contextualized word representations. In: Proceedings of NAACL (2018)
10. Schuster, M., Paliwal, K.K.: Bidirectional recurrent neural networks. IEEE Trans. Signal Process. **45**(11), 2673–2681 (1997)
11. Schwartz, R., Tsur, O., Rappoport, A., Koppel, M.: Authorship attribution of micro-messages. In: Proceedings of the 2013 Conference on Empirical Methods in Natural Language Processing, Seattle, Washington, USA, pp. 1880–1891. Association for Computational Linguistics, October 2013. https://www.aclweb.org/anthology/D13-1193
12. Silessi, S., Varol, C., Karabatak, M.: Identifying gender from SMS text messages. In: 15th IEEE International Conference on Machine Learning and Applications (ICMLA), pp. 488–491 (12 2016)
13. Thomale, J.: Interpreting MARC: where's the bibliographic data? Code4Lib J. (11) (2010). https://journal.code4lib.org/articles/3832
14. Torkkola, K.: Discriminative features for textdocument classification. Formal Pattern Anal. Appl. **6**(4), 301–308 (2004). https://doi.org/10.1007/s10044-003-0196-8
15. Walkowiak, T., Datko, S., Maciejewski, H.: Bag-of-words, bag-of-topics and word-to-vec based subject classification of text documents in polish - a comparative study. In: Zamojski, W., Mazurkiewicz, J., Sugier, J., Walkowiak, T., Kacprzyk, J. (eds.) Contemporary Complex Systems and Their Dependability, pp. 526–535. Springer, Cham (2019)

16. Walkowiak, T., Datko, S., Maciejewski, H.: Low-dimensional classification of text documents. In: Zamojski, W., Mazurkiewicz, J., Sugier, J., Walkowiak, T., Kacprzyk, J. (eds.) Engineering in Dependability of Computer Systems and Networks, pp. 534–543. Springer, Cham (2020)
17. Walkowiak, T., Piasecki, M.: Stylometry analysis of literary texts in polish. In: Rutkowski, L., Scherer, R., Korytkowski, M., Pedrycz, W., Tadeusiewicz, R., Zurada, J.M. (eds.) Artificial Intelligence and Soft Computing, pp. 777–787. Springer, Cham (2018)
18. Yeo, I.K., Johnson, R.A.: A new family of power transformations to improve normality or symmetry. Biometrika **87**(4), 954–959 (2000)
19. Zou, H., Hastie, T.: Regularization and variable selection via the elastic net. J. Roy. Stat. Soc. B **67**, 301–320 (2005)

Searching Algorithm for an Optimal Location of Warehouses in a Distribution Network for Predicted Order Variability

Henryk Piech and Grzegorz Grodzki(✉)

Czestochowa University of Technology,
Dabrowskiego 69, 42-201 Czestochowa, Poland
{Henryk.piech, grzegorz.grodzki}@icis.pcz.pl

Abstract. In reality, from the point of view of a logistics company, we have little control over the location of customers and producers. As a logistics company, we can, however, situate warehouses [2, 5–7]. The construction and location of warehouses is aimed at achieving organizational independence from production activities, to reduce transport costs, and better adapt to the temporary and overarching supply policy of recipients [1, 9, 11]. So there is an organizational buffer between the producer and the recipient. As often is the case in real-life situations, the producer or recipient has logistics departments creating so-called stocks of raw materials and distribution inventory [3, 4], but nowadays, this is not the norm. Getting back to the topic of warehouses, it is extremely important to have an optimum location. Here, we encounter a number of restrictions that take into account both infrastructural and personnel matters. Nevertheless, we can still search for an optimum location while taking restrictions into account. In reality, when looking for an optimum location one must also take into account the number of deliveries, the size, direction and delivery rate. The main purpose is to assess the usefulness and structure of network connections using the rough set theory, more specifically, several of its parameters, such as strength, degree of certainty and coverage ratio. Parameters used in works [8, 10, 12] were also used to assess risk, reliability, security, etc. Based on the estimators of selected parameters, one can create a standard set of rules and use them in the inference mechanism to categorize network connections and determine their usefulness. The proposed concept is multi-tiered, which corresponds to the individual sections of the publication. The final goal consists in using the rough set theory to make predictions (and thus a structural proposal) for a treated (functionally better than the current one) configuration of the distribution network with an optimum (or near optimum) warehouse location.

Keywords: Storage location optimization · Rough set theory · Structural forecasts

W. Zamojski et al. (Eds.): DepCoS-RELCOMEX 2020, AISC 1173, pp. 482–492, 2020.
https://doi.org/10.1007/978-3-030-48256-5_47

1 Introduction

The following considerations start from obvious, heuristic assumptions such as: routes passed most frequently should be the shortest, the largest deliveries should also be realized within the shortest connections, one should consider the possibility of combining deliveries – it is associated with time and with the capacity of containers, the predictions of the size and, range of supplies assortment and their directions, which should be realistic and carefully analysed. In this case, however, the occurrence of errors is possible to a large extent. Hence, the possibilities of dislocation or creating new warehouses should be predicted. Nonetheless, by assuming constancy of supplies, within a specified time (month, quarter, etc.), it is possible to accept that the cost of transportation along one section is proportional to its length, transportation intensity and the size of deliveries:

$$c_{i,j} = \sum\nolimits_{k=1}^{l_{i,j}} n_{i,j,k} v_{i,j,k} d_{i,j} \tag{1}$$

where:

i,j – the codes of each section border node,
k – the number of the delivery type,
$l_{i,j}$ – the number of supply types at a given section i,j,
$n_{i,j,k}$ – the intensity of a delivery type k, along the section i,j,
$v_{i,j,k}$ – the size of the delivery type k, along the section i,j,
$d_{i,j}$ – the length of the section i,j.

The dependency of the cost on the size of the delivery cannot be regarded as proportional to its value. In order to simplify, a loading factor is introduced, and then it includes a cost change factor when changing the degree of loading, so instead $v_{i,j,k}$ the parameter cfl_k is used:

$$cfl_k = fl_k(1 - (1 - u_k)d_k) \tag{2}$$

where:

fl_k – the unitary (for unit distance) costs of fuel consumption at full loading of the delivery vehicle for the delivery type k.
d_k – the factor of transport costs reducing resulting from the under loading of the delivery vehicle for the delivery k.

This task is reduced to the creation of an optimization criterion and a system of limitations related to the real situation.

2 The Task of the Storage Location Optimization of the Logistics Company

Each node, which corresponds to the warehouse or the place of manufacturing, has its coordinates, which will be interconnected for the full network. Some of the nodes are already created; therefore, there is no influence on their location. Connections between some nodes exist, however, elsewhere they do not. Supply routes use some connections, while others do not. The current status of the connections does not mean maintaining it in the time perspective. Therefore, the coefficients of the existence of connections $eg_{i,j} \in \{0, 1\}$ are introduced. The coefficient takes binary values, that is the value 1 if the connection between nodes i, j exists or the value 0 if it does not, and also when $i = j$. The inclusion coefficient of the route connection is not necessary, since it is regulated by the parameter $n_{i,j,k}$.

It is suggested that the criterion of minimizing the length of segments including the potential and currently used connections is:

$$gc = min\left\{\sum\nolimits_{i-1}^{n}\sum\nolimits_{j=1}^{n} eg_{i,j}\sum\nolimits_{k=1}^{l_{i,j}} n_{i,j,k}cfl_k d_{i,j}\right\} \tag{3}$$

that means

$$gc = min\left\{\sum\nolimits_{i-1}^{n}\sum\nolimits_{j=1}^{n} eg_{i,j}\sum\nolimits_{k=1}^{l_{i,j}} n_{i,j,k}cfl_k\sqrt{\left(x_i - x_j\right)^2 + \left(y_i - y_j\right)^2}\right\} \tag{4}$$

where x_i, y_i – the coordinates of the i-th node.

This criterion is complemented with a set of limitations $x_i \in [a_i, b_i], y_i \in [c_i, d_i]$, $i = 1, 2, \ldots, n$. It is assumed that what is sent from the node i reaches the node j, so for each i it is possible to write the partial criterion:

$$gc_i = min\left\{\sum\nolimits_{j=1}^{n} eg_{i,j}\sum\nolimits_{k=1}^{l_{i,j}} n_{i,j,k}cfl_k\left(\left(x_i - x_j\right)^2 + \left(y_i - y_j\right)^2\right)\right\} \tag{5}$$

Partial derivatives are compared to zero or to minimal ranges:

$$\frac{\partial gc_i}{\partial x_i} = 2x_i\sum\nolimits_{j=1}^{n} eg_{i,j}\sum\nolimits_{k=1}^{l_{i,j}} n_{i,j,k}cfl_k - 2\sum\nolimits_{j=1}^{n} eg_{i,j}\sum\nolimits_{k=1}^{l_{i,j}} n_{i,j,k}cfl_k x_j \tag{6}$$

$$\frac{\partial gc_i}{\partial y_i} = 2y_i\sum\nolimits_{j=1}^{n} eg_{i,j}\sum\nolimits_{k=1}^{l_{i,j}} n_{i,j,k}cfl_k - 2\sum\nolimits_{j=1}^{n} eg_{i,j}\sum\nolimits_{k=1}^{l_{i,j}} n_{i,j,k}cfl_k y_j \tag{7}$$

For all nodes the system of criterion equations is obtained:

$$\frac{\partial gc_n}{\partial x_n} = 2x_n\sum\nolimits_{j=1}^{n} eg_{n,j}\sum\nolimits_{k=1}^{l_{n,j}} n_{n,j,k}cfl_k - 2\sum\nolimits_{j=1}^{n} eg_{n,j}\sum\nolimits_{k=1}^{l_{n,j}} n_{n,j,k}cfl_k x_j = \delta x_n \tag{8}$$

and

$$\frac{\partial gc_n}{\partial y_n} = 2y_n \sum\nolimits_{j=1}^{n} eg_{n,j} \sum\nolimits_{k=1}^{l_{n,j}} n_{n,j,k} cfl_k - 2\sum\nolimits_{j=1}^{n} eg_{n,j} \sum\nolimits_{k=1}^{l_{n,j}} n_{n,j,k} cfl_k y_j = \delta y_n \quad (9)$$

where $|\delta x(y)_i|$ should be as small as possible and is caused by given limitations corresponding to real information $|\delta x(y)_i| = 0$ and if the location of the nodes is not limited.

By solving the system of Eq. (1) and taking into account given limitations, the solution was obtained regarding the location of the nodes which is close to optimal.

The problem is mainly limited to the creation of coefficients of the unknown x_i and y_i. Therefore, the preparation and initial processing of data is in the center of attention. With the maximal number of unknowns equal to $2n$, and the same number of equations, the optimal location of nodes can be obtained. When the location of some nodes is given in advance, limits can be presented as follows $x_i \in [a_i, a_i], y_i \in [c_i, c_i]$, for nodes that already exist. The system of equations can be reduced or simplified (in terms of the number of unknowns in the equations) when the nodes do not participate in the description of supply routes.

3 The Categorization of Connections in the Distribution Network

As an initial step one should determine routes by using algorithms: Dijkstry, Kruskal, Floyd-Fulkerson, algorithm for the traveling salesman and others. This study is realized for current levels of supplies, predicted deliveries, for current node structures and the predicted structures of nodes. The following data are obtained:

$ld_{i,j}, lu_{i,j}$ – the minimal and maximal number of supply types at a given section i,j in the supply forecast,
$nd_{i,j,k}, nu_{i,j,k}$ – the minimal and maximal intensity of the delivery type k, along the $ld'_{i,j}, lu'_{i,j}$ section i,j in the supply forecast,
$vd_{i,j,k}, vu_{i,j,k}$ – the minimal and maximal size of the delivery type k, along the section i,j in the supply forecast,
$ld'_{i,j}, lu'_{i,j}$ – the minimal and maximal number of supply types at a given section i,j with the forecast structure,
$nd'_{i,j,k}, nu'_{i,j,k}$ – the minimal and maximal intensity of the delivery type k, along the section i, j with the forecast structure,
$vd'_{i,j,k}, vu'_{i,j,k}$ – the minimal and maximal size of the delivery type k, along the section i,j with the forecast structure,
$egd'_{i,j}, egu'_{i,j}$ – connection activities for the maximal and minimal extended structure.

Therefore, obtained solutions concern the type of supplies in relation to their sizes and structures. The magnitude of supplies is multiplied by the intensity parameter. The control of the analysis process is realized by three parameters: i, j, k. The analysis leads, inter alia, to the estimation of accumulated parameters referring to their usage to assign their validity scales to connections, corresponding to load scales. Hence, they may be the following:

$sld_{i,j}, slu_{i,j}$ – the minimal and maximal total number of supply types after a given section i,j in the supply forecast,
$svd_{i,j,k}, svu_{i,j,k}$ – the minimal and maximal total size of the delivery type k, along the section i,j in the supply forecast,
$sld'_{i,j}, slu'_{i,j}$ – the minimal and maximal total number of delivery types after a given section i,j with the forecast structure,
$svd'_{i,j,k}, svu'_{i,j,k}$ – the minimal and maximal total size of the delivery type k, along the section i,j with the forecast structure.

The complex structure is formed by taking into account control parameters i, j, k and the attributes of connections defined by them: $svd, svu, svd', svu', sld, slu, sld', slu'$. This structure can be presented as the structure describing rough sets:

$i = 1, j = 2,\ k = 1 \quad svd_{1,2,1},\ svu_{1,2,1},\ svd'_{1,2,1},\ svu'_{1,2,1},\ sld_{1,2},\ slu_{1,2},\ sld'_{1,2},\ slu'_{1,2}$
$i = 1,\ j = 3,\ k = 1 \quad svd_{1,3,1},\ svu_{1,3,1},\ svd'_{1,3,1},\ svu'_{1,3,1},\ sld_{1,3},\ slu_{1,3},\ sld'_{1,3},\ slu'_{1,3}$
$i = n-1,\ j = n,\ k = m \quad svd_{n-1,n,m},\ svu_{n-1,n,m},\ svd'_{n-1,n,m},\ svu'_{n-1,n,m}$

The categorization process conducted on the basis of the normalized values of accumulated attributes:

$$nsvd_{i,j,k} = svd_{i,j,k}/max_{i=1,2,\ldots n-2;j=i+1,\ldots,n;k=1,2,\ldots,m}\left\{svd_{i,j,k}, svd'_{i,j,k}\right\}$$
$$nsvu_{i,j,k} = svu_{i,j,k}/max_{i=1,2,\ldots n-2;j=i+1,\ldots,n;k=1,2,\ldots,m}\left\{svd_{i,j,k}, svd'_{i,j,k}\right\}$$
$$nsvd'_{i,j,k} = svd'_{i,j,k}/max_{i=1,2,\ldots n-2;j=i+1,\ldots,n;k=1,2,\ldots,m}\left\{svd_{i,j,k}, svd'_{i,j,k}\right\}$$
$$nsvu'_{i,j,k} = svu'_{i,j,k}/max_{i=1,2,\ldots n-2;j=i+1,\ldots,n;k=1,2,\ldots,m}\left\{svd_{i,j,k}, svd'_{i,j,k}\right\}$$

where $max_{i=1,2,\ldots n-2;j=i+1,\ldots,n;k=1,2,\ldots,m}\left\{svd_{i,j,k}, svd'_{i,j,k}\right\}$ – the largest sum of supplies realized by the connection i,j for both types of forecasts (supply and structure). And

$$nsld_{i,j,k} = sld_{i,j,k}/max_{i=1,2,\ldots n-2;j=i+1,\ldots,n}\left\{sld_{i,j}, sld'_{i,j}\right\}$$
$$nslu_{i,j,k} = slu_{i,j,k}/max_{i=1,2,\ldots n-2;j=i+1,\ldots,n}\left\{sld_{i,j}, sld'_{i,j}\right\}$$
$$nsld'_{i,j,k} = sld'_{i,j,k}/max_{i=1,2,\ldots n-2;j=i+1,\ldots,n}\left\{sld_{i,j}, sld'_{i,j}\right\}$$
$$nslu'_{i,j,k} = slu'_{i,j,k}/max_{i=1,2,\ldots n-2;j=i+1,\ldots,n}\left\{sld_{i,j}, sld'_{i,j}\right\}$$

where $max_{i=1,2,\ldots n-2;j=i+1,\ldots,n}\left\{sld_{i,j}, sld'_{i,j}\right\}$ – the largest number of supply types realized by the connection i,j for both types of forecasts (supply and structure).

Algorithmically this aspect of the analysis can be presented as follows:

```
for k := 1 to m do
  for i := 1 to n-2 do
    for j := i+1 to n do
    begin
      svd_{i,j,k} := svu_{i,j,k} := 0;
      svd'_{i,j,k} := svu'_{i,j,k} := 0;
    end;
    for i := 1 to n-2 do
      for j := i+1 to n do
      begin
        sld_{i,j} := slu_{i,j}:= 0;
        sld'_{i,j} := slu'_{i,j}:= 0;
      end;
        for k := 1 to m do
          for I := 1 to n-2 do
            for j := i+1 to n do
            begin
              if eg_{i,j} = 1 then
              begin
                svd_{i,j,k} := svd_{i,j,k} + vd_{i,j,k};
                svu_{i,j,k} := svu_{i,j,k} + vu_{i,j,k};
                svd'_{i,j,k} := svd'_{i,j,k} + vd'_{i,j,k};
                svu'_{i,j,k} := svu'_{i,j,k} + vu'_{i,j,k};
                if vd_{i,j,k} > 0 then sld_{i,j} := sld_{i,j} + 1;
                  if vu_{i,j,k} > 0 then slu_{i,j} := slu_{i,j} + 1;
                    if vd'_{i,j,k} > 0 then sld'_{i,j}:= sld'_{i,j} + 1;
                      if vu'_{i,j,k} > 0 then slu'_{i,j} := slu'_{i,j} + 1;
              end;
            end.
```

Let us introduce category thresholds:

$$tv = \frac{1}{lv} and\ tl = \frac{1}{ll} \tag{10}$$

where

lv – the number of categories in estimating the size of supplies,
ll – the number of categories in estimating delivery types (the assortment of goods), and then for each attribute value, categories are defined (with integer values) as follows:

$cvd_{i,j,k} = round\left(nsvd_{i,j,k}/tv\right) + 1,$
$cvu_{i,j,k} = round\left(nsvu_{i,j,k}/tv\right) + 1,$

$$cvd'_{i,j,k} = round\left(nsvd'_{i,j,k}/tv\right) + 1,$$
$$cvu'_{i,j,k} = round\left(nsvu'_{i,j,k}/tv\right) + 1,$$

and

$$cld_{i,j} = round(nsld_{i,j}/tl) + 1,$$
$$clu_{i,j} = round(nslu_{i,j}/tl) + 1,$$
$$cld'_{i,j} = round\left(nsld'_{i,j}/tl\right) + 1,$$
$$clu'_{i,j} = round\left(nslu'_{i,j}/tl\right) + 1.$$

One can enter the total categorization of the supply value aggregating the values (magnitude) of supplies for different types:

$$ckvd_{i,j} = \sum\nolimits_{k=1}^{m} cvd_{i,j,k} \tag{11}$$

$$ckvu_{i,j} = \sum\nolimits_{k=1}^{m} cvu_{i,j,k} \tag{12}$$

$$ckvd'_{i,j} = \sum\nolimits_{k=1}^{m} cvd'_{i,j,k} \tag{11}$$

$$ckvu'_{i,j} = \sum\nolimits_{k=1}^{m} cvu'_{i,j,k} \tag{12}$$

Global categorization table has dimensions $\frac{n}{2}(x8)$ and has the form:

$i = 1, j = 2 \quad ckvd_{1,2},\ ckvu_{1,2},\ ckvd'_{1,2},\ ckvu'_{1,2},\ cld_{1,2},\ clu_{1,2},\ cld'_{1,2},\ clu'_{1,2}$
$i = 1, j = 3 \quad ckvd_{1,3},\ ckvu_{1,3},\ ckvd'_{1,3},\ ckvu'_{1,3},\ cld_{1,3},\ clu_{1,3},\ cld'_{1,3},\ clu'_{1,3}$
$i = 1, j = n \quad ckvd_{1,n},\ ckvu_{1,n},\ ckvd'_{1,n},\ ckvu'_{1,n},\ cld_{1,n},\ clu_{1,n},\ cld'_{1,n},\ clu'_{1,n}$
$i = n - 1, j = n$
$ckvd_{n-1,n},\ ckvu_{n-1,n},\ ckvd'_{n-1,n},\ ckvu'_{n-1,n},\ cldn_{n-1,n},\ clu_{n-1,n},\ cld'_{n-1,n},\ clu'_{n-1,n}.$

4 The Analysis of the Algorithm Use of Delivery Route Creation and the Suitability of Connections

Let us introduce the attributes of decision-making concerning the membership of the section (connection to) in the route. Specific connections are used for different routes. The route of the specific delivery, according to different algorithms, can have a different shape. In this case, some connections may overlap, while another part may be mutually exclusive for different algorithms. The membership of the connection to a particular delivery will have a certain or probable character. The problem of estimation of probabilities for the situation of selection the connection remains. Here one can use both values and the numbers of supply types and their corresponding categories.

Delivery can be identified, in general, with the initial and final place of transportation procedures. For example, the distribution network shown in Fig. 1 is disposed.

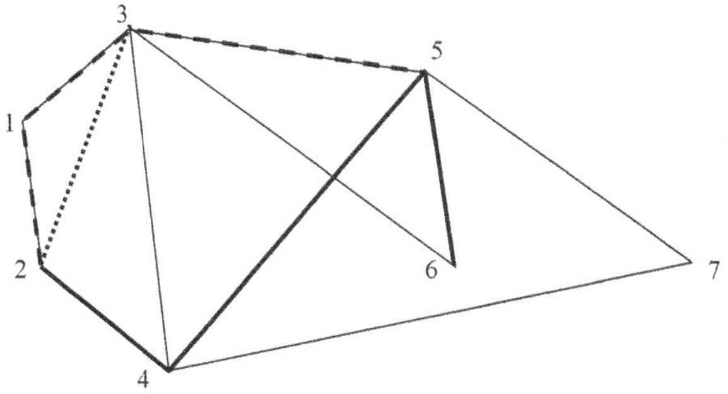

Fig. 1. Delivery between nodes 2 and 6 can be realized in two routes: 2-1-3-5-6 and 2-4-5-6, and in the forecast of structural changes also in the rout 2-3-5-6.

In the abovementioned example one uses connections (1,2), (1,3), (2,4), (3,5), (4,5), (5,6). In addition, the forecast of the structure change foresees the creation of connections (2,3). The segments of connections that are the selected data on the probability of their use have the form of Table 1.

Table 1. Categorized connection parameters

Connection	*ckvd*	*ckvu*	*ckvd'*	*ckvu'*	*ckld*	*cklu*	*ckld'*	*ckvu'*
1,2	5	4	0	0	4	3	0	0
1,3	5	4	0	0	4	3	0	0
2,3	0	0	5	4	0	0	4	3
2,4	4	4	4	4	6	3	6	3
3,5	6	5	6	5	5	3	5	3
4,5	5	4	5	4	6	5	6	5
5,6	10	10	10	10	10	10	10	10

The connection that regardless of the forecast will be included in the route of delivery is included to the area of the lower approximation [6]:

$$LA = \bigcup\nolimits_{x \in U} \{CR(x) : CR(x) \subseteq T\} \tag{13}$$

where

x – the number of the connection,
U – universe, the set of all values of category connections,
$CR(x)$ – connection which fulfils criteria for the membership in the route,
T – the set of all routes of delivery.

The remaining connections of the given delivery belong to the area of the upper approximation because with the probability greater than zero they are included in one of its routes:

$$LA = \bigcup_{x \in U} \{CR(x) : CR(x) \cap T \neq 0\} \tag{14}$$

For the analysis of the strength, certainty and coverage one introduces sums in the rows of the Table 2.

Table 2. The indicators of the connection use in forecasts expressed as the sum of category values.

Connection	$N(\pi)$
1,2	16
1,3	16
2,3	16
2,4	34
3,5	38
4,5	40
5,6	80

The number of elements of the universe $|U|$ is the sum of all $N(\pi)$ that is equal to 240. The decision of the connection belonging to the route of delivery can be certain as for the connections 5, 6, and uncertain for other connections. It is possible to estimate the strength of the decision of the connection belonging on the basis of values of conditional attributes C and decision-making D, and practically on the basis of their corresponding values: $S(x) = suppx(C, D)/|U| = |C(x) \cap D(x)|/|U|$, where $suppx(C, D)$ - support for decision-making rule D by attributes C.

The degree of certainty is evaluated by separating the probability of belonging according to the sums of categories: $DC(x) = suppx(C, D)/|C(x)| = |C(x) \cap D(x)|/|C(x)|$, that is $DC(x) = S(x)/\pi\ (C(x))$, where $\pi(C(x)) = |C(x)/|U||$.

Moreover, the degree of coverage is related to various decisions including the probabilities of belonging attributes:

$DCV(x) = suppx(C, D)/|D(x)| = |C(x) \cap D(x)|/|D(x)|$ or

$DC(x) = S(x)/\pi(D(x))$ where $\pi(D(x)) = |D(x)|/|U|$

The variant containing the parameters of strength, certainty and decision coverage is shown in Table 3.

Decision making rule $C(x) \rightarrow D(x)$ is a decision on the certainty and means that $0 < DC(x) < 1$. Inverse decision making rule $D(x) \rightarrow C(x)$ is used to explain reasons of the decision. Routes 1, 2, 3 are mutually independent and can be realized this way.

Therefore, the sum of coverage for individual routes is equal to 1, however, if one assumes the mutual exclusion of paths *t1*, *t2*, then the conditional probability is used to describe the degree of connection utilization:

$$DCV(x) = p(t1)DC(x/t1) + p(t2)DC(x/t2) \quad (15)$$

where $p(t1), p(t2)$ - the probability of route selection 1 (2), $DC(x/t1), DC(x/t2)$ – the coverage of the decision on the route 1 (2) by including the connection x.

Table 3. The parameters of decision rules

Connection	Strength	Certainty	Route 1	Route 2	Route 3	Coverage t1	Coverage t2	Coverage t3
1,2	0,07	0,1	1	0	0	0,11		
1,3	0,07	0,1	1	0	0	0,11		
2,3	0,07	0,1	0	1	0		0,12	
2,4	0,14	0,21	0	0	1			0,22
3,5	0,16	0,24	1	1	0	0,25	0,28	
4,5	0,17	0,25	0	0	1			0,26
5,6	0,33	1	1	1	1	0,53	0,60	0,52

The reorganization of the route configuration as seen from the data did not affect the efficiency of transportation in terms of size and the order type. Hence, it is assumed that $p(t1) = p(t2) = 0,5$. The form of filling the table will be different (Table 4).

Table 4. The parameters of decision rules for exclusive routes t1 and t2.

Connection	Strength	Certainty	Route 1	Route 2	Route 3	Coverage t1	Coverage t2	Coverage t3
1,2	0,07	0,1	1	0	0	0,05		
1,3	0,07	0,1	1	0	0	0,05		
2,3	0,07	0,1	0	1	0		0,06	
2,4	0,14	0,21	0	0	1			0,22
3,5	0,16	0,24	1	1	0	0,27	0,27	
4,5	0,17	0,25	0	0	1			0,26
5,6	0,33	1	1	1	1	0,57	0,57	0,52

In cases of exclusive or dependent decisions one introduces probabilistic apparatus for determining the degree of certainty and decision coverage.

5 Conclusions

The proposed analytical model refers to the search for warehouse locations and tracking the efficiency of deliveries. It is carried out in three stages, i.e. the solution of the system of partial equations (for minimizing the distance of connections), categorization of connections (for creating connection attributes, which is the basis for describing rough sets) and analyzing the usefulness of routes based on the Pawlak rough set theory parameters. This corresponds to how work is divided out and can be used to forecast logistics processes related to deliveries and transport route structures.

An interval (with the option of using fuzzy) way was used to present data. Using the rough set theory allows us to categorize and evaluate the connection rate, routes and deliveries. In such analyses, we use only selected parameters of the rough set theory, which in the proposed approach means strength, degree of certainty and coverage. To select parameters for a rough analysis, forecasts for supply and structural changes in the distribution network were used. The optimum warehouse location depends on the range and volume of deliveries, which is included in Sect. 2. The categories and how often connections and routes are used depend on the location of warehouses, which can be included in structural forecasts. Obtained solutions most likely do not provide optimal solutions but they allow us to optimize prognostic heuristics. The more data on a diversified structure, the greater the contribution to the optimization process, which will improve organizational and financial results. The process model shows that performance indicators improve in the range of a few to several percent.

References

1. Baumgarten, H.: Logistik-Management, vol. 12. Technische Universitaet Berlin, Berlin (2004)
2. Blackstock, T.: Keynote speech. In: International Association of Food Industry Suppliers, San Francisco, CA, March 11 (2005)
3. Chen, I.J., Paulraj, A., Lado, A.: Strategic purchasing, supply management, and firm performance. J. Oper. Manag. **22**(5), 505–523 (2004)
4. Cohen, M.A., Huchzermeir, A.: Global supply chain management: a survey of research and applications. In: Tayur, S., Ganeshan, R., Magazine, M. (eds.) Quantitative Models for Supply Chain Management, Kluwer, Boston, pp. 669–702 (1999)
5. Lambert, D.M., Cooper, M.C.: Issues in supply chain management. Ind. Mark. Manag. **29** (1), 65–83 (2000)
6. Lambert, D.M., Garca-Dastugue, S.J., Croxton, K.L.: An evaluation of process-oriented supply chain management frameworks. J. Bus. Logist. **26**(1), 25–51 (2005)
7. Gattorna, J.: Supply chains are the business. Supply Chain Manag. Rev. **10**(6), 42–49 (2006)
8. Croxton, K.L., Garca-Dastugue, S.J., Lambert, D.M., Rogers, D.S.: The supply chain management processes. Int. J. Logist. Manag. **12**(2), 13–36 (2001)
9. Sawicka, H., Zak, J.: Mathematical and simulation based modeling of the distribution system of goods. In: Proceedings of the 23rd European Conference on Operational Research, Bonn, 5–8 July (2009)
10. Simchi-Levy, D., Kaminski, P., Simchi-Levy, E.: Designing and Managing the Supply Chain: Concepts, Strategies, and Case Studies, Boston. Irwin/McGraw Hill, MA (2000)
11. Straka, M., Malindzak, D.: Distribution Logistics. Express Publicity, Kosice (2008)
12. Wisner, J.D., Keong Leong, G., Tan, K.-C.: Supply Chain Management: A Balanced Approach. Thomson South-Western, Mason (2004)

Tackling Access Control Complexity by Combining XACML and Domain Driven Design

Paweł Rajba(✉)

Institute of Computer Science, University of Wrocław, Joliot-Curie 15, 50-383 Wrocław, Poland
pawel@cs.uni.wroc.pl
http://pawel.ii.uni.wroc.pl/

Abstract. In the paper we propose an approach for designing software architecture with the access control solution which reflects the real business needs in a consistent, maintainable and complete way. We identify and analyze key drivers and requirements for access control, show the complexity of authorizations and propose an approach based on XACML and Domain Driven Design.

Keywords: Access control · Authorization · XACML · Domain-Driven Design · Application architecture · Hexagonal architecture

1 Introduction

Access control is a security service [5] to protect information from unauthorized access. It executes authorization rules (also referred as policies or just authorizations) to support basic security requirements [6] like confidentiality, integrity and availability. One of the areas where access control is widely adopted is the development of information systems for different industries (like banking, healthcare, automotive) to which we will refer further as business applications. During the last two decades we can observe a strong focus on making sure that the software reflects the real business needs and that is maintainable over time, especially when it comes to applying changes and new requirements. As a result a number of papers [3,9,15] has been published in the field of Model Driven Engineering and related field of Model Driven Security [1,11,14].

Domain-Driven Design, firstly described by Eric Evans in [4], is a philosophy of application development where attention is focused on the complexity in the business domain itself and can be considered as one of the model-driven development methodologies. One of the key elements is the development of a domain model where business and IT experts work together and where all the business logic is implemented. Domain-Driven Design introduces strategic and tactical levels where complexity is handled and structured on different levels of details.

W. Zamojski et al. (Eds.): DepCoS-RELCOMEX 2020, AISC 1173, pp. 493–502, 2020.
https://doi.org/10.1007/978-3-030-48256-5_48

In the paper we propose an access control solution based on XACML and Domain-Driven Design. Firstly we analyze the complexity of authorizations' decisions in the large business applications as being a part of the overall business logic. Then we enumerate common business requirements towards access control and proposal a solution which based on XACML and aligned with Domain-Driven Design. Finally, we show how the proposed approach supports specified requirements and reflects on the consequences in case key principles are not followed.

The paper is structured as follows. In the next section we provide a basics of Domain-Driven Design. In Sect. 3 we define the access control model we are going to analyze and propose as a part of the solution. Section 4 shortly explains the complexity of authorization logic and in Sect. 5 we describe architectural building blocks influencing the access control consideration. Finally in Sect. 6 we describe the proposed solution and conclude in Sect. 7.

2 Basic Domain Driven Design Concepts

There are many publications where Domain-Driven Design is described and analyzed [4,17], however, there are not so many where security aspects are considered and discussed. In this section we describe the main concepts relevant to our further access control considerations. As indicated in the previous section, Domain-Driven Design introduces strategic and tactical levels. Strategic DDD is about the (a) splitting the business into *domains* and *subdomains* (what business does), (b) identifying *bounded contexts* where all operations are being executed, and map them to *subdomains*, (c) establishing relationships and interactions between bounded contexts and create a *context map*.

Within every bounded context there is one *domain model* essential to all the business logic (data and operations). This model is developed and built by joint cooperation between business and IT experts. To make sure the cooperation results in a solution that meets business needs, both teams need to communicate using a language that is understood by both sides in the same way. This *ubiquitous language* is not given, but rather is being developed over time and it is a very core concept in Domain-Driven Design.

Domain-Driven Design defines a number of elements on the tactical level that can be used to create a single bounded context: services, aggregates, entities, value objects and factories. In our consideration we exclude factories as they play more syntactic role, however we differentiate services into application services, domain services and infrastructure services. In order to represent application architecture in the further consideration, even though very often the layered architecture is used, we consider the hexagonal architectural pattern [2] as more expressive and flexible one.

3 Access Control Model

There are many approaches how access control model can be structured or organized [16]. In our consideration we hire XACML OASIS Standard [18] where we can distinguish roles like Policy Enforcement Point (PEP), Policy Decision Point (PDP), Policy Information Point (PIP) and Policy Administration Point (PAP) which provide a very good separation of different responsibilities.

In a common identity and access management scenario we can split activities into configuration and operation parts. In the configuration part the main activities are identity registration and provisioning as well as obtaining and provisioning permissions a.k.a. authorizations (provisioning is delivered in Policy Administration Point). In the configuration part we can also include identity removal and access revocation – important and very often underestimated activities. The operation part contains identification, authentication and access control. Moreover, in the access control usually we distinguish the following constituents: subject, resource, policy enforcement point, policy decision point and policy information point. The complete access management scenario (without the identity management part) is presented on the Fig. 1.

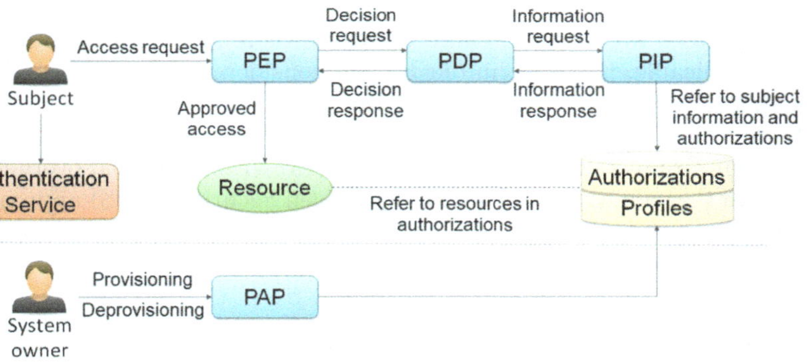

Fig. 1. Access management scenario

In the further consideration we assume that the configuration part has been already delivered and we focus on the operation part. Moreover, we assume that the authentication has been also executed and as a result we get a proof like an ID token with attributes describing the subject.

4 Complexity of Authorization

Authorizations can be described in many different ways, also including many details [7,10,12,18], but a general definition is as follows.

Definition 1. *A function (subject, object, action, context) → {authorized, denied} defines authorizations (or authorization rules, or policies). A subject (a user, a system, etc. with a set of attributes including roles) wants to perform an action (e.g. edit, read) on an object (e.g. an order or a list of products) in a specific context (e.g. the current time, a location, related objects).*

For the simplicity of our consideration, let's narrow down the definition and assume that *context* = *related objects* which are necessary to evaluate the policy.

Next, we can quickly come to the conclusion that authorization rules might be very simple, based on e.g. basic subject's attributes like role or age, but in most of the cases business objects not being part of the user's profile like ownership of an order or amount of transactions from the last month are needed to execute the rules which may become very complicated relatively quickly (e.g. to show the right discount offer). That lead us to the conclusion that authorization rules might be (usually quite often) a part of the business logic what is well known fact in the Model Driven Engineering area. From the Domain-Driven Design perspective that means that authorizations rules may become a part of the domain model. Let's formulate that as a statement:

Remark 1. Authorization rules may be considered as a business logic, i.e. as a part of the domain model. As a consequence, the PDP logic may become a part of the domain model.

In both cases we use word "may", because there are simple solutions where it is not applicable. However, in our further consideration we put our attention to the systems where it is applicable, so the word "may" can be replaced by "is".

On the other hand there is a well recognized design principle to distinct features or main components that are loose-coupled with a minimal overlap [8]. We can observe that a majority of the Model-Driven Security (MDS) approaches are following that principle what result in frameworks where security and business are separated [13]. That obviously is in conflict with Remark 1.

5 Architectural Building Blocks Influencing the Access Control Solution

Now let's investigate the following aspects which have a significant impact on designing the access control solution: trust boundaries, models and types of operations.

5.1 Trust Boundaries

The scope for access control solution vary a lot and depends on many factors. It can be considered in a simple web site, but also in a complex system with many integrated components. No matter of the scope, we can always distinguish several trust boundaries, i.e. distinct boundary within which a system trusts all sub-systems and including data. To simplify our further consideration w.l.o.g. let's define an example solution where we will apply presented concepts, the example being a typical business application which usually consists of the following components: C_B – a business logic, C_D – a database, C_A – an API (including integration capabilities), C_W – a web UI and a C_M – a mobile UI. Each of those distinguish their own trust boundary.

5.2 Types of Operations

We are going to follow a very common Command-Query Separation (CQS) principle where every operation is classified as a command (which is changing the state of the model) or a query (which is only reading). As we will see the classification (command vs. query) also reflects the hardness level in the authorization area.

5.3 Models

Each trust boundary or context may introduce its internal model (language) and it is very important that authorization rules are expressed in that model. Based on our example we can easily enumerate the following models:

- M_B. Business logic model (domain model),
- M_D. Database model,
- M_A. API (REST) public model,
- M_W. Web UI presentation model,
- M_M. Mobile application model.

Referring to the previous section we can quickly conclude that as, in many cases, authorization rules are being part of the business logic, they are also a part of the domain model.

6 Proposed Access Control Solution

In this section we provide a description on how to design an access control solution based on the XACML concept in a business application based on Domain-Driven Design. We consider every access control function (PDP, PEP, PIP, PAP) one by one and show how to implement it within the Domain-Driven Design and the hexagonal architecture mindset.

Before we go into description let's examine a common set of business requirements towards access control in most of the business applications.

- R_0. Trusted identification and authentication process.
- R_1. Proper authorization rules reflecting business needs.
- R_2. Correct implementation of authorization rules.
- R_3. Completeness. All information is protected appropriately.
- R_4. Consistency. The same protection level of information in all contexts.
- R_5. Maintainability and adaptability, especially in the area of meeting new requirements and handling changes.
- R_6. Re-usability of the authorization rules description in different places.

According to our assumption, the R_0 requirement is out of scope of our consideration, but we assume the architecture is designed in a way to support that.

6.1 Policy Decision Point

PDP is the most challenging responsibility in supporting enumerated security requirements and as we have already previously concluded that authorization logic is scattered in different components, there are going to be more than on PDP in our solution. To handle the relationship between different PDPs let's introduce the concept of *master PDP* and *shadow PDP* as well as a PDP hierarchy.

Definition 2. *Let M_C be the core model, A_{M_C} be the complete set of authorization rules expressed in the model M_C and $M_1, \ldots, M_j$ be other models where we need some part of authorization decisions. Let a PDP based on A_{M_C} be the master PDP. Then, in order to apply the correct protection in models $M_1, \ldots, M_j$ we need to define projections $\Pi_1, \ldots, \Pi_j$ (to scope the needed protection) and transformation functions $T_1, \ldots, T_j$ which are making transformations A_{M_C} into authorization rules expressed in the target model. Combining all together we get the set of authorization rules $T_1(\Pi_1(A_{M_C})), \ldots, T_j(\Pi_j(A_{M_C}))$. Let PDPs based on transformed authorization rules be shadow PDPs. The structure constitutes a tree with levels. Ideally should have max. 2 levels, but for obvious reasons this is not always possible. An example tree is presented on the Fig. 2.*

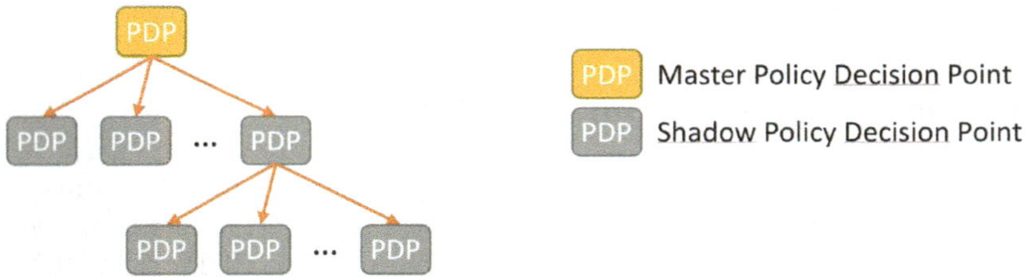

Fig. 2. Hierarchy of policy decision points

Applying the above to our example we obviously conclude that M_B is the core model and by that *master PDP* is based on A_{M_B}. All other models (M_D, M_A, M_W, M_M) are non-core models and PDPs based on them – shadow ones.

Let's now review how PDP is implemented in different components.

- C_B with M_B and A_{M_B}. As master PDP is here, A_{M_B} is the main source of authorization rules. From a design perspective we strongly recommend creating appropriate domain services to capture authorization rules, moreover, if relevant, part of that logic can be moved to the respective entities and value objects, according to the Domain-Driven Design principles.
- C_D with M_D and A_{M_D}. PDP is reflected in different ways here. **(1)** Appropriate authorization on database objects, it can be more strict approach where one can find users in a database corresponding to application users, including different levels of permissions' granularity, however, most often it is less

strict approach where authorizations are defined based on a single database account. **(2)** Authorizations expressed in parts of queries making sure that a user can see only relevant data. Again, the whole setup can expressed as an appropriate projection Π_D and model translation T_D applied on A_{M_B}. Moreover, part (2) can be supported by ORM system and the right implementation of *specification pattern*.

- C_A with M_A and A_{M_A}. Scope for PDP may differ a lot. From having $\Pi_A(A_{M_B}) = \emptyset$ where all requests are passed further to having projection quite extensive, however limited by the scope of public model M_A offered by API. A common solution is that translation is executed "on the fly" by sending respective authorization requests from C_A directly to A_{M_B}.
- C_W with M_W and A_{M_W}. As C_W very often is placed in an untrusted domain, usually the purpose of applying authorization rules is an adjustment of presentation layer to reflect the actual permissions of a user and by that to increase the user experience, so projection Π_W and translation T_W are to support implementation of showing/enabling and hiding/disabling appropriate functions, presenting correct messages, etc.
- C_M with M_M and A_{M_M}. Partly the same applies as for the component C_W, however as mobile OS usually offers trusted enclaves, there is a possibility to execute actual access control, so appropriate authorization rules need to be hired here what needs to be reflected in projections and translations.

Let's revisit how the above approach meet the business requirements stated as R_1–R_6. Starting from putting the domain model in the center we strongly support R_1. Doing correctly projections and transformations, we meet R_1–R_4. The best approach would be to have patterns and libraries which are providing all P_i and T_i automated, then we would contribute to R_5 and R_6. In addition, by setting A_{M_B} as a single master PDP, we additionally contribute to R_6.

Now let's consider the selected consequences of not following the proposed approach.

- Multiple master PDPs with edge case where there is a master PDP per model or trust boundary. As master PDPs are based on requirements, it is very likely that **(a)** different teams will understand requirements in different ways, **(b)** over the time changes applied in one master PDP will not be populated to other master PDPs. As a consequence R_1, R_2 and R_4 are affected.
- Master PDP outside the M_B. There is a quite common approach to consider security as a part of the cross-cutting component (CCC) within the software design. Usually it results in putting all authentication and authorization related matters in this component. Even though apparently R_6 is supported, there are many ways how things may go wrong.
  - Simplifying the authorization logic to accommodate the lack of access to the business objects. Then at least R_1 and R_3 are affected.
  - Designers are trying to meet the business requirements. However, then there are the following risks. **(1)** Lifecycles of domain model and CCC might be different, so changes in domain model might not be reflected in CCC what affects R_4. **(2)** Let's recall an example of the relationship

between an application and a database. It is a common pattern that filtering and other detailing in queries in the object model is translated down to the data source (because otherwise in most of the cases the performance would be not acceptable). In most of the complex systems queries are rather complex and they are being constructed in many places of the application as well as by many developers. Part of those queries is authorization logic and an attempt to extract those in order to put it in a separate, loose-coupled component will (most likely) fail, if not at the beginning, it is almost certain that after applying several change requests. If it fails, then R_2 is affected, and in the consequence, all other requirements as well. **(3)** Due to the complexity of referencing business objects (including details which potentially shouldn't be exposed externally) from an external component, R_5 is affected.

To sum up, the proposed PDP setup is supporting all the stated requirements, moreover not following the proposed setup leads to risks.

6.2 Policy Enforcement Point

Enforcements points are very much dependent on trust boundaries and the recommended approach is quite straightforward: each component where we want to execute authorization rules to protect data requires PEP. Referring to our example it means that PEP is required in C_B, C_D, C_A, potentially in C_M depending on the requirements and application features. By the correct trust modeling we can contribute to R_3 and enable R_1, R_2 and R_4.

Obviously, if we don't follow the recommendation, authorization logic is not executed and the whole access control collapses.

6.3 Policy Information Point

An information point is to support PDP with the appropriate data to take the right decisions. By that, we need to secure the right data transformations and projections to ensure they are aligned with PDPs' authorization rules projections and transformations. However, if we combine PDP and PIP in a good way, we contribute to $R_1 - R_4$.

6.4 Policy Administration Point

As stated in the Sect. 3 introduction responsibility of PAP is out of scope of this paper. Nevertheless it is worth to mention that even if at first glance PAP seems to be quite straightforward, it complicates very quickly, especially when we consider: many different platforms and COTS/SaaS based solutions, roles descriptions (when we want to have roles on a company or even portfolio level), automated provisioning processes, temptation to move parts of PDP to PAP.

6.5 Complete Solution

Combining all the functions described in the previous section we get a complete solution. We can see that applying Domain-Driven Design thinking combined with right trust modeling and automation, increases the probability of delivering all the expected requirements. Nevertheless we would like to emphasize several aspects. First, as we discussed previously there is strong push towards defining security as a separate concern from the business logic. However, as we have seen already, this is a very risky approach and not according to the Domain-Driven Design philosophy. Nevertheless when designing a system the right balance of centralization and decentralization must be taken very consciously.

Another aspect is that in literature (including famous Vaugn book [17]) very often Identity and Access Management bounded context is introduced as a separate one. That might be considered as a way for centralization, but we must be very careful again what we put there. We recommend to consider that context more as a PIP and PAP, but much less (if at all) as PDP. For sure it can't be considered as PEP for obvious reasons.

Finally, the architecture diagram with all the policy points applied may look as on the Fig. 3.

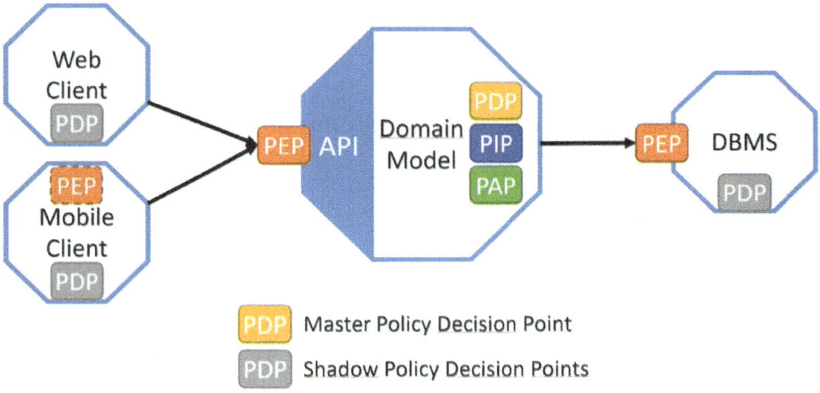

Fig. 3. Proposed access control solution

7 Conclusions

In the paper we proposed a solution for designing software architecture with the access control based on XACML and Domain Driven Design. We showed that many authorization rules are indeed a part of the business logic and investigated consequences of that fact. We reviewed several common requirements and architectural building blocks influencing and driving the access control function and analyzed how those can be supported by the proposed approach.

References

1. Basin, D., Clavel, M., Egea, M.: A decade of model-driven security. In: Proceedings of the 16th ACM Symposium on Access Control Models and Technologies, pp. 1–10. ACM, June 2011
2. Cockburn, A.: Hexagonal Architecture: Ports and Adapters ("Object Structural"), 19 June 2008
3. Cysneiros, L.M., do Prado Leite, J.C.S.: Non-functional requirements: from elicitation to modelling languages. In: Proceedings of the 24th International Conference on Software Engineering, pp. 699–700. ACM, May 2002
4. Evans, E.: Domain-Driven Design: Tackling Complexity in the Heart of Software. Addison-Wesley Professional, Boston (2004)
5. ISO 7498-2:1989. https://www.iso.org/standard/14256.html. Accessed 24 Mar 2019
6. ISO/IEC 27000:2018. https://www.iso.org/standard/73906.html. Accessed 24 Mar 2019
7. Jiang, H., Bouabdallah, A.: A Lightweight JSON-based Access Control Policy Evaluation Framework (2017)
8. Jurjens, J.: Sound methods and effective tools for model-based security engineering with UML. In: 2005 Proceedings of the 27th International Conference on Software Engineering. ICSE 2005, pp. 322–331. IEEE, May 2005
9. Kleppe, A.G., Warmer, J., Bast, W.: MDA Explained: The Model Driven Architecture: Practice and Promise. Addison-Wesley, Boston (2003)
10. Lobo, J., Bhatia, R., Naqvi, S.: A policy description language. In: AAAI/IAAI 1999, pp. 291–298 (1999)
11. Lucio, L., Zhang, Q., Nguyen, P.H., Amrani, M., Klein, J., Vangheluwe, H., Le Traon, Y.: Advances in model-driven security. In: Advances in Computers, vol. 93, pp. 103–152. Elsevier (2014)
12. Margheri, A., Masi, M., Pugliese, R., Tiezzi, F.: A rigorous framework for specification, analysis and enforcement of access control policies. IEEE Trans. Softw. Eng. **45**, 2–33 (2017)
13. Nguyen, P.H., Klein, J., Le Traon, Y., Kramer, M.E.: A systematic review of model-driven security. In: 2013 20th Asia-Pacific Software Engineering Conference (APSEC), vol. 1, pp. 432–441. IEEE, December 2013
14. Nguyen, P.H., Kramer, M., Klein, J., Le Traon, Y.: An extensive systematic review on the model-driven development of secure systems. Inf. Softw. Technol. **68**, 62–81 (2015)
15. Schmidt, D.C.: Model-driven engineering. Comput.-IEEE Comput. Soc. **39**(2), 25 (2006)
16. Uzunov, A.V., Fernandez, E.B., Falkner, K.: Security solution frames and security patterns for authorization in distributed, collaborative systems. Comput. Secur. **55**, 193–234 (2015)
17. Vernon, V.: Implementing Domain-Driven Design. Addison-Wesley, Boston (2013)
18. OASIS XACML Technical Committee: "eXtensible access control markup language (XACML) Version 3.0. Oasis Standard, OASIS (2013). http://docs.oasis-open.org/xacml/3.0/xacml-3.0-core-specos-en.html. Accessed 24 Mar 2019

Large Scale Attack on Gravatars from Stack Overflow

Przemysław Rodwald(✉)

Polish Naval Academy, ul. Śmidowicza 69, 81-127 Gdynia, Poland
p.rodwald@amw.gdynia.pl

Abstract. Stack Overflow is a globally recognizable service features questions and answers on a wide range of topics in computer programming. Even though stackoverflow is not an anonymous service, users posting comments hope to not reveal personal email. Email, which according to EU laws is consider as a personal information. Unfortunately, emails are possible to recover, because stackoverflow provides them in the obfuscated form in source code. This article explains all stages of the real large-scale attack on Stack Overflow user emails. Those stages could be easily adopted to any website with comments system based on the Gravatar service. In our attack we crawled and were able to recover more than (1,25 M) 20% real emails of the stackoverflow users.

Keywords: Gravatar · Deanonymization · Computer security · MD5 hash function

1 Introduction

Stack Overflow is a service created in 2008 and available on the website under www.stackoverflow.com. It is one of the most popular platform for computer programmers, where one can ask and answer questions. Behind the popularity of this service state some statistics: as of June 2018 Stack Overflow has over 8.6 million users [1] and 9.3 million visits per day [2]. In mid 2018 it exceeded 16 million questions.

Gravatar, a webservice available at www.gravatar.com, provides globally unique avatars. Avatar is an individual image uploaded by a user and linked with his email address. Those avatars are provided in obfuscated form as a hash calculated by the cryptographic hash function. A cryptographic hash function could be defined as an algorithm that maps data of arbitrary size to a fixed size string of bits (called hash). A strong hash function must be designed in such a way to be a one-way function. The function which is easy to calculate in one direction but infeasible to invert. Gravatar as a unique identifier for user's emails uses MD5 hash algorithm, designed by Ronald Rivest in 1991 and produces a 128-bit hash value [3]. From cryptographic point of view MD5 hash is consider as a broken algorithm. In 2004 Wang et al. presented the first collision attack [4]. From this time real MD5 collisions for: X.509 certificates with different public keys [5], executable program files [6] pdf files [7], jpeg files [8] have been presented. So far, no real MD5 email addresses collision is known. Even though MD5 hash algorithm should be avoided and is unsuitable for further use [9] it is still quite popular and widely used. It is particularly popular among developers as a

W. Zamojski et al. (Eds.): DepCoS-RELCOMEX 2020, AISC 1173, pp. 503–512, 2020.
https://doi.org/10.1007/978-3-030-48256-5_49

password storage mechanism, even though it is obvious that they should stop using fast cryptographic hash functions (like MD5, SHA-1) and replace it by computation-hard or memory-hard functions (like for example bcrypt, ARGON2, Balloon) [10]. The reason does not lie in collision resistance vulnerability. Instead, the reason lies in the "speed" of MD5 hash function. A feature desirable for cryptographic hash algorithms is a nightmare in password storage or obfuscated email storage. Attacker can brute-force billions of hashes per second what makes it quite feasible. We prove this in our article.

The rest of this article the is organized as follows: in Sect. 2 existing literature on that topic is briefly reviewed; in Sect. 3 our attention is focused on two web-services: stackoverflow and gravatar from users' perspective. The attack methodology, which was divided into preparation phrase and attack itself is shown in Sects. 4 and 5. Next section introduces two different hardware platforms. Section 7 presents the results of our real attacks, where besides statistics, randomly chosen images (in obfuscated form) of re-identified users are presented as well. And finally, in the last section, conclusions are presented.

2 Current Work

The results of recovering email addresses from Gravatar hashes was previously demonstrated at least three times. First, in 2008, a user nicknamed Abell [11] crawled 80000 MD5 Gravatar hashes from Stack Overflow webpage and was able to recover 10% of email addresses. Second, in 2013, Bongard [12] acquired 2400 MD5 Gravatar hashes from French political blog Fdesouche and was able to recover 70% of the email addresses. Third, in 2019, Rodwald [13] showed how to prepare for the attack on Gravatar MD5 hashes and how to carry out it step by step. He showed results of a real attack on two polish language webservices and revealed 65% of real email addresses. Cited attacks indicate that a significant difference between attacks on global websites (like StackOverflow.com, GitHub.com) with millions of users around the globe and national-base websites (e.g.: fdesouche.com, jakoszczedzacpieniadze.pl) where users are mainly from one country. In this article we adopt a small-scale approach for the large-scale attack.

3 Web-Services Description

StackOverflow from user's point of view works as follows. A user creates an account on stackoverflow.com website. One can use existing Google or Facebook account or create a new account with email address. Choosing Google/Facebook login option a user's picture is provided by those services. Choosing email login option and after verifying an email address user can log in and setup his/her picture. There are three options: Identicon, Gravatar or upload a new picture. Identicon is a geometric pattern based on an email hash. Gravatar is a picture provided by a Gravatar webservice. Both options use Gravatar webservice as a source of a picture. Only third option, upload a new picture, is not linked with Gravatar webservice.

Gravatar is a Globally Recognized Avatar, as stated on the gravatar.com website. An avatar is an image that represents a user online. To be more precisely, it is a little picture that appears next to the user's email address (or user name or user nickname) when a user interacts with websites by adding comments. From the user's perspective Gravatar works in the following way: a user creates an account on gravatar.com website; then uploads an image (avatar) and assigns it to his own email address. From this moment this avatar will be visible on all websites that use Gravatar.

From developer's perspective to activate Gravatar on the website, that is to retrieve avatars of users, the web service must make a request to gravatar.com webservice. This request is based on HTTP GET request, so no authentication is required. The website must first generate the MD5 hash of the user's email address (HASH) and then request for the avatar using a particular URL address in the form: https://www.gravatar.com/avatar/HASH. For example, if the user's email is p.rodwald@amw.gdynia.pl, then the request looks like https://www.gravatar.com/avatar/0aec9b599eeb18ae640234f62683104b. Developers have some flexibility to adjust gravatars of users. They might configure a few parameters such as: size in pixels (e.g. s = 96), rating (e.g. r = PG) or default image (e.g. d = identicon). Such settings are used by Stack Overflow.

Let us focus the attention for a moment on the theoretical question, if gravatar is a bijection (one to one function). From the mathematical point of view, one can compute the number of possible email addresses. An email address consists of three elements: username, @ (at sign) and domain. According to RFC's [14, 15] username has a maximum length of up to 64 characters, and domain of up to 253 characters. Assuming (which is not fully true, according to standards but is not significantly important) that username and domain are composed only of lower case Latin letters, digits, and '.', '-', '_', one could roughly estimate the total possible number of email addresses as $39^{64} \times 39^{253} \approx 2^{1676}$ distinct values (let's mark it as e). From the cryptographic point of view, one knows that the length of MD5 hash function is 128 bits, what gives us 2^{128} possible hash values (marked as g). One can notice that $e \gg g$ what could lead us to a theoretical conclusion that Gravatar is a many to one function. However, assuming that all email addresses are in use and are evenly distributed - which in fact is not valid. Firstly, the domain part in the real world should not be approximated to 39^{253} but rather to only 350 million registered domains [16]. Secondly the estimated number of email accounts in 2018 is only 6690 million [17], calculated as Worldwide Email Users (3823 million) $\times$ the average number of email accounts per user ratio (1.75). It could be roughly compared to the number of people alive today ($1.1 \times 6.5 \times 10^9$). In fact, the number of real email addresses is smaller than 2^{33}, and as a consequence $2^{33} \ll g$, which leads us to the conclusion of one-to-one mapping between email addresses and gravatar addresses. This conclusion leads us to a uselessness of collision attack on gravatar MD5 hashes. Only preimage attack is in the area of our interest in this article.

It is worth mentioning that gravatar is not the only one service that offering similar functionality in the Internet. A few examples are presented in Table 1.

Table 1. Sample webservices offering avatars.

Website	Algorithm	Sample link	Sample image
gravatar.com	MD5	gravatar.com/avatar/ 547d20f2c04a3dc4838aae94b1ff06e1	
evatar.io	SHA256	evatar.io/ 8d5218972a45ba5309db7d70f3373d4b fdbae040a336df4091cab8b66f642149	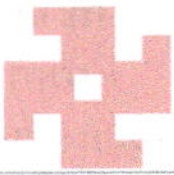
dicebear.com	plaintext	avatars.dicebear.com/v2/male/art@wp.pl.svg	
robohash.org	plaintext	robohash.org/art@wp.pl	

Some of the services (row 3 and 4 in Table 1) offer only the service of generating a certain personalized graphic (avatar) for a given input string (not only email). Another services (row 1 and 2), in addition to generating a graphic element, allow also to assign own graphic (photo) to an email address, and call an avatar by providing hashed email (MD5 or SHA-256).

4 Preparation to the Attack

As one of the first steps in this survey, an analysis of known global websites was done, sites which use gravatar as a service to present users' avatars. As a result, we decided to choose Stack Overflow as source of Gravatars for two reasons: the number of Gravatar users (6M+ in the first quarter 2019) and geographical distribution of users. The preparation phase was divided into 3 parts.

4.1 Part 1 – Crawl (Target Identification)

The structure of the HTML code of the Stack Overflow website has been analyzed. As a list of users, the subpages https://stackoverflow.com/users?page=pageid were crawled, where: $1 \leq pageid \leq 264200$. At the time of crawling time (January 2019) a number of users is estimated to the value 9.511.200 (264.200 [quantity of sub-pages] $\times$ 36 [users per subpage]). A dedicated web-crawler was implemented. The aim of this crawler was searching all mentioned subpages with users available on the Stack Overflow website and then extracting nicknames (file *nicknames.txt*) provided by users with Gravatar MD5 hash. As a result, our webcrawler has extracted 6016434 unique MD5 Gravatar hashes of users' emails along with additional data (id, nickname, location). All data were stored in the MySQL database. Stackoverflow is protected by a security mechanism which forbid to many requests per particular period of time. Because of this mechanism our web-crawler must have been slow down. This slow-down has been used to check how many users use individual picture in the Gravatar account (not a standard picture provided by the Gravatar service). To show that some users use they real photos, a picture constructed from 40 randomly chosen Gravatar images was prepared and is presented as Fig. 1.

Fig. 1. Randomly chosen Gravatar images (obfuscated form).

As a first result of this survey we identified that only 8,74% of Stack Overflow users has individual picture in the Gravatar account.

4.2 Part 2 – Stats of User Localizations and Email Domains

The second step in preparation phase was identification of the most frequent user localizations. We identified (based on location results of webcrawler) that 8 the most popular user countries are: United States, India, United Kingdom, Germany, Canada, France, Russian Federation, Poland. Our results are similar to researches: [18, 19]. For each country statistics about the most popular email domains have been found in the Internet and saved in files *domains_xx.txt* where *xx* indicates the country. For example for Poland the following domains: @wp.pl, @poczta.onet.pl, @o2.pl, @interia.pl, @op.pl, @tlen.pl, @gmail.com, @poczta.fm, @gazeta.pl, @yahoo.com are the most popular free webmail providers and most popular ISP domains and those list was saved in *domains_pl.txt* file.

4.3 Part 3 – Username Patterns

As a third step a username patterns analysis has been done. Our analysis was based on the polish emails. But our results, presented in Table 2, could also be used for other nationalities. In the table the most popular notation is used, where: `?d` – any digit, `?l` – any letter, `?s` – one of the sign {, ., _}.

Table 2. Username patterns.

Pattern	Examples
`[lastname][?s][?d]{0,4}`	rodwald, rodwald_1990
`[firstname][?s][?d]{0,4}`	Paul, paul07
`[firstname][?s][lastname][?d]{0,4}`	paulrodwald, paul_rodwald2017
`[lastname][?s][firstname][?d]{0,4}`	rodwald.paul, rodwald_paul77
`[?l][?s][lastname][?d]{0,4}`	p.rodwald, p_rodwald33
`[lastname][?s][?l][?d]{0,4}`	rodwald_p, rodwald_p01

To prepare for the hybrid attack, all the national most popular domain files (*domains_xx.txt*) have been extended with the digit-based masks identified in this step (*domains_xx.hmask*). For example a content of the *domains_pl.hmask* file contains 50 rows and looks like: @wp.pl, …, @yahoo.com, ?d@wp.pl, …, ?d@yahoo.com, ?d?d@wp.pl, …, ?d?d@yahoo.com, … …, ?d?d?d?d@wp.pl, …, ?d?d?d?d@yahoo.com.

4.4 Part 4 – National Dictionaries

As a final step of preparation phase searching for lists of national names and national surnames for the eight countries identified in part 2 has been done. To prepare for the hybrid attack, all the patterns identified in Table 1 have been created and saved as *usernames_xx.txt* files. This process was preceded by merging, cleaning and removing duplicates from downloaded national dictionaries (i.e.: replacement of national letters, like: ą - a, ä – a, ß – ss, б – b, œ – oe; removing of unneeded chars, like: ‘, ’).

5 Attack Overview

After preparatory stages mentioned above, the attack itself could be carried out. The attack itself has been carry out in the *hashcat* software [20]. The attack is divided into three main approaches: dictionary attack based on leaked email addresses, hybrid attack and finally brute force attack.

5.1 Dictionary Attack

Dictionary attack, by definition, uses a precompiled list of words [21]. As the dictionaries we decided to use two sources of real leaked emails. The first one is known as a Exploit.in leakage. This source of emails could be identified and download by any torrent program. The unique ident of the file is C9D10AB5F3D7504978C5AFE2-CA7BD68FF131E9BF. Size of the zipped file is larger than 10 GB. The Exploit.in leakage has 805,499,391 rows of email address and plain text password pairs, but actually has 593,427,119 unique email addresses. The second one is more up to date source of emails, dated January 2019, called Collection #1-5 [22]. Size of the compressed files is larger than 870 GB.

5.2 Hybrid Attack

Hybrid attack combines a dictionary and mask attack by taking input from the dictionary and adding mask placeholders [21]. In our attack a dictionary is a list of usernames (i.e. *usernames_pl.txt*) or a list of nicknames (*nicknames.txt*) and masks are composed of digits and domains (i.e. *domains_pl.hmask*).

5.3 Brute Force Attack

Brute force attack, by definition, attempts every possible combination of a given character set up to a certain length. We decided to check all possible usernames from 1 up to 8 chars long. As a set of possible chars an expression `?l?d_.` was used. As a domain part all domains identified in Part 2 of preparation process were used. For example, for gmail.com domain the following eight batch hashcat commands were used:

hashcat.exe -a 3 -m 0-1 ?l?d._ "hash.txt" ?1@gmail.com
hashcat.exe -a 3 -m 0-1 ?l?d._ "hash.txt" ?1?1@gmail.com
hashcat.exe -a 3 -m 0-1 ?l?d._ "hash.txt" ?1?1?1@gmail.com
...

6 Hardware

Cracking time is strongly connected with attacker hardware capabilities. Author decided to present two real different hardware platforms, platforms which correspond to two types of attackers, marked as: a *GAMER* and a *MINER*.

The first one is a homemade attacker is a hacker with a typical gaming PC with good, modern Graphic Card. For the purpose of this article a benchmark of a platform with one overclocked MSI GeForce GTX 1060 6 GB was run. The result was 6063.7 MH/s. During a process of hash calculations an energy consumption was measured: 200 W.

Recent popularity of bitcoin and other cryptocurrencies have lead to a situation where private rigs with 6 or even 12 GPUs are not so rare. For the purpose of this research a rig (called hashkiller [23]) with six overclocked GPU's MSI GeForce GTX 1080 8 GB was used. The benchmark result was 100GH/s [24] and a measured energy consumption is about 1000 W. This particular platform was used in our attack and marked as MINER.

7 Results

With the power of the first dictionary (Expoit.in) 442270 (7,35%) Gravatar's emails were discovered. With the power of the second dictionaries (Collection #1-5) 885690 (14,72%) Gravatars' emails were revealed. 438890 emails were discovered by both sources. Total number of revealed emails by dictionary attack is 14,78% (889070 from 6016434).

The speed of dictionary attack is mainly dependent of files (dictionaries) processing, and in our case (hashcat, hashkiller) was between 500 kH/s and 1000 kH/s. So, the attack for Expoit.in dictionary took about 40 min, for Collection #1 – about 80 min. Much more time consuming was the process of extracting emails from files belonging to mentioned leakages.

The results of the hybrid attack are as follows: 406490 (6,8%) Gravatar's emails were discovered where as a dictionary the national dictionaries were used; 357744 (5,9%) when a list of nicknames was used; 169149 emails were revealed by both approaches. Total number of revealed emails by hybrid attack is 9,89% (595085 of 6016434). The whole hybrid attack took more than 12 days.

With the brute force approach, we were able to crack 539178 (8,96%) Gravatar's emails. The time of brute-forcing one domain took about 9 h. We have investigated 43 most popular national domains.

The summary of our attacks: 20,88% (1256315 of 6016434) of all Gravatar MD5 hashes were broken and emails revealed; 8,74% (526070 of 6016434) of emails have individual Gravatar picture; the effectiveness of individual attacks: dictionary - 14,78%, hybrid – 9,89%, brute force – 8,96%. The Venn diagram of all three types of attack is presented on Fig. 2.

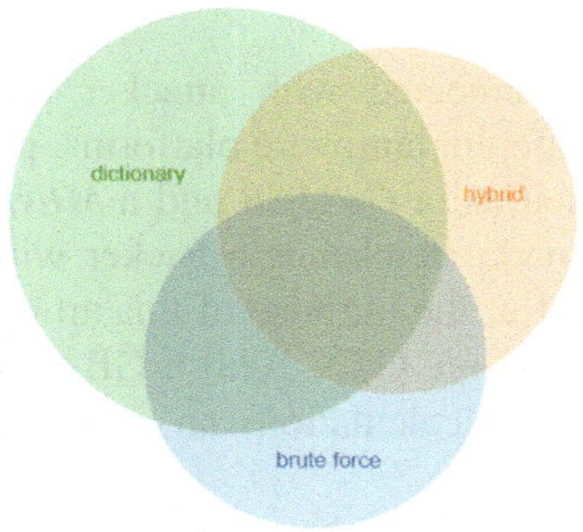

Fig. 2. Venn diagram for all types of attacks.

Randomly chosen e-mail addresses recovered from Gravatars on the website stackoverflow.com are presented, in anonymized form, on the dedicated website [25].

One could ask what kind of email addresses are hidden behind other Gravatar MD5 hashes – those not recovered. Some of the possible answers: emails from personal (rodwald.pl) domains; emails from regional (krakow.pl) domains; emails from company (amw.gdynia.pl) or government (mf.gov.pl) domains; emails from rarely used (t-net.com.pl) email providers; emails from mistyped domains (gmial.com); emails from anonymizing services (jetable.org); emails from disposable services (mailinator.com); randomly typed, fake emails (hflksfdssa@cjdsalf.oj) or more complex email formats.

8 Conclusions

The primary contribution of this work is a providing a step-by-step procedure of recovering user's email addresses from global website which use Gravatar. Authors describe every single step in such a way that an attack could be adopted to any Gravatar based website. Authors estimate cost and time effectiveness for two different hardware platform. Finally, based on presented approach, authors recovered 1.25 million real email addresses (20.88% of all MD5 Gravatar hashes).

It is worth pointing out that in many jurisdictions an email address is consider as a sensitive private data. For example: in the USA the National Institute of Standards and Technology considers an email address to be Personally Identifiable Information [26]; in European Union an email address such as name.surname@domain is an examples of personal data. Such a data in many circumstances must be anonymized. For data to be truly anonymized, the anonymization must be irreversible [27]. We prove in this article that the usage of MD5 hash function as an anonymization technique for email addresses is not always irreversible.

A usage of gravatar in places where deanonymization could be a problem must be stop. For example, Disqus (a worldwide blog comment hosting service for web sites) disabled the use of the Gravatar service after revealing security breach. And services like Gravatar (a few examples in Table 1) should stop using fast cryptographic hash functions (like MD5, SHA-1, and even SHA-2) and replace it at least by computation-hard or memory-hard functions like for example bcrypt, ARGON2, Balloon. This piece of advice, widely known for password protection [10], should be adopted to other domains where security has a significant role.

Acknowledgments. Author thanks Michał Raczyński, a student from Polish Naval Academy in Gdynia, for his help in searching, identifying, providing and cleaning of national dictionaries.

References

1. StackOverflow Users. https://stackoverflow.com/users. Accessed 01 June 2018
2. StackExchange.com. https://stackexchange.com/sites#traffic. Accessed 01 June 2018
3. Rivest, R.: The MD5 Message-Digest Algorithm. RFC 1321 (1992). https://tools.ietf.org/html/rfc1321. Accessed 23 Nov 2019
4. Wang, X., Yu, H.: How to Break MD5 and other hash functions. In: Cramer, R. (eds.) Advances in Cryptology – EUROCRYPT 2005. Lecture Notes in Computer Science, vol. 3494, pp. 19–35 (2005). https://doi.org/10.1007/11426639_2
5. Lenstra, A., Wang, X., de Weger, B.: Colliding X.509 Certificates. Cryptology ePrint Archive Report 2005/067 (2005). https://eprint.iacr.org/2005/067. Accessed 23 Nov 2019
6. Selinger, P.: MD5 Collision Demo (2006). https://mathstat.dal.ca/~selinger/md5collision/. Accessed 23 Nov 2019
7. Stevens, M., Lenstra, A., de Weger, B.: Predicting the winner of the 2008 US Presidential Elections using a Sony PlayStation 3 (2007). https://www.win.tue.nl/hashclash/Nostradamus/. Accessed 23 Nov 2019
8. McHugh, N.: Create your own MD5 collisions (2015). https://natmchugh.blogspot.com/2015/02/create-your-own-md5-collisions.html. Accessed 23 Nov 2019

9. Dougherty, C.R.: Vulnerability Note VU#836068 MD5 vulnerable to collision attacks. Vulnerability notes database. CERT Carnegie Mellon University Software Engineering Institute (2008). https://www.kb.cert.org/vuls/id/836068. Accessed 23 Nov 2019
10. Rodwald, P., Biernacik, B.: Password protection in IT systems. Bull. Mil. Univ. Technol. **67** (1), 73–92 (2018). https://doi.org/10.5604/01.3001.0011.8036
11. Abell. Gravatars: why publishing your email's hash is not a good idea (2009). http://www.developer.it/post/gravatars-why-publishing-your-email-s-hash-is-not-a-good-idea. Accessed 23 Nov 2019
12. Bongard, B.: De-anonymizing Users of French Political Forums. Technical report, 0xcite LLC, Luxembourg (2013). http://archive.hack.lu/2013/dbongard_hacklu_2013.pdf. Accessed 23 Nov 2019
13. Rodwald, P.: E-mail recovery from websites using Gravatar. Bull. Mil. Univ. Technol. **68**(2), 59–70 (2019). https://doi.org/10.5604/01.3001.0013.3003
14. Yao, J., Mao, W.: RFC 6531 - SMTP Extension for Internationalized Email, IETF (2012). http://www.ietf.org/rfc/rfc6531.txt. Accessed 23 Nov 2019
15. Mockapetris, P.: RFC 1035 - Domain Names - Implementation and Specifications, IEFT (1987). http://www.ietf.org/rfc/rfc1035.txt. Accessed 23 Nov 2019
16. Verisign: The Domain Name Industry Brief. **16**(1) (2019). https://www.verisign.com/assets/domain-name-report-Q42018.pdf. Accessed 23 Nov 2019
17. The Radicati Group: Email Statistics Report 2018-2022 (2018). www.radicati.com/wp/wp-content/uploads/2017/12/Email-Statistics-Report-2018-2022-Executive-Summary.pdf. Accessed 23 Nov 2019
18. StackOverflow: Developer Survey Results 2019 (2019). https://insights.stackoverflow.com/survey/2019/. Accessed 23 Nov 2019
19. Robinson, D.: A Tale of Two Industries: How Programming Languages Differ Between Wealthy and Developing Countries (2017). https://stackoverflow.blog/2017/08/29/tale-two-industries-programming-languages-differ-wealthy-developing-countries/. Accessed 23 Nov 2019
20. Hashcat. https://hashcat.net/hashcat/. Accessed 23 Nov 2019
21. Picolet, J.: Hash Crack: Password Cracking Manual v3 (2019)
22. RAID Forum (2019). https://raidforums.com/Thread-Collection-1-5-Zabagur-AntiPublic-Latest-120GB-1TB-TOTAL-Leaked-Download. Accessed 23 Nov 2019
23. Rodwald, P.: Hashkiller 1080 – hardware spec (2017). https://www.rodwald.pl/blog/1156/. Accessed 23 Nov 2019
24. Rodwald P.: Hashkiller 1080 benchmark (2017). https://www.rodwald.pl/blog/1161/. Accessed 23 Nov 2019
25. Rodwald, P.: Randomly chosen e-mail addresses recovered from Gravatars on the website stackoverflow.com. https://www.rodwald.pl/blog/1201/. Accessed 23 Nov 2019
26. McCallister, E., Grance, T., Scarfone, K.: Guide to Protecting the Confidentiality of Personally Identifiable Information (PII), NIST Special Publication (NIST SP) - 800-122 (2010). https://csrc.nist.gov/publications/detail/sp/800-122/final. Accessed 23 Nov 2019
27. European Commission: What is personal data? https://ec.europa.eu/info/law/law-topic/data-protection/reform/what-personal-data_en. Accessed 23 Nov 2019

Safety Analysis for the Operation Process of Electronic Systems Used Within the Mobile Critical Infrastructure in the Case of Strong Electromagnetic Pulse Impact

Adam Rosiński(✉), Jacek Paś, Jarosław Łukasiak, and Marek Szulim

Faculty of Electronics, Military University of Technology, Gen. Sywestra Kaliskiego 2, 00-908 Warsaw, Poland
adam.rosinski@wat.edu.pl

Abstract. The concept of electronic equipment and systems operation means all of the activities to be conducted by an operator after accepting a given technical object from a manufacturing plant and incorporating it into the operation process as per the intended use. The process of operating an electronic system shall be stable and the operating states as foreseen in the range of their permissible consequences – changes over the operation period. Electronic systems shall be maintained in constant readiness throughout an assumed period of time, especially in the case of technical objects used in buildings falling within the group of the so-called critical infrastructure (CI) structures. Electronic equipment and systems in CI structures operate in specific environmental conditions. However, their operating process shall also take into account extreme conditions of external and internal environmental interference, changes in the electro-climate and deliberately generated high-power electromagnetic fields. All activities aimed at maintaining the constant readiness of electronic equipment and systems are of technical and organizational nature. This elaboration discusses issues in the field of a reliability and operational analysis of electronic equipment and systems, which can be operated in an environment exposed to the impact of strong electromagnetic pulses. A relationship graph was developed for specific operating states.

Keywords: Electromagnetic interference · Model · Electronic system

1 Introduction

The impact of operation conditions (usage, maintenance and power supply) on damage intensity λ, electronic equipment and systems in the case of action from external and internal factors adverse to the process can be expressed with a damage intensity coefficient k_λ [6, 22, 23]. This coefficient indicates how much the actual damage intensity λ_{rz} in given environmental conditions (temperature, humidity, pressure, present electromagnetic interference) is higher compared to the damage intensity λ in laboratory or nominal conditions – Fig. 1 [4, 20, 25]. This can be expressed using Expression 1.

W. Zamojski et al. (Eds.): DepCoS-RELCOMEX 2020, AISC 1173, pp. 513–522, 2020.
https://doi.org/10.1007/978-3-030-48256-5_50

$$k_{\lambda} = \frac{\lambda_{rz}}{\lambda} \tag{1}$$

where: k_{λ} – damage intensity coefficient, λ_{rz} – actual damage intensity of electronic systems, elements or devices in given environmental conditions, λ – damage intensity in laboratory or nominal conditions.

Electronic devices and systems function in different, often extreme conditions [1, 8, 12]. One of the basic issues to be considered when using electronic equipment and systems involves the conditions of their proper functioning within a given electromagnetic environment [28–30]. It is an issue associated with ensuring compatibility [11, 15]. Due to the high saturation of CI structures with electronic equipment and systems, which execute all functions associated with management, operation and maintenance of the resources necessary for the functioning of a given organization [2, 26, 27], determining the strength, susceptibility and resistance to electromagnetic interference generated intentionally or unintentionally is a very important issue [5, 10, 21]. The commonly available literature sources lack data on the impact of strong electromagnetic pulses (EMP) on electronic systems or devices [1, 4, 8]. Some research papers and elaborations contain only data regarding the impact of EMP on electronic elements [1, 8]. The objective of the studies conducted by the authors is to determine

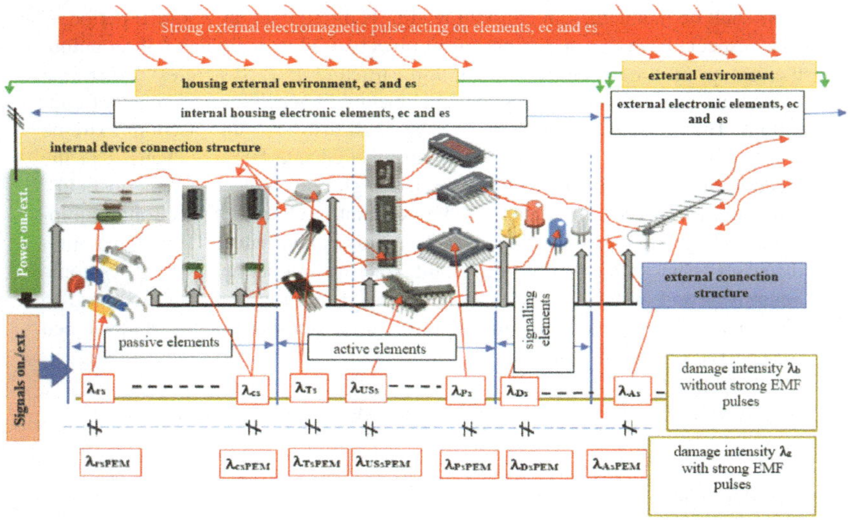

Fig. 1. Damage intensities λ in given environmental conditions and in the case of acting strong pulses of an electromagnetic field (EMF), Designations in the figure: ec – electronic circuits, es – electronic systems, $\lambda_{rs},\ldots,\lambda_{cs}$ – damage intensities λ for passive electronic elements, $\lambda_{Ts},\ldots,\lambda_{Ps}$ – damage intensities λ for active electronic elements, λ_{As} – damage intensities λ for external electronic elements coupled with an es, $\lambda_{rsPEM},\ldots,\lambda_{csPEM}$ – damage intensities λ for passive electronic elements in the case of EMF impact, $\lambda_{TsPEM},\ldots,\lambda_{PsPEM}$ – damage intensities λ for active electronic elements in the case of EMF impact, λ_{AsPEM} – damage intensities λ for external electronic elements coupled with an es in the case of EMF impact

the impact of EMP on electronic systems and devices. In the light of the EMP generator technical parameters and an electromagnetic wave propagation route, it is possible to work out guidelines and standards to determine the methods of protection against the effects of unintended pulse impact [11, 16, 18].

2 Reliability and Operational Modelling of Electronic Equipment and Systems Used Within the Critical Infrastructure, Taking into Account the Impact of Strong Electromagnetic Pulses

The development of reliability and operational models for electronic systems used within a CI in terms of their exposure to the impact of strong electromagnetic pulses shall take into account the technical parameters of such sources [17, 24]. Determining the guidelines concerning the values of indicators for a strong EMF impact on electronic elements, equipment and systems is a rather complex issue [4, 20, 21]. The analysis of the indicators regarding the impact on electronic elements, equipment and systems operated in a CI shall involve all technical and organization aspects, as well as attack tactics – including wave propagation using strong EMF pulses – Fig. 2.

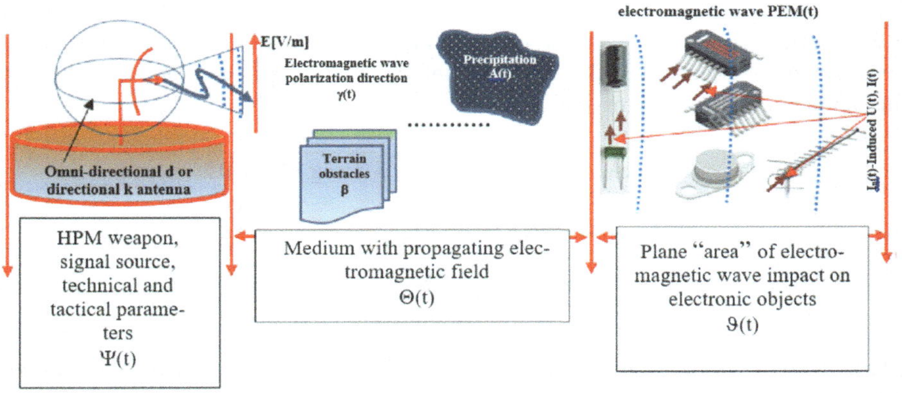

Fig. 2. Determining the parameters for an indicator characterizing strong EMF pulse impact on equipment and systems operated in CI structures

The parameters for an indicator characterizing strong EMF pulse impact on electronic elements, equipment and systems vary over the electromagnetic wave propagation period, and additionally are not subject to changes over a given pulse impact time – Fig. 2. The calculating indicators shall also include all technical parameters present within this system [6–8]. Due to the fact that strong EMF pulses use high frequencies, the conditions of the electromagnetic wave propagation "environment" shall also be taken into account – i.e., wave propagation path within a given medium. An electromagnetic wave reaching a given technical structure penetrates given electronic elements, equipment or systems, and is subject to dispersion, reflection, interference, etc. [7, 18, 24]. Electronic objects are constructed using various materials. Some of these materials constitute a peculiar shield, e.g. a metal housing of an active element or an entire device – e.g. field radio [10, 12, 16]. The determination of the indicator characterizing the impact of a strong electromagnetic pulse shall also take into account the shielding effect of such a housing. Also, the range of used frequency is decisive in terms of damage to electronic structures [7, 18, 24]. Metal connections, integrated circuit connectors and connection tips form specific antenna systems, which draw electromagnetic energy from the environment and, jointly with the "desirable" signal, direct it to a given electronic element, system or circuit. The dimensions of these leads or connectors (length, width, thickness), and their impedance for a given frequency range are crucial in terms of determining the amount of interfering energy generated at a certain distance from a device entering the desirable signal, which are intentionally generated by electronic structures and ensure the proper operation of electronic elements, equipment or systems. The key issues in terms of determining these indicators also include the properties of the very electromagnetic wave – i.e. polarization, interfering pulse duration, pulse processing time, pulse rise time, decay time, overshooting and oscillations, as well as their carrier frequency [4, 6, 7]. The aforementioned parameters affect the coverage of strong EMF pulses. The determination shall also take into account that the coverage of a "high-power pulse – called incapacitating" decreases exponentially to the increasing distance between signal sources and a structure under attack. In order to determine the damage intensity coefficient for a selected electronic element, in the case of strong electromagnetic field pulse impact, one should take into account the following indicators characterizing interference impact – expression 2.

actual damage intensity | strong EMF pulse impact indicators

$$\mathbf{k}_{\lambda E} = \frac{\lambda_{rz}}{\lambda} = \frac{\lambda_L \cdot \frac{\pi_{ST}}{\pi_{STL}} \cdot \pi_k \cdot \pi_A \cdot \pi_U \cdot \pi_R \cdot \therefore(t) \cdot T(t) \cdot M(t)}{\lambda} \tag{2}$$

where: $\Psi(t)$ – technical and tactical parameters of EMF pulses treated as an incapacitating signal source – such as: pulse power, duration, repetition period, pulse increment time, frequency, wave polarization, transmitting antenna gain, antenna characteristics, etc.; $\Theta(t)$ – using a medium (environment) for the wave propagation; $\vartheta(t)$ – Plane "area" of electromagnetic wave impact on electronic objects elements, equipment or systems; λ_L – laboratory damage intensity, $\pi_{STL} = 0.048$ for transistors; π_{ST} – coefficient value in laboratory load conditions; π_K – coefficient including the impact of environmental factors; π_A – coefficient taking into account the application; π_U – coefficient including power grid load, and also correcting the model relative to the electrical load included in π_{ST}; π_R – coefficient including the maximum permissible device parameters.

The following protection by electronic security systems will be utilized to ensure appropriate protection of a CI structures based on the conducted safety hazard analyses [10, 15, 19]:

- external – peripheral on long connections to the object [10],
- internal – using accessible electronic security systems and other measures, which can provide the required risk level – Fig. 3. From the point of view of reliability and operation, taking into account the impact of strong electromagnetic pulses, electronic equipment and systems used at CI structures can remain in the following safety states:
  - full safety (fitness) [13, 15],
  - safety hazard (partial fitness) [3, 9],
  - safety unreliability (unfitness) [3, 13, 14].

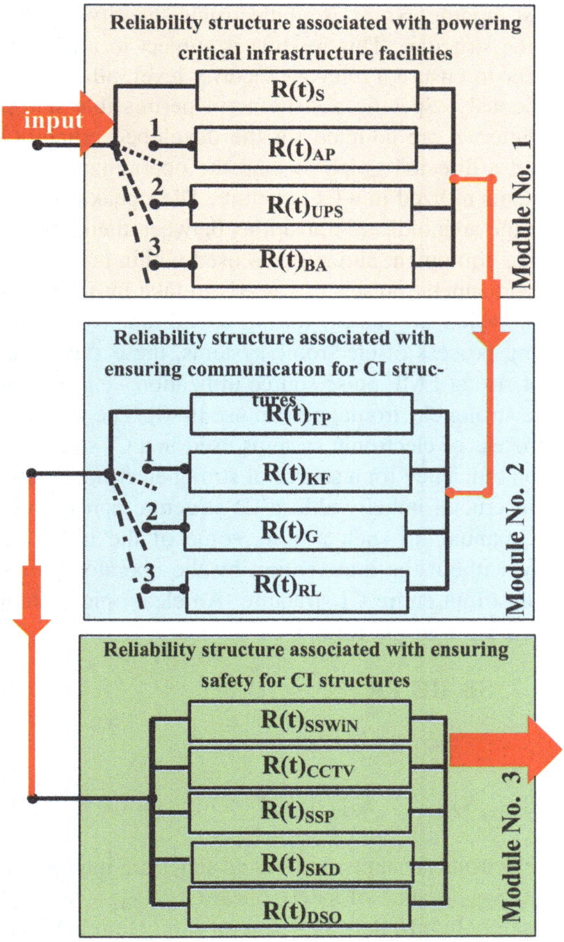

Fig. 3. Reliability structure for the CI safety system – modules M1 – M3

Table 1. Designations for Fig. 3

No.	Designation	Name in figure
1	R(t)S	mains power reliability
2	R(t)AP	power generator supply reliability
3	R(t)UPS	UPS power reliability
4	R(t)BA	battery bank power reliability
5	R(t)TP	hardwired network communication reliability
6	R(t)KF	wireless network communication reliability
7	R(t)G	reliability of communication using runners
8	R(t)RL	radio link network communication reliability
9	R(t)SSWiN	ESS reliability - the intrusion detection system
10	R(t)CCTV	ESS reliability - CCTV
11	R(t)SSP	ESS reliability - fire alarm system
12	R(t)SKD	ESS reliability - access control system
13	R(t)DSO	ESS - sound alarm system

ESS – electronic safety systems

The deliberations regarding the reliability and operational analysis of electronic equipment and systems used in IT buildings and technical facilities are abundant and concern various technical aspects [10, 12, 15]. They take into account different reliability structures, redundancy and the impact of electromagnetic interference. However, available publications do not contain an assessment of strong electromagnetic pulse impact and their effect on damage intensity λ. The reliability structure in terms of operating a field CI structure is shown in Fig. 3. Module 1 within a technical security system is responsible for the power supply of all devices [10, 19, 22]. Its reliability structure is parallel unloaded. Mains power damage results in subsequent, automatic switching onto other available back-up power types – item 1, 2, 3. Module 2 is responsible for the CI structure external and internal communication with managed entities, e.g. business. Damage to the hardwired communication systems results in a transition to other solutions (1, 2, 3), which ensure information exchange – parallel unloaded reliability structure. The last module, No. 3, is an electronic security system, which is responsible for protecting the structure. This module is subject to a parallel reliability– loaded – structure. In order to ensure a relevant security level, all systems must execute their function simultaneously. Specific transitions of permissible states, undergoing over a considered time interval are adopted for the developed reliability model. The authors of the paper used a directed graph to map the operating process model for electronic devices and systems utilized in a CI structure. Their peaks are the reliability and operating states, while the arcs indicate transitions between them. When considering the behaviour of electronic equipment and systems used within a structure and under the impact of strong electromagnetic pulses, one needs to take into account the presence of safety hazard states, which are absent during a "normal" operating process. In the course of the operating process of electronic systems, these states can appear depending on various cases of strong EMF pulse source utilization - e.g. partial or full coverage of a CI structure by a strong electromagnetic pulse [6–8]. The graph in Fig. 4, which shows the operating process of electronic systems used in a CI structure, also takes into account the η utilization efficiency for a source of strong electromagnetic field pulses. The execution of certain crucial items within a CI structure can be previously stipulated in the form of a manual. In such a case, some of the tasks are conducted manually or automatically without the intervention by the operator (s) responsible for the technical functioning of an entire CI structure. An electronic system operating process model is an ordered triple in the form:

$$\mathrm{M} = \langle \mathbf{SB}, \mathbf{RE}, \mathbf{FR} \rangle \tag{3}$$

where:

$$SB = \{\mathrm{S_{PZ}}, \mathrm{S_{ZB1}}, \mathrm{S_{ZB2}}, \ldots, \mathrm{S_{ZBn\text{-}1}}, \mathrm{S_Z}\} \tag{4}$$

SB is a set of operating states of electronic systems. Individual states are interpreted as follows: S_{PZ} – state of full fitness, S_{ZB1} – state of safety hazard 1, S_{ZB2} – state of safety hazard 2, …, $S_{ZBn\text{-}1}$ – state of safety hazard n-1, S_Z – state of safety unreliability. The second **RE** element of the ordered M triple is a set of pairs with elements interpreted as follows: (S_{PZ}, S_{ZB1}) indicates a possibility of a system transition from state

S_{PZ} to state S_{zB1} resulting from a "normal" wear process present in electronic equipment and systems, (S_{ZB1}, S_Z) indicates a possibility of a system transition from state S_{ZB1} to state S_Z resulting from the impact of a strong electromagnetic field source,…, (S_{ZB2}, S_Z) indicates a possibility of a system transition from state S_{ZB2} to state S_Z resulting from the impact of a strong EMF source. **FR** is a set of functions, each of which is determined based on the **RE** set and adopts values from a set of positive real numbers, i.e. R^+.

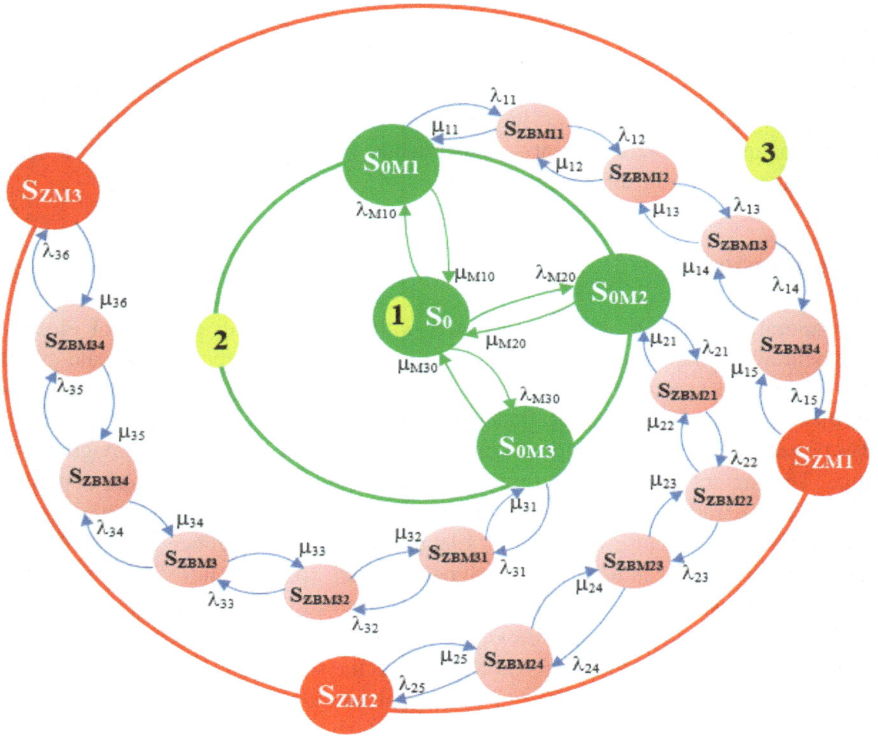

Fig. 4. Operating safety process graph for a CI structure – modules M1 - M3, where: S_0, S_{OM3}, S_{ZM3},… – model safety state, μ_{31}, μ_{M30}, …, μ_{11} – restoration intensity, λ_{11}, λ_{32}, …, λ_{M30} – damage intensity, 2, 3 – CI structure protection system potential circles, 2 – circle for the permissible safety decay resulting from "manual" control of a CI structure safety, 3 – CI structure safety unreliability potential circle, S_0 – full safety fitness state, λ_{M10}, λ_{M20}, λ_{M30} – intensity of damage to signal transmission devices – modules – alarm receiving centre, μ_{M10},μ_{M20},μ_{M30} – restoration intensity for signal transmission devices, circle No. 2 – permissible safety potential decrease within a system, resulting from damaged alarm signal transmissions, S_{ZM31}, S_{ZM32}, S_{ZM33}, S_{ZM34}, S_{ZM35} – safety hazard states for module No. 3, resulting from damage to successive electronic safety systems – IDS, CCTV, FAS, ACS and AWS, S_{ZM21}, S_{ZM22}, S_{ZM23}, S_{ZM24} – safety hazard states for module No. 2 resulting from damage to subsequent signal transmission subsystems – Table 1, S_{ZM11}, S_{ZM12}, S_{ZM13}, S_{ZM14} – safety hazard states for module No. 1 resulting from damage to subsequent power supply subsystems.

Unloaded back-up, where it is technically and financially feasible, is used in order to improve the reliability level of a CI structure protection system [19, 20, 30]. IT structures are also characterized by particular tactical and technical requirements – they must be deployed and removed within a short time. It is also very important to ensure continuity – reliability of the power supply system over a vast area [10, 13, 17]. Computer simulation and calculation of reliability and operating parameters for the graphs, as in Fig. 4, is a complex process. A CI safety system reliability structure can be treated as three independent M1 – M3 modules – Fig. 3. The reliability parameters of a CI system can then be determined successively, for each individual module, regardless of the initial point on circle No. 2 - Fig. 4. Then all M1 - M3 modules are fit, are in the states of S_{0M1}, S_{0M2}, S_{0M3}, respectively. Damage, e.g., to the M1 module means a transition from circle No. 2 to circle No. 3, hence, reaching a safety unreliability potential for the CI object. Individual M1-M3 modules of the CI system can be treated separately, as independent – not mutually burdened. Modules execute various tasks associated with ensuring object safety.

3 Conclusion

The elaboration presents issues associated with the operating process of electronic equipment and systems used within a CI structure. Electronic equipment and systems operated in a CI structure shall meet the external and internal electromagnetic compatibility requirements [4, 8, 24]. Satisfying the aforementioned requirements does not mean protecting the used structures against the impact of a strong electromagnetic field source. The impact of strong electromagnetic field pulses on a CI structure depends on the technical parameters of the interference source and the technical systems for the protection of the structure [5, 11, 16]. Electromagnetic interference reaches CI structures through direct, indirect and induced impact (power and signalling cables). A safety relationship graph was developed and conventionally divided into two circles. The first permissible decrease of the operating process safety values is associated with manually controlling modules 1-3. Circle No. 3 is the safety unreliability potential for a CI structure in the case of damage to individual modules 1-3.

Acknowledgments. The work was supported by the Polish National Centre for Research and Development within the project "Methods and ways of protection and defence against HPM impulses" pending within strategic project: "New weaponry and defense systems of directed energy".

References

1. Benford, J., Swegle, J.: High Power Microwaves. Taylor & Francis Group, New York (2007)
2. Burdzik, R., Konieczny, Ł., Figlus, T.: Concept of on-board comfort vibration monitoring system for vehicles. In: Mikulski, J. (ed.) Activities of Transport Telematics, pp. 418–425. Springer, Heidelberg (2013)

3. Caban, D., Walkowiak, T.: Dependability analysis of hierarchically composed system-of-systems. In: Zamojski, W., Mazurkiewicz, J., Sugier, J., Walkowiak, T., Kacprzyk, J. (eds.) Contemporary Complex Systems and Their Dependability. DepCoS-RELCOMEX 2018, pp. 113–120. Springer, Heidelberg (2018)
4. Charoy, A.: Interference in electronic devices. WNT, Warsaw (1999)
5. Chen, S., Ho, T., Mao, B.: Maintenance schedule optimisation for a railway power supply system. Int. J. Prod. Res. **51**(16), 4896–4910 (2013)
6. Chernikih, E.V., Didenko, A.N., Gorbachev, K.V.: High Power Microwave pulses generation from Vircator with inductive storage. In: EUROEM (Electronic Environments and Consequences), Bordeaux (1994)
7. Chmielińska, J., Kuchta, M., Kubacki, R., Dras, M., Wierny, K.: Selected methods of electronic equipment protection against electromagnetic weapon. Przegląd elektrotechniczny **1**, 1–8 (2016)
8. Dras, M., Kałuski, M., Szafrańska, M.: HPM pulses – disturbances and systems interaction – basic issues. Przegląd elektrotechniczny **11**, 11–14 (2015)
9. Duer, S., Zajkowski, K., Płocha, I., Duer, R.: Training of an artificial neural network in the diagnostic system of a technical object. Neural Comput. Appl. **22**(7), 1581–1590 (2013)
10. Dyduch, J., Paś, J., Rosiński, A.: The Basic of the Exploitation of Transport Electronic Systems. Publishing House of Radom University of Technology, Radom (2011)
11. Dziubinski, M., Drozd, A., Adamiec, M., Siemionek, E.: Electromagnetic interference in electrical systems of motor vehicles. In: Scientific Conference on Automotive Vehicles and Combustion Engines (KONMOT 2016), Book Series: IOP Conference Series-Materials Science and Engineering, vol. 148, pp. 1–11 (2016)
12. Dziula, P., Paś, J.: The impact of electromagnetic interferences on transport security system of certain reliability structure. In: 12th International Conference on Marine Navigation and Safety of Sea Transportation TransNav 2017, pp. 185–191. Gdynia, Poland (2017)
13. Jin, T.: Reliability Engineering and Service. Wiley, New York (2019)
14. Jodejko-Pietruczuk, A., Werbińska-Wojciechowska, S.: Analysis of maintenance models' parameters estimation for technical systems with delay time. Eksploatacja i Niezawodnosc – Maint. Reliab. **16**(2), 288–294 (2014)
15. Klimczak, T., Paś, J.: Selected issues of the reliability and operational assessment of a fire alarm system. Eksploatacja i Niezawodnosc – Maint. Reliab. **21**(4), 553–561 (2019)
16. Lheurette, E. (ed.): Metamaterials and Wave Control. ISTE/Wiley, London/Hoboken (2013)
17. Loeffler, C., Spears, E.: Uninterruptible power supply system. In: Hwaiyu Geng, P.E. (ed.) Data Center Handbook, pp. 495–521. Wiley, New York (2015)
18. Ogunsola, A., Mariscotti, A.: Electromagnetic Compatibility in Railways. Analysis and Management. Springer, Heidelberg (2013)
19. Paś, J., Rosiński, A., Wiśnios, M., Majda-Zdancewicz, E., Łukasiak, J.: Electronic Security Systems. Introduction to the Laboratory. Military University of Technology, Warsaw (2018)
20. Paś, J.: Shock a disposable time in electronic security systems. J. KONBiN **2**(38), 5–31 (2016)
21. Reddig, K., Dikunow, B., Krzykowska, K.: Proposal of big data route selection methods for autonomous vehicles. Internet Technol. Lett. **1**(36), 1–6 (2018)
22. Rosiński, A., Paś, J., Szulim, M., Lukasiak, J.: Determination of safety levels of electronic devices exposed to impact of strong electromagnetic pulses. In: Beer, M., Zio, E., (eds.) Proceedings of the 29th European Safety and Reliability Conference (ESREL), pp. 818–825. Research Publishing, Singapore (2019)

23. Siergiejczyk, M., Krzykowska, K., Rosiński, A.: Reliability assessment of integrated airport surface surveillance system. In: Zamojski, W., Mazurkiewicz, J., Sugier, J., Walkowiak, T., Kacprzyk, J. (eds.), Proceedings of the Tenth International Conference on Dependability and Complex Systems. DepCoS-RELCOMEX, pp. 435–443. Springer (2015)
24. Siergiejczyk, M., Paś, J., Rosiński, A.: Issue of reliability–exploitation evaluation of electronic transport systems used in the railway environment with consideration of electromagnetic interference. IET Intell. Trans. Syst. **10**(9), 587–593 (2016)
25. Skorupski, J., Uchroński, P.: A fuzzy reasoning system for evaluating the efficiency of cabin luggage screening at airports. Transp. Res. Part C Emerg. Technol. **54**, 157–175 (2015)
26. Stawowy, M., Kasprzyk, Z.: Identifying and simulation of status of an ICT system using rough sets. In: Zamojski, W., Mazurkiewicz, J., Sugier, J., Walkowiak, T., Kacprzyk, J. (eds), Proceedings of the Tenth International Conference on Dependability and Complex Systems. DepCoS-RELCOMEX 2015, pp. 477–484. Springer (2015)
27. Stawowy, M.: Comparison of uncertainty models of impact of teleinformation devices reliability on information quality. In: Nowakowski, T., Młyńczak, M., Jodejko-Pietruczuk, A., WerbińskaWojciechowska, S. (eds.), Safety and Reliability: Methodology and Applications - Proceedings of the European Safety and Reliability Conference ESREL 2014, pp. 2329–2333. CRC Press/Balkema, London (2015)
28. Suproniuk, M., Skibko, Z., Stachno, A.: Diagnostics of some parameters of electricity generated in wind farms. Przegląd Elektrotechniczny **95**(11), 105–108 (2019)
29. Weintrit, A.: Technical infrastructure to support seamless information exchange in e-Navigation. In: Mikulski, J. (ed.) Activities of Transport Telematics. CCIS, vol. 395, pp. 188–199. Springer, Heidelberg (2013)
30. Zajkowski, K., Rusica, I., Palkova, Z.: The use of CPC theory for energy description of two nonlinear receivers. In: MATEC Web of Conferences, vol. 178, pp. 1–6 (2018)

Job Scheduling with Machine Speeds for Password Cracking Using Hashtopolis

Jarosław Rudy[1](✉) and Przemysław Rodwald[2]

[1] Department of Computer Engineering, Wrocław University of Science and Technology, Wybrzeże Wyspiańskiego 27, 50-370 Wrocław, Poland
jaroslaw.rudy@pwr.edu.pl

[2] Department of Computer Science, Polish Naval Academy, Śmidowicza 69, 81-127 Gdynia, Poland
p.rodwald@amw.gdynia.pl

Abstract. Due to the current challenges in computer forensics and password cracking, a single GPU is no longer sufficient. Thus, distributed password cracking platforms with dozens of GPUs become a necessity in the race against criminals. In this paper, we show a multi-GPU cracking platform build on Hashcat-based open-source distributed tool Hashtopolis for use in password cracking and computer forensics. We present a mathematical model of the problem, formulating it as a specific case of the problem of scheduling independent jobs on parallel machines with machines speed and makespan criterion. We propose two metaheuristic algorithms based on the Simulated Annealing and Genetic Algorithm method to solve this problem. We employ the algorithms in a computer experiment using real-life password cracking instances and hash functions. The results indicate moderate (4–8%) to considerable (14–38%) improvement in makespan compared to default schedule for most instances. We also show that hash function type affects the improvement. SA and GA show little difference in quality, with SA being slightly better.

1 Introduction

A password, a secret string, still remains the widely used method for user authentication in the majority of IT systems. The methods of securing passwords stored in IT systems have evolved over the years [13], starting from storing passwords as plaintext, through ciphering passwords and to using cryptographic hash functions. The security of the last approach mainly depends on the hash function used. Fast cryptographic hash functions (like MD5 or SHA-1) offer considerably lesser protection than adaptive password algorithms: computation-hard ones (like bcrypt or PBKDF2) or memory-hard ones (like Argon2 or Balloon).

Password cracking is the process of recovering passwords (getting the plaintext) from hashes created through hash functions. The purposes of password cracking differ: from helping users to recover forgotten passwords, through gaining unauthorized access to a system to, in our case, retrieve some encrypted data

W. Zamojski et al. (Eds.): DepCoS-RELCOMEX 2020, AISC 1173, pp. 523–533, 2020.
https://doi.org/10.1007/978-3-030-48256-5_51

during criminal investigation. The perspective of our study is that of the digital forensic investigator or examiner. Law enforcement faces the task of accessing password protected information. Some cases concern full disk encryption, other file/directories encryption or gaining access to user account. All those examples have one thing in common: the need of password cracking.

This paper is a case study, presenting preliminary research on modeling a specific real-life problem of password cracking on a computer cluster through the use of theory of job scheduling. We propose two metaheuristic solving algorithms and report the obtained improvements in makespan for several real-life problem instances and various hash functions. As such, the results can be a reference for future research of problem properties and more advanced solving algorithms.

2 Password Cracking

In this section we present a brief overview of approaches to password cracking than can be found in the literature. In many papers the idea of distribution of cracking tasks is similar: the server divides the keyspace (a set of all password candidates) to a pre-defined number of chunks with each client receiving an equal chunk to solve. Pippin et al. proposed a parallel dictionary attack for cracking multiple hashes [11]. Bengtsson described a feasibility study in building a Linux-based high-performance computing (HPC) cluster and the development of a parallelized password cracker [4]. Apostal et al. implemented a divided dictionary algorithm and used it in HPC-based dictionary password cracking [3]. Marks et al. designed a hybrid CPU/GPU cluster formed by devices from different vendors (Intel, AMD, NVIDIA). Mentioned solutions work well for a homogenous cluster with a static set of nodes. For non-homogenous environments, Zonenberg created a distributed solution for cracking MD5 hashes using a brute-force attack [15]. Crumpacker came up with the idea of using BOINC [2] to distribute chunks, and implemented a tool for distributed cracking with Jack the Ripper (JtR) [8]. This approach was later improved by Hranicky et al.: first by using a password recovery application developed by his team and later by adopting Hashcat as a replacement of custom password cracking software [9].

Among open-source password-cracking tools Hashcat [1] dominates in: the number of supported hash formats, updates frequency, community support and, finally, in hash calculation speed. It supports both Windows and Linux and various hardware platforms (CPUs, GPUs, FPGAs). Because of this, we assume in our research Hashcat as a standalone agent. For task distribution among agents, we use Hashtopolis as the only well-known open-source solution: it is built using client-server model, allows to create and manage cracking tasks, handles benchmarking, divides keyspace in chunks and distributes them to agents.

In our study we adopted a cluster system made up of the following GPU-based platforms: Hashkiller 4 x GeForce 2080 Ti FE (2 PCs), Hashkiller 4 x GeForce 2080 (5 PCs), Hashkiller 6 x GeForce 1080 (1 PC) and PC 1 x GeForce 1060 (17 PCs). Time benchmarks (hash processing speed) for the cluster system and several hash functions is shown in Table 1.

Table 1. Benchmark of the platform for a few example hash functions (values are H/s)

Hardware platform name	quant.	Hash algorithm			
		MD5	SHA1	NTLM	bcrypt
Hashkiller 4x2080 Ti FE	2	2.19×10^{11}	7.15×10^{10}	3.70×10^{11}	1.13×10^{5}
Hashkiller 4x2080	5	1.63×10^{11}	5.22×10^{10}	2.66×10^{11}	7.77×10^{4}
Hashkiller 6x1080	1	6.45×10^{10}	2.91×10^{10}	9.29×10^{10}	7.08×10^{4}
PC 1x1060	17	1.0×10^{10}	3.83×10^{9}	1.72×10^{10}	6.90×10^{3}
Total	25	1.49×10^{12}	4.98×10^{11}	2.46×10^{12}	8.02×10^{5}

Various methods ("attacks") of password cracking were reviewed in many papers, for example see [10, 12]. Dictionary, brute-force, rule, and mask attacks are the most popular techniques. A dictionary attack, known as a wordlist attack, uses a precompiled list of words (file of common or known passwords) to attempt to match a password. Rule attack generates permutations against a given dictionary by trimming, extending, expanding, reversing, lower/upper casing, etc. Brute-force attack, called exhaustive key search, tries every possible combination of a given keyspace or character set for a given length. Finally, mask attack is a form of brute-force attack which uses placeholders for characters in certain position (i.e. `?u?l?l?l?l?d?d`). The particular symbol pairs[1] mean:

`?l` – lower-case letters ([a-z] regex),
`?u` – upper-case letters ([A-Z] regex),
`?d` – digits ([0-9] regex),
`?s` – special characters (!"#$%&'()*+,-./:;<=>?@[\]^_'{}~| and space),
`?a` – all character sets `?l?u?d?s`.

In this work, attention is focused on the search for the optimal strategy for breaking a single password with the power of mask attack, assuming that we do not have any knowledge about the structure of the password itself (length, complexity). In Table 2 an estimated time for cracking sample masks with our Hashtopolis-based computer cluster is presented.

Hashcat community provide sets of popular masks. A Hashcat mask file (*.hcmask) is a set of masks stored in a single plain text file. Masks are calculated and sorted based on the passwords coming from large data breaches (i.e. RockYou).

3 Problem Formulation

Job scheduling is a broad subfield of discrete optimization with various applications. It had been used to model various industrial processes ranging from

[1] Supported in various password-cracking software (Hashcat, JtR).

Table 2. Time of mask attack for two hash functions MD5 and SHA1

Mask	Number of hashes	Time to crack the mask	
		MD5	SHA1
?l?l?l?l?l?l?l?l?l	1.41×10^{14}	1 m 35 s	4 m 43 s
?l?l?l?l?l?l?l?l?l?l	3.67×10^{15}	41 m 06 s	2 h 02 m 51 s
?a?a?a?a?a?a?a?a	6.63×10^{15}	1 h 14 m 18 s	3 h 42 m 03 s
?a?a?a?a?a?a?a?a?a	6.30×10^{17}	4 d 21 h 38 m 18 s	14d 15 h 34 m 06 s

manufacturing [5] and transport [6] to project management [7] and software testing [14]. In this section we will formulate the problem of password retrieval as a problem of job scheduling on parallel machines with machine speeds and makespan criterion and present its formal mathematical model.

Let $\mathcal{J} = \{1, 2, \ldots, n\}$ be a set of n jobs, each corresponding to some hash mask, for example `u?l?l?l?d?`. For each job $j \in \mathcal{J}$ let t_j be the size of that job, understood as the number of hashes the job contains. Thus, a job with hash mask `u?l?l?l?d?` will have size of 4569760. Next, let $\mathcal{M} = \{1, 2, \ldots, m\}$ be a set of non-identical machines, each corresponding to a single node in a computer cluster. For each machine $i \in \mathcal{M}$ let s_i be the speed of that machine, understood as the number of hashes the machine can check in one second. The number of seconds machine i will take to process job j will be denoted p_j^i and defined as:

$$p_j^i = \frac{t_j}{s_i} \quad j \in \mathcal{J}, i \in \mathcal{M}. \tag{1}$$

Thus, machine with speed 1000000 will be able to check all hashes in job with mask `u?l?l?l?d?` in approx. 4.57 s.

The task is to determine a schedule of processing jobs on machines, while meeting the following two conditions: (1) each machine can process up to one job at any given time, (2) if job j is assigned to machine i then it has to be processed on i for p_j^i seconds without interruption. Any job can be processed on any machine, although the processing time will vary depending on the machine.

Let us start with the assignment of jobs to machines. This can be represented by a sequence (or a tuple) $\pi = (\pi^1, \pi^2, \ldots, \pi^m)$ of m elements. Each element π^i is a (possibly empty) sequence of jobs that will be processed on machine i. The jobs from π^i will be processed on i in the order of their appearance in π^i. Thus, π_j^i is the job that will be processed as j-th on machine i. From now on, the sequence π will be called a processing order.

A schedule is described by a pair (S, M) of sequences (vectors) S and M of j elements, where S_j is the starting time of job j and $M_j \in \mathcal{M}$ is the machine that will process that job. The schedule (S, M) can be determined from the processing order π as follows:

$$M_{\pi_j^i} = i, \quad i \in \mathcal{M}, j \in \pi^i, \tag{2}$$

$$S_{\pi_j^i} = S_{\pi_{j-1}^i} + p_{\pi_{j-1}^i}^i, \quad i \in \mathcal{M}, j \in \pi^i, \tag{3}$$

$$S_{\pi^i_1} = 0. \tag{4}$$

Due to (2) each job will be assigned to exactly one machine. Due to (3) and (4) jobs assigned to the same machine will not overlap, will be processed by their order in π^i and will be processed by the necessary time without interruption.

For a given schedule (S, M) we can define the makespan $C_{\max}(S, M)$ as the completion time of the job that finished last:

$$C_{\max}(S, M) = \max_{j \in \mathcal{J}} S_j + p_j^{M_j}. \tag{5}$$

The goal is to choose a schedule (S^*, M^*) as to minimize the makespan:

$$C_{\max}(S^*, M^*) = \min_{(S,M)} C_{\max}(S, M). \tag{6}$$

The schedule (S^*, M^*) is called the optimal schedule.

4 Algorithms Description

In this section we will describe the algorithms used for solving the problem of scheduling jobs for password cracking. Due to the way hash masks files are structured, it is not uncommon to encounter files with thousands of hash masks. Because of that we have decided to use metaheuristic algorithms. Such algorithms do not guarantee obtaining of optimal solution, but are able to perform controlled search of the solution space within a reasonable polynomial time complexity. We employed two such algorithms: Simulated Annealing method (SA) and Genetic Algorithm (GA). In both algorithms the solution was represented as the processing order π (*i.e.* sequence of sequences of jobs).

Let us start with SA method, which is a well-known iterative local-search metaheuristic based on imitating the annealing process in metallurgy. Our implementation is as follows:

1. The initial solution is chosen as the best one from 1000 random solutions.
2. The initial temperature T_0 was set to the difference between the best and worst of those 1000 solutions.
3. An exponential cooling scheme was used with temperature T_i on iteration i given as $T_i = \alpha T_{i-1}$. Parameter α was set to 0.9.
4. The neighbor solution π' is obtained by applying the insert move on the current solution π *i.e.* $\pi' = \text{ins}(\pi)$. The move is defined by four parameters: j_1, i_1, j_2 and i_2 and works by removing the j_1-th job from machine i_1 and putting it as j_2-th job on machine i_2. Some parameters values are forbidden in order to avoid creating infeasible solutions or making moves that do not change the solution. The size of the insert neighborhood for this problem is $O(n(n+m))$. The best of 20 random neighbors was selected in each iteration.
5. The probability of accepting worse solution on iteration i was given as:

$$\exp\left(\frac{C_{\max}(\pi) - C_{\max}(\pi')}{T_i}\right). \tag{7}$$

6. If no improvement of the globally best solution was found in 200 iterations, then temperature was raised such that $t_i = \frac{T_0}{2}$.
7. The halting condition was performing 100 000 iterations.

Let us now move onto the GA method, which is population metaheuristic based on the processes of species evolution and natural selection. The following implementation was used:

1. The initial population was made up of random solutions (specimens).
2. The population size was set to 100 specimens.
3. Roulette wheel parent selection was used with 50 pairs of parents, each yielding 2 offspring (for a total of 100 offspring).
4. The crossover operator was implemented as follows. In the first stage, for each element (machine) π^i we choose a cutting point $c \in \{1, 2, \ldots, |\pi^i|\}$ at random. Then the first c elements of π^i are copied from the first parent to the offspring, in that order. In the second stage, each element π^i of the second parent is examined and every element $\pi^i_j \in \pi^i$ that has not yet appeared in offspring is pushed at the end of π^i in offspring.
5. The mutation operator was implemented by applying a single random insert move. The probability of mutation was 3%.
6. The current population and offspring population are merged and the best 100 specimens become the population for the next generation.
7. The halting condition was performing 600 generations.

The running time of both SA and GA is $O(nm)$, which makes them applicable even for large problem instances (thousands of jobs).

5 Computer Experiment

In this section we will describe the results of computer experiment involving applying the proposed algorithms for solving the considered problem.

We will start with description of the instances (hcmask files). We considered 8 basic instances: 6 from the `rockyou` set (excluding `rockyou-1-60`, which is too small to be considered here) as well as instances `8char-1l-1u-1d-1s-compliant` and `pathwell`. The basic information for each instance (number of jobs (masks), number of hashes and the instance shorthands), is presented in Table 3.

The final processing time of each job depends on: (1) the number of hashes in a job, (2) the number of hashes a machine can check in a given unit a time. The number of hashes is known from the mask itself. However, the number of hashes that can be checked in a second depends not only on the machine itself, but also on the particular hash function used. 15 such hash functions were considered. Relative speeds of machines for each hash function are given in Table 4. The values in the table were obtained through Hashcat benchmark tests (computed using real-life computer cluster). In each row the slowest machine has speed shown as 1.000. The average is presented in the last row.

We can notice that the relative speed of machine is partially dependent on hash function type. In extreme cases (hash function types 3 and 12 for machine

Table 3. Summary of information on the 8 considered basic problem instances

Shorthand	Filename	No. of jobs	No. of hashes
r2	rockyou-2-1800.hcmask	2968	1.99×10^{13}
r3	rockyou-3-3600.hcmask	3971	5.91×10^{13}
r4	rockyou-4-43200.hcmask	7735	4.94×10^{14}
r5	rockyou-5-86400.hcmask	10613	9.03×10^{14}
r6	rockyou-6-864000.hcmask	17437	8.70×10^{15}
r7	rockyou-7-2592000_cleaned.hcmask	20560	2.60×10^{16}
p	pathwell.hcmask	40824	8.27×10^{16}
8c	8char-1l-1u-1d-1s-compliant.hcmask	100	3.03×10^{15}

Table 4. Relative speed of processing hashes for all 4 machine types

No.	Hash type description	4x2080	4x2080Ti FE	6x1080	1060
1	MD5	16.266	22.230	6.436	1.000
2	SHA1	13.640	18.960	7.596	1.000
3	SHA2-256	16.573	23.149	6.256	1.000
4	SHA2-512	14.802	20.901	9.900	1.000
5	WPA-EAPOL-PBKDF2	13.260	18.732	9.680	1.000
6	NTLM	15.498	21.790	5.413	1.000
7	LM	14.633	20.148	3.948	1.000
8	NetNTLMv1/NetNTLMv1+ESS	15.430	21.247	5.882	1.000
9	NetNTLMv2	13.557	20.373	8.464	1.000
10	descrypt, DES (Unix), Traditional DES	13.477	18.494	8.663	1.000
11	md5crypt, MD5 (Unix), Cisco-IOS 1	11.773	18.865	9.468	1.000
12	bcrypt 2*, Blowfish (Unix)	11.145	16.332	10.155	1.000
13	sha512crypt 6, SHA512 (Unix)	15.274	18.579	3.895	1.000
14	Kerberos 5 AS-REQ Pre-Auth etype 23	13.984	20.818	7.939	1.000
15	Kerberos 5 TGS-REP etype 23	13.575	20.230	5.841	1.000
	Average	14.192	12.057	7.302	1.000

type 4x2080) the difference can be as big as 48%. However, for machine type 4x2080Ti_FE the same hash function types (3 and 12) show difference of only 41%. Thus, we conclude that the quality of solutions obtained by solving algorithms can be dependent on the particular hash function, even if the same machines are used. Due to the above, we decided to test each of the 8 instances for each of the 15 hash functions, yielding 120 actual tested instances in total.

Both the SA and GA methods are probabilistic, meaning the quality of their solutions can vary between executions. In order to alleviate this, each of those algorithms was run for each of the 120 instances 10 times. The best result out of those 10 was chosen. That should mean that the reported running times of algorithms should be multiplied by 10, however, we can easily execute the algorithm 10 times in parallel with little to no time increase compared to a single run. All experiments were conducted as simulations using a Dell Latitude 5590 laptop with Intel Core i7-8650U 1.9 GHz CPU, running 64-bit Linux Mint 19.2. The reason for running the experiment as a simulation instead of running the experiments on the actual cluster system was because the cluster system was at the time used for password cracking. The algorithms were implemented in C++.

Table 5. Summarized normalized results for the SA algorithm

Instance	Minimum	Average	Maximum	St. dev.
r2	7.10	8.95	12.00	1.19
r3	28.10	28.47	29.00	0.30
r4	8.60	9.98	14.30	1.48
r5	17.70	23.73	38.40	6.42
r6	13.80	14.70	14.80	0.25
r7	8.60	12.11	13.90	1.62
p	4.00	4.45	10.70	1.73
8c	0.00	0.09	0.10	0.04
Average	10.99	12.81	16.65	1.63

Table 6. Summarized normalized results for the GA algorithm

Instance	Minimum	Average	Maximum	St. dev.
r2	7.10	8.95	12.00	1.19
r3	28.10	28.47	29.00	0.30
r4	8.60	9.98	14.30	1.48
r5	17.20	22.52	35.60	5.28
r6	13.80	14.70	14.80	0.25
r7	7.20	10.65	12.50	1.67
p	3.90	4.44	10.70	1.73
8c	0.00	0.07	0.10	0.05
Average	10.74	12.47	16.13	1.49

The reported results of each algorithm are normalized. This is done by comparing the makespan obtained from the schedule returned by the SA (or GA) algorithm to the makespan obtained from "default" schedule. The default schedule is obtained by a greedy algorithm that schedules jobs one by one in the order of their appearance in the instance file. The current job is scheduled on the machine that will guarantee the smallest makespan in such partial solution. Thus if $C_{\max}(A, I)$ is the makespan obtained by algorithm A on problem instance I, then the reported normalized result for instance I is:

$$\frac{C_{\max}(\text{greedy}, I)}{C_{\max}(\text{SA}, I)}. \tag{8}$$

For the GA method the definition is similar. The summarized normalized results for the SA and GA method are shown in Tables 5 and 6 respectively.

First we notice that the results obtained by both metaheuristics are nearly identical. The SA method obtains slightly better results than GA overall at the cost of higher standard deviation. This effect is minor, but consistent. The second observation is that the improvement compared to the default greedy algorithm is considerable, with the exception of instance denoted as 8c. This particular instance is either very difficult or the default schedule is already close to optimum. This can be also caused by this instance size (only 100 jobs). On the other hand, for other instances we obtain improvement from 4% to 38%. This improvement is thus considerable, but not very stable and is dependent on instance type. The best improvement occurs for instances r3, r5 and r6. We also see that the improvement does not depend on instance size in any obvious way and the smallest instance (8c) is associated with least improvement.

Next we notice that the particular hash function used affects the results as well. The effect is the most visible in the case of instance denoted p, where for one "best" hash function the resulting makespan improvement was 2.7 times larger than for the "worst" hash function. Similar, but less pronounced effect occurs with instances r2, r4, r5 and r7. However, there are also instances (like r3, r6 and 8c), where the hash function used had little effect on the improvement.

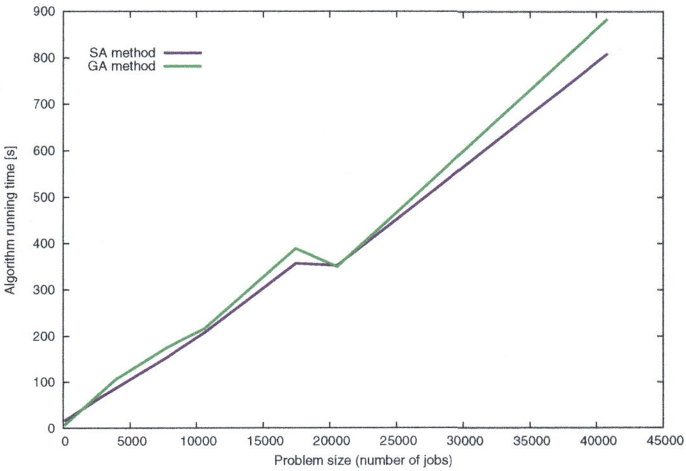

Fig. 1. Running time of algorithms with regards to problem size

We have also researched the running time of both algorithms with regards to problem size which is shown in Fig. 1. The results confirm the linear dependency on the number of jobs (the number of machines does not vary). We conclude that the proposed algorithms are able to obtain their results in under 15 min for the considered instances. This can be further improved with parallel computing (especially in the case of the GA method). We also notice that GA requires a little more computation time in general. Thus, the SA method slightly wins in both solution quality and running time.

6 Conclusions

In this paper we modeled the problem minimizing the time of password cracking in a multi-machine multi-GPU environment as a problem of job scheduling on parallel machines with machine speeds and makespan criterion. We proposed two linear-time metaheuristic algorithms to solve the problem. The results of a computer experiment using real-life hash mask files indicate moderate to considerable improvement in makespan for most of the considered instances compared to default schedule. Moreover, the results for the same instance can vary significantly depending on the hash function used. The SA method is also slightly better than the GA method (both quality and running time).

References

1. Hashcat. https://hashcat.net/hashcat. Accessed 13 Jan 2020
2. Anderson, D.P.: BOINC: a system for public-resource computing and storage. In: Proceedings of the 5th IEEE/ACM International Workshop on Grid Computing, pp. 4–10. IEEE Computer Society (2004)
3. Apostal, D., Foerster, K., Chatterjee, A., Desell, T.: Password recovery using MPI and CUDA. In: 19th International Conference on High Performance Computing, pp. 1–9. IEEE (2012)
4. Bengtsson, J.: Parallel password cracker: a feasibility study of using Linux clustering technique in computer forensics. In: Second International Workshop on Digital Forensics and Incident Analysis (WDFIA 2007), pp. 75–82. IEEE (2007)
5. Bożejko, W., Gnatowski, A., Idzikowski, R., Wodecki, M.: Cyclic flow shop scheduling problem with two-machine cells. Arch. Control Sci. **27**(2), 151–167 (2017)
6. Bożejko, W., Grymin, R., Pempera, J.: Scheduling and routing algorithms for rail freight transportation. Procedia Eng. **178**, 206–212 (2017). https://doi.org/10.1016/j.proeng.2017.01.098
7. Bożejko, W., Hejducki, Z., Wodecki, M.: Applying metaheuristic strategies in construction projects management. J. Civ. Eng. Manage. **18**(5), 621–630 (2012). https://doi.org/10.3846/13923730.2012.719837
8. Crumpacker, J.R.: Distributed password cracking. Naval Postgraduate School, Monterey, California (2009)
9. Hranický, R., Zobal, L., Ryšavý, O., Kolář, D.: Distributed password cracking with BOINC and hashcat. Digit. Invest. **30**, 161–172 (2019)
10. Picolet, J.: Netmux LLC: Hash Crack: Password Cracking Manual. CreateSpace Independent Publishing Platform, Scotts Valley (2016)

11. Pippin, A., Hall, B., Chen, W.: Parallelization of John the Ripper using MPI (2006)
12. Rodwald, P.: Choosing a password breaking strategy with imposed time restrictions. Bull. Mil. Univ. Technol. **68**, 79–100 (2019). https://doi.org/10.5604/01.3001.0013.1467
13. Rodwald, P., Biernacik, B.: Password protection in IT systems. Bull. Mil. Univ. Technol. **67**, 73–92 (2018). https://doi.org/10.5604/01.3001.0011.8036
14. Rudy, J.: Algorithm-aware makespan minimisation for software testing under uncertainty. In: Zamojski, W., Mazurkiewicz, J., Sugier, J., Walkowiak, T., Kacprzyk, J. (eds.) Engineering in Dependability of Computer Systems and Networks, pp. 435–445. Springer, Cham (2020)
15. Zonenberg, A.: Distributed hash cracker: a cross-platform GPU-accelerated password recovery system. Rensselaer Polytech. Inst. **27**(395–399), 42 (2009)

Standard Dropout as Remedy for Training Deep Neural Networks with Label Noise

Andrzej Rusiecki(✉)

Department of Computer Engineering, Wroclaw University of Science and Technology, Wybrzeże Wyspiańskiego 27, Wrocław, Poland
andrzej.rusiecki@pwr.edu.pl

Abstract. Deep neural networks, trained on large annotated datasets, are often considered as universal and easy-to-use tools for obtaining top performance on many computer vision, speech understanding, or language processing tasks. Unfortunately, these data-driven classifiers strongly depend on the quality of training patterns. Since large datasets often suffer from label noise, the results of training deep neural structures can be unreliable. In this paper, we present experimental study showing that simple regularization technique, namely dropout, improves robustness to mislabeled training data, and even in its standard version can be considered as a remedy for label noise. We demonstrate it on popular MNIST and CIFAR-10 datasets, presenting results obtained for several probabilities of noisy labels and dropout levels.

Keywords: Neural networks · Deep learning · Dropout · Label noise · Categorical cross-entropy

1 Introduction

Deep neural networks attract nowadays much attention, mainly because of their impressive, and very often state-of-the-art performance, obtained for many computer vision, speech recognition, or natural language processing tasks [3]. Availability of well-annotated large data collections is also a reason for their popularity [26], because such networks trained on large supervised datasets can potentially represent high-level abstractions [4,23].

Unfortunately, as in many other data-driven approaches, models we obtain can be only as reliable as our training data [13,22]. In particular, when training patterns have uncertain labels, a network may learn incorrect classification rules for a given task. This phenomenon is even more prominent for deeper than shallow networks, simply because they posses much more parameters, giving them more degrees of freedom and higher probability to overfit erroneous data. It is clearly evident that large datasets suffer from label noise, which is usually introduced by the way they are collected. Annotating data by many different human

W. Zamojski et al. (Eds.): DepCoS-RELCOMEX 2020, AISC 1173, pp. 534–542, 2020.
https://doi.org/10.1007/978-3-030-48256-5_52

annotators, with no precise distinction between classes, using search engines, or data mining algorithms analyzing social media websites, results in noise that can degrade model performance [11]. In this paper, we present a simple way to partially overcome this problem, demonstrating that popular dropout regularization may be considered as a tool to make deep network training robust to label noise. To our knowledge, it is the first experimental study on how noisy labels affect network performance for different dropout rates.

2 Learning from Noisy Labels and Standard Dropout

For shallow neural networks, noise in the training data has been often looked at from the point of view of learning in the presence of outliers [9]. Such outlying data points, defined as observations distant from the majority of data, have been usually considered for regression-like tasks, where network outputs are continuous. For classification tasks, however, reliability of training data is usually perturbed by erroneous labels.

2.1 Learning with Label Noise

In the field of learning with noisy labels, two basic groups of methods exist. In the first group of approaches, efforts are directed to clean data by removing or correcting noisy patterns. The label noise can be considered as conditionally independent from the input [18,25], or the models are image-conditional [28,29]. In the latter case, training process is designed to be robust against noisy data, and the methods aim to learn directly from noisy labels with slightly modified algorithms [8,12,17,20,27], or corrected loss functions are applied [7,19,22].

2.2 Dropout Methods

As dropout methods we describe a wide range of stochastic techniques applied to artificial neural network training and/or simulation [6]. The term itself refers to dropping out units (neurons) or connections. Dropout was firstly introduced in [10] as a simple method to avoid overfitting and improve generalization ability of deep networks. It was further described in [24] and successfully applied in [15]. Following [6], sometimes we refer to the original version of dropout technique as *standard dropout* to make it distinguishable from other variations. Its main idea is based on omitting randomly chosen neurons or weights in each training iteration. In original approach, single neurons are turned off with probability 0.5 in each step of the training phase, and during testing phase all neurons are taken into account.

Many techniques, exploiting similar idea, has been recently proposed in the field of network weights regularization, measuring uncertainty of network outputs, or in model compression. The dropout-inspired approaches are e.g. fast, variational or Monte Carlo dropout (a good survey of such methods can be found in [6]).

The whole procedure of dropout technique is as follows: during each iteration of training algorithm, each single neuron is omitted by making its output equal to 0, with certain probability p_d. After the training phase is finished, all the neurons are used with outputs multiplied by p_d to compensate for the larger network size. Note, that during training, the network structure has, on average, approximately $(1 - p_d)\%$ of all existing neurons.

A single layer output, after applying standard dropout, can be written as:

$$y = f(\mathbf{W}x) \circ m, \quad m_i \sim \text{Bernoulli}(p_d), \tag{1}$$

where x is layer input, $f(\cdot)$ is the activation function, and $\mathbf{W}$ is the matrix of layer weights. For the layer dropout mask m, its elements are equal 1 with given probability p_d. When a network is simulated, the layer output is scaled as:

$$y = p_d f(\mathbf{W}x). \tag{2}$$

Dropout can prevent neural nets from overfitting and improve their generalization abilities because it regularizes a single model by approximating process of averaging an exponential number of learned models that share parameters [24]. From another point of view, such averaging can be considered as efficient way of combining many different neural networks architectures. It can be described as sampling thinned networks consisting of units that survived dropout.

2.3 Dropout for Noisy Labels

If applying dropout to regularize the network training allows for effective learning for longer periods without overfitting, it can potentially reduce the effect of noise on the whole training process. It is well known that dropout can improve generalization but its influence on training with label noise has not been extensively studied yet. Some results can be found in [2], where dropout memorization reduction ability is shown, and in [11] where dropout is used to learn a non-trivial noise model.

In the next section, we present experimental results of training deep models on data containing different amounts of noisy labels, for several dropout rates. The dropout is applied in its standard form, with the same dropout rate (probability p_d) in each dropout mask.

Label Noise. In our experiments we considered only the basic uniform noise model. This type of label noise is applied by random generating correct or incorrect labels for all the training patterns. In this approach, for each element of a training set, belonging to one of C classes, its label is flipped and uniformly sampled within the set of all available incorrect labels with certain probability μ. Hence, the label can be correct with probability $1 - \mu$. The noisy training data available to the learner are $\{(x_i, \hat{y}_i), i = 1, \ldots, N\}$, where:

$$\hat{y}_i = \begin{cases} y_i & \text{with probability } 1 - \mu \\ k, \quad k \in [C], k \neq y_i, & \text{with probability } \frac{\mu}{C-1} \end{cases} \tag{3}$$

In Eq. 3, y_i denotes true label, while k is incorrect label uniformly drawn from the set $k \in [C], k \neq y_i$.

Categorical Cross-entropy. As our previous efforts have demonstrated [21,22], proper choice of a loss function to be minimized in network training has its obvious impact on robustness to label noise, and modifying such functions can improve performance on testing data. Hence, to make our investigation reliable, we decided to use standard, non-modified error measure. In the case of classification problems, the most popular loss is definitely so-called categorical cross-entropy (CCE) that can be technically defined as:

$$E_{CC} = -\frac{1}{N}\sum_{i=1}^{N}\sum_{c=1}^{C}(p_{ic}\log(y_{ic})), \tag{4}$$

where p_{ic} is a binary function indicating whether the ith training pattern belongs to cth category. Target p_{ic} and output y_{ic} can be thought of as true probability, and predicted probability distribution for ith observation belonging to cth class.

3 Experimental Results

Two well-known classification datasets were chosen to conduct our experiments. MNIST [16] and CIFAR-10 [14] are widely used also in the field of testing robustness to label noise [7]. These, relatively small image classification sets, are especially convenient to simulations where some averaging should be made, because good-performing deep architectures for these task are also of reasonable number of parameters.

3.1 Testing Methodology

We decided to perform our experiments on deep convolutional neural networks (CNN) with architectures described in [7] and used previously in training with noisy labels. The details of CNN structures and basic description of MNIST and CIFAR-10 datasets were gathered in Table 1.

Label Noise Level. The noise was introduced into training sets by flipping labels with preset probability, following Eq. 3. For each training pattern its label was sampled within all the remaining classes with probability μ. The noise level was varied in the range from $\mu = 0$ up to $\mu = 0.6$, which is equivalent, on average, to 60% of training patterns with incorrect labels in the training data.

Dropout Rate. We tested standard dropout version defined by Eqs. 1 and 2. In Table 1 dropout layers positions were specified for both tested CNN architectures. As dropout rate we denote probability p_d which was naively set equal for each neural network layer. In the experiments, the rate was varied in the range $p_d = 0$ up to $p_d = 0.8$.

Table 1. Network architectures and dataset characteristic

Dataset description	Network architecture
MNIST dataset: Input 28×28, 10 classes, 60k/10k training/test	Convolutional layer → max pooling → **dropout** → fully connected 1024 neurons → **dropout** → fully connected 1024 neurons → **dropout** → softmax
CIFAR-10 dataset: Input $32 \times 32 \times 3$, 10 classes, 50k/10k training/test	2 Convolutional layers → max pooling → **dropout** → 2 Convolutional layers → max pooling → **dropout** → fully connected 512 neurons → **dropout** → softmax

Training Algorithm. CNN used in our simulations were implemented with Python 3.6 in TesorFlow environment [1]. To speed up network training, all the experiments were run on GTX 1080Ti GPU. To train deep neural networks we used popular and robust to noisy gradients Adam algorithm [5]. The training parameters were set as follows: learning rate $lr = 0.001$, $\beta_1 = 0.9$ and $\beta_2 = 0.999$, and the networks were trained for 200 epochs. Test accuracies were averaged over only 6 runs of simulations because one training took approximately 30 min on 1080Ti GPU (obtaining one data point in the Fig. 2 took about 3 h).

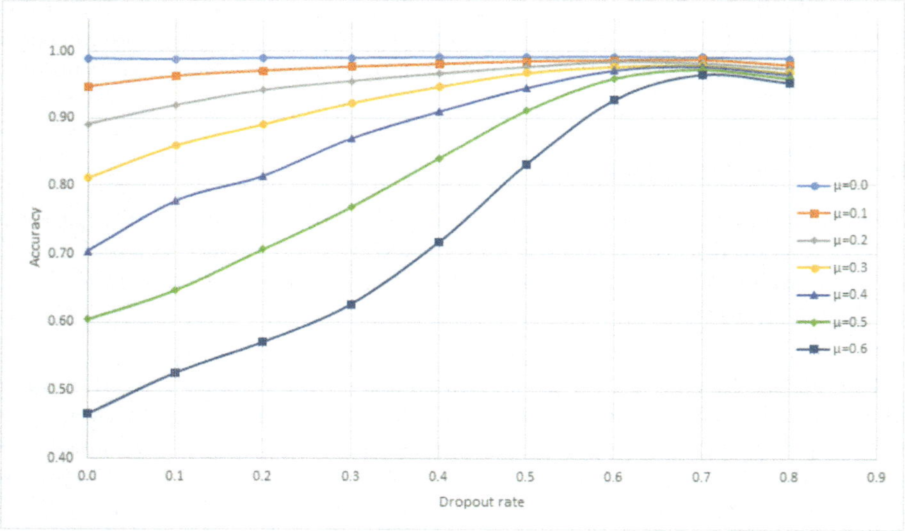

Fig. 1. Averaged test accuracies for several levels of dropout rate for MNIST dataset: noise level is varied in range $\mu = 0.0 - 0.6$.

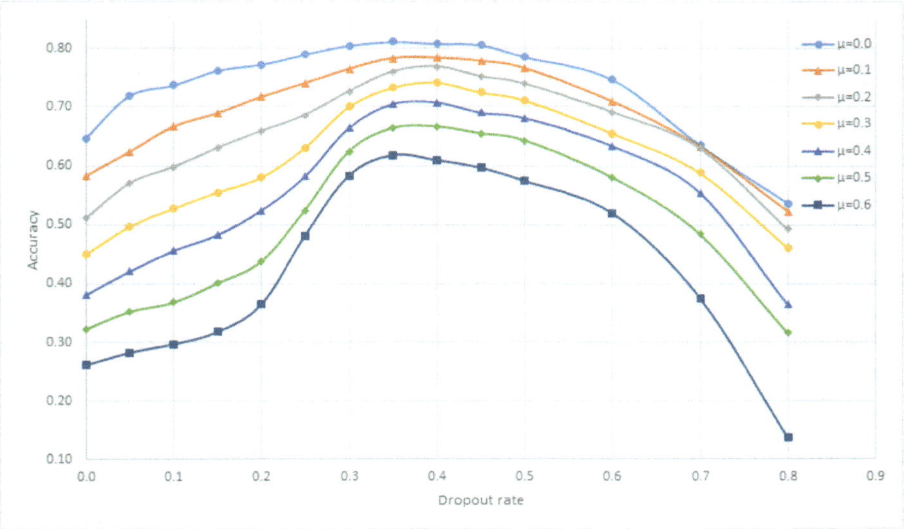

Fig. 2. Averaged test accuracies for several levels of dropout rate for CIFAR-10 dataset: noise level is varied in range $\mu = 0.0 - 0.6$.

3.2 Simulation Results

Averaged test accuracies (calculated on clean data), for several probabilities of noisy labels and several dropout rates, were presented in figures. Shapes of resulting curves are as expected: increasing to maximal accuracy until they reach optimal dropout rate, and then decreasing. Looking at the Figs. 1 and 2, one may notice, that applying even naive version of dropout regularization can indeed increase generalization ability of deep CNN trained on data containing label noise. For MNIST dataset (Fig. 1), when networks were trained on data not containing noise, using dropout doesn't influence CNN performance. However, when label noise appears, dropout improves generalization. For example, when probability of noisy labels $\mu = 0.6$, accuracy without regularization is less than 50% but can reach over 95% when proper dropout rate is applied. Another interesting phenomenon is that optimal dropout rate seems to be equal (close to 0.7) for each amount of noise.

Analysing results of naive dropout for CIFAR-10 dataset, one may observe very similar properties. However, the variability of obtained accuracies is much faster in Fig. 2, so we decided to use denser sampling, which allows us to identify optimal rate close to 0.35–0.40 (which is less than typical rates 0.5–0.8 proposed in original paper [24]). Moreover, while for clean training data there exist a small range, where the accuracy curve is flat, for data with contaminated labels, these curves go up below, and down, over optimal dropout rate value. In this case, choosing too large dropout rate may result in performance poorer than without applying such regularization. This could mean that CNN architecture, though considered suitable for CIFAR-10 task, should be in fact more sophisticated.

Based on the results described in the previous paragraphs, we may formulate four general observations:

1. Using even naive standard dropout improves network performance when training dataset contains label noise.
2. Choosing optimal dropout rate is crucial for obtaining good performance (small changes in p_d can degrade accuracy).
3. Optimal dropout rate does not depend on the probability of noisy labels (it is equal for each probability of noise).
4. Optimal dropout rate seems to be rather a single number, not an interval providing similar performance (which is common for clean data).

4 Conclusions

In this paper, we presented preliminary experimental study showing that applying standard dropout regularization to deep neural network learning process, improves robustness to mislabeled training data and can be considered as a remedy for label noise. Based on the results obtained for MNIST and CIFAR-10 datasets, we formulated several observations on how dropout rate influences network generalization abilities for training with noisy labels. To our knowledge it is the first such study, focused on training with dropout on data with label noise. It is especially important to stress that applying this regularization technique, with properly chosen dropout rate, can dramatically improve performance for highly contaminated data. However, keeping the rate in reasonable range can be still considered as an efficient way to deal with noisy labels.

Future research should be directed towards more extensive experimental examination, taking into account variability of network architectures (especially deeper ones), different dropout rates in each layer, larger datasets, and more sophisticated models of label noise. This could result in some rules on how to effectively choose dropout parameters, in order to make network training robust to label noise. Studying other popular regularization techniques may also contribute in this field.

References

1. Abadi, M., et al.: TensorFlow: large-scale machine learning on heterogeneous systems (2015). Software available from https://www.tensorflow.org/
2. Arpit, D., Jastrzebski, S., et al.: A closer look at memorization in deep networks. arXiv preprint arXiv:1706.05394 (2017)
3. Erhan, D., et al.: Why does unsupervised pre-training help deep learning? J. Mach. Learn. Res. **11**, 625–660 (2010)
4. Bengio, Y., et. al.: Greedy layer-wise training of deep networks. In: Advances in Neural Information Processing Systems, vol. 19, pp. 153—160. MIT Press (2007)

5. Kingma, D., Ba, J.: Adam: a method for stochastic optimization. arXiv preprint arXiv:1412.6980 (2014)
6. Labach, A., Salehinejad, H., Valaee, S.: Survey of dropout methods for deep neural networks. arXiv preprint arXiv:1904.13310 (2019)
7. Ghosh, A., Kumar, H., Sastry, P.S.: Robust loss functions under label noise for deep neural networks. arXiv:1712.09482v1 (2017)
8. Guan, M.Y., Gulshan, V., Dai, A.M., Hinton, G.E.: Who said what: modeling individual labelers improves classification. arXiv:1703.08774 (2017)
9. Hampel, F.R., Ronchetti, E.M., Rousseeuw, P.J., Stahel, W.A.: Robust Statistics: The Approach Based on Influence Functions. (Wiley Series in Probability and Statistics), revised edn. Wiley, New York (2005)
10. Hinton, G.E., Srivastava, N., Krizhevsky, A., Sutskever, I., Salakhutdinov, R.R.: Improving neural networks by preventing co-adaptation of feature detectors. arXiv preprint arXiv:1207.0580 (2012)
11. Jindal, I., Nokleby, M., Chen, X.: Learning deep networks from noisy labels with dropout regularization. In: IEEE 16th International Conference on Data Mining (ICDM), pp. 967–972. IEEE (2016)
12. Joulin, A., van der Maaten, L., Jabri, A., Vasilache, N.: Learning visual features from large weakly supervised data. In: European Conference on Computer Vision (ECCV). Springer (2016)
13. Korodos, M., Rusiecki, A.: Reducing noise impact on MLP training. Soft Comput. **20**(1), 49–65 (2016)
14. Krizhevsky, A.: Learning multiple layers of features from tiny images, Technical report (2009)
15. Krizhevsky, A., Sutskever, I., Hinton, G.E.: ImageNet classification with deep convolutional neural networks. In: Advances in Neural Information Processing Systems, vol. 25, pp. 1097–1105 (2012)
16. LeCun, Y., Cortes, C.: MNIST handwritten digit database. http://yann.lecun.com/exdb/mnist/
17. Misra, I., Lawrence, Z.C., Mitchell, M., Girshick, R.: Seeing through the human reporting bias: visual classifiers from noisy human-centric labels. In: Computer Vision and Pattern Recognition (CVPR) (2016)
18. Natarajan, N., Inderjit, S.D., Ravikumar, P.K., Tewari, A.: Learning with noisy labels. In: Advances in Neural Information Processing Systems (NIPS) (2013)
19. Patrini, G., Rozza, A., Menon, A., Nock, R., Qu, L.: Making neural networks robust to label noise: a loss correction approach. In: Computer Vision and Pattern Recognition (2017)
20. Reed, S., Lee, H., Anguelov, D., Szegedy, C., Erhan, D., Rabinovich, A.: Training deep neural networks on noisy labels with boot-strapping. arXiv preprint arXiv:1412.6596 (2014)
21. Rusiecki, A.: Robust learning algorithm based on LTA estimator. Neurocomputing **120**, 624–632 (2013)
22. Rusiecki, A.: Trimmed categorical cross-entropy for deep learning with label noise. Electron. Lett. **55**(6), 319–320 (2019)
23. Salakhutdinov, R., Hinton, G.E.: Semantic hashing. In: Proceedings of the Workshop on Information Retrieval and Applications of Graphical Models (SIGIR 2007). Elsevier, Amsterdam (2007)
24. Srivastava, N., Hinton, G., Krizhevsky, A., Sutskever, I., Salakhutdinov, R.: Dropout: a simple way to prevent neural networks from overfitting. J. Mach. Learn. Res. **15**, 1929–1958 (2014)

25. Sukhbaatar, S., Bruna, J., Paluri, M., Bourdev, L., Fergus, R.: Training convolutional networks with noisy labels. arXiv preprint arXiv:1406.2080 (2014)
26. Vahdat, A.: Toward robustness against label noise in training deep discriminative neural networks. In: Neural Information Processing Systems (NIPS) (2017)
27. Van Horn, G., Branson, S., Farrell, R., Haber, S., Barry, J., Ipeirotis, P., Perona, P., Belongie, S.: Building a bird recognition app and large scale dataset with citizen scientists: the fine print in fine-grained dataset collection. In: Computer Vision and Pattern Recognition (CVPR) (2015)
28. Veit, A., Alldrin, N., Chechik, G., Krasin, I., Gupta, A., Belongie, S.: Learning from noisy large-scale datasets with minimal supervision. In Computer Vision and Pattern Recognition (CVPR) (2017)
29. Xiao, T., Xia, T., Yang, Y., Huang, C., Wang, X.: Learning from massive noisy labeled data for image classification. In: Proceedings of the IEEE Conference on Computer Vision and Pattern Recognition (CVPR), pp. 2691–2699 (2015)

State Assignment of Finite-State Machines by Using the Values of Output Variables

Valery Salauyou(✉) and Michal Ostapczuk

Faculty of Computer Science, Bialystok University of Technology, Bialystok, Poland
valsol@mail.ru

Abstract. Structural models of finite-state machines (FSMs) that make it possible to use the values of the output variables for encoding the internal states are studied. To minimize the area (the parameter *area* is used to denote cost in the context of this paper) of FSM implementation, it is proposed to use the structural model of the class D FSM. A method for the design of the class D FSM in FPGA is proposed. This method involves two phases—splitting the internal states of the FSM (to satisfy the necessary conditions for the construction of the class D FSM) and encoding the internal states (to ensure that the codes are mutually orthogonal). It is shown that the proposed method reduces the area of FSM implementation for all families of FPGAs of various manufacturers by a factor of 1.85–2.67 on average and by a factor of 5.7 for certain cases. Practical issues concerning the method and the specific features of its use are discussed, and possible directions of the elaboration of this approach are proposed.

Keywords: Finite state machine (FSM) · Field programmable gate array (FPGA) · State assignment · Area minimization · State splitting · Synthesis · Look up table

1 Introduction

In the general case, a digital system can be represented by a set of combinational circuits and finite state machines (FSMs). FSMs are also widely used as individual units as controllers and control devices. Usually, when working on a project, the designer has to develop new FSMs each time. It is clear that the parameters of FSMs used in a digital system to a large extent determine the success of the whole project. For this reason, the issue of minimization of FSMs is very important. As the FSM optimization criteria, one typically uses area, delay, and power consumption. Presently, field programmable gate arrays (FPGAs) are widely used in digital systems; for this reason, many FSM optimization methods are designed for the implementation of FSMs based in FPGAs.

The idea of using the values of the input and output variables of the FSM for encoding its internal states was first proposed in [1]. Later, this approach was elaborated in [2], where various combinations of the input and output variables that can be used for encoding the internal states are considered. The choice of the minimum number of input and output variables for encoding is an NP-hard problem. In [3], it was

W. Zamojski et al. (Eds.): DepCoS-RELCOMEX 2020, AISC 1173, pp. 543–553, 2020.
https://doi.org/10.1007/978-3-030-48256-5_53

proposed to use the values of the output variables of the Moore FSM as the codes of the internal states. In [4], structural models of FSMs based on the architectural capabilities of FPGAs were proposed; these models make it possible to use the values of the FSM input and output variables as internal state codes. In [5], the values of input variables are used for state assignment of finite-state machines.

The analysis of the known methods for designing FSMs shows that these methods are still being developed intensively for optimization an area [6, 7], a power [8, 9], and a performance [10, 11]. However, recently there has been no progress in using the values of input and (or) output variables for encoding the internal state. At the same time, modern FPGAs provide means for implementing such encoding because their input and output buffers, as well as logical elements, contain flip-flops. In addition, the logical elements of FPGAs admit the implementation on their outputs of both combinational and register functions, as well as both types of functions simultaneously, which has opened new possibilities for designing FSMs.

In practice, two FSM structural models have received the most widespread use—these are the Mealy and Moore FSMs. In [1–3], only the Moore FSM is considered. However, such an FSM cannot be always used in real-life projects. The application of the Moore FSM model often requires the transition from the Mealy to the Moore FSM. Compared with the Mealy FSM, the Moore FSM operates with one clock signal cycle delay at power up and at reset in the initial state, which is inadmissible in some projects. In addition, the Moore FSM typically has more states than the similar Mealy FSM. In [4], a structural model of the Mealy FSM, the class D FSM was proposed. This model allows one to use the values of the output variables as codes of the internal states of the FSM.

In the present paper, it is proposed to use the structural model of the class D FSM to minimize the cost of implementing FSMs in FPGAs. The distinction of the proposed approach from the known approaches lies in the fact that the transition from the Mealy to the Moore FSM is not required.

The paper is organized as follows. We present structural models of the Mealy FSMs in Sect. 2. The synthesis method of the class D FSM is considered in Sect. 3. Splitting of internal states to satisfy the necessary conditions for synthesis of the class D FSM is discussed in Sect. 4. The algorithms for the state assignment of the class D FSM is described in Sect. 5. Experimental research results are analyzed in Sect. 6. Finally, concluding remarks and feasible future trends of research are discussed in Conclusions.

2 Structural FSM Models

The most general model of the Mealy FSM can be described by means of following equations:

$$a_{t+1} = \Phi(z_t, a_t);$$
$$w_t = \Psi(z_t, a_t),$$

where Φ is the transition function, Ψ is the output function, a_t is the present state of the FSM at time t (t = 1, 2, 3, …), a_{t+1} is the next state of the FSM, z_t is a collection of values of the input variables (the input vector) on the FSM input at time t, and w_t is a collection of values of the output variables (the output vector) formed at time t.

The Mealy FSM in classification [4] received a title the *class A FSM*. The structural model of the Mealy FSM show on Fig. 1,*a*, where CL_Φ is the combinational circuit forming the values of the transition functions, CL_Ψ is the combinational circuit forming the values of the output functions, and RG is the FSM's memory.

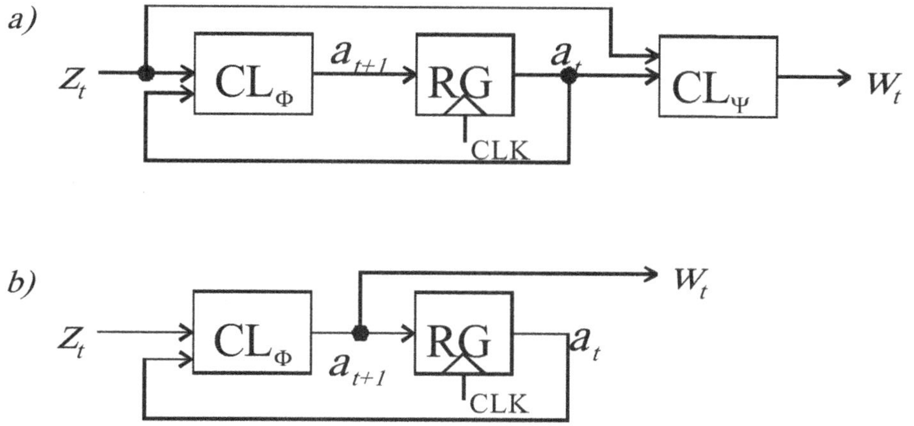

Fig. 1. The structural models of FSMs: *a* – the class A FSM; *b* – the class D FSM

In the *class D FSM* the code of the next state $a_{t+}1$ determines the value of the output vector w_t; therefore, the equations of functioning of the class D FSM have the following view:

$$a_{t+1} = \Phi(z_t, a_t);$$
$$w_t = a_{t+1}.$$

In contrast to the Mealy FSM of the class A, the structure of the class D FSM does not include the combinational circuit CL_Ψ (Fig. 1,*b*) that allows to build FSMs of a low cost (an area) and a high-speed performance.

3 The Synthesis Method of the Class D FSM

We will describe each FSM by the number L of input variables in the set $X = \{x_1, \ldots, x_L\}$, the number N of the output variables in the set $Y = \{y_1, \ldots, y_N\}$, the number M of the internal states in the set $A = \{a_1, \ldots, a_M\}$, and the parameter R equal to the number of the state encoding bits sufficient for encoding of the FSM internal states. In the traditional approaches to the design of FSMs, we have $R \geq \text{intlog}_2 M$.

The basic idea underlying the proposed method is the use of the output vector of the Mealy FSM formed at the transition to a state as the code of this state. It is clear that the output vector suitable for encoding an internal state must be formed on *all* transitions into this state. In addition, the output vectors used as the codes of internal states must be mutually orthogonal.

A necessary condition for the construction of a class D FSM is that the same output vector must be formed at all transitions to the same sate. Let $Y(a_s, a_i)$ be the output vector formed at the transition from the state a_s to the state a_i ($a_s, a_i \in A$) and $B(a_i)$ be the set of states the transitions from which end at the state a_i; then, a necessary condition for the construction of the class D FSM can be written as

$$Y(a_s, a_i) = Y(a_t, a_i) \forall a_i \in A : a_s, a_t \in B(a_i) \wedge s \neq t \quad (1)$$

If condition (1) is not satisfied for a certain state $a_i \in A$, we propose to split this state into several states.

A sufficient condition for the construction of a class D FSM is the mutual orthogonality of all codes of its internal states. It is clear that at transitions to different states of the FSM arbitrary output vectors can be formed that are not necessarily mutually orthogonal. To ensure the orthogonality of the state codes, the proposed approach introduces additional bits into the state codes. As a result, the structural model of the FSM shown in Fig. 2 is obtained, where $U = \{u_1, \ldots, u_R\}$ is the set of additional transition functions that ensure the orthogonality of the state codes.

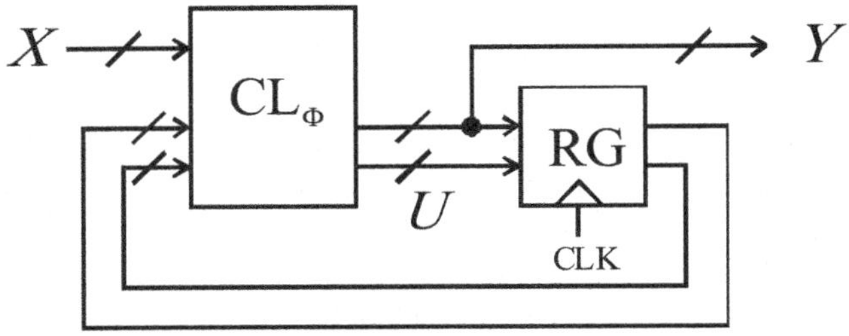

Fig. 2. The structure of the class D FSM with additional transition function in *U*

Thus, the design of class D FSMs consists of two phases - splitting the internal states to satisfy conditions (1) and a special state assignment to guarantee the mutual orthogonality of the codes.

4 Splitting of Internal States to Satisfy the Necessary Conditions for Synthesis of the Class D FSM

Note that splitting the internal states is an equivalent transformation of the FSM and it does not change the operation algorithm of the FSM. Let $C(a_i)$ be the set of transitions to the state a_i, and $P(a_i)$ be the set of transitions from the state $a_i \in A$; then, the algorithm for splitting the internal states with the aim to satisfy conditions (1) is as follows.

Algorithm 1

1. In the set A, find a state a_i for which conditions (1) are not satisfied. If such a state is found, then go to Step 2; otherwise, go to Step 7.
2. Determine the output vectors $Y_1, \ldots, Y_Q$ formed at the transitions to the state a_i, $Y_k \neq Y_h$, $k \neq h$, $k, h = \overline{1, Q}$.
3. Introduce Q new states $a_{i\_1}, \ldots, a_{i\_Q}$.
4. Determine the subsets $C(a_{i\_1}), \ldots, C(a_{i\_Q})$ of transitions to the states $a_{i\_1}, \ldots, a_{i\_Q}$. Each subset $C(a_{i\_q})$ is assigned transitions on which the output vector Y_q, $q = \overline{1, Q}$, $C(a_{i\_q}) \subseteq C(a_i)$ is formed.
5. The subsets $P(a_{i\_1}), \ldots, P(a_{i\_Q})$ of transitions from the states $a_{i\_1}, \ldots, a_{i\_Q}$ are determined in the following way: $P(a_{i\_q}) := P(a_i)$ for all $q = \overline{1, Q}$.
6. Set $A := A \backslash \{a_i\}$, $A := A \cup \{a_{i\_1}, \ldots, a_{i\_Q}\}$, and $M := M+Q-1$; go to Step 1.
7. Stop.

5 State Assignment of the Class D FSM

Note that in the general case both the input and output vectors of the FSM are ternary; i.e., each component takes the value of 0, 1, or −1, where the dash denotes an undetermined value.

The main purpose of encoding the internal states when designing the class D FSMs is to ensure the mutual orthogonality of these codes. To encode the internal states of a class D FSM, a ternary matrix W is constructed in which the rows correspond to the internal states and the columns correspond to the output variables of the FSM. The rows of W are filled by the values of the output vectors that are formed on the transitions to the corresponding states. Later, the rows of W will determine the codes of the internal states of the class D FSM.

To ensure the orthogonality of the rows of W, additional columns are introduced that correspond to the additional transition functions in the set U created to ensure the mutual orthogonality of the state codes (Fig. 2). Thus, we want to create the minimum number R of additional columns in W and set their values in such a way that the extended rows of W are mutually orthogonal.

Since the matrix W is ternary, we can try to orthogonalize the rows of W by replacing the undetermined values of the output vectors by specific values. Taking into account the reasoning above, the algorithm for encoding the internal states of a class D FSM is as follows.

Algorithm 2

1. Construct the ternary matrix W for encoding the internal states.
2. The undetermined elements in W are replaced by concrete values using Algorithm 3.
3. The graph H for the orthogonalization of the rows of the matrix W is constructed. The nodes of H correspond to the rows of W (internal states of the FSM). Two nodes of H are connected by an edge if the corresponding rows of W are orthogonal.
4. The nodes connected to all other nodes (the rows of W corresponding to these nodes are orthogonal to all other rows) are removed from H.

5. The graph H is decomposed into the minimum number of complete subgraphs $H_1, \ldots, H_T$ using Algorithm 4.
6. The subgraphs $H_1, \ldots, H_T$ are encoded by binary codes of the minimum length $R = intlog_2T$ using Algorithm 6.
7. R columns that correspond to the positions of the codes of the subgraphs $H_1, \ldots, H_T$ are added to the matrix W. In row i of W, the positions of the additional columns are filled by the code of the subgraph H_t, $t = \overline{1, T}$ containing the node a_i, $i = \overline{1, M}$. The other positions of the additional columns in W are filled by zeros.
8. The contents of the row i of W is used as the code of the internal state a_i, $i = \overline{1, M}$.
9. Stop.

The undetermined values in the matrix W are replaced at Step 2 of Algorithm 2 using the following algorithm.

Algorithm 3

1. Consider the undetermined elements in the matrix W one by one. If all such elements have been examined and modified, then go to Step 4.
2. Let the current undetermined element in W be at the position (i, j), $i = \overline{1, M}$, $j = \overline{1, N}$. For row i of W, two values k_0 and k_1 are found, where k_0 is the number of rows orthogonal to row i in which the undetermined element is replaced by 0 and k_1 is the similar number of rows with the undetermined element replaced by1.
3. If $k_1 > k_0$, then the current undetermined element in W is replaced by 1; otherwise, it is replaced by 0; then, return to Step 1.
4. Stop.

The decomposition of the graph H into the minimum number of complete graphs $H_1, \ldots, H_T$ (at Step 5 of Algorithm 2) is solved approximately by the following algorithm.

Algorithm 4

1. Set $T: = 0$.
2. Set $T: = T + 1$. In the graph H, find a complete graph H_T with the maximum number of nodes.
3. Remove the vertices of H_T from the graph H.
4. If the set of nodes of H is not empty, the go to Step 2; otherwise, go to Step 5.
5. Stop.

The maximal complete subgraph H_t, $t = \overline{1, T}$ at Step 2 of Algorithm 4 can be found approximately using the following algorithm.

Algorithm 5

1. Find a node a_i in H with the greatest local degree.
2. Include a_i into the graph H_t.
3. Among all the nodes of H not included in H_t, find a node a_i connected to all the nodes of the subgraph H_t. If several such nodes are found, choose a node with the greatest local degree among them.

4. If a node connected to all the nodes of the subgraph H_t was found at Step 3, then go to Step 2; otherwise, go to Step 5.
5. Stop.

To encode the subgraphs $H_1, \ldots, H_T$ (Step 6 of Algorithm 2) the following algorithm is used to minimize the area of implementing the transition functions.

Algorithm 6

1. Calculate the length R of the codes of the subgraphs $H_1, \ldots, H_T$: $R = intlog_2T$.
2. Form the set K of binary codes of length R.
3. The subgraph containing the initial state a_1 is encoded by the zero code from K.
4. If all the subgraphs $H_1, \ldots, H_T$ are encoded, then go to Step 5; otherwise, find among the not yet encoded subgraphs $H_1, \ldots, H_T$ a subgraph H_t for which

$$\sum_{a_i \in H_t} |C(a_i)| = \max,$$

where $|A|$ is the cardinality (the number of elements) of the set A. To encode the subgraph H_t, the code with the minimum number of unities is chosen in the set K. Go to Step 4.
5. Stop.

Example. Let us apply the proposed method for designing the FSM described by the state diagram shown in Fig. 3. The state diagram nodes correspond to the internal states $a_1, \ldots, a_5$ of this FSM, and its edges correspond to the FSM transitions. Beside each arc, the value of the input vector that triggers the transition and, separated by a slash, the value of the output vector formed on this transition are indicated. In this example, the FSM has five states, one input variable, and three output variables.

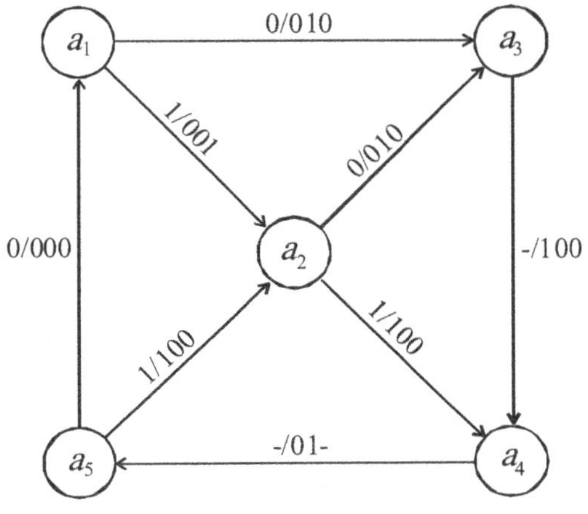

Fig. 3. The state diagram of a Mealy FSM

In this example, conditions (1) are violated for the state a_2 because $Y(a_1, a_2) \neq Y(a_5, a_2)$; therefore, a_2 is split into two states a_{2_1} and a_{2_2}. The state diagram of the FSM obtained upon splitting the state a_2 is shown in Fig. 4.

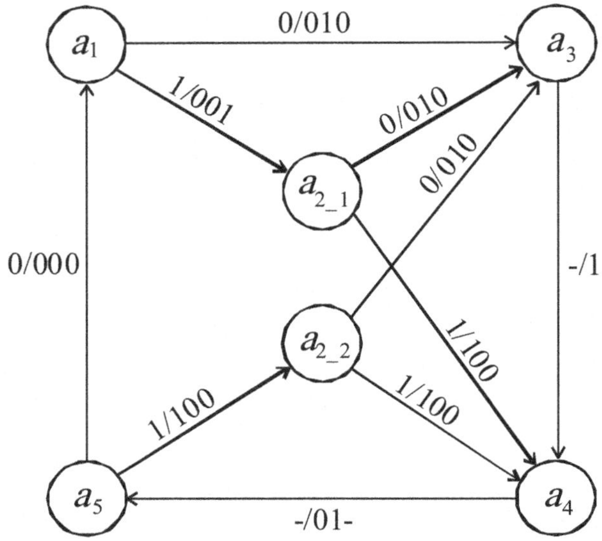

Fig. 4. The state diagram of a D class FSM upon splitting the state a_2 into two states a_{2_1} and a_{2_2}

The encoding of the internal states begins with constructing the matrix W (Table 1). The matrix W contains one undetermined element "-" at the position corresponding to the state a_5 and the output variable y_3. For this element, we have $k_0 = 4$ and $k_1 = 5$; since $k_1 > k_0$, the undetermined element in W is replaced by unity.

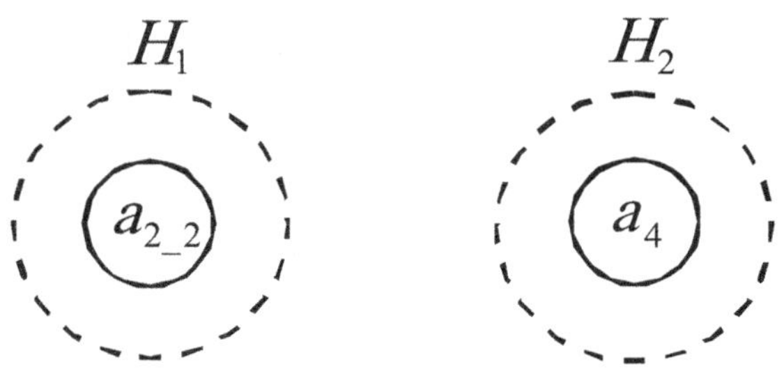

Fig. 5. The graph H of orthogonality of the rows of the matrix W and its decomposition into the subgraphs H_1 and H_2

Figure 5 shows the orthogonality graph *H* of the rows of *W* after the nodes connected to all other nodes are deleted. The graph *H* contains only two nodes a_{2_2} and a_4 that are not orthogonal to each other. For this reason, *H* is decomposed into two complete subgraphs H_1 and H_2, where H_1 contains the node a_{2_2}, and H_2 contains the node a_4. For our example, $T = 2$; therefore, the number of state encoding bits $R = intlog_2 2 = 1$. Since $|C(a_{2\_2})| = 1$ and $|C(a_4)| = 3$, the subgraph H_1 is encoded by 1, and H_2 is encoded by 0.

The matrix *W* together with the additional column u_1 for the orthogonalization of the rows is shown in Table 2.

Table 1. The matrix *W* for state assignment of a class D FSM

a_i	y_1	y_2	y_3
a_1	0	0	0
a_{2_1}	0	0	1
a_{2_2}	1	0	0
a_3	0	1	0
a_4	1	0	0
a_5	0	1	–

Table 2. The matrix *W* with the additional column u_1 for the orthogonalization of the rows

a_i	y_1	y_2	y_3	u_1
a_1	0	0	0	0
a_{2_1}	0	0	1	0
a_{2_2}	1	0	0	1
a_3	0	1	0	0
a_4	1	0	0	0
a_5	0	1	1	0

6 Experimental Study

The efficiency of the proposed method for designing the class D FSMs was tested on the FSM benchmarks MCNC [12]. For this purpose to each benchmark of the FSM the considered synthesis method was applied. Both finite state machines, the initial class A FSM and synthesized the class D FSM, were described in language Verilog. Then standard implementation on FPGA of FSMs by means of CAD Quartus II version 18.1 was fulfilled. It is the synthesis parameters by default of CAD Quartus were used. As criteria of optimization the implementation cost (*C*), defined by the number of used logical elements LUT was considered.

Table 3 shows the results of experiments for various FPGA families, where C_A and C_D are the numbers of the logical elements used in the implementation of the A and D class FSMs, C_A/C_D is the ratio of the corresponding parameters, and *mid* is the mean value of the parameter. The data in Tables 3 show that the proposed method designing the D class FSM reduced the implementation cost of the FSM by a factor of 1.85–2.67 on average and by a factor of 5.7 for certain cases.

Table 3. Implementation of classes A and D FSMs in Intel FPGAs

	Cyclone III			MAX II			Arria GX, Stratix III		
	C_A	C_D	C_A/C_D	C_A	C_D	C_A/C_D	C_A	C_D	C_A/C_D
Keyb	70	86	0.81	66	77	0.86	90	49	1.84
Lion	10	8	1.25	10	8	1.25	5	4	1.25
S1	118	40	2.95	137	37	3.70	171	30	5.70
S27	22	9	2.44	22	8	2.75	14	6	2.33
shiftreg	9	5	1.80	9	5	1.80	9	4	2.25
mid			1.85			2.07			2.67

7 Conclusions

In the description of the FSM, the possible undetermined values of the output vectors should be indicated. This helps better minimize the internal states and state assignment of the class D FSM, i.e., minimize the area of implementation.

In this work, we used the values of the output values of the Mealy FSM for state assignment; however, we could also use the values of the input variables for the same purpose. Therefore, it would be interesting to investigate the possibility of using the values of the input variables as the codes of the internal states of Mealy FSMs and to combine these two approaches.

Acknowledgements. The present study was supported by a grant S/WI/3/2018 from Bialystok University of Technology and founded from the resources for research by Ministry of Science and Higher Education.

References

1. McCluskey, E.J.: Reduction of feedback loops in sequential circuits and carry leads in iterative networks. Inform. Control **6**, 99–118 (1963)
2. Pomeranz, I., Cheng, K.T.: STOIC: state assignment based on output/input functions. IEEE Trans. Comput. Aided Des. Integr. Circuits Syst. **12**(8), 1123–1131 (1993)
3. Forrest, J.: ODE: output direct state machine encoding. In: Proceedings of European Design Automation Conference, EURO-DAC 1995, pp. 600–605. IEEE, Brighton (1995)
4. Klimovicz, A.S., Solov'ev, V.V.: Structural models of finite-state machines for their implementation on programmable logic devices and systems on chip. J. Comput. Syst. Sci. Int. **54**(2), 230–242 (2015)
5. Salauyou, V., Ostapczuk, M.: State assignment of finite-state machines by using the values of input variables. In: Saeed, K., Homenda, W., Chaki, R. (eds.) CISIM 2017. LNCS, vol. 10244, pp. 592–603. Springer, Cham (2017). https://doi.org/10.1007/978-3-319-59105-6_51

6. Barkalov, A., Titarenko, L., Chmielewski, S.: Mixed encoding of collections of output variables for LUT-based Mealy FSMs. J. Circuits Syst. Comput. **28**(8), 1950131 (2019)
7. Klimowicz, A.: Area targeted minimization method of finite state machines for FPGA devices. In: Saeed, K., Homenda, W. (eds.) CISIM 2018. LNCS, vol. 11127, pp. 370–379. Springer, Cham (2018). https://doi.org/10.1007/978-3-319-99954-8_31
8. Nag, A., Das, S., Pradhan, S.N.: Low-power FSM synthesis based on automated power and clock gating technique. J. Circuits Syst. Comput. **28**(5), 1920003 (2019)
9. Tao, Y.Y., Zhang, L.J., Wang, Q.Y., Chen, R., Zhang, Y.Z.: A multi-population evolution strategy and its application in low area/power FSM synthesis. Nat. Comput. **18**(1), 139–161 (2019)
10. Salauyou, V., Bulatowa, I.: Synthesis of high-speed ASM controllers with Moore outputs by introducing additional states. In: Saeed, K., Homenda, W. (eds.) CISIM 2018. LNCS, vol. 11127, pp. 405–416. Springer, Cham (2018). https://doi.org/10.1007/978-3-319-99954-8_34
11. Klimowicz, A.: Performance targeted minimization of incompletely specified finite state machines for implementation in FPGA devices. In: Proceedings on 20th Euromicro Conference on Digital System Design (DSD), pp. 145–150. IEEE, Vienna (2017)
12. Yang, S.: Logic synthesis and optimization benchmarks user guide. Version 3.0. Microelectronics Center of North Carolina (MCNC), North Carolina, USA (1991)

Integration of Enterprise Resource Planning (ERP) System in Value Based Management of the Corporation

Elena V. Savenkova, Alexander Y. Bystryakov, Oksana A. Karpenko(✉), Tatiana K. Blokhina, and Andrey V. Guirinsky

Economics Department, Peoples' Friendship University of Russia (RUDN University), Miklukho-Maklaya Street 6, Moscow 117198, Russia
{savenkova-ev, bystryakov-aya, karpenko-oa, blokhina-tk, guirinsky-av}@rudn.ru

Abstract. The use of the automated systems in the joint stock company allows to achieve compliance of internal business processes of activity and safety of information. Besides Enterprise resource planning (ERP) system can be used in value based management. In the modern Russian market technological and commercial stability is based on two major products 1C: ERP and SAP ERP. This research is focused on the study of necessity, shortcomings and advantages of use of ERP SAP and 1C:ERP by the enterprise. The object of the research is JSC Acron. The group Acron is one of the leading vertically integrated producers of mineral fertilizers in Russia and over the world. The authors suggest to introduce the ERP system in the Acron company on the basis of 1C the platform because the company is focused on the Russian market and considers the Russian features of business. The effect of ERP system implementation in JSC Acron was investigated. The carried out calculation allows to draw conclusions that introduction of Enterprise resource planning at above mentioned JSC allows to increase activity efficiency of the company by increasing its value.

Keywords: Value based management · Cash flows of firm discounted to the present moment · Modeling of operating activities of the company · Enterprise resource planning (ERP) system

1 Introduction

The experience of joint stock companies in many countries confirms high efficiency of value based management. It may become one of the tools of increasing attractiveness in their investment. The Russian companies use the value based management, increasing the competitiveness and provide the most effective use of all factors of production, reaching the leading positions in the industry.

Increase the value of the corporation in the long perspective satisfies the interests of shareholders. They can define optimal strategy of business and develop ways of its implementation and chose the most perspective investment projects to start.

W. Zamojski et al. (Eds.): DepCoS-RELCOMEX 2020, AISC 1173, pp. 554–564, 2020.
https://doi.org/10.1007/978-3-030-48256-5_54

From the point of view of income approach in value based management the price of the business is the sum of the discounted cash flows taking into account risks and costs of the capital. Therefore, the size of cash flow of the company is influenced by profit, taxes and the sum of capital investments, changing the working capital, expenses of the company and cost of the production.

So as the value of the company is calculated on the basis of future cash flows of firm discounted to the present moment, this indicator is becoming the base of value-based management.

The most effective method of the analysis of the factors influencing on the value of the company is modeling of operating activities of the company. The creation of the control system of business based on information technologies may improve value based management. The information system of the firm should meet requirements of the company and promote increase in production efficiency. Therefore use of the automated systems in the joint stock company allows to achieve the compliance of internal business processes with activity and safety of information. It can reduce cost and increase attractiveness of the company for external investors and it is one of the main facilities to increase the enterprise value. The most popular system is ERP (Enterprise Resource Planning) in technology of the enterprise resource management.

2 Methodology

Enterprise resource planning is the resource management system in the company which allows to store and process the majority of data, crucial for work of the company. ERP provides collaboration of all business processes of the company in one system, provides the leadership for the company in expeditious obtaining information of all kinds of activity, allows to plan and control the operations of the organization. There are following modules in Enterprise resource planning system:

- cash flow management;
- stockpile management;
- steering purchases and orders;
- planning sales;
- system of monitoring of transport
- interactions with debtors and creditors;
- accounting bank and cash operations;
- steering contracts;
- tax accounting;
- accounting;
- means of analytical financial accounting;
- financial analysis;
- financial planning;

ERP system is generally used for planning resources. These system has to answer questions not only "as was" and "as is" in the company using, but also "as will be", "as has to be", i.e. have pins and needles in the set parameters in purchases and production [1].

The main feature of this system is that it represents "identification and planning of all resources of the enterprise which are used at implementation the whole cycle of production, realization and interaction with clients; effective planning, conducting operational accounting, monitoring procedure and implementation of the analysis when using resources of the enterprise which are necessary for purchases, production and realization [2]. Thus, when we consider the automated system ERP we bear in mind that the automated systems allows to solve effectively difficult complex tasks including optimum distribution of business resources, ensuring fast and effective delivery of goods and services to the consumer [3].

In the modern Russian market the technological and commercial stability is making by two major products 1C:ERP and SAP ERP. Export of products by 1C company is fulfilling in the localized versions for the following countries: Russia, Belarus, Ukraine, Kazakhstan, Kyrgyzstan, Tajikistan, Georgia, Moldova, Uzbekistan, Romania, Latvia, Lithuania, Estonia, Azerbaijan. SAP ERP is international systems which is used all over the world [4]. The main aim of our research is to consider necessity, shortcomings and advantages for use of ERP SAP and 1C:ERP in the enterprise and to take a decision about implementation one of them at the chemical joint stock company.

3 Literature Review

The researches dedicated to Enterprise Resource planning systems stated that these integrated systems allow managers to share information, and this information can be used to monitor firm performance [5, 6]. Poston and Grabskistate that one of the two chief expectations of ERP system implementations is enhanced managerial decision making via the provision of accurate and timely enterprise-wide information [7]. Hunton et al. posit that these potential advantages allow ERP adopters to financially outperform non-adopting firms and their results support this hypothesis [8].

ERP systems integrate different areas among the organization [9]. Enterprise Resource planning could be a great tool that builds sturdy capabilities, improves performance, supports higher cognitive process, and provides competitive advantage for businesses by providing management with right and updated information [10]. Its main task is to manage all the resources, data, and actions needed to complete business processes such as manufacturing, sales, finance, marketing, human resources and accounting [11]. Literature stresses the importance of ERP benefits. Unfortunately there is not any research concerning the shortcomings and advantages of use of ERP systems based on SAP or 1C.

4 Analysis

Many researchers have studied ERP systems, their implementation and the success factors. In our research we will analyze the possibility to introduce the automated control system which is the component of value based management in PJSC Acron.

The main requirements to the system: it has to be the smallest at cost and promote increase in value of the company as main indicator of efficiency in its activity.

Automation of this sort are quite expensive, each project is individual and the price depends on the performed works, the scale of the organization. Besides purchase of the license we will consider the cost of introduction, version and support of all services. The acquisition of this system can bring considerable benefit of the company.

The group Acron is one of the leading vertically integrated producers of mineral fertilizers in Russia and the world. The company unites two chemical plants and mining and processing works on extraction of phosphatic raw materials in Russia, develops potash fields in Russia and Canada [12].

From 2005 to 2018 the Acron Group realized the strategy of vertical integration based on own production of all three main input products for production of complex fertilizers – nitrogen, phosphorus and potassium. The effective chain of the interconnected business segments is the cornerstone of business model of the Acron Group: extraction of raw materials, chemical production, logistics and distribution. Vertical integration allows to control all chain of creation of value added and provides efficiency and competitiveness.

The chemical companies of Group are located in Veliky Novgorod (PJSC Acron) and in the Smolensk region (PJSC Dorogobuzh), the main office is situated in Moscow. The group conducts own extraction of phosphatic raw materials in Murmansk region (JSC SZFK) and realizes the project on development of the potash field in Perm region (CJSC VKK), has own transport and logistic infrastructure which is turning on three port terminals on Baltic and marketing networks in Russia and China.

First we will calculate the value of the company within model of discounting of cash flows without possible introduction of ERP system. We forecasted FCFF for 2019,2020,2021,2022 years (Table 1)

Table 1. Calculation of value of the PJSC Acron

Indicator	2018	2019F	2020F	2021F	2022F
Free cash flow (FCFF)	−4795 113	5 280 411	5 84 447	6418 482	6987 518
Discount rate	7,26%	7,26%	7,26%	7,26%	7,26%
The discounted cash flow (DCFF)	−4 469 718	4 588 072	4 737601	4845 709	4 917 327
The discounted cash flow (DCFF) in 5 years	14 618 991				

Source: calculated by the authors on financial statements of PJSC Acron.

Further it is necessary to determine the cost of the company at post-forecast period – the terminal value (Terminal Value). For calculation of terminal value of the company the model of constant growth of M.J. Gordon Growth Model assuming that the further growth of business will go stable rates is used. Calculation is carried out on formula (1):

$$TV = \frac{FCF}{WACC - g} \tag{1}$$

where TV – terminal value;
$FCF(1+g)$ – cash flow in the first year of post-forecast period;
WACC – weighted average cost of the capital;
g – long-term gain rates of cash flow.

The average annual growth rate (CAGR) was 4,0%, cash flow in the first year of post-forecast period – 7 267 019 thousand rubles, and at last, the terminal value (TV) – 221 555 458 thousand rubles. Thus, the value of the PJSC Acron company calculated by method of the discounted cash flows was 236 174 million rubles.

Now we will try to raise the value of the PJSC Acron by implementation of ERP system. ERP the automated systems store the historical (saved-up) data about occurring and occurring (at the moment time) in the company. Modules of planning and optimization of the following resources enter the automated system using the methodology of ERP: finance, material stocks, shots; and functions of operational and financial accounting which are realized in systems and directed for fixing of economic operations.

The degree of applicability of information defines type of administrative information or financial information. If information is used only in firm, it is possible to speak about private (internal) administrative information". As for external consumers, the regulated (financial) information is used. In the registration module of the automated ERP system the current data the companies, (operational) about economic activity are fixed, in this module there are no functions of automation in planning and comparison "the plan - the fact". Thus, in the registration module records of the current events are carried out, selection of records on the periods which provide various options of formation of the reporting of the public companies is carried out and allow to open information on segments.

The methodology of creation of the automated ERP systems is based on use of the uniform database which contains all corporate business information, the system provides access to data of employees according to their roles in the company. The choice of the platform is a complex process. It takes place within the compliance of system to functional requirements, such as possibility of system in standard configuration. There are special technical requirements – infrastructure power, authorization, safety, integration and other aspects. Also there is a cluster of nonfunctional requirements – scalability, reliability, fail-safety, reliability of vendors, existence of examination and other parameters.

Statistics in introduction terms of 1C: on average prior to trial operation there pass about 3 months, and industrial—8, 5 months. Average project cost is 53 thousand rubles counting on one automated workplace. The SAP is more expensive.

Functionality of system allows to make change of data to users with certain roles. The opportunities of Enterprise resource planning allow to solve problems of automation in the company and to carry out activity in the difficult automated Enterprise resource planning areas of all functional divisions of the company.

At 1C there is system of the class ERP – "1C: UPP" which is intended for automation of planning and resource management of the enterprise. This decision allows to organize the complex information system conforming to corporate, Russian and international standards and providing financial and economic activity of the enterprise without specialized software solutions use.

Let's consider shortcomings and advantages of use of ERP on SAP and on 1C.

We suppose that, 1C has number of the following advantages in the Russian software market in comparison with other ERP-systems:

The SAP and Oracle experience has big difficulties at adaptation of decisions to the Russian conditions of business. It is connected with the fact that business processes at the Russian enterprises and are not accurately recorded, and is often changing under requirements of business and the external environment. In the Russian firms it is difficult to reconstruct business processes under requirements accepted for the western ERP systems.

The built-in 1C language has many common features with other languages, such as Pascal, Java Script, Basic that facilitates its development by the beginning developers. However, it is not direct analog any of the listed languages. In turn, ABAP with which SAP developers work significantly differs from them, however, it not so important as the ecosystem has already developed and there are competences in the market. At the same time SAP has expanded tool kit of development. The platform of the user Web Dynpro interface can use Java, and the Fiori platform – HTML5.SAP offers significantly ampler opportunities, and 1C partially compensates it by existence of partner decisions in the field of, for example, TORO or warehouses, but the maturity level of such decisions is lower. Advantage of 1C is openness for standard decisions and simplicity of their modernization and at the same time emphasizes that any introduction means completion of the decision under requirements of the specific customer.

The extensive experience accumulated by SAP for many years of existence in the market is its main advantage. Therefore the SAP company has huge advantage, for example, the correct introduction of software products programming. As for 1C, according to experts, it is very well elaborated at the functional level. The interface of system is ergonomic and modern. The 1C architecture of system was very successful and develops in high gear. The scalability, productivity and fail-safety of the 1C platform allows to do large projects on thousands and tens of thousands of jobs.

One more indicator of quality of the software product is its convenience. 1C represents strong basic level from the point of view of ergonomics, however for business modules of the logician is less harmonious. As for SAP ERP, in spite of the fact that initially the system was under construction as the engineering decision with the minimum adaptation under the user, in recent years in new products there was Fiori interface, more progressive and friendly to the user. SAP provides decisions at the level of the best samples. Let's state all advantages and shortcomings of systems, based on 1C and SAP in the Table 2.

1C was initially created for automation of procedures and jobs, SAP is intended for resource management of the enterprise – ensuring continuity of deliveries, uninterrupted production, timely correction of the production plan, change-over of capacities, etc. Basic purpose of SAP is not automation of jobs, and the organization of performance of difficult production and logistic functionality. The decisions 1C are developed together with development of business in Russia. Therefore they comprise the Russian experience of business management, unlike SAP which has the western traditions. So, we suggest to introduce in the Acron company the ERP system on the basis

Table 2. Advantages and shortcomings of the SAP ERP systems and 1C:ERP.

Group of parameters	SAP ERP	1C:ERP
Price	The price of SAP became just very high for most of the Russian companies	1C offers cost-effective product which in many cases will be effective on combination of the price and quality
Maturity	The decision of SAP is based on thousands of introductions in the large companies which were created by the best managers	Decisions 1C developed together with development of business in Russia therefore they comprise the Russian business management
	The system is working for more than 20 years at the world market. Many largest companies use it because of the world practices in business management	1C is a local specialized program for Russia and neighbour countries
	SAP is really international standard of Enterprise resource planning. The vast majority of multinational corporations which work in our country apply SAP	There was no case when the client would not receive as a result functionality and necessity
Functionality	SAP has functionality under large number of verticals in all kinds of businesses. The functionality peculiar to concrete vertical is activated under the specific customer. There is an opportunity to use the best practicing who are put in SAP. Nothing needs to be thought out. It helps to avoid mistakes	1C can have advantage over SAP in the Russian account and the expanded functionality of ERP is not required
	Uniform working environment for all blocks of the company, including production, logistics, sales, the international operations, efficiency (data from transaction level on the level of the reporting are broadcast daily)	Lack of the opportunity checked and confirmed with experience of introduction to provide through support of chain of process of chemical production
	Variability of processes and extensive set of functional areas with possibility of expansion of automation on processes, adjacent to ERP, within the uniform integrated platform on the basis of SAP. For example, in the field of purchases, steering of warehouses, steering of transport	Existence of significant functional gaps in support of production, transportation, sales and international operations

(*continued*)

Table 2. (*continued*)

Group of parameters	SAP ERP	1C:ERP
	The business content allows to reduce time at introduction of the reporting	There is a convenient system of the analytical reporting
	In the new solution of SAP S/4HANA implementation of such new concepts as In Memory, big data, the Internet of things. It is declared that will it give the chance to the enterprise to receive all range of the decisions integrated with each other from one producer	It combines complex of ready decisions for business with the flexible tools allowing to realize quickly know-how of the concrete enterprise
Scalability	Proposes solutions of the maximum scalability	1C has limit of scalability at the level of 1000–2000 users
	Productivity questions extremely seldom are critical for SAP within big installations	The scalability, productivity and fail-safety of the platform allow to do large projects on thousands and tens of thousands of jobs
Workmanship	The logic and workmanship are significantly higher, than in 1C	In modern branch of development 1C - 1C:ERP is significant shortcomings of structure, of stability of core of system so far
	There is no opportunity to change documents backdating. The so-called "auditor trace" allows to look and understand origin of any figure in the final reporting. SAP gives transparent picture of situation both for owners of the company, and for tax authorities	1C supports recarrying out documents. Thus, at each reconducting of operation there is recalculation of the price of materials according to all documents of system. It leads to vital issues with productivity. The architecture 1C is not calculated on processing of large number of documents
	The methodology of introduction of SAP is focused on large projects and allows if it to adhere to realize projects of any scale with the minimum risks	Approaches of 1C first of all are focused on small and medium business
Quality of technical support	The centralized support which is provided by vendor and also the partners having the certified escort service. The quality of this support is regulated by standards of SAP, it identical to all clients	The quality of support 1C depends entirely on the integrator or even on the specific person on the party of the integrator which introduced 1C at the client

(*continued*)

Table 2. (*continued*)

Group of parameters	SAP ERP	1C:ERP
Efficiency of updating	SAP, in whatever country it would work, meets requirements of the legislation	The efficiency of updating of products 1C on changes of requirements of the legislation of the Russian Federation became the inaccessible standard long ago, and can cause even feeling that these changes are developed with direct participation of representatives “1C
Possibility of control and completion	In SAP the core is significantly better isolated that allows to carry out updating of versions with smaller labor costs	There is a wide partner network, huge number of introductions therefore it is possible to find something similar in the project for any situation
	SAP has expanded tool kit of development. The platform of the user Web Dynpro interface can use Java, and the Fiori platform uses HTML5	Tools of control and completion of products “1C are the most convenient and effective
		1C allows to use the written applications in different options – local, client-server, distributed, cloudy, in different operating systems, with various DBMS. Besides, the platform allows to develop mobile applications for iOS and Android

of the platform 1C so as the company is focused on the Russian market and considers the Russian features of business. Further we will estimate as it will affect the listed measures in the price of the company. For this purpose we will substitute the corrected data in model of cost of the company (Table 3).

Thus, the value of the company calculated by method of the discounted cash flows was 277 251 957 thousand rubles. Use of business value allowed to see visually that optimization of expenses and assets of the PJSC Acron company, by optimization of expenses as a result of introduction of new technologies – introductions of modern Enterprise resource planning can positively affect creation of free cash flows of firm. It can increase the cost of the company by 17% in comparison with that would be without introduction of system.

Table 3. The calculation of the value of the PJSC Acron with ERP:1C.

Name	2018	2019F	2020F	2021F	2022F
Revenue	59542997	63971211	68399426	72827641	77255856
Cost of sales	35205725	38539988	41874251	45208514	48542777
Gross profit	24337272	25431224	26525175	27619127	28713079
Business expenses	3765035	4065859	4 366682	4667505	4968328
Management expenses	3725818	3964892	4 203966	4443040	4682114
Operating profit	16846418	17400473	17954528	18508583	19062637
Interest receivable	164811	0	0	0	0
Interest payment	10860356	12330114	13799872	15269630	16739388
Other income	12007894	13166747	14325600	15484453	16643306
Other expenses	7927050	6798167	5669284	4540400	3411517
Profit before the taxation	12911830	14572639	16398258	18223878	20049498
Income tax (20%)	2582366	2914528	3279652	3644776	4009900
Net profit	10329464	11658111	13118607	14579103	16039598
Depreciation	1894813	2035730	2176648	2317565	2458482
Capital expenditure (CAPEX)	9870915	10605016	11339116	12073216	12807317
Change of net working capital	6995832	−2510656	−2510656	−2510656	−2510656
Free cash flow (FCFF)	−4510622	5 599482	6 466794	7334107	8201420
The discounted cash flow (DCFF)	−4204532	4 865308	5 237605	5536970	5771587
The discounted cash flow (DCFF) in 5 years	17 206 937				
Terminal value	260 045 020				
Value of the company	277 251 957				

Source: calculated by the authors.

5 Conclusion

It is necessary to underline that the choice of integrated management system for the enterprise is not simple action. It is not only question of money. We have to find out if it is necessary or it is not necessary to invest in Enterprise resource planning introduction. It is question of competitiveness maintenance and leadership of the company in the market. Return from investments into ERP system goes from ability of the company to be the best with new business processes. The costs of ERP possession should be planned and considered. The carried out calculation allows to draw conclusions that introduction of Enterprise resource planning at manufacturing enterprise allows to increase efficiency of activity of the company by increase in its value. The moment of ERP system introduction is a competitive step on the way of increase management efficiency for any company.

Acknowledgements. The publication was prepared with the support of the "RUDN University program 5-100".

References

1. Review: IT market [An electronic resource]. http://www.cnews.ru. Accessed 20 Nov 2019
2. The certified franchisees 1C [An electronic resource]. https://1c.ru. Accessed 20 Nov 2019
3. Enterprise management systems (ERP) market of Russia [An electronic resource]. http://www.tadviser.ru. Accessed 20 Nov 2019
4. Review of the Russian market of ERP [An electronic resource]. http://www.tadviser.ru. Accessed 20 Nov 2019
5. Oliver, R.: ERP is dead: long live ERP. Manag. Rev. **88**, 12–13 (1999)
6. Davenport, T.: Mission critical: recognizing the promise of enterprise systems. Harvard Bus. Rev. **76**, 121–131 (2000)
7. Poston, R., Grabski, S.: Financial impacts of enterprise resource planning implementations. Int. J. Account. Inf. Syst. **2**, 271–294 (2001)
8. Hunton, J.E., Wright, A.M., Wright, S.: Are financial auditors overconfident in their ability to assess risks associated with enterprise resource planning systems? J. Inf. Syst. **18**, 7–28 (2004)
9. Rajan, C.A., Baral, R.: Adoption of ERP system: an empirical study of factors influencing the usage of ERP and its impact on end user. IIMB Manag. Rev. **27**(2), 105–117 (2015)
10. Ahmed, Z., Zbib, I., Arokiasamy, S., Ramayah, T., Chiun, L.M.: Resistance to change and ERP implementation success: the moderating role of change management initiatives. Asian Acad. Manag. J. **11**(2), 1–17 (2006)
11. Nawaz, M.N., Channakeshavalu, K.: The impact of enterprise resource planning (ERP) systems implementation on business performance. Asia Pac. J. Res. **2**(4), 13 (2013)
12. Financial Statements: Acron. Dorogobuzh. IFRS Reporting

Landscape Imaging of the Discrete Solution Space

Czeslaw Smutnicki(✉)

Department of Computer Engineering, Wroclaw University of Science and Technology, Wrocław, Poland
czeslaw.smutnicki@pwr.edu.pl

Abstract. We consider a general class of NP-hard combinatorial optimization problems with solutions represented by permutations. Assuming that each problem is solved then by suitable approximate method based on local search or population based approaches, the landscape analysis of the solution space is carried out. Trajectories performed by various algorithms differ each other and can be controlled by parameters derived from the image. The topic of space images appears seldom in the literature, so at the beginning we refer shortly to the state-of-the-art on this subject. Then we propose our own original concepts of space imaging. As an image we are taking the two-dimensional and three-dimensional Euclidean cube and the sphere. Necessary distance measures in the solution space and images space are provided. The new method of visualization is proposed based on reference points derived from the specific auxiliary optimization task. The method is used to track manifold search trajectories performed in the solution space. Variability of the images is illustrated by few experimental research.

Keywords: Scheduling · Space landscape · Optimization

1 Introduction

Metaheuristic algorithms dedicated for practical combinatorial optimization problems, scheduling, routing, timetabling, balancing, warehousing, and so on, usually use certain combinatorial objects (as an example permutations, set partitions, sequences) to represent the solution. Many well-known in the literature cases, as an example the Traveling Salesman Problem (TSP), single-machine scheduling (SP), Quadratic Assignment Problem (QAP), operate on permutations on the set $N = \{1, 2, \ldots, n\}$, which represent the huge set of possible solutions. We consider this case as the challenge for the space landscape analysis. Thus, in this paper we operate on the general class of NP-hard problems

$$f(\pi^A) = \min_{\pi \in A \subseteq \Pi} f(\pi), \tag{1}$$

National Science Centre, grant OPUS 8 no. DEC 2017/25/B/ST7/02181.

W. Zamojski et al. (Eds.): DepCoS-RELCOMEX 2020, AISC 1173, pp. 565–574, 2020.
https://doi.org/10.1007/978-3-030-48256-5_55

where π is the permutation on N, $f(\pi)$ is the goal function value defined for each given $\pi \in \Pi$, Π is the set of all permutations on N, A is the subset of solutions searched by an approximate algorithm A, $|A| \ll |\Pi|$. Our overall aim is to display graphical representation of A, because of the numerousness of Π.

Mentioned the above problems, being instantiations of the formula (1), cause significant troubles in finding solution as close as possible to the optimal one, [11]. There have been already identified several fundamental factors responsible for this phenomenon: huge number of haphazardly distributed local extremes, roughness, curse of dimensionality, NP-hardness, balance between running time and quality of $f(\pi^A)$, [11]. That is why in recent years parallel approaches dominate traditional sequential methods, [1,2,12]. In order to observe the behavior of the solution algorithms (sequential as well as parallel) many researchers uses various charts, plots and graphics, [8,9]. Visualization can be applied for: (1) graphical representation of the solution space, (2) detecting features of the solution space, (3) analyzing local extremes distribution, (4) detection of valleys (attractors), (5) visualization of the crucial elements of the algorithm as an example neighborhood, population, swarm, (6) trajectory visualization and examination, and (7) guiding the search. Because solution space has unknown in advance landscape and the trajectory tracking has a dynamic unpredictable character, a self-adapting, low-costly method of visualization is highly desirable.

Actually, thinking about the visualization depends on the philosophy of used metaheuristics which define set A. The approach is extremely different in cases: (a) local search approaches (LS) with single search trajectory or multiple parallel independent/cooperative search trajectories based on local neighborhoods, and in (b) the population-based (PB), distributed population-based or swarm approaches. Both approaches consider distance measures, however in the different context and for different purposes: compact local neighborhoods versus dispersed population. The excellent review of space features oriented on PB-type algorithms one can find in [9]. Different measures are used for the LS-type methods (e.g. simulated annealing, simulated jumping, tabu search, random search, greedy random search, variable neighborhood search) where the given solution is modified slightly step-by-step using so-called moves made in the neighborhood. Hence, this paper presents an alternative to [9] stream of the landscape analysis.

2 Landscape Analysis

Approximate LS-type algorithms use several notions in order to justify the choice of the subset A in (1) and the searching strategy. A few of these notions will be elucidated next. The overall aim of the landscape imaging is to define the transformation from Π on the plane or cube such that the "intuitions" of these notions remain unchanged. The example question is "whether 2D image of the simulated annealing trajectory in the space Π suggests convergence of the method?".

2.1 Local Optima, Plateau, Modality

Local optimum π_* is defined only in the context of a local neighborhood $N(\pi_*) \subseteq A \subset \Pi$. The formal definition is mathematically obvious

$$f(\pi_*) < f(\pi), \quad \forall \pi \in N(\pi_*) \subset \Pi, \tag{2}$$

and depends on the notion $N(\pi)$. Neighborhood $N(\pi)$ contains solutions $\sigma \in N(\pi)$ "close" to π, i.e.

$$D(\pi, \sigma) \leq d, \tag{3}$$

where $D(\pi, \sigma)$ denotes the distance between solutions (permutations) discussed in detail in Sect. 3. Since one admits only integer values $d = 0, 1, \ldots$, the most common approach is to set $d = 1$ in order to ensure a small computational complexity of the neighborhood search. Indeed, neighborhood $N(\pi) = \{\sigma : \ D(\pi, \sigma) = 1\}$ generated by N-moves contains $O(n^2)$ solutions, whereas for $N(\pi) = \{\sigma : \ D(\pi, \sigma) = 2\}$ already $O(n^4)$ solutions. Thus, checking whether π_* is the local optimum may cause different amounts of computations.

Plateau is the region of the solution space with the same goal function value. Allowing operator "$\leq$" in the definition (2) we can perceive plateau as the subset of solutions $\sigma \in B \subseteq A$ such as $f(\sigma) = const$. Such phenomenon is observed, for example, in scheduling problems with tardiness criterion.

Modality evaluates the number of local optima (and their distribution) with respect to the size of the solution space. It can be analyzed either experimentally (dominated approach) as well as theoretically (seldom approach).

2.2 Trajectories

Landscape can be examined by performing the *trajectories* (single, parallel, multiple) passing through the solution space Π. Formally, we define individual trajectory as the sequence of solutions $\pi_0, \pi_1, \pi_2, \ldots \in \Pi$, such that π_0 is given (a starting solution) and $\pi_{i+1} \in N(\pi_i)$, $i = 0, 1, 2, \ldots$. This definition is suitable for LS-type methods. Clearly, we have $D(\pi_{i+1}, \pi_i) \leq d$, $i = 0, 1, \ldots$. We expect that the trajectory ensures theoretical possibility of reaching any solution in the space Π or at least an optimal solution. This feature follows from the proper definition and properties of $N(\pi)$. In the literature there are exist two notions: "strong connectivity property" and "weak connectivity property" tested for $N(\pi)$. Strong connectivity for neighbourhood $N(\pi)$ means that for any starting solution π_0 and any solution $\sigma \in \Pi$ exists a trajectory leading from π_0 to σ. Weak connectivity means that for any given π_0 exists a trajectory leading to the optimal solution (or one of the optimal solutions if many of them exist). One distinguishes several types of trajectories:

1. random: we choose $\pi_{i+1} \in N(\pi_i)$ randomly;
2. adaptive: we choose $\pi_{i+1} \in N(\pi_i)$ using a priority rule, e.g. any descent, steepest descent, greatest descent or similar,
3. reverse adaptive: we choose $\pi_{i+1} \in N(\pi_i)$ using some inverse rule as in p. 2;

4. downhill-uphill: at first we generate an adaptive trajectory until we cannot attain lower goal function value than we perform reverse adaptive trajectory until we cannot obtain the higher goal function value;
5. neutral: we choose $\pi_{i+1} \in N(\pi_i)$ so that $f(\pi_{i+1}) = f(\pi_i)$, trying to increase adaptive distance to the starting point.

Referring to PB-based approaches, the above definition of the trajectory have to be modified. Actually, in each iteration i we have a subset $A_i \subset \Pi$ of dispersed solutions called the population. Sequence $A_0, A_1, A_2, \ldots$ we call the stream (generalized trajectory). One can analyze the behavior of stream (average, dispersion, envelope, and so on). To complete considerations we define the distance between sets of solutions A and B

$$D(A, B) = \min_{\pi \in A,\ \sigma \in B} D(\pi, \sigma). \tag{4}$$

To ensure the convergence to the optimal solution we need a generalization of the mentioned already connectivity property. It means that components of the algorithm should be designed so that there exists the sequence $\pi_0, \pi_1, \pi_2, \ldots$, such that $\pi_i \in A_i$, $i = 0, 1, 2, \ldots$, leading from the initial population A_0 to the population A_s containing the optimal solution. Some authors call this feature evolvability. Notice $D(\pi_{i+1}, \pi_i)$ varies in iterations and may not be bounded.

2.3 Basins of Attraction

The basin of attraction means the subset of solutions in the solution space, starting from which the algorithm performs a trajectory to the same local optimum located in this basin. Convergence to the best solution in the basin depends on the basin size and its structure. Since discrete problems own many local extremes, identification of all basins of attractions is computationally very expensive. The most common approach is to store attractors found during the search introducing certain classification between them (e.g. the distance).

2.4 Ruggedness

One among the most important features of the space landscape is the ruggedness. Intuitively, for small differences between neighboring solutions we observe the flat landscape and thus LS-type algorithms behave quite efficient. The measure of the space ruggedness is the auto-correlation function

$$\rho(d) = 1 - \frac{AVE((C(\pi) - C(\sigma))^2)|_{D(\pi,\sigma)=d}}{AVE((C(\pi) - C(\sigma))^2)} \tag{5}$$

where $AVE((C(\pi) - C(\sigma))^2)$ denotes the average value of $(C(\pi) - C(\sigma))^2$ calculated on the set of pairs $(\pi, \sigma) \in \Pi$, and $AVE((C(\pi) - C(\sigma))^2)|_{D(\pi,\sigma)=d}$ denotes the average value of $(C(\pi) - C(\sigma))^2$ calculated on the set of pairs $(\pi, \sigma) \in \Pi$ so that the distance $D(\pi, \sigma)$ between solutions π and σ is exactly d.

Measure $\rho(d)$ defines the correlation between solutions located on the distance d. The most important for LS-type algorithms is $\rho(1)$. Value $\rho(1)$ close to one means the flat landscape. Value $\rho(1)$ close to zero means the rough landscape.

Because of the huge size of Π landscape roughness is evaluated on the base of random trajectory $\pi_0, \pi_1, \pi_2, \ldots$, redefining (5) with the help of the autocorrelation function

$$r(s) = 1 - \frac{AVE((C(\pi_i) - C(\pi_{i-s}))^2)}{2 \cdot AVE((C(\pi_i) - C(\pi_{i-s}))^2)} \tag{6}$$

where $\pi_1, \pi_2, \ldots, \pi_k$ is a random trajectory. Using $r(1)$ we define the autocorrelation coefficient $\xi = 1/(1 - r(1))$. Greater ξ means flat landscape. It has been verified experimentally that value ξ depends, among others, on the problem size, problem type, constraints, neighborhood type.

2.5 Barriers

Barrier is defined as the expected deterioration of the goal function value necessary to reach one optimum from another through an arbitrary trajectory. It evaluates possibility of going out from the local optimum to continue the search. It has significance for Tabu Search method (length of tabu list), Simulated Annealing (temperature), Simulated Jumping (temperature). Barriers are identified by target-oriented search trajectories and can be collected in the memory. There is no relation between barrier height, slope and the distance to nearest local extreme. This feature depends on the problem type.

3 Solution Space

LS-type metaheuristics uses the notion "move" to define certain sligth modification of the current solution π and to define the local neighborhood $N(\pi) \subset \Pi$ of the current solution. For given $\pi \in \Pi$ it corresponds to various technologies of generating other permutations from π. At least three types of moves have been commonly distinguished in the literature: A (adjacent swap), N (non-adjacent swap), I (insert), see Table 1. Symbol $\circ$ denotes superposition. Denoting by $D^Z(\pi, \sigma)$ $(Z \in \{A, S, I\})$ the distance between permutations $\pi, \sigma \in \Pi$, we have $D(\pi, \sigma) = 1$ for move $\pi \to \sigma$ as well as for any $\sigma \in N(\pi)$. The measures were already analyzed in the literature, see e.g. [4]. Beside Table 1, the collection of the remain properties (mean, max, variance, complexity) one can find also in [5].

4 Image

Imagining of the solution space can be perceived as a mapping from Π into a plane, cube or other more complex surfaces like sphere, ellipsoid, disc. Actually, only a sample B of permutations $A \subset B \subseteq \Pi$ with moderate cardinality of B is

Table 1. Distance $D(\pi, \sigma)$ between permutations π and σ

Move type	Adjacent swap	Swap	Insert
Source	Kendall's tau	Cayley's	Ulam's, Floyd's game
Algorithm	Number of inversions in $\pi^{-1} \circ \sigma$	n minus the number of cycles in $\pi^{-1} \circ \sigma$	n minus the length of the maximal increasing subsequence in $\pi^{-1} \circ \sigma$
Mean	$\frac{n(n-1)}{4}$	$n - \sum_{i=1}^{n} \frac{1}{i}$	$n - O(\sqrt{n})$
Maximum	$\frac{n(n-1)}{2}$	$n - 1$	$n - 1$
Variance	$\frac{n(n-1)(2n+5)}{72}$	$\sum_{i=1}^{n} (\frac{1}{i} - \frac{1}{i^2})$	$\Theta(n^{\frac{1}{3}})$
Complexity	$O(n^2)$	$O(n)$	$O(n \log n)$

visualized because of huge space size. We expect that the mapping has several advantageous features: (a) cost of calculation is reasonable small, (b) distance in the space is invariant on the image, (c) it fits to the local search trajectories performed by common LS metaheuristics. Below, we discuss briefly two classes of the distance measures on the image.

Hypercube Model. We assume that mapping T has the form $T : B \rightarrow R^m$, where $B \subseteq \Pi$, R^m is the plane ($m = 2$) or cube ($m = 3$), R is the set of real numbers. Value $f(\pi)$ can represented by the color $C(\pi)$, $C(\pi) = (f(\pi) - f_*)/(f^* - f_*) \cdot RGB$, where $RGB = 65536 \cdot 255 + 256 \cdot 255 + 255$ is the range of RGB colors, $f_* = \min_{\pi \in B} f(\pi)$, $f^* = \max_{\pi \in B} f(\pi)$ or f_*, f^* are set as their lower and upper evaluations, respectively. Approximation of values f_* and f^* can be made also off-line or on-line by limited sampling of the space.

Let us discuss the topic of distance between points on the image. For two points $x = (x_1, \ldots, x_m)$ and $y = (y_1, \ldots, y_m)$, $x, y \in R^m$, the *Minkowski distance* is provided by the fundamental definition

$$L_p(x, y) = (\sum_{i=1}^{m} |x_i - y_i|^p)^{\frac{1}{p}}. \tag{7}$$

Using selected values of p in (7) we get a few special cases. For $p = 1$ we have the *Manhattan distance*, for $p = 2$ we get the *Euclidean distance*

$$L_1(x, y) = \sum_{i=1}^{m} |x_i - y_i|, \quad L_2(x, y) = \sqrt{\sum_{i=1}^{m} (x_i - y_i)^2}, \tag{8}$$

and taking the limit we obtain *Chebyshev distance*

$$L_\infty(x, y) = \lim_{p \rightarrow \infty} L_p(x, y) = \max_{1 \leq i \leq m} |x_i - y_i|. \tag{9}$$

Despite the general definitions, only $m = 2, 3$ will be considered next.

Spherical Model. We assume that mapping Q has the form $Q : B \rightarrow S(n)$, where $B \subseteq \Pi$, $S(n)$ is the sphere with radius $R = D^Z_{\max}(n)/\pi$, where $D^Z_{\max}(n)$ is the solution space diameter, i.e. the maximum value of the metric $D^Z(..)$, for chosen $Z \in \{A, N, I\}$, see Sect. 3 and [5] for detail. Thus, solution π is transformed in the point (θ, ϕ) on the sphere $S(n)$, where $-\pi \le \theta \le \pi$ is the geographical lenght, $-\pi/2 \le \phi \le \pi/2$ is the geographical width. Let us consider the distance measure between images $x = (\theta_x, \phi_x)$ and $y = (\theta_y, \phi_y)$. The commonly used in the literature is the shortest path on the sphere (an arc) between x and y, namely $d(x, y) = R \arccos[\sin\theta_x \sin\theta_y \cos(\phi_x - \phi_y) + \cos\phi_x \cos\phi_y]$. The goal function value can be transformed into colors in the way already given or in the height with respect to the reference sea level. Justification of the spherical model follows from several facts: (1) hypergraph defining the neighboring permutations becomes for large n similar to the sphere, see Fig. 1; (2) distance between any two points on the sphere is less than the space diameter; (3) distribution of distances from any fixed permutation roughly approximates the normal distribution, see Fig. 2 observed in computer experiments [7]; (4) distribution of distances does not depend on the reference permutation, [7]; (5) search trajectory has an analogy to the path on the Earth surface, [5]. Lack of normal distribution of distances

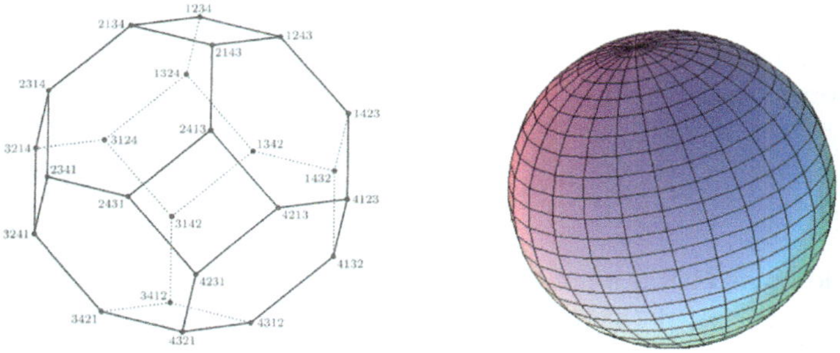

Fig. 1. Hypergraph of neigboring permutations (left) and sphere model (right).

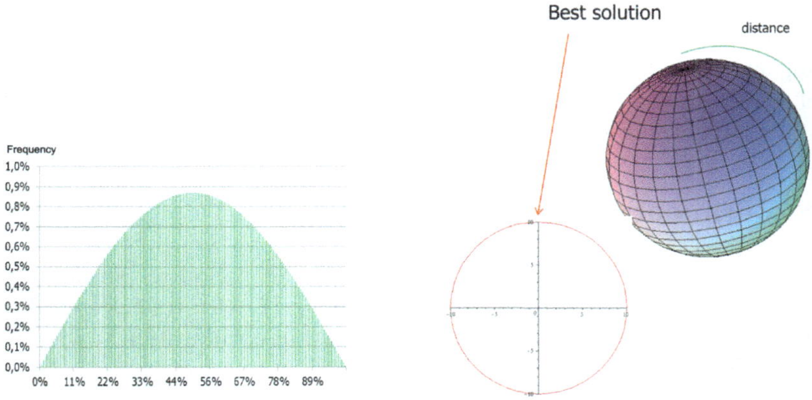

Fig. 2. Simulation of distribution distances in sphere model.

in the basic sphere model incline us to introducing modifications called shortly "disturbances", namely (A) relocation of the center of the sphere or (B) changing the shape of the image (disc, ellipsoid).

5 Mappings

In order to reduce calculation we propose the transformation T called next the *reference points* mapping. The basic definition deals with the image on hypercube R^m. The method consists of two phases. In the first phase we transform r permutations $\sigma_1, \ldots, \sigma_r$ from Π into r points $e_1, \ldots, e_r$ in R^m called reference images. They remain unchanged next and create the reference set. In the second phase, we transform step-by-step each individual permutation τ into its image t in R^m. Distance in Π is represented by a measure $D(..)$ and distance in R^m by a measure $d(..)$. Images are set by solving the following task

$$\sum_{i=1}^{r}\sum_{j=1}^{r} ||d(e_i, e_j) - D(\sigma_i, \sigma_j)|| \rightarrow \min_{e_1,\ldots,e_r} \tag{10}$$

where $||..||$ denotes a certain norm, e.g. $(..)^2$ or $|..|$. Mapping for any individual permutation τ assigns point $t \in R^m$ so that

$$\sum_{i=1}^{r} ||d(e_i, t) - D(\sigma_i, \tau)|| \rightarrow \min_{t} \tag{11}$$

Optimization task (10) is a nonlinear case with mr continuous decision variables, whereas (11) has m decision variables and is also nonlinear. The first phase, computationally rather expensive (depends on r) is performed once. The second phase performed multiple times is appreciably cheaper. Notice, the visualization is used for tuning the approximate algorithm, so rarely. For $m = 2$ and $r = 3$ we can use instead of (10)–(11) certain fast geometric approach.

The reference points mapping can be applied in the sphere model as well. Leaving measure $D(..)$ and using definition of $d(..)$ dedicated for the sphere model, we can create suitable images in two phases. From the methodology point of view nothing will change.

6 Experimental Research

Distribution of images depends on: (a) number of reference permutations r; (b) selected distance measure in R^m, from among (7)–(9); (c) selected distance measure in Π, see Table 1; (d) selected norm $||..||$ in optimization tasks (10) and (11); (e) selected $m \in \{2, 3\}$ in R^m; (f) technology of selecting $\sigma_1, \ldots, \sigma_r$. The comprehensive experimental research seems are vast and exceed the volume of this paper. Indeed, at least a few values $r = 3, 4, 5, \ldots$ should be considered in (a), $r \geq 3$ for $m = 2$ and $r \geq 4$ for $m = 3$; 4 measures in (b); 3 measures in (c); 4

norms in (d); 2 in (e); and 2,3 in (f). This implies for single problem and single instance over 500 runs. Skipping consciously the excess of experimental research we provide only comparison between various search trajectories, Figs. 3–4, for an instance of the flow-shop scheduling problem: random, made by simulated annealing method, made by tabu search method on the background of the same B and three reference points, see Fig. 3 left.

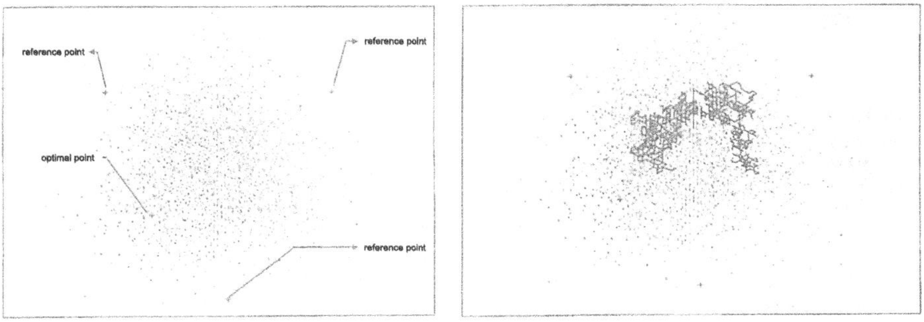

Fig. 3. The flow-shop scheduling instance TA51. Three reference points (left), their images and images of certain set $B \subseteq \Pi$. Random search trajectory on the background of B (right).

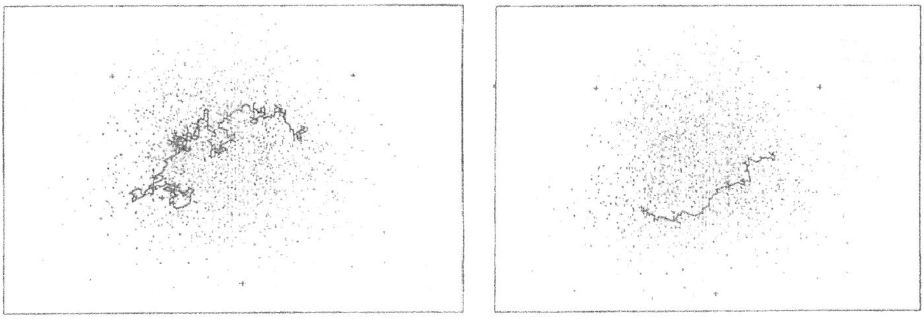

Fig. 4. The flow-shop scheduling instance TA51. Simulated annealing trajectory on the background of B (left). Tabu search trajectory on the background of B (right).

7 Remarks and Comments

If we consider visualization process as a mapping $R^n \rightarrow R^2$ then we can apply methods from the exploratory pattern analys, see the survey in [10]. Skipping consciously further details, we only pinpoint valid components: (a) methods (linear/nonlinear mappings), (b) complexity of mapping calculation, (c) usage of the increased history (incremental calculation) and (d) non-standard approaches. Among linear mappings we refer to: principal components, generalized decluttering, last squares, projection pursuit. Among nonlinear mappings we refer to: Sammon's, triangular, distance from two means, k-nearest neighbor. Such an approach defines further research stream. Additionally findings of this paper can be applied in the study of algorithms for more advanced problems [3, 6, 13–15].

References

1. Bożejko, W., Pempera, J., Smutnicki, C.: Parallel simulated annealing for the job shop scheduling problem. In: Allen, G., et al. (eds.) ICCS 2009. LNCS, vol. 5544, pp. 632–640. Springer, Heidelber (2009)
2. Bożejko, W., Smutnicki, C., Uchroński, M.: Parallel calculating of the goal function in metaheuristics using GPU. In: Allen, G., et al. (eds.) ICCS 2009. LNCS, vol. 5544, pp. 1014–1023. Springer, Heidelberg (2009)
3. Jain, A.S., Rangaswamy, B., Meeran, S.: New and "stronger" job-shop neighbourhoods: a focus on the method of Nowicki and Smutnicki (1996). J. Heuristics **6**(4), 457–480 (2000)
4. Knuth, D.E.: The Art of Computer Programming. Addison Wesley, Longman, Boston (1977)
5. Nowicki, E., Smutnicki, C.: Some new ideas in TS for job-shop scheduling. In: Rego, C., Alidaee, B. (eds.) Adaptive Memory and Evolution: Tabu Search and Scatter Search. Kluwer Academic Publishers, Dordrecht (2004)
6. Nowicki, E., Smutnicki, C.: An advanced tabu search algorithm for the job shop problem. J. Sched. **8**(2), 145–159 (2005)
7. Nowicki, E., Smutnicki, C.: Some aspects of scatter search in the flow-shop problem. Eur. J. Oper. Res. **169**(2), 654–666 (2006)
8. Nowicki, E., Smutnicki, C.: 2D and 3D representations of solution spaces for CO problems. In: Bubak, M., et al. (eds.) ICCS 2004. LNCS, vol. 3037, pp. 483–490. Springer, Heidelberg (2004)
9. Pitzer E., Affenzeller M.: A comprehensive survey on fitness landscape analysis. In: Recent Advances in Intelligent Engineering Systems, pp. 161–191. Springer, Heidelberg (2012)
10. Siedlecki, W., Siedlecka, K., Sklansky, J.: An overview of mapping techniques for exploratory pattern analysis. Pattern Recogn. **21**(5), 411–429 (1988)
11. Smutnicki, C.: Optimization technologies for hard problems. In: Fodor, J., Klempous, R., Araujo, C.P.S. (eds.) Recent Advances in Intelligent Engineering Systems, pp. 79–104. Springer, Heidelberg (2011)
12. Smutnicki, C., Bożejko, W.: Parallel and distributed metaheuristics. In: Moreno-Díaz, R., Pichler, F., Quesada-Arencibia, A. (eds.) Computer Aided Systems Theory - EUROCAST 2015. LNCS, vol. 9520, pp. 72–79. Springer, Cham (2015)
13. Smutnicki, C., Bożejko, W.: Tabu search and solution space analyses. The job shop case. In: Moreno-Díaz, R., Pichler, F., Quesada-Arencibia, A. (eds.) Computer Aided Systems Theory - EUROCAST 2017. LNCS, vol. 10671, pp. 383–391. Springer, Cham (2017)
14. Watson, J.P., Beck, J.C., Howe, A.E., Whitley, L.D.: Problem difficulty for tabu search in job-shop scheduling. Artif. Intell. **143**, 189–217 (2003)
15. Watson, J.: An introduction to fitness landscape analysis and cost models for local search. In: International Series in Operations Research and Management Science, pp. 599–623 (2010)

Smart Services for Improving eCommerce

Andrzej Sobecki[1](✉), Julian Szymański[1], Henryk Krawczyk[1], Higinio Mora[2], and David Gil[2]

[1] Faculty of Electronic Telecommunications and Informatics, Gdańsk University of Technology, Gdańsk, Poland
{andsobec,julian.szymanski,hkrawk}@pg.edu.pl
[2] Department of Computer Science Technology and Computation, University of Alicante, Alicante, Spain
{hmora,david.gil}@ua.es

Abstract. The level of customer support provided by the existing eCommerce solutions assumes that the person using the functionality of the shop has sufficient knowledge to decide on the purchase transaction. A low conversion rate indicates that customers are more likely to seek knowledge about the particular product than finalize the transaction. This is facilitated by the continuous development of customers' digital competencies, resulting in the increasing popularity of web services enabling the exchange of information, e.g. through social networks. Currently the user act with eCommerce platform like a source of information. At the same time, he or she usually use more than one source of information e.g., web portals, social networks, etc. The existing online shops seem unsuited to these trends because they remain simple trading platforms without integration with external web services and sources of knowledge. New categories of smart services are suggested, enabling the newly implemented eCommerce network platform to enhance the offered knowledge and reduce the abandonment of the platform by the user. Our empirical studies show an increase in the conversion rate in the case of shops which increased the level of customer support using the proposed model of integration.

Keywords: Electronic commerce · Smart Services · Transaction scenarios · User knowledge development · Integrated platforms

1 Introduction

In recent years the rapidly increasing customer interest in shops operating only online has been observed. One of the reasons for the popularity of this method of selling is customer convenience related to the ability to view and purchase goods using web services. The eCommerce solutions are becoming increasingly popular also due to the development of customers' digital skills and related lifestyle changes, including the performance of more and more tasks remotely over the

W. Zamojski et al. (Eds.): DepCoS-RELCOMEX 2020, AISC 1173, pp. 575–584, 2020.
https://doi.org/10.1007/978-3-030-48256-5_56

Internet or unwillingness to visit traditional shops. The customers of online shops can use their functionality to familiarize themselves with their offer—they can compare selected products, negotiate the terms of purchase, place orders or use the automated order payment process.

In general, customers do not have enough knowledge to independently decide on the purchase of goods. The standard eCommerce solutions support such customers as regards comparing the attributes of alternative products and presenting lists of the most frequently purchased products or goods that arouse most interest among the clientele. Such functionality of the shop is often sufficient when the customer has sound enough preconceptions about the product and their motivation to make the purchase is high. In other cases, they are forced to use different types of web pages describing in detail the properties of the required goods, and often avoid making a purchasing decision. This thesis is confirmed by the low average conversion rate in many of the world's online shops, that is reported 2.95% in the fourth quarter of 2016 [1]. It indicates i.a. how many shop visitors were successfully transformed into customers within a specified time. Based on the average conversion rate it can be said that most customers (97.05% of online shop sessions) are interested in information about the goods, whereas transactions are rare (2.95% of cases).

The existing heuristic optimization methods [2] are intended to increase the conversion rate through testing further hypotheses based on well-defined changes in the offer, and the tools for measuring the effects of such modifications. This approach is aimed only at the optimization of the result, and not necessarily, providing a potential customer-oriented online shop.

The customers differ by their level of digital skills. The first group includes the occasional customers who shop on the Internet while following information from traditional sources, such as the press or friends. They buy a narrow range of products online because their purchasing decisions depend mostly on talking with an assistant in a traditional shop. They are not familiar with the available online services and use only basic services for simple orders. The second group of customers is commonly referred to as Millennials. They are familiar with mobile technology, available on-line services and the culture of sharing knowledge.

The existing eCommerce solutions do not provide effective tools that would support knowledge management [3] and using this knowledge to help customers make choices. As a consequence, the process of selecting products by the customer becomes prolonged. This results in prolonging the sessions, reducing the number of conversions, increasing the probability of customer fatigue and, as a consequence, a decrease in customer satisfaction.

The proposed solution combines selected advantages of traditional and online shops, enriching them with the opportunity to support the customer in the process of goods selection. Thus, the description of the customer's decision-making process is proposed to be expressed in the form of a service purchase scenario which will include services corresponding to the functionalities of traditional online shops (Basic Services) and new smart services (Smart Services), gathering and exploiting the knowledge available in many online shops in order to e.g.

formulate an adequate purchase proposition and provide the customer with a more substantive rationale of the product value.

The paper presents examples of services that support the proposed model, and indicate its usefulness in many areas of trade. The research related to the proposed model was carried out in a test environment created by 34 modified online shops offering an assortment from different fields, such as construction, medical services, children's articles, bookstores, electronics shops etc., which enabled the implementation of integrated shopping related to mutually complementary products.

2 Service-Centric eCommerce

The process of selling goods in shops is accomplished through a set of functions that can be grouped by the scope of their use by shop users (customers and vendors): 1) Providing product information—presenting a catalog offer in the form of letters, e.g. categories, products, as well as a collection of information pages detailing the products. 2) Support for the customer related to make decisions as regards the choice of a shop or a product—a collection of tools aimed at complementing product descriptions and supporting decision making—including goods popularity rankings, reviews of other customers, mutual comparisons of alternate product values, presenting lists of alternative products, as well as support for the processes of order placing and payment execution. 3) Gathering customer information—the possibility to define ratings and comments by the customers and monitor customer behavior using external tools such as Google Analytics (GA) or PIWIK.

Such services, implemented in the SOA (Service Oriented Architecture) standard [4], shall be called Basic Services here, as they are available in almost any online shop. It is possible to distinguish system services that support the functioning of the shop and usability services that directly support the customer. The fixed packages of services offered by the shop shall be called the 'shop configuration' here.

The proposed service model of a shop can also be applied to the modeling of traditional shops—in such cases services represent the actions performed by the customer or the seller on their own. One of these new capabilities in traditional shops is increasing sales through appropriate organization of the exhibition of the goods in the shop [5,6] or enabling the customers to make purchases without waiting at the checkout [7]. The introduction of such services in an online shop will, on the one hand, increase the satisfaction of its customers, and on the other hand, increase the income of the entity that offers them.

Popular eCommerce solutions, like Magento [8] or Prestashop [9], offer a range of basic services. The services provided to customers in the popular eCommerce solutions, such as a list of available categories, a list of products similar to the selected one, a 'wish list' or a comparison of the features of selected products fail to assist the customer in fulfilling an order related to a comprehensive project. The fulfillment of such orders most frequently involves the purchase of many different products at the same time, while taking into account the specific purpose of a particular project.

3 Customer-Oriented Shopping Scenarios

Each transaction results directly from the customer's needs and motivations, which can be described as a project to execute. The customer's choice depends on their initial knowledge and the product preferences resulting from it. The development and usage of knowledge by the customer of an online shop can be presented in the form of a process that we will called shopping scenario. It describes their activity expressed in terms of the services used by them to make purchases in the web market. We can distinguish four types of shopping scenarios: 1) SPSS (Single product - Single Shop)—refers to ordering a single product by the customer based on the data derived from using one shop; the SPSS scenario can use available services comparing the selected products in a single shop or recommendation services using the database of the shop. 2) SPMS (Single product - Many Shops)—describes the order of a single product based on the information presented by multiple shops; SPMS scenario can use available services comparing the range of products from multiple online shops, and dedicated services for users following the SPSS scenarios. 3) MPSS (Many products - Single Shop)—refers to the history of user activity related to ordering multiple products related to a specific project based on the information available in a single shop; MPSS scenario most frequently use only a list of product proposals linked to the product added to the cart or proposals of other similar products from the same category. The customers also use services supporting other types of scenarios, i.e. SPSS and MPSS. 4) MPMS (Many products - Many Shops)—refers to the customer's ordering multiple products related to a specific project based on the information available in multiple shops. MPMS scenario can use an additional set of services aimed at assisting the customer in addition to those offered for the SPSS, SPMS, MPSS scenarios.

4 Smart Services for eCommerce

The personalization of the offerings according to the customer's characteristics is more and more frequently based on recommendation systems [10]. The recommendation methods are increasingly used by web providers offering various types of products or services, including Netflix [11], clothing shops (Levi [12]), large retail shops (Amazon [12]) or auction sites such as eBay [12,13]). Currently, the recommender systems used in eCommerce platforms are focused on products rather than services or sources of knowledge.

The general model of the proposed solution, based on the new type of services called Smart Service. Smart Services fall under the category of recommendation services and can be globally used in any shop. It is a collection of knowledge-based services that allow collecting, processing and using knowledge to support the decision-making process executed by the customer and the seller. We propose to use recommender systems that can help not only in find appropriate product but also the needed service or source of knowledge. The proposed model of an eCommerce solution using Smart Services facilitates the use of knowledge collected by customer monitoring tools and external web services, such as discussion forums, social networks or information portals (articles, rankings, videos). This allows the process of making purchases in an online shop to be supported by additional knowledge, which is unavailable in existing eCommerce solutions.

Supporting the exchange of information on the offered goods can have a significant impact on the customer's decisions; by enhancing their knowledge and clarifying the vision of the expected product. The following Smart Service categories are proposed: 1) Intelligent product recommendation for the customer using feedback from various customers in the eCommerce network – estimating customer needs by building product or service selection rules based on customer behaviour in the eCommerce network, and recommending the right products from any of the shops, as long as they match the discovered needs of the customer. 2) Extending information on the proposed products using the new capabilities of external web services e.g., Wikipedia, YouTube.com, TomsHardware.com, thespruce.com etc. – providing information relevant to the current customer context based on the knowledge of available web services (e.g. social networking information, information derived from promotional materials, videos) and discussions with other customers considering the purchase of the selected product. 3) Monitoring current user context and behaviour based on distributed customer activity – discovering relevant knowledge based on the information about customer activity (e.g. services used, products selected, requests made), enabling the identification of customer needs and classifying customer behaviours into groups of similar customers based on their level of expertise, inspirations, trends, promotions or goals.

Ultimately, Smart Services offered by different vendors should enable the interoperability of online shops using such services. This will increase of dependability of the shopping scenarios which could be performed in more than one e-commerce platform.

The proposed expansion of the shopping scenario by Smart Services may affect the customer in all phases of this scenario. Selecting the right service for the scenario should be supported by a recommendation system that tailors the list of proposed services to the current context of customer activity. Such a recommendation system should enable monitoring of the activity of customers in multiple collaborating shops in order to precisely define the context and provide opportunities for customer integration. It is proposed to extend the set of collected information by data that enable the representation of the customer's behavior in the form of a shopping scenario consisting of services.

In order to increase the dependability and proper support for the customer executing a shopping scenario, the knowledge discovery process should be implemented at a higher level of generality, i.e. above the individual eCommerce solutions. This will allow for specifying a single profile of an individual viewing the offerings of multiple shops, as well as for detecting the current context associated with the relevant stage of a shopping scenario.

5 Test Environment and the Experiment

The environment for the experiments was created using the computing cloud provided by CI TASK[1]. The participants of the experiment were 120 programmers divided into teams of three to five people, each of which was responsible for developing and managing a single online shop. Each of the shops was created on the basis of existing eCommerce solutions, i.e. Magento or Prestashop. Each of the created online shops had to offer at least 100 products matching the offering of a selected actual shop. In total, 3911 goods were included in the offering. The created eCommerce solutions were embedded in the computing cloud and integrated with other cloud-based systems, i.e. the global repository (SaaS) of Smart Services, the PIWIK event logging tool and the external Google Analytics tool [14,15]. The team identified the Smart Services that will be used during the experiment, and defined certain shopping scenarios based on a selected actual shop (e.g. IKEA or MediaMarkt). There were many possible configurations of services and knowledge available to the online shops. During the experiment, the authors focused on configurations using knowledge related to i.a.: 1) the offerings of similar shops participating in the eCommerce network, 2) the collaboration of virtual customer groups exchanging information through a social network; 3) specialized advice, product tests and marketing information.

The basic eCommerce software was extended to include the capability to log users' decisions related i.a. to used services, ordered products and defined ratings. The logged user decisions supported complex shopping scenarios that were recorded in a central database. Based on these scenarios, customers were recommended to purchase products tailored to their needs to use online services supporting the identification of the buyer's needs. The sellers were provided with information on the prediction of demand for goods, and the trends observed in the established market of shops. The architecture of the environment corresponded to the proposed Smart Shop network. The role of an intermediary providing access to the offerings of established online shops was played by the Smart Service repository made available for this purpose.

At the next stage, each participant of the experiment was required to execute a shopping scenario based on the received description of the requirements, related i.a. to: 1) shopping scenario description; 2) definition of a user project (a mountain trip, flat renovation, party organization) required for buying some products; 3) the list of products for preparation of the user project (5–10 items); 3) the maximum user budget for product shopping; 4) the maximum time devoted by the user for buying products; 5) categories of shopping scenarios: SPSS, MPSS, SPMS, MPMS.

[1] Academic Computer Centre in Gdansk (http://task.gda.pl/).

Each of the customers, in order to choose products for their project, used i.a. a global catalogue of online shops, basic services provided by the creators of solutions such as Magento or Prestashop, as well as additional Smart Services. The set of basic services included the following service categories: available products and their rankings, monitoring client behaviors. The smart services set included: understanding customer plans and shopping needs, coordination of co-operation in eCommerce network, monitoring shopping scenarios. In addition, they had to execute three shopping scenarios. The customer behavior was monitored by PIWIK (locally running) and Google Analytics (remotely) systems.

The last stage of the experiment was the analysis of the results and drawing conclusions regarding the usefulness of the proposed service model as regards supporting the purchase process from the point of view of meeting user needs and the satisfaction of customers.

6 The Results

During the experiment, users executed 340 shopping scenarios that included i.a. using product search services 13, 940 times and using services providing information on the selected goods or groups of goods 37, 433 times. The average response time of the services made available was 0.53 s. Upon the completion of the experiment, the repository contained 12, 873 users' evaluations of search services, 9, 764 evaluations related to the recommended product and service information, and 8, 688 ratings related to the level of customer support offered by the shops. The products, services, shops and search results were evaluated according to a four-level scale: (1) – poor, (2) – acceptable, (3) – good, (4) – excellent. Two similar experiments were carried out, one concerning the online shop configuration using only Basic Services, and the other using both Basic Services and Smart Services. The results of experiments was showed in Table 1.

The satisfaction with Smart Services offered by the shops was rated at 3.58. In the case of shops offering traditional services to their customers this rating was 2.22. The customers used the services integrating customers executing similar projects based on the Facebook social network, as well as the services recommending products sources of product knowledge (e.g. specialist portals, Wikipedia or product rankings) most frequently.

Another aspect rated by the customers was the relevance of recommendations regarding product information. In the case of shops using basic services, recommendations were limited to a narrow list of related products based on the similarity of customer feedback. In this case, the relevance of the recommended information was evaluated on average as 2.97. With the use of Smart Services, customer support was extended to include recommendations from external sources of information. In this case, the relevance of the received information was evaluated on average as 3.84. Most of the recommendations matched the expectations of the users and the weak ratings were related to the lack of knowledge about some of the products available in the market or to erroneously given recommendations.

Table 1. List of parameters acquired through monitoring tools integrated with the test environment – BS - Basic Sercices; BSS - Basic and Smart Services

Parameter name	BS	BSS
Client satisfaction (average score) [User evaluation: poor (1.0), acceptable (2.0), good (3.0), excellent (4.0)]	2.22	3.58
Adequacy of product recommendation to user needs (average score) [User evaluation as above]	2.97	3.84
Shops visited by a user during searching for product adequate to their plan [Avg value]	6	2
Session duration for one shopping scenario [Average value in minutes]	41′	14′
The transaction rate [Avg value]	10%	47.5%
The percentage of available budget spent for buying products [Avg value]	100%	89%
The percentage of time used by user to realizing their plan [Avg value]	100%	79%
The number of used basic services by user during searching for product adequate to their plan [Avg value]	81	34
The number of used smart services by user during searching for product adequate to their plan [Avg value]	0	7
The number of used basic knowledge source (product description, comparison etc.) by user during searching for product adequate to their plan [Avg value]	74	29
The number of used additional knowledge source (product reviews, social media etc.) by user during searching for product adequate to their plan [Avg value]	0	3

Prior to providing the Smart Services, customers searched through 6 shops on average in order to find the right product. Providing the support of a recommendation system made it possible to find a product by looking at an average of two shops. The average session time before the provision of Smart Services amounted to 41 min, and was due to the need to identify a shop's offering, organize goods and compare information about products found independently on the Internet and in multiple stores in the created market. When the recommendation system was launched, the average session time was reduced to 14 min. The conversion rate, i.e. the proportion of sessions ended with a transaction, has also improved. Prior to launching the recommendation system, it amounted to approximately 10% and, after providing a recommendation system, it ranged from 35% to 60% (47.5% on average) for the entities having the offering most suited to the needs of customers and providing all the Smart Services.

A relationship connected to the degree of budget utilization identified in the executed scenarios was also observed. The budget of the scenario was determined by the designer by summing up the cost of the independently chosen products alone, and the costs of delivery. The average budget utilization rate in the case of individuals using Smart Service support amounted to 89%. This means saving about 11% in comparison to the budget planned by the requirements designer. The average time used for ordering decreased to 79% of the time spent while using basic services. Most individuals executing the scenarios chose products faster than anticipated by the designer.

Lastly, the individuals executing shopping scenarios for specific projects used most Smart Services when choosing the first few products (3–5). Subsequent shopping scenarios were executed with minimal support from such services. The reason for this was probably the fact that the customers were provided with all the required information regarding the project already, when selecting the first project-related products. The shopping scenarios recorded for the same projects indicated high similarity of subsequent selections as regards products chosen and services used.

7 Discussion and Conclusions

The popularity of online shops will most likely lead to the unification of shop offers and, as a result, the competitive advantage of particular entities will be due to the services offered to support the execution of the purchase scenario and the knowledge that is being used and offered.

The experiment was conducted on individuals aged 23–25, being conscientious customers of online shops, who use them frequently to order various types of goods. Their digital competences are high, as they have the knowledge and experience related to using a variety of technologies, including programming skills. The test group was selected in such a way as to show certain trends in the habits of shoppers who know the capabilities of web services, and know how to use them. The experimental shops presented a range of products corresponding to real market institutions, i.a. Tesco, OBI, IKEA and MediaMarkt. The aim was to reflect the real market of products and services as closely as possible. During the experiments, specific service configurations that use different types of knowledge were selected. The proposed variety of test environments was purposeful, and intended to show how the proposed approach works in different contexts of user activity.

The contribution is a usage of Smart Services model which support the customers in the execution of complex shopping scenarios through recommending them first a knowledge sources and available services before particular products. The second contribution is proposition of extending description the user behavior through including information about used services and knowledge sources.

Enriching the offer of shops with the knowledge provided by external entities has resulted in an increase in the number of transactions per started session. This also allowed for the reduction of the number of shops that a user had to search in order to find the right product. The analysis of customer behavior in the shop network has also enabled predicting the demand for products and accelerating the response to inventory shortages.

The service model for the eCommerce solutions described in the article, along with the proposed Smart Service extensions, provides an opportunity to expand customer support by the stage of gathering the knowledge required to initiate a transaction. The service model of an online shop, and supporting shopping scenarios with web services, enable the creation of information hubs that connect information, products and customers in one place. Such an approach provides new development opportunities for eCommerce solutions in the context of appropriate management of information and services provided to potential customers.

The observed effects of the proposed change include i.a. shortening of the transaction time and increasing the average conversion rate. Changing the model involves modifying the perception of the customers of online shops, and the role of these shops in the global information exchange network. By using Smart Services, online shops can act as nodes linking various information and trade services, thus becoming knowledge-based transactional platforms. It is also important to retain flexibility in designing such shops using the available independently developed basic and Smart Services.

References

1. Insight, S.: Ecommerce conversions rates (2019)
2. Phillips, J.: Ecommerce Analytics: Analyze and Improve the Impact of Your Digital Strategy. FT Press, Upper Saddle River (2016)
3. Helms, M.M., Ahmadi, M., Jih, W.J.K., Ettkin, L.P.: Technologies in support of mass customization strategy: exploring the linkages between e-commerce and knowledge management. Comput. Ind. **59**, 351–363 (2008)
4. Erl, T., Gee, C., Kress, J., Chelliah, P.R., Normann, H., Maier, B., Shuster, L., Trops, B., Utschig-Utschig, C., Winterberg, T., et al.: Next Generation SOA: A Concise Introduction to Service Technology & Service-Orientation. Pearson Education, London (2014)
5. Chen, Y.L., Chen, J.M., Tung, C.W.: A data mining approach for retail knowledge discovery with consideration of the effect of shelf-space adjacency on sales. Decis. Support Syst. **42**, 1503–1520 (2006)
6. Aloysius, G., Binu, D.: An approach to products placement in supermarkets using prefixspan algorithm. J. King Saud Univ.-Comput. Inf. Sci. **25**, 77–87 (2013)
7. Amazon: Description of the amazon go! (2019)
8. Magento: Magento ecommerce software (2019)
9. Prestashop: Prestashop ecommerce software (2019)
10. Aggarwal, C.C.: Recommender Systems. Springer, Heidelberg (2016)
11. Gomez-Uribe, C.A., Hunt, N.: The netflix recommender system: algorithms, business value, and innovation. ACM Trans. Manag. Inf. Syst. (TMIS) **6**, 13 (2016)
12. Akshita, S.: Recommender system: review. Int. J. Comput. Appl. **71**, 38–42 (2013)
13. Li, H., Zhang, S., Wang, X.: A personalization recommendation algorithm for e-commerce. JSW **8**, 176–183 (2013)
14. García, M.D.M.R., García-Nieto, J., Aldana-Montes, J.F.: An ontology-based data integration approach for web analytics in e-commerce. Expert Syst. Appl. **63**, 20–34 (2016)
15. Gerrikagoitia, J.K., Castander, I., Rebón, F., Alzua-Sorzabal, A.: New trends of intelligent e-marketing based on web mining for e-shops. Procedia-Soc. Behav. Sci. **175**, 75–83 (2015)

Probabilistic Modelling of Reliability and Maintenance of Protection Systems Incorporated into Internal Collection Grid of a Wind Farm

Robert Adam Sobolewski(✉)

Faculty of Electrical Engineering, Bialystok University of Technology, Wiejska 45D Street, 15-351 Bialystok, Poland
r.sobolewski@pb.edu.pl

Abstract. The capacity factor (performance) of wind farms is quite unsatisfactory and is much lower as compared to conventional power generation units. The performance refers, among others, to 'electric and electronic components' of wind farms, i.e. generators, transformers, cables, busbars, protection systems, power electronic units, and many more. Assuring a high availability of the components can require their high both reliability and quality of preventive and corrective maintenance strategies. Reliability and maintenance of protection systems (one or more protective relays, circuit breaker, wiring, and other components) can result in major wind farms operation upsets and substantially influence their performance. Recently, microprocessor-based electronic relays have been developed and are being applied at increasing rate. They are usually equipped with self-monitoring and checking module. Some of the protection system failures can be hidden ones, i.e. to be detected: within planned maintenance, by self-monitoring and checking module or while fault or failure of protected component occurred. Availability rate can be a main criterion of system' reliability and maintenance. It is a probability that a system occupies up state within a long time, and can be calculate relying on semi-Markov model presented in the study. One of the application of the model can be valuable feedback and recommendations on time interval of periodic planned maintenance that maximizes the system' availability. As an example, the optimal time interval of planned maintenance along different both failure rates and probabilities of detecting the failures, is calculated relying on the model and presented in the case study.

Keywords: Protection system · Reliability · Maintenance · Wind farm · Semi-Markov model

1 Introduction

Cumulative capacity of both onshore and offshore wind farms (WF) is still growing very extensively around the world. Independently on the location of WF their capacity factor (performance) is quite unsatisfactory and is much lower as compared to conventional generation units. Technical and economic performance of WF depends on:

W. Zamojski et al. (Eds.): DepCoS-RELCOMEX 2020, AISC 1173, pp. 585–595, 2020.
https://doi.org/10.1007/978-3-030-48256-5_57

availability of their components, availability of an internal electrical collection grid arrangement, WF layout, wind resources, maintenance (planned and corrective) strategies, and many others. The performance refers, among others, to ‘electrical and electronic components’ of WF, such as: generators, transformers, cables, busbars, protection systems, power electronic units, and so on. Assuring a high availability of the components can require their high both reliability and quality of preventive and corrective maintenance strategies. Moreover, the components could be equipped with continuous self-monitoring and checking modules to detect their hidden failures immediately, if possible. The maintenance strategies should allow: restriction of imperfect maintenance (including imperfect detecting the failures), limitation of maintenance and repair time, optimization of interval time of periodic planned maintenance.

The protection system (PS) failure to operate and incorrect operation (unfolded tripping) can result in major WF operation upsets involving increased components damage, increased personnel hazards, and possible long interruption of WF operation. Some of these failures can be hidden ones. They can be detected in following way: within planned maintenance, by self-monitoring and checking module or while fault or failure of protected component occurred. Reliability and maintenance strategies of PS can be one of the major concerns about assuring the performance of WF. Good reliability and maintenance strategies of PS can serve an important contribution in: (i) preventing WF components’ faults/failures and related outages, (ii) achieving a high level of continuity of WF service, and (iii) when intolerable conditions occur – minimizing the extent and time of WF components outage. Concerning the preventive maintenance (essentially the optimal intervals of periodic maintenance) of PS the ambient wind conditions should be taken into account while making decision about shutting down the turbine, feeder section or whole wind farm. Partial or complete outages in WF to perform preventive maintenance can cause energy loses (energy not served) while the favourable wind conditions enable the reasonable power generation.

Each PS to be used in wind energy projects, consists of: one or more protective relays, circuit breaker and wiring. Moreover, it can be equipped with: current and voltage transformers, ancillary power supply, communication units, and so on. Depending on the size of WF (number of wind turbines, their rated capacity) and WF internal collection grid arrangement, one WF can be equipped with more than one PS, i.e. PS of individual wind turbine generators, PS of section feeders, and PS of inter-tie link that interconnects collector bus of WF and point of common coupling (PCC). They act as primary protection function and some of them – remote back-up protection function while primary PS are unavailable.

In recent years, microprocessor-based electronic relays have been developed and are being applied in wind energy sector at an increasing rate. Such relays sometimes refer to numerical type relays since the analogue inputs are converted to digital numbers that are then processed within the relay. With electronic relays, the protection principles and fundamentals are essentially unchanged as they are the issues regarding protection reliability. Microprocessor type relays do provide many benefits such as higher accuracy, reduced space, lower equipment and installation costs, wider application and setting capabilities, self-monitoring and checking, plus various other desirable supplemental features (e.g. control logic, remote and peer-to-peer communications, data acquisition, event recording, fault location, remote setting). Microprocessor-based relays technology enables detecting the relay failures both within planned maintenance (out of service of one or more WF components is mandatory) and relying on continuous self-monitoring and checking (usually without shouting down WF components protected). The latter ability can refer to a few tens percent's of PS internal failures.

Availability rate can be taken as a main criterion of PS reliability and maintenance. This rate is a probability that PS occupies the up state within a long time, and can be calculate relying on reliability and maintenance of PS model.

The relevant literature offers some approaches that can used for WF reliability analysis [1–3]. In [1] the method of component outage leading to WF outage is applied. The methodology relies on Reliability Block Diagram technique and reliability model consists of components in series. In [2] a logical diagram construction method is applied with a sequential grouping of the model components in series or in parallel arrangement. In [3] a step-by-step procedure is developed for calculation of WF reliability using combinatorial algorithm. All these models either incorporate simplified representation or neglect availability of protection systems and their impact on WF performance.

In the paper the approach to availability of PS calculations is presented relying on semi-Markov reliability and maintenance model. The model enables obtaining the relationship among PS availability and: (i) failure rate, (ii) time interval of periodic planned maintenance, (iii) imperfect failure detection within planned maintenance and thanks to self-monitoring and checking module), (iv) duration of planned and corrective maintenance, (v) and time to repair. One of the application of the model can be valuable feedback and recommendations on time interval of periodic planned maintenance that maximizes the PS availability. As an example, the optimal time interval of PS planned maintenance along different both failure rates and probabilities of detecting the failures, is calculated relying on the model and presented in the case study.

2 Internal Collection Grid Topology of Onshore Wind Farm and Protective Systems Operation

Figure 1 shows the example of internal collection grid topology of onshore wind farm [4]. Five power feeder sections (PFS1, …, PFS5) are interconnected with collector bus (BUSBAR). This bus is linked with the PCC through the inter-tie line and main transformer. The PCC consists of double busbar system. Each busbar system is equipped with disconnector and grounding switch (there are not any protection systems there).

One can point out three locations of protective systems inside internal collection grid, i.e.: (i) either outcome terminal of each WTG or high voltage side of each MV/LV transformer (depending on the design in question) interconnected with WTG (not depicted in Fig. 1), (ii) on each feeder section (PS0, …, PS5 in Fig. 1), and (iii) at both sides of inter-tie line (PS00 and PS01 in Fig. 1). The main role (primary protection function) of WTG (or WTG and MV/LV transformer) protection system is to clear the faults that occurred in a WTG or MV/LV transformer.

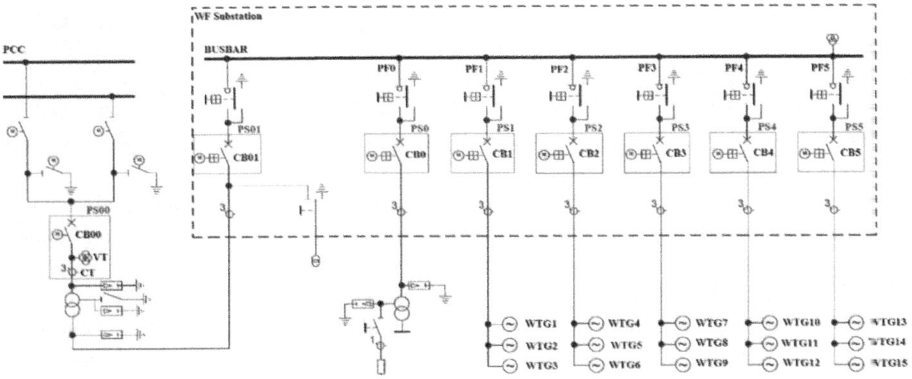

Fig. 1. Example of internal collection grid of onshore WF [4]

Usually, such protection system consists of simple overcurrent relays and LV circuit breaker. A protection system of the feeder section should be able to clear the faults that can occur in the cable that links WTGs of the section and collector bus (primary protection function) and comprise the remote back-up protection function while the fault occurred in WTG or MV/LV transformer while their protection system is broken (on outage). Such protection system is equipped with the relay, current and voltage transformers and MV circuit breaker (see circuit breakers CB0, …, CB5, in Fig. 1). Finally, the main role of protection systems installed at both sides of inter-tie line is to clear the faults that can occur in the line (primary protection function) and in feeders sections interconnected to collector bus. Each of these systems is equipped with relays, current and voltage transformers (see CT and VT respectively, in Fig. 1), and MV (or MV and HV, depending on voltage level of both sides of main power transformer) circuit breakers (see CB00 and CB01, in Fig. 1).

3 Reliability and Maintenance Model

The availability rate of PS A_{PS} is a probability that PS is in the up state(s) within the long time t, i.e.

$$A_{PS} = \lim_{t \to \infty} A(t), \tag{1}$$

where $A(t)$ is the PS availability within time t.

Calculation of A_{PS} can be performed based on semi-Markov model. The model refers to PS of feeders (e.g. see PS0, …, PS5 in Fig. 1) and inter-tie line (e.g. see PS00 and PS01 in Fig. 1). Its transition diagram is depicted in Fig. 2. The reliability and maintenance states are as follows:

S1 – PS in service while there are not any failures,
S2 – planned maintenance of PS started off while there are not any failures,
S3 – planned maintenance of PS started off while there is a failure,

S4 – PS under repair,
S5 – checking PS after false failure detecting by self-monitoring and checking module,
S6 – checking PS after failure detecting by self-monitoring and checking module,
S7 – PS in service while there is a failure (dormant failure).

The states S1, S2, S3 take into account condition that there are not any failure (false failure) detected by continuous self-monitoring and checking module of PS. Then availability rate of PS can be derived relying on a general formula [5]

$$A_{PS} = \frac{\pi_1(T) \cdot \mu_1(T)}{\sum_{i=1}^{7} \pi_i(T) \cdot ET_i(T)} \tag{2}$$

where: $\pi_i(T)$ is a stationary probability of state Si, $\mu_1(T)$ is a mean time of PS in service, $ET_i(T)$ is a mean waiting time in state Si, and T is a set time interval of periodic planned maintenance.

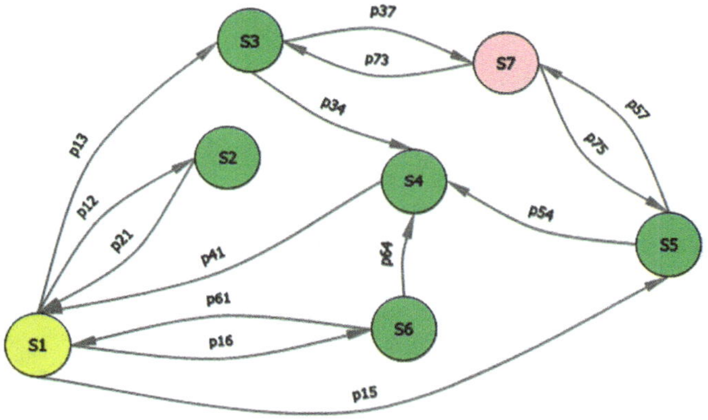

Fig. 2. Transition diagram for reliability and maintenance of protection system model

Let assume: time to failure of PS, time to false detection of the failure by self-checking and monitoring module, and time to detection of the failure not terminated after previous maintenance procedure (both planned and corrective one) are random variables η, φ and v respectively. These variables are exponentially distributed, i.e. $F_{PS}(t) = P(\eta \leq t) = 1 - e^{-\lambda \cdot t}$, $F_{SC}(t) = P(\varphi \leq t) = 1 - e^{-\gamma \cdot t}$ and $F_{DF}(t) = P(v \leq t) = 1 - e^{-\theta \cdot t}$ respectively, where: λ is failure rate of PS, γ is intensity of false detection of failure, and θ is intensity of lack of failure termination after previous maintenance service. Since the random times both to detect the failure by self-monitoring and checking module and to detect the false failure by self-monitoring and checking module, are independent to each other the mean time of PS in service is following

$$\mu_1(T) = \int_0^T e^{-(\lambda + \gamma)\cdot T} dt = \frac{1}{\lambda + \gamma} \cdot \left(1 - e^{-(\lambda + \gamma)\cdot T}\right). \quad (3)$$

The unique stationary distribution of the embedded Markov chain satisfies system of equations $\pi(T) = \pi(T) \cdot P(T)$ and $\sum_{i=1}^{7} \pi_i(T) = 1$, where $\pi(T) = [\pi_1(T), \pi_2(T), \ldots, \pi_7(T)]$, and $P(T)$ is the matrix of transition probabilities among the states (matrix size is 7×7) [6]. To sake the simplicity the attribute T of transition probabilities depicted in the diagram in Fig. 2 is omitted. Moreover, the probabilities pij depicted in Fig. 2 refer to the transition probabilities $p_{ij}(T)$ in analytical representation of the model. The transition probabilities ($p_{ij}(T) > 0$, $\sum_{j=1}^{7} p_{ij} = 1$, $i,j = 1, 2, \ldots, 7$, $i \neq j$) are as follows:

$$p_{12}(T) = e^{-(\lambda+\gamma)\cdot T}, \quad p_{13}(T) = (1 - \rho) \cdot \frac{\lambda}{\lambda + \gamma} \cdot \left(1 - e^{-(\lambda+\gamma)\cdot T}\right), \quad (4)$$

$$p_{15}(T) = \rho \cdot \frac{\lambda}{\lambda + \gamma} \cdot \left(1 - e^{-(\lambda + \gamma)\cdot T}\right), \quad p_{16}(T) = \frac{\gamma}{\lambda + \gamma} \cdot \left(1 - e^{-(\lambda + \gamma)\cdot T}\right), \quad (5)$$

$$p_{21}(T) = p_{41}(T) = 1, \quad p_{34}(T) = \alpha, \quad p_{37}(T) = 1 - \alpha, \quad p_{54}(T) = \beta, \quad (6)$$

$$p_{57}(T) = 1 - \beta, \quad p_{61}(T) = \delta, \quad p_{64}(T) = 1 - \delta, \quad (7)$$

$$p_{73}(T) = e^{-\theta\cdot T}, \quad p_{75}(T) = 1 - e^{-\theta\cdot T} \quad (8)$$

where ρ, α, β and δ are the probabilities of: detecting the failure by self-monitoring and checking module, detecting the failure within planned maintenance, confirmation the failure after its detection by self-monitoring and checking module, and confirmation the false failure after its detection by self-monitoring and checking module, respectively.

Stationary probabilities of the states S1, …, S7 are following:

$$\pi_1(T) = \frac{A(T)}{M(T)}, \quad \pi_7(T) = \frac{B(T)}{M(T)}, \quad (9)$$

$$\pi_2(T) = p_{12}(T) \cdot \pi_1(T), \quad \pi_3(T) = p_{13}(T) \cdot \pi_1(T) + p_{73}(T) \cdot \pi_7(T), \quad (10)$$

$$\pi_4(T) = (1 - p_{12}(T) - p_{16}(T) \cdot p_{61}(T)) \cdot \pi_1(T), \quad (11)$$

$$\pi_5(T) = p_{15}(T) \cdot \pi_1(T) + p_{75}(T) \cdot \pi_7(T), \quad \pi_6(T) = p_{16}(T) \cdot \pi_1(T), \quad (12)$$

$$A(T) = 1 - p_{37}(T) \cdot p_{73}(T) - p_{57}(T) \cdot p_{75}(T), \quad (13)$$

$$B(T) = p_{13}(T) \cdot p_{37}(T) + p_{15}(T) \cdot p_{57}(T), \quad (14)$$

$$M(T) = A(T) \cdot [2 + p_{13}(T) + p_{15}(T) + p_{16}(T) \cdot (1 - p_{61}(T))] + 2 \cdot B(T). \quad (15)$$

Mean waiting time in state Si (i = 1, ..., 7) can be calculated relaying on general formula [6]

$$ET_i(T) = \sum_{j=1}^{7} p_{ij}(T) \cdot ET_{ij}(T) = \sum_{j=1}^{7} p_{ij}(T) \cdot \int_0^{\infty} \tau dF(\tau, T), \qquad (16)$$

where $F(\tau, T)$ is a probability distribution of time transition from Si up to Sj within the time T.

The mean waiting time in states S1, ..., S7 are as follows:

$$\begin{aligned} ET_1(T) = T \cdot \left(e^{-(\lambda+\gamma)\cdot T} + (1-\rho) \cdot \frac{\lambda}{\lambda+\gamma} \cdot \left(1 - e^{-(\lambda+\gamma)\cdot T}\right)\right) \\ + \frac{\rho \cdot \lambda + \gamma}{(\lambda+\gamma)^2} \cdot \left(1 - ((\lambda+\gamma)\cdot T + 1) \cdot e^{-(\lambda+\gamma)\cdot T}\right) \end{aligned} \qquad (17)$$

$$ET_2(T) = t_1 + t_2, \qquad ET_3(T) = t_1 + t_2(1 - \alpha), \qquad ET_4(T) = t_3 \qquad (18)$$

$$ET_5(T) = ET_6(T) = t_4, \qquad ET_7(T) = \frac{1}{\theta} \cdot \left(1 - e^{-\theta \cdot T}\right) \qquad (19)$$

where: t_1 is a duration of PS testing while planned maintenance is being performed, t_2 is a planned maintenance duration while there are not any failures (or the failure is undetected during planned maintenance), t_3 is a time to repair, t_4 is a corrective maintenance duration while the failure (false failure) is detected by self-monitoring and checking module. In general, the times t_1, t_2, t_3 and t_4 can be random variables, but to the sake of simplicity we assumed their mean values.

4 Case Study

The reliability and maintenance model is applied to each protective system introduced in wind farm internal collection grid depicted in Fig. 1. Let assume, the parameters of the model are the same for each PS of the feeder sections (PS0, ..., PS5) and inter-tie line (PS00, PS01). The investigation aims at finding the best (optimal) time interval of periodical planned maintenance T^{BEST} and maximum availability assured by the best time interval A_{PS}^{BEST}, given the parameters of the model: λ, γ, θ, ρ, α, β, and δ. Their values for all variants considered in the study are provided in Table 1. The variants involve different combinations of ρ and α. The values of the rest parameters t_1, t_2, t_3 and t_4, are the same for each variant and are following: 5 h, 10 h, 100 h and 3 h, respectively. All the values provided are of order expected/confirmed in real PS operation conditions. The best time interval is the maximum time to planned maintenance that assures A_{PS} as much as possible. Let assume the resolution of T^{BEST} is 10 h.

Table 1. The parameters of semi-Markov reliability and maintenance model

Number of Fig.	Parameter							
	λ [1/h]	γ [1/h]	θ [1/h]	ρ	α	β	δ	T [h]
3	10^{-7}, …, 10^{-4}	10^{-4}	10^{-6}	0.65	0.95	0.95	0.95	100…5•10^4
4	10^{-7}, …, 10^{-4}	10^{-4}	10^{-6}	0.35	0.95	0.95	0.95	100…5•10^4
5	10^{-7}, …, 10^{-4}	10^{-4}	10^{-6}	0.65	0.75	0.95	0.95	100…5•10^4
6	10^{-7}, …, 10^{-4}	10^{-4}	10^{-6}	0.35	0.75	0.95	0.95	100…5•10^4

Results of A_{PS} along T given different both λ and combinations of ρ and α are presented in Fig. 3, …, Fig. 6, whereas T^{BEST} and A_{PS}^{BEST} given $\lambda = 10^{-5}$ 1/h and all variants of ρ and α are provided in Table 2. Figure 3, …, Fig. 6 show, the higher failure rate of PS the shorter T^{BEST} and lower A_{PS}^{BEST} can be expected. In the figures, a color of the frame of each λ corresponds to the color of A_{PS} curve obtained given λ. Moreover, they prove the following: (i) the essential impact of both ρ and α on A_{PS}, and (ii) similar A_{PS} given $\lambda = 10^{-5}$, …, 10^{-7} 1/h, different variants of ρ and α, and short T (<1000 h).

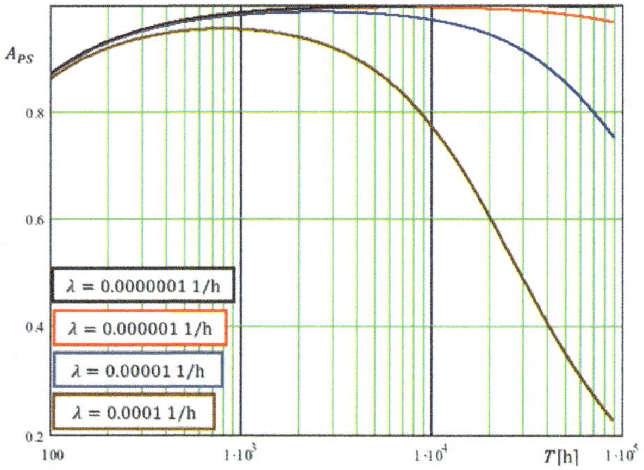

Fig. 3. A_{PS} against the time interval of periodic planned maintenance T, given $\rho = 0.65$, $\alpha = 0.95$, and $\lambda = (10^{-7}$ 1/h (black), 10^{-6} 1/h (red), 10^{-5} 1/h (blue), 10^{-4} 1/h (brown))

According to Table 2, if $\lambda = 10^{-5}$ 1/h, $\alpha = 0.95$ and it is possible (technically/practically) to increase ρ by 86%, i.e. from 0.35 (in Fig. 4) up to 0.65 (in Fig. 3), then T^{BEST} will be longer by 42% (from 2440 h up to 3470 h, i.e. by 1030 h).

However, if $\lambda = 10^{-5}$ 1/h, $\alpha = 0.75$ the growth of ρ by 86%, i.e. from 0.35 (in Fig. 6) up to 0.65 (in Fig. 5) can cause the extending of T^{BEST} by 37% (according to Table 2 from 1880 h up to 2580 h, i.e. by 600 h).

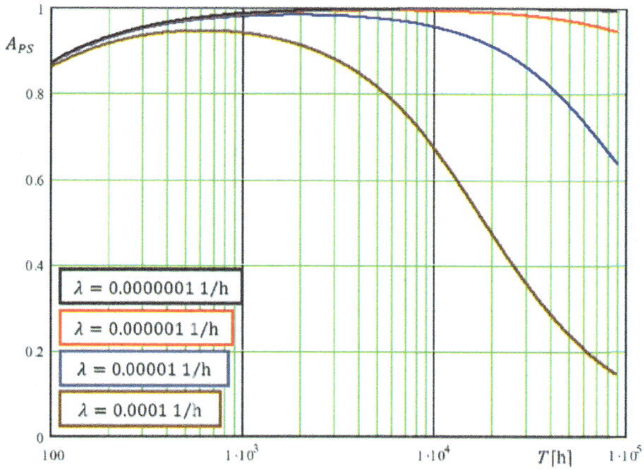

Fig. 4. A_{PS} against the time interval of periodic planned maintenance T, given $\rho = 0.35$, $\alpha = 0.95$, and $\lambda = (10^{-7}$ 1/h (black), 10^{-6} 1/h (red), 10^{-5} 1/h (blue), 10^{-4} 1/h (brown))

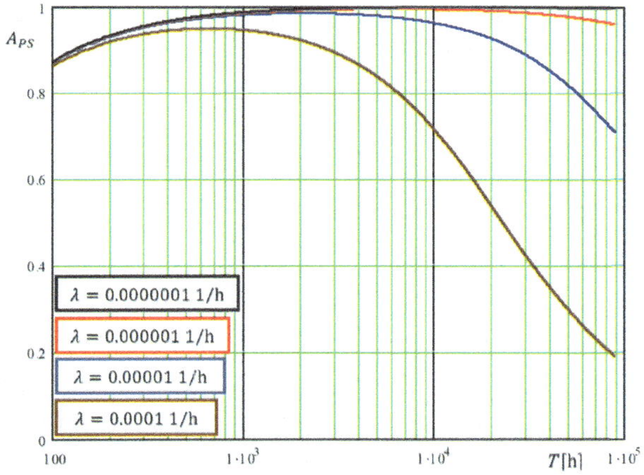

Fig. 5. A_{PS} against the time interval of periodic planned maintenance T, given $\rho = 0.65$, $\alpha = 0.75$, and $\lambda = (10^{-7}$ 1/h (black), 10^{-6} 1/h (red), 10^{-5} 1/h (blue), 10^{-4} 1/h (brown))

If $\lambda = 10^{-5}$ 1/h, $\rho = 0.65$ and it is possible (technically/practically) to go up α by 27%, i.e. from 0.75 (in Fig. 5) up to 0.95 (in Fig. 3), then (according to Table 2) T^{BEST} will be longer by 34% (from 2580 h up to 3470 h, i.e. by 890 h), whereas assumption of $\lambda = 10^{-5}$ 1/h and $\rho = 0.35$, and growth α from 0.75 (in Fig. 6) up to 0.95 (in Fig. 4) can extend T^{BEST} by 30% (from 1880 h up to 2440 h, i.e. by 560 h).

According to Table 2, A_{PS}^{BEST} is almost the same for all variants investigated. The maximum variation for extreme variants (in Fig. 3 and Fig. 6) is around 6% only.

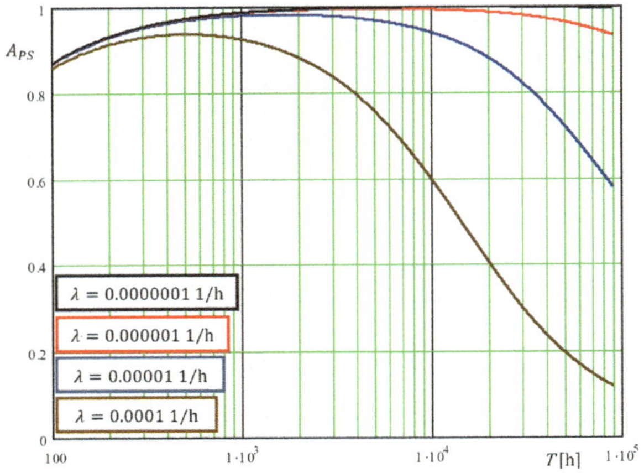

Fig. 6. A_{PS} against the time interval of periodic planned maintenance T, given $\rho = 0.35$, $\alpha = 0.75$, and $\lambda = (10^{-7}$ 1/h (black), 10^{-6} 1/h (red), 10^{-5} 1/h (blue), 10^{-4} 1/h (brown))

Table 2. Results of the best planned maintenance period and maximum availability assured by the best period given $\lambda = 0.00001$ 1/h

Number of Fig.	Best planned maintenance period T^{BEST} [h]	Max availability A_{PS}^{MAX} given T^{BEST}
3	3470	0.987
4	2440	0.984
5	2580	0.985
6	1880	0.981

The results of case study prove that growth of both ρ and α can extend T^{BEST} in essential way. Moreover, improving the λ (especially from of 10^{-4} 1/h up to 10^{-5} 1/h or more) can produce some benefits, i.e. extending T^{BEST} and going up of A_{PS}^{BEST}, and finally the performance of the WF.

5 Conclusions

The reliability and maintenance model of protection system of WF is presented. The model relies on semi-Markov approach. The model takes into account all important features that determine correct and incorrect operation of protection system, i.e. failures occurrence (including hidden ones), planned maintenance, detecting the failures under planned maintenance service, detecting failures/false failures by self-monitoring and checking module, repair, and holding the failure after preventive maintenance service (failure undetected). The case study prove, the model is able to obtain the best interval time of preventive maintenance and the best availability of PS given the quantitative parameters that represent the features mentioned above. Improving the probability of both detecting the failure within planned maintenance and detecting the failures by self-monitoring and checking module can extend the best time interval of planned maintenance substantially.

Acknowledgment. The research has been financially supported by Ministry of Science and Higher Education, Poland, under the grant of Institute of Control Engineering, Electronics and Electrical Engineering, Department of Electrical Engineering, Power Electronics and Power Engineering.

References

1. Bahirat, H.J., Mork, B.A., Hoidalen, H.K.: Comparison of wind farms topologies for offshore applications. In: IEEE Power and Energy Society General Meeting (2012)
2. Segura-Heras, I., Escriva-Escriva, G., Alcazar-Ortega, M.: Wind farm electrical power production model for load flow analysis. Renew. Energy **36**, 1008–1013 (2011)
3. Ali, M., Matevosyan, J., Milanovic, J.V.: Probabilistic assessment of wind farm annual energy production. Electr. Power Syst. Res. **89**, 70–79 (2012)
4. Lubośny, Z.: Farmy wiatrowe w systemie elektroenergetycznym. Wydawnictwa Naukowo-Techniczne, Warszawa (2009). (in Polish)
5. Sinica M.A., Jablonskij P,M.: Ob odnoj modeli obsluziwanja techniczeskich obiektov z pomoszcziu dwoch sistem kontrolia. Osnownyje woprosy teori i praktyki nadeznosti. Moskva 'Sovetskoje Radio', pp. 342–351 (1975). (in Russian)
6. Grabski, F.: Semi-Markowskie modele niezawodności i eksploatacji. Instytut Badań Systemowych PAN, Warszawa (2002). (in Polish)

On the Influence of the Coding Rate and SFN Gain on DAB+ Coverage

Kamil Staniec(✉), Sławomir Kubal, Michał Kowal, and Piotr Piotrowski

Wroclaw University of Science and Technology, Wyb. Wyspiańskiego 27, 50-370 Wrocław, Poland
{kamil.staniec,slawomir.kubal,michal.kowal, piotr.piotrowski}@pwr.edu.pl

Abstract. The ubiquitous digitization which also affects the audio broadcasting technologies, has spawned multiple standards, such as IBOC/HD Radio, ISDB-TSB, DRM, DBM or DAB+ , out of which the latter has gained the wideset endorsement, especially among European countries. Despite requiring a ten-fold wider channel, it offers multiple advantages over the traditional analog (FM) broadcasting counterpart. These are enumerated and evaluated in terms of financial benefits, enhanced/new technological functionalities or even ecological aspects. Some of the highlights include the possibility of attaching synchronized images (even video clips), website broadcasting, journaline service, traffic messaging etc. from multiple broadcasters within a single multiplex (MUX) occupying a single frequency channel emitted from multiple sites simultaneously, forming so called a Single Frequency Network (SFN). The research part of the paper regards simulations obtained for a DAB+ SFN network, with a purpose of demonstrating how coverage will differ the coding rate which is the only variable parameter at an operator's disposal, by means of which they can scale the audibility and data transmission range in digital audio broadcasting systems. A separate analysis is devoted to the SFN gain; an effect observed by users caused by the transmit diversity from multiple (three in the analyzed case) transmitters.

Keywords: Sensitivity · Fixed · Portable · Mobile

1 Introduction

1.1 Analog vs. Digital Audio Broadcasting

The ubiquitous digitization, so well observed in multiple domains of information and communication technologies (ICT), has also affected the broadcasting domain. At first, the digital television DVB-T (Digital Video Broadcasting - Terrestrial) replaced its analog counterpart. A similar process was inevitable in the audio broadcasting as well, although in this matter there appears to be far less international agreement than with TV, which results in the digital audio switchover proceeding at much slower pace. The advantages, however, leave no doubts as to whether this direction is or is not correct. On the end-user's side, the digital transmission, first of all, opens up a plethora of

W. Zamojski et al. (Eds.): DepCoS-RELCOMEX 2020, AISC 1173, pp. 596–605, 2020.
https://doi.org/10.1007/978-3-030-48256-5_58

services unavailable in the analog FM broadcasting which allowed only for the audio signal (assisted by functionally-restrained the Radio Data System, RDS, messaging), as will now be enumerated. Another distinguishing feature of the digital broadcasting is its ability to "pack" several broadcasters into a single channel, called a multiplex (MUX). From the users' point of view, the constituent sub-channels are discoverable on their radio receivers as separate radio stations.

- Dynamic Label Segment (DLS) – sending long text messages;
- Slide Show (SLS) – sending graphical messages (images;
- Electronic Program Guide (EPG) – information (listed and/or detailed) on the program timetable;
- Broadcast Website (BWS) – broadcasting web sites to receivers with LCD displays;
- Journaline – an e-newspaper service;
- Traffic Message Channel/Traffic and Travel Information (TMC/TTI) – broadcasting local, regional or national information on a current traffic situation;
- Emergency Warning System (EWS) – broadcasting local, regional or national information in case of e.g. natural disasters;
- Surround Sound over DAB+ – multiple audio channels transmission for superb hearing effects.

1.2 Highlights of the Digital Broadcasting

A practical question that now remains be answered is why not simply use cellular technologies to convey all this information, including radio stations, to end-users. The answer is: due to the lack of available transmission capacity if one considers the total amount of traffic generated by dozens of millions of users every day just on listening to the Internet radio. Taking Norway as an example, estimates indicate that an average listening time per day equals 92 min (mainly while driving cars), which is equivalent to the daily/monthly data transfer of 190 MB/5.7 GB, which is close to the estimated monthly data transfer planned by operators per a statistical user in total (i.e. including all other network services) also for USA and Great Britain. This gives the total of 130 000 TB of data in a single year, which exceeds the total capacity available, for instance, to all four Norwegian telecom operators. It is therefore obvious that some services – with the audio broadcasting opening this list – should always be broadcast to preserve the operators' resources, for at least the following reasons:

- economic – the total cost of transferring 130 000 TB of data in the cellular way would cost 860 mln EUR a year, whereas achieving the same using the DAB+ standard would cost 10–20 mln EUR per year. Also, the access to the mobile cellular Internet is mostly paid by subscribers; the access to the broadcast services, on the other hand, cost either nothing at all or just the general annual TV/radio subscription fee which is usually on the order of a few tens of EUR different by a country.
- technical – an average smartphone batteries usually run out after c.a. 7 h of constant playout of the Internet radio whereas the uptime for listening to the broadcast radio prolongs to 40–50 h.

The other advantages, as enumerated below [1], will only concern advantages of the digital vs. analog (FM) audio broadcast. The basic argument for digitization lies in the nature of the audio multiplex (MUX), which allows to pack several (up to 30) radio stations into a single physical channels, which has the following ramifications in terms of up to 30 less:

- transmitters, resulting in proportionally fewer masts, antennas and emitter locations;
- emission licenses from national telecom regulators;
- physical spectral channels to coordinate;
- distribution links by which signals carrying separate "radio stations" are fed from the broadcaster to the transmitter.

Another group of advantages stems from major technical differences resulting from the digital transmission, which include [1]:

- a few decibels lower minimum electric field strength (E_{min}) necessary to maintain a comparable coverage, in terms of subjectively evaluated audio quality (e.g. 66 dBμV/m in FM vs. 58 dBμV/m in DAB+), at the cost of slightly increased feeder and adapter losses due to higher operating frequencies used in digital transmission;
- the above argument implies less equivalent isotropic radiated power (EIRP) necessary to meet the minimum E_{min} criterion (e.g. 2.5 kW_{rms} in DAB+ can be equivalent to 10 kW in FM);
- the total price for a DAB+ transmitter (emitting, for instance, 15 channels in a MUX) will reach c.a. 120 kUSD whereas the purchase of 15 separate transmitter for emitting the same number of FM channels will cost c.a. 750 kUSD altogether;
- with the above assumptions in mind, the total energy consumed per year by a single DAB+ transmitter (requiring lower EIRP, as stated earlier), will generate a cost of 7.3 kUSD, whereas that for 15 transmitters (each additionally requiring a greater EIRP) will be c.a. 182 kUSD. This argument can also be viewed as one benefiting ecology due to a 25-fold lower demand for power supply (and hence less CO_2 emission etc.).

Lastly, one should mention that as opposed to the analog transmission, the digital signals yield perfect hearing quality as long as some threshold carrier-to-noise ratio (C/N) is maintained. It means that while a listener keeps moving away from the transmitter, their observed audio experience does not gradually deteriorate as is the case with the analog signal. Instead, it upholds high quality until abruptly ending as the minimum C/N has been met.

2 The Digital Audio Broadcasting (DAB+) System – A Technical Specification and Current Research

The DAB+ standard, that appears to be the most popular for digital broadcasting, has been developed by ETSI (European Telecommunications Standards Institute) and described in [2–4]. According to the GE06 Agreement [5] (Geneva, 2006), it operates in 174–230 MHz VHF III band which causes a few-decibel greater losses incurred during propagation (c.a. 3 dB in free space) as well as in signal feeders or adapters than its FM counterpart. The DAB system uses differential QPSK modulation coupled with a multi-carrier scheme: known as Orthogonal Frequency Division Multiplexing (OFDM), with 1536 orthogonal, individually modulated, subcarriers, forming approximately rectangular, Gaussian noise-like spectrum occupying a bandwidth of c.a. 1.54 MHz [6], allowing the maximum of 2432 kb/s of throughput per channel and multiplexing up to 30 different audio stations. The basic principle consists in dividing the information to be transmitted into a large number of bit-streams, having low bit-rates individually, which are then used to modulate individual orthogonal carriers, such that the corresponding symbol duration becomes larger than the delay spread of the transmission channels. By inserting a temporal guard interval (GI) between successive symbols, the channel selectivity and multi-path propagation will not (to a large extent) cause inter-symbol interference. As in [2], the system provides four transmission mode options which allow the use of a wide range of transmitting frequencies, up to 3 GHz for mobile reception. These transmission modes have been designed to cope with Doppler spread and delay spread, for mobile reception in the presence of multi-path echoes. Since the project referred to in this paper concerns the DAB+ system operating in the SFN (Single Frequency Network), the most suitable of these is mode I, well-fitted for a terrestrial single-frequency network (SFN) in the VHF range because it allows the greatest transmitter separations (up to 96 km). When in this mode, it is recommended that the operational frequency is no greater than 375 MHz with GI = 246 μs.

The audio codec applied in DAB+ is HE-AAC v2 [7] (or AAC++), deliberately chosen due to its specific features outperforming the popular MPEG4 codec, used in DAB, for instance. AAC++ is a solution specifically optimized for work at lower frequencies, i.e. better adapted to the human psychoacoustic curve centered at c.a. 2–4 kHz. It has been found that its subjectively evaluated performance at 40 kb/s is comparable to that achievable at 128 kb/s with MPEG4. In other words, the same audio quality is attained at c.a. 3 times lower bit rate.

Since the modulation in DAB+ is DQPSK (i.e. 2 bits per a modulation symbol), the effective reception range is only dependent on the coding rate, which comes in the Equal Error Protection (EEP) variant, with the coding rate R – depending on which of the four protection levels (PL) has been endorsed: 0.25 (1A), 0.375 (2A), 0.5 (3A) and 0.75 (4A). With EEP, all signal parts are equally protected.

Despite its invention in 2007 not a lot of attention has been paid to the matter of SFN networks thus far, treating it as mainly an engineering issue. Some items, however, do require scientific effort to answer, these are: provision of limited-range services in wide-area SFN networks or coping with the power and multipath imbalance and finally – which is treated in detail in this paper – systematic statistics of the expected SFN gain in terms of range and coverage. Some of the most prominent writings on the former aspects include: e.g. [8] where the average SFN gain of 1.1 dB and its deviation of 3.3 dB were obtained for an exemplary DVB-H network in Ghent, Belgium, or [9, 10] where authors propose a Local Service Insertion (LSI) method allowing to provide local services in wide-area SFN networks (impossible in the original system version) by means of inserting an additional phase reference symbol to the DAB transmission frame. In [11] experimental SFN DVB-T2 network trials were examined for the SFN gain leading to updating empirical minimum C/N (Carrier/Noise) thresholds and providing a thorough discussion on the influence of power imbalance and relative delays occurring in SFN networks on performance in the Rayleigh and Rician channel. Analytical approach to the fading phenomenon, intrinsic to single-frequency architectures, has been provided in papers [12] and [13] wherein a propagation model was developed that allows to determine both the signal level at the reception site as well as the phase of this signal, which – applied in a precise interference analysis of a real-life case study (i.e. a DAB+ 'LokalDAB' network in Wrocław, Poland) – have led to some non-obvious conclusions regarding the strength of fades. These turned out to be deep for receivers placed equidistantly to a pair of transmitters and far from the third one, while being relatively flat when placed at similar distances to each of the three transmitters.

3 Definitions of the Effective Maximum Range and the Coverage

Calculating the median electric field strength E_{med} involves several steps. Firstly, in accordance with Eq. (1), one shall calculate the receiver noise power, where: *NF* is the noise factor [dB], k is the Boltzmann's constant ($1.38 \cdot 10^{-23}$ *K*), T_0 is the reference temperature (290 *K*) and B is the receiver noise bandwidth [Hz].

$$P_n = NF + 10\log_{10}(kT_0B) \tag{1}$$

Knowing the receiver noise power one can calculate its sensitivity with the use of Eq. (2), wherein *C/N* [dB] is the required signal-to-noise ratio. Thereupon it is also necessary to determine the receiving (Rx) antenna aperture, as in Eq. (3) in which G [dBd] and λ represent, respectively, the Rx antenna gain and the signal wavelength.

$$P_{s\,min} = \frac{C}{N} + P_n \tag{2}$$

$$A_a = G + 10\log_{10}\left(1.64\lambda^2/4\pi\right) \tag{3}$$

The minimum power density at the reception point shall be found from Eq. (4), where L_f denotes the feeder loss in decibels. The last step towards obtaining E_{med} consists in converting φ_{min} to the electric field strength domain, as by Eq. (5).

$$\varphi_{min} = P_{s\,min} - A_a + L_f \tag{4}$$

$$E_{med} = \varphi_{min} + 120 + 10\log_{10}(120\pi) = \varphi_{min} + 145.8 \tag{5}$$

$$C_l = \mu\sqrt{\sigma_b^2 + \sigma_m^2} \tag{6}$$

$$E_{min} = E_{med} + C_l + L_h + L_b + L_{freq} \tag{7}$$

Table 1. Standard parameters used in sensitivity determination for DAB+ [14–16].

Parameter	Fixed reception	Portable reception	Mobile reception
L_{freq} *[dB]*	$20\log_{10}(f_c/200$ MHz)	$30\log_{10}(f_c/200$ MHz)	
NF	7 dB		
C/N	Depends on PL (1A–4A), see [10] (Table 3 therein)		
G	7 dBd	−2 dBd	
Percentage of locations	95%		99%
C_l	1.64·5.5 dB (=1.64·σ_m dB)		2.33·5.5 dB (=2.33·σ_m)
L_h	0 dB (for h_{Rx} = 10 m)	12 dB (for h_{Rx} = 1.5 m)	
L_b	9 dB		
σ_b	3 dB		
σ_m	5.5 dB		

The effective range, or coverage, will however be determined not by the median electric field but rather by its minimum value E_{min} that accounts for the location correction factor C_l [dB] meant to assure that along the maximum range line delineating the effective coverage, the required percentage of users will enjoy reception at a satisfactory level (usually 95% or 99%), assuming a log-normal EM signal distribution. The parameter C_l, defined by Eq. (6), considers therefore: the normal distribution factor μ and standard deviations associated with the radio signal variation outdoors (σ_m) and indoors (σ_b). Moreover, it should also account for corrections due to other factors, such as: the Rx antenna height h_{Rx} other than the referential 10 m (L_h correction), building penetration losses for indoor reception (L_b correction) and the center frequency (f_c) other than the reference 200 MHz (L_{freq}). The resultant and final value of the minimum electric field strength is therefore given by Eq. (7), which constitutive parameters are provided in Table 1, with discrimination of three reception modes: - fixed, where a stationary Rx antenna is assumed, - portable, where the Rx antenna can be moving slowly at a pedestrian's pace and – mobile, with the Rx antenna used in or mounted upon a moving vehicle. The final magnitudes of the minimal electric field strength E_{min} calculated for all possible combinations of code rates (determined by a respective *PL* as defined in Sect. 2) and division into the outdoor and indoor reception, are provided in Table 2.

Table 2. Minimum electric field strength E_{min} [dBμV/m] for different modes in DAB+.

PL	Fixed reception (outdoor/indoor)	Portable reception (outdoor/indoor)	Mobile reception (outdoor/indoor)
1A	46.7/57.0	70.9/80.8	74.7/85.2
2A	47.3/57.6	73.2/83.1	77.0/87.5
3A	48.6/58.9	75.7/85.6	79.5/90.0
4A	51.5/61.8	81.2/91.1	85.0/95.5

4 Effective Range/Coverage Calculations for DAB+

The results of analyses presented in this chapter refer to a DAB+ network deployed in the SFN configuration, operating in Wrocław and consisting of three transmitters (which sites are marked as 'WUST', 'IŁ-PIB' and 'Radio Wrocław') located with respect to each other as shown in Fig. 1 (left).

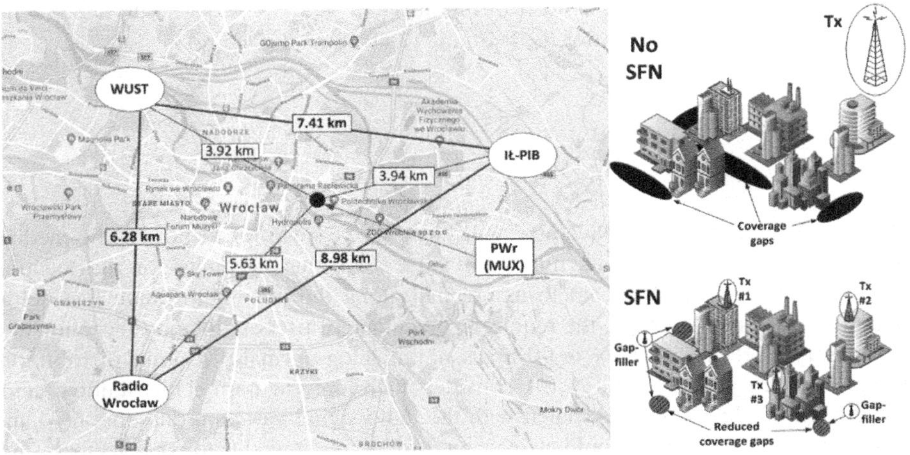

Fig. 1. (Left) A geometry of the LokalDAB network; (right) SFN principle explanation.

As can also be noticed in Fig. 1 (right), with a single transmitter, there may occur spots ("black holes" in the upper right corner of the figure), especially in high urban environments, where the electric field strength drops below E_{min} leading to sudden lack of reception. The idea behind the SFN approach consists in transmitting the same information (MUX) simultaneously from several transmitters, each emitting at proportionally lower EIRP to achieve the same effective range as would be obtained with a single high-power transmitter, as shown in the lower part of Fig. 1. This is happening unbeknownst to end users yet considerably reducing the number of possible blackouts. The remaining ones can further be effectively coped with by means of using, so called 'gap-fillers' (or 'DAB-repeaters'), which receive and then retransmit the signal at much lower power than the basic transmitters forming the SFN, thus reducing or eliminating areas deprived of a proper signal quality.

The maximum effective range is defined as a maximum distance originated at the center of the triangle formed by the three transmitters and stretched out to the farthest end of coverage, as shown in Fig. 2 (right). The coverage, in turn, is the total area covered by the signal of electric field strength $E > E_{min}$, with E_{min} thresholds as defined in Table 2 for various combinations of: Protection Layer (*PL*), reception mode (fixed, portable, mobile) and the environment type (indoor/outdoor).

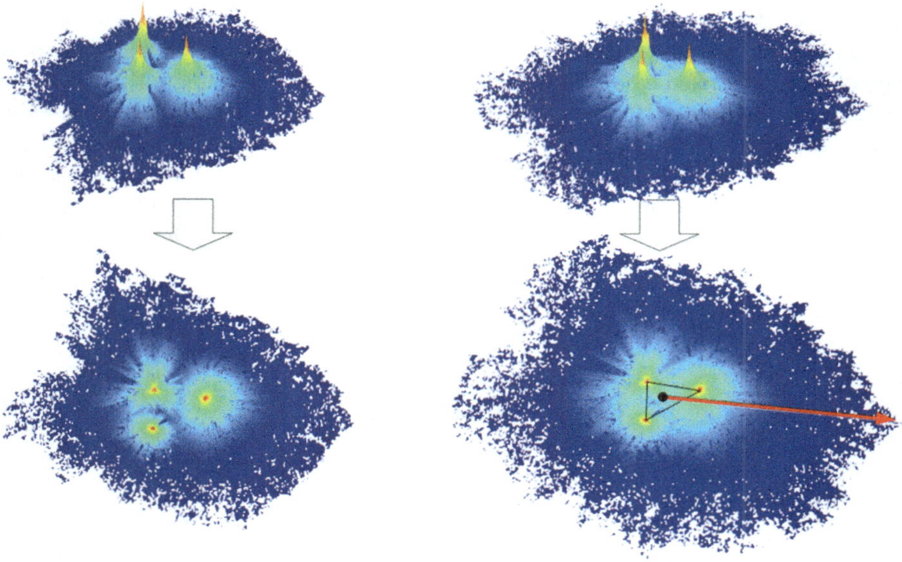

Fig. 2. A coverage and a maximum range definitions: without (Left) and with SFN (Right) solution (calculated for PL = 1A, fixed, outdoor).

The results were obtained with the use of a software simulator called "Piast" [17] created by the National Institute of Telecommunications, based on ITU-R P.1546 method [18] for point-to-area predictions, currently in force. Figure 3 allows one to draw two major takeaways. Firstly, the maximum reduction between extreme *PL*'s (i.e. *PL* = 4A relative to *PL* = 1A) oscillates around 50% for both outdoors and indoors. Secondly, while the range reductions with increasing *PL* are comparable in the portable and the mobile mode, reductions in the fixed mode are more than twice greater.

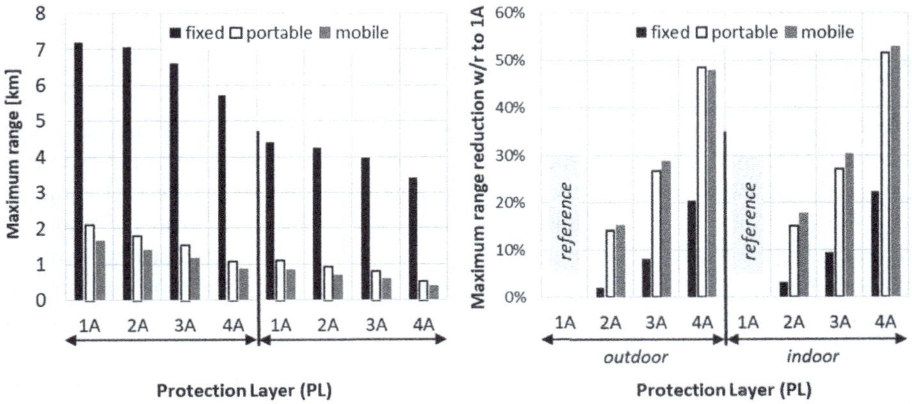

Fig. 3. The maximum range w/r to *PL*: in the absolute measure (Left); relative to *PL* = 1A (Right).

As for the coverage area, it was analyzed with respect to the third parameter, namely the 'SFN gain' which can be defined as the surplus electric field strength stemming from simultaneous reception from multiple signal sources (the three transmitters in the presented case). In the following analyses E at any point of the reception plane (located either at 10 m or 1.5 m, as set out in Sect. 3) was calculated in either case – i.e. with and without the SFN configuration – using formulas (8) and (9), respectively. Similarly to the previous maximum range analysis, significant degradation in coverage occurs when considering the receiver placement inside buildings (indoor reception), as well as a notable difference between absolute coverage areas between the fixed mode and the portable/mobile modes, as presented in Fig. 4.

$$E_{SFN} = \sum_i E_{Txi} \tag{8}$$

$$E_{NoSFN} = \max\{E_{Tx1}; E_{Tx2}; E_{Tx3}\} \tag{9}$$

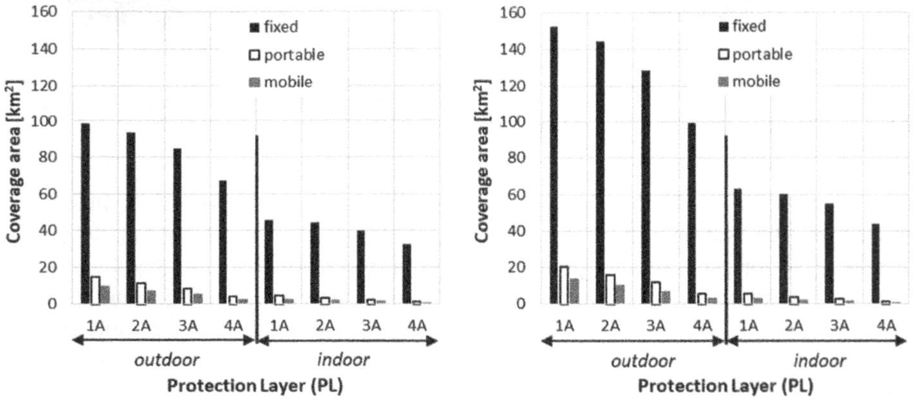

Fig. 4. The coverage area: without SFN (Left); with SFN (Right).

5 Conclusions and Further Research

The paper is meant to provide answers regarding the influence of the Protection Layer and the use of the Single Frequency Network solution on the maximum range and the coverage of DAB+ service. The former parameter affects the received signal immunity to noise (with 1A being the most and 4A the least robust) but also the multiplex capacity, i.e. the maximum bit rate available per sub-channel/a radio station (with 1A allowing for the smallest capacity and 4A for the greatest). Thus, the broadcast network planner is forced to make a trade-off between the multiplex capacity (i.e. the number of radio stations that can be included in a MUX) and the effective coverage. It was shown that the strongest downgrade occurs upon transition from 3A to 4A which renders the latter setting not recommendable for practical use. It was also demonstrated, the number of population covered by the DAB+ network service drops a few times for portable and mobile users with respect to those receiving signals in stationary localizations (i.e. in the fixed mode).

References

1. Stockmann, J.: DAB+ The cost effective Radio transmission. Harris Broadcast www.harrisbroadcast.com
2. ETSI TR 101 496-1 V1.1.1. Digital Audio Broadcasting (DAB); Guidelines and rules for implementation and operation; Part 1: System outline
3. ETSI TR 101 496-2 V1.1.2. Digital Audio Broadcasting (DAB); Guidelines and rules for implementation and operation; Part 2: System features
4. ETSI TR 101 496-3 V1.1.2. Digital Audio Broadcasting (DAB); Guidelines and rules for implementation and operation; Part 3: Broadcast network
5. ITU. FINAL ACTS of the Regional Radiocommunication Conference for planning of the digital terrestrial broadcasting service in parts of Regions 1 and 3, in the frequency bands 174–230 MHz and 470–862 MHz (RRC-06)
6. ETSI EN 300 401 v2.1.1 (2017-01), "Radio Broadcasting Systems; Digital Audio Broadcasting (DAB) to mobile, portable and fixed receivers)", January 2017
7. ETSI TS 102 563 V1.2.1: Digital Audio Broadcasting (DAB); Transport of Advanced Audio Coding (AAC) audio (May 2010)
8. Plets, D., et al.: On the methodology for calculating SFN gain in digital broadcast systems. IEEE Trans. Broadcast. **56**(3), 331–339 (2010)
9. Schrieber, F.: A backward compatible local service insertion technique for DAB single frequency networks: first field trial results. In: 2018 IEEE International Symposium on Broadband Multimedia Systems and Broadcasting (BMSB), Valencia, pp. 1–5 (2018)
10. Schrieber, F.: A differential detection technique for local services in DAB single frequency networks. In: 2019 IEEE International Symposium on Broadband Multimedia Systems and Broadcasting (BMSB), Jeju, Korea (South), pp. 1–7 (2019)
11. Morgade, J., et al.: SFN-SISO and SFN-MISO gain performance analysis for DVB-T2 network planning. IEEE Trans. Broadcast. **60**(2), 272–286 (2014)
12. Zieliński, R.: Fade analysis in DAB+ SFN network in Wroclaw. In: Zieliński, R. (ed.) Proceedings of the 2019 International Symposium on Electromagnetic Compatibility (EMC Europe 2019), Barcelona, Spain, 2–6 September 2019
13. Zielinski, R.J.: Analysis and comparison of the fade phenomenon in the SFN DAB + network with two and three transmitters. Int. J. Electron. Telecommun. **66**(1), 85–92 (2020)
14. ITU: ITU Final Acts of the Regional Radiocommunication Conference for planning of the digital terrestrial broadcasting service in parts of Regions 1 and 3, in the frequency bands 174–230 MHz and 470–862 MHz (RRC-06)
15. ITU: ITU-R BS.1660-7 (10/2015), "Technical basis for planning of terrestrial digital sound broadcasting in the VHF band"
16. European Broadcasting Union (EBU): TR 025, "Report on frequency and network planning parameters related to DAB+", version 1.1, Geneva, October 2013
17. The "Piast" program official website, hosted by the National Institute of Telecommunications: http://www.piast.edu.pl/About
18. ITU: ITU-R P1546-6. Method for point-to-area predictions for terrestrial services in the frequency range 30 MHz to 4000 MHz (August 2019)

Intra-round Pipelining of Keccak Permutation Function in FPGA Implementations

Jarosław Sugier(✉)

Faculty of Electronics, Wrocław University of Science and Technology, Janiszewskiego St. 11/17, 50-372 Wrocław, Poland
jaroslaw.sugier@pwr.edu.pl

Abstract. The Keccak permutation function constitutes the essential part of the computations in the SHA-3 (Secure Hash Algorithm-3) standard. Its fast implementation in hardware is crucial for efficient operation of many contemporary ICT systems which commonly apply hashing e.g. in data storage and transmission or in security protection. This paper analyzes potential improvements in computation speed of the function if its hardware implementation uses a pipelined organization where the single Keccak round is divided into two or three pipeline stages. The discussion starts with examination of various options for such pipelining and then the proposed architectures are implemented in a Spartan-7 FPGA device from Xilinx. Estimations of their maximum frequencies of operation illustrate speed gains (in terms of total throughput) which can be accomplished by the increased parallelization achieved through fine-grained pipelining. The results indicate that after careful tuning of the overall control framework the complete module calculating the 1600-bit permutation can operate at frequencies exceeding 500 MHz even in a device from this economy-grade FPGA family.

Keywords: Cryptographic hash function · SHA-3 · Hardware implementation · Pipelining

1 Introduction

The cryptographic hash algorithms for computer applications have been developed since 1990s but new research activities intensified in the new century with novel advances in cryptanalysis. Probably the most significant progress in this area was made as a result of a contest for the new SHA-3 algorithm which was announced by the U.S. National Institute of Standards and Technology in 2007. Its idea was to utilize positive momentum created in cryptographer's society by a successful competition for the AES block cipher (finalized in 2001) when the new block cipher was selected in an open and intensive debate of expert researchers and practitioners from all over the world. In an analogous manner, a public discussion of the SHA-3 competition ended with selection of the Keccak algorithm for the new hash standard. This paper explores possible hardware implementations of the Keccak permutation function in a Field Programmable Gate Array (FPGA) for a specific architecture option: when the single cipher round is divided into multiple pipeline stages.

W. Zamojski et al. (Eds.): DepCoS-RELCOMEX 2020, AISC 1173, pp. 606–615, 2020.
https://doi.org/10.1007/978-3-030-48256-5_59

1.1 Motivation and Scope of This Work

KECCAK's high cryptographic strength and resistance to attacks is achieved thanks to its involved and massive computational scheme which transforms a bulky 1600-bit state data – hence its efficient implementation is a task which requires careful consideration. In hardware architectures pipelining is a classic method of achieving high data throughput and this direction was considered from the very beginnings as one of the evident options in implementation of this cipher.

The particular topic of this paper – intra-round pipelining, i.e. an organization where the single cipher round is split into multiple stages which allows to process multiple sets of data in parallel – was e.g. examined in [9] where all the 4 elementary Keccak transformations were implemented as separate stages. As not all of them are equally complex, in order to balance the load the θ transformation was split into two stages but authors do not present details of the split and of the final organization neither discuss viability of other options – apart from reporting maximum operational frequency of their circuit in a Virtex-5 FPGA chip. A solution where the entire round is divided into two pipeline stages with registers between π and χ transformations was proposed in [1] where it was verified across Virtex-5, 6 and 7 devices. In [7] authors presented an implementation with two computational cores operating in a sequence where each core computed half of the round iterations – which facilitates processing of multi-block messages – and this was also only tested in different Virtex chips. In all these papers (and numerous other ones about FPGA implementations of various cryptographic algorithms, e.g. in [4–6]) the problem of optimization was perceived as a task of finding only the best (with regard to the criteria adopted by the authors) design which operates by some margin faster than competition, without presentation of other feasible choices.

More comprehensive discussion including additional, probably relatively close alternatives would enable taking an informed decision about choosing maybe some sub-optimal solution but a preferable one for a reason other than raw performance. This can be especially significant in FPGA practice where the cryptographic unit must co-exist with other parts of the system realized in the same chip and with the same implementation process so that its optimization cannot govern the overall implementation course. In this context, the character of this work is somewhat different and the attention is more broadly put on examination of how the idea of intra-round pipelining works in the case of this particular algorithm including evaluation of benefits and costs of various design alternatives. The study comprises a total of 15 analyzed designs and they all were verified by implementation in a Spartan-7 FPGA device from Xilinx.

The paper is organized as follows. The next section analyzes operations of the KECCAK round in order to identify viable pipelining options and introduces specific cases proposed in this work. The third section reports the implementation results and discusses optimizations in the overall control scheme which were necessary to keep its speed up with the fast pipeline stages. Section 4 evaluates the numerical results and is followed by conclusions.

2 Hardware Implementation of the Permutation Function with Intra-round Pipelining

2.1 Implementing the KECCAK Function in Hardware

The algorithm was defined in [2] as a family of seven different size variants: with parameter $l = 0 \ldots 6$, each variant operated on words of bit length $w = 2^l$ ($w = 1 \ldots 64$) called lanes. A 5×5 array of lanes formed a state A; its total bit size was thus given as $b = 25 \cdot w = 25 \ldots 1600$. The last parameter derived from l was a number of rounds in the permutation function: $n_r = 12 + 2\,l = 12 \ldots 24$. The SHA-3 standard [8] employs the largest 1600 bit version where $l = 6$, $w = 64$, $b = 1600$ and $n_r = 24$. This variant is the sole interest of our work.

The overall hashing scheme in KECCAK is based on a simple but effective concept of a cryptographic sponge ([3]) which operates by repeated application of the permutation function KECCAK-*f*() iteratively to consecutive chunks of the input stream or recursively to the state itself – during, respectively, actual hashing (so called absorbing phase) or generation of the output (squeezing phase). With straightforward implementation of the sponge scheme, the challenge of realization of the whole algorithm consists entirely in efficient implementation of the permutation function.

The function is computed by application of the round transformation n_r times to the 1600-bit state which at first is initialized directly with the input bits. Operation of a single round is a sequence of 5 elementary transformations of the state: θ, ρ, π, χ and ι. Every transformation takes as its input the state produced by the preceding one.

Figure 1 presents different hardware organizations of the complete permutation module which are considered in this paper. After common taxonomy of various architectural options available for any round-based cipher which is used e.g. in [5] or [6] we will use notation X*n*P*m* to describe an architecture with a cascade of *n* round instances which are split into *m* pipeline stages.

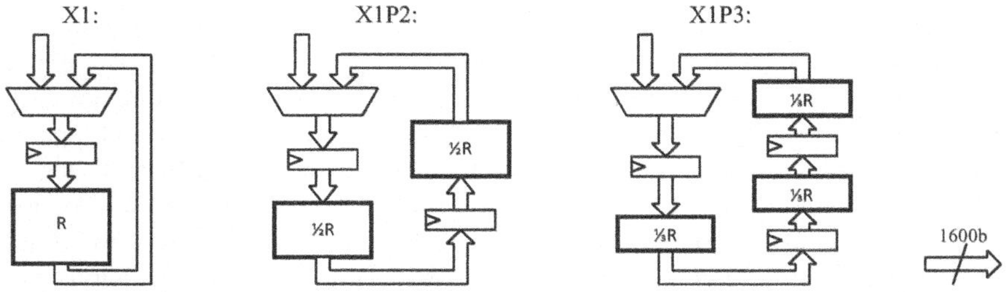

Fig. 1. Implementations of the complete permutation function considered in this paper, from left to right: iterative one and with round divided into two or three pipeline stages.

As the starting point in preparation of the pipelined designs (and as a reference solution used for their evaluation) the standard iterative case X1 was used where the state A is repeatedly processed by the only round instance n_r times (in n_r clock cycles) without pipelining. Such an implementation of the KECCAK permutation was already examined in our previous works in [10] and [11] where it was used as a reference for comparison with loop unrolled cases (Xn where $n > 1$) optionally with extra-round pipelining ($n \geq m$). The two remaining organizations shown in Fig. 1 were created from the X1 case by application of intra-round pipelining and they constitute the actual subject of the study in this work.

2.2 Organization of the Intra-round Pipelining

Decision how to split the round logic into two or three pipeline stages depends on its internal structure in hardware. When analyzed form the point of view of data propagation in a digital circuit, the round code [8] can be visualized as in Fig. 2.

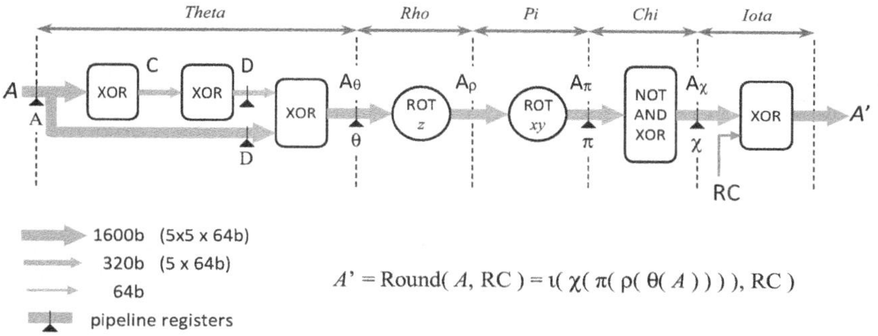

Fig. 2. Structure of internal data processing inside the round; markers identify locations of the pipeline registers tested in this paper.

Every elementary transformation in KECCAK can be seen as a block of hardware (logic gates) which alters value of the A vector. Most of the transformations involve XOR gates; only the χ step uses also NOT and AND operators but still it can be expressed with a relatively simple equation where each new state bit depends on three old ones. The data flow is strictly linear with the only minor aberrance in the θ step: first an auxiliary signal C and then D need to be computed from A and only afterwards the state is bitwise xor'ed with D. It should be noted that from the point of view of a hardware implementation the ρ and π transformations are empty because they only re-order bits in the state vector. In software such re-ordering can be quite time consuming (loading the word to a CPU register, performing the rotation, saving the result) but in a digital circuit such re-ordering is implemented entirely in routing of the state vector when it is transmitted from θ to χ logic. Nevertheless, because the A vector is 1600 bit wide, in FPGA devices such re-routing can lead to complicated and lengthy – i.e. slow – propagation paths. Their successful implementation heavily depends on efficient optimizations performed by both placement and routing tools.

The figure also shows locations of registers introduced to create pipeline stages in the tested architectures. Most of them appear exactly at transformation boundaries with the only exception of θ: its operation is expressed by a 3-level function so the D signal (at 2/3 of θ) was also considered as a candidate location. Moreover, in this particular case also the *A* input needs to be registered in order to delay its arrival at the final xor operation synchronously with D (i.e. one clock period later).

In this study the pipelining concept was tested in the following configurations:

- X1P2 (two pipeline stages, one intra-round register) – three variants with the round split at D, θ or π locations;
- X1P3 (three pipeline stages, two intra-round registers) – also three variants with registers at θ and π, θ and χ, or D and π.

Placement of the pipeline registers will be denoted with lower indices in the architecture symbol, e.g. $X1P2_D$ or $X1P3_{\theta\pi}$. The ρ & π logic is actually empty so these transformations were not treated as a candidate for a separate stage but their allocation to either one or another stage could make a difference – thus the option $X1P2_\theta$ should be considered different than $X1P2_\pi$, for example.

3 Implementation Results

3.1 Application of the Pipelining

All the six pipelined architectures were implemented in a Spartan-7 xc7s50fgga484-2 device from Xilinx using the Xilinx Synthesis Tool (XST) and the latest Vivado 2019.2 software suite. The module of the permutation function was extended with basic serial-to-parallel input/output logic so that the design was completely functional. Table 1 lists parameters of the implementations with the X1 case added for reference. The group of the first four columns describe performance that was determined by the longest propagation path: minimum clock period which leads to the maximum operating frequency, location of the critical path (components it runs from : to), percentage of its delay generated by logic resources (i.e. excluding routing) and number of logic levels i.e. the function generators. All these parameters were reported by the implementation tools for the final, completely routed design. Additionally, in column 3 some particular components of the overall control infrastructure are introduced as sources/destinations of the critical paths: CR is a round counter (mod 24) which holds current number of iteration and E is a general enable (active) flag controlling globally operation of all mayor elements. The last three columns give sizes of the designs in, respectively, slices, LUTs (logic generators) and registers of the FPGA array.

In the ideal case, in any X1P*m* architecture the T_{clk} period should be reduced approximately to 1/*m* of the value of the X1 case (ignoring in this crude approximation delays of the register flipflops – this will be considered in the next section). It is clearly seen that the actual T_{clk} values are far from this reduction and, to make the situation even worse, all the P3 cases are actually *slower* than any of the P2 one – which indicates some serious implementation problems caused by an increased total number of individual paths to be routed in P3 organizations vs in P2 ones. The reason is indicated in the column giving the location of the critical path: only in the X1 case it runs as expected through the round logic (from and to register A located at the round entrance; compare Fig. 1) while in the P2 & P3 cases the limits come from the control infrastructure: the round counter or the global active flag rather than from the in-round

Table 1. Parameters of the first X1P2 and X1P3 implementations.

Architecture	Critical path				Slices		LUTs		Registers	
	T_{clk} [ns]	Src : Dst[(a)]	Logic delay	Logic levels	Tot.	%	Tot.	%	Tot.	%
X1	3,67	A:A	19%	3	1332	16,3%	5080	15,6%	4806	7,4%
$X1P2_D$	2,86	CR:A	28%	3	1412	17,3%	4628	14,2%	6739	10,3%
$X1P2_\theta$	2,87	CR:A	22%	2	1573	19,3%	5460	16,7%	6418	9,8%
$X1P2_\pi$	2,84	CR:A	24%	3	1574	19,3%	5459	16,7%	6416	9,8%
$X1P3_{\theta\pi}$	3,17	CR:A	19%	2	1762	21,6%	5464	16,8%	8028	12,3%
$X1P3_{\theta\chi}$	3,15	CR:A	19%	2	1677	20,6%	5604	17,2%	8034	12,3%
$X1P3_{D\pi}$	3,11	E:A	25%	3	1608	19,7%	5460	16,7%	8353	12,8%

(a) CR – 4-bit round counter (0...23); E – global enable (active) signal

propagation. This shows that analogously to a situation when very fast DES crackers were developed in [12] the high reductions in propagation delays of the pipelined round logic made the overall control setup too slow.

3.2 Optimizations in the Iteration Scheme

To remedy this problem two modifications in the overall X1 control scheme were introduced. The first one addressed problem of slow interpretation of the CR state: it was identified that excessive delay was generated by a comparator signaling maximum value CR = 23 used to stop the loop, so it was replaced with a new dedicated flipflop (denoted here as F23) which was set by a rising clock edge with a condition CR = 22. In this way operation of the comparator was moved to the preceding clock period. Such a modification will be marked with a subscript $X1_a$. Results after its application to the three $X1P2_z$ cases are shown in Table 2.

The presented performance parameters indeed confirm that the comparator was no longer generating the longest propagation but – while the best $X1_aP2_\theta$ case started to operate 13% faster than before – the critical delay still was not caused by the round logic. Another bottleneck exposed this time was the global enable flipflop E used to control operation of all mayor components of the X1 infrastructure. Such a flipflop was a simple and convenient method for overall control of the design and was successfully used up to this point (including all our previous KECCAK implementations in [10] and [11]) but now it could not keep up with the high speed of the intra-pipelined rounds. To remove it an entirely different control framework was applied: in place of a global flag only one Clock Enable input was used which did not enter any other logic other than CE inputs at the registers so that it could be implemented in one control set. The designs with this second modification (added to the first one) are denoted with a subscript $X1_b$. The modification changed mode of module operation (continuous enable instead of a start pulse) but, as the results in Table 2 reveal in the column Src:Dst, finally the designs were limited by the propagation of round logic i.e. by organization of the pipeline. This modification reduced the clock period by another 13% compared to the $X1_aP2$ cases to the total of 34% vs the unoptimized X1P2 organizations.

Table 2. The pipelined architectures after optimizations in the overall control circuitry.

Architecture	Critical path				Slices		LUTs		Registers	
	T_{clk} [ns]	Src : Dst[(a)]	Logic delay	Logic levels	Tot.	%	Tot.	%	Tot.	%
$X1_aP2_D$	2,64	F23:A	19%	1	1388	17,0%	4641	14,2%	6740	10,3%
$X1_aP2_\theta$	2,49	E:A	24%	2	1517	18,6%	5140	15,8%	6418	9,8%
$X1_aP2_\pi$	2,54	E:A	21%	1	1611	19,8%	5139	15,8%	6417	9,8%
$X1_bP2_D$	2,36	D:A	28%	2	1432	17,6%	4643	14,2%	6737	10,3%
$X1_bP2_\theta$	2,25	F23:A	25%	1	1570	19,3%	5071	15,6%	6415	9,8%
$X1_bP2_\pi$ [(b)]	2,16	A:π	28%	2	1582	19,4%	5074	15,6%	6420	9,8%
$X1_bP3_{\theta\pi}$	2,10	A:θ	29%	2	1627	20,0%	5077	15,6%	8019	12,3%
$X1_bP3_{\theta\chi}$	2,24	A:θ	28%	2	1809	22,2%	5904	18,1%	8019	12,3%
$X1_bP3_{D\pi}$ [(b)]	1,93	CR:F23	36%	2	1671	20,5%	5460	16,7%	8336	12,8%

[(a)] F23 – dedicated flipflop signaling condition CR = 23
[(b)] The best case of all two- or three-stage pipelines

The $X1P3_{zz}$ organizations were also implemented with this modification and in the further discussion only results received on the $X1_b$ platform (the last 6 rows in Table 2) will be considered as the representative applications of the pipelined architectures.

4 Evaluation of the Results

The two basic speed and size attributes of all 6 pipelined architectures – the maximum frequency of operation and the number of occupied slices in the FPGA array – are shown in Fig. 3. In all types of architectures considered here (X1, X1P2 and X1P3) the operational frequency is proportional by the same factor to the data throughput expressed e.g. in permutations per second so the shape of the left chart can serve also for comparison according to this criterion.

Among the P2 cases, the fastest design was the one with the furthest pipeline register i.e. located at the π output but the second fastest – with stage boundary at θ – was slower only by 4% (2.25 vs 2.16 ns). Both designs occupied very similar number of slices and were by ~10% larger than the $X1_bP2_D$ case but this increase was compensated by better speed. In the P3 architectures, the verdict is unquestionable: the winner is the $X1_bP3_{D\pi}$ variant which offers clearly the best speed and good size.

Overall, adding the pipeline to the pure iterative X1 case did not significantly increase size of the implementations: if subtract 2 × 1600 input/output storage flipflops from the number of registers in Tables 1 and 2, their increases (by 2 or 3 times) correspond closely to the new pipeline flipflops yet such large growth is not seen in slice occupancy. This confirms that pipelining was very well absorbed by the slices

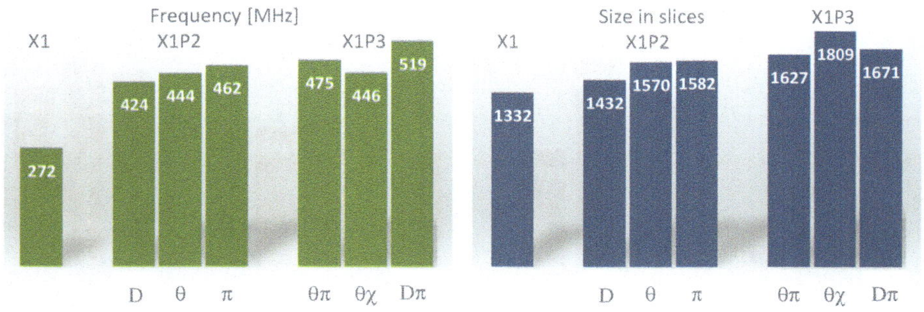

Fig. 3. Speed and size of the pipelined designs vs the basic X1 architecture.

already used for round logic and provides another example of the fact that FPGA arrays can efficiently adopt this technique without engaging extra logic cells.

Apart from identifying the best designs, another (and more interesting) question is whether the pipeline concept really brings the expected improvements when applied to KECCAK processing. The speed progress will be evaluated in the following analysis. In every digital design the minimum clock period comes from the delay of the critical path which is a sum of three components: switching time of a source flipflop, delay of the propagation path to the destination flipflop and signal setup time required at its input:

$$\mathrm{T}_{clk} = t_{CQ} + t_P + t_{SU} \tag{1}$$

When dividing the paths into m stages in a pipelined design only the middle component is reduced – in an ideal case proportionally to m – hence:

$$\mathrm{T}_{clk}\big|_{\mathrm{X1P}m} = t_{CQ} + t_P\big|_{\mathrm{X1}}/m + t_{SU} = t_{CQ} + \left(\mathrm{T}_{clk}\big|_{\mathrm{X1}} - t_{CQ} - t_{SU}\right)/m + t_{SU} \tag{2}$$

Actual values of T_{clk} can be compared to these estimations and such results are in the left chart of Fig. 4. They indicate that the best effect of the two-stage pipelining is actually very close to the theoretical optimum with its clock slower by only 5%. The result of the three-stage pipelining is somewhat worse but still not so distant (+27%). These effects should be considered very good especially in such a complex and structurally involved circuit as the KECCAK core. It was proved in our other studies [10, 11] that especially cryptographic algorithms – with their dense, complicated and purposely awkward propagation paths – can use capacities of FPGA arrays up to the limits, in particular saturating their routing resources. No signs of such saturation are seen in case of these pipelined designs on the Spartan-7 platform.

Nevertheless, it should be noted that the final raw speed of the modules is not high enough to keep the latency – the calculation time for one set of input data – on the same level as in the X1 architecture. In any m-stage pipeline the data must go through m times more steps so keeping latency down is possible only if the clock period is divided exactly by the m factor. This can be easier to reach (or to approach) in a design with longer propagation paths but when they are – as in our case – single nanoseconds

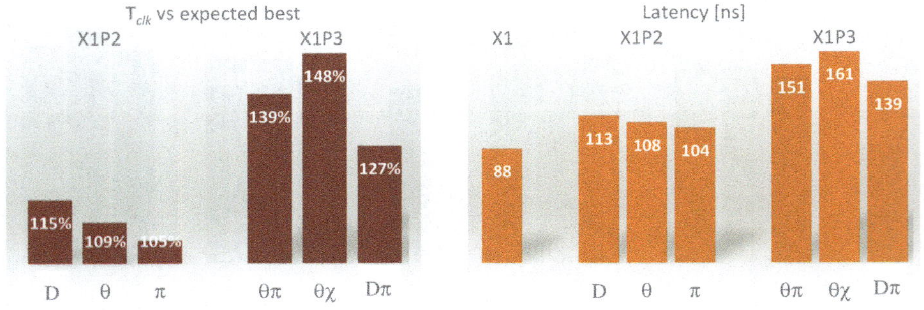

Fig. 4. Actual values of the clock periods expressed as percentages of the theoretical best values (left) and latencies of calculation in the Keccak modules (right).

long and comprise only one or two levels of logic then the $t_{CQ} + t_{SU}$ component in Eq. (2) is relatively too high. As a result, actual latencies in the pipelined designs (as shown in the right chart of Fig. 4) are longer by 18% vs the X1 architecture in the best P2 case and by 57% in the P3 one – so this slowdown is inevitable.

5 Conclusions

The paper presented studies on efficiency of intra-round pipelining applied to FPGA implementations of the Keccak permutation function. Starting from the standard iterative organization with one cipher round instantiated in hardware 6 candidate designs with two and three pipeline stages were proposed after analysis of the round logic. Very high speed of operation of the stages forced additional optimizations in the iteration control logic after which the complete permutation modules could reach frequencies of 462 MHz (two stages) and 519 MHz (three stages) in a Spartan-7 FPGA device from Xilinx. These frequencies correspond to effective speed of, respectively, 19.3 and 21.6 million permutations per second which translates to hashing throughput of e.g. 22.2 and 24.9 Tbps in the SHA3-224 standard. Compared to the ordinary iterative case the pipelining increased the throughput by 70% (2 stages) or 91% (3 stages) although at the cost of latencies extended by 18% and 57%.

The results prove also good implementation efficiency of the pipelined architectures in the target FPGA family especially regarding the speed of the two-stage cases where the results deviate within single percents from theoretically best achievable values. Implementation tools were also effective with regard to design sizes which in terms of slice utilization increased acceptably by 19% and 25%.

References

1. Athanasiou, G.S., Makkas, G., Theodoridis, G.: High throughput pipelined FPGA implementation of the new SHA-3 cryptographic hash algorithm. In: 2014 6th International Symposium on Communications, Control and Signal Processing (ISCCSP), Athens, pp. 538–541 (2014)

2. Bertoni, G., Daemen, J., Peeters, M., Van Assche, G.: The KECCAK reference. PDF file. http://keccak.noekeon.org. Accessed March 2020
3. Bertoni, G., Daemen, J., Peeters, M., Van Assche G.: The KECCAK sponge function family. http://keccak.noekeon.org. Accessed March 2020
4. Gaj, K., Homsirikamol, E., Rogawski, M., Shahid, R., Sharif, M.U.: Comprehensive evaluation of high-speed and medium-speed implementations of five SHA-3 finalists using Xilinx and Altera FPGAs. In: The Third SHA-3 Candidate Conference. Available: IACR Cryptology ePrint Archive, 2012, p. 368 (2012)
5. Gaj, K., Kaps, J.P., Amirineni, V., Rogawski, M., Homsirikamol, E., Brewster B.Y.: ATHENa – automated tool for hardware EvaluatioN: toward fair and comprehensive benchmarking of cryptographic hardware using FPGAs. In: 20th International Conference on Field Programmable Logic and Applications, Milano, Italy (2010)
6. George Mason University: ATHENa - Automated Tools for Hardware EvaluatioN. http://cryptography.gmu.edu/athena. Accessed March 2020
7. Ioannou, L., Michail, H.E., Voyiatzis, A. G.: High performance pipelined FPGA implementation of the SHA-3 hash algorithm. In: 4th Mediterranean Conference on Embedded Computing (MECO), Budva, pp. 68–71 (2015)
8. National Institute of Standards and Technology: SHA-3 Standard: Permutation-Based Hash and Extendable-Output Functions. http://dx.doi.org/10.6028/NIST.FIPS.202. Accessed March 2020
9. Pereira, F., Ordonez, E., Sakai, I., Souza, A.: Exploiting parallelism on KECCAK: FPGA and GPU comparison. Parallel Cloud Comput. **2**(1), 1–6 (2013)
10. Sugier, J.: Efficiency of Spartan-7 FPGA devices in implementation of contemporary cryptographic algorithms. J. Pol. Saf. Reliab. Assoc. **9**(3), 75–84 (2018)
11. Sugier, J.: Low cost FPGA devices in high speed implementations of KECCAK-*f* hash algorithm. In: Zamojski, W., Mazurkiewicz, J., Sugier, J., Walkowiak, T., Kacprzyk, J. (eds.) Proceedings of the Ninth International Conference on Dependability and Complex Systems DepCoS-RELCOMEX. Advances in Intelligent Systems and Computing, Brunów, Poland, 30 June–4 July 2014, vol. 286. Springer, Cham (2014)
12. Sugier, J.: Optimizing the pipelined DES cracker implemented in contemporary popular-grade FPGA devices. In: Kabashkin, I., et al. (eds.): RelStat 2019, LNNS 117 (to be published)

Investigation and Detection of GSM-R Interference Using a Fuzzy Hierarchical Model

Marek Sumiła(✉)

Railway Research Institute, 04275 Warsaw, Poland
msumila@ikolej.pl

Abstract. The paper presents the application of the well-known fuzzy logic theory to help investigate places in the railway network where GSM-R interference could exist. In the introduction to the article the reasons for implementation GSM-R in UE are explained as well as the need to undertake research on the subject of interference in the GSM-R network. After the review of the current state of the knowledge, the author presents today's methods suggested to investigate interference in the network. The following section presents fuzzy logic in the process of identifying places exposed to interference and the most important stages of building the fuzzy model. The final part of the article contains a summary of the work and directions for further research.

Keywords: Fuzzy logic · GSM-R interference · Expert systems

1 Introduction

Currently we can observe renaissance of the popularity of railways in Europe. It is a consequence of the actions undertaken by the European Union over twenty years ago. These activities were aimed at creating a modern, uniform network of transport corridors in the European Union enabling the fast and safe transport of people and goods. The main key to implementing these assumptions was to develop new standards for all EU countries as an element unifying technical solutions on the path of evolution of existing, hermetic standards used by each of the European countries. These decisions were made on the threshold of a rapid development of Internet networks and launching of a fully functional mobile GSM cellular radio network. The development of these technologies has finally sealed the direction and shape of a huge progress currently being observed in the field of information technologies used in the field of transport.

In the case of rail transport, this development was to be implemented through Technical Specifications for Interoperability (TSI). In the area of rail traffic control, this task was evident in the form of European Rail Traffic Management System (ERTMS), which according to common EU-wide standards was to "open the borders" of countries enabling trains to pass without the time and effort-consuming adjustments each time a train passed the national borders. Achieving rail interoperability in the EU therefore meant implementing uniform TSI standards in each country.

Until the idea of European rail interoperability was introduced, each country had its own analogue radio communication system, operating differently and incompatible

W. Zamojski et al. (Eds.): DepCoS-RELCOMEX 2020, AISC 1173, pp. 616–625, 2020.
https://doi.org/10.1007/978-3-030-48256-5_60

with the systems of other countries. Therefore, the train radio communication system has become one of the important elements requiring harmonization. An example of such a network could be the Polish VHF 150 MHz rail radio network having not only a different broadcast band compared to its neighbors, but also specific functional features that the networks of other railway operators do not have. In the nineties of the last century, as a result of consultations, it was decided to choose the GSM 2G network as the future, uniform standard of railway radio throughout the European Union. The choice of this standard at that time seemed to be the most rational and the range of adaptation to railway needs was considered acceptable. Since then, we have seen an increase in the number of railway lines having access to the interoperable GSM for Railway (GSM-R) standard across Europe. It is worth adding that this standard [6] is unchanged despite the significant progress of public cellular networks. This is due to the need to ensure a uniform technical solution in the main corridors of the trans-European rail network.

Here we come to the main problem, because after the appearance of the GSM-R network in Western Europe, individual railway managements began to report to UIC problems with the functioning of the GSM-R network in selected network locations [13]. The problem was so serious that it contributed to disturbances in train traffic and the ineffectiveness of the train control system [3]. The investigation conducted in this case led to the identification of the causes of GSM-R network interference and methods of dealing with it [1, 2, 9]. The conclusion from this work was that the source of interference was public cellular networks operating in the 900 MHz public band. Technologically more developed networks (GSM 3G and 4G) turned out to be particularly dangerous due to the phenomenon of intermodulation (IM3) appearing in these networks. Concerned by reports [1, 9, 14], railway managers expected the International Union of Railways (UIC) to present a method for solving the GSM-R interference problem. Therefore, UIC in cooperation with European Conference of Postal and Telecommunications Administrations (CEPT) led to the creation of Report 229 [2] and O-8740 [15] which present a set of guidelines to help railway managers in limiting the effects of emerging radio interference.

Unfortunately, the proposed solutions are not easy to implement and do not always give a clear result. This is due to many ambiguous factors that can contribute to interference in various ways. Therefore, it is worth to propose a unconsidered previously method of investigating inference using fuzzy logic theory [16]. The method may be relevant during the testing of thousands of kilometers of railway line because the experts' knowledge can be implemented in automation algorithms.

2 Problem Identification

Known and researched problems related to the availability of GSM-R networks do not help to identify places in risk of interference. Proposals for such identification in the form of a general procedure are presented in Report 229 [2]. Other proposals based on accumulated experience can be found in [10–12]. Sources of problems should be sought in:

- according to ETSI TR 103 134 [8], in 2013 it covered about 68,000 from over 220,000 km of railway lines (the planned target is GSM-R coverage is about 150,000 km of railway lines),
- public networks that are sources of interference cover railway areas, base stations (BS) are located in places unpredictable from the point of view of the GSM-R network and they need to be analyzed individually and in groups.

As a consequence, we receive for analysis a huge area of the railway network with a thousand of public base stations working around in different bands, different technologies and with different Equivalent Isotropical Radiated Power (EIRP).

In this situation, identifying locations at risk of interference requires one of the following actions to be taken:

1. conducting surveys among railway employees (especially train drivers) to obtain information on observed events and places with failed radio communication in the GSM-R network,
2. manual analysis of radio maps taking into account the location of base stations of public networks and the railway network,
3. conducting measurement campaigns of GSM-R network[1].

Each of the presented actions has its good and bad sides. The advantage of the first solution is a quick and undoubtedly the cheapest method of identifying the places of disturbances based on the employee's notification. The disadvantage of this method is the lack of complete information about the source of the interference. Such information could help in the selection of necessary ways to eliminate or mitigate the effects of adverse radio phenomena. The second solution does not have the disadvantage of the first, because the analysis is carried out based on a wide set of sources technical data. A person performing such a task must have appropriate expert knowledge in the radio propagation field and the problem of interference because it requires considering many factors that have implication in the process of identification phenomena. In this case, it is not difficult to make a mistake, because it is a monotonous and time-consuming solution. The last solution is the most effective method of identifying the places of interference in the GSM-R network and one giving a full radio picture of the radio environment at a given location of the railway line. Albeit effective, the method also has disadvantages. They are primarily high research costs and time consumption.

In practice, many railway managers decide to carry out radio tests of their GSM-R networks periodically with the effectiveness of the method in view. Many of them also launched internal programs and procedures for rail employees to report communication problems in GSM-R [15]. The second of the proposed solutions is a compromise between extreme solutions and is not very popular. However, the transfer of expert knowledge to computer tools enabling process automation could significantly increase the potential of the method by achieving good results of carrying out the research task in a short time.

[1] The number of methods that can be used is higher, but due to the size of the article they are not mentioned here. More methods can be found in the article [12].

3 Fuzzy Logic Modeling

The theory of fuzzy sets has been well known since the sixties of the last century [16]. The fuzzy set theory is therefore an extension of the classical set theory, where the true values of variables may be any real number between 0 and 1 both inclusive (many-valued logic). It is particularly useful in systems where there are many complex dependencies, multi-dimensional, hierarchical, often with internal feedback loops. Generally, fuzzy logic can be applied in situations where the quantitative computational techniques reveal its weaknesses, and where the basis are imprecise ideas. Nowadays this theory is widely used in solving many decision-making, not necessarily technical, problems.

In general fuzzy logic modelling involve three stages: fuzzification, inference (fuzzy rule) and defuzzification. The fuzzification is the process of assigning input variables of the system to fuzzy sets. Fuzzification consists of transforming individual input signals into qualitative quantities representing fuzzy sets, i.e. Membership Functions (MF). These fuzzy sets are typically described by a number but also by words, and that's why it is easy to assign the system input to fuzzy sets. These functions can have different shapes. The inference stage should include such elements as the rule base, inferential mechanism and membership functions of exiting the model. The rule database contains logical rules defining the cause-effect relationships existing between the fuzzy sets of input and output. An important problem is how to create inference rules. The modelling of fuzzy systems is carried out where using two basic methods based either on expert knowledge, as so-called linguistic modelling, either on analytical data derived either from measurement systems or mathematical models.

The final step is the defuzzification. It involves the process of quantifiable result in Crisp logic given fuzzy sets and corresponding membership degrees. It is a process transforming the result of fuzzy set into a sharp real value that is the response of the system, which is e.g. a control signal or other decision variable. For this purpose many methods of defuzzification can be used such as: Center of Area (COA), First Maximum (FOM), Last Maximum (LOM), Center of Gravity (COG), etc. Among the many different rules, the Center of Gravity Method is very popular.

4 Model Development

4.1 The Source of Input Data and Fuzzification

The development of a model supporting the process of identifying places susceptible to interference from public cellular networks requires the identification of factors affecting such interference. Extensive research in this area with the participation of technical experts and network managers provide many valuable tips for selecting factors and levels (ranges) that affect interference in the GSM-R network. Currently, the sources of knowledge are:

- CEPT ECC technical reports (among them [1, 2]),
- UIC documents (for example [13–15]),
- dedicated to this topic workshops [3–5],
- ETSI technical specifications [7].

Based on this, the following factors can be identified for the fuzzy model:

- exact location (latitude, longitude),
- distance from rail tracks,
- antenna in line-of-sight of the rail tracks,
- public BS carrier,
- carrier types (2G, 3G, 4G) and channel/block bandwidth,
- antenna height, azimuth, tilt and transmitted EIRP power,
- distance from GSM-R sites and their azimuths,
- identification of the suspected neighbouring sites,
- known issues of problems in the GSM-R network (e.g. poor voice quality, drop call, no network connection, etc.) and frequency of the problem (permanent or temporary),
- occurrence of intermodulation frequencies.

A closer analysis of these factors allows them to be ordered in terms of belonging to the group, but also to indicate places in the hierarchy of factor importance in the occurrence of interference. The results of this work are presented in Table 1.

Table 1. The list of features for the fuzzy model.

Feature	Symbol	Factor influence	
		Small	Large
Distance from railroad tracks	D_y	Long (>1 km)	Short (<700 m)
Direct visibility	W_i	Poor	Good visibility
Carrier type	G_x	2G	3G/4G
Carrier frequency from GSM-R	C_r	Distant (>10 MHz)	Near (0,1÷10 MHz)
Channel width	B_c	Narrow (200 kHz)	Wide ($\geq$ 1,4 MHz)
Power emitted by the public BS	P_n	Weak	Strong
The height of the public BS antenna	H_n	Does not affect direct visibility	Affects direct visibility
Azimuth BS public antenna	A_n	Does not aim at railway area	Aims at the railway area
Tilt public BS antenna	T_n	Does not aim at railway area	Aims at the railway area
Number of nearby BS public transmitters	L_n	1	>1
GSM-R signal level	P_g	Strong	Weak (<−92 dBm)
The distance of the active public channel/block from GSM-R	S_b	Wide (>10 MHz)	Narrow (0,1÷10 MHz)
Appling ETSI TS 102 933-1 [7]	K_t	Fulfilled	Unfulfilled
Level of foreign signals in the railway area	P_{onx}	Weak (<−90 dBm)	Strong (>−40 dBm)
Out-of-band interference	O_{oob}	Weak	Strong
Occurrence of intermodulation	I_{m3x}	Weak	Strong
GSM-R interference reports	Z_{int}	0	≥ 1
Threat of GSM-R communication interference	Y_{dang}	Little	High

Each of the features can be assigned in the process of fuzzification to appropriate membership functions. The easiest way to describe each variable in the form of membership function is divide it into two or three functions describing the effect of the factor: large, medium and small. Membership functions can take different shapes. The most commonly used is the trapezoidal, triangular and so-called s-function. For selected input variables, e.g. support type, the function has a rectangular shape, due to the Boolean nature of this feature. An example of such a feature is "Carrier type" and its value: 2G and 3G/4G. An exemplary graphic illustration of a linguistic variable "Direct visibility" and its corresponding fuzzy sets along with membership functions are shown in the figure below (Fig. 1).

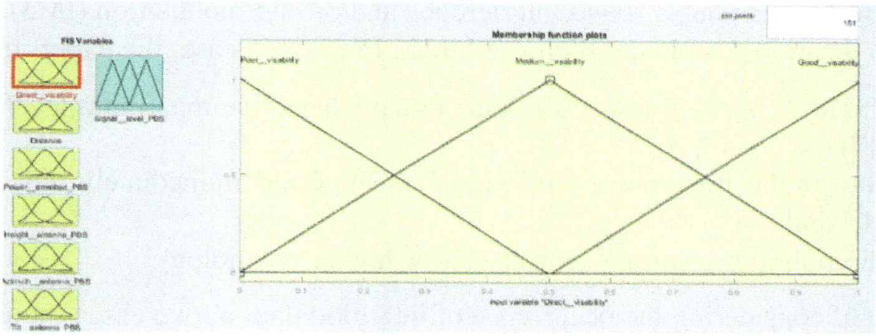

Fig. 1. Assumed membership functions for fuzzy sets describing in the Mamdani model the linguistic variable "Direct visibility".

4.2 Model Description and Fuzzy Rules

The use of fuzzy logic to build a decision model for radio hazard assessment requires a new approach to the underlying problem. In the classic approach, decisions are made based on the radio link budget, where all variables for the equation are known. This model should reflect the thought process that the expert must follow when deciding on the likely threat of interference. For this purpose, an expert has to take a decision based on the following guidelines:

1. Are there any public BS near the railway line?
2. Do they work in the 900 MHz band?
3. Do they use 2G or higher technology?
4. Do the parameters of the transmitters increase the potential threat?
5. What is the distance between public BS and the railway area?
6. Are there any obstacles in the space between the public BS and the railway area and how big are they?

The result of data processing is the P_{onx} parameter, which will be used in the further inference process. In the fuzzy logic model, this can be represented as a "Public Base Station" block (see Fig. 2).

In the process of inference, the degrees of belonging fuzzy sets representing the values of input variables should be assigned to rules that meet the condition of the premises. As example it can be develop in the fuzzy model necessary rules as follow:

Rule 1: IF(D_y IS SM) AND (W_i IS SM) AND (G_x IS LG) AND (C_r IS LG) THEN (P_{onx} IS MD)
Rule n: IF(D_y IS SM) AND (W_i IS LG) AND (G_x IS LG) AND (C_r IS LG) THEN (P_{onx} IS LG)

Where: SM means small, MD means medium and LG means large. For a full description of the proposed model is necessary to create all rules occurring between input and output values.

In further proceedings, the expert must consider the possibility of two possible cases of interference: out-of-band interference and/or intermodulation (IM3). Each case requires to develop a separate inference block. In the first case, the expert must check:

1. Does the level of the foreign BS radio signal exceed the maximum level of TS 102 933-1 [7]?
2. Does the public transmitter emit signal in the band immediately adjacent to the GSM-R band?
3. Does the public transmitter emit in 2G or higher technology?

In case of considering the occurrence of intermodulation, two cases should be taken into account:

- narrow-band intermodulation occurs for several transmitters operating in the entire GSM 900 MHz band in 2G technology,
- for 3G/4G transmitters working in bands adjacent to GSM-R, broadband intermodulation occurs (the band is considered 925–935 MHz).

This means that in the Intermodulation block there must be two embedded blocks describing each case may have impact on GSM-R users terminals. The inference process for blocks out-of-band interference and intermodulation assumes the use of knowledge about every public Base Station in the analyzed area and considered in the previous block of the model.

The last element of the fuzzy model is the block enabling final evaluation of data based on earlier blocks and two additional critical input data:

1. compliance to the needs of technical specification TS 102 933-1 [7],
2. registered cases of problems with GSM-R communication in a given railway area.

At the output of the fuzzy model we get the result Y_{dang} describing the overall threat status for a given area of the railway line. Figure 2 shows the block diagram of the elaborated fuzzy model.

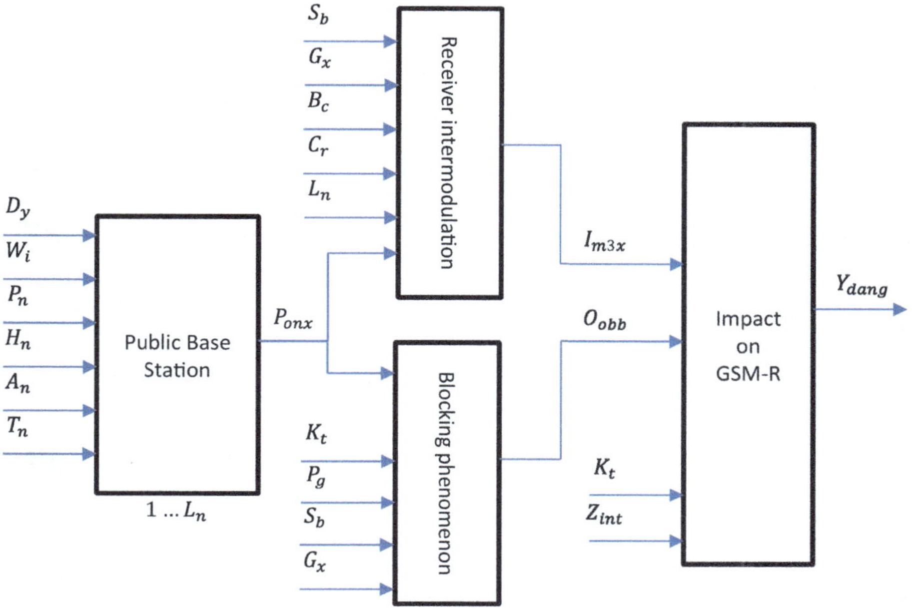

Fig. 2. A block diagram of the fuzzy model.

4.3 Defuzzification

The final stage of inference is the defuzzification process. Its purpose is to obtain a numerical synthetic value. The defuzzification operation is performed using publicly available methods. In engineering practice, this type of calculations is performed in computer technology due to the rapidly increasing computational difficulties in more complex fuzzy models. This postulate is met, among others, by the standard Fuzzy Logic Toolbox package found in the Matlab engineering system, allowing for effective use of all potential possibilities of Fuzzy theory and systems.

Figure 3 presents a fragment of the visualization of the probability of GSM-R interference at the ending block “Impact on GSM-R”.

Launching the model in a simulating environment generates a set of results for all possible combinations of data used in the model and their impact on the probability of GSM-R radio interference. Figure 3 simulated a state in which there is a high probability of intermodulation (0.8), the radio terminal meets the requirements of ETSI TS 102 933-1 (1.0) and there are known cases of network interference in this area (1.0). The overall probability of interference in considered place of railway network is also high.

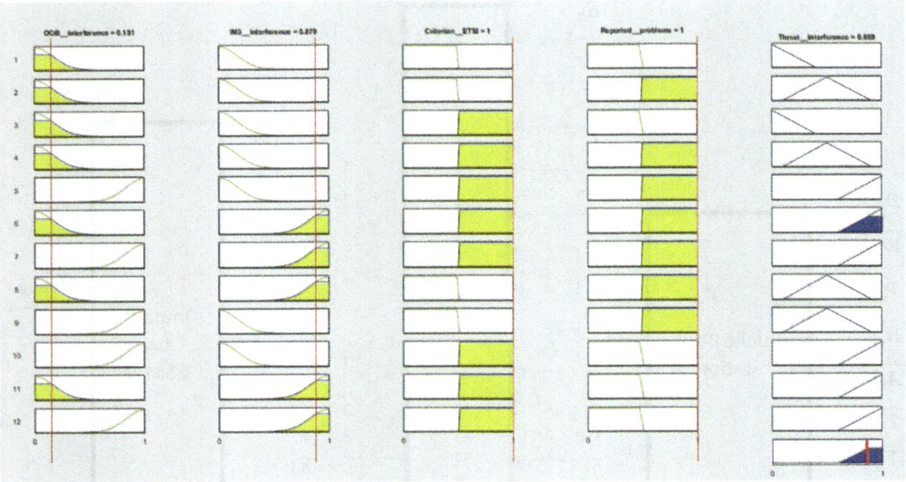

Fig. 3. Visualization the fuzzy model rules.

5 Summary and Further Actions

Heuristic methods, including fuzzy sets, allow us to build models based on intuition, expert knowledge and verbal description, i.e. wherever quantitative methods fail. The article presents a proposal to create an expert tool based on fuzzy logic theory to identify places covered by GSM-R network susceptible to interference. The model uses a number of factors that can have impact on radio communication. Their parameters are based on the radio equipment catalog cards, propagation models, reports and technical recommendations of international organizations. The model is a tool enabling expert analysis of many factors having a direct or indirect impact on the occurrence of radio interference in the GSM-R network. Nevertheless, it should be clearly stated that its full usefulness will be achieved at the time of linking it with Geographic Information System (GIS) tools. Therefore, further development activities should focus on improving the fuzzy model and creating a link to the extra data source.

An important test for the model would consist of comparing its results with the results of field tests in the particular area of the railway network. However, performing such research requires greater organizational and financial expenditure.

References

1. CEPT ECC Report 162: Practical mechanism to improve the compatibility between GSM-R and public mobile networks and guidance on practical coordination (2011)
2. CEPT ECC Report 229: Guidance for improving coexistence between GSM-R and MFCN in the 900 MHz band (2015)
3. First GSM-R Interference ERA workshop. http://www.era.europa.eu/Document-Register/Pages/Presentation-workshop-5.11.2012.aspx. Accessed 16 Nov 2016

4. Second GSM-R Interference ERA workshop. http://www.era.europa.eu/Document-Register/Pages/Presentations-2ndWorkshop-11.11.2013.aspx. Accessed 16 Nov 2016
5. Third GSM-R Interference ERA workshop. http://www.era.europa.eu/Document-Register/Pages/3rd-Interference-Workshop.aspx. Accessed 16 Nov 2016
6. ETSI EN 301 515 (V2.3.0): Global System for Mobile communication (GSM); Requirements for GSM operation on railways (2004)
7. ETSI TS 102 933-1 (V2.1.1): Railway Telecommunications. GSM-R improved receiver parameters. Part 1: Requirements for radio reception (2015)
8. ETSI TR 103 134 (V1.1.1): Railway Telecommunications (RT); GSM-R in support of EC Mandate M/486 EN on Urban Rail (2013)
9. FM(13)134 GSM-R Measurement Report - BNetzA Germany, Radio Spectrum Committee. Updated version of the working document on GSM-R interferences and coexistence with public mobile networks, taking into consideration observations made by RSC delegations. Working Document. RSCOM15-60 rev3. Brussels (2016)
10. Sumiła, M., Miszkiewicz, A.: Analysis of the problem of interference of the public network operators to GSM-R. In: Mikulski, J. (ed.) Tools of Transport Telematics. TST 2015. Communications in Computer and Information Science, vol. 531, pp. 76–82. Springer, Cham (2015)
11. Sumiła, M.: Risk analysis of interference railway GSM-R system in Polish conditions. In: Proceedings of the Eleventh International Conference on Dependability and Complex Systems DepCoS-RELCOMEX, Advanced in Intelligent Systems and Computing, Brunów, Poland, 27 June–1 July 2016, vol. 470. Springer, Heidelberg (2016)
12. Sumiła, M.: Methodical approach to assessing interference in GSM-R network. In: Siergiejczyk, M., Krzykowska, K. (eds.) Research Methods and Solutions to Current Transport Problems. ISCT21 2019. Advances in Intelligent Systems and Computing, vol. 1032, pp. 420–429. Springer, Cham (2020)
13. UIC O-8700-2.0: Report on interferences to GSM-R. UIC (2012)
14. UIC O-8736-2.0: UIC assessment report on GSM-R current and future radio environment. UIC (2014)
15. UIC O-8740: Report on the UIC interference field test activities in UK (2013)
16. Zadeh, L.A.: Fuzzy sets. Inf. Control **8**, 338–353 (1965)

Using Convolutional Network Visualisation to Determine the Most Significant Pixels

Tomasz Szandała(✉)

Wroclaw University of Science and Technology, Wroclaw, Poland
Tomasz.Szandala@pwr.edu.pl

Abstract. Over the last years, Deep Neural Network models have been recognized as successful in solving many complex problems. However, these methods are mostly focused on the efficiency of final results and rarely provide sufficient evidence and details on factors that contribute to their outcomes This is why a growing demand for analysis techniques appeared. Thanks to visualisation techniques we can if network works as expected or even improve output of given model if possible. Moreover we can use these methods as optimization technique to boost network's performance but pruning less important neurons. Finally, if we know how a given model works we can prepare a disruption to its work process. This paper shows how we can combine Class Activation Map with feature map to determine a few of the most contributing pixels for given input and modify them to perform an adversarial attack.

Keywords: Deep neural networks · Convolutional networks · Visualisation · Deep learning, adversarial attacks

1 Introduction

Recent years have generated great advances in training large, deep neural networks (DNNs), where the most notable successes occured in the field convolutional neural networks (CNNs) to classify natural images [1, 2]. This work tackles an important problem related to classification of images using CNNs: deep networks trained to successfully recognize some classes may often focus on some parts of the image. These features are often not the same characteristics which humans take into consideration when recognizing a given class. CNN may hence classify based on some image artifacts or even some few pixels on the image which happen to correlate with class labels therefore "discovering" very spurious, counter-intuitive features.

However, our knowledge on how these models work, what makes them give certain output is still lackinging behind. Progress in the field will be further accelerated by the development of better tools for visualizing and interpreting neural nets. As for now a few methods have been proposed. We have studied a few of the most promising techniques. The two of them: Class Activation Map (CAM) [3] and Guided Backpropagation [4] can be combined to obtain Guided CAM, which displays the features that contributed the most to the network answer. These features can be further processed by thresholding, which finally leaves the most important pixels. This outcome proves we can dampen network's efficiency in recognizing chosen object. The method

W. Zamojski et al. (Eds.): DepCoS-RELCOMEX 2020, AISC 1173, pp. 626–632, 2020.
https://doi.org/10.1007/978-3-030-48256-5_61

appears to work satisfyingly for any image recognition network, but in this paper we aim to focus in detail on one example.

2 Methodology

In order to interact with the most important pixels on the picture we need to determine them. There are many methods that provides insight into networks behaviour [12, 13], but for this paper we have chosen only three of them.

2.1 CNN Visualisation Technique: Class Activation Map (CAM)

CNN networks utilizes technique known as Global Average Pooling (GAP) [14]. Global Average Pooling is an operation that calculates the average output of each feature map in the previous layer. This fairly simple operation reduces the data, extracts the leading features and thus prepares the model for the final classification.

In the experiments [3], it has been found that the advantages of GAP layer extend beyond simply acting as a regularizer. Moreover, with a little modification, the network can retain its remarkable localization ability until the final layer. This tweaking allows identifying easily the discriminative image regions in a single forward pass for a wide variety of tasks, even those that the network was not originally trained for.

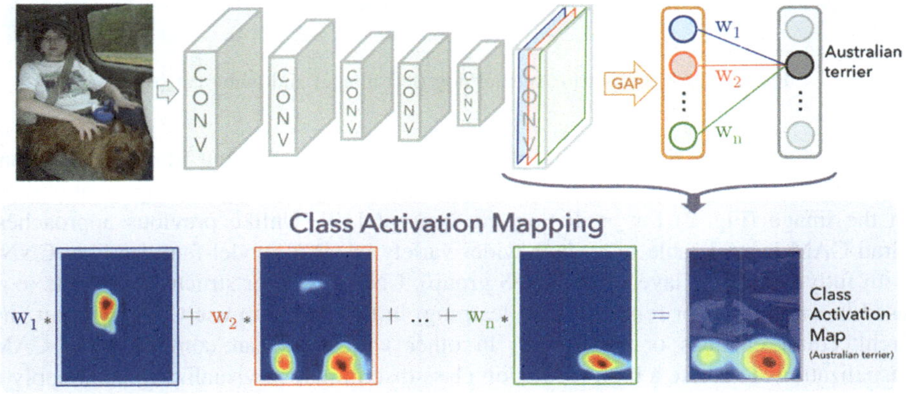

Fig. 1. Schematic for obtaining CAM for a sample image

A class activation map for a particular category indicates the discriminative image regions used by the CNN to identify that category. As illustrated in Fig. 1, GAP outputs the average of the feature map of each unit at the last convolutional layer. A weighted sum of these values is used to generate the final output in fully connected layer. Similarly, we compute a weighted sum of the feature maps of the last convolutional layer to obtain our class activation maps.

A drawback of CAM is that it requires feature maps to directly precede softmax layers, so it is only applicable to a particular kind of CNN architectures performing global average pooling over convolutional maps immediately prior to prediction (i.e. conv feature maps → global average pooling → softmax layer). Such architectures may achieve inferior accuracy compared to general networks on some tasks (e.g. image classification) or may simply be inapplicable to any other tasks (e.g. image captioning). For a fully-convolutional architecture an enhancement known as Gradient-weighted Class Activation Mapping (Grad-CAM) can be used.

Fig. 2. CAM output for spaniel dog (left image) and tabby cat (right)

Grad-CAM uses the gradients of any target concept, flowing into the final convolutional layer to produce a coarse localization map highlighting the important regions in the image (Fig. 2) for predicting the concept [10]. Unlike previous approaches, Grad-CAM is applicable to a much wider variety of CNN model-families like: CNNs with fully-connected layers (e.g. VGG group), CNNs used for structured outputs (e.g. used in captioning) or reinforcement learning. In these we can use CAM without any architectural changes or re-training. In other words we can combine Grad-CAM visualizations to create a high-resolution class-discriminative visualization and apply it to off-the-shelf image classification networks, including ResNet-based architectures [9], which we use in this paper.

2.2 CNN Visualisation Technique: Guided Backpropagation

Guided backpropagation computes the gradient of the target output with respect to the input, but backpropagation of ReLU functions is overridden so that only non-negative gradients are back-propagated (any negative gradients are set to 0). Guided backpropagation was proposed in the context of an all-convolutional network and is generally used for convolutional networks, although the approach can be applied generically [4].

Fig. 3. Guided Backpropagation result showing the features that contributed for network's output

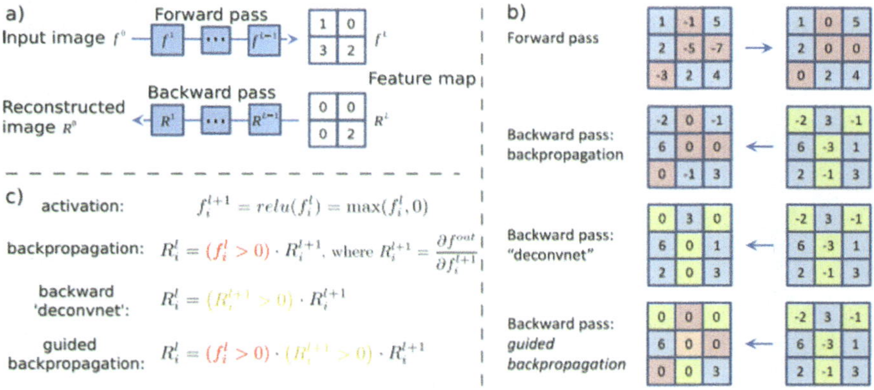

Fig. 4. Family of backpropagation related techniques

We call this method guided backpropagation, because it adds an additional guidance signal from the higher layers to usual backpropagation [5]. In vanilla backpropagation we use derivatives that are calculated using backpropagation to determine saliency map.

DeconvNets are the same as the "Gradients" approach except for a difference in backpropagation through the ReLU nonlinearity. It propagates back all positive error signals. Note that this is equivalent to applying the linear rectifier to the error signal.

Guided Backpropagation basically combines vanilla backpropagation and DeconvNets when handling the ReLU nonlinearity. Like DeconvNets, in Guided Backpropagation we only backpropagate positive error signals, while negative are set to zero. This is the application of the ReLU to the error signal itself during the backward pass. And like vanilla backpropagation, we also restrict ourselves to only positive inputs (see Fig. 4, section c)

While simple backpropagation tends to look messy, the guided one displays all features detected on given network (Fig. 3). Nonetheless it is without respect to the importance of specific votes.

2.3 CNN Visualisation Technique: Guided Class Activation Map

To gain an insight into intermediate convolutional layers, the Deconvolution method was introduced by Zeiler and Fergus [5]. WhileCAM visualizations are class-discriminative and localize relevant image regions well, they lack the ability to show fine-grained importance like pixel-space gradient visualization methods (Guided Backpropagation and Deconvolution). In order to combine the best aspects of both, we can fuse Guided Backpropagation and the Grad-CAM visualizations via a pointwise multiplication [6].

The feature maps of the last convolutional layer are used to obtain the Grad-CAM heatmap, while Guided Backpropagation is applied in the backward pass of the CNN to obtain features from the last fully-connected layer. Then, the Grad-CAM heatmap is upsampled to the size of the input image and element-wisely multiplied with the Guided Backpropagation features. Figure 5, below, illustrates the result of this fusion.

Fig. 5. Combination of CAM and Guided Backpropagation for spaniel and cat detected by network on an image.

3 Results

In the last step we have obtained the output of Guided-CAM. This outcome will be used to find the most important pixels for given classification. We assume those will be the ones with color about certain threshold. We have adjusted the threshold so that only

14 pixels stayed above it. Now we used obtained the mask to cut out those pixels from the original image and once again put it through the networks (Fig. 6).

Fig. 6. Mask obtained by thresholding dog's Guided-CAM output (14 pixels chosen) and the application of mask to the original image

In the experiment we define as success a decreased probability of recognition of previously detected class. Now the vote by ImageNet pre-trained Resnet50 [9] for spaniel was significantly decreased by over 25% (confidence dropped from 0.209 to 0.150). Also interesting were the results for the cat: these increased only by 3%: 0.195 to 0.201. Afterwards we have repeated the process for few state-of-the-art networks AlexNet [10], GoogleNet [11], VGG-19 [8] and the results can be found in the table below (Table 1).

Table 1. Probability of dog class detection on original and modified image

Network	Original image dog probability	Modified image dog probability	Percentage confidence drop
AlexNet	0.865	0.691	20.1%
Resnet50	0.209	0.150	28.2%
VGG-19	0.467	0.396	15.2%
GoogleNet	0.637	0.499	21.6%

4 Conclusions

Deep neural network visualisation techniques can be crucial in the development of deep neural networks. They can be useful in investigating network efficiency, correctness, but also the vulnerabilities as we have proven above. By changing only 14 pixels (out of 224 $\times$ 224 image) we decreased the confidence level of class by approximately 20%.

To sum up: this method can test whether our network is immune to small image perturbations during classification. Furthermore this case can be considered as non-targeted adversal attack, which just reduces probability of recognition of given class. Lastly it is worth pointing out that such small modification, approximately over dozen pixels, can be equal to an eyelash sticked to the camera [7]. If our network is vulnerable to it, it might be a signal to enhance it.

References

1. Boureau, Y-L., Ponce, J., LeCun, Y.: A theoretical analysis of feature pooling in visual recognition. In: Proceedings of the 27th International Conference on Machine Learning (ICML-10) (2010)
2. Szyc, K.: An impact of different images color spaces on the efficiency of convolutional neural networks. In: International Conference on Dependability and Complex Systems. Springer, Cham (2019)
3. Selvaraju, R.R., et al.: Grad-cam: visual explanations from deep networks via gradient-based localization. In: Proceedings of the IEEE International Conference on Computer Vision (2017)
4. Springenberg, J.T., et al.: Striving for simplicity: The all convolutional net (2014). arXiv preprint arXiv:1412.6806
5. Zeiler, M.D., Graham W.T., Rob F.: Adaptive deconvolutional networks for mid and high level feature learning. In: ICCV, vol. 1, No. 2 (2011)
6. Chattopadhay, A., et al.: Grad-cam ++: generalized gradient-based visual explanations for deep convolutional networks. In: 2018 IEEE Winter Conference on Applications of Computer Vision (WACV). IEEE (2018)
7. Nguyen, A., Yosinski, J., Clune, J.: Deep neural networks are easily fooled: High confidence predictions for unrecognizable images. In: Proceedings of the IEEE conference on computer vision and pattern recognition (2015)
8. Simonyan, K., Zisserman, A.: Very deep convolutional networks for large-scale image recognition (2014). arXiv preprint arXiv:1409.1556
9. He, K., et al.: Deep residual learning for image recognition. In: Proceedings of the IEEE conference on computer vision and pattern recognition (2016)
10. Krizhevsky, A., Sutskever, I., Hinton, G.E.: Imagenet classification with deep convolutional neural networks. In: Advances in neural information processing systems (2012)
11. Szegedy, C., et al.: Going deeper with convolutions. In: Proceedings of the IEEE conference on computer vision and pattern recognition (2015)
12. Yu, W., Yang, K., Bai, Y., Yao, H., Rui, Y.: Visualizing and Comparing Convolutional Neural Networks (2014)
13. Evans, R., Grefenstette, E.: Learning explanatory rules from noisy data. J. Artif. Intell. Res. **61**, 1–64 (2018)
14. Lin, M., Chen, Q., Yan, S.: Network in network (2013). arXiv preprint arXiv:1312.4400

Android Methods Hooking Detection Using Dalvik Code and Dynamic Reverse Engineering by Stack Trace Analysis

Michał Szczepanik(✉), Michał Kędziora, and Ireneusz Jóźwiak

Faculty of Computer Science and Management, Wroclaw University of Science and Technology, Wroclaw, Poland
{michal.szczepanik,michal.kedziora, ireneusz.jozwiak}@pwr.edu.pl

Abstract. This research paper is focused on the issue of method hooking detection which in most of the existing solutions is focused on detection of the framework itself without checking of its real usage. The Authors proposed an algorithm that compares Dalvik bytecode from the application file with the stack trace of the executed code. Detection of the flow which is not a part of the app can be application method hooking. In this paper, the Authors ty to confirm it the concept can be used as a practical solution for the detection of this dynamic application tampering.

Keywords: Mobile application security · Method hooking · Malware detection

1 Introduction

1.1 Method Hooking and Malware

Mobile applications are nowadays much more popular in use than laptops or desktop computers. They are also the most popular type of device for web browsing and online services usage [1]. This is the reason why malicious creators are currently focused on Android and iOS platforms and implement malware that is dedicated to these technologies. The most common way of integrating malicious code inside an application is the modification of an apk file. This file is not easy to distribute to many users as all applications distributed by the official channels are verified [2], so the only way is to use alternative sources. To analyze Android application according to malicious code detection two technics can be used [3]:

- dynamic analysis techniques which rely on monitoring the application in real time in an isolated specially prepared environment [4],
- static analysis which works on decompiled source code, without launching applications [5].

The malicious code can be injected in the application not only by modification of the code or resources in apk file, but also by method hooking which allows modifying

W. Zamojski et al. (Eds.): DepCoS-RELCOMEX 2020, AISC 1173, pp. 633–641, 2020.
https://doi.org/10.1007/978-3-030-48256-5_62

behavior of specific application call [6–11]. It can be used to modify specific methods form the application code but also to modify methods from system API which are used for data storing, in-app purchases, network communication, and many others.

1.2 Risk of Method Hooking

The most popular tool is Xposed Framework [12]. It allows to create modules dedicated for a specific application or system API that modifies the code of listed methods and changes the behavior of them. The framework works on the main Zygote process level (see Fig. 1). Zygote is a special process in Android which handles the forking of each new application process, so it is controlling virtual machine and application sandbox. The */init.rc* script starts Zygote during booting time, the maim flow and configuration are defined in */system/bin/app_process* file, which loads the all needed classes and invokes methods. The Xposed framework provides extended *app_process* executable which is copied to*/system/bin*. This extended startup process adds an additional jar to the classpath and calls methods from there at certain places.

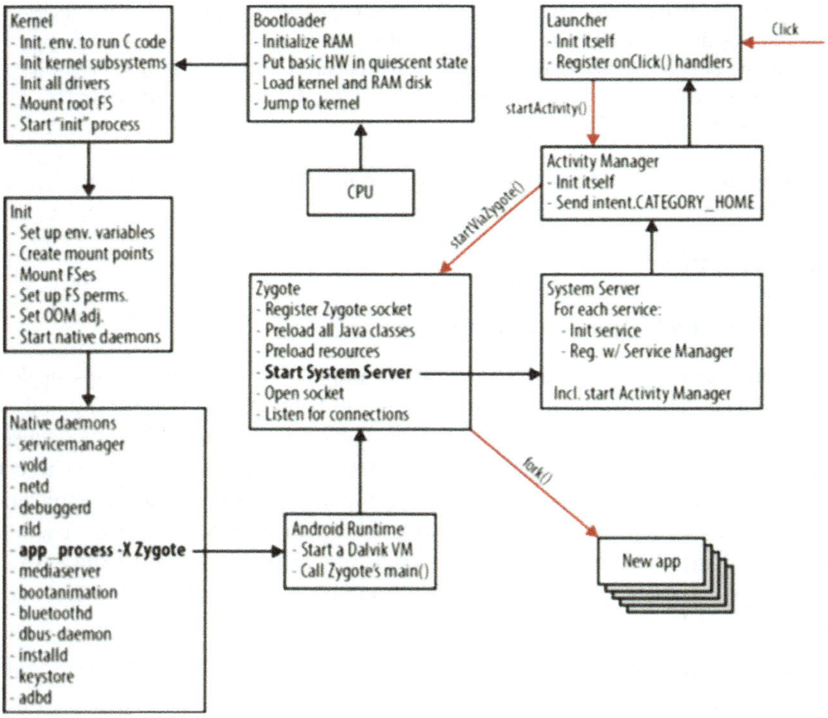

Fig. 1. Zygote process model.

Xposed Framework is popular because its setup is very easy and can be done by Android users even without any specific technical knowledge. The only requirement is that the phone needs to be rooted, then a dedicated application setup framework environment. The Pokemon GO release in 2016 was the main reason why a lot of people install this framework without real knowledge of which type of cases it can be used. One of the Xposed framework modules was used to simulate user movement which allows collecting Pokemons [13]. After that, the number of modules for other applications also grows and started to be popular. The typical cases of usage were: unlock of paid features, simulation of some user behavior or additional features to applications that were not provided in the original program.

The malware software can be also run as a module. Potentially its code can be used to collect data from a specific application like an email client, social media, banking, messaging apps or photo gallery. The module with malicious code can be distributed as a solution for games or other popular applications.

There are also other solutions similar to Xposed framework like edxposed (modification of original Xposed dedicated for Android OS 7-9), magisk (modifying the boot partition), Frida (required Frida server to work and need to be connected with the computer). In this paper, the Authors focus on the solution which detects method hooking done by most popular frameworks.

1.3 Existing Solutions

The most popular tool which detects tampering and method hooking is Dexguard [14]. It needs to be integrated with the specific application during the building process and the detection is limited for only this application. It provides information about framework detection even when it is not used. The detection based on specific fingerprints of frameworks so it may not detect the new version or not supported frameworks. Additionally, smartphone manufacturers frequently customize Android distributions, which provide false positives for the detector.

2 Proposed Solution

2.1 Dalvik Code

The code of the application and all other resources are stored in the Android Package (apk) file. The most important from the solution perspective are dex flies which store Dalvik bytecode. The code of the application is usually obfuscated by tools like Dexguard or Proguard [14]. To get the Java like code of the application the tools like dex2jar and JD-jar need to be used, but they required a lot of resources to work, so they cannot be used on the device itself. The Dalvik bytecode structures cannot be eastly converted back to Java code and usually different tools use different patterns for this type of conversion. The other way of analysis is to use Dalvik bytecode. Table 1 presents listing of Hello World application and its Dalvik code represented in Smali The Smali code include list of methods, used objects and current package which are needed for comparison with application code which can be collected by stack traces.

Table 1. Java and Dalvik code of "Hello World" application.

Hello World Java code	Smali

package com.test.helloworld;
 	
import android.app.Activity;
import android.os.Bundle;
import android.widget.TextView;
 	
public class HelloWorldActivity
extends Activity {
    @Override
    public void onCreate(Bundle
savedInstanceState) {
        super.onCreate(savedIn-
stanceState);
 	
        TextView text = new
TextView(this);
        text.setText("Hello
World");
        setContentView(text);
    }
}
	.class public Lcom/test/hel-
	loworld/HelloWorldActivity;
	.super Landroid/app/Activity;
	.source "HelloWorldActivity.java"
	
	# direct methods
	.method public constructor ()V
	    .locals 0
	    .prologue
	    .line 7
	    invoke-direct {p0},
	Landroid/app/Activity;->()V
	    return-void
	.end method
	
	# virtual methods
	.method public onCre-
	ate(Landroid/os/Bundle;)V
	    .locals 2
	    .parameter "savedInstanceS-
	tate"
	    .prologue
	    .line 11
	    invoke-super {p0, p1},
	Landroid/app/Activity;->onCre-
	ate(Landroid/os/Bundle;)V
	    .line 13
	    new-instance v0,
	Landroid/widget/TextView;
	
	    invoke-direct {v0, p0},
	Landroid/widget/TextView;-
	>(Landroid/content/Context;)V
	    .line 14
	    .local v0,
	text:Landroid/widget/TextView;
	    const-string v1, "Hello World"
	
	    invoke-virtual {v0, v1},
	Landroid/widget/TextView;-
	>setText(Ljava/lang/CharSe-
	quence;)V
	    .line 15
	    invoke-virtual {p0, v0},
	Lcom/test/helloworld/Hel-
	loWorldActivity;->setCon-
	tentView(Landroid/view/View;)V
	    .line 17
	    return-void
	.end method

2.2 Extracting Stack Trace

The stack traces of the code can be collected by injecting exception which returns current stack trace of the application code. The stack trace can include information about hooking (see Table 2) and in that case it stores specific methods calls:

- call to the de.robv.android.xposed.XposedBridge.main method after the dalvik.system.NativeStart.main
- the calls of de.robv.android.xposed.XposedBridge.handleHookedMethod and de.robv.android.xposed.XposedBridge.invokeOriginalMethodNative methods.
- the hooked method can appear twice.

Table 2. Stack trace of normal execution and exposed execution

Stack trace of normal execution	Stack trace of exposed execution
com.example.testapp.sampleClass- >getData com.example.testapp.MainActivity- >onCreate android.app.Activity- >performCreate android.app.Instrumentation- >callActivityOnCreate android.app.ActivityThread- >performLaunchActivity android.app.ActivityThread- >handleLaunchActivity android.app.ActivityThread- >access$800 android.app.ActivityThread$H- >handleMessage android.os.Handler- >dispatchMessage android.os.Looper->loop android.app.ActivityThread->main java.lang.reflect.Method- >invokeNative java.lang.reflect.Method->invoke com.android.internal.os.ZygoteInit $MethodAndArgsCaller->run com.android.internal.os.ZygoteInit ->main dalvik.system.NativeStart->main	com.example.testapp.sampleClass- >getData com.example.testapp.MainActivity- >onCreate **de.robv.android.xposed.XposedBridge** **-> invokeOriginalMethodNative** **... // alternative (hooked) code** **de.robv.android.xposed.XposedBridge** **->handleHookedMethod** com.example.hookdetection.DoStuff- >getSecret com.example.hookdetection.MainActi vity->onCreate android.app.Activity- >performCreate android.app.Instrumentation- >callActivityOnCreate android.app.ActivityThread- >performLaunchActivity android.app.ActivityThread- >handleLaunchActivity android.app.ActivityThread- >access$800 android.app.ActivityThread$H- >handleMessage android.os.Handler- >dispatchMessage android.os.Looper->loop android.app.ActivityThread->main java.lang.reflect.Method- >invokeNative java.lang.reflect.Method->invoke com.android.internal.os.ZygoteInit $MethodAndArgsCaller->run com.android.internal.os.ZygoteInit ->main **de.robv.android.xposed.XposedBridge** **->main** dalvik.system.NativeStart->main

The injected code can be also listed by the stack trace and it is really difficult to remove it from there. Unfortunately, the stack trace of it cannot be determine for all of the devices, because of native execution and device specific call. Most of the malware software hide the basic information by hooking the stack trace and removing information about framework and duplicated calls from stack trace. The code which is executed by hook method stay in stack trace and removing of it is complex process. The anomalies which occur in the stack trace can be detected by comparison of the stack trace methods with the Dalvik bytecode.

2.3 Malicious Code Detection

Malicious code features analysis was done during researches presented in the paper [15]. The Table 3 presents percentage of code occurrence in malware and safe applications. Modules with are usually created to extract data from the application, so all file operations related to database or file access which are not part of current code of program can be malicious code.

Table 3. Percentage of occurrences of features derived from java code, which are in the top features chosen by 3 selection algorithms in the malware and safe application sets. Source: [15].

Feature	Percentage in malware	Percentage in safe apps
startService	0.34809	0.00118
getString	0.35412	0.04471
setPackage	0.25553	0
putExtra	0.27767	0.01059
startActivity	0.19215	0.01882
getSystemService	0.17907	0.02000
append	0.20121	0.04588
indexOf	0.11268	0.00941
getInputStream	0.09558	0.00235

3 The Method Hooking Detection

3.1 Code Comparison Process

The application code is stored in dex files in the apk repository. Based on packages and methods in the dex files the map of program structure and dependencies is created. This structure is used to find code related to the current stack trace and compare Dalvik bytecode from dex files with it by the classifier.

3.2 Classification Process

The classification process is done by TensorFlow text classification module. TensorFlow Lite is an open-source deep learning framework for on-device inference, dedicated to the Android operating system. It is optimized for this type of devices, so it is not affecting the device performance as any other external solution Code stack trace and methods lists are usually less than <1500 words, based on that to analysis process the data tokenize as n-grams was used. For the classification, simple multi-layer perceptron (MLP) model is used. Both these algorithms are predefined by TensorFlow text classification and additionally optimized for typical hardware like arm processors. For the learning process, top 20 native applications from 10 most popular Play Store categories were selected. According to different code base and structure any cross-platform applications which were written in React Native, Xamarin or Flutter were excluded from the analysis. The first part of the learning process was focused on code structure and its similarity to the stack trace. It was mandatory because of Android API dependencies and code optimizations done by Proguard or R8. The next part of the learning process was related to the detection of method hooking. For learning 100 Xposed and 20 EdXposed modules were selected from modules repository [16] and XDA Developer Portal [17]. Data were collected from Motorola Moto G6 Plus and Google Pixel 3. Devices were rooted as this was mandatory for method hooking frameworks to be run.

3.3 Testing

The solution was tested on 200 modules available for EdXposed framework and 1000 modules available for Xposed framework on modules repository [16] and via XDA Developer Portal [17]. The number of hooked methods was 4628 For Xposed framework and 458 for EdXposed framework. Applications without Xposed or EdXposed frameworks installed were used for false-positive detection. The results are presented in Table 4. Tested applications were grouped into three categories based on their code structure:

- Not obfuscated – code of the application is not obfuscated, and similar to the original code which is written by the developer.
- Obfuscated – code of application includes obfuscated names, for example, a.a, a.b. The code can be also optimized during the building process on the bytecode level.
- Secured – code is obfuscated, strings or classes encryption was detected in the application file.

Table 4. Results of method hooking detection.

	Xposed framework		EdXposed framework	
Application type	False negative	False positive	False negative	False positive
Not obfuscated	6	7	2	0
Obfuscated	137	34	8	2
Secured	243	237	21	15

For applications without obfuscation, the method hooking was detected correctly in most of the cases. The code doesn't have any modification or hidden calls, so the analysis is easy. The only issue which occurs was related to the reflection mechanism which breaks the standard flow of the application. For obfuscated applications, the mechanism of code optimization provides some structures which were not correctly recognized by the classifier. Applications with additional security levels like class encryption do not include the whole code in the dex files, as some part of it is encrypted outside and it cannot be extracted from the apk in unencrypted form.

4 Conclusions and Future Work

The proposed solution can detect 91,8% of method hooks, additionally, the level of false positives reports is very low. The experiment confirms that the proposed solution is good and can be used for method hooking detection for all installed applications. It does not require to be integrated within the application code as market solutions. As future works Authors plan to test other classification models and compare them according to levels of method hooking detection. Additionally, Authors plan to provide solution which detects class encryption in apk file.

References

1. Szczepanik, M., Jóźwiak, I.J., Jamka, T., Stasiński, K.: Security lock system for mobile devices based on fingerprint recognition algorithm. In: Świątek, J., Borzemski, L., Grzech, A., Wilimowska, Z. (eds.) Proceedings of 36th International Conference on Information Systems Architecture and Technology – ISAT 2015 – Part III. Advances in Intelligent Systems and Computing, vol 431. Springer, Cham (2016)
2. Elenkov, N.: Android Security Internals: An In-Depth Guide to Android's Security Architecture, 1st edn. No Starch Press, San Francisco (2014)
3. Vidas, T., Nicolas C.: Evading android runtime analysis via sandbox detection. In: Proceedings of the 9th ACM Symposium on Information, Computer and Communications Security. ACM. (2014)
4. Lim, K., Jeong, Y., Cho, S., Park, M., Han, S.: An android application protection scheme against dynamic reverse engineering attacks. JoWUA **7**, 40–52 (2016)
5. Lim, K., Jeong, J., Cho, S. J., Choi, J., Park, M., Han, S., Jhang, S.: An anti-reverse engineering technique using native code and obfuscator-LLVM for android applications. In: Proceedings of the International Conference on Research in Adaptive and Convergent Systems (RACS 2017). Association for Computing Machinery, New York, NY, USA, pp. 217–221 (2017)
6. Park, J.K., Choi, S.Y.: Studying security weaknesses of android system. Int. J. Secur. Appl. **9**(3), 7–12 (2015)
7. Costamagna, V., Zheng, C.: ARTDroid: a virtual-method hooking framework on android ART runtime. In: IMPS@ ESSoS, pp. 20–28 (2016)
8. Mitul, B., Hinaxi, P., Swati, K.: A survey permission based mobile malware detection. Int. J. Comput. Technol. Appl. **6**, 2 (2015)

9. Jang, W.J., Cho, S.W., Lee, H.W., Ju, H.I., Kim, J.N.: Rooting attack detection method on the Android-based smart phone. In: Proceedings of 2011 International Conference on Computer Science and Network Technology, Harbin, pp. 477–481 (2011)
10. Szczepanik, M., Jóźwiak, I.: Security of mobile banking applications. In: Kościelny, J., Syfert, M., Sztyber, A. (eds.) Advanced Solutions in Diagnostics and Fault Tolerant Control. DPS 2017. Advances in Intelligent Systems and Computing, vol 635. Springer, Cham (2018)
11. Gallo, R., Hongo P., Dahab, R., Navarro, L., Kawakami, H., Galvão, K., Junqueira, G., Ribeiro, L.: Security and system architecture: comparison of Android customizations. In: Proceedings of the 8th ACM Conference on Security & Privacy in Wireless and Mobile Networks (2015)
12. Xposed Framework. https://xposed.info/. Accessed 10 Jan 2020
13. Gotta hack‘em all, Meet_Mobile Group meetup (2016). https://www.meetup.com/meet-mobile/events/232943997/. Accessed 10 Jan 2020
14. Guardsquare Homepage. https://www.guardsquare.com/en. Accessed 10 Jan 2020
15. Kedziora, M., Gawin, P., Szczepanik, M., Jozwiak, I.: Malware detection using machine learning algorithms and reverse engineering of android java code (2019)
16. Xposed Module Repository. https://repo.xposed.info/. Accessed 10 Jan 2020
17. XDA Developers Portal. https://www.xda-developers.com/. Accessed 10 Jan 2020

Reliability of Ultrasonic Distance Measurement in Application to Multi-Rotor MAVs

Boguslaw Szlachetko(✉) and Michal Lower

Faculty of Electronics, Wroclaw University of Science and Technology, Wroclaw, Poland
{Boguslaw.Szlachetko,Michal.Lower}@pwr.edu.pl

Abstract. An ultrasonic sensor with the use of a modulated waveform is presented in the paper. This proposition makes it possible to increase the reliability of the measurement and to substantially shorten the period between consecutive distance readings. The distance measurements are commonly made using ultrasonic sensors in multi-rotor micro aerial vehicles (MAVs). The distance measurement is a complex issue. In practice, the distance is usually calculated through a data fusion of measurement of several different types of sensors. However, due to their irreplaceable advantages, very often one of the basic sensor that MAVs are equipped with is US sensor. Ultrasonic sensors are readily used in take-off and landing and for obstacle detection. Therefore, multi-rotor MAVs are equipped with two or more such sensors because of the need to measure distance in several directions (e.g. forward and down).

Keywords: Ultrasound · Distance measurement · Intelligent sensor · MAV

1 Introduction

Ultrasonic measurement is often used by unmanned micro aerial vehicles (MAV) to measure distances up to 5 m [2,5]. Due to the sound's propagation speed in the air, the time needed to measure a distance of 5 m for a classic ultrasound sensor equals about 30 ms. A classic ultrasound sensor generates a bunch of square impulses - five to eight - and wait for an echo which comes back after bouncing off an obstacle. Because each of successive bunch of pulsed is identical, the sensor cannot make another measurement until the echo returns. It is because a sensor can not differentiate between received echo signals. The only solution is to define the time interval long as much as a flight time of bunch of pulses to the furthest possible distance sum of forward and back. Also, taking measurements simultaneously by two ultrasound sensors is inadvisable, because the signals of both sensors can interfere with each other - they are identical. In the case of a single MAV, the

W. Zamojski et al. (Eds.): DepCoS-RELCOMEX 2020, AISC 1173, pp. 642–651, 2020.
https://doi.org/10.1007/978-3-030-48256-5_63

activation of individual sensors can be controlled at the expense of the measurement frequency. However, the problem becomes more serious if another independent MAV close to it uses the same type of measurement. Such MAV becomes a source of interference and ultrasound measurement becomes worthless.

To increase the reliability of ultrasound measurement, the authors propose a sensor using modulated waves. Modulation of ultrasonic waves is a method mainly used in medical diagnostics, but this is another case because the ultrasound wave propagates in soft tissues. However, the use of a modulated ultrasound wave which propagates in the air is also considered in the literature in the context of communication systems [4,7,8]. Thanks to wave modulation, each generated bunch of pulses can be tagged with a unique label. This solution makes it possible to distinguish between received echo signals. It is because each returning echo signal can be associated with the moment in time it was generated. Therefore, there is no need to wait for the returning echo and one can transmit further probe pulses in short intervals. In addition, the signal given by the foreign sensor will be rejected because they will not contain the appropriate identification code/label. This method increases the reliability of the measurement and significantly speeds up the distance readings.

The other idea of using ultrasound signal is positioning system which relies on the measurement of the angle of arrival of the ultrasound wave instead of the time of flight of the wave to the obstacle and back [1,3,6]. But this solution entails a build of external to MAV system of many receivers. Each of them measures an angle of arrival and next external computer can solve a triangulation problem to find a source of ultrasound waves e.g. the MAV which emits it.

An ultrasound distance measurement is widely used in the context of MAV. These vehicles can move in all directions simultaneously. Usually forward and downward movements are considered, which forces two parallel measurements in these directions. Therefore, it is expected that the proposed solution using a modulated waveform will improve the reliability of distance measurement by MAV and increase the frequency of measurement pulses. The authors present a detailed analysis of the problem of classic solutions in combination with their own proposal based on modulated impulses. It is worth noting that the proposed solution has been implemented in the form of software for the embedded SMT32 Cortex M4 microprocessor. Also, due to the specific properties of the chosen microprocessor and due to the intensive use of hardware clock generators, the library functions for processing ultrasonic signals can not be utilized [9].

2 Ultrasound Distance Measurements Dependency

Ultrasonic sensors available on the market are not able to generate a modulated ultrasound waveform due to their closed architecture composed of transducer (one or two), amplifier microprocessor and some less important parts. Because hardware and software specifications are unknown, any modifications are highly difficult. Thus, it was necessary to design a dedicated electronic device which enables a modulation and demodulation of the ultrasound wave. The implementation of some Digital Signal Processing algorithm is also a very important achievement.

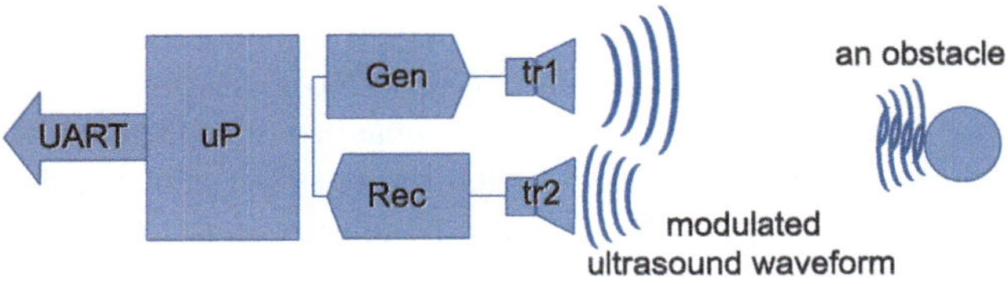

Fig. 1. A diagram of the ultrasonic sensor design and measurement principles.

Low cost ultrasound transducers demonstrate necessary properties to be used as an emitter and as a receiver/microphone of an ultrasound signals. The scheme of the proposed ultrasonic sensor is shown in Fig. 1. It is composed of:

- **tr1, tr2** - two low cost ultrasonic transducers TCF40-18TR1 playing roles of emitter and microphone
- **Gen** - electronic generator and amplifier designed for the ultrasonic band about 48 kHz
- **Rec** - electronic receiver designed for ultrasound band
- **uP** - microprocessor STM32F4 Cortex M4 used software modulator and demodulator of ultrasound signals

The proposed device has been built and tested at Wroclaw University of Science and Technology. A communication with the device is possible through serial interface UART.

The microprocessor uP is responsible for modulation/demodulation of a probing pulse and also for calculation of time delay between the pulse and its echo. Note that each probing pulse gives a unique label which, thanks to the modulation process, can be amplified and emitted by P1 transducer. In accordance with the transducer characteristics, an ultrasound wave propagates through the surrounding space and returns to the P2 transducer after reflection from an obstacle. Than echo is received by the P2 transducer. The microprocessor uP demodulates the signal, compares the label against the sent one and calculates a time delay, which is equivalent to giving a distance to the obstacle.

Binary Phase Shift Keying (BPSK) modulation has been utilized in our design, due to its simplicity. In this type of modulation the phase of a carrier frequency is used to encode the value of the currently transmitted bit. Because of specific nature of ultrasonic transducers they are controlled by a square wave generator. Thanks to this simplification, it was decided to use a hardware timer (a device implemented in microprocessor structure) that generates a rectangular carrier wave. At the right time, polarization of a timer is changed according to the binary information and this way a modulated signal is generated. Next, the square carrier wave goes to an analogue signal amplifier circuit that allows achieving an optimal ultrasonic transducer operation voltage. It is worth note, that low cost ultrasonic transducers utilize an piezoelectric disc which acts as an strong band-pass filter. Therefor, even if a transducer is excited by square carrier wave it emits nearly sinusoidal waveform.

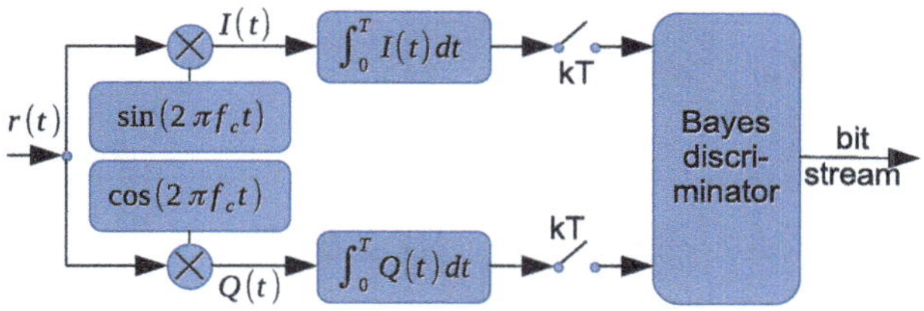

Fig. 2. A diagram of the noncoherent BPSK receiver.

The demodulator works based on incoherent reception. According to the theoretical foundation of incoherent reception, the received signal is multiplied by an artificially generated wave with a frequency equal to the carrier frequency of 48 kHz. Multiplication takes place in two channels, I (In-phase) and Q (quadrature) as shown in Fig. 2. The received signal is multiplied by the corresponding sine and cosine functions. Then the I and Q signals are integrated over the duration of a single bit and the integral value goes to the discriminator, which decides whether the received bit is a logical state of "0" or "1". The discriminator makes this decision using the highest probability rule.

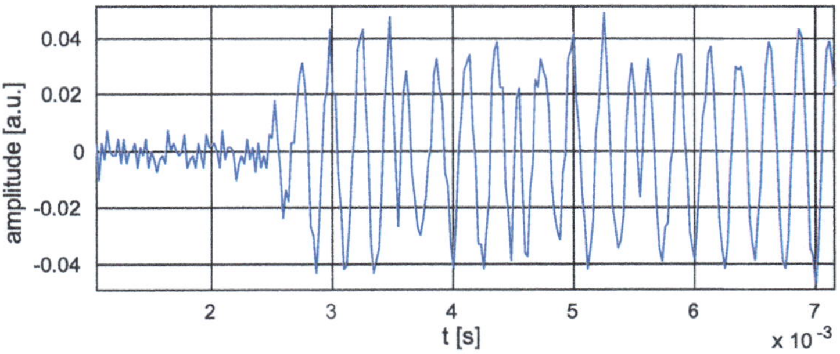

Fig. 3. Echo signal registered by ultrasonic transducer after transmitting a probing pulse with BPSK modulation.

In our device, an adaptive discriminator was implemented, which utilizes known training sequence of bits sent at the beginning of each frame. The sequence is used for transmission clock synchronization. Also, it makes possible to identify the phase of the received symbols. The graph of the fragment of the received signal recorded by an oscilloscope is shown in Fig. 3. It clearly shows the moments of phase change of the carrier wave at the time limits of the subsequent symbols/bits. The graph of I and Q signals is shown in Fig. 4. These signals after integration over the duration of the symbols allow to determine the logical values of subsequent bits.

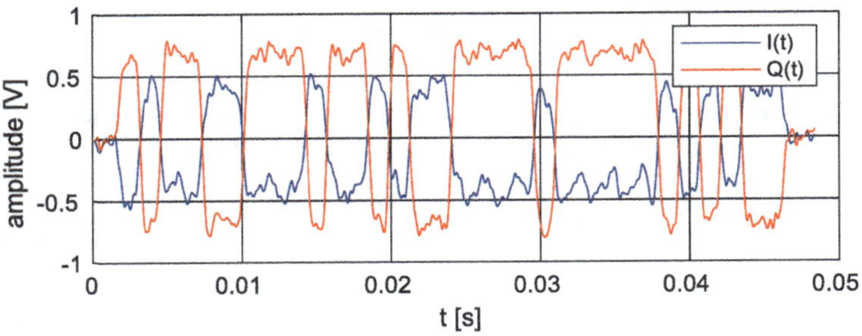

Fig. 4. $I(t), Q(t)$ signals used in a demodulation procedure.

2.1 The Use of a Microcontroller's Embedded Hardware

The use of the largest possible number of hardware components instead of software is the main factor to increase the efficiency and speed of the entire conversion system. Thanks to the microcontroller from the Cortex M4 family, which has hardware timers, PWM mechanism and built-in ADC converters using DMA channels, it was possible to generate, receive and demodulate a modulated ultrasonic wave in real time. It is worth to note that for selected piezoelectric transducers carrier frequency equals 48 kHz, which gives only about 2000 clock cycles of the microcontroller. During such a short period microcontroller has to catch samples of received signal, multiply it by $\sin(.)/\cos(.)$ functions, calculate integral of I/Q signals and distinguish logical levels. In addition to these operations, the conversion system should also realize other functions like communication with the main computer.

The generation of the modulated waveform has been performed using the embedded timer. This task is relatively simple because the transducer must be excited by a square wave. So, in registers of Cortex microcontroller a PWM waveform with a frequency of 48 kHz and a duty cycle of 50 % has been configured. Square carrier waveform generated in this way flows to a specially designed amplifier working in class D. In the case of (BPSK) the carrier frequency has got only two phases. Moreover, if the baud rate is adopted, which is a multiple of carrier frequency, then keying the phase of the carrier frequency can be transformed into a simple sign change of the generated waveform. Therefore, the key element is to switch the hardware output of the timer between positive/negative states. This can be done by changing a single bit of the timer register even without stopping the timer.

The next hardware tricks are used in demodulation chain. Embedded Analog-to-Digital Converter (ADC) is used to sample and hold received signal. To left a core of microcontroller free of transmitting data between ADC hardware register and microcontroller memory a Direct Memory Access (DMA) channel has been configured. In this case a conventional double-buffer mechanism is utilized. The DMA transfers a data from ADC to the buffer after each conversion, then after collecting half portion of data hardware interrupt is generated. Interrupt procedure changes only a current address for data in a buffer and gives a semaphore to a thread which waits on it. The thread starts the detection and eventual demodulation procedure while DMA still transfers data to a second half of the buffer. This way the thread gains more time for processing data and time constrains become slightly less restrictive.

3 Results

Classic ultrasonic sensors generate identical rectangular bunch of impulses and wait until echo return after bouncing off the obstacle. In the proposed solution, based on the modulated wave, pulses are individually labeled. For research purposes, adverse situations have been simulated that show the increased reliability of measurement based on modulated signals. Two tests were carried out.

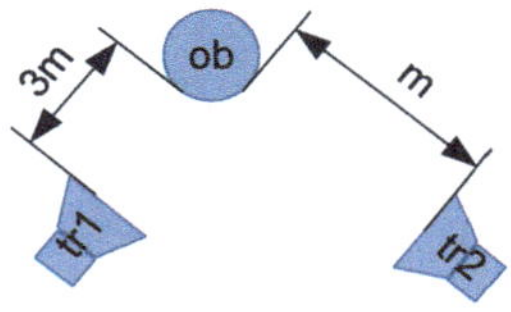

Fig. 5. Test 1 situational diagram. Two independent ultrasonic sensors illuminate a pole with a diameter of about 1 m.

3.1 Test 1

Test 1 consists of setting two ultrasonic sensors of one type, but working independently of each other, i.e. the measuring systems did not know that another system was operating nearby, and therefore measurement systems were not adapted to such a situation in any way. First, this test was performed for classical sensors, and then for sensors with modulated wave. The situational diagram is shown in Fig. 5. The object Ob was a concrete pole with a smooth wall with a diameter of about 1 m. The sensors were set at a distance of 3 m (sensor S1) and 4 m (sensor S2) at an angle of about 90°. So an echo reflected from the pole went to both sensors regardless of which sensor emitted a probing pulse. The times of emission of a probing pulse were random at both sensors.

In the Table 1 results of measurements in case of use of classical sensors are presented. Values in bold emphasise faulty measurements. It is clearly visible that the faulty measurement happens quite often, which means that this type of sensors is not resisting to situation when more than one sensor works nearby. In the Table 2 results of measurements in case of using the designed sensors are presented.

Table 1. Measurements results for classical ultrasonic sensors.

Sensor S1 - dist. 3 m		Sensor S2 - dist. 4 m	
Delay [ms]	Distance [m]	Delay [ms]	Distance [m]
17.435	2.997	**9.011**	**1.549**
17.444	2.999	**6.415**	**1.103**
7.886	**1.356**	23.270	4.000
5.774	**0.992**	23.250	3.997
17.442	2.998	23.262	3.999
17.462	3.002	**10.183**	**1.750**
10.299	**1.770**	23.281	4.002
17.456	3.001	23.259	3.998
17.463	3.002	23.247	3.996
8.492	**1.460**	23.266	3.999
12.437	**2.138**	23.257	3.998
17.451	3.000	**2.745**	**0.472**
7.887	**1.356**	23.272	4.001
17.460	3.001	**1.131**	**0.194**

Table 2. Measurements results for ultrasonic sensors with a modulated signal.

Sensor S1 - dist. 3 m		Sensor S2 - dist. 4 m	
Delay [ms]	Distance [m]	Delay [ms]	Distance [m]
17.48	3.005	23.28	4.002
17.42	2.994	23.28	4.002
17.47	3.003	23.27	4.000
17.44	2.998	23.28	4.002
17.52	3.012	23.30	4.005
17.46	3.001	23.23	3.993
17.46	3.001	23.28	4.002
17.45	3.000	23.30	4.005
17.44	2.998	23.28	4.002
17.44	2.998	23.26	3.998

The results of distance measurements in the Test 1 are given in tables Table 1 and Table 2. It can be seen that when traditional sensors are used, the measurements are random in nature. One can meet a correct measurements - i.e. a distance of about 3 m to the sensor S1 and 4 m to the sensor S2 - but there are also many random measurements. Unfortunately, there are no effective methods to distinguish between normal and disturbed measurements. The interference observed in tables come from the fact that the sensor receives an echo signal, but it is an echo of the probing pulse generated by the neighboring sensor. For calculations it is necessary to measure the time of flight of the ultrasound wave to the obstacle and back. In the opposite, the use of a modulated ultrasound wave allows to clearly distinguish between own and foreign probing pulses. What is more important, this method allows to distinguish several subsequent own impulses as well. Because each probing pulse contains unique label, the calculation of the distance is not burdened with fatal errors resulting from confusion between the measuring pulses.

3.2 Test 2

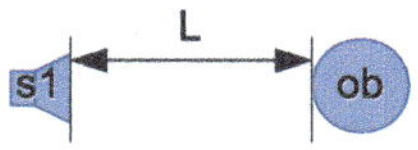

Fig. 6. Test 2 situational diagram. The ultrasonic sensor which illuminates an obstacle moves away from it at a constant speed.

The situational diagram of the Test 2 is shown in Fig. 6. The sensor was tested by measuring the distance from the obstacle Ob - a concrete pole with smooth walls and a diameter of about 1 m - while the distance to the obstacle has been increasing at a constant speed. The frequency of the probing pulses was constant. In the case of the classical ultrasonic sensors it was 100 ms, which results from the design of the sensors. For the proposed ultrasonic sensor with a modulated wave it was set up to 10 ms.

It is worth noting that measuring the distance at probing pulse period 100 ms is too slow to be useful in anti-collision systems. This is due to several facts. Firstly, during 100 ms MAV moving at a high speed is able to fly a considerable distance, which in effect makes it difficult to calculate the actual distance to the obstacle and additionally threatens to run into an obstacle. Therefore, using traditional sensors, the usual assumption is made that a MAV should fly with a low speed and far away from obstacles. Secondly, reducing the measurement period, e.g. up to 20 ms makes it impossible to perform measurements for greater distances, because the echo's time-of-flight is received after a time longer than 20 ms. It makes problems because measurements are unambiguous. Thus, the sensor can not distinguish whether the echo comes from the current measurement pulse or from the previous one. However, using the proposed ultrasonic sensors with modulated wave there is no such a threat, because each consecutive probing pulse has got its own unique number which can be decoded from the received echo signal. As a consequence, with probability one it is possible to measure the delay time correctly and thus the distance from the obstacle.

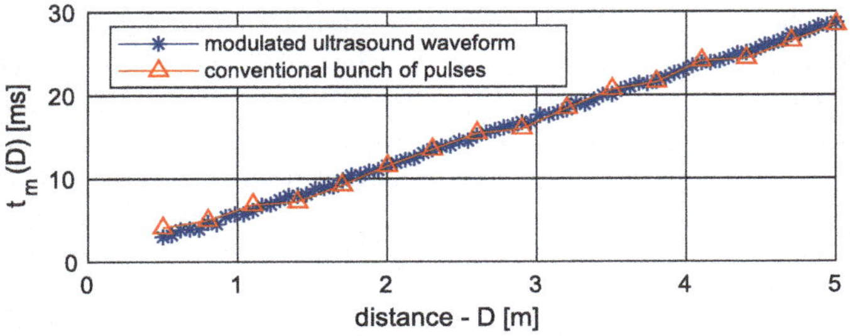

Fig. 7. Results of time-of-flight measurement as a function of distance from an obstacle Ob during constant speed of MAV. Blue stars mark results given with the proposed method, red triangles mark results given with the classical ultrasound sensor.

In Fig. 7 a graph of the echo delay time measurements recorded during Test 2 is presented. Asterisks indicate the measurements made by the proposed ultrasonic system. Triangles represent times measured by classical COTS sensors. It can be seen that the frequency of measurements done by proposed system is much higher, which means that MAV obtains more measurements during flight time, thus, information about the distance to obstacle is more reliable. This is especially important for small distances because the MAV autopilot can check the distance to the obstacle more often. At the same time, it is also an advantage in the case of long distance measurement - let's say 5 to 8 m - because MAV can continue to carry out measurements at a high frequency without waiting for the echo to arrive. Such properties gives hope to develop a 3D spatial model using ultrasonic sensors without the need for vision sensors. The similar method is used by bat in a dark cave.

4 Summary

The proposed technical solution allow us to increase the reliability of the measurement and to substantially shorten the period between consecutive distance readings.

The proposed measurement method and designed device allow to increases the reliability and effectiveness of the measurement relative to a classic solution. Additionally, the proposed method shortens substantially period between consecutive reading. Thanks to the use of unique pulse labeling, the problem of mutual interference of sensors has ceased to be noticeable. Also the ambiguity problem has been solved. Therefore, by labeling each subsequent pulse, the measurement frequency can be increased. The proposed construction uses measuring pulses lasting a little longer than in a traditional solution. As a result, the so called dead zone of the measurement is about 70 cm. For this reason, further work will focus on the use of faster modulation and shortening of the measuring pulse.

Finally our device would be subjected to miniaturization process which allows to reduce the size of our device up to the conventional ultrasonic sensors.

References

1. Barbagallo, M., Kleiner, M., Sarti, A., da Vinci, P.L.: Modulation and demodulation of steerable ultrasound beams for audio transmission and rendering templates. In: Proceedings of the 11th International Conference on Digital Audio Effects, Espoo, Finland (2008)
2. Holm, S.: Ultrasound positioning based on time-of-flight and signal strength. In: 2012 International Conference on Indoor Positioning and Indoor Navigation (IPIN), pp. 1–6 (2012). https://doi.org/10.1109/IPIN.2012.6418728
3. Holm, S., Nilsen, C.C.: Robust ultrasonic indoor positioning using transmitter arrays. In: 2010 International Conference on Indoor Positioning and Indoor Navigation, pp. 1–5 (2010). https://doi.org/10.1109/IPIN.2010.5646198
4. Hosman, T., Yeary, M., Antonio, J.K., Hobbs, B.: Multi-tone fsk for ultrasonic communication. In: 2010 IEEE Instrumentation Measurement Technology Conference Proceedings, pp. 1424–1429 (May 2010). https://doi.org/10.1109/IMTC.2010.5488066
5. Kelemen, M., Virgala, I., Kelemenova, T., Mikova, L., Frankovsky, P., Liptak, T., Lorinc, M.: Distance measurement via using of ultrasonic sensor. J. Autom. Control **3**(3), 71–74 (2015). https://doi.org/10.12691/automation-3-3-6
6. Kreczmer, B.: Azimuth angle determination for the arrival direction for an ultrasonic echo signal. J. Autom. Mob. Rob. Intell. Syst. **11**(2), 31–41 (2017)
7. Li, C., Hutchins, D.A., Green, R.J.: Short-range ultrasonic communications in air using quadrature modulation. IEEE Trans. Ultras. Ferroelectr. Freq. Control **56**(10), 2060–2072 (2009). https://doi.org/10.1109/TUFFC.2009.1289
8. Lopes, C.V., Aguiar, P.M.Q.: Aerial acoustic communications. In: Proceedings of the 2001 IEEE Workshop on the Applications of Signal Processing to Audio and Acoustics (Cat. No.01TH8575), pp. 219–222 (2001). https://doi.org/10.1109/ASPAA.2001.969582
9. Robledo, E., Hinojosa, R., de Boer, M.: A C++ development platform for real-time audio processing and synthesis applications. In: Proceedings of the 5th International Conference on Digital Audio Effects (2002). Software implementations

Determining the Minimal Number of Images Required to Effectively Train Convolutional Neural Networks

Kamil Szyc(✉)

Wrocław University of Science and Technology, Wrocław, Poland
kamil.szyc@pwr.edu.pl

Abstract. Determining the minimal number of images required to effectively train convolutional neural networks (CNNs) is not trivial and no simple solution exists. In this paper, we propose a method that helps to determine that number of images and checks if sufficient data is collected. We did a series of experiments using different CNN architectures (classic one with 3 convolutional layers, ResNet18, and Inception v3) and images data sets with different complexity (MNIST, CIFAR-10, CIFAR-100, and STL-10). These experiments allowed us to propose guidelines for the discussed problem. We also proposed a method for establishing the complexity of image sets.

Keywords: Convolutional Neural Networks · Training neural network · Images data sets

1 Introduction

Deep learning methods are commonly used nowadays. Even beginners can use complex methods to solve non-trivial problems using high-level programming libraries. Many people use deep neural networks as a black-box, without understanding what is happening inside the network. Nevertheless they can still achieve good (or even the best possible) results.

People can now use deep-learning methods as tools. There are many ready to use programming libraries, pre-training models, numerous scientific papers and other solutions. In order to solve even complex problems it is often required just to do a few steps: preparing data, choosing a model, choosing hyper-parameters, training model and testing it. Choosing a model is a really important step but usually comes down to using the state-of-the-at model or just a popular one. The training and testing model is just a few lines of code when we use popular deep learning libraries like PyTorch or Keras. So in practice, the most important steps are choosing hyper-parameters and preparing the input data. There exist methods like Random Search [1] or Bayesian Optimization [2] to help correctly choose hyper-parameters. In this paper, we would like to focus on preparation

W. Zamojski et al. (Eds.): DepCoS-RELCOMEX 2020, AISC 1173, pp. 652–661, 2020.
https://doi.org/10.1007/978-3-030-48256-5_64

of the data. The rule on how to do that seems to be simple: preparing as much data as possible with correct marked labels.

We decided to focus on image classification problem using Convolutional Neural Networks (CNNs). To solve that kind of problem we can usually use the same method as before: prepare lots of images, label them and use some good and popular model like VGG [3], Inception [4,5], ResNet [6] or DenseNet [7]. If we use enough images and set good hyper-parameters we can achieve satisfactory results.

Our contribution is to define what "enough images" means. The general rule is preparing as many images as possible. However, building data sets is a time-consuming task (we need to choose images, label them and pre-process them), so it is also expensive. We can use online crowd-sourcing services like Amazon Mechanical Turk, but we don't know if we can trust the results [8]. On the other hand, CNNs are able to generalize what an object looks like, so more images help them to achieve better accuracy. So there is ambiguity - we would like to prepare a small number of images (because it is expensive), but to achieve better results we need an increasing number of them. We proposed a method to help estimate a satisfactory number of images.

There are methods that help to increase final accuracy when we don't have enough images. The most important one is for sure data augmentation [9]. The idea is simple - increasing the number of images by changing them a little. We can achieve that by for example flipping, changing colors or brightness, rotating, using GAN. Thanks to that we can simulate how our object can look like with different scenarios. Another one is using transfer-learning [10,11] method or fine-tuning. The idea of both is similar - we would like to use the network which was trained on big data-sets (like the ImageNet [12]). We know that first layers in CNN can recognize simple patterns, so usually, we can use them to recognize images on different subsets too. The latest layers on CNN can be replaced by others (transfer learning) or retrain (fine-tuning). These methods will not work, when images are completely different from the ones used for training - like for example comics or x-ray images. There is also a method called a few shot learning [13] which enables training deep networks using only single examples from each classes. There are also other less known methods - for instance, we can change color space [14] and expect that some features will be able to recognize better or use Semi-Supervised Learning method [15] and collect unlabeled images and used them. In this paper, we decided not to focus on these methods and to not use them.

We did a series of experiments using different deep learning models for image classification problems and different images of data sets. These tests allow us to propose a rule on how to estimate the number of images required to achieve satisfactory accuracy.

2 Experiments

We decided to do a series of experiments to estimate the number of images in the training set. We are aware that final accuracy depends on many different factors like quality of input data, labels, used CNN model architecture or hyper-parameters. We decided to use images from different data sets and model architectures in our tests. We choose data sets with different complexity of images, so different numbers of channels, resolutions, number of classes and number of images in sets. We also choose models with different architectures: classic one with 3 convolution layers, ResNet-18 and Inception v3.

The main idea of our tests was training models using parts of a chosen data sets. The images were split to train and test subsets by default. We used naturally only training subset to train, and whole test subset to determine final accuracy. However, the idea was to use only part of the training data. We trained different models showing them only 1%, 5%, 10%, 20%, 25%, 30%, 40%, 50%, 60% 70%, 75%, 80%, 90% or 100% of available images in training subset. We picked up chosen images randomly.

Based on these experiments we were able to present a method that helps to determine the minimal number of images.

2.1 Data Sets

We decided to use four popular data sets: MNIST, CIFAR-10, CIFAR-100, and STL-10. For STL-10 we used the only train and test images - originally it contains also not labeled data. We also created a new one - CIFAR-5 - by using all images from CIFAR 10 but chosen from only 5 classes. We compare all of it in Table 1. We defined a new parameter and we called it *complexity*. We calculated it by using the formula:

$$complexity = \frac{w * h * d}{\sqrt{c} * n} \tag{1}$$

where w is width, h is height, d is depth (number of channels - for instance 3 in RGB color space), c is number of classes and n is number of images per class in training sub-set. Using the square root of c in the equation is used to reduce differences when they are more spread out. It is a common technique in the statistic. Complexity determines how difficult it is to learn patterns from an image by deep neural networks. We used image sets with different complexity and it allowed us to draw more general conclusions.

Proposing *complexity* we assumed that width, height, and depth are determined by set itself or determined by chosen model architecture. We assumed that all parameters except a number of images per class should not be modified when using our method.

2.2 Models

We tested three different models: classic/simple CNN, ResNet-18 and Inception v3. The simple CNN is a network created by us and has only 3 classic convolutions layers. The ResNet-18 and the Inception v3 are popular architectures that use popular techniques. The ResNet-18 uses skip connections method which

Table 1. Comparison of used images data sets (and ImageNet)

Name	Width	Height	Depth	Classes	Training images per class	Complexity
MNIST	28	28	1	10	6000	0.04
CIFAR-5	32	32	3	5	5000	0.27
CIFAR-10	32	32	3	10	5000	0.19
CIFAR-100	32	32	3	100	500	0.61
STL-10	96	96	3	10	500	17.49
ImageNet	256	256	3	500	21841	2.66

keeps gradient stronger and consequently helps to avoid vanishing gradient problem. The Inception v3 uses special modules with filters with multiple sizes that operate on the same level. That helps with better features extraction.

We defined hyper-parameters for all models based on the CIFAR-10 data set and used them in all other experiments. We set the epochs number to 50 and batch size to 128. We used Cross-Entropy loss and Stochastic Gradient Descent optimizer. We had to modify the input of networks for each set (based on width, height, and depth on image) and final classifier (based on the number of classes).

Table 2. Number of images per class and complexity depends on percent of training data used in our experiments

	MNIST		CIFAR-5		CIFAR-10		CIFAR-100		STL-10	
Percent	n	c	n	c	n	c	n	c	n	c
1	60	4.13	50	27.48	50	19.43	5	61.44	5	1748.61
5	300	0.83	250	5.50	250	3.89	25	12.29	25	349.72
10	600	0.41	500	2.75	500	1.94	50	6.14	50	174.86
20	1200	0.21	1000	1.37	1000	0.97	100	3.07	100	87.43
25	1500	0.17	1250	1.10	1250	0.78	125	2.46	125	69.94
30	1800	0.14	1500	0.92	1500	0.65	150	2.05	150	58.29
40	2400	0.10	2000	0.69	2000	0.49	200	1.54	200	43.72
50	3000	0.08	2500	0.55	2500	0.39	250	1.23	250	34.97
60	3600	0.07	3000	0.46	3000	0.32	300	1.02	300	29.14
70	4200	0.06	3500	0.39	3500	0.28	350	0.88	350	24.98
75	4500	0.06	3750	0.37	3750	0.26	375	0.82	375	23.31
80	4800	0.05	4000	0.34	4000	0.24	400	0.77	400	21.86
90	5400	0.05	4500	0.31	4500	0.22	450	0.68	450	19.43
100	6000	0.04	5000	0.27	5000	0.19	500	0.61	500	17.49

n - Training images per class

c - Complexity

2.3 Experiments

We did a series of experiments. For each image sets we choose 1%, 5%, 10%, 20%, 25%, 30%, 40%, 50%, 60% 70%, 75%, 80%, 90% or 100% images from training set. We randomly choose these images, but in that way to always keep the same number of images per class. We present the number of used images and complexity (Eq. 1) in Table 2. We trained chosen networks and we measured final accuracy based on the whole testing subset. We repeated each experiment 30 times to also consider the impact of randomness.

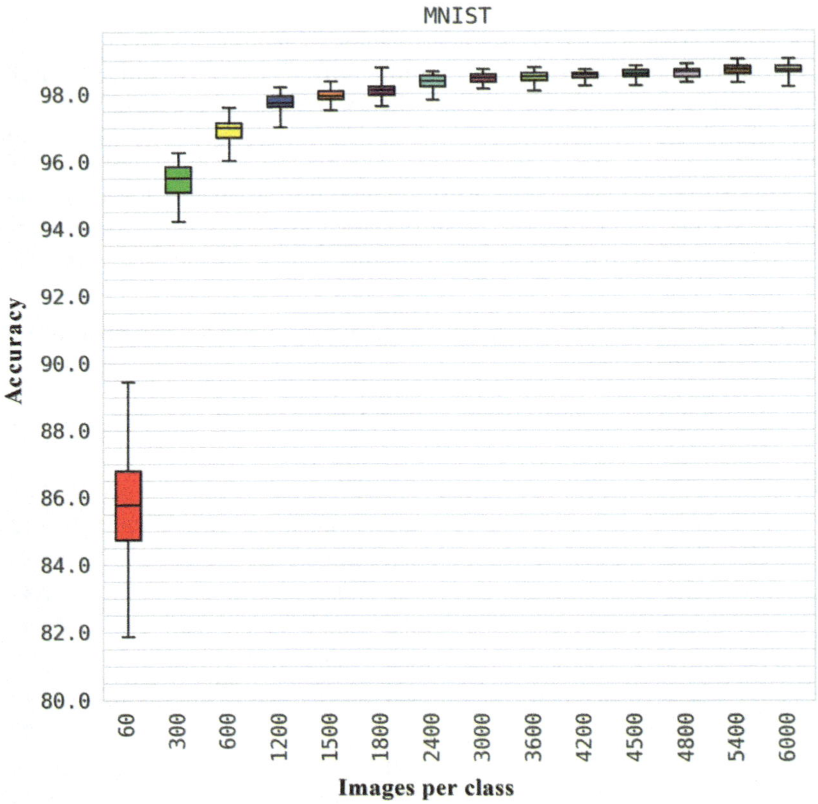

Fig. 1. Final accuracy when used only percentages of images per class from MNIST - used classic CNN.

2.4 Results

We collected the results of our experiments. In Figs. 1, 2, and 3 we can see how final accuracy was changing - we present only examples of the graphs, but the

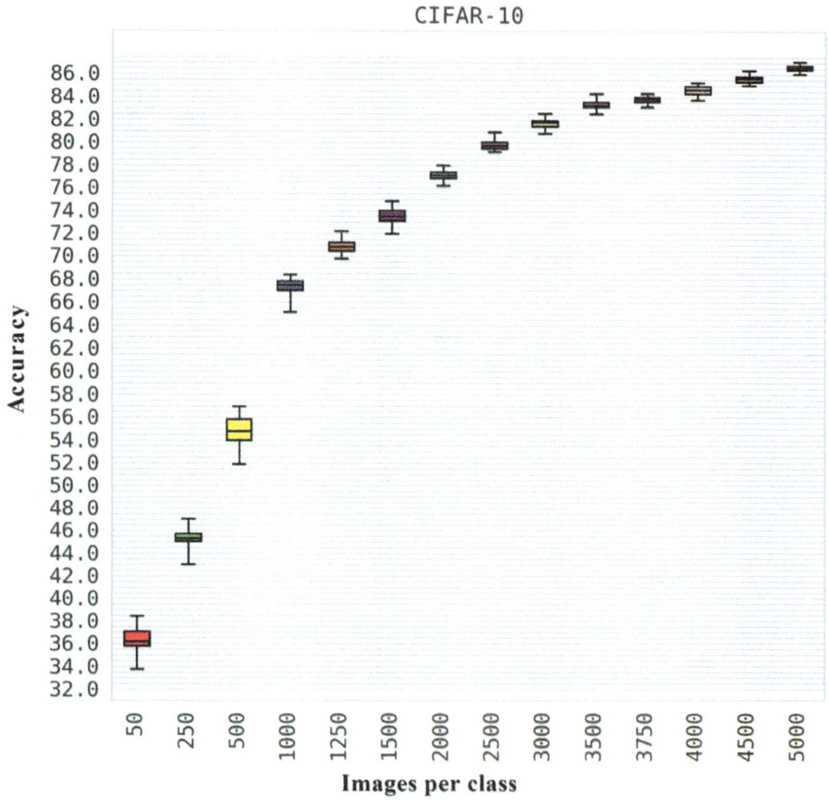

Fig. 2. Final accuracy when used only percentages of images per class from CIFAR-10 - used ResNet18.

trend is similar. Accuracy depends on model architecture and percentage of used training images.

We can notice that for not complex images sets like MINST or CIFAR-5 all tested CNN architectures achieved satisfactory accuracy - even with a small number of images. Adding more images doesn't improve the result.

In CIFAR-10 image set situation is different. Although we achieved good accuracy, we can see that adding new images could help a little. All graphs were becoming more and more flattened, so it couldn't help with an as long solution. In this case, we should use the data augmenting method.

For CIFAR-100 and STL-10, where complexity is relatively higher and we can see that we didn't use enough images. The networks could better generalize what each class looks like if we trained the network with more images.

In Fig. 4 we can see how complexity influences accuracy. These graphs are key to this paper. Based on these graphs we proposed a method that helps to determine a sufficient number of images.

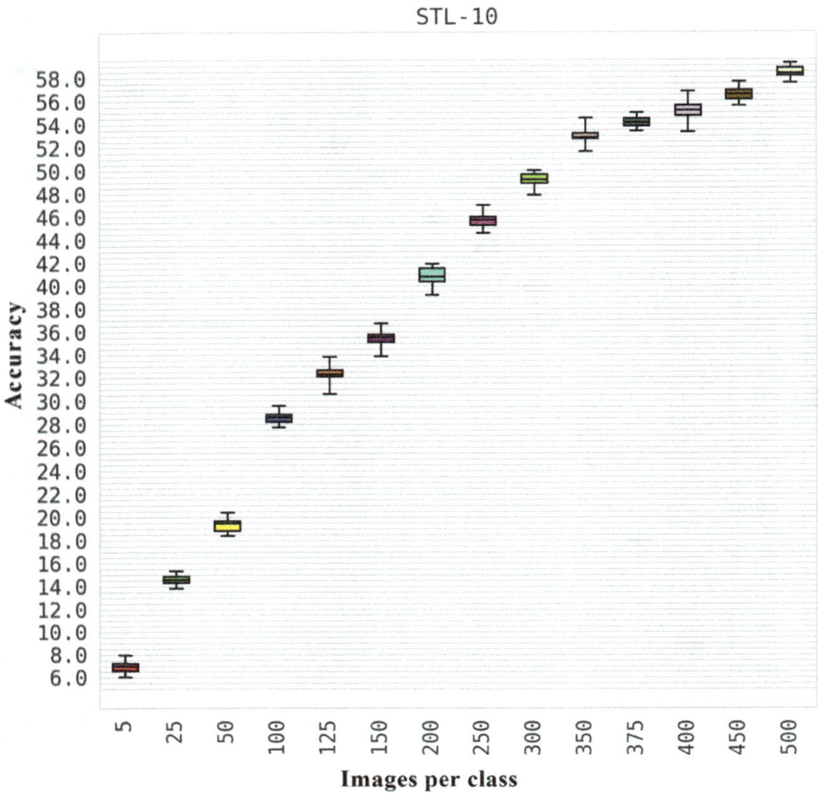

Fig. 3. Final accuracy when used only percentages of images per class from STL-10 - used ResNet18.

3 Conclusions

3.1 Proposed Method

Based on our experiments we can propose a method that helps to determine if sufficient data has been collected. We assumed that we have selected the best possible model for a given problem and with correct hyper-parameters. First, we should collect some images, calculate complexity based on Eq. 1 and train our model. Then we should repeat that process over and over by adding new images and training our model. We should observe how complexity(it will decrease as the number of images increases) and accuracy changed in a line chart. We suggest using a log scale for complexity. The graph should look similar to those shown in Fig. 4. We should stop adding new images when the graph will become more flattened. Then we suggest using data augmenting methods.

We are aware that our method looks similar to a simple rule - add more images and stop when it doesn't help anymore. However, based on our method

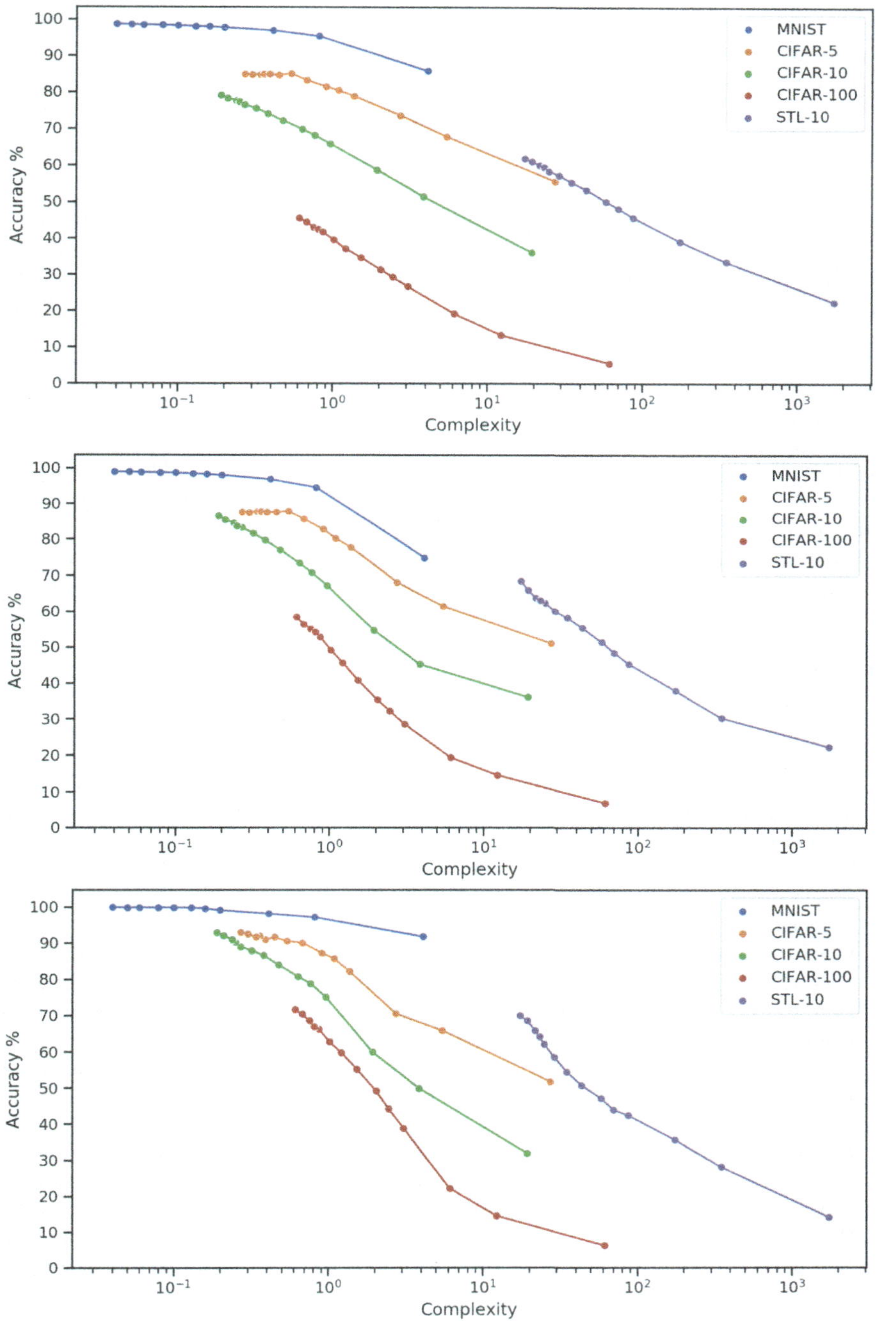

Fig. 4. Comparison of accuracy (the arithmetic mean from 30 same parameters experiments represented as single point) and complexity of set (when using (top) the classic CNN, (middle) the ResNet18, (bottom) the Inception v3).

we can easily compare different data sets and models. Thanks to that fact we can deduct what the initial number of images should be.

We propose use this equation:

$$n = \frac{w * h * d}{\sqrt{c} * complexity} \quad (2)$$

to estimate required images. We should set *complexity* to 1 on thread that as good start. We can use then data augmentation method. If we have enough resources then we should try to collect images given by *complexity* equal to 0.1. We should try to keep *complexity* below 10 or try using different methods like transfer learning.

3.2 Summary

We proposed a method to calculate complexity from the given data set - see Eq. 1. We did a series of experiments which showed that we can deduct an optimal number of images required to effectively train convolutional neural networks. We propose using our method for people who treat deep-learning just as a tool and do not have sufficient experience to estimate that number on their own.

References

1. Bergstra, J., Bengio, Y.: Random search for hyper-parameter optimization. J. Mach. Learn. Res. **13**, 281–305 (2012)
2. Snoek, J., Larochelle, H., Adams, R.P.: Practical Bayesian optimization of machine learning algorithms. In: Advances in Neural Information Processing Systems, pp. 2951–2959 (2012)
3. Simonyan, K., Zisserman, A.: Very deep convolutional networks for large-scale image recognition. *arXiv preprint* arXiv:1409.1556 (2014)
4. Szegedy, C., et al.: Going deeper with convolutions. In: Proceedings of the IEEE Conference on Computer Vision and Pattern Recognition, pp. 1–9 (2015)
5. Szegedy, C., Vanhoucke, V., Ioffe, S., Shlens, J., Wojna, Z.: Rethinking the inception architecture for computer vision. In: Proceedings of the IEEE Conference on Computer Vision and Pattern Recognition, pp. 2818–2826 (2016)
6. He, K., Zhang, X., Ren, S., Sun, J.: Deep residual learning for image recognition. In: Proceedings of the IEEE Conference on Computer Vision and Pattern Recognition, pp. 770–778 (2016)
7. Huang, G., Liu, Z., Van Der Maaten, L., Weinberger, K.Q.: Densely connected convolutional networks. In: Proceedings of the IEEE Conference on Computer Vision and Pattern Recognition, pp. 4700–4708 (2017)
8. Welinder, P., Perona, P.: Online crowdsourcing: rating annotators and obtaining cost-effective labels. In: 2010 IEEE Computer Society Conference on Computer Vision and Pattern Recognition-Workshops, pp. 25–32. IEEE (2010)
9. Perez, L., Wang, J.: The effectiveness of data augmentation in image classification using deep learning. *arXiv preprint* arXiv:1712.04621 (2017)
10. Pan, S.J., Yang, Q.: A survey on transfer learning. IEEE Trans. Knowl. Data Eng. **22**(10), 1345–1359 (2009)

11. Szyc, K.: Comparison of different deep-learning methods for image classification. In: 2018 IEEE 22nd International Conference on Intelligent Engineering Systems (INES), pp. 000341–000346. IEEE (2018)
12. Deng, J., Dong, W., Socher, R., Li, L.-J., Li, K., Fei-Fei, L.: ImageNet: a large-scale hierarchical image database. In: 2009 IEEE Conference on Computer Vision and Pattern Recognition, pp. 248–255. IEEE (2009)
13. Snell, J., Swersky, K., Zemel, R.: Prototypical networks for few-shot learning. In: Advances in Neural Information Processing Systems, pp. 4077–4087 (2017)
14. Szyc, K.: An impact of different images color spaces on the efficiency of convolutional neural networks. In: International Conference on Dependability and Complex Systems, pp. 506–514. Springer (2019)
15. Wang, X., Kihara, D., Luo, J., Qi, G.-J.: EnAET: self-trained ensemble autoencoding transformations for semi-supervised learning. *arXiv preprint* arXiv:1911.09265 (2019)

Heuristic Allocation Strategies for Dependable Scheduling in Heterogeneous Computing Environments

Victor Toporkov(✉) and Dmitry Yemelyanov

National Research University "MPEI", ul. Krasnokazarmennaya, 14, Moscow 111250, Russia
{ToporkovVV, YemelyanovDM}@mpei.ru

Abstract. In this work, we consider heuristic resources allocation strategies for parallel jobs execution in heterogeneous computing environments. Existing modern job-flow execution features and realities impose many restrictions for the resources allocation procedures and applicable scheduling criteria. Subject to these restrictions a special dynamic programming scheme is proposed to select resources depending on their meta-features. When applied to secondary scheduling criteria this procedure may be used to generate job flow scheduling strategies while maintaining the same primary scheduling criterion outcomes. Such secondary criteria optimization is commonly referred as *breaking a tie* rules. Based on a conservative backfilling scheduling procedure we study how different breaking a tie heuristics affect integral job-flow scheduling characteristics in a dedicated simulation environment.

Keywords: Distributed computing · Grid · Coordination · Resources allocation · Slot · Job · Backfilling · Breaking a tie rules

1 Introduction and Related Works

Some of the most important efficiency indicators of distributed computational environments include both system resources utilization level and users' jobs time and cost execution criteria [1–5]. In distributed environments with non-dedicated resources, such as utility Grids, the computational nodes are usually partly utilized and reserved in advance by jobs of higher priority [6]. Thus, the resources available for use are represented with a set of slots - time intervals during which the individual computational nodes are capable to execute parts of independent users' parallel jobs. These slots generally have different start and finish times and a performance difference.

The presence of a set of independent slots complicates the problem of coordinated selection of the resources necessary to execute jobs from computational environment users. Resource fragmentation also results in a decrease of the total computing environment utilization level [7, 8]. High-performance distributed computing systems (HPCS), including Grid, cloud and hybrid infrastructures provide access to large amounts of resources [9, 10]. These resources are typically required to execute parallel

W. Zamojski et al. (Eds.): DepCoS-RELCOMEX 2020, AISC 1173, pp. 662–671, 2020.
https://doi.org/10.1007/978-3-030-48256-5_65

jobs submitted by HPCS users and include computing nodes, data storages, network channels, software, etc.

HPCS organization and support bring certain economical expenses: purchase and installation of machinery equipment, power supplies, user support, maintenance works, security, etc. Thus, HPCS users and service providers usually interact in economic terms, and the resources are provided for a certain payment. In such conditions, resource management and job scheduling based on the economic models is considered as an efficient way to coordinate contradictory preferences of computing system participants and stakeholders [7, 11–13].

Metaschedulers and metabrokers are considered as intermediate links between the users, local resource management and job batch processing systems [7, 14] based on uniform rules of a resource sharing and consumption to improve the overall scheduling efficiency [6–8].

As an example, this approach allows improving the job-flow level scheduling and resource distribution efficiency in a virtual organization (VO) to satisfy both users' and VO global preferences [15]. The VO scheduling problems may be formulated as follows: to optimize users' criteria or utility function for selected jobs, to keep resource overall load balance, to have job run in strict order or maintain job priorities, to optimize overall scheduling performance by some custom criteria, etc.

The main contribution of this paper is heuristic strategies for a coordinated and hindsight resources allocation and parallel jobs execution.

Coordinated resources selection algorithms take into account available slots configuration as well as individual jobs' meta-features: their size, runtime, cost, etc. When used in HPCS metaschedulers, it is sometimes possible to affect global resources utilization efficiency by choosing the appropriate secondary optimization criteria.

Based on a conservative backfilling procedure we study how different resources allocation heuristics affect integral job-flow scheduling characteristics.

The rest of the paper is organized as follows. Section 2 introduces a job-flow scheduling model, backfilling algorithm and main ideas behind the secondary criteria optimization strategies. Section 3 contains algorithms implementation details along with simulation results and analysis. Finally, Sect. 4 summarizes the paper.

2 Resources Allocation for Job-flow Scheduling

2.1 Basic Components of Computing Model

In order to cover a wide range of computing systems we consider the following model to describe a heterogeneous resource domain.

The resource domain consists of a list of computing nodes with different usage costs and performance levels.

A space-shared resources allocation policy is implemented for a local queuing system like in SimGrid [5] or CloudSim [12, 13]. Each node can process only one task at any given time.

Users and resource owners operate within an economic model by using some currency or it's analog to coordinate resources allocation transactions. This model allows to regulate interaction between different organizations and to settle on fair equilibrium prices for resources usage.

Thus, we consider a set R of heterogeneous computing nodes with different performance p_i and price c_i characteristics. Each node has a local utilization schedule known in advance for a considered scheduling horizon time L. This schedule corresponds to job reservations, maintenance works and other resources utilization events.

A node may be turned off or on by the provider, transferred to a maintenance state, reserved to perform computational jobs. Thus, it is convenient to represent all available resources as a set of slots. Each slot corresponds to one computing node on which it's allocated and may be characterized by its performance and price.

The execution cost of a single job depends on the allocated node's price c_i and execution runtime.

The runtime is proportional to the node's performance level p_i. In order to execute a parallel job one needs to allocate the specified number of simultaneously idle nodes (execution window) ensuring user requirements from the resource request. Common start and finish times for each slot in the window ensures the possibility of inter-node communications during the whole job execution.

The resource request specifies a number n of nodes required simultaneously and their minimum performance level p, job's computational volume V and a maximum available resources allocation budget C. These parameters constitute a formal generalization for resource requests common among distributed computing systems and simulators [7, 12, 16].

In heterogeneous computing environment a parallel job execution time is defined based on a computing node with the minimum performance. For example, if the execution window consists of two slots with performances $p \in \{p_i, p_j\}$ and $p_i < p_j$, then the job execution time is calculated as $T = \frac{V}{p_i}$. In this way V really defines a computational volume for each single job subtask. The total cost of the computing nodes allocation is then calculated as $C_W = \sum_{i=1}^{n} T * c_i$.

These parameters constitute a formal generalization for resource requests common among distributed computing systems and simulators.

2.2 Backfilling Scheduling Procedure

The most straightforward way to schedule a job-flow is by using the First-Come-First-Served (FCFS) procedure. FCFS executes jobs one by one in an order of arrival. However, this approach proved inefficient in terms of a global resources usage and backfilling [2, 4, 6, 8] was proposed to improve computing system utilization.

Backfilling procedure makes use of advanced resources reservations in order to prevent starvation of jobs with a relatively large resource request requirements. During FCFS job-flow execution a number of nodes may stay idle waiting for the next jobs in the queue. The main idea behind the backfilling procedure is to use those idle slots for lower-priority jobs execution and improve the overall system utilization. Thus, backfilling may execute some smaller jobs from the back of the queue ahead of the general priority order.

There are two common variations of backfilling - conservative and aggressive (EASY). In conservative variation the execution of jobs submitted later (from the back of the queue) cannot delay start of the jobs arrived earlier. EASY backfilling aggressively backfills smaller jobs as long as they do not delay the start of a single job next in the queue. Conservative backfilling considers jobs in the order of their arrival and either immediately starts a job or makes an appropriate reservation upon the arrival.

Jobs priority in backfilling queue may be additionally modified in order to improve system-wide job-flow execution efficiency metrics. Under default FCFS policy the jobs are arranged by their arrival time. Some other priority reordering-based policies like Shortest Job First or expansion Factor were proposed to further improve the overall resources utilization efficiency [2, 6, 8].

Another backfilling variation separates jobs into multiple queues based on their resource requirements size: small, medium, large, etc. The main idea behind this approach is that earlier arriving jobs and smaller-sized jobs are given the highest execution priority [8]. The look-ahead algorithm [6] implements dynamic programming scheme to find a combination of jobs for backfilling that together will maximize the resources utilization.

2.3 Resources Selection Algorithms

Backfilling scheduling approach defines a general procedure determining high level policies and rules for jobs prioritization and resources reservations. However, the resources selection step usually remains sidelined since its more system specific nature. On the other hand, applying different resources allocation policies for parallel jobs may affect scheduling results not only for individual jobs but for a whole job-flow.

In [7], we presented a Slot Subset Allocation (SSA) dynamic programming scheme for resources selection optimization in heterogeneous computing environments. Each computing resource has a local utilization schedule known in advance and it's convenient to represent available resources as a set of time-slots. SSA algorithm takes these time slots as input and performs resources selection for a specified parallel job in accordance with the computing model described in Sect. 2.1.

SSA performs window search optimization by a general additive criterion $Z = \sum_{i=1}^{n} c_z(s_i) \rightarrow \max$, where $c_z(s_i) = z_i$ is a target optimization characteristic value provided by each slot of a target window W. These criterion values z_i may represent different slot characteristics: time, cost, power, hardware and software features, etc. SSA optimization is implemented based on a special knapsack problem solution and its total computational complexity for a single parallel job may be estimated as $O(|R| * n * C)$.

2.4 Resources Allocation Strategies

One important factor for an efficient resources usage in distributed and parallel computing is a job placement metric. By the job placement we mean its timeline position and proximity to other jobs running or reserved on the same resources in regard to the utilization schedule.

For example, Fig. 1 shows Gantt chart for a 3 slots co-allocation problem in a sample computing environment with resources pre-utilized with jobs of a higher priority (filled rectangles). Slots 1–4 (hollow rectangles on Fig. 1) represent a candidate window of slots suitable for a parallel job execution.

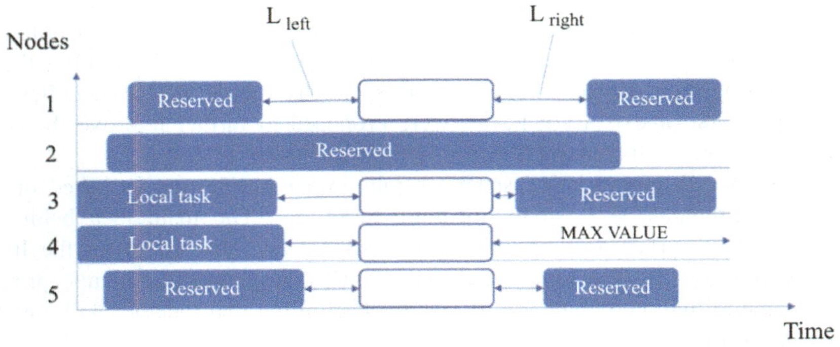

Fig. 1. Coordinated window co-allocation and placement metrics.

Firstly, there are many options to position this window on the considered timeline by shifting its start time left or right. The start time minimization problem usually corresponds to a primary optimization criterion [2, 6, 8, 16]. However, any starting position additionally may be described by L_{lefti} and L_{righti} distances to the neighbor jobs for every slot s_i (see Fig. 1).

Secondly, as the job in our example requires only three nodes there are four different options to allocate the resources. Important to note, that each such option will provide the same start and finish time, but different sets of L_{lefti} and L_{righti}. Thus, it is possible to form a set of *breaking a tie* rules to choose between slots providing the same start time. The rules proposed in [8] for Picking Earliest Slot for a Task (PEST) procedure may be summarized as following.

- Minimize number of idle slots left after the window allocation; slots adjacent to the already reserved slots should have higher priority.
- Maximize length of idle slots left after the window allocation; the main point is to leave longer slots available for the subsequent jobs in the queue.

By analogy, we propose the following Coordinated Placement (CoP) heuristic rules [4].

- Prioritize slots allocated on nodes with lower performance. When deciding between two slots providing the same primary criterion value it makes sense to leave higher performance node available for the subsequent jobs.
- Prioritize slots with relatively small distances to the neighbor jobs: $L_{left\ i} \ll T$ or $L_{right\ i} \gg T$, where T is a current job runtime. The general idea is similar to the first rule in PEST, but CoP does not require perfect adjacency and uses threshold values for a satisfactory resources usage.

- Penalize slots leaving significant, but insufficient resources to execute a full job. For example, when $T/3 > L_{right\ i} > T/8$, the resulting slot may be too short to be used by any of the subsequent jobs. Thus, it will remain idle and underutilized.
- On the other hand, the slots leaving sufficient resources compared to the current job's runtime T should be prioritized. One example of maintaining sufficient resources amount is $L_{left\ i} > T$.

An important feature of CoP and similar breaking a tie approaches [4, 8] is that they do not affect the primary scheduling criterion and do not change a base scheduling procedure.

However there are many options to add diversity during the resources allocation breaking a tie step. This allows to precalculate and to choose between several scheduling scenarios obtained with different resources allocation rules. The resulting schedule will represent the same baseline algorithm outcome and maintain all its required and special features. Such *Hindsight approach* may be used when it's possible to consider job queue execution on some scheduling interval or horizon.

For this purpose we consider *a family of backfilling-based algorithms* with different breaking a tie rules: *Random*, *Greedy*, *CoP* and *an empty ruleset*. After the scheduling step is over the best solution obtained by these algorithms is chosen as a Hindsight result. Except for CoP, this family is chosen almost arbitrary in order to evaluate how each rule will contribute to the final solution.

Implementation details for each algorithm are provided in Sect. 3.

3 Simulation Study

3.1 Scheduling Algorithms Implementation Details

Based on heuristic rules described in Sect. 2 we implemented the following scheduling algorithms and strategies for SSA-based resources allocation.

- Firstly, we consider conservative backfilling *BF* procedure. For a finish time *minimization*, BF criterion for *i*-th considered slot has the following form: $z_i = -s_i.finishTime$. As there are no secondary criteria, BF generally selects a random subset of slots providing the earliest finish time for a job execution.
- *Rand* algorithm uses SSA algorithm for a fully random resources selection over a conservative backfilling: $z_i = -s_i.finishTime + r_i$. Here r_i is a small random value uniformly distributed on interval $[0; 01]$ representing the secondary criterion.
- *Greedy* backfilling-based algorithm performs resources selection based on the following greedy metric of the resources *profitability*: $\frac{p_i}{c_i}$. Thus, the resulting SSA criterion function is: $z_i = -s_i.finishTime + \alpha\frac{p_i}{c_i}$. Here α defines the weight of secondary criteria and is supposed to be much less compared to a primary BF criterion in order to break a tie between otherwise even slots. In the current study we used $\alpha = 0.1$ value.
- *CoP* resources allocation algorithm for backfilling is implemented in accordance with rules and priorities described in Sect. 2.4. More details were provided in [4].

3.2 Experiment Setup

The experiment was prepared with the following setup of a custom distributed computing environment simulator [4, 7]. For our purpose, simulator implements a heterogeneous resource domain model: nodes have different utilization costs and performance levels. The job-flow processing and resources allocation policies simulate a local queuing system: each node can process only one task at any given simulation time. The execution of a single job requires parallel execution of all its tasks. The execution cost of each task depends on its execution time, which is proportional to the dedicated node's performance level. More details regarding the simulation computing model were provided in Sect. 2.1.

During each simulation experiment a new instance for the computing environment is automatically generated. Node performance level is given as a uniformly distributed random value in the interval [2, 16]. This configuration provides a sufficient resources diversity level while the difference between the highest and the lowest resource performance levels will not exceed one order.

The jobs were generated with the following resources request requirements: number of simultaneously required nodes is uniformly distributed as $n \in [1; 8]$ and computational volume is $V \in [60; 1200]$, which also contribute to a wide diversity in user jobs.

3.3 Simulation Results

The experiment setup included a job-flow of $N = 50$ jobs in a domain consisting of 42 heterogeneous computing nodes. The jobs were accumulated at the start of the simulation and no new jobs were submitted during the queue execution. Such problem statement allows us to statically evaluate algorithms' efficiency and simulate high resources load.

Table 1 contains simulation results obtained from 500 scheduling experiments for a family of breaking a tie heuristics contributing to the Hindsight solution.

Table 1. Breaking a tie heuristics scheduling results

Characteristic	BF	CoP	Rand	Greedy	Hindsight
Number of experiments	500	500	500	500	500
Average makespan	677	670	683	680	662
Average finish time	254	250	257	257	246
Earliest finish number	114	238	62	86	500
Earliest finish number, %	22.8	47.6	12.4	17.2	100%
Algorithm working time, s	0.01	52.6	54.4	53.2	160.2

We consider the following global job-flow execution criteria: a makespan (finish time of the latest job) and an average jobs' finish time.

Without taking into account the Hindsight solution the best results were provided by CoP: nearly 1% advantage over BF and 2% over both Rand and Greedy algorithms.

Hindsight approach reduces makespan and average finish time even more: 1% advantage over CoP, 2% over BF and 3% over Rand and Greedy.

Although these relative advantage values do not look very impressive, an important result is that they were obtained almost for free: CoP or Hindsight represent *the same baseline backfilling procedure*, but with a more efficient resources usage.

CoP made the largest contribution to the Hindsight solution: in 238 experiments (47.6%) CoP provided the earliest job-flow finish time. Baseline BF contributed to Hindsight in 114 experiments (22.8%). Greedy provided the earliest finish time in 86 experiments (17.2%), Rand – in 62 experiments (12.4%).

In this way, CoP ruleset actually implements heuristics which allow better resources allocation for parallel jobs compared to other considered approaches. At the same time even algorithm with a random tie breaking procedure outperformed BF, Greedy and CoP in 17.2% of experiments. Thus, by combining a larger number of random algorithms in a single family may result in comparable or even better Hindsight solution.

However, the major limiting factor for the Hindsight approach is SSA's actual working time. Baseline BF with a single criterion implements a simple procedure with almost a linear computational complexity over a number of available resources $O(|R|)$. Consequently, its working time is relatively short: only 10 ms for a whole 50 jobs queue scheduling. SSA computational complexity is $O(|R| * n * C)$ and it required almost a minute to perform the same job-flow scheduling in each experiment. Hindsight approach requires all completion of all the component algorithms. Thus, in our experiment setup Hindsight algorithm was executed for almost 3 min to obtain the resulting scheduling solution.

For a more detailed CoP and BF comparison we additionally considered job queues with $N \in [1; 5; 10; 15; 20; \ldots; 100]$ jobs. 250 independent scheduling scenarios were simulated for each size N of jobs in the queue. As a general result, CoP provided at average 1–2% earlier jobs finish times compared to the baseline BF, but not in every single simulation. Figure 2 presents median, 25% and 75% percentiles for relative difference (%) between average jobs' finish time provided by CoP and BF: $(BF.finishTime - CoP.finishTime)/CoP.finishTime$. Positive values represent scenarios when earlier job-flow finish time was provided by CoP, while negative value – scenarios when BF provided better solution.

As it can be observed, CoP generally provides better scheduling outcomes for all considered jobs number except for $N = 1$. In this case, both BF and CoP by design provide the same finish time for the single job in each experiment, resulting in a zero difference between them on Fig. 2 when $N = 1$.

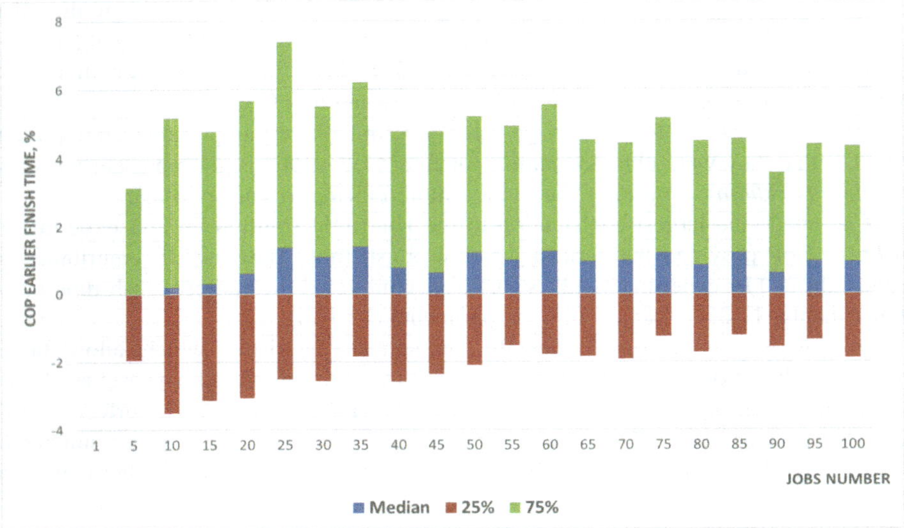

Fig. 2. Average job finish time difference between CoP and BF.

4 Conclusion

In this work, we address the problem of a resources allocation for parallel jobs in distributed and heterogeneous computing environments.

The main idea is to use a hierarchy of scheduling criteria and strategies after the primary scheduling algorithm, for example, backfilling. Thus, it is possible to construct the *Hindsight* solution by choosing the best over a family of backfilling solutions obtained with different strategies and secondary criteria.

Without any changes in the baseline backfilling algorithm, the simulation study showed 2–4% Hindsight strategy advantage against average job-flow finish time and makespan criteria.

The major limiting factor for the composite optimization approach is its actual working time, so further research will be focused on reducing its computational complexity.

Acknowledgments. This work was partially supported by the Council on Grants of the President of the Russian Federation for State Support of Young Scientists (YPhD-2979.2019.9), RFBR (grants 18-07-00456 and 18-07-00534) and by the Ministry on Education and Science of the Russian Federation (project no. 2.9606.2017/8.9).

References

1. Kurowski, K., Nabrzyski, J., Oleksiak, A., Weglarz, J.: Multicriteria aspects of grid resource management. In: Nabrzyski, J., Schopf, J.M., Weglarz, J. (eds.) Grid Resource Management. State of the Art and Future Trends, pp. 271–293. Kluwer Academic Publishers (2003)

2. Srinivasan, S., Kettimuthu, R., Subramani, V., Sadayappan, P.: Characterization of backfilling strategies for parallel job scheduling. In: Proceedings of the International Conference on Parallel Processing, ICPP 2002 Workshops, pp. 514–519 (2002)
3. Menasce, D.A., Casalicchio, E.: A framework for resource allocation in Grid computing. In: The 12th Annual International Symposium on Modeling, Analysis, and Simulation of Computer and Telecommunications Systems (MASCOTS 2004), Volendam, The Netherlands, pp. 259–267 (2004)
4. Toporkov, V., Yemelyanov, D.: Coordinated resources allocation for dependable scheduling in distributed computing. In: Zamojski, W., et al. (eds.): DepCoS-RELCOMEX 2019. AISC, vol. 987, pp. 515–524. Springer, Cham (2020)
5. Nazarenko, A., Sukhoroslov, O.: An experimental study of workflow scheduling algorithms for heterogeneous systems. In: Malyshkin, V. (ed.) Parallel Computing Technologies, pp. 327–341. Springer (2017)
6. Shmueli, E., Feitelson, D.G.: Backfilling with lookahead to optimize the packing of parallel jobs. J. Parallel Distrib. Comput. **65**(9), 1090–1107 (2005)
7. Toporkov, V., Yemelyanov, D.: Dependable slot selection algorithms for distributed computing. In: Advances in Intelligent Systems and Computing, vol. 761, pp. 482–491. Springer (2019)
8. Khemka, B., Machovec, D., Blandin, C., Siegel, H.J., Hariri, S., Louri, A., Tunc, C., Fargo, F., Maciejewski, A.A.: Resource management in heterogeneous parallel computing environments with soft and hard deadlines. In: Proceedings of 11th Metaheuristics International Conference (MIC 2015) (2015)
9. Lee, Y.C., Wang, C., Zomaya, A.Y., Zhou, B.B.: Profit-driven scheduling for cloud services with data access awareness. J. Parallel Distrib. Comput. **72**(4), 591–602 (2012)
10. Bharathi, S., Chervenak, A.L., Deelman, E., Mehta, G., Su, M., Vahi, K.: Characterization of scientific workflows. In: 2008 Third Workshop on Workflows in Support of Large-Scale Science, pp. 1–10 (2008)
11. Rodriguez, M.A., Buyya, R.: Scheduling dynamic workloads in multi-tenant scientific workflow as a service platforms. Future Gener. Comput. Syst. **79**(P2), 739–750 (2018)
12. Calheiros, R.N., Ranjan, R., Beloglazov, A., De Rose, C.A.F., Buyya, R.: CloudSim: a toolkit for modeling and simulation of cloud computing environments and evaluation of resource provisioning algorithms. J. Softw.: Pract. Exp. **41**(1), 23–50 (2011)
13. Samimi, P., Teimouri, Y., Mukhtar, M.: A combinatorial double auction resource allocation model in cloud computing. J. Inf. Sci. **357**(C), 201–216 (2016)
14. Rodero, I., Villegas, D., Bobroff, N., Liu, Y., Fong, L., Sadjadi, S.: Enabling interoperability among grid meta-schedulers. J. Grid Comput. **11**(2), 311–336 (2013)
15. Toporkov, V., Toporkova, A., Yemelyanov, D.: Global and private job-flow scheduling optimization in grid virtual organizations. In: Kotenko, I., et al. (eds.), IDC 2019. SCI, vol. 868, pp. 160–169. Springer, Cham (2020)
16. Epema, D., Iosup, D.: Grid computing workloads. J. IEEE Internet Comput. **15**(2), 19–26 (2011)

Prediction of Selected Personality Traits Based on Text Messages from Instant Messenger

Marek Woda[1(✉)] and Jakub Batogowski[2]

[1] Department of Computer Engineering, Wroclaw University of Technology, Janiszewskiego 11-17, 50-372 Wrocław, Poland
marek.woda@pwr.edu.pl
[2] Wrocław, Poland

Abstract. The aim of the work was to check effectiveness of machine learning to predict personality traits based on technical parameters of text messages. At the beginning, personality traits of test group (based on Big Five model) were determined. Each person provided a collection of text messages from the instant messenger. The authorial analyzer was used to aggregate the technical parameters: (number of emoji used, average message length, number of punctuation characters used, etc.) During the analysis, the most important parameters of text messages were identified, with the help of which it was possible to predict personality traits. In addition, based on the collected data and conducted analysis, a proprietary system to predict personality traits was created. To this end, methods of supervised machine learning were used. Finally, tests were carried out on the implemented solution, its prediction effectiveness was verified, and conclusions were drawn.

Keywords: Personality traits · Prediction · Machine learning

1 Introduction

The high popularity of social media has contributed to many studies related to content analysis of content posted on the Internet. It is worth mentioning the study in which the existence of a relationship between personality and emoji used was checked - namely in Twitter messages [5]. Additionally, posts on internet messengers were also frequently researched [6, 13]. A few studies highlighted elements that could affect written text messages: gender, context of speech, relationship between the interlocutors or, examined in this work - personality traits [8, 10, 16]. It can be said with certainty that personality influences many aspects of written content. One of them is punctuation, on that basis it is possible to predict certain personality traits [9, 11]. For example, one who frequently uses "full-stops", is a person who has usually a strong inclination towards latest technologies and who has no problems with self-limits. On the other hand, people who frequently use commas are more peaceful and more willing to help others in need [2]. Each punctuation mark and its frequency can have a real relationship with the personality.

Emoticons were invented in early 1970s to express emotions. Since then, they have appeared more and more often in short text messages. Today, several years after first use, they are an indispensable element of almost every message sent via instant

W. Zamojski et al. (Eds.): DepCoS-RELCOMEX 2020, AISC 1173, pp. 672–685, 2020.
https://doi.org/10.1007/978-3-030-48256-5_66

messengers [13, 14]. Due to the increasing popularity of emoticons, scientists began to consider their function and importance in media communication [9, 10]. Emoticons, in electronic communication, fill the gap created by the lack of non-verbal gestures used by the interlocutors during a normal conversation. As is well known, non-verbal messages contain much more information than the content of the words spoken, and often reduce the ambiguity of the statement [3]. It is common knowledge that non-verbal messages are directly related to body language, and this difference depends on the person's character [1, 4, 7].

Along with the development of emoticons, emoji was also created - pictograms used mainly to express feelings and emotions. Since most current instant messengers automatically replace many emoticons with their graphic counterparts, emoji are used much more often than emoticons. In addition, with the help of emoji, it is possible to express feelings more accurately [1, 7, 15], and therefore there is a higher probability (than in the case of emoticons) of a strong correlation between emojis used and selected personality traits.

The literature analysis [5, 12, 18] shows that the relationship between personality traits and the style of writing text messages is not trivial. In addition to the character of the interlocutor, there are many different factors that affect how text messages are written [13, 16]. For example, conversations of people who know each other well contain much more emoticons [15]. It is assumed that this is strongly associated with the freedom to write messages to loved ones, people feel much more comfortable among friends, and thus behave there more naturally. Therefore, it should be expected that when writing messages from a well-known person we will write much more, and a message will be enriched more with emoticons that reflect well our emotions and character traits.

On the other hand, in [15] is stated that there is no significant connection between the interlocutors and the emoticons used. This may be because during conversations between poorly known interlocutors, unconsciously or specifically use more emoticons to be better understood and more favorably received by the other party. Based on the above premise, it was decided to choose FB Messenger as the source of text messages. Messenger is mainly used in chat with friends. As expected, messages sent via this instant messenger were often enriched by the presence of emoticons and emoji.

The influence of gender on the style of writing text messages cannot be ignored. As one may presume, women write more messages than men, and each of them contains statistically more emoticons used. In addition, the style of messages written by both sexes should be quite different. This was partly confirmed in one study on the relationship with personality, social context and style of writing text messages [6]. In this study it was found that women more often than men use personal pronouns and words related to emotions and society. On the other hand, men use more swear words and words related to frustration, failure and anger [18]. It is worth mentioning that many other studies were also carried out, which not only confirmed the conclusions from the previous paragraph, but also indicated the existence of a larger number of dependencies between different situations and phrases used by women and men [7, 10]. As it can be deduced from the analyzed research, gender significantly affects the content of text messages. Nevertheless, it was decided not to analyze the content of the message, but only to use emoticons, emojis and punctuation marks. However, information on the sex of people from the research group was included at the stage of data analysis and preparation of the model used to predict personality traits.

It is well known that emotions significantly influence human behavior. Careless conversation with an irritated interlocutor can turn into a quarrel, and this looks much different than normal conversation. This bright, well-known example seems to confirm the existence of the influence of emotions on the way of communication during conversations. Therefore, it should be assumed that, just like in a real-world conversation - emotions can have a major impact on the style of written text messages. In connection with the above conclusion, it was decided to examine only those conversations that contained enough messages in order to minimize the impact of strong, momentary emotions on the analyzed technical parameters of text messages.

In [10], authors found out - how the context of written messages affects the emoticons used in them. To this end, two groups were distinguished, according to which the analyzed conversations were divided. The first, a task-oriented group contained those conversations whose priority was to achieve a specific, most often professional goal. Messages in this group were often conducted carefully by the interlocutors and contained a lot of substantive content. Nevertheless, they did not show many emotions in the form of emoticons, which appeared to hinder conversation. The second group, into which the interviewed interviews were divided - a social-emotional one, contained conversations conducted without a clear purpose in a relaxed atmosphere. The conversations belonging to this group were saturated with emotional messages, even in the form of emoticons, and the interlocutors were more focused on talking than on what she was concerned with. The detailed analysis described in the cited study clearly confirmed the relationship between the context of the message and their content. Therefore, it was decided to analyze text messages originating from a messenger, which is more likely to have very informal and loose conversations. So, this was another reason to use Messenger as the source of the text messages tested.

Emoji is not only popular in messages sent via instant messengers. They are also very popular on social networks. Therefore, most of the research on the impact of personality on the content style sent via the Internet concerned only used emojis. Among the surveys reviewed as part of the thesis, those that were mainly focused on the content of posts posted on social networks [16] and those in which the emoji contained in text messages were analyzed [15]. Nevertheless, regardless of the source of the studied content, the personality model most often used in the analyzed studies was the Big Five model [19] which is a taxonomy, or grouping, for personality traits. In connection with the great popularity of model, it was decided to use it in the conducted research. Referring to research carried out to verify the relationship between emojis and personality [3, 5], it is worth mentioning some of the most noteworthy details. People with a high level of *openness* do not show any preferences regarding the most commonly used emojis. In addition, no positive results were clearly found, and negative correlations associated with any emoji. Therefore, one should expect that openness will be a feature whose prediction can be significantly more difficult. People characterized by a high *conscientiousness* are distinguished by above-average vigilance and self-control. They use less emoji associated with negative emotions. Instead, they prefer those that express joy, happiness and satisfaction. *Extroverts* prefer emoji associated with positive emotions. In addition, they less frequently use negative emojis and those that have mixed feelings. People with a high level of *agreeability* use many emoji related to love and those expressing positive feelings. In addition, they insert less emoji associated with negative feelings in their messages.

In contrast to others, people with high levels of *neuroticism* use more emoji representing extreme, mostly negative emotions. This is the only feature of the *Big Five* model that stands out for its positive correlation with negative emojis. It turns out that there are many other parameters for text messages related to personality traits. For example, people with a high *extraversion* ratio use more personal pronouns, including women writing messages containing noticeably more words, and men are much less likely to use negatively marked words. In addition, *extroverts* use words that strengthen their message much more often in their statements. People with significant *neuroticism* often use negatively marked phrases, and women have a greater tendency to use pronouns. Women and men with a high *agreeableness* rate have very little use of swear words, and most of their statements are not negative [5].

2 Research Methodology

In the *first stage*, an initial interpretation of personality traits from the persons included in the research group were collected. For this purpose, Polish version of the IPIP-NEO-FFI-100 [17] questionnaire, containing 100 questions, was used. Gender information of people completing the prepared personality test was included. Thanks to this, it was possible to carry out gender-specific analysis. A research group consisting of 16 people (9 men, 7 women) in the age group from 21 to 25 years old was assembled. Data on technical parameters of text messages (in Polish) and personality traits were obtained from all persons. In order to obtain data on personality traits, each person in the group was asked to complete the IPIP-NEO-FFI-100 test. Collected answers were converted into numerical values corresponding to personality traits according to *Big Five* model. *Second stage* consisted of collecting selected technical parameters of text messages, based on which it was possible to predict personality traits. According to the literature review, personality can have a real impact on many technical parameters of text messages, so it was decided to analyze the emojis used and their relationship with selected personality traits of this model. In addition, the impact of punctuation marks was considered. The analysis of technical parameters of text messages was carried out using the authorial parser implemented for this purpose. During the implementation of the parser, several issued were encountered. One of them was the need to correctly interpret emoticons and emoji. The problem was the ambiguous coding of the emoji used in the file provided from Facebook. The emoji used was saved differently depending on how they were inserted into the message. Emoji inserted directly from the so-called "emoji keyboards" were recorded according to the officially used standard. For example, emoji was clearly represented by means of \UD83D\uDE00. Thanks to this, capturing such encoded emojis was not a problem. On the other hand, emoji created from emoticons were written in exactly the way the interlocutor wrote them. For example, emoji 😀 was mapped from the emoticon *:D*, and could be saved as *:)* or *:]*. In connection with this problem, during the implementation of the parser, a module was created to find emoticons in the text. The emoticons found this way were mapped to the corresponding emojis, which were later analyzed. Mapping emoticons on emoji allowed to unify all text and graphic representations of emotions captured by the parser. Acquiring technical parameters of text messages, on the other hand, consisted in implementing a parser, which was made available to the research group. In this way, the surveyed persons were able to independently analyze their own messages without disclosing their confidential data to anyone. The results of each person from the research group were collected and prepared for further analysis in following stages. The *third stage* of the study consisted in analyzing data collected from the research group regarding personality traits and parameters of text messages. The results of personality traits test for the research group are presented in Table 1. As it can be seen, for most women the neuroticism rate was noticeably higher than for men. Nevertheless, none of the other personality traits of Big Five model seemed to correlate with gender. In addition, due to the small amount of data, most relationships between gender and personality may have gone unnoticed.

Table 1. The results of the personality test based on Big Five model.

	Gender	Openness	Extraversion	Agreeableness	Conscientiousness	Neuroticism
1.	M	63	50	84	50	63
2.	M	64	49	78	66	35
3.	M	63	50	65	37	70
4.	M	68	82	72	75	56
5.	M	77	73	87	65	44
6.	M	62	66	78	71	51
7.	M	70	72	75	59	51
8.	M	58	74	67	79	48
9.	*M**	*66*	*72*	*86*	*51*	*55*
10.	F	69	54	60	73	90
11.	F	63	54	53	57	72
12.	F	67	52	71	77	45
13.	F	83	62	83	74	61
14.	F	69	72	71	65	73
16.	F	73	57	95	76	74
17.	*F**	*69*	*55*	*78*	*67*	*66*

It is also worth notice to max. and min. values of the results obtained for each personality trait. In the case of neuroticism, the largest possible range of values was obtained. A huge difference between the extreme values was also observed for conscientiousness. Unfortunately, in the case of the remaining personality traits, the collected data did not contain such a wide range of achievable values, which could affect the quality of predictions. At this stage, it was also decided to select two people whose data was intended for testing of predictions of personality traits. These people (marked with * in Table 1) were selected in such a way that the results of the personality test for each of them did not include an extreme value for any personality trait.

Further analysis included the average number of points scored and their median, first and third quartiles. The summary is presented in Table 2. Neuroticism turned out to be the most evenly represented personality trait. Among surveyed the people, two of them had values close to the range limits (from 20 to 100) that could be scored. One of them obtained an extremely high result - 90, the other very low - 35. Considering that the other people evenly represented all intermediate values, it could be assumed that neuroticism could be the most effectively predicted personality trait. The other three features of *Big Five* model were not very evenly distributed. This should have been considered at the stage of drawing conclusions regarding the effectiveness of predicting personality traits. Increasing the size of the research group could affect a more even distribution of people among each of the personality traits, so that the obtained predictions could be more effective.

Table 2. Summary of the personality test

	Openness	Extraversion	Agreeableness	Conscientiousness	Neuroticism
avg	67.75	62.13	75.19	65.13	59.63
sigma	6.16	10.77	10.8	11,7	14.16
min	58	49	53	37	35
25%	63	53.5	70	58.5	50.25
50%	67.5	59.5	76.5	66.5	58.5
75%	69.25	72	83.25	74.25	70.5
max	83	82	95	79	90

The next step was to analyze the collected data on the technical parameters of text messages. As mentioned earlier, among the collected data were such message features as: the number of messages written, the number of characters and emoji used, and punctuation used. In total, *4508* conversations were collected from *17* people from the research group. A fragment of the summary is presented in Table 3.

Table 3. Excerpt from the conversation summary

	Messages	Characters	Emojis	Dots	Commas
avg	472.28	12609.98	180.03	69.67	64.83
sigma	4369.8	101121.5	1543.17	554.35	518.28
min	1	0	0	0	0
25%	2	115	1	1	1
50%	9	429	4	3	3
75%	50	1938.25	20	13	14
max	216942	4570151	67131	17027	14921

After a separate analysis of data on personality traits and parameters of text messages, merger of all collected data began. Then data was prepared so that it can be used in the machine learning process. The preparation of the data consisted in combining the collected conversation parameters with the results of a personality test of selected people. The features of text messages only from persons from the research group were considered. Then the conversations in which the respondents sent less than 1000 messages were filtered out. As described in earlier, people feel much more comfortable chatting with friends, resulting in a more natural message tone. Therefore, separation of conversations containing less than 1000 messages sent by each of the interlocutors allowed to leave for analysis only those in which the participants knew each other well. The next stage of initial data preparation was their appropriate transformation into a form in which they could be processed by machine learning algorithms. To this end, the process of coding non-numeric data was performed, and the numerical scale was appropriately scaled. The method used for scaling was standardization, consisting of scaling the data so that their mean was 0 and the standard deviation 1. Such prepared data was used in the prediction of personality traits.

Table 4. Spearman's correlation between personality traits

	Neuroticism	Extraversion	Openness	Agreeableness	Conscientiousness
Neuroticism	1	−0.1818	0.1673	−0.2496	−0.1487
Extraversion	−0.1818	1	0.2333	0.1497	0.2897
Openness	0.1673	0.2333	1	0.4113	0.1658
Agreeableness	−0.2496	0.1497	0.4113	1	−0.0303
Conscientiousness	−0.1487	0.2897	0.1658	−0.0303	1

Spearman's correlation between personality traits for the test group was examined. The results obtained are shown in Table 4. The most correlated personality traits were *openness* and *agreeableness*. However, their lack of dependence could have influenced the final quality of the predictions made. No significant correlations among the remaining personality traits were noted. As part of the overall data analysis, the relationships between individual personality traits and those used in emoji text messages were established. Results are presented in Tables 5 and 6. Carrying out this stage of research enabled better understanding and revealing relationship between personality traits and the parameters of text messages. The knowledge gathered at this stage turned out to be very useful in the last, *fourth* stage of the research.

Table 5. Emoji most correlated with personality traits

Neuroticism	Extraversion	Openness	Agreeableness	Conscientiousness
0.445788	0.393166	0.417176	0.268414	0.623730
0.405648	0.386792	0.407604	0.197708	0.488593
0.388073	0.331287	0.358588	0.174558	0.457863
0.367115	0.288534	0.278788	0.144394	0.313716
0.351853	0.254617	0.269506	0.143295	0.296929
-0.178794	-0.263813	-0.399244	-0.366951	-0.330288
-0.196128	-0.276124	-0.413846	-0.411893	-0.347913
-0.222111	-0.280581	-0.474326	-0.445431	-0.354661
-0.231399	-0.323209	-0.693649	-0.446668	-0.391609
-0.236293	-0.619338	-0.722552	-0.661374	-0.42965

The last, *fourth stage* was focused on the preparation of software for the prediction of personality traits. To this end, algorithms of supervised machine learning were used. Experiment aimed at choosing the most effective way to predict personality traits. Several different methods of supervised machine learning were tested. Experiment included: scrutiny of the impact of selected regression models on the quality of prediction, along with impact of selected hyper-parameters for each model, examination the effectiveness of prediction with/without dimensional reduction and verification the prediction efficiency for: complete dataset containing all appearances of emoji and dataset containing only a number of occurrences of emoji categories. During research it was decided to solve the regression problem, in which the predicted value was the number of points obtained by the subjects in the personality test. The features based on which the predictions were made emoji used, their categories and punctuation marks used.

Table 6. Emoji categories most correlated with personality traits

Neuroticism		Extraversion		Openness	
love	0.410908	face unwell	0.398096	love	0.275520
face affection	0.354491	face smiling	0.306121	face hand	0.265716
face sleepy	0.223923	face costume	0.152051	face costume	0.261514
hands	0.214006	face hand	0.127376	face concerned	0.152478
face neutral sceptical	0.185728	face neutral sceptical	0.120464	emotion	0.149802
face glasses	-0.023447	face affection	-0.119202	monkey face	-0.134467
face costume	-0.045120	plant	-0.131246	face negative	-0.216205
family	-0.048582	hand single finger	-0.135272	face glasses	-0.222880
hand fingers closed	-0.076683	face negative	-0.202628	hand fingers closed	-0.455591
face unwell	-0.083044	face tongue	-0.259843	face tongue	-0.784293

Agreeableness		Conscientiousness	
face negative	0.133340	face tongue	0.588411
hand single finger	0.084422	hand fingers closed	0.448162
plant	0.083949	face negative	0.224062
hands	0.051306	face glasses	0.217912
person sport	0.046778	monkey face	0.216448
hand fingers closed	-0.216297	face unwell	-0.153591
face neutral sceptical	-0.244597	face costume	-0.204213
love	-0.362593	face concerned	-0.213134
face unwell	-0.460666	face hand	-0.232489
face smiling	-0.479109	love	-0.340421

There were five, separate methods prepared to predict values of all five personality traits from *Big Five* model. The use of this approach allowed for an easy verification of quality of the prediction for any personality trait and allowed for selection of different prediction methods for each of them. For the prediction of values of personality traits, a vector machine and three linearized linear models were used. The first method used was *SVM* regression (with three different kernels: *linear*, *rbf* and *sigmoid*). The second variation of linearized linear regression was the *LASSO* method. The third method was the *Elastic Net* method. Another method used was *Ridge* regression. In addition to teaching selected regression models and their tuning using hyper-parameters, it was decided to check to what extent the representation of data used to teach the model affects the quality of predictions.

3 Results

Table 7 presents a complete summary of the results obtained for the training and test data sets. This made it possible to compare the prediction efficiency of all regression models while trying to find corresponding value of personality traits: who were included in the machine learning process and those whose data did not appear in the learning process. As it can be deduced, predictions for unknown data turned out to be on average 2–3 times worse than those performed on the data used during learning process. However, this is a normal phenomenon, and the quality of all obtained prediction results turned out to be at a surprisingly high level.

As expected, it turned out that more accurate predictions were obtained for models trained using data containing the frequency of appearance each emoji separately. For the training set, among 60 pairs of compared regression model configurations - 45 of them obtained a smaller error in predicting the value of personality traits when using frequency of appearance of individual emoji. It constituted 75% of all verified pairs. However, for the test set - only in 32 cases - use of data containing frequency of appearance individual emoji led to more effective predictions. As it was mentioned, two different methods were tested, which were used to describe the emoji used by the subjects. The *first method* was to count all instances of written emoji. The *second* one boiled down to counting the occurrences of the emoji category according to the division established by the authors. Each group contained emoji with a similar emotional message. Thanks to this, when analyzing all the collected data, it was possible to consider, for example, the relationship with the selected personality trait and all the emoji used depicting a smiling face with a drawn tongue.

4 Discussion of the Prediction Results

The results of the experiment exposed that the most effective prediction of *neuroticism* is given using tuned SVM regression with the *rbf* kernel. In order to minimize prediction errors, an accurate data representation was necessary in order to minimize prediction errors - including the frequency of appearance of individual emojis. Multidimensional reduction was needless. The best prediction was noted for values: $C = 100$, $\varepsilon = 0.1$ and $\gamma = 0.0001$. The model with these parameters returned the mean prediction error equal to 8.664565. Despite the relatively average quality of the prediction, the obtained results were considered satisfactory. Unfortunately, trained data set was too small to be considered representative. Based on the analysis of the literature, and received results, it was found that *neuroticism* could be the best predicted personality trait. In fact, it turned out that *neuroticism* was the second most-effectively

Table 7. Results of RMSE for the best-tuned regression models

PCA		Lasso		Elastic net		Ridge		SVM linear		SVM rbf		SVM sigmoid	
		Y	N	Y	N	Y	N	Y	N	Y	N	Y	N
	emoji cat.	*Training set results*											
N	Y	10.18	7.33	10.18	7.58	10.24	7.45	7.78	–	4.61	3.25	8.44	8.31
	N	10.18	6.64	9.40	6.64	9.66	6.45	10.86	10.85	7.14	6.81	7.19	6.66
E	Y	6.78	5.45	6.74	5.45	6.74	5.46	5.68	–	5.54	2.39	11.25	5.19
	N	6.87	4.50	6.69	4.50	6.69	4.51	4.81	3.64	4.70	4.69	5.25	4.20
O	Y	4.48	5.78	4.47	5.76	4.48	5.76	4.41	–	3.89	4.04	4.43	4.13
	N	5.81	6.37	5.81	6.37	5.93	6.37	4.02	3.07	3.29	3.01	4.08	3.59
A	Y	6.70	9.28	6.85	9.28	6.75	9.28	9.16	–	9.28	9.28	9.28	9.28
	N	9.28	9.28	9.28	9.28	9.28	9.28	8.68	8.67	4.77	5.49	7.99	6.98
C	Y	9.44	9.44	9.44	9.44	9.44	9.44	10.19	–	10.29	10.29	10.29	10.28
	N	9.44	9.44	9.44	8.64	9.44	8.20	6.17	4.35	3.61	3.17	6.60	4.35
Test set results													
N	Y	12.94	9.85	12.94	12.92	12.23	12.70	1014.14	–	16.34	14.91	25.94	11.76
	N	40.14	18.47	148.53	18.47	103.78	74.86	30.57	30.84	13.10	8.67	63.42	21.44
E	Y	20.22	11.58	97.40	11.58	97.40	11.34	1033.89	–	11.67	6.97	15.98	97.45
	N	263.07	53.52	248.16	53.52	248.16	45.26	297.73	236.40	10.81	11.50	123.88	19.41
O	Y	345.64	5.20	205.87	4.45	271.41	4.45	13.07	–	3.82	7.02	75.71	16.40
	N	57.95	1.78	57.95	1.78	41.28	1.78	117.65	110.85	2.35	3.59	28.09	23.46
A	Y	131.66	7.06	60.36	7.06	93.81	7.06	13.88	–	7.03	7.03	7.04	7.04
	N	7.06	7.06	7.06	7.06	7.06	7.06	84.12	74.22	11.60	9.03	86.79	110.97
C	Y	9.63	9.63	9.63	9.63	9.63	9.63	17.69	–	12.52	12.52	12.52	12.52
	N	9.63	9.63	9.63	90.81	9.63	40.36	296.85	54.61	9.32	9.77	56.95	24.50

predicted after *conscientiousness*. It is presumed that this was due to the insufficient amount of data analyzed and the fact that most subjects with high neuroticism were women.

As expected, *extraversion* turned out to be an effectively predicted personality trait. Among the tested methods, SVM regression using the *rbf* kernel proved to be the most accurate again. Giving up on multidimensional reduction at the expense of learning speed of the model – it allowed for a smaller prediction error. It was on average 4.523 points smaller than that obtained with the help of reduced data. In contrast to *neuroticism* - *extraversion* predictions were most effectively carried out using the trained model based on data containing information on the frequency of emoji categories. The best results were obtained for: $C = 1000$, $\varepsilon = 0.1$ and $\gamma = 0.01$. This trained model had the second highest prediction efficiency among five personality traits. An average prediction error score of 6.972 was obtained for the test set data. Given the aforementioned factors preventing effective prediction of personality traits, the obtained prediction error was considered acceptable. As for *openness*, it was not possible to clearly determine positively correlated emoji. Literature studies indicated on lack of connection between high *openness* and emojis. As a result of the experiment, turned out that the number of *openness* points obtained in the personality test was the most effectively predicted value among all five traits. Unlike other personality traits - the

Table 8. Best configurations for the most effective predictions of personality traits.

	Neuroticism	Extraversion	Openness	Agreeableness	Conscientiousness
Regression	SVM				
Kernel	rbf				
C	100	1000	10	0.001	100
ε	0.1	0.1	0.1	0.1	0.1
γ	0.0001	0.01	0.01	0.0001	0.01
Data	Emoji	Emoji categories	Emoji	Emoji categories	Emoji
PCA	No	No	Yes	Yes	Yes

value of *openness* was most effectively predicted using each of the three methods: *LASSO*, *Ridge* and *Elastic Net*. Obtaining the same good, identical results by three different regression methods prompted reflection on the effectiveness of each of them. As a result, the above three models were ultimately replaced by SVM regression. This model provided with effective prediction with an average error of just 2.349 points. To obtain such good results, an accurate data representation was used, considering the frequency of appearance of individual emojis. In addition, there was no a reduction in dimensionality, which further improved the quality of the prediction by just over one point. The best results were obtained for: C = 10, ε = 0.1 and γ = 0.01. During general data analysis, *no significant* correlation between people with high *agreeableness* rates and emoji was observed. In addition, the obtained results of analysis mostly turned out to be contrary to the assumptions generated during the work on the project. Despite this, the quality of the prediction of agreeableness turned out to be at an acceptable level. SVM regression with *rbf* kernel was used to train the most effective model (C = 0.001, ε = 0.1 and γ = 0.0001). The most effective data representation turned out to be the one in which the frequency of emoji categories was considered. The use of dimensional reduction for the data representation did not affect the quality of the predictions made. The model allowed for prediction of agreeableness with an accuracy of 7.030 points. Even though the results obtained were not very accurate, but given the potential obstacles encountered during data analysis - were considered acceptable. Contrary to expectations, *conscientiousness* turned out to be the least effectively predicted value among all personality traits. The lowest average error of predictions was 9.315 points. Considering the range of values possible to obtain, which ranged from 20 to 100 - the obtained results of the prediction were considered lower than satisfactory. Yet again, SVM regression using the *rbf* kernel (C = 100, ε = 0.1 and γ = 0.01) proved to be the best model. The best representation of the data turned out to contain accurate information on all emoji graphics used, while the use of dimensional reduction only slightly improved the results.

The final step was to check the actual effectiveness of the trained regression models. To this end, for each of the five personality traits, the most effective prediction method was chosen (Table 8). Then again, using the training set, each model was properly trained and predictions of the value of personality traits were made for the data from the test set. The obtained prediction results were in line with the conclusions described earlier. The value of *neuroticism* for person #17 was most often predicted with an accuracy of less than 3 or just over 11 points, so that the average error remained at 7 points. Whereas for the person #9 the prediction error was in the range from −18.89 to −0.75 points. Several identical values, among the obtained prediction results, were noted. It is presumed that this was due to a poorly learned model due to the small amount of data collected for its training. The obtained results of *extraversion* prediction were at a satisfactory level. The smallest obtained prediction error was 0.25, while the largest was 15.27 points. Both extreme results were obtained for the data from conversation of the person #9. In addition, no significantly better quality of extraversion predictions for any of the checked persons were noted. According to the expectations - *openness* was the most effectively predicted value among all five personality traits. Most predictions did not exceed the error of 2 points. The worst result of the prediction differed from the actual state by only 5.7 points. All *agreeableness* predictions carried out as part of the tests provided identical results. A score of 75.10 was obtained for any prediction. which raised doubts about the effectiveness of the trained model. It is assumed that much more data should be used to prepare a working model. Unfortunately, due to the inability to obtain them - it was impossible to train a new, effective model. As expected, the predicted value of *conscientiousness* was only average. Nevertheless, as noted - for person #16, the obtained prediction results turned out to be accurate and much better than those obtained for person #9. The difference in average prediction errors for both people was as much as 11.16 points. Presumably, the actual value of conscientiousness, as well as any other personality traits from *Big Five* model, may depend on many different, previously unknown factors. Therefore, the effectiveness of prediction can be different for different people and one should be aware of this when conducting further thematically similar research.

5 Conclusions

The presented results allowed to state that the prediction of personality traits based on the technical parameters of text messages is possible. The results obtained were not considered accurate, but nevertheless given the modest set of data used, they were finally found promising. As it turned out - even an unrepresentative data set allowed to train regression models whose average prediction error, in the worst cases, did not exceed 20 points. Considering the range of achievable values - from 20 to 100 points - from the mathematical point of view the obtained results of the prediction could not be considered precise. However, human personality does not depend only on a few numerical values. In fact, this is a much more complex issue, and the personality test result - in the form of five numbers - only partially reflects the facts. In addition, the exact numerical values of the five personality traits from the Big Five model are not necessary to effectively predict a person's personality. As it turns out, only the approximate values of the personality test are enough for quite effective personality assessment. Therefore, the recognized prediction results were considered satisfactory.

From a technical point of view - the results of experience and tests have provided a lot of valuable information. First, it was learned that the prediction of personality traits based on text messages can be more effective than expected. As already mentioned in first chapter - emoticons along with emoji supplemented the content of text messages with an emotional message that dominates in non-verbal communication and is unique to each person. It should therefore be assumed that such good quality predictions were made due to the large number of information-rich emoji contained in the analyzed messages. Nevertheless, one should know for more detailed research into the possibility of predicting personality traits, much more data should be collected. Carried out research confirmed the possibility of predicting personality traits using text messages. However, for a more thorough verification of the problem, a much more extensive dataset would be needed for many different people across a broader age range. An additional aspect that is worth paying attention to, is the accuracy of the research. Despite their wide range including many factors such as data representation, regression models, hyper parameter values - the accuracy of the tests could also be improved. As part of future work, it would be worth checking: more regression models, more configurations of hyper-parameter, and other data representations. In addition, it would be worth testing methods based on an artificial neural network, which nowadays often give high-accuracy results. To sum up - prediction of personality traits based on collected technical parameters of text messages is possible. However, more research would be needed to improve its effectiveness.

References

1. Aldunate, N., González-Ibáñez, R.: An integrated review of emoticons in computer-mediated communication. Front. Psychol. **7**, 2061 (2017)
2. Burke, K.: 107 Texting Statistics That Answer All Your Questions (2016). https://www.textrequest.com/blog/texting-statistics-answer-questions/. Accessed 20 Jan 2020
3. Derks, D., Bos, A.E., Von Grumbkow, J.: Emoticons and social interaction on the Internet: the importance of social context. Comput. Hum. Behav. **23**(1), 842–849 (2007)
4. Giannoulis, E., Wilde, L.R.: Emoticons, Kaomoji, and Emoji: The Transformation of Communication in the Digital Age. Routledge (2019)
5. Golbeck, J., Robles, C., Edmondson, M., Turner, K.: Predicting personality from Twitter. In 2011 IEEE Third International Conference on Privacy, Security, Risk and Trust and 2011 IEEE Third International Conference on Social Computing, pp. 149–156. IEEE, October 2011
6. Holtgraves, T.: Text messaging, personality, and the social context. J. Res. Pers. **45**(1), 92–99 (2011)
7. Kaye, L.K., Malone, S.A., Wall, H.J.: Emojis: insights, affordances, and possibilities for psychological science. Trends Cognit. Sci. **21**(2), 66–68 (2017)
8. Li, W., Chen, Y., Hu, T., Luo, J.: Mining the relationship between emoji usage patterns and personality. In: Twelfth International AAAI Conference on Web and Social Media, June 2018
9. Lund, M.G.: Punctuation and personality: why 50 pages of rules? Clearing House: J. Educ. Strat. Issues Ideas **27**(9), 541–543 (1953)
10. Paul, K.J.: Text messaging and personality (2011)
11. Petersen, L.: Your Punctuation Personality Type by Leah Petersen. http://bryanthomasschmidt.net/guest-post-your-punctuation-personality-type-by-leah-petersen/. Accessed 20 Jan 2020
12. Rothmann, S., Coetzer, E.P.: The big five personality dimensions and job performance. SA J. Ind. Psychol. **29**(1), 68–74 (2003)

13. Różańska, A.: Kultura korzystania ze znaków graficznych w komunikacji medialnej. Funkcja emotikonów z perspektywy społeczno-psychologicznej. (The culture of using graphic signs in media communication. The function of emoticons from a socio-psychological perspective). Zeszyty Naukowe Państwowej Wyższej Szkoły Zawodowej im. Witelona w Legnicy, vol. 2, no. 27, pp. 195–202 (2018)
14. Tankovska, H.: Perception of people who never use emojis in Sweden 2018, 16 May 2018. https://www.statista.com/statistics/860085/perception-of-people-who-never-use-emojis-in-sweden/. Accessed 20 Jan 2020
15. Xu, L., Yi, C., Xu, Y.: Emotional expression online: the impact of task, relationship and personality perception on emoticon usage in instant messenger. In: PACIS 2007 Proceedings, p. 79 (2007)
16. Li, X., Chan, K.W., Kim, S.: Service with emoticons: how customers interpret employee use of emoticons in online service encounters. J. Consum. Res. **45**(5), 973–987 (2019)
17. International Personality Item Pool Polish Version IPIP-NEO-FFI-100. http://www.ipip.uksw.edu.pl/test.php?id=36. Accessed 20 Jan 2020
18. Walther, J.B., D'addario, K.P.: The impacts of emoticons on message interpretation in computer-mediated communication. Soc. Sci. Comput. Rev. **19**(3), 324–347 (2001)
19. De Raad, B.: The Big Five Personality Factors: The Psycholexical Approach to Personality. Hogrefe & Huber Publishers (2000)

Computer Aided Urban Landscape Design Process

Tomasz Zamojski(✉)

Wrocław University of Science and Technology, Wybrzeże Wyspiańskiego 27, 50-370 Wrocław, Poland
tomasz.zamojski@pwr.edu.pl

Abstract. In the article, special interests are focused on evaluation of harmony, composition and order of vertical buildings and their urban environment. Contemporary, dynamic development of XR technologies (VR/AR/MR) allows to create digital applications targeted on mobile devices and define evaluation processes based on quantitative values. Urban planners and architects need measures to facilitate assessment of specific solution of a given district and urban landscapes but tools allowing quantitative evaluation are still limited.

Following some comments on the use of XR technology (Virtual Reality and Augmented Reality) in urban landscape design, a problem was formulated and measures of landscape assessment were proposed enabling quantitative assessment of its quality based on the luminance of the XR scene. The procedure of designing changes in the existing buildings was analyzed and the concept of a computer program using XR technologies for numerical evaluation of the modified landscape was presented.

Keywords: Virtual Reality · Augmented Reality · Mixed reality · Technology · Architecture · Urban · Composition · Planning · Proportions · Harmony · Balance · 3denviroment · Landscape · Fitness · Design · Application · Quantitative · Evaluation · Concept · Smart phone · Mobile

1 Introduction

The atmosphere of the urban space, especially in highly urbanized areas, is a consequence of many factors affecting perception and feeling of a given place by man. Dense and high urban development limit possibilities of assessing results of architects and urban planners' work (a frog's perspective and a limited number of observation points), and yet the perception of the landscape in which we live has a significant influence on our feeling of the surrounding world [7, 10, 11].

'For a long time, civilization has been trying not only to define beauty, but also to assign a numerical measure to it, in order to be able to answer the question concerning its core (see the golden ratio). There is no definite answer to this question because it refers to issues of a qualitative character and is connected with subjective feelings of an observer. It is even more difficult to solve the problem of choosing a 'nicer' solution when we want to use the computer help and then we also need to move our considerations of 'beauty'

W. Zamojski et al. (Eds.): DepCoS-RELCOMEX 2020, AISC 1173, pp. 686–699, 2020.
https://doi.org/10.1007/978-3-030-48256-5_67

into the sphere of quantitative considerations, so that the computer can choose a solution closer to the value of the introduced criterion of 'beauty' [20].

Architects of modern urban agglomerations often solve problems of land development so that the introduced development do not violate the existing traditional harmony of the urban landscape. The new development is the body of a building (shape, volume, height), its location in relation to the existing development and visual interference with the surroundings. Visual changes in the landscape constitute primarily a change in the horizon created by a city's overall structure, i.e. the City Skyline, a violation of the existing lighting distribution (shading of the neighbouring lower development during the day and the introduction of intense luminance at night – LEDs!) as well as a change in the colour of the development landscape resulting from the application of new building coating technologies (texture, luminescence and LEDs) [21].

For many years architects and urban planners have been looking for appropriate criteria to facilitate the adaptation of designed objects to the existing development, landscape and social conditions, e.g. Kazimierz Wejchert [18] and hence it seems purposeful to adapt their results to computer-processed (3D) models.

Computer tools and systems supporting the work of urban planners, architects and constructors focus primarily on the automation of activities connected with the preparation of design documentation, visualization of architectural and urban solutions, optimization of the organization of the design and implementation process, etc. However, despite the observed flourishing of XR[1] technology, the number of computer tools supporting the process of composing beauty and harmony of urban and architectural solutions is still limited.

This study is an attempt at formulating the premises, dependencies, conditions and regulations to create a computer tool to aid the process of designing the urban environment with high-rise buildings with the basic assumption that the introduced object should not disturb the harmony of the already existing landscape.

After comments concerning the application of XR (Virtual Reality and Augmented Reality) technology in urban landscape design (Sect. 2), a research problem was formulated and pattern metrics of landscape were proposed, hereinafter referred to as fitness, enabling quantitative assessment of the landscape quality (Sect. 3). In terms of applying the proposed pattern metrics, the procedure for designing modifications of the existing development was analyzed (Sect. 4) and the concept of a computer program using XR technologies for numerical evaluation of the modified landscape was presented (Sect. 5).

[1] In the VR Focus article by Editor-in-Chief Kevin Joyce which states: 'A new emerging trend is to define any content using these emerging technologies as 'XR'. However, XR is also used to define 'cross reality', which is any hardware that combines aspects of AR, MR and VR; such as Google Tango' [19].

2 State of the Art

Technologies which are based on virtual reality image processing (Virtual Reality) are gradually implemented into the architect's workshop. At present, they are widely used in interior architecture, where, on the one hand, they make it easier to visualize the designer's proposals, and on the other, they enable a dialogue, almost on-line, between an architect and a client. It is worth emphasizing that a client-architect dialogue is already based on Augmented Reality technologies. The application of these two technologies (VR and AR) in the design of urban architecture solutions encounters many difficulties resulting primarily from the complexity and uniqueness of landscapes as well as the complexity of tasks set for a designer, more precisely a design studio. Virtual Reality and Augmented Reality technologies are expected to support and greatly accelerate decision-making processes connected with the location of architectural objects and the selection of their form in the context of the existing urban landscape, in particular the one with high-rise buildings [1, 2, 5, 6, 8, 20, 21].

Virtual Reality of Landscape images are relatively easy to create, e.g. on the basis of digital photographs and it is relatively easy to process and analyse them for the variability of selected parameters, e.g. sizes, coloration or exposure. More difficult issues are connected with the analysis of the image content, for example, distinguishing fixed landscape elements from variables (plantings and their seasonal variation or also over time). An architect often has to determine the building coating (material, texture, etc.) or to assess the quality of soil on which a building is to be constructed – how to implement it on the basis of X Reality images? The indicated issues are connected with the selection of pattern metrics which unambiguously and numerically describe a given phenomenon and proposing methods for determining the values of these metrics on the basis of the analysis of optical parameters which can be read from a digital image. Therefore, in many cases complex physical and mathematical models of the analysed phenomena visible in digital images are built and sophisticated mathematical methods are applied.

In [4] the research directions connected with pattern metrics and methods of processing Landscapes of ecology were discussed and in [3] problems of processing geographical landscapes were dealt with. Architectural Landscapes are equally complex, they lack unequivocal pattern metrics and their processing includes transforming quality criteria (beauty, harmony) into digital quantities. Further considerations put forward pattern metrics which are useful for assessing modified architectural landscapes, a mathematical model based on image luminescence and assumptions of a software system supporting the architect's work.

3 Problem and Model

The three-dimensional landscape (3D Model) illustrated in two dimensions on the computer screen (2D Model) is covered by elements of different luminance which in combination with the edges of the development create the illusion of perspective and depth of the scene. The luminance intensity of an element depends on lighting, a texture of its façade as well as on the colour – the colour spectrum of the element is modified by solar radiation directly illuminating it and reflected from other landscape elements.

The assessment of the aesthetics of the incorporated object depends on the geographical location of the scene under consideration, existing landscape and climate conditions (e.g. air cleanness and humidity), a type of development, sunlight as well as the time of day and year.

We consider an urban situation (Fig. 1), in which in a 'free' area (a plot of land limited by dashed lines) it is planned to build an object of such a shape and luminance (coatings, texture, colouration and location in relation to the sun) in order to modify the current landscape. A degree of modification of the view will be assessed on the basis of the proposed pattern metric called the coefficient of change in the aesthetics of the architectural solution and marked with symbol ε.

By definition, coefficient ε describes fitness changes of the landscape between two views p and r determined at moment t determined on a 24-hour scale and on an annual scale (sun position) as

$$\varepsilon_{p,r}(t) = \frac{F_r(t) - F_p(t)}{F_p(t)} \tag{1}$$

where
$F_p(t)$ – global fitness value (of luminescence) p^{th} – landscape,
$F_r(t)$ – global fitness value (of luminescence) r^{th} – landscape.

The energetic value of luminance, hereinafter referred to as fitness, provided by kth – element of the scene (or through a distinguished cluster with homogeneous luminance) at moment t is estimated as the product of its area and the sum of the components of luminance, i.e.

$$f_k(t) \approx c_k(t)\left[l_B(t) \cup l_{CF}(t) \cup l_S(t)\right] = c_k(t) l_\Sigma(t) \tag{2}$$

where
$c_k(t)$ – area of kth – cluster,
$l_B(t)$ – luminance of the background,
$l_{CF}(t)$ – chrominance of colour and texture of the cluster,
$l_S(t)$ – luminance of shadow,
$\cup$ - summation luminance operator

The instantaneous fitness value for the scene in question is

$$f_{scene}(t) \cong \sum\nolimits_{k=1}^{K} f_k(t) \tag{3}$$

and the average energetic value of luminance of the scene in the time interval [0, T] can be estimated as

$$F_{scene,T} = \frac{1}{T}\int_0^T f_{scene}(t)dt \tag{4}$$

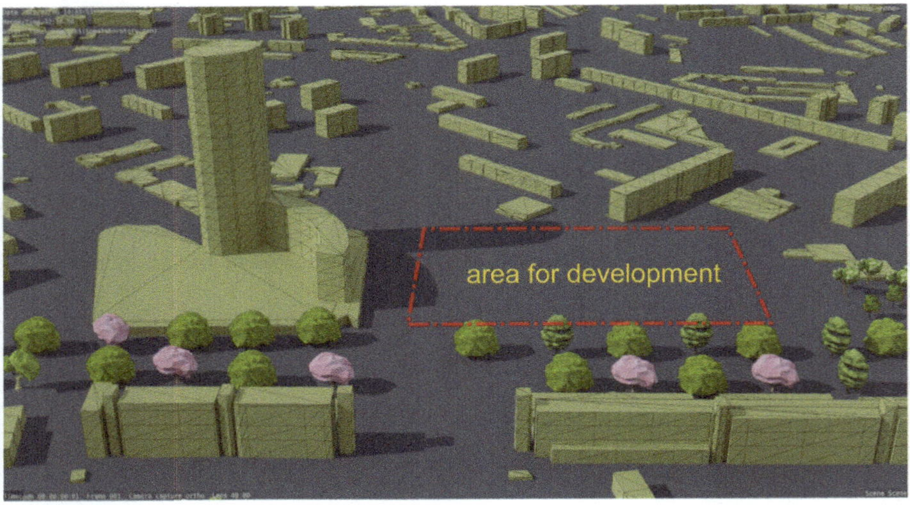

Fig. 1. 3D landscape with a marked plot for the development – geographical location: Wroclaw, Poland, latitude: 57.08; Longitude: 17.01 [4, 9, 13, 16, 17]

Dependencies (3) and (4), after appropriate modifications, enable us to evaluate the aesthetics of the architectural solution and carry out research on changes in the existing landscape according to the procedure described below (algorithm).

Normalized values of ε(p,r) will allow a designer to choose a solution with parameters which are closest to the conditions set before him, e.g. if an object is to dominate the surroundings, then value ε(p,r) should significantly exceed 1, and when it is to be an object which is harmonized with the existing surroundings, then ε will assume values near one.

Further considerations assume that the assessment of the aesthetics of the architectural solution will be based on the assessment of luminescence of the scene. The objects introduced into the landscape change the lighting of the scene by, among other things, introducing additional shadows with a length depending on the height of a development element and the location of the sun in the sky (local time).

The shadow length of building B with height h_B at moment t is determined on the basis of the following relationship

$$d_S(t) = \frac{h_B}{\tan \varphi(t)} \tag{5}$$

where

$d_S(t)$ – length of shadow at moment t,
h_B – height of building B,
$\varphi(t)$ – angle of incidence of sunlight at moment t,

whereas a direction of the shadow as $180 + \delta\ (t)$, where $\delta\ (t)$ corresponds to the local position of the sun in the sky, i.e. the astronomical angle of the movement of the sun in the sky for moment t (compare Fig. 2).

4 The Procedure of Architectural Modification of the Urban Landscape

The computer-aided architectural landscape modification process includes a number of stages of a traditional character such as the preparation of design assumptions, construction conditions, architectural conditions, etc. as well as activities implemented by means of computer support or even by the computer with the support of a human being (architect). The computer aids the preparation of Virtual Reality of Landscape images for the planned area of development, including planned observation sites. On the basis of the architect's instructions and inventions, the computer will generate Augmented Reality scenes for which it will calculate fitness values connected with incorporating the proposed bodies of development into the existing landscape, which may constitute the basis for the architect's choice of the appropriate (or optimal) solution.

Step 1 – Design assumptions

- plot intended for development (landscape, location and borders)
- assumptions regarding the development planned
  - location of an 'embedded' object in the current landscape,
  - boundary requirements of the object's size – minimum and maximum 'sizes',
  - urban conditions, e.g. a built-in object should not violate the current harmony of the scene or should dominate it by its size of the body, height, colours, luminance, etc.,
  - time of exposure of the urban solution, e.g. assumed urban conditions should be met in an all-year cycle or only in the summer period (resort on the beach), winter period (skiing) or for a certain period[2].

Step 2 – Urban Landscape Conditions

- defining evaluation criteria for fitness of Landscape
- determining a set of location points of a landscape observer (scenes) – the basis for evaluation of the architectural solution proposed

[2] Technical conditions to be met by buildings and their location. Section 3. Buildings and rooms. Chapter 2. Lighting and sunlight [https://www.muratorplus.pl/biznes/prawo/nowe-warunki-techniczne-jakim-powinny-odpowiadac-budynki-i-ich-usytuowanie-aa-nAPR-CdXw-wB4V.html].

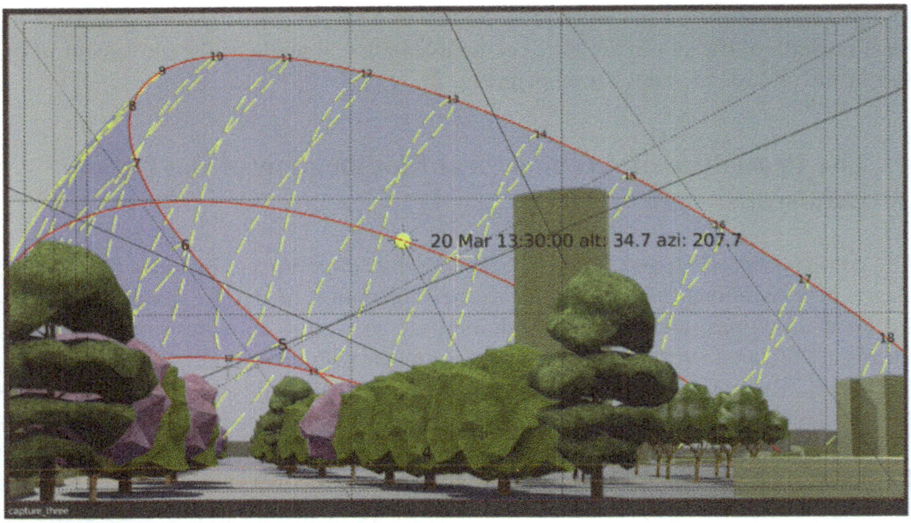

Fig. 2. Solar ruler on the spring equinox (March 20, 2020) at 1:30 pm for the analyzed development location [4, 9, 13, 16, 17]

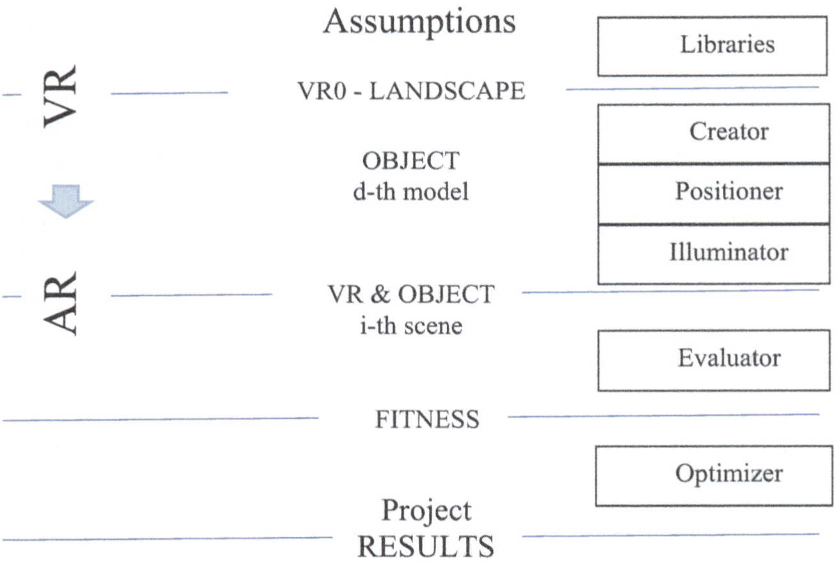

Fig. 3. Landscape design procedure

Step 3 – Primary Virtual Reality of Landscape (VR0) with indicating the boundaries of the expected development location

- generating $VR0^m(t)$ scenes for the particular points of an observer's location and observation times
- determining (calculating) the original value of the aesthetics of the architectural solution, defined e.g. by the fitness scene $(\varepsilon_0^m \stackrel{\text{def}}{=} f_{scene}^m(t))$

Step 4 – 3D model of the in-built object

- sketch definition of the form and parameters of an object (body, dimensions, types of surfaces, luminescence, etc.)
- defining a set of considered models $d = \{1, 2, \ldots\}$

Step 5 – Augmented Reality of Landscape

- building in d^{th} –object model in VR0,
- calibration (scaling),
- interaction of the object's lighting with the existing scene lighting – shadows, shades, reflections, interferences, etc.

Step 6 – Fitness of Landscape

- determination of the fitness value for Augmented Reality of Landscape of scene $AR^{d,m}$ for m^{th} – location of an observer $XR^{d,m}$ and set time conditions for the observation of Landscape with a built-in object model
- checking compliance with the project's landscape requirements (1)

Step 7 – Project optimization

- generating more objects (Step 4) and their building-in in Landscape (Step 5)
- determination of the fitness scene (Step 6)
- selection of the optimal solution – a form and parameters (dimensions, surface types), location.

Selection of the optimal solution – a form and parameters (dimensions, surface types), location.

5 Program Urban Landscape Design (Comments)

Below there are some comments on the future implementation of the computer system (application for a laptop/smart phone/tablet) supporting the work of an architect modifying the existing landscape by introducing appropriate development.

5.1 Libraries

Libraries collecting knowledge about used (imported, generated or designed) components of the scene;

- solids and their components (they can be treated as Lego blocks) described by physical parameters and visualizations,
- catalogues of coatings (textures, colouration, luminescence) applied for covering surface of the existing or planned buildings,
- images/VR scenes.

5.2 Tools

VR Creator – a set of tools enabling generation of the original landscape scene (VR0) into which a designed object will be built. In fact, VR0 will be a set of photographs taken from selected observation points for determined lighting and time conditions. The quality of images should enable determination of the baseline of building development along with their physical parameters (size, distance), City Skyline and luminance.

Object Creator – a set of tools enabling the creation (sketching) of graphic 3D models of objects then incorporated into Virtual Reality landscapes.

A designer, by means of a graphic menu, builds, using three-dimensional blocks (cubes, cuboids, cones, cylinders, spheres, etc.), the body of a designed building by choosing its shape, dimensions and location in the landscape (VR scene). The next step is to choose colours and textures of the building's surface to achieve the desired luminance.

The work of OBJECT CREATOR to a large extent reflects a preliminary design phase of an architect, during which sketches of future solutions are created. Significant difficulties are connected with the development of appropriate menus and interfaces which allow the selection of 'Lego blocks' (appropriate library), which are easy to manipulate and calibrate (scale). Decisions and assessments are subjective here and belong to a designer. The computer 'connects' (integrates) blocks, stains them and gives them the desired luminance, and following the designer's acceptance, the computer generates a preliminary 3D model of an object. The results obtained (the body of an object, its colour and luminance and 3D model) are stored in the Library.

Augmented Reality Creator – a set of tools enabling fixing a generated OBJECT into the selected VR0 scene. In order to do it, OBJECT should be located on the scene while maintaining appropriate distances and proportions between the existing elements of the landscape and the introduced body – the POSITIONER tool. After locating OBJECT, it is necessary to identify changes in lighting of the scene (shadows, luminances and chrominances) with a possible consideration of design requirements for dates (calendar), daytimes or globe position (tool: LUMINATOR).

Evaluator – a set of tools enabling determination of the fitness value for Landscape with a built-in (designed) object; the choice of the adopted solution which meets the assumed design requirements is based on a numerical assessment of the difference between the fitness value for VR0 and the particular implementations of Augmented Reality scenes.

Table 1. Fitness calculation of a scene

```
d = [1:D]   % matrix of generated objects
m = [1:M]   % location matrix of observers
t = [1:T]   % observation moments matrix

for d = 1: D   % next versions of built-in objects
  for m = 1: M   % observer's locations
  for t = 1: T   % observation moments
```

1.1.1 segmentation of scene – determining clusters K understood as disjoint sets of points with the same luminance

1.1.2 $f_t^{D,m} = \sum_{k=1}^{K} c_k^{D,m,t}\, l_k^{D,m,t}$ % instantaneous fitness scene values

1.1.2.1 $F_{t_2-t_1}^{D,m} = \boldsymbol{E}_t\left(f_t^{D,m}\right) \cong \frac{1}{t_2-t_1}\left(\sum_t \left(\sum_{k=1}^{K} c_k^{D,m,t}\, l_k^{D,m,t}\right)\right)$ % average fitness value over a period of time $[t_1, t_2]$

end

1.1.3 $F_m^{D,t_2-t_1} = \boldsymbol{E}_m\left(F_{t_2-t_1}^{D,m}\right) = \frac{1}{m}\left(\sum_m F_{t_2-t_1}^{D,m}\right)$ % average fitness value for location m

or

1.1.4 $F_{m,w}^{D,t_2-t_1} = \boldsymbol{E}_{m,w}\left(F_{t_2-t_1}^{D,m}\right) = \frac{1}{m}\left(\sum_m w_m F_{t_2-t_1}^{D,m}\right)$ % weighted average – w_m value of weight assigned to a given observer's location

end

Depending on the formulated design requirements, global fitness values are determined, e.g. an instantaneous average for a given observer's location or a weighted average of fitness determined for different locations of an observer moving along the assumed observation path (compare Table 1). The choice of the right solution (meeting adopted design assumptions) can be based on a mean square error describing changes between luminescence of the examined landscapes, e.g. the original VR0 landscape and a modified development body introduced $AR^{d,m}$.

Optimizer – a set of tools enabling optimization of the adopted design solution [12].

6 Example

To verify the proposed method of assessing changes in the urban landscape under the influence of changes in its development, the comparison of how modifications of the development affect relative changes in the fitness value of the urban landscape was performed.

For the established observation point, a visualization of the existing urban development dominated by a high residential skyscraper (hereinafter referred to as Sky Tower), constructed among a relatively low development (Fig. 4) was made. The existing solution was marked with letter A. In the area indicated in Fig. 1, development of the marked plot with a complex of four high-rise buildings (located between the green belt and the existing Sky Tower building) was developed. Two variants of solving this problem were considered, i.e. concept B (Fig. 5) - a complex of four equal height skyscrapers with parameters similar to the existing Sky Tower high building. This layout clearly dominates over the existing urban development. However, concept C (Fig. 6) with a newly designed complex of high-rise buildings of various heights fits harmoniously into the described fragment of the urban landscape, using for this purpose the principles of perspective, the golden ratio and a harmony of triangles [18].

Fig. 4. Urban Landscape A. Visualization of the existing state; the observation was made from direction NE towards SW at 1:30 pm on the spring equinox 20.03.2020 [4, 9, 13, 16, 17]

For both proposed solutions (A → B; A → C) an assessment of luminescence changes for a given moment and location of the observer was carried out. Changes in luminance values were assessed on the basis of a mean square difference in luminance in the particular image elements. And thus, it turned out that the modification of A → B landscape causes a relative change in the fitness value greater than in the case of A → C (compare Fig. 7), which confirms the aesthetic feelings of man mentioned above. Figure 7 also indicates changes in the landscape fitness in the case of transformations B → C and C → B.

Fig. 5. Visualization of variant B. The concept of a newly designed complex of skyscrapers with a height similar to the existing skyscraper Sky Tower located on the neighboring plot [4, 9, 13, 16, 17].

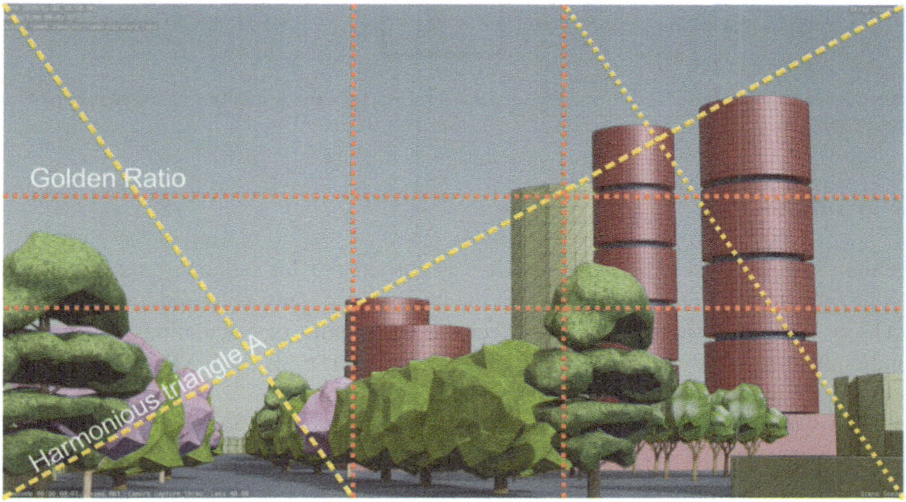

Fig. 6. Visualization of variant C – the scene after modifying the height of the newly designed complex of skyscrapers using the principles of composition [4, 9, 13, 16, 17]

Experiment. In Unity3d [16], a 3D scene was built on the basis of maps and location data from Maps SDK [13]. For the purposes of the experiment, the geographical location of Wroclaw (Poland) was selected, referring to the close context of Sky Tower skyscraper. The 3D model (.obj file) was then exported to Blender3d software [4]. The generated scene of the existing fragment of the city is the original Virtual Reality scene (VR0), in which both appropriately parameterized variants of the development modification were built-in using the add-on sverchok3d [15]. Two scenes were obtained in 3D (Augmented Reality) environment. For each of the scenes, using the add-on VI-Suite [17] to Blender [4], lighting/luminance conditions were calculated (for simplification in a linear gray scale). The obtained luminance maps of scenes enabled simulation calculations in Matlab [14], the results of which were presented in Fig. 7.

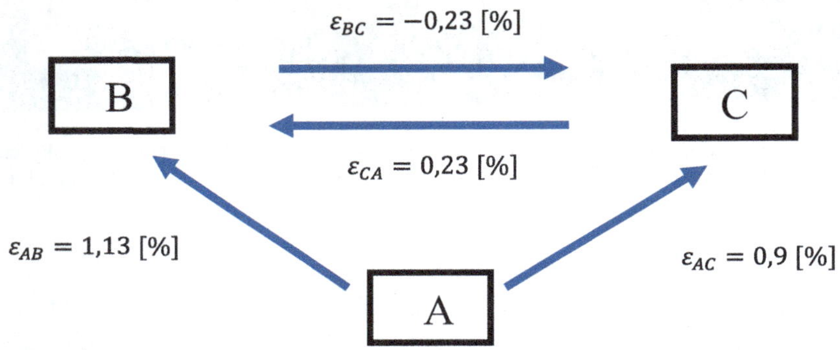

Fig. 7. Fitness changes for two variants of land development modifications of the plot in front of Sky Tower (Fig. 4)

7 Conclusions

In the example presented, the research was carried out for monochromatic luminescence (grayscale), but we can hope that the transition to multi-coloured luminance, e.g. RGB or CMYK systems will only affect the complexity of calculations and will not undermine the proposed idea of applying computer tools and XR (VR and AR) technology in aiding the architect's decision-making process.

The basic difficulties are connected with the implementation of computer applications (smart phone, mobile) indicated in Fig. 4 and mentioned in point 5 (CREATOR, POSITIONER or LUMINETOR tools), whereas for the architect, the two most interesting issues are connected with the computer (numerical) assessment of the landscape aesthetics and the application of the golden ratio idea in the implementation of VR and AR images (scaling of objects, their location deep in the scene and luminescence variability).

References

1. 9 Augmented reality technologies for architecture and construction. ArchDaily. https://www.archdaily.com/914501/9-augmented-reality-technologies-for-architecture-and-construction
2. Abboud, R.: Architecture in an age of augmented reality: opportunities and obstacles for mobile ar in design. Construction, and Post-Completion. https://www.academia.edu/14677741/Architecture_in_an_Age_of_Augmented_Reality_Opportunities_and_Obstacles_for_Mobile_AR_in_Design_Construction_and_Post-Completion
3. Baker, W.L., Cai, Y.: The r.le programs for multiscale analysis of landscape structure using the GRASS geographical information system. Landscape Ecol. **7**, 291–302 (1992). https://doi.org/10.1007/BF00131258
4. Blender Foundation. Blender 2.80 [blender.org]
5. Costanza, J.K., Riitters, K., Vogt, P., et al.: Describing and analyzing landscape patterns: where are we now, and where are we going? Landscape Ecol. **34**, 2049–2055 (2019). https://doi.org/10.1007/s10980-019-00889-6
6. Deng, J., Desjardins, M.R., Delmelle, E.M.: An interactive platform for the analysis of landscape patterns: a cloud-based parallel approach. Annals of GIS. **25**, 99–111 (2019). https://doi.org/10.1080/19475683.2019.1615550
7. Eco, U.: Historia Piękna. REBIS Publishing House Ltd., Poznań (2005)
8. Filipowiak, J.: Augmented Reality (AR) In Architecture. https://virtualist.app/augmented-reality-ar-in-architecture/
9. Google Poly: [poly.google.com]
10. Greiner, P., Specjalisty, A.B.C.: Technologia VR. Architektura. Murator. Warsztat Architekta Programy, technologie. 1/19. Wydawnictwo TIME S.A. Warszawa (2019)
11. Konopacki, J.: Rozszerzona rzeczywistość – jako narzędzie wspomagające procesy analityczno-decyzyjne w architekturze i planowaniu przestrzennym. przestrzeń i FORMA, **21**, 89–108 (2014)
12. Pitzer, E., Affenzeller, M.: A comprehensive survey on fitness landscape analysis. Josef Ressel Center "Heureka!", School of Informatics, Communications and Media, Upper Austria University of Applied Sciences (2011)
13. Mapbox SDK: [mapbox.com]
14. Matlab – : [Matlab R2019b]
15. Sverchok3d for Blender. http://nikitron.cc.ua/sverchok_en.html
16. Unity 3d: [Unity3d.com]
17. VI-Suite for Blender3d. [http://arts.brighton.ac.uk/projects/vi-suite/downloads]
18. Wejchert, K.: Elementy kompozycji urbanistycznej. Arkady, Warszawa (1984)
19. X Reality. [https://en.wikipedia.org/wiki/X_Reality_(XR)]
20. Zamojski T.: Elementy kształtowania architektury małych siedlisk. (Elements of forming architecture of small habitats). Habitaty. Oficyna Politechniki Wrocławskiej (2014)
21. Zamojski, T.: Wieżowce mieszkalne w Europie w latach 2000–2019. Ph.d. Dissertation. Wydział Architektury. Politechnika Wrocławska (2020)

Choosing Exploration Process Path in Data Mining Processes for Complex Internet Objects

Teresa Zawadzka(✉) and Wojciech Waloszek

Gdansk University of Technology, Narutowicza 11/12, Gdansk, Poland
{tegra, wowal}@eti.pg.edu.pl

Abstract. We present an experimental case study of a novel and original framework for classifying aggregate objects, i.e. objects that consist of other objects. The features of the aggregated objects are converted into the features of aggregate ones, by use of aggregate functions. The choice of the functions, along with the specific method of classification can be automated by choosing of one of several process paths, and different paths can be picked for different parts of the domain. The results are encouraging and show that our approach allowing for automated choice, can be beneficial for the data mining results.

Keywords: Data mining · Complex objects

1 Introduction

In this paper we present application our novel framework for classification of complex internet object, namely Web sites and Web pages.

Our work was a part of a larger project conducted jointly by academic and industrial partners. The industrial partners were using contents of Web pages harvested from the Internet to perform various analyses, especially from the field of sentiment analysis [1], and therefore had a large corpus of pages at their disposal.

The task we focus on in this paper consisted in classifying Web sites to one of functional categories on the basis on the features of the pages being their components.

Classification of Web sites posed a challenge because the objects that actually made the dataset were posts and Web pages coming from the sites. Consequently, we have faced the problem of categorizing aggregate, complex objects (sites) basing on the processing of aggregated, simpler objects (posts and pages).

We approached the problem from meta learning positions. Our primary observation was that there was no single method of contriving the features of the complex object from the simple ones. That observation led us to develop a new framework in which we called this step an aggregation, and included it in our workflow, at the same time allowing it to be variable. As a result we obtained a flow description which allowed for different methods of aggregating features. Each of the methods determined one path in a multi-path workflow, and the choice of the specific aggregation method (and so the path) has been automated. The further part of the paper describes the details of our approach (workflow, choosing mechanism, and the results of our experiments).

W. Zamojski et al. (Eds.): DepCoS-RELCOMEX 2020, AISC 1173, pp. 700–710, 2020.
https://doi.org/10.1007/978-3-030-48256-5_68

2 The Outline of the Solution

The proposed solution is based on the observation that the process of data classification in this task is in fact multi-staged: to properly classify a web site (aggregate object) it is necessary to use the features of web pages (aggregated objects). Basing on this observation we approached the problem systematically by creating means for describing such kinds of processes.

The processes in this approach are multi-path: a single object may be subjected to different steps of data mining process (a single exploration path), basing on the results of previous classification. The method is accompanied by procedures which supported automated choice of the single exploration path.

Therefore, for the design of the process we adopted the following desiderata:

1. It should be possible to define diverse exploration paths for aggregate Internet objects and describe them in computer readable form.
2. It should be possible to decompose the set of complex Internet objects into subsets, and map those subsets to specific process paths.

We set our primary research goal to verify if it is beneficial to process specific subsets of aggregate Internet objects with specific methods of data mining. Our further work in the project consisted in creating a framework within which it is possible to systematically describe exploration path and to choose one basing on object characteristics and to verify both the soundness of the framework and desiderata, along with the hypothesis, by an experiment. Within the experiment we used data provided by our industrial partner. The subsequent chapters describe the framework and the experiment.

3 The Framework

3.1 Assumptions

Within our project we faced the challenge of classifying complex Internet objects acquired from various sources. The process of classification was multi-staged and required us to use data describing different structures: Web sites, Web pages, and internet posts. The character of data was also specific; they were characterized by high volume and velocity, allowing us to include them under the topic of "Big Data".

As a consequence of such a setting, we decided to use formalized description of tasks undertaken during the data mining processes. It would allow us to test and store descriptions of different possible course of actions (path of a workflow), along with the results of their application. With the growing number of possibilities this idea evolved into automated choice of exploration process path.

The pattern for describing a workflow from the technical point of view was taken from the tool of our choice. Within our project we chose Spark ML [2, 3]. The specific requirements that supported the choice were the need for scalability of the solution for larger datasets, the object-oriented model approach to modelling the flow, and the availability of required data mining algorithm implementations.

SparkML standardizes APIs for machine learning algorithms to make it easier to combine multiple algorithms into a single pipeline, or workflow. Machine learning can be applied to a wide variety of data types, such as vectors, text, images, and structured data. This API adopts the DataFrame from Spark SQL in order to support a variety of data types.

A DataFrame is a Dataset organized into named columns. It is conceptually equivalent to a table in a relational database or a data frame in R/Python. Usually, these data are loaded from csv files and then transformed, such as for example changing a column with a long text into a column with a list of words from that text.

Machine learning operations performed on data sets are organized into a Pipeline. A Pipeline is an ordered collection of subsequent elements, which may be estimators or transformers (pipelines can also be nested to create more complex flows).

A transformer is an object that represents an algorithm that converts one dataframe into another. Typically, the conversion is based on the addition of a single column or set of columns. An estimator is an abstraction which, basing on the data produced by transformers, usually returns the learned model that implements a transformer interface (which it is able to return a result, e.g. of classification, for every data instance, in the form of an additional column). Evaluator in turn, assesses the obtained dataframe according to a specific set of criteria.

Our system used those standard Spark ML components for data transformation and evaluation of results (resp. Transformer and Evaluator) to match the specifics of a multi-track process (including alternate flows) described by the flow ontologies. The stress was put to enable the evaluation of the results of the exploration process according to different paths, and this task has been carried out by implementation of custom evaluators.

In addition to the technical dimension of flow components, we considered also the functional one. In general we assumed that every flow consists of tasks. But among the tasks we distinguished those that are the focal point of data preparation and modeling phases of data mining process. The main categories for the tasks are:

- data loading tasks - due to specificity of the data being process, we assume that additional actions may be needed to prepare a vector of features. Those task may include text processing, parsing the contents of a Web page etc.
- data transforming tasks - embracing tasks that perform transformations on data sets expressed as vectors of features. The tasks may be standard, like feature selection, but we also allow for more complex ones like aggregating several instances into a new one represented by a vector consisting of a different set of features.
- data modelling tasks - tasks for creating the models.

3.2 The Process Organization

Since we have chosen an object-oriented framework as a basis for designing the data mining process, it seemed suitable to use UML for more detailed description.

Figure 1 presents the pipeline elements we borrowed from SparkML along with our extension, which focuses on aggregating the metrics of smaller objects (Web pages) into the features of larger ones (Web sites). According to our discussion, there can be several ways of aggregating the features, which is reflected in the figure.

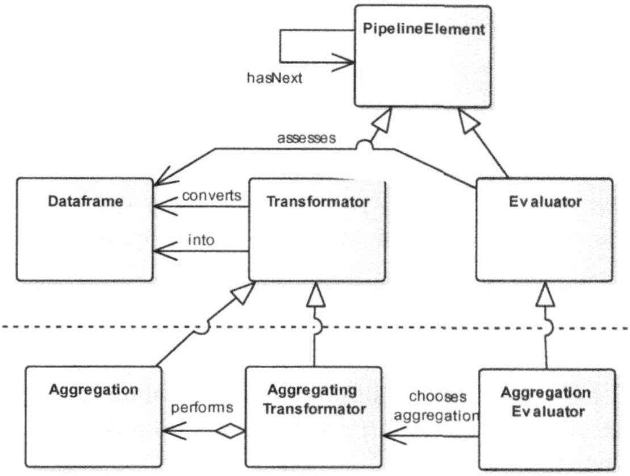

Fig. 1. Pipeline elements used in our framework. Below the dashed line, custom elements that focus on aggregation.

This kind of workflow construction comes with the assumption that several competing procedures of aggregation can be used. In the typical use of such a framework, we can distinguish the learning phase and the deployment phase. Within the phase of learning (when a smaller sample dataset is used) the method of aggregation can be picked, and then used during the deployment.

In such an arrangement *Aggregating Transformator* performs a set of aggregations during the learning phase, and *Aggregation Evaluator* chooses the best one from among them. After deployment, *Aggregating Transformator* performs only the aggregation that has been chosen, and *Aggregation Evaluator* is inactive.

While it is perfectly valid to pick one aggregation method for the whole workflow, we decided to explore the possibilities that can be introduced by diverting the aggregations used. To pursue this, we introduced the preliminary phase of clustering the dataset of Web sites. Then, for each of the clusters, we picked its own best aggregation method, therefore making our data exploration process multi-path. In the experiment described in the following sections we compared the obtained results to those achieved when using a single aggregation method.

4 Experiment Description

The aim of the experiment is to prove that it is beneficial to process specific complex Internet objects with specific methods of data mining by decomposing the set of complex Internet objects into subsets, and map those subsets to specific process paths (second desideratum).

The experiment consisted of three stages. The first stage concerned data preparation. In the second stage data was clustered using k-means algorithm and the rules classifying data into those clusters were defined. In the third stage the optimized path was assigned to the specified subset of data.

4.1 Data Preparation

Data sets are specific websites provided by business partners. By a website we mean the view visible under a single URL. The website has a collection of various page elements. Each page element may have certain features defined. The set of features describing page elements were extracted. The following ones are those taken into analysis:

- *has-link-to-page-h* – feature indicates that an anchor node has a *href* attribute and the text content of anchor can also be found in a header on the page the *href* leads to.
- *siblings-similar-ids* – feature indicates that a set of siblings all have similar values in their id attribute. Similarity is determined by removing or digit characters and then comparing the attribute values.
- *leaf* – feature indicates the place of the node in a DOM tree (is it a leaf or is it close to the leaf).
- *dummy-tree-similar* – feature indicates that a set of siblings share a similar tree.
- *has-author-names* – feature indicates that the text content of a node contains commonly used names from different languages.
- *stop-words-count* – feature indicates the number of stop words (for given language) found in the text content of a node.
- *token-count* – feature value equals the number of tokens in text nodes directly inside of the node.
- *has-date* – feature indicates that the node has an extractable date string in the text or attribute.
- *contains-date* – feature counts the number of nodes with has-date feature that are anywhere inside the checked node.
- *has-emoticons* – feature indicates that the text content of a node contains a number of emoticons commonly used on the Internet.
- *has-IP* – feature indicates that a node has IP. On many sites IP of author is displayed next to comment or post. IP can be directly detected using regex rule.
- *contains-IP* – feature counts the number of nodes with has-date feature that are anywhere inside the checked node.
- *has-pagination* – feature indicates that page contains pagination feature (links to previous, next and numbered page.

One sample means one Web page with aggregated attribute values for individual page elements. It is possible to use different aggregates to determine aggregated value of the specified feature for the specified sample: maximum, average, median and count. Each sample is thus described by 14 aggregated features in 4 different ways. Column names are built according to the principle: `features_[word representing aggregation function]/[name of the feature]`. The word `representing aggregation function` is occurrences for count function, max for maximum function, mean for average function and median for median function. For example, `features_max/contains_date` is an aggregated value of all values of `contains_date` feature, occurring on the website and the aggregation is done using maximum function.

For the purpose of the experiment, 649 samples (Web pages) were selected from 8 Web sites: ceneo.pl, dynatrace.com, epam.com, eti.pg.edu.pl, goyello.com, jakubrozalski.artstation.com, jetbrains.com and rubik.pl. The samples were prepared as described above.

In addition, the samples have been appropriately tagged by the authors during brainstorming sessions to indicate what category the page is. The list of website categories with number of samples is depicted in Table 1.

Table 1. Website categorization.

Category	Category description	No of samples
Article	Pages presenting articles concerning a topic	67
Booklet	Pages describing a single product, company, movie, etc.	145
Catalog	A page displaying a list of elements of the same type	39
Contact	Pages showing a contact or providing a contact form	22
Dictionary	Pages similar to a wiki page, but have a specific format for presenting word-expression information and its definition or translation	1
Event calendar	Pages presenting events by date	36
Forum	Forums	1
Information	Sites that provide information on a topic that cannot be classified into other categories	184
News	Pages displaying news	39
Price comparator	Pages displaying product or product prices in selected stores	62
Price list	Pages containing the product price list	5
Q&A	Ages containing questions like "questions and answers"	5
Registration	Pages with registration forms	1
Terms & conditions	Pages containing information on terms of use, copyrights, privacy	8
User profile	Pages containing user information (more than just the identifier or nickname), possible statistics, and often contact details	20
Work offer	Pages presenting job offers or cooperation offer	14

4.2 Data Clusterization

The set of samples has been divided into the predefined number of clusters (4 or 5) – the maximal number of clusters is connected with the number of samples. The higher number of clusters caused creation of clusters with too small number of samples. The k-means algorithm has been applied with 0–1 normalization of features' values.

For both experiments there were some common features affecting clusterization results. These are `contains_IP`, `token-count` and `leaf`. However, in the process of clusterization for 4 clusters also `contains-adjectives` feature is regarded whereas in the process of clusterization for 5 clusters the values of `has-link-to-page-h` and `stop-words-count` are analyzed. It also can be noticed that count aggregation function is mostly used. The process of clusterization shows that it is possible to decompose the set of complex Internet objects into subsets, while Sect. 4.3 shows the possibility of mapping those subsets to specific process paths.

4.3 Optimized Path Selection

The experiment was conducted according to the following rules:

1. The number of clusters has been chosen.
2. The set of samples is divided into the training and test set. We assumed that the size of training set should be at least 70% of the whole set and should contain at least 100 web pages per cluster. The cardinality of both sets is depicted in Table 2, respectively for the two experiments with 4 and 5 clusters.
3. For each chosen number of clusters *n* the following steps are conducted:
   a. The optimized path is determined twice: for the whole training set (the model is built for the training set consisting of the samples from all clusters), and for the subset of the training set assigned to the cluster (the model is built for the training set consisting of the samples from the specified cluster only).

Table 2. Cardinality of training and test sets.

4 clusters	Training set	Test set	Both sets
0	24	4	28
1	236	82	318
2	23	11	34
3	183	86	269
	Total: 466	**Total**:183	**Total**:649
5 clusters	**Training set**	**Test set**	**Both sets**
0	204	45	249
1	16	3	19
2	144	45	189
3	7	1	8
4	142	42	184
	Total:513	**Total**:136	**Total**:649

b. For each optimized path the validation is done for each cluster.

The optimized path is determined using the designed framework. The path consists of two steps. In the first step the aggregation function is settled: mean (*mn*) or max (*mx*) can be chosen. In the second step the classification algorithm is chosen. There are two possibilities: decision tree (*dt*) and logistic regression (*lr*) algorithms.

Validation has been done using cross-validation with 4 folds. The following points show the results of the experiment (the experiment shows the process of mapping clusters to specific process paths, which follows the second desideratum).

4.4 Discussion of Experiment Results

For both experiments the training sets were created according to the steps described in Sect. 4.3. Experiment for 5 clusters, which detailed results are presented in Table 3, showed that it is possible to choose a path for a cluster that gives better results than the path chosen globally. This effect is visible for cluster 2 (the middle column of Table 3) where the choice of max aggregation function and decision tree algorithm led to accuracy increased from 71.11% to 77.8%.

Table 3. Experiment results for 5 clusters (chosen aggregation and classifier; percentage value for accuracy)

	Cluster 0	Cluster 1	Cluster 2	Cluster 3	Cluster 4
Trained on the cluster	*mx, dt* (44.4%)	*mx, dt* (100%)	*mx, dt* (77.8%)	*mx, dt* (0%)	*mx, dt* (66.67%)
Trained on the whole set	*mn, dt* (60%)	*mn, dt* (100%)	*mn, dt* (71.11%)	*mn, dt* (100%)	*mn, dt* (66.67%)

Nevertheless the combined accuracy (accuracy weighted by the number of samples in each cluster) for this experiment was better for the single path setting (66.91% comparing to 63.24% for multi path processing).

It is worth noting that the result of this experiment was specific because for every path (cluster) the same combination of aggregation function and classification algorithm has been chosen. In such a situation one can probably expect that the global classifier trained on the whole set might give better results. This observation also underlines the importance of the clustering phase.

In the second experiment we applied different segmentation (for 4 clusters). The results are presented in Table 4. First of all we can notice that a different aggregation methods clusters were picked. Consequently, the multipath setting allowed as to obtain better combined accuracy result (51.91% comparing to 47.54%). In this experiment the classification accuracy for 3 out of 4 clusters was better the single path classification. The only cluster with worse accuracy was also the smallest one and contained exactly 4 samples. The seemingly large difference in accuracy (25% vs. 50%) was therefore the result of just one misclassification.

Table 4. Experiment results for 4 clusters (chosen aggregation and classifier; percentage value for accuracy)

	Cluster 0	Cluster 1	Cluster 2	Cluster 3
Trained on the cluster	*mx, dt* (25%)	*mx, dt* (58.54%)	*mn, dt* (81.82%)	*mn, dt* (43.02%)
Trained on the whole set	*mn, dt* (50%)	*mn, dt* (54.88%)	*mn*, dt (72.72%)	*mn, dt* (37.21%)

The results of these experiments can be treated as partial confirmation of our hypothesis. They show that careful selection of clusterization method can result in that the multipath process can provide better results than one global classifier.

5 Related Work

When comparing to the other works, it is worth noticing that the aim of our approach is not to be an alternative to the existing classification algorithms. Our work presents the framework for enhancing existing solutions by allowing for automatic selection of best workflows, especially algorithms and type of aggregations, depending on categories (clusters) the sample being classified belongs to. Due to that fact, this section is mostly concentrated on works on choosing the best methods of classification, rather than comparing the classification accuracy. Therefore, the next paragraphs present the various perspectives of the problems stated above (so workflows, aggregations etc.).

One of the perspectives of our proposal is to use the description of workflow. Such descriptions exist, and the most sophisticated assume the form of a formal ontology [4]. The most prominent ontology in the area of describing data mining processes is OntoDM [5]. The ontology embraces much broader set of problems, covering the goals of the process, mining algorithms, and the space of data types [6].

While our solution resembles more an *ad-hoc* ontological approach well-suited for a specific problem (like, e.g. [7]), it in fact may also be perceived as a step towards describing alternative flows which use different aggregating functions. As such, during further works, it might be integrated with OntoDM as specialized types of data transformation and mining algorithms.

Another perspective that one may apply to see our results bases on the observation that with use of pre-clustering we in fact obtain a form of an aggregate or ensemble classifier [8, 9]. However, in contrast to typical ensemble classifiers [10], we use here clustering for partitioning of the input data set, in order to use different aggregated features (selection is done on per cluster basis). This approach ensures that only one classifier is finally used for each partition (cluster), and follows the assumption that calculation of aggregates may be generally computationally expensive as the size of the dataset of Web Pages (smaller objects) might be considerable.

Problem of analyzing the features of Web Sites by examining Web Pages has lots in common with analyzing graphs (as we can perceive a HTML page as a sophisticated multi-labeled graph). One of the possible approaches to this problem could be use of graph mining methods [11]. Here, however, we wanted a method which will be useful for improving performance of Internet bots, which can focus simply on basic calculations made during shallow analysis of DOM graphs.

6 Conclusions

In this paper we presented a proposal of a novel and original framework for automated choose of one of several process paths in order to find those optimized for specific parts of the domain. The system we created has been built following the two desiderata that assumed the use of a description of multi-track data mining process and choosing the best path for distinguished subsets of the domain.

For describing diverse exploration paths we used the Spark ML engine, and decomposed each path into the steps which associate the basic flow elements of the engine. These expanded elements have been used to choose the best possible path, but not for the whole domain but for its part. It allowed us to determine specific methods of data mining for specific fragments of the domain.

The obtained results show the potential of the method. Throughout the experiments we were able to achieve better accuracy for the segmented domain than for the domain treated as a whole. This effectively supported the thesis, that the data could be gathered in aggregated form in different ways for different web pages.

Our plans for future work embrace the expansion of the framework by the new tools of aggregating features. But the most promising direction seems to be shifting towards the more precise description of both the domain and selected process paths in order to create a more systematic view of the domain. We plan to use ontologies for this task, more specifically to create our own extension of OntoDM.

References

1. Liu, B.: Sentiment Analysis and Opinion Mining. Morgan & Claypool, San Rafael (2012)
2. Meng, X., et al.: MLlib: machine learning in apache spark. J. Mach. Learn. Res. **17**(34), 1–7 (2016)
3. Zaharia, M., et al.: Apache Spark: a unified engine for big data processing. Commun. ACM **59**, 56–65 (2016)
4. Dou, D., Wang, H., Liu, H.: Semantic data mining: a survey of ontology-based approaches. In: IEEE 9th International Conference on Semantic Computing (2015)
5. Panov, P., Dzeroski, S., Soldatova, L.: OntoDM: an ontology of data mining. In: IEEE International Conference on Data Mining Workshops. IEEE (2008)
6. Panov, P., Soldatova, L., Dzeroski, S.: Representing entities in the OntoDM data mining ontology. In: Džeroski, S., Goethals, B., Panov, P. (eds.) Inductive Databases and Constraint-Based Data Mining. Springer, New York (2010)
7. Euler, T., Scholz, M.: Using ontologies in a KDD workbench. In: Workshop on Knowledge Discovery and Ontologies at ECML/PKDD 2004 (2004)
8. Breiman, L.: Bagging predictors. Mach. Learn. **24**(2), 123–140 (1996)

9. Louppe, G., Geurts, P.: Ensembles on random patches. In: Flach, P.A., De Bie, T., Cristianini, N. (eds.) Machine Learning and Knowledge Discovery in Databases, vol. 7523, pp. 346–361. Springer, Heidelberg (2012)
10. Nalwoga-Lutu, P.E.: Dataset selection for aggregate model implementation in predictive data mining (2010)
11. Rehman, S., Fong, S.: Graph mining: a survey of graph mining techniques. In: Seventh International Conference on Digital Information Management (ICDIM) (2012)

Author Index

W. Zamojski et al. (Eds.): DepCoS-RELCOMEX 2020, AISC 1173, pp. 711–713, 2020.
https://doi.org/10.1007/978-3-030-48256-5

The manufacturer's authorised representative in the EU is Springer Nature Customer Service Centre GmbH, Europaplatz 3, 69115 Heidelberg, Germany. If you have any concerns regarding our products, please contact ProductSafety@springernature.com

Printed and bound by CPI Group (UK) Ltd, Croydon, CR0 4YY
30/09/2026
02225643-0017